Fossil Vertebrates of Greece Vol. 1

Evangelos Vlachos
Editor

Fossil Vertebrates of Greece Vol. 1

Basal Vertebrates, Amphibians, Reptiles, Afrotherians, Glires, and Primates

Springer

Editor
Evangelos Vlachos
Museo Paleontólogico Egidio Feruglio
CONICET
Trelew, Argentina

ISBN 978-3-030-68397-9 ISBN 978-3-030-68398-6 (eBook)
https://doi.org/10.1007/978-3-030-68398-6

This Springer imprint is published by the registered company Springer Nature Switzerland AG
The registered company address is: Gewerbestrasse 11, 6330 Cham, Switzerland

This book is dedicated
to the memory of
our friend and colleague
Katerina Vasileiadou

Foreword

When I first started working on fossil mammal material from Greece, the one thing I wanted most was something that did not exist: a comprehensive guide to the region's fossil vertebrates. *The Fossil Vertebrates of Greece* fills this gap admirably, marking a major and timely milestone in the synthesis of Greece's natural heritage. For the first time, an account of almost all of the fossil vertebrate species known from Greece – from the smallest conodonts to the Titans of the Proboscidea – can be found in one place. By bringing these data and knowledge together, *The Fossil Vertebrates of Greece* is more than a regional record of past biodiversity. It makes the clearest statement yet that Greece is a treasure-trove of vertebrate fossil history, and – thanks to its synthetic nature and the inclusion of vital geographical, geological and historical context as well as collections information — it is also a tool that will help to unlock a new wave of high-quality, globally important research.

The table of contents for Volume 1 immediately highlights just how remarkable the Greek Neogene-Quaternary fossil record is, and in particular its fossil mammals. We are offered glimpses of the Paleozoic and Mesozoic through conodonts (Chap. 3) and actinopterygian fishes (Chap. 4), but the richest data is found in Neogene deposits, including for fishes. Similarly, for non-mammalian taxa, the Greek occurrences of a whole Class or Order can be summarized in a single chapter (with Chaps. 6, 7, 8 and 9 covering lizards and snakes, amphibians, crocodilians, and turtles and tortoises, respectively). But from here on in, the story is one of mammals from – with the exception of some Oligocene rodents from Kyprinos (Chap. 15) – the Neogene onwards.

Greece sits at the crossroads to Europe, and this is as relevant to its mammalian palaeobiogeography as it is to geopolitics today. From the first European appearance of the deinotheres following the establishment of a land bridge connection between Africa and Eurasia in the early Miocene (Chap. 12, to the first appearance of our own species, *Homo sapiens,* in Europe 210 thousand years ago (Chap. 19), Volume 1 traces key moments in the emergence of Greece and the impacts of major climatic and geological events on the biotic exchanges between Europe, Asia and Africa. At the same time, Volume 1 provides clear pointers for future work: the need for more and better dating across the board (how many more first appearances are to be discovered, given Greece's strategic location?); the opportunities for high-resolution studies on otoliths (Chap. 5), especially in freshwater setting (the incredible palaeoenvironmental record of Lake Ohrid cries out for just such work); and the relative lack of study on basal vertebrate groups, to name but a few examples.

Works like this are a way for a discipline to showcase both its treasures and the community that studies them. Our colleagues are saying, 'We have incredible material, and we want you to come and work with us on it'. In this context, the forthright look at the history and wider colonial influence on Vertebrate Palaeontology in Greece taken by Vlachos and colleagues in these volumes provide powerful lessons for new collaborations. The accompanying unflinching but vital assessment of the current status of scholarship and research in Greece shows that there is much to be proud of, but also much that could be improved. Together, *The Fossil Vertebrates of Greece* becomes more than a major work of scholarship, it is a catalyst for change.

The birth and development of our discipline coincided with the birth and development of the modern Greek nation, and the influence of the Western powers that shaped and buffeted nineteenth and early twentieth century Greek politics is apparent in the flow of fossil vertebrates from Greece to collections in London, Paris, Vienna and Munich. The international scattering of these specimens – many of them illegally, like the fossils of *Mammuthus creticus* from Cape Maléka that Dorothea Bate deliberately and knowingly smuggled out of Crete (they are now in the NHM, London) – is emblematic of the struggle Greek palaeontologists have faced from the outset. Vlachos highlights the lack of both a national museum of natural history and a national palaeontological society as key hurdles to the progression of Greek palaeontology, as evidenced by the lag in scientific milestones for first Greek men, and then Greek women. He does not need to join the dots between this and the historical legacy of foreign research in Greece; it is obvious.

I am proud to have a research connection to Dorothea Bate, the first woman to have described a Greek vertebrate fossil species. I have benefitted not only from easy access to her Greek specimens, but also the strong personal connection to Greek palaeontology that Bate's story gave me. This helped engender a feeling of 'belonging' to a place I had never visited, giving me the confidence and desire to pursue research there. I am so grateful for this last aspect especially, and the adventures, discoveries and friendships it spurred. But by not troubling myself to learn about, let alone face up to, the full colonial legacy on the history of Greek palaeontology, I did not fully appreciate just how generous, kind and open my Greek colleagues have been towards collaboration, when they must surely have felt some degree of chagrin at yet another foreigner trying her luck in their backyard. My gratitude to them is now even more profound.

Science is at its best when it is collaborative, open and international, and this is exactly what synthetic works like *The Fossil Vertebrates of Greece* foster as they facilitate access to knowledge, sites and material. But we should all be alert to the imbalance in advantage that the hand of history has dealt and ensure we don't repeat the same mistakes. I can't help wondering, for example, how the course of Greek palaeontological history might have run differently had the pioneering discoverer of *Mammuthus creticus* someone whose surname ended in –ακη? Might she have inspired and supported a whole host of other Greek female pioneers? These connections do matter and are wonderful to trace – like the pioneering women linked through Petralona Cave – from Anna Petrochilou's first exploration of the site through the first description of a new Greek fossil species – a horse from Petralona Cave – by a Greek woman, Evangelia Tsoukala, to Katerina Harvati's pioneering work on the Petralona hominin skull. And rather neatly, Spyridoula Pappa's work with Tsoukala on the ursids from Petralona Cave (Chap. 21 in Vol. 2) joins these pioneers back to Dorothea Bate: Pappa is now curator of fossil mammals at the NHM, London, with Bate's fossils (including *M. creticus*) under her care.

Looking to the future, *The Fossil Vertebrates of Greece* will be the benchmark for both the state of knowledge, and the demographics, of Vertebrate Palaeontology in Greece, and for Greek vertebrate palaeontologists globally. It will be the baseline from which progress can be measured, and even greater strides can be made. The Who's Who of authors in these volumes, from emeritus professor to up-and-coming student, is testament to a time of unprecedented activity and scholarship. The outstanding effort to reflect the current diversity of Greek vertebrate palaeontology researchers, especially women active in the field, in the author list regardless of seniority sews the seed for change, and the hope for a more equitable future.

It is always impossible in a foreword to do justice to a work of scholarship, especially one of such breadth at this, encompassing a fossil record that stretches from the Palaeozoic to the Holocene, and that showcases the discoveries and achievements of so many dear colleagues. But I predict that, in 10 years' time, there will be ground-breaking projects, research networks and collaborations – and the research positions that stem from them – that will trace their origins back to these two volumes, and the names therein. In that sense, *The Fossil Vertebrates of Greece* will speak for itself in the legacy it leaves.

Victoria Herridge

Department of Life Sciences
Natural History Museum
London, UK

Preface

Each young student of vertebrate paleontology in Greece quickly learns about two important dreams of every Greek paleontologist: the creation of a National Museum of Natural History and the necessity of a volume that summarizes the fossil record of vertebrates from Greece. My case was not different. Fueled by the dreams and ideas of my former supervisor, Prof. Evangelia Tsoukala, those dreams became my dreams as well. To my mind, these two important goals are connected: a National Museum of Natural History would lead to the publication of such a book eventually. Similarly, the existence of a book about the fossil record of Greek vertebrates would highlight the necessity of a National Natural History Museum and would assist in its creation.

The fossil record of Greek vertebrates preserves important information that help formulating several ideas on the evolution of life in the area, especially for the last 25 million years in the case of terrestrial environments. In the case of marine environments, however, Greek fossils trace back to the Mesozoic and Paleozoic as well. Some Greek localities, such as Pikermi near Athens, Samos Island in the Aegean, and Petralona Cave near Thessaloniki are among the most famous fossil localities in the world. Greek fossils form important parts of the collections and exhibitions of numerous international museums of natural history.

As soon as I was given the chance, I decided to pursue one of these two dreams, the publication of a high-quality volume that would review the fossil record of Greek vertebrates. It was February 2017 when the original plan for this project was made, and immediately the challenges of this task were clear. The need to gather most of the active paleontologists with first-hand experience and research on these fossils, organize the chapters, design numerous high-quality figures depicting the most important fossils, go through an extensive literature search, and bring all this heterogeneous information together into a book. Fortunately, my teachers, professors, colleagues, and friends answered the call and blessed me with their trust to organize this project. Almost 4 years later, this project grew into a two-volume book with a combined length of more than 1300 pages with more than 300 figures and maps, arranged in 42 chapters written by 31 different scientists.

This project is divided into two volumes, the first one covering basal vertebrates, amphibians, reptiles, afrotherians, glires, and primates, whereas the second deals with laurasiatherians, artiodactyles, perissodactyles, carnivorans, and island endemics. Each volume is organized in chapters per taxonomic group (in most cases per family), containing the most up-to-date information on the systematics, taxonomy, distribution, and evolution of each clade. Each chapter contains maps and tables with all known vertebrate occurrences from Greece. The most important and well-preserved fossils are illustrated. Our aim is to present detailed information of all valid vertebrate species with special emphasis on taxa whose type locality is in Greece. Some chapters are quite extensive, whereas others are quite brief, reflecting the alpha diversity of the given group in the fossil record of the country. Even so, this project is not complete: for different reasons, chapters on some important clades are missing. But given the circumstances during the last 4 years, we have done our utmost best to present the most complete and inclusive result possible.

The documented vertebrate assemblage has a long and complex evolutionary history. Since its appearance in the geological record, Greece stands in between three continents, Europe, Asia, and Africa, playing an important role on our understanding of the events that shaped the biota in the Eastern Mediterranean through time. As such, Greek vertebrates have multiple origins that can be inferred by known dispersals from the surrounding areas, as well as some vertebrates that represent native, endemic, forms of life. The geological and tectonic complexity of the region has further affected the evolution of Greek vertebrates leading to its unique extant fauna.

This book primarily addresses all those who are looking for a detailed account of the evolution of life in Greece: the scientific community, paleontologists and neontologists, teachers and professors, students, and researchers from Greece and abroad. However, because of the numerous figures and illustrations, this book could be useful for a wider audience as well. It summarizes work done in Greece by many generations of Greek and foreign paleontologists, providing a state of the art of the taxonomy and systematics of fossil vertebrates from Greece. This monumental task also has pointed out several areas of potential future research. As such, I hope that this book will prove to be a useful reference and research tool for future generations of paleontologists working with the amazing fossils from the country. I do not pretend that this project will be the final word on the fossil record of vertebrates in Greece. To the contrary, I view this project as the starting point for similar efforts in the future, which will fill the gaps, correct the mistakes, and expand the information presented herein.

Trelew, Argentina Evangelos Vlachos

Spellings of Toponyms and Personal Names

The reader should understand that the spelling of most of the Greek localities with fossil vertebrates varies in the literature and might vary in the chapters of this volume as well. In many cases, some localities are known from two or more completely different names. Whereas we opted for the most uniform result, in some cases the choice between these different spellings and/or names depends on personal preferences of the authors and historical usage, which are respected herein. To avoid confusion and ambiguity, the great majority of the localities in the tables herein is accompanied by the collection number in the Paleobiology Database. To some extent, the same problem exists with some Greek personal names.

Acknowledgments

First and foremost I would like to express my sincere thanks to all contributing authors of this two-volume project. They are beloved teachers and professors, esteemed collaborators and colleagues, respected scientists, and dear friends who entrusted me to edit and coordinate this project. I deeply thank them for their collaboration and for giving their best scientific efforts to meet the goals of this huge effort. Among them, I feel obligated to specially mention E. Tsoukala, G.D. Koufos, D.S. Kostopoulos, A. Athanassiou, and, of course, K. Vasileiadou for their crucial help and support in matters exceeding their respective chapters, including helping to organize the list of chapters and their main authors. The completion of this project would not have been possible without the help, support, and understanding of Springer Nature and especially the Senior Publishing Editors Zachary Romano, Aaron Schiller, Dr. Johanna Schwarz, and Claudia Mannsperger. I am indebted to many wonderful scientists that reviewed the chapters, namely (in alphabetical order): S. Bartolini, G. Baryshnikov, G. Carnevale, M. Cherin, A. Collareta, M. Coombs, R. Croitor, J. de Vos, E. Delson, D. Geraads, L. Hristova, Ch. Kevrekidis, J. Klietmann, T. Lehmann, F. Lihoreau, V. Popov, R. Rabinovich, M. Roksandic, G. Roessner, C. Romano, M. Salesa, R. Sardella, G.L. Stringer, W. Schwarzhans, E. Syromyatnikova, M. Uhen, L. van den Hoek Ostende, J. van der Made, A. Villa, A. Xafis, D. Youlatos, and three anonymous reviewers; most of the chapters have been additionally reviewed by other contributors of this project. Juliana Sterli accompanied me along this journey in any way possible.

Contents

List of Abbreviations

Symbols

★ The presence of a star next to a taxon name denotes that the type of this taxon (e.g., type specimen in the case of species-group name and type species in the case of a genus-group name) is from Greece. These type localities are marked also with boldface in the tables of the various chapters of this book.

⊙ The presence of the two concentric circles next to a taxon name denotes that it is an extant species. Extant species are also followed by their vernacular common name.

Institutional Abbreviations

AMNH	American Museum of Natural History, New York
AMPG	Department of Palaeontology and Historical Geology, University of Athens, Greece
BMNH	British Museum of Natural History (= Natural History Museum), London
BSPM	Bayrische Staatssamlung für Paläontologie und historische Geologie, Munich, Germany
CM	Carnegie Museum of Natural History, Pittsburgh, Pennsylvania
FSL	Univ. Claude-Bernard, Lyon, France
GPMM	Geologisch-Paläontologisches Museum (Geomuseum) der Westfälischen Wilhelms-Universität Münster
GPIH	Geologisch-Paläontologisches Institut der Universität Hamburg (Centrum für Naturkunde, Hamburg)
HLMD	Hessisches Landesmuseum, Darmstadt, Germany

IGF	Museum of Natural History of the University of Florence, Italy
IPS	Institut de Paleontologia Miquel Crusafont de Sabadell, Barcelona, Spain
IPUW	Institut für Paläontologie der Universität Wien, Austria
ISERB	Institute of Speleology "Emil Racoviţă," Bucharest, Romania
LAY	Lucas Mallada Institute, Spain
LGPUT	Laboratory of Geology and Palaeontology, School of Geology, Aristotle University of Thessaloniki, Greece
MAFI	Magyar Állami Földtani Intézet, Budapest
MCGL	Musée Cantonal de Géologie Lausanne
MCM	Montevarchi Paleontological Museum, Tuscany, Italy
MNH.MG	Museum of Natural History of Milia, Grevena, Greece
MGPP	Museo di Geologia e Paleontologia dell'Università di Padova
MGPT-PU	Museum of Geology and Paleontology of Torino University, Italy
MNHB	Museum der Naturkunde für Humboldt Universität zu Berlin
MNHN	Museum National d'Histoire Naturelle, Paris, France
MNM	Museum für Naturkunde, Magdeburg, Germany
MPUR	Museum of Palaeontology, Department of Earth Sciences of Sapienza University, Rome, Italy
MSZ	Strasbourg University Zoological Museum
NHMA	Natural History Museum of Aegean, Mytilinii, Samos
NHMB	Naturhistorisches Museum, Basel
NHMC	Natural History Museum of Crete, Herakleion, Crete, Greece
NHMUK or NHML	Natural History Museum, London, UK
MHNO	Musée des Sciences Naturelles, Orleans
NHMW	Naturhistorisches Museum Wien, Austria
NKUA	National and Kapodistrian University of Athens, Greece
NMBE	Naturhistorisches Museum Bern
NMP	Natural History Museum of Nostiko, Kastoria, Greece
NRM	Naturhistoriska Riksmuseet Stockholm
PIN	Palaeontological Institute of the Russian Academy of Sciences
PMMS	Paleontological Museum of Mytilinii, Samos
RIM	collection of B. von Freyberg, Erlangen, Germany
SAM	Natural History Museum of Stuttgart, Germany
SMF	Senckenbergisches Naturhistorisches Museum-Frankfurt, Germany
SMNK	Staatliches Museum für Naturkunde Karlsruhe
SMNS	Staatlisches Museum für Naturkunde, Stuttgart, Germany

STIPB	Steinmann-Institut für Geologie, Mineralogie und Paläontologie (Goldfuss Museum), Rheinische Friedrich-Wilhelms-Universität Bonn
UCMP	University of California Museum of Paleontology (Vertebrate Collection)
UGR	Ungarische Geologische Reichsanstalt, Budapest, Hungary
UMSL	University of Montpellier, France
UU	Department of Earth Sciences of the University of Utrecht, The Netherlands
YPM	Yale Peabody Museum of Natural History

Other Abbreviations

GPTS	Geomagnetic Polarity Time Scale
Ma	million years ago
MN	Mammal Neogene Zones
MNQ	Quaternary Mammal Zones

Introduction to the Study of Greek Fossil Vertebrates

Evangelos Vlachos

1 Vertebrates

Vertebrates (Chordata: Vertebrata) are animals closely related to other animals with a notochord, namely, the lancelets (Cephalochordata) and ascidians (Urochordata). Although traditionally considered as subphyla of the phylum Chordata, a recent review by Irie et al. (2018) proposes the recognition of Vertebrata as a separate phylum, because of new evidence from comparative genomic, gene-expression, and morphological studies. Chordates form a natural or monophyletic group defined by several shared derived characters (or synapomorphies) which separate them from other deuterostomians: the presence of an endostyle, the presence of a dorsal and hollow nerve cord, the presence of a notochord and myomeres, and the presence of a postanal tail (de Iullis and Pulerá 2019). Vertebrates are, of course, animals that have a backbone; however, this simplistic definition does not hold against modern definitions of this clade. During the last two decades, extensive discussions on the definition of vertebrates took place, especially considering the relationships between vertebrates without jaws (e.g., lampreys and hagfishes) and vertebrates with jaws (i.e., gnathostomatans). Lampreys possess only rudimentary vertebrae, termed arcualia, considered as precursors of true vertebrae (de Iullis and Pulerá 2019). Solving the relationships between these three groups was crucial to define Vertebrata: the currently accepted scheme groups lampreys and hagfishes (= cyclostomatans) which form the sister group of jawed vertebrates (= gnathostomatans) (de Iullis and Pulerá 2019 and references therein). Vertebrates are defined by several synapomorphies: two rounds of genome-wide duplication, full repertoire of neurogenic placodes, a neural crest, the formation of tripartite brain, at least one set of semicircular canals in the ear, and the presence of arcualia (de Iullis

E. Vlachos (✉)
Museo Paleontológico Egidio Feruglio, CONICET, Trelew, Chubut, Argentina
e-mail: evlacho@mef.org.ar

E. Vlachos (ed.), *Fossil Vertebrates of Greece Vol. 1*,
https://doi.org/10.1007/978-3-030-68398-6_1

and Pulerá 2019 and references therein)—the last character is the modern expression of the classical phrase "vertebrates are animals with backbone."

As vertebrates are composed of several and many hard parts, they have an important potential for fossilization in terrestrial and marine environments. There are more than 50,000 collections with more than 245,000 occurrences of vertebrates around the world, covering the entire Phanerozoic (Paleobiology Database 2020). Greece holds more than 3000 occurrences from more than 500 collections of fossils. The actual number of fossil specimens is impossible to calculate precisely at the moment. Considering each individual as a single specimen, even my most conservative and preliminary estimations suggest that 50,000 fossil individuals is a realistic starting point—the actual number of fossil specimens is higher as many of these individuals are represented by several specimens of their skeleton. Considering that Solounias (1981) calculates that only from Samos more than 30,000 specimens are dispersed around the world, my estimation herein is certainly a quite conservative one.

Data Statement The data and information used to perform the basic analyses that support the statements in the following sections are based on the list of references of the various chapters in this book and the Paleobiology Database.

2 The Fossil Treasures of Greece

The history of vertebrate fossils in Greece starts in the 1830s with Pikermi (Attica, near Athens), arguably the most famous and one of the most important fossiliferous localities of the country. The discovery of the locality is credited to the Scottish historian G. Finley who spotted bone accumulations in the Pikermi ravine, during prospection of the area in 1836. Together with the help of the German ornithologist A. von Lindermayer, they continued exploring the area and made the first excavations there (Symeonidis and Theodorou 1989; Roussiakis et al. 2019). The next part of the story is a classic one: the diamonds of the Bavarian soldier! The tale goes that in 1838 a Bavarian soldier, who presumably participated in the excavations of Finley and Lindermayer, took to Munich some fossils from Pikermi (Woodward 1901). Some of them contained calcite crystals in their cavities, which he thought were diamonds (Gaudry 1862–67). The zoologist A. Wagner immediately refuted the treasure find of the soldier, only to suggest that the real "treasure" was something else: amazing fossils of extinct animals, including the maxilla of the cercopithecid *Mesopithecus* Wagner, 1839.

The publication of Wagner (1839) should be considered as the official beginning of vertebrate paleontological research in Greece. There are some earlier mentions of fossil discoveries from Greece by Finley, Jameson, and Goldfuβ (for further information, see Athanassiou (n.d.) volume 2; Giaourtsakis (n.d.) volume 2), but these works contained simple mentions without any actual study of the fossils.

Almost immediately, several excavations took place in Pikermi, enriching collections in the country and abroad: J. R. Roth in 1852–1853, H. Mitzopoulos in 1853, A. Chairetis in 1854, A. Gaudry in the winter of 1855–1860 and summer of 1860, W. Dames in the 1880s and 1890s, M. Neumayer and L. von Tausch in 1885, the Prince of Orleans in 1888, M. from Dijon in 1895, A. S. Woodward and Th. Skouphos in the turn of the century, and O. Abel in 1912 (Roussiakis et al. 2019).

Around the same time, another famous paleontological area from Greece is discovered: the Adrianos ravine, in Mytilini Basin of Samos Island. The discovery of these localities is credited to the C.J. Forsyth Major who collected specimens in 1885–1887 and also returned later in 1887 and 1889 to search for more fossils (Forsyth Major 1894; Koufos 2009). However, fossil mammals from Samos were already known by accidental findings given to Italian travelers, now housed in Padova (Koufos 2009 and references therein). In the case of Samos, it is likely that the fossils were also known much earlier, hidden inside the ancient Greek myths of monsters, gods, and amazons (Mayor 2000; Koufos 2009; and references therein). Just like in Pikermi, many more excavations took place afterward and by numerous people, also enriching collections in European and North American museums: T. Stutzel in the end of the nineteenth century, B. Sturtz in 1889–1900, A. Hentschel in 1901–1902, E. Fraas in 1901, K. Acker, Th. Skouphos in 1903, and B. Brown in 1921–1924 (see Koufos 2009 and references therein).

At the beginning of the nineteenth century, the trio of the classical vertebrate paleontological localities from Greece is completed. During the World War I (and particularly in 1915–1916), the French naturalist C. Arambourg was serving in the 156th French Infantry Division and discovered fossils in various localities of the Axios valley and Thessaloniki area, joined under the term "Salonique": Ravin X, Ravin des Zouaves, Ravin R, Ravin C, Ravin de Vatilük (= Vathylakkos), Ravin de Konikovo (= Dytiko), and Falaise de Karabouroun (= Megalo Emvolon). Additional collections were made later in the area by Dr. Puyhaubert. These fossils were packed and sent to the MNHN of Paris (via Algeria) and formed the subject of the monumental monograph of Arambourg and Piveteau (1929).

These three, in the broad sense, localities (Pikermi, Samos, and "Salonique") form the basis not only of the mammalian paleontology in Greece but also that of other vertebrate groups: the legendary Owen (1857) named the first snake from Greece, based on some vertebrae from Megalo Emvolon; Gaudry (1862) named the first tortoise from Greece, based on some shells from Pikermi; Forsyth Major (1891) named the first avian species from Greece and Samos; and Woodward (1901) mentioned the first fossils of giant tortoises, from Pikermi.

The leading Greek figure of that time is, without any doubt, Th. Skouphos, who not only excavated in both Pikermi and Samos but also initiated the investigations in another important paleontological area: the Megalopolis basin in Peloponnesus peninsula (see Athanassiou this volume). Skouphos also started the first university position on paleontology in 1906, in the Faculty of Natural Sciences of the University of Athens. At the same time, he founded the Laboratory and Museum of Geology and Paleontology, which are the "ancestors" of the present-day paleontological institutions of the National and Kapodistrian University of Athens.

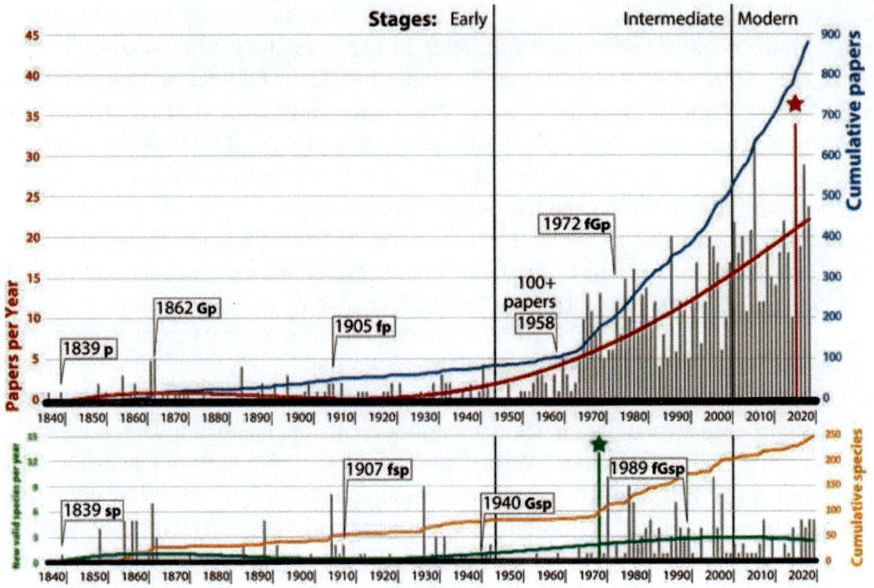

Fig. 1 Counts of published papers (red line) and new species (green line) based on Greek fossils per year vs. their cumulative growth of knowledge on Greek fossils (blue and orange lines, respectively). The diagram is divided in the three stages that are defined in the text. Some milestones are indicated, whereas stars indicate the most prolific years. Abbreviations: **f**, female; **G**, Greek author; **p**, published paper; **sp**, new species (considered as valid at present)

During this *Early Stage* of vertebrate paleontological research in Greece—and amidst the destruction of the World Wars—paleontological research is mostly focused on Pikermi, Samos, and "Salonique," although fossils from other localities from Greece begin to appear. Perhaps the most notable cases include the early investigations of the endemic vertebrates from Crete (e.g., Simonelli 1894, 1907; Bate 1907, 1912) and the Early Miocene fossils from Chios (e.g., Paraskevaidis 1940). The great majority of the published research during this stage (which contains roughly 90 published papers, books, and monographs; Fig. 1) is made by foreign scientists and deals with Greek fossils housed in foreign institutions. The most important collections are located in BSPM (Germany), MNHN (France), NHMUK (UK), and NHMW/IPUW (Austria), including more than 75 holotypes of species named from Greek type localities.

Following the end of the World War II, the *Intermediate Stage* of vertebrate paleontological research in Greece commences. This period is marked by the specific focus of enriching or creating fossil collections in national institutions, mostly Greek universities and their associated museums. During the *Intermediate Stage*, the number of trained Greek paleontologists with permanent positions gradually increases and expands geographically. As a result, Greek paleontologists gradually emerge as important and leading authorities on several topics, with an important component of foreign collaboration as well. Paleontological research continued in

Athens after the times of Skouphos, with P. Psarianos, M. Mitzopoulos, G. Marinos, and N. Symeonidis. During this period, J. Melentis, who started working in Athens, moves to Thessaloniki and the School of Geology (former Faculty of Natural Sciences) and impulses paleontological research there. The Museum of Geology and Paleontology of the University of Thessaloniki opens in 1940 and operates officially since the 1990. Melentis and Symeonidis are, without any doubt, two leading figures of the first part of the *Intermediate Stage*, forming two poles of paleontological research in Thessaloniki and Athens, respectively. They carried out numerous excavations in areas like Samos, Attica, Megalopolis, Axios valley, and Crete that have been accompanied in many cases by the creation of museums and exhibitions and numerous key publications that remain relevant to date. Also, both professors invested in training new paleontologists, creating a solid basis for paleontological research in the country. As a result, during the last third of the twentieth century, additional vertebrate paleontologists joined, both in Athens (M. Dermitzakis, G. Theodorou, C. Doukas, P. Pavlakis) and Thessaloniki (G. Koufos, E. Tsoukala) with permanent positions of teaching and research. They continued this legacy, with excavations, development of museums and exhibitions, publications and studies, formation of new students, and national and international collaboration, causing a sharp increase and subsequent accelerated growth in the paleontological knowledge of the country after the 1970s (Fig. 1).

During the *Intermediate Stage*, the classic, now, localities of Pikermi, Samos, and "Salonique" are revisited, with new excavations, revisions of the previously published materials, discovery of new materials, and applications of new techniques and methodologies (see the relevant chapters of this book). New paleontological sites are discovered, focusing the interest in adjacent areas (e.g., Western Macedonia, Chalkidiki peninsula, Mygdonia basin, Peloponnese, Thessaly, Euboea Island, other Aegean Islands). Numerous new exhibitions emerge, holding and preserving important new collections of fossils (see next section). The core of permanent paleontologists and researchers train new doctoral students who will lead paleontological research during the transition to the *Modern Stage*; there is no need to mention their names, as they are many of the authors of the various chapters in this book.

During the second half of the twentieth century, the scientific production has soared compared to the Early Stage, with more than 411 additional publications (Fig. 1). This accumulative information led to amazing new discoveries and the descriptions of many new taxa of vertebrates from Greece. These taxa and the localities are extensively covered in the following chapters, and they will not be repeated herein.

It is difficult to define a boundary between the *Intermediate* and the *Modern Stage* of paleontological research, which I consider to be defined by the presence of permanent working paleontological teams in Greece and leading studies on vertebrate paleontology. Also, this varies per taxonomic group and locality. For most mammalian families, this transition was achieved already in the 1980s, whereas in reptiles, for example, this is a recent phenomenon and occurred in the twenty-first century. The Geological Department of the Athens University, containing the oldest paleontological research position in the country, has a headstart of almost half a century over the School of Geology of the University of Thessaloniki; the Department of Geology of the University of Patras only recently gained a member specializing on vertebrate paleontology.

The *Modern Stage* of paleontological research in Greece—roughly defined herein from the beginning of the twenty-first century—finds an unprecedented group of highly qualified Greek vertebrate paleontologists, who lead the scene in the national and international level. There are at least 20 active researchers with a PhD and approximately 10 PhD candidates; more than a third of them are women, which have been significantly underrepresented in the previous stages. Adding to the names mentioned before, more Greek paleontologists hold now permanent positions in Greek universities and abroad (e.g., D.S. Kostopoulos, S.R. Roussiakis, N. Solounias, K. Harvati, G. Iliopoulos, G. Lyras). They have formed teams with researchers, permanent staff members, postdocs, and PhD/Master candidates. Additionally, permanent or long-term research positions are occupied by vertebrate paleontologists in museums (e.g., K. Vasileiadou) and other national governmental institutions (e.g., A. Athanassiou). This pool of scientists is actively discovering, preserving, and describing the Greek fossil record of vertebrate animals. Almost every year, new fossil localities are discovered. During the last decades, even more museums and exhibitions have been developed (see below), bringing paleontology closer to the public. And only in the last two decades, more than 381 new scientific works have been published (Fig. 1); this is nearly as much as the published record during the *Intermediate Stage*, especially if we consider that additional works published in 2020 are not counted herein. In several years during the *Modern Stage*, more than 20 new papers per year have been consistently published by Greek vertebrate paleontologists, marking unprecedented scientific production records. These last two decades also saw an increased national collaboration, both intra- and inter-institutionally, producing comprehensive and integrated works of global interest and outreach. This modern community of Greek paleontologists is much more specialized and diverse than before, with many scientists specializing in single family group taxa and/or to new analytical techniques.

As a result of the growth during the *Intermediate* and *Modern Stages*, the fossil collections in Greek institutions have grown, now hosting more than 100 different holotypes of vertebrate species. LGPUT (Thessaloniki; 52+ holotypes) and AMPG (Athens; 41+ holotypes) have been transformed as the two most important institutions in terms of the number of type specimens preserved.

3 Milestones

The purpose of this short section is to mention and draw attention to some important milestones related to the published record of the fossil vertebrates from Greece (Fig. 1). As it has been mentioned above, the first publication to include studies on vertebrate fossils from Greece is that of Wagner (1839), presenting fossils from Pikermi. In the same publication, the first ever genus and species is named from

Greek fossils: *Mesopithecus pentelicus* Wagner, 1839, from the Late Miocene of Pikermi. Thus, the famous type specimen (BSPM-PIK-AS II.11) should be considered—objectively, based on the published record—as the first vertebrate fossil to be formally described and published from Greece.

The first publication written (at least partly) by Greek authors and dealing with some vertebrate fossils is the one of Mitzopoulos et al. (1862), dealing with a general report of the specimens in the Physiographic Museum of Athens. Another alternative is the work of Skouphos (1905), presenting important elephant fossils from Megalopolis. The first new and valid species erected by a Greek, though, came much later by Paraskevaidis (1940): *Georgiomeryx georgalasi*. The first publication dealing with vertebrate fossils from Greece and written by a female author is by the British pioneer Dorothy Bate (Bate 1905), who also named few years later the first species from Greece named by a female author: *Elephas* (now *Mammuthus*) *creticus* from Crete (Bate 1907). The first paleontological publication by a Greek female author came 60 years later, when Helene Sakellariou-Mane published some gomphothere remains from Axios valley, near Thessaloniki (Sakellariou-Mane 1972). However, a special mention must be made to Anna Petrochilou, one of the founders of speleological research and exploration in Greece. Although she never studied or published fossilized remains in detail, she mentioned fossil vertebrates found in many caves that she explored, presented in a series of scientific works in the *Bulletin of the Hellenic Speleological Society* in the 1960s and 1970s. Some of these caves described by her would become some of the most prolific fossil vertebrate localities in the country. For example, she remarked that the study of the paleontological remains from Petralona Cave is mandatory, immediately after the first explorations in the cave (Petrochilou 1964; Petrochilos 1965). The first new species named from Greece by a Greek female author came even later, when E. Tsoukala (1989) named *Equus petraloniensis* from Petralona Cave. During the last few decades, the participation of female paleontologists in Greece has been rising, as is clearly evident in the list of authors in this volume.

Based on the cited publications herein, the top ten of authors—in terms of total number of contributions—have participated in more than half of the available literature dealing with Greek vertebrate fossils. This list includes (in descending order based on the number of their contributions) G. D. Koufos, D.S. Kostopoulos, E. Tsoukala, N. Symeonidis, A. Athanassiou, J.K. Melentis, L. de Bonis, G. Theodorou, S. Roussiakis, and G. Bouvrain.

In these nearly 900 publications published so far dealing primarily with Greek vertebrate fossils, paleontologists have identified and described more than 1100 different species of vertebrates preserved as fossils in the sediments and rocks of Greece, collected and excavated in more than 500 different localities in the country (Fig. 2). One out of four of these species (at least 254 and counting only those considered valid herein) was new to science and has been named based on unique type fossil specimens from Greece. The extreme majority of these new species (at least 219 spp.) were mammals; mammalian species represent half of the identified species (at least 538 spp.) in the country.

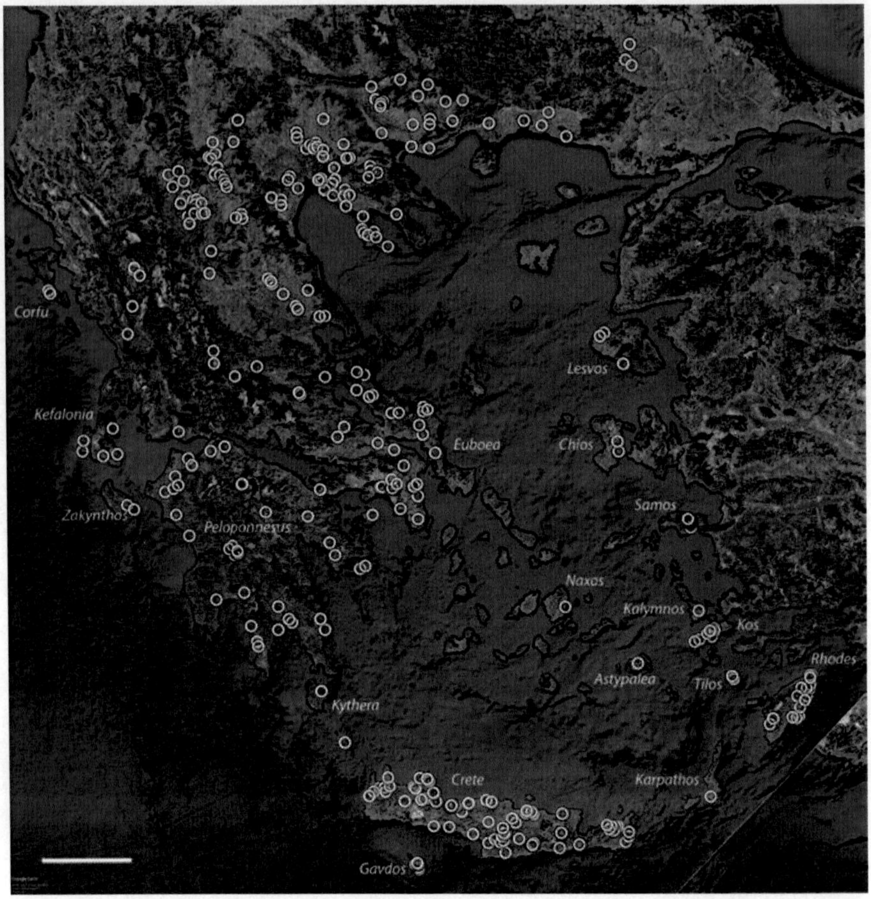

Fig. 2 Map of Greece showing the geographic distribution of the most important localities with fossils of vertebrates. (Data from the various chapters of this book and the Paleobiology Database. Image exported from Google Earth Pro © 2019, map data from US Dept. of State Geographer, SIO, NOAA, US Navy, NGA, GEBCO, image from Landsat/Copernicus. Scale bar equals 80 km, North faces upward)

4 Future Perspectives

The current status is most certainly positive and allows projecting the future of paleontology in the country with optimism. But we shouldn't forget to evaluate the context in which the modern paleontological community has evolved; if anything, this has been a tough struggle against extinction. We need to keep in mind, and compared to other countries with similarly important vertebrate fossil record, that paleontological research in Greece lacks many of the advantages that are frequently found abroad. For example, there is no specialized National Paleontological Society, which could define and preserve our rights and trigger communication, divulgation, and growth of the community. This society could also assist in the creation of laws specific to fossils and excavations that are placed outside the archaeological context of existing legislation; these laws would cover and protect the extreme majority of the fossils presented herein. The lack of a National Paleontological Society is also partially responsible for the lack of a specialized scientific journal of paleontology with a concrete regional focus. In many cases, important works and discoveries are rejected from top international journals because of their regional interest, whereas at the same time this publishing niche is empty in Greece.

For a long time, and especially during the *Intermediate Stage*, this niche was partly filled with the *Annales Géologiques des Pays Hélléniques* (1942–2004), a journal founded by M. Mitzopoulos and succeeded by the *Hellenic Journal of Geosciences*—the last volume of this journal was published in 2010. To place it in context, and although *Annales* had a broader earth sciences scope, more than 90 papers related to Greek fossil vertebrates have been published there, with a publishing frequency that diminished through time, in stark contrast to the overall publishing record that experienced an accelerated growth. It seems to me that a modern, open access, regional paleontological journal is essential for the continuous scientific growth of the Greek paleontological community. Finally, a major obstacle on the development of paleontological research in Greece is the lack of a clear growth plan and associated funding. Only very few of the completed PhD studies have been supported by adequate scholarships. Very few of the paleontological excavations and field collections received funding from local, national, or international entities. Permanent research positions are almost exclusively linked to teaching positions in universities. Additional research positions that will not be linked with teaching need to be created, otherwise all this recent growth in human resources remains unused. Proper and professional preparation laboratories are virtually nonexistent, and the preparation of fossils is mainly done through volunteering work by the researchers themselves and students, without the assistance of professional preparators. Collections are seriously understaffed and poorly financed, and researchers devote a lot of their time for curatorial work; the preservation of important type specimens and other unique fossils is left to the responsibility of the scientists alone. At least, many regional museums and exhibitions have been developed (Table 1) and in several cases provide high-quality exhibits and fulfill an important educational and touristic role. Even so, in many cases they lack a clear plan of viable development which would permit the protection of the fossils and the sustainability of these institutions. But in numerous cases, these facilities are understaffed as well as their development has been done with poor funding and limited long-term planning. All these famous fossiliferous localities and many new ones that have been discovered along the way remain unprotected as well. At the same time, Greece still lacks a National Museum of Natural History, a prerequisite for paleontological growth in many countries of the world. And last but not least, during the *Modern Stage* (so far the most fruitful period from a scientific point of view), research has been carried out under a severe financial crisis and brain-drain to other countries.

Table 1 List of major museums, institutions, and collections containing fossils of vertebrates from Greece

Name	City	Acronym
Museum of Paleontology and Geology of the National and Kapodistrian University of Athens	Athens	AMPG
Palaeontological Exhibition of Pikermi	Pikermi	–
Palaeontological Museum of Megalopolis	Issoma Karyon	–
Palaeontological Museum of Rethymnon, Crete	Rethymnon, Crete	–
Vrisa Natural History Museum	Vrisa, Lesvos	AMPG-VM
Dwarf elephant's Exhibition	Tilos	
Museum of Fossil Mammals of Kerasia	Kerasia, Euboea	AMPG-K
Museum of Geology—Palaeontology—Palaeonthropology, School of Geology, Aristotle University of Thessaloniki	Thessaloniki	LGPUT
Aristotle Museum of Natural History	Thessaloniki	ΑΜΦΙΘ
Palaeontological Exhibition of Milia, Grevena	Milia, Grevena	LGPUT-MIL
Palaeontological Exhibition of Kalamoto	Kalamoto, Langadas	LGPUT-KAL
Museum of Loutra Almopias Cave	Aridea	LGPUT-LAC
Palaeontological Exhibition of Agios Georgios Hill	Kilkis	LGPUT-SGK
Palaeontological Exhibition of Thermopigi	Thermopigi	LGPUT-SIT
Historical and Palaeontological Collection of Siatista and Kaloneri Elephant	Siatista	LGPUT-SIA LGPUT-AKL
Palaeontological Exhibition of Neokaissareia	Neokaissareia, Pieria	LGPUT-NKP
Aegean Museum of the Natural History	Samos	PMMS
Natural History Museum of Crete	Heraklion, Crete	NHMC
Goulandris Natural History Museum	Athens	GNHM
Natural History Museum of the Lesvos Petrified Forest	Sigri, Lesvos	–
Petralona Museum	Petralona, Chalkidiki	–
Palaeontological and Historical Museum of Ptolemaida	Ptolemaida	PHP
Natural History Museum of Volos	Volos	–
Perdikkas' Elephant Exhibition	Perdikkas, Eordaia	–
Museum of Palaeontology and Palaeobotany of Nostimo	Nostimo, Kastoria	NMP
"Anna Petrochilou Center," Perama Cave	Perama, Ioannina	–
Historical—Folklore and Natural History Museum of Kozani	Kozani	–

It wouldn't be a stretch to claim that the current status of paleontological research in the country is based on two pillars: the quality and importance of the Greek fossils and the selfless work of all the professors and researchers during the *Intermediate Stage*. Nearly two centuries after the discoveries in Pikermi, many more "diamonds" have been discovered in the country, treated beyond doubt as scientific "treasures" by the Greek and foreign paleontologists who discovered, preserved, and studied them. But if we are to build upon this momentum and treat this amazing paleontological heritage accordingly, serious investment and changes need to be made.

Acknowledgments I first and foremost thank E. Tsoukala for the continuous and selfless guidance and inspiration, and for comments that improved this short account. I also deeply thank G.D. Koufos, G. E. Theodorou, D.S. Kostopoulos, K. Vasileiadou, A. Athanassiou, and S.R. Roussiakis for all the help in completing this project and this manuscript. Finally, I deeply thank all the authors of this book who provided excellent and detailed accounts, allowing the preparation of this short summary herein.

References

Arambourg C, Piveteau J (1929) Les vertébrés du Pontien de Salonique. Ann Paleontol 18:57–139

Athanassiou A (this volume) The fossil record of non-endemic elephants and mammoths (Mammalia: Proboscidea: Elephantidae) in Greece. In: Vlachos E (ed) The fossil vertebrates of Greece Vol. 1—Basal vertebrates, amphibians, reptiles, afrotherians, glires, and primates. Springer – Nature Publishing Group, Cham

Athanassiou A (n.d.) (volume 1) The fossil record of non-endemic fossil deer (Mammalia: Artiodactyla: Cervidae) in Greece. In: Vlachos E (ed) The fossil vertebrates of Greece Vol. 2—Laurasiatherians, artiodactyles, perissodactyles, carnivorans, and island endemics. Springer – Nature Publishing Group, Cham

Bate DMA (1905) Four and a half months in Crete in search of Pleistocene mammalian remains. Geol Mag 2:193–202

Bate DMA (1907) On elephant remains from Crete, with description of *Elephas creticus* sp. n. Proc Zool Soc Lond 1907:238–250

Bate DMA (1912) On a New Species of Mouse and other Rodent Remains from Crete. Geol Mag 9:4–6

De Iuliis G, Pulerà D (2019) The dissection of vertebrates. Academic, Tokyo

Forsyth Major CI (1894) Le gisement ossifère de Mytilini et catalogue d'ossemens fossiles recueillis à Mitylini, île de Samos, et déposés au Collège Galliard, à Lausanne. G Bridel et Cie (éditeurs), Lausanne

Forsyth Mayor CI (1891) Considérations nouvelles sur la faune des vertébrés du Miocène supérieur dans l'île de Samos. C R Hebd Séances Acad Sci 113:608–610

Gaudry A (1862) Résultats des fouilles exécutées en Gréce sous les auspices de l'Académie. C R Acad Sci Paris 52:297–300

Gaudry A (1862–67) Animaux fossiles et géologie de l'Attique. Savy, Paris

Giaourtsakis IX (n.d.) (volume 1) The fossil record of rhinocerotids (Mammalia: Perissodactyla: Rhinocerotidae) in Greece. In: Vlachos E (ed) The fossil vertebrates of Greece Vol. 2—Laurasiatherians, artiodactyles, perissodactyles, carnivorans, and island endemics. Springer – Nature Publishing Group, Cham

Irie N, Satoh N, Kuratani S (2018) The phylum Vertebrata: a case for zoological recognition. Zool Lett 4:32. https://doi.org/10.1186/s40851-018-0114-y
Koufos GD (2009) The Late Miocene Mammal Faunas of the Mytilinii Basin, Samos Island, Greece: new collection. History of the Samos fossil mammals. Beitr Paläontol 31:1–12
Mayor A (2000) The first fossil hunters: paleontology in Greek and Roman times. Princeton University Press, Princeton
Mitzopoulos H, Orphanides TG, von Heldreich T (1862) General report on the accessions in the collections of the Physiographic Museum in Athens during the academic year 1860–61. Athens (in Greek)
Owen R (1857) On the fossil vertebræ of a Serpent (*Laophis crotaloïdes*, Ow.) discovered by Capt. Spratt, R.N., in a tertiary formation at Salonica. Q J Geol Soc Lond 13:196–199
Paleobiology Database (2000) Navigator. Accessed 7 Jan 2020
Paraskevaidis I (1940) Eine obermiocäne Fauna von Chios. N Jb Min Geol Paläont 83:363–442
Petrochilos J (1965) Découverte de restes de Mammifères du Quaternaire moyen dans la région de Petralona en Chalkidiki. Troisième Congr int Spéléol Wien
Petrochilou A (1964) Spilaio Kokkinon Petron Petralonon Chalkidikis Arith. 1044. Bull Hell Speleol Soc 7:160–166
Roussiakis S, Filis P, Sklavounou P, Giaourtsakis I, Kargopoulos N, Theodorou G (2019) Pikermi: a classical European fossil mammal geotope in the spotlight. Eur Geologist 48:28–32
Sakellariou-Mane H (1972) Presence d'*Anancus arvernensis* dans les couches supérieures du Pliocène de la Valée du Vardar. Folia Biochim Biol Graeca 9:31–36
Simonelli V (1894) Appunti sopra i terreni neogenici e quaternari dell'Isola di Candia. Atti Accad Naz Lincei Cl Sci Fis 2:265–268
Simonelli V (1907) Mammiferi quaternari dell'isola di Candia I. Memorie, Accademia delle Scienze. Ist Bologna, Cl Sc Fis 4:455–471
Skouphos TG (1905) Über die palæontologischen Ausgrabungen in Griechenland in Beziehung auf das Vorhandensein des Menschen. In: Comptes Rendus du Congrès International d'Archéologie, Athènes, pp 231–236
Solounias N (1981) The Turolian fauna from the island of Samos, Greece. Contr Vertebr Evol 6:1–232
Symeonidis N, Theodorou G (1989) History of fossil life in Pikermi, Attica. Attiko Topio kai Perivallon, Athina
Tsoukala ES (1989) Contribution to the study of the Pleistocene fauna of large mammals (Carnivora, Perissodactyla, Artiodactyla) from Petralona cave Chalkidiki (N. Greece). PhD thesis, Aristotle University of Thessaloniki
Wagner A (1839) Fossile Ueberreste von einem Affenschädel und andern Säugethierreste aus Griechenland. Gel Anz Bayer Akad Wiss 38:301–312
Woodward AS (1901) On the bone-beds of Pikermi, Attica and on similar deposits in Northern Euboea. Geol Mag 4:481–486

The Sedimentary Basins from the Miocene to the Present in Greece: Examples for the Most Studied Basins from North Greece

Angelos G. Maravelis, Nicolina Bourli, Evangelos Vlachos, and Avraam Zelilidis

1 Introduction

There are many post-Oligocene basins in Greece, where both terrestrial, lacustrine-lagoonal, or marine deposits have been accumulated (Fig. 1). These correspond to the main areas of targeted field research to find fossil vertebrates, although in some exceptional cases such fossils could be found in older rocks as well. For the purpose of this book, which concerns the fossil vertebrates from Greece ranging mainly from the Miocene to present time, we present an introduction on three major basins, situated in central and north Greece, Mesohellenic, Axios-Thermaikos, and Strymonikos basins, and their associated minor ones. These are some of the basins where many vertebrate fossils have been discovered, including also more detailed studies on the stratigraphy and paleoenvironment (Fig. 1).

These basins are situated in central and north Greece and have been developed on the Internal Hellenic Units (Fig. 2). The Mesohellenic basin is situated in the contact between External and Internal Hellenides, the Axios-Thermaikos basin on Axios-Vardar unit, whereas the Strymonikos basin on the contact between Serbo-Macedonian with Rhodope Massif.

It seems that these three major basins were active from late Eocene to the present, during which clastic deposits were accumulated. However, all of them were accompanied by three minor, and younger, basins, which were developed east of the

A. G. Maravelis
Department of Geology, Aristotle University of Thessaloniki, Thessaloniki, Greece
e-mail: angmar@geo.auth.gr

N. Bourli · A. Zelilidis (✉)
Department of Geology, University of Patras, Patra, Greece
e-mail: n_bourli@upnet.gr; a.zelilidis@upatras.gr

E. Vlachos
Museo Paleontológico Egidio Feruglio, CONICET, Trelew, Chubut, Argentina
e-mail: evlacho@mef.org.ar

E. Vlachos (ed.), *Fossil Vertebrates of Greece Vol. 1*,
https://doi.org/10.1007/978-3-030-68398-6_2

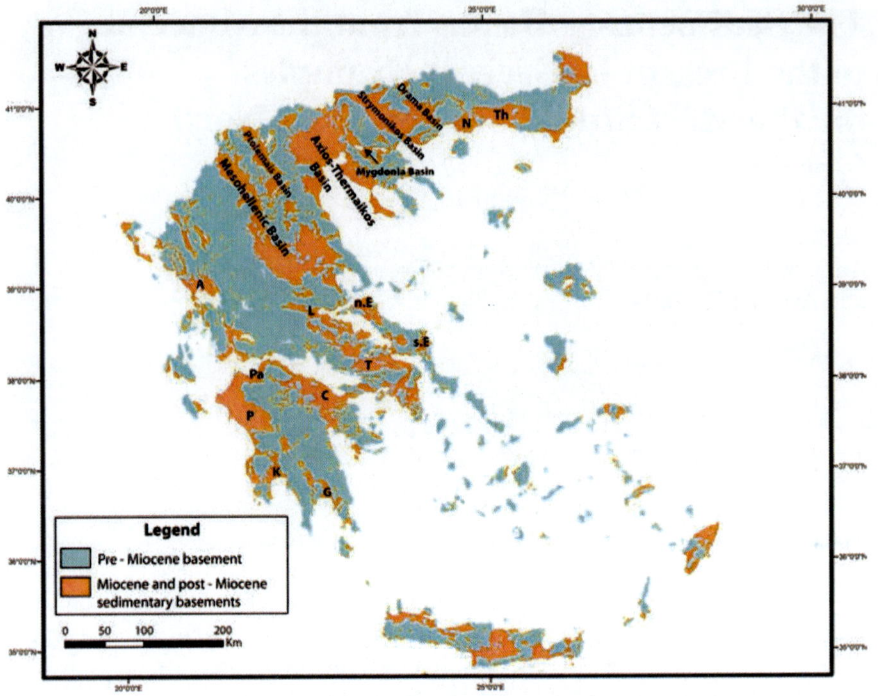

Fig. 1 The post-Oligocene basins of Greece, where the studied Mesohellenic, Ptolemais, Axios-Thermaikos, Mygdonia, and Strymonikos basins were showed. Additional, smaller, basins not presented in detail in the text: K, Kalamata; G, Gytheion; P, Pyrgos; Pa, Patras; C, Corinth; T, Thiva; L, Lamia; n.E, North Evia; s.E, South Evia; A, Amvrakikos Gulf; N, Nestos; and Th, Thrace

major basins from the Miocene onwards. The Mesohellenic major basin is accompanied by the Ptolemais minor basin, the Axios-Thermaikos major basin is accompanied by the Mygdonia minor basin, and finally, the Strymonikos major basin is accompanied by the Drama minor basin (see Figs. 3, 4 and 5).

The remaining post-Alpine deposits from Greece are located in other, smaller, basins, which nevertheless show important fossil vertebrate potential. These basins were strongly influenced by normal faults producing asymmetrical grabens.

At least four basins can be distinguished in Crete Island (Kasteli-Chania, Rethymnon, Irakleion-Messara, Ag. Nikolaos-Siteia basins), being characterized by strong synsedimentary tectonic influence with normal faults, where sedimentation started during late Miocene time (Tortonian) and continues with some interruptions as far as during the Pleistocene (Kontopoulos et al. 1996; Pasadakis et al. 2012; Moforis et al. 2013; Zidianakis et al. 2015; Maravelis et al. 2016; Zelilidis et al. 2016). Compared to other post-Oligocene basins from Greece that are predominantly terrestrial, the Neogene deposits from Crete contain several marine sediments. As such, they play an important role to our knowledge of the fossil record of vertebrates adapted to marine niches, including a diverse record of ray-finned fishes (see Argyriou this volume, and references therein) and the entire Greek fossil record of sea cows (see Iliopoulos et al. this volume, and references therein).

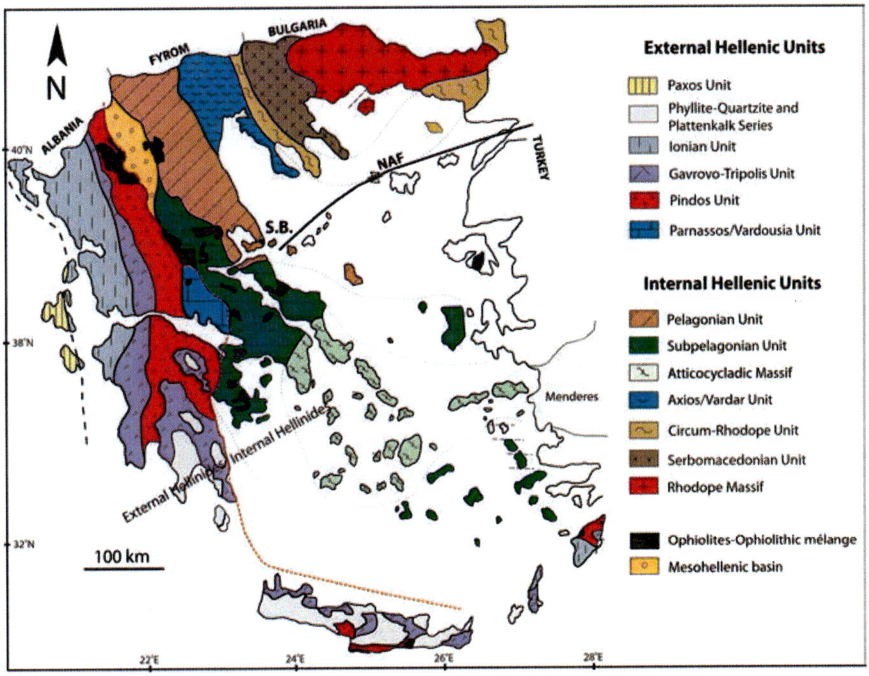

Fig. 2 Geological map of Greece showing the External and Internal Hellenides with their Units, Series, and Massifs; S.B., Sporades Basin. (Modified from Koukouvelas 2019)

The remaining Aegean islands also contain smaller-scale Neogene deposits. The most interesting are situated in Rhodes Island, where sedimentation during the Neogene covered a very high percentage of the present-day island. These deposits have not been extensively explored for fossil vertebrates, but they most certainly show an important potential, as evidenced by the rich diverse micromammalian faunas of the island (see Vasileiadou and Sylvestrou this volume-a, b; Vasileiadou and Doukas volume 2; and references therein).

In Peloponnesus, also there are many quite large basins, like these of Kalamata, Gytheion, Pyrgos, Patras, and Corinth (see Fig. 1), where mostly sedimentation took place in deltaic environments from early Pliocene to present, with many cycles of sedimentation, introducing regression and transgression events, owed mostly to tectonic activity and less to eustatic sea-level changes (Kontopoulos and Zelilidis 1992; Poulimenos et al. 1993; Zelilidis and Kontopoulos 1994; Zelilidis 2000; Zelilidis and Kontopoulos 2001). Many fossil vertebrates have been discovered in these deposits, including mainly large mammals like elephants and hippos (Athanassiou this volume-a, volume 2) or turtles (Vlachos this volume).

There are also smaller basins, characterized by Neogene deposits, like these of Thiva, Lamia, north Evia, central Evia in eastern part of central Greece, Amvrakikos gulf in the western Greece, and Nestos and Thrace basins in North Greece (see Fig. 1). With the exception of Evia island, the other basins mentioned above contain only some few known localities with fossil vertebrates.

Although there are many basins with great interest, as mentioned above (Fig. 1), we focused in three (3) synthetic basins (a major with an accompanied minor) as their complicated evolution seems to present the higher interest. Moreover, as this chapter represent the introductory chapter for all other chapters, with these three-basin analyses, a detailed description for post-Oligocene evolution in Greece is presented. Finally, all detailed maps were organized with unified basement and different lithologies and environments with chronologically divisions and not per formation, in order for readers to understand and follow the remaining chapters.

2 The Mesohellenic Basin (MHB)

Geological Setting

The Mesohellenic Basin (MHB) corresponds to a Late Eocene/Oligocene–Middle Miocene thrust-top basin that was developed because of the westwards progradation of the Pindos Orogen (Avramidis et al. 2000; Zelilidis et al. 2002). The MHB is approximately 150 km long and 30 km wide (Figs. 1 and 3) and is positioned at the boundary between the Apulian Platform to the west and the Pelagonian Microplate to the east (Doutsos et al. 1994).

The basement of the MHB is represented by the ophiolite complex, remnants of the older subducted oceanic crust (of the so-called Pindos Ocean) that was subsequently emplaced over the margin of the continental crust (Apulian Platform) during compression (Moores 1969), along with Upper Cretaceous limestones of the western Pelagonian margin (Killias et al. 2015). The dip directions of the sedimentary succession at the western margin of the MHB (towards the ENE) and the eastern margin (towards the WSW) define a large-scale asymmetrical syncline (Killias et al. 2015), whereas to the north, the basin is further subdivided into two narrower synclines, separated by a region of uplift (Doutsos et al. 1994). Structural analyses reveal that during the Middle Eocene to Quaternary, the MHB displays a complex tectonic pattern (Vamvaka et al. 2010) that occurred in semi-ductile to brittle conditions (Killias et al. 2015).

Contrasting scenarios have been proposed to account for the origin of the MHB, with the most possible being (1) a multi-story strike-slip and piggyback-type basin from Middle Eocene to the present, above the westward emplacing Neotethyan ophiolites and Pelagonian units on the cold Hellenic accretionary prism (Killias et al. 2015); (2) a foreland type basin developed in response to eastwards back-thrusting from Eocene to Miocene, because of the growth of the Pindos Orogen (Doutsos et al. 1994); (3) a forearc type basin that developed during the early stages of a Mid–Late Eocene subduction (Pindos Ocean) and turned into a piggyback basin as a result of Oligocene underthrusting of the Gavrovo-Tripolis domain (Ferriere et al. 2004).

Since the Late Miocene, the MHB exhibits gentle strike-slip deformation along major faults, uplift, and erosion, with the oldest rocks being exposed along the western basin margin (Doutsos et al. 1994).

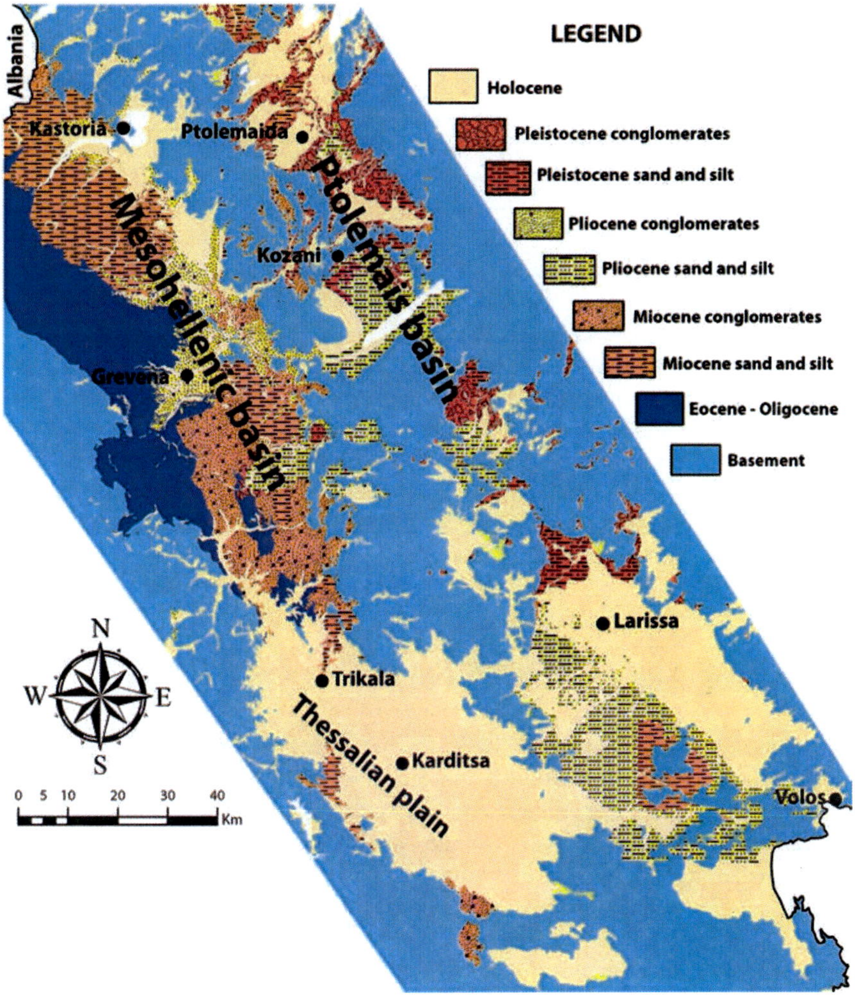

Fig. 3 Geological map of the Mesohellenic and accompanied Ptolemais basins

Stratigraphic Evolution

The MHB receives sediments that have been accumulated in a variety of depositional environments from the Early Oligocene to the Middle Miocene. The stratigraphic record of the MHB initiates with the Eptahori Formation (1000 m thick, Lower Oligocene) that comprises conglomerates and interbedded sandstones that accumulated on a fan-delta depositional environment (Zelilidis and Kontopoulos 1996; Zelilidis et al. 2002; Avramidis et al. 2002). These deposits evolve abruptly upward into fine-grained submarine fan sediments. At the southwestern margin of the MHB, the Eptahori Formation unconformably overlays the Middle to Upper Eocene shales and sandstones of the Kranea Formation (Bizon et al. 1968). The Kranea Formation (<2 km thick) evolves up-sequence from fan-delta conglomerates and shales to turbidite sandstones and shales and finally to deltaic and flood plain deposits (Zelilidis and Kontopoulos 1996; Zelilidis et al. 2002). During the Late Oligocene to Middle Miocene, the depositional environments are represented by fan-delta deposits that evolve into shelf deposits around the uplifted areas to the north and south of the MHB. Submarine fan deposits were continuously deposited in the central part of the MHB. Over the Eptahori Formation, the Pentalofos Formation (2500 m thick, Upper Oligocene–Lower Miocene) comprises submarine fan deposits that evolve to the southeast into shelf deposits. The overlying Tsotili Formation (1500 m thick, Lower to Middle Miocene) is also composed of submarine fan deposits, which in the northern part of the basin are overlain by Middle Miocene sandy shelf deposits (Ondria beds, 350 m thick). Finally, in the eastern parts of the basin fluvial deposits have been accumulated. South of the MHB, the Thessalian plain is characterized by coarse-grained shallow water deltaic deposits that formed after the separation of the MHB into three-distinct sub-basins (according to Zelilidis 2003), during early Miocene.

Vertebrate Fossils in the MHB Basin

The majority of the vertebrate fossils from the MHB basin area are known actually from the more-recent, Plio–Pleistocene deposits. These sediments contain perhaps the most important records of derived proboscidean specimens, including elephants and mammoths (see Athanassiou volume 2, and references therein), and mastodons (see Konidaris and Tsoukala this volume, and references therein). The Miocene deposits are much poorer in fossil vertebrate remains, given that they mostly represent more coastal or littoral depositional environments; they are, however, quite rich in invertebrate fossil remains. As such, Miocene vertebrate localities from the MHB basin are few, including some aquatic turtle remains (see Vlachos this volume, and references therein). These few occurrences, however, are of great importance for their respective clades and highlight the important potential of vertebrate fossil findings in the Miocene and older deposits of the MHB.

3 The Ptolemais Basin (PTB)

Geological Setting

The Ptolemais Basin (PTB) is a northwest-southeast trending sedimentary basin and is located in the Pelagonian Zone (Figs. 1 and 3). The study region was affected by at least six deformational stages from the Carboniferous to the Miocene (Mountrakis 1983). This tectonic activity triggered uplift and deformation of the basement rocks, along with the development of large-scale synclines and anticlines (Mountrakis 1983). The PTB basin was developed along a pre-existing syncline, located within the regional mountain ranges. The study area is regarded as a rift-type basin developed because of extensional tectonic activity, which followed the Alpine orogenesis in Greece (Anastasopoulos and Koukouzas 1972; Koukouzas et al. 1979, 1981, 1984, 1985). During the Pliocene, NE-SW to NNE-SSW directed normal faults affected the Neogene sedimentary succession of the PTB and controlled topography, forming grabens and horsts. Tectonic analysis studies suggest that at the early stages of basin development, the PTB was controlled by NW-SE directed normal faults, whereas the later stages are impacted by NE-SW normal faults (Pavlides 1985; Pavlides and Mountrakis 1987). The fault activity is subdivided into two stages that were active during the Late Miocene-Pleistocene and post-Early Pleistocene, respectively (Pavlides and Mountrakis 1987).

Stratigraphic Evolution

In the PTB, Neogene in age sediments (Miocene to Pliocene) are deposited, which come from the erosion of the pre-existing basement rocks. These basement rocks belong to the Pelagonian Zone and consist of Pre-Cambrian to Paleozoic metamorphic rocks, Carboniferous granites, Triassic to Lower Jurassic limestones, and Middle Jurassic ophiolites that are unconformably overlaid by Upper Cretaceous limestones and submarine fan deposits (Mountrakis 1983). The Neogene (Late Miocene to Early Pleistocene) deposits have been interpreted as lagoonal in origin and are divided into two discrete members, based on the age, composition and type of lignite-bearing deposits (Koukouzas et al. 1979; Steenbrink 2001). The lower member was accumulated during the Late Miocene to possibly earliest Pliocene age. It is composed of yellow to green, fine-grained silty sandstone and sandy claystone. The upper member was deposited during the Pleistocene and is composed of repetitions of silty sandstone and sandy claystone, along with sandstone and mudstone. During the Early to Middle Pleistocene, fluvial in origin sediments consisting of loose conglomerate, sandstone and red in colour mudstone accumulate in the PTB. The fluvial deposits overlay the pre-existing lagoonal deposits, and their boundary is represented by an erosional unconformity.

Vertebrate Fossils in the PTB Basin

The PTB basin hold important occurrences of vertebrate fossils, many of which have been discovered during prospecting or mining activities related to the lignite exploration and research in the basin. The most important vertebrate fossils from the PTB basin belong to two different categories. On one hand, there are plenty and diverse localities with numerous micromammalian fossils (e.g., see Vasileiadou and Sylvestrou this volume-a, this volume-b; Vasileiadou and Doukas volume 2; and references therein). These occurrences mostly come from the lower member, as described above. On the other hand, there are several occurrences of elephants and mammoths, including some partial skeletons with exceptional preservation (see Athanassiou volume 2). Most elephant fossils come from the Pleistocene deposits of the basin.

4 The Axios-Thermaikos Basin (ATB)

Geological Setting

The Axios-Thermaikos Basin (ATB) (Figs. 1, 2 and 4) corresponds to a fault-bounded sedimentary basin (Ferentinos et al. 1981) and forms part of the North Aegean region, an area with complex geological history (Kilias et al. 2013; Maravelis et al. 2015). The ATB overlies the eastern margin of the Pelagonian Zone

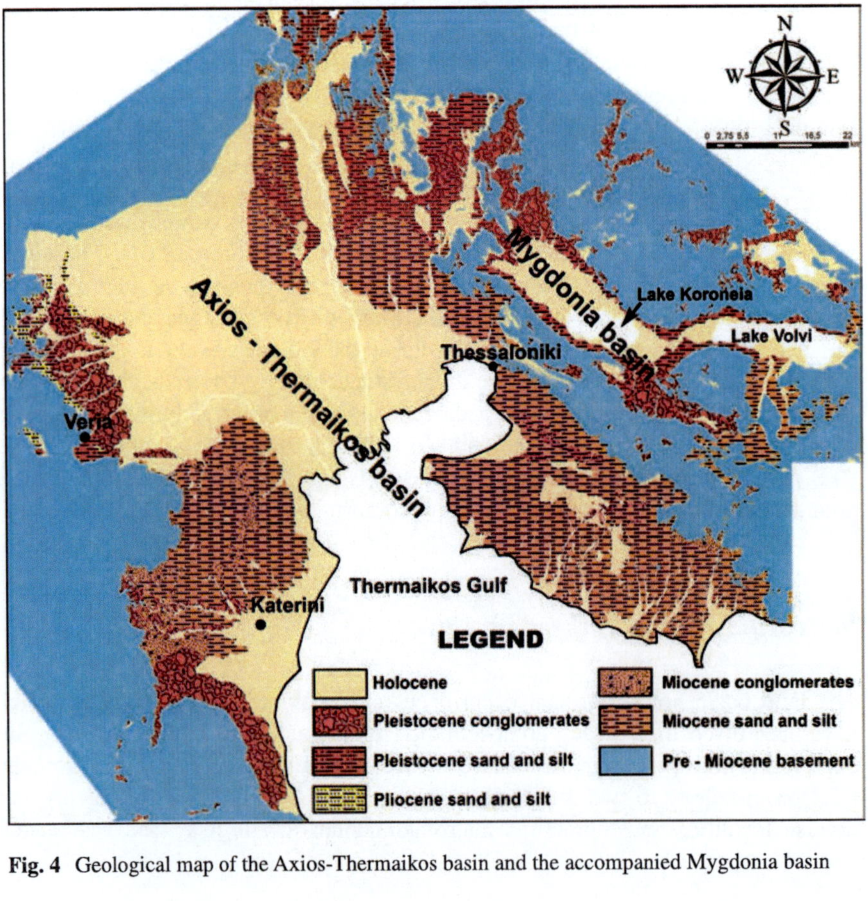

Fig. 4 Geological map of the Axios-Thermaikos basin and the accompanied Mygdonia basin

(the so-called Vardar-Axios Zone), a fault-related crystalline massif and part of the Internal Hellenides (Brooks and Ferentinos 1980). The Vardar-Axios Zone is NNW-SSE directed and extends through Bulgaria, Northern Crete, and Western Turkey. An important regional feature is the North Aegean Trough (NAT), a 300 km long trough along the Tethyan Ocean suture zone (Mountrakis 2006). The western margin of the NAT is the Sporades Basin, which in turn corresponds to the offshore extension of the ATB (Brooks and Ferentinos 1980). The development of the NAT is ascribed to late collisional processes that occurred between the Apulian and Eurasian continental lithosphere during the Late Oligocene to Middle Miocene (Tranos 2009), probably combined by the westward progradation of the North Anatolia Fault (McKenzie 1972; Pavlides et al. 1990).

The study area exhibits a complex tectonic regime characterized by periods of both extensional, compressional, and strike-slip tectonics since the Oligocene (Pavlides et al. 1990; Sokoutis et al. 1993; Kilias et al. 1999; Tranos 2009; Koukouvelas and Aydin 2002). Tectonic analyses suggest that the NE-SW Sporades Basin was initiated either during the (1) Early–Middle Miocene under regional contraction and strike-slip to transpressional deformation and subsequently involved into extensional basin (Tranos 2009) and (2) Middle–Late Miocene WNW-ESE extension (Tranos 2009). In both scenarios, the Sporades Basin was initiated because of late collisional processes, rather than the westward propagation of the North Anatolia Fault into the North Aegean Sea (Tranos 2009). Strike-slip tectonic activity played a key role in the development of the study area since the seismic reflection data suggest that basin-bounding faults display strike-slip architecture (Ferentinos et al. 1981; Roussos and Lyssimachou 1991). Normal and strike-slip interaction is also supported by fault pattern analyses, bathymetric data, and onshore and offshore seismic profiles (Koukouvelas and Aydin 2002). Since the Early Pleistocene, the deformation of the NAT is related to a NNE-SSW back-arc extension of the present-day Hellenic subduction zone (Tranos 2009).

Stratigraphic Evolution

The Neogene sedimentary succession in the ATB is ~1000 m thick (Kalkreuth et al. 1991). The ATB stratigraphy is subdivided into three major units with sediments that belong to a terrestrial to low salinity lacustrine depositional environment (Benda and Steffens 1981; Kalkreuth et al. 1991; Kotis and Papanikolaou 1999). The lower unit is ~400 m thick and is exposed in the northern margin of the ATB because of extensive erosion of the overlying strata. It includes conglomerates of terrestrial origin and sandstone beds (up to 10 m thick). Silt and sandy clay beds are also present. The middle part is over 200 m thick and is composed of sandstone, mudstone, and thick coal beds. The age of this unit, as determined by palynological data, is Lower Miocene (Benda and Steffens 1981), and it has been interpreted as being deposited in a swamp to limnic lower delta environment (Kalkreuth et al. 1991). Unconsolidated sandstones, silt, and clay follow the thick-bedded coals (up to 1 m thick). The upper unit consists of sandstone and mudstone beds that accumulated during the Late Miocene age ("Pontian"). In this unit, lignite beds occur but they are of no economic importance (Kalkreuth et al. 1991). The sediment flow direction is generally to the south and east further feeding the Sporades Basin. During the Late Quaternary, deposition is controlled primarily by prodeltaic and delta plain dynamics and, secondarily, by relative sea level changes (Lykousis et al. 2005). Three major river systems (Axios, Aliakmon, and Pinios), along with minor ones (Loudias and Gallikos), contribute freshwater and sediment to the sea. All but Pinios debouch in the north part of the Thermaikos Gulf (Lykousis et al. 2005). The excess in river-borne sediment is responsible for the development of an extensive bird-foot type delta (Poulos et al. 1994). In the eastern margins of the ATB (western Chalkidiki), the basin was formed during early Miocene and was filled with Neogene–Quaternary deposits that are over 5 km in total thickness (Syrides 1990). The deposits are grouped into six formations: (1) Antonios Formation was deposited during the Early to Middle Miocene and Late Miocene, (fluvial deposits), (b) Triglia Formation was deposited during the Late Miocene (Vallesian-Lowermost Turolian, continental deposits and red-beds), (c) Trilophos Formation was deposited during the Latest Miocene (Pontian or Turolian, brackish-lacustrine deposits), (d) Gonia Formation was deposited during the Pliocene (Ruscinian, fluvio-lacustrine deposits), and (e) Moudania Formation was deposited during from the Villafranchian onwards (continental sediments). To the west (Katerini sub-basin), the Neogene sedimentation is represented by eight Formations (Sylvestrou 2002): (1) Elatochori (alluvial fans, Early Miocene), (2) Moschopotamos (continental to lagoonal-lacustrine, Early to Middle Miocene), (3) Sykea (meandering fluvial, Middle Miocene), (4) Ryakia (meandering fluvial, Middle Miocene), (5) Lagorachi (braided fluvial, Late Miocene), (6) Sfendami (lagoonal, Late Miocene), (7) Makrygialos (braided fluvial, Late Miocene to Early Pleistocene), and (8) Lofos Formation (fluvial and lagoonal, Pleistocene to Quaternary).

Vertebrate Fossils in the ATB Basin

The ATB Basin is one of the most important regions that contain vertebrate fossils. The first collections of vertebrate fossils in the area took place in the beginnings of the twentieth century by C. Arambourg (Arambourg and Piveteau 1929) and continue until present with new collections and extensive exploration. In the northern part of the basin, on each margin of the Axios river, numerous and famous localities with vertebrate fossils have concentrated enormous interest during the last 100 years. The majority of these localities are quite diverse, containing important primate (Koufos this volume-a, b), bovid (Kostopoulos volume 2), equid (Vlachou et al. volume 2), carnivoran (Koufos volume 2-c, d, e, f), and proboscidean (Konidaris and Tsoukala this volume) fossils. The western Chalkidiki sub-basin contains several vertebrate localities as well, mostly discovered along the coastline of the eastern border of Thermaikos gulf. These localities contain important non-mammalian fossils, including snakes (Georgalis and Delfino this volume), turtles, and giant tortoises (Vlachos this volume). The Katerini sub-basin is just as promising as the western Chalkidiki one, but focused fieldwork and exploration targeting vertebrate remains only recently took place (Sylvestrou 2002).

5 The Mygdonia Basin (MB)

Geological Setting

The Mygdonia Basin (MB) (Fig. 4) is situated in central Macedonia, to the northeast of Thessaloniki city, and is characterized by active seismic activity (Papazachos et al. 1979; Martinod et al. 1997). The MB is an E-W directed sedimentary basin (graben) that was developed during the Early-Middle Miocene because of extensional tectonic activity (Le Pichon and Angelier 1981). The MB is nowadays characterized by N-S extension (Martinod et al. 1997) that has been interpreted as the result of the spreading of the Aegean region over the subducted oceanic Mediterranean plate (Le Pichon and Angelier 1981). The extension principally influences a narrow (less than 2 Km), E-W oriented zone at the southern edge of the basin (Martinod et al. 1997). The integration of seismological and neotectonic data indicate that normal faults prevail the tectonic features in the MB (Papazachos et al. 2001). Despite the variety in fault azimuths observed in some small faults, the majority of the faults and the major-magnitude seismic events are in agreement with the average N-S extension (Papazachos et al. 2001). The MB exhibits S-shape geometry, with its edges being NW-SE oriented and the central parts being E-W oriented (Papazachos et al. 2001). The central part has drawn the scientific attention because of the seismic activity of the Mw = 6.5 magnitude that affected the city of Thessaloniki (Papazachos et al. 1979; Pavlides et al. 1988). The basin is situated within the Serbo-Macedonian massif and is surrounded by a several sediment depocenters (Axios Basin, Strymon Basin, and North Aegean Trough). The MB receives sediments from Neogene to Quaternary that uncomformably overlay the pre-Neogene basement. Gneisses and schists represent the basement rocks and amphibolites, metamorphic rocks that form parts of the Serbo-Macedonian Massif (in the central and eastern part of the MB, Kockel et al. 1977). The basement rocks in the western part of the basin correspond to phyllites, limestone, and sandstone that belong to the Circum-Rhodope Belt and have experienced low-grade metamorphism (Kockel et al. 1977).

Stratigraphic Evolution

The sedimentary pile in the MB is divided into two units: (1) the Pre-Mygdonian Group and (2) the Mygdonian Group. The Pre-Mygdonian Group encompasses Neogene and Early Pleistocene in age sediments whereas the Mygdonian Group was deposited during the Middle Pleistocene to the Holocene and overlies uncomformably the Pre-Mygdonian Group (Koufos et al. 1995; Konidaris et al. 2015). The

Pre-Mygdonian Group is further subdivided into three formations. The oldest Chrysavgi Formation (Middle Miocene, 40–50 m thick) uncomformably overlies the basement and is composed of laterally discontinuous conglomerates interbedded with sandstones. The conglomerate is unconsolidated and consists of well-rounded gravels (up to 40 cm in diameter). The gravels consist of mica schist, gneiss, granite, quartzite, and pegmatite. Finer-grained deposits are rare and form thin-bedded lenses of silt and clay. The conglomeratic deposits mainly occupy the lower parts of the Chrysavgi Formation and exhibit an upward decrease in the gravel diameter within the formation. The upper parts of the Formation are sand-dominated and the finer-grained deposits (siltstones, silty sandstones, and silty claystones) are sporadically present. The overlying Gerakarou Formation (Early Pleistocene, over 100 m thick) is represented by red-beds consisting of repetitions of unconsolidated gravels, sand, silt, and clay. The gravelly beds are lenticular in shape, laterally discontinuous, and they have been interpreted as being accumulated in a fluvio-terrestrial environment of deposition. These red-beds erode significantly the underlying deposits and develop steep, valley-type landforms. The overlying Platanochori Formation (Early Pleistocene, 10–20 m thick) consists of sandstones and conglomerates, interbedded with silty sandstones, silty claystones, marlstones, and marly limestones. The limit between the Gerakarou and Platanochori Formations is transitional, consisting of laterally discontinuous sandstones, sandy marlstones, and red beds. The Platanochori Formation is thought as being accumulated in fluvial, fluvio-lacustrine depositional environment. This Formation represents the early stages of evolution between the continental Gerakarou Formation and the overlaying lacustrine sediments of the Mygdonian Group.

Vertebrate Fossils in the MB

Targeted exploration in the MB focusing on fossil vertebrates has contributed greatly to our knowledge of the stratigraphy of the basin; it is not a coincidence that some of the most important localities share the same name with the formations mentioned above. Vertebrate fossils are found mainly in the Pre-Mygdonian Group and are encountered in all formations. Occurrences from the great majority of mammalian families covered in this volume are found in the MB (e.g., Koufos 2006; Koufos et al. 1995; Tsoukala and Chatzopoulou 2005; Konidaris et al. 2015; and references therein).

6 The Strymonikos Basin (SB)

Geological Setting

The Strymonikos Basin (SB) is positioned over the boundary between the Rhodope massif and Serbo-Macedonian massif (Fig. 5). The boundary between the basement massifs is documented by the overthrusting of biotitic gneiss and marbles

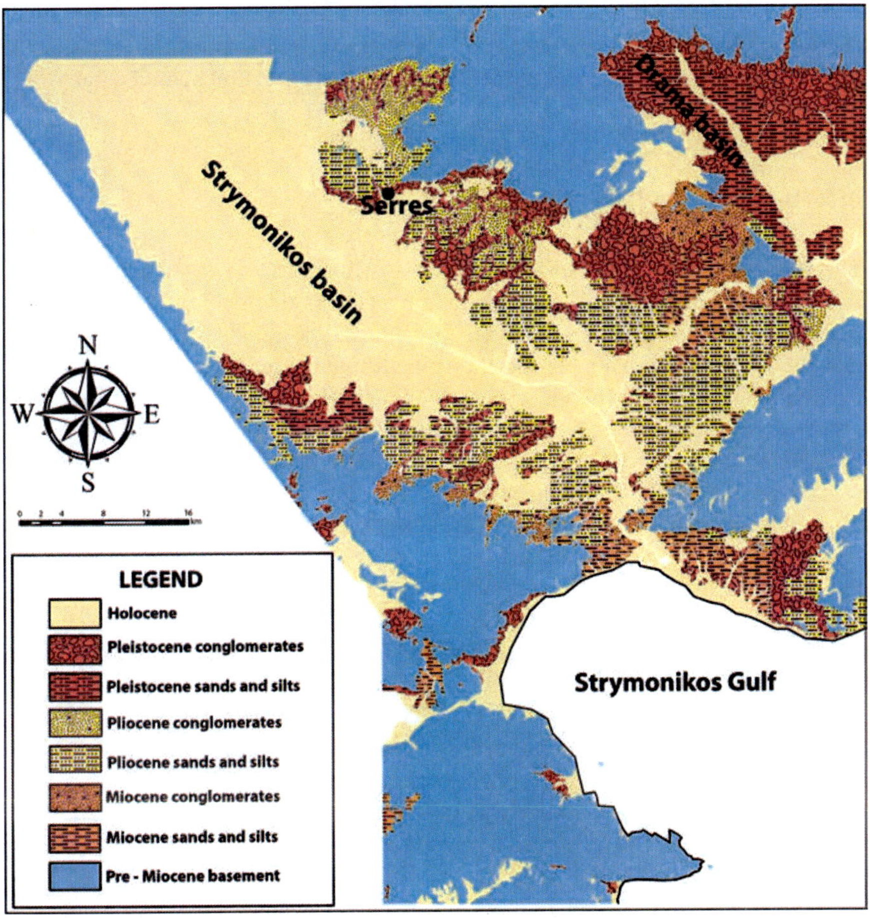

Fig. 5 Geological map of the Strymonikos basin and the accompanied Drama basin

(Serbo-Macedonian) over the marbles of the Pangaion Mt. (Rhodope). The SB is thought as a NNW-SSE oriented graben-type basin and is arcuate in shape (Psilovikos 1994). The SB is interpreted to have been developed in response of a low-angle normal boundary fault (Dinter and Royden 1993). The fault action triggered collapse of the material and slumping towards the southwest. This tectonic event is the result of the extensional regime that controlled the basin evolution during the Neogene and extends from the Aegean Sea to Bulgaria (Dinter and Royden 1993).

Gravity faults affected the development of the SB, particularly the central parts, triggering post-depositional sediment deformation and anticline formation (Lalechos 1986). The uplift of the surrounding regions leads to the erosion of the basement rocks that provided the detritus for the sediment in the SB. The basin includes evidence of relative sea-level fluctuations, with stages of lagoonal-lacustrine sedimentation and stages of marine sedimentation (Lalechos 1986). At the latest stages of evolution, sedimentation in the SB is entirely continental.

Stratigraphic Evolution

The SB receives Neogene in age sediments (Miocene to Pleistocene), stemming from the erosion of the pre-existing basement rocks that belong to the Rhodope massif and Serbo-Macedonian massif (Lalechos 1986). Sub-surface data indicate that the Miocene deposits correspond to both marine and lacustrine depositional environments and are subdivided into three units (Lalechos 1986). The basal unit comprises brecciated deposits, with the internal clasts being composed of fine-grained sandstones and conglomerates.

The overlying unit is interpreted as lacustrine deposits and is composed of repetitions of sandstones, mudstones, and dark brown marlstones that evolve upsection into petroliferous limestones. The younger deposits represent marine deposits and develop repetitions of sandstones, siltstones, and claystones that accumulate along with micro-brecciated deposits and lignite layers. The Pliocene sediments are interpreted to represent both brackish and continental environments of deposition (Lalechos 1986). The brackish sediments include sandstones, mudstones, and siltstones, along with travertines and lignites. The continental deposits are composed of alternations of reddish sandstones and mudstones. Intercalations of sandstones, siltstones, and micro-conglomerates are common. These sediments also include travertines, lignite-bearing limestones, and chalk limestones. The Pleistocene deposits are composed of quartz gravels, coarse-grained sandstones, siltstones, and claystones. The sedimentary succession suggests that the sedimentation in the SB was dominated by lacustrine conditions and low-energy sedimentation on continental deposits (Lalechos 1986). The travertinic and lignitic deposits represent material accumulated at the basin margin, adjacent to the basement rocks (Lalechos 1986).

Vertebrate Fossils in the SB

The majority of the vertebrate fossil-bearing localities of the SB are located in the northern parts of the basin. These localities are mainly quite rich and diverse in micromammalian (e.g. see Vasileiadou and Sylvestrou this volume-a, b; Vasileiadou and Doukas volume 2; and references therein) and squamate (Georgalis and Delfino this volume) remains, including some of the most diverse vertebrate localities in Greece (Schmidt-Kittler 1995). These localities document the transition from the Miocene to the Pliocene in great detail. Large vertebrate remains are fewer but include important chalicothere specimens together with associated Late Miocene remains (Tsoukala volume 2 and references therein). Stratigraphically younger vertebrates are known from the central and southern parts of the SB and the associated Drama basin, including mainly proboscidean remains (Athanassiou volume 2, Konidaris and Tsoukala this volume).

References

Anastasopoulos I, Koukouzas K (1972) Geology and economy geology in the south part of coal bearing basin of Ptolemais. IGEY, Athens

Arambourg C, Piveteau J (1929) Les Vertébrés du Pontien de Salonique. Ann Paleontol 18:57–139

Argyriou T (this volume) The fossil record of ray-finned fishes (Actinopterygii) in Greece. In: Vlachos E (ed) The fossil vertebrates of Greece vol. 1 – basal vertebrates, amphibians, reptiles, afrotherians, glires, and primates. Springer, Cham

Athanassiou A (volume 2) The fossil record of non-endemic elephants and mammoths (Mammalia: Proboscidea: Elephantidae) in Greece. In: Vlachos E (ed) The fossil vertebrates of Greece vol. 1 – basal vertebrates, amphibians, reptiles, afrotherians, glires, and primates. Springer, Cham

Athanassiou A (this volume-b) The fossil record of non-endemic hippopotamids (Mammalia: Artiodactyla: Hippopotamidae) in Greece. In: Vlachos E (ed) The fossil vertebrates of Greece vol. 2 – laurasiatherians, artiodactyles, perissodactyles, carnivorans, and island endemics. Springer, Cham

Avramidis P, Zelilidis A, Kontopoulos N (2000) Thrust dissection control of deep-water clastic dispersal patterns in the Klematia-Paramythia foreland basin, western Greece. Geol Mag 137:667–685

Avramidis P, Zelilidis A, Vakalas I, Kontopoulos N (2002) Interactions between tectonic activity and sea level changes in the Pindos and Mesohellenic Basins, NE Greece: basin evolution and hydrocarbon potential. J Pet Geol 25:53–82

Benda L, Steffens P (1981) Aufbauy and Alter des Neogens von Katerini, Griechenland. Geol Jahrb B 42:93–103

Bizon JJ, Lalechos G, Savoyat E (1968) Présence d'Eocene transgressif en Thessalie. Bull Soc Geol Fr 10:36–38

Brooks M, Ferentinos G (1980) Structure and evolution of the Sporadhes basin of the North Aegean trough, Northern Aegean Sea. Tectonophysics 68:15–30

Dinter DA, Royden LH (1993) Late Cenozoic extension in northeastern Greece: Strymon Valley detachment and Rhodope metamorphic core complex. Geology 21:45–48

Doutsos T, Koukouvelas I, Zelilidis A, Kontopoulos N (1994) Intracontinental wedging and post-orogenic collapse in the Mesohellenic trough. Geol Rundsch 83:257–275

Ferentinos G, Brooks M, Collins M (1981) Gravity-induced deformation on the north flank and floor of the Sporades Basin of the North Aegean Sea Trough. Mar Geol 44:289–302

Ferriere J, Reynaud JY, Pavlopoulos A, Bonneau M, Migiros G (2004) Geologic evolution and geodynamic controls of the Tertiary intramontane piggyback Meso- Hellenic Basin, Greece. Bull Soc Geol Fr 175:361–381

Georgalis GL, Delfino M (this volume) The fossil record of lizards and snakes (Reptilia: Squamata) in Greece. In: Vlachos E (ed) The fossil vertebrates of Greece vol. 1 – basal vertebrates, amphibians, reptiles, afrotherians, glires, and primates. Springer, Cham

Iliopoulos G, Svana K, Roussiakis S (this volume) The fossil record of sea cows (Mammalia: Sirenia) in Greece. In: Vlachos E (ed) The fossil vertebrates of Greece vol. 1 – basal vertebrates, amphibians, reptiles, afrotherians, glires, and primates. Springer, Cham

Kalkreuth W, Kotis T, Papanikolaou C, Kokkinakis P (1991) The geology and coal pe-trology of a Miocene lignite profile at Meliadi Mine, Katerini, Greece. Int J Coal Geol 17:51–67

Kilias A, Falalakis G, Mountrakis D (1999) Cretaceous–Tertiary structures and kinematics of the Serbomacedonian metamorphic rocks and their relation to the exhumation of Hellenic hinterland (Macedonia, Greece). International Journal of Earth Sciences 88:513–531

Kilias A, Falalakis G, Sfeikos A, Papadimitriou E, Vamvaka A, Gkarlaouni C (2013) The Thrace basin in the Rhodope province of NE Greece – a tertiary supradetachment basin and its geodynamic implications. Tectonophysics 595–596:90–105

Killias A, Vamvaka A, Falalakis G, Sfeikos A, Papadimitriou E, Gkarlaouni CH, Karakostas B (2015) The mesohellenic trough and the paleogene thrace basin on the rhodope massif, their structural evolution and geotectonic significance in the Hellenides. J Geol Geosci 4:198(2):17

Kockel F, Mollat H, Walther HW (1977) Erläuterungen zur geologischen Karte der Chalkidiki und angrenzender Gebiete 1:100.000 (Nord-Griechenland). Bundesanst Geowiss Rohstoffe, Hannover, p 119

Konidaris G, Tsoukala E (this volume) The fossil record of Neogene Proboscidea (Mammalia) in Greece. In: Vlachos E (ed) The fossil vertebrates of Greece vol. 1 – basal vertebrates, amphibians, reptiles, afrotherians, glires, and primates. Springer, Cham

Konidaris GE, Tourloukis V, Kostopoulos DS, Thompson N, Giusti D, Michailidis D, Koufos GD, Harvati K (2015) Two new vertebrate localities from the Early Pleistocene of Mygdonia Basin (Macedonia, Greece): preliminary results. C R Palevol 14:353–362

Kontopoulos N, Zelilidis A (1992) Upper Pliocene lacustrine environments in the intramontane Rio graben basin, NW Peloponnesus, Greece. N Jb Geol Paläont (Monatsh) 2(102):114

Kontopoulos N, Zelilidis A, Frydas D (1996) Late Neogene sedimentary and tectonostratigraphic evolution of southwestern Crete island, Greece. N Jb Geol Paläont (Monatsh) 202:287–311

Kostopoulos DS (volume 2) The fossil record of bovids (Mammalia: Artiodactyla: Pecora: Ruminantia: Bovidae) in Greece. In: Vlachos E (ed) The fossil vertebrates of Greece vol. 2 – laurasiatherians, artiodactyles, perissodactyles, carnivorans, and island endemics. Springer, Cham

Kotis T, Papanikolaou K (1999) Depositional environment of lignite bearing sediments in the Neogene basin of Moschopotamos Pieria, Greece. Bull Geol Soc Greece 33:17–28

Koufos GD (2006) The Neogene mammal localities of Greece: faunas, chronology and biostratigraphy. Hell J Geosci 41:183–214

Koufos GD (this volume-a) The fossil record of the Old World monkeys (Mammalia: Primates: Cercopithecidae) in Greece. In: Vlachos E (ed) The fossil vertebrates of Greece vol. 1 – basal vertebrates, amphibians, reptiles, afrotherians, glires, and primates. Springer, Cham

Koufos GD (this volume-b) The fossil record of the Miocene hominoids (Mammalia: Primates: Hominoidea) in Greece. In: Vlachos E (ed) The fossil vertebrates of Greece vol. 1 – basal vertebrates, amphibians, reptiles, afrotherians, glires, and primates. Springer, Cham

Koufos GD (volume 2-a) The fossil record of felids (Mammalia: Carnivora: Felidae) in Greece. In: Vlachos E (ed) The fossil vertebrates of Greece vol. 2 – laurasiatherians, artiodactyles, perissodactyles, carnivorans, and island endemics. Springer, Cham

Koufos GD (volume 2-b) The fossil record of hyaenids (Mammalia: Carnivora: Hyaenidae) in Greece. In: Vlachos E (ed) The fossil vertebrates of Greece vol. 2 – laurasiatherians, artiodactyles, perissodactyles, carnivorans, and island endemics. Springer, Cham

Koufos GD (volume 2-c) The fossil record of canids (Mammalia: Carnivora: Canidae) in Greece. In: Vlachos E (ed) The fossil vertebrates of Greece vol. 2 – laurasiatherians, artiodactyles, perissodactyles, carnivorans, and island endemics. Springer, Cham

Koufos GD (volume 2-d) The fossil record of mustelids (Mammalia: Carnivora: Mustelidae) in Greece. In: Vlachos E (ed) The fossil vertebrates of Greece vol. 2 – laurasiatherians, artiodactyles, perissodactyles, carnivorans, and island endemics. Springer, Cham

Koufos GD, Syrides GE, Kostopoulos DS, Koliadimou KK (1995) Preliminary results about the stratigraphy and the palaeoenvironment of Mygdonia Basin, Macedonia, Greece. Geobios 18:243–249

Koukouvelas IK (2019) Geology of Greece. Liberal Books, Athens

Koukouvelas IK, Aydin A (2002) Fault structure and related basins of the North Aegean Sea and its surroundings. Tectonics 21(5):1046. https://doi.org/10.1029/2001TC901037

Koukouzas K, Kotis Th, Ploumidis M, Metaxas A (1979) Coal exploration of Anargiri-Amynteon area, I.G.M.E. Miner Deposit Res 9:1–69 (in Greek)

Koukouzas K, Kotis Th, Ploumidis M, Metaxas A (1981) Coal exploration of "Apophisis" field of Anargiri-Amynteon area, IGME, Res Energy Resour 1:1–52. (in Greek)

Koukouzas K, Kotis Th, Ploumidis M, Metaxas A, Dimitriou D (1984) Coal exploration of Komnina area, Ptolemais district (W. Macedonia). IGME, Res Energy Resour 2:1–103 (in Greek)

Koukouzas K, Kotis Th, Ploumidis M, Metaxas A, Dimitriou D (1985) Lignite deposits of Ptolemaida trench. Proceedings of the International meeting for the exploitation of low energy solid fuel, Ptolemaida (in Greek)

Lalechos N (1986) Correlations and observations in molassic sediments in onshore and offshore areas of northern Greece. Miner Wealth 42:6–11

Le Pichon X, Angelier J (1981) The Aegean Sea. Philos Trans R Soc Lond A 300:357–372

Lykousis V, Karageorgis AP, Chronis GT (2005) Delta progradation and sediment fluxes since the last glacial in the Thermaikos Gulf and the Sporades Basin, NW Aegean Sea, Greece. Mar Geol 222–223:381–397

Maravelis AG, Pantopoulos G, Tserolas P, Zelilidis A (2015) Accretionary prism forearc interactions as reflected in the sedimentary fill of southern Thrace Basin (Lemnos Island, NE Greece). Int J Earth Sci 104:1039–1060

Maravelis A, Panagopoulos G, Piliotis I, Pasadakis N, Manutsoglou E, Zelilidis A (2016) Pre-Messinian (sub-Salt) Source-rock potential on Back-stop Basins of the Hellenic Trench system (Messara Basin, Central Crete, Greece). Oil Gas Sci Technol-Rev IFP Energies nouvelles 71:6

Martinod J, Hatzfeld D, Savvaidis P, Katsambalos K (1997) Rapid N-S extension in the Mygdonian graben (Northern Greece) deduced from repeated geodetic surveys. Geophys Res Lett 24:3293–3296

McKenzie DP (1972) Active tectonics of the Mediterranean region. Geophys J R Astron Soc 30:109–185

Moforis L, Kostopoulou S, Panagopoulos G, Pyliotis J, Triantaphyllou M, Manoutsoglou E, Zelilidis A (2013) Sedimentation processes and palaeogeographic evolution of Makrilia Pliocene deposits, SE Crete. Bull Geol Soc Greece 47:216–225

Moores EM (1969) Petrology and structure of the Vourinos ophiolitic complex of northern Greece. Geol Soc Am Spec Pap 118:1–74

Mountrakis D (1983) Structural Geology of the North Pelagonian Zone s. l. and Geotectonic Evolution of the Internal Hellenides (Macedonia, Greece). Habilitation Scholarship, University of Thessaloniki, Greece [in Greek with English abstract]

Mountrakis D (2006) Tertiary and Quaternary tectonics of Greece. In: Dilek Y, Pavlides S (eds), Postcollisional tectonics and magmatism in the Mediterranean region and Asia. Geol Soc Am Spec Pap 409:125–136

Papazachos B, Mountrakis D, Psilovikos A, Leventakis G (1979) Surface fault traces and fault plane solutions of the May–June 1978 major shocks in the Thessaloniki area, Greece. Tectonophysics 53:171–173

Papazachos B, Vamvakaris DA, Vargemezis GM, Aidona EV (2001) A study of the active tectonics and deformation in the Mygdonia Basin (N. Greece) using seismological and neotectonic data. Bull Geol Soc Greece 34:303–309

Pasadakis N, Dagounaki V, Chamilaki E, Vafidis A, Zelilidis A, Piliotis I, Panagopoulos G, Manoutsoglou EM (2012) Organic geochemical evaluation of Neogene formations in Messara (Heraklion, Crete) basin as source rocks of biogenetic methane. Mineral Wealth 166:8–26

Pavlides S (1985) Neotectonic evolution of the Florina-Vegoritis-Ptolemais basins, PhD thesis, University of Thessaloniki Greece, 256

Pavlides SB, Mountrakis DM (1987) Extensional tectonics of northwestern Macedonia, Greece, since the late Miocene, J Struct Geol 9(4):385–392

Pavlides SB, Kondopoulou DP, Kilias AA, Westphal M (1988) Complex rotational deformations in the Serbo-Macedonian massif (north Greece): structural and palaeomagnetic evidence. Tectonophysics 145:329–335

Pavlides S, Mountrakis D, Kilias A, Tranos M (1990) The role of strike-slip movements in the extensional area of Northern Aegean (Greece): a case of transtensional tectonics. Ann Tecton 4:196–211

Poulimenos G, Zelilidis A, Kontopoulos N, Doutsos T (1993) Geometry of trapezoidal fan deltas and their relationship to extensional faulting along the south-western active margins of the Corinth rift, Greece. Basin Res 5:179–192

Poulos S, Papadopoulos A, Collins MB (1994) Deltaic progradation in Thermaikos bay, northern Greece and its socio-economic implications. Ocean Coast Manag 22:229–247

Psilovikos A (1994) Study - Research of environmental impacts from protected areas constructions at upstream and downstream of Strymonas River at Lake Kerkini at torrents and at the plain area of Serres. 2nd Volume. Res Comm Aristotle University Thessaloniki (in Greek)

Roussos N, Lyssimachou T (1991) Structure of the Central North Aegean Trough: an active strike-slip deformation zone. Basin Res 3:39–48

Schmidt-Kittler N (ed) (1995) The vertebrate locality Maramena (Macedonia, Greece) at the Turolian–Ruscinian boundary (Neogene). Pfeil

Sokoutis D, Brun JP, Van den Driessche J, Pavlides S (1993) A major Oligocene-Miocene detachment in southern Rhodope controlling north Aegean extension. J Geol Soc Lond 150:243–246

Steenbrink J (2001) Orbital signatures in lacustrine sediments -The Late Neogene intramontane Florina-Ptolemais-Servia Basin, northwestern Greece, Faculteit Aardwetenschappen Universiteit Utrecht, Utrecht, 167

Sylvestrou I (2002) Stratigraphic, palaeontological and palaeogeographical study of the Neogene-Quaternary deposits of Katerini basin, Northern Greece. PhD Thesis, Department of Geology, Aristotle Univ of Thessaloniki, Greece

Syrides G (1990) Lithostratigraphic, biostratigraphic and palaeogeographic study of the Neogene - Quaternary sedimentary deposits of Chalkidiki Peninsula, Macedonia, Greece, PhD Thesis, School of Geology, Aristotle Univ of Thessaloniki, Greece

Tranos MD (2009) Faulting of Lemnos Island; a mirror of faulting of the North Aegean Trough (Northern Greece). Tectonophysics 467:72–88

Tsoukala E (volume 2) The fossil record of chalicotheres (Mammalia: Perissodactyla: Chalicotheriidae) in Greece. In: Vlachos E (ed) The fossil vertebrates of Greece vol. 2 – laurasiatherians, artiodactyles, perissodactyles, carnivorans, and island endemics. Springer, Cham

Tsoukala E, Chatzopoulou K (2005) A new Early Pleistocene (latest Villafranchian) site with mammals in Kalamotó (Mygdonia Basin, Macedonia, Greece) – preliminary report. Mitt Komm Quartärforsch Österr Akad Wiss 14:213–233

Vamvaka A, Spiegel C, Frisch W, Danisik M, Kilias A (2010) Fission track data from the Mesohellenic Trough and the Pelagonian zone in NW Greece: cenozoic tectonics and exhumation of source areas. Int Geol Rev 52:223–248

Vasileiadou K, Doukas CS (volume 2) The fossil record of insectivores (Mammalia: Eulipotyphla) in Greece. In: Vlachos E (ed) The fossil Vertebrates of Greece vol. 2 – laurasiatherians, artiodactyles, perissodactyles, carnivorans, and island endemics. Springer, Cham

Vasileiadou K, Sylvestrou I (this volume-a) The fossil record of rodents (Mammalia: Rodentia) in Greece. In: Vlachos E (ed) The fossil vertebrates of Greece vol. 1 – basal vertebrates, amphibians, reptiles, afrotherians, glires, and primates. Springer, Cham

Vasileiadou K, Sylvestrou I (this volume-b) The fossil record of hares, rabbits and pikas (Mammalia: Lagomorpha) in Greece. In: Vlachos E (ed) The fossil vertebrates of Greece vol. 1 – basal vertebrates, amphibians, reptiles, afrotherians, glires, and primates. Springer, Cham

Vlachos E (this volume) The fossil record of turtles and tortoises (Reptilia: Testudines) in Greece. In: Vlachos E (ed) The fossil vertebrates of Greece vol. 1 – basal vertebrates, amphibians, reptiles, afrotherians, glires, and primates. Springer, Cham

Vlachou T, Koufos GD, Gkeme A (volume 2) The fossil record of equids (Mammalia: Perissodactyla: Equidae) in Greece. In: Vlachos E (ed) The fossil vertebrates of Greece vol. 2 – laurasiatherians, artiodactyles, perissodactyles, carnivorans, and island endemics. Springer, Cham

Zelilidis A (2000) Drainage evolution in a rifted basin, Corinth graben, Greece. Geomorphology 35:69–85

Zelilidis A (2003) The geometry of fan-deltas and related turbidites in narrow linear basins. Geol J 38:31–36

Zelilidis A, Kontopoulos N (1994) Pliocene-Pleistocene fluvial/wave dominated deltaic sedimentation: the Pamisos delta in SW Peloponnesus, Greece. Geol Mag 131:653–668

Zelilidis A, Kontopoulos N (1996) Significance of fan deltas without toe-sets within rift and piggy-back basins: examples from the Corinth graben and the Mesohellenic trough, Central Greece. Sedimentology 43:253–262

Zelilidis A, Kontopoulos N (2001) Post-Miocene sedimentary evolution of south Peloponnesus, Greece. Gaia 16:1–12

Zelilidis A, Piper DJW, Kontopoulos N (2002) Sedimentation and basin evolution of the Oligocene -Miocene Mesohellenic basin, Greece. AAPG 86:161–182

Zelilidis A, Tserolas P, Chamilaki E, Pasadakis N, Kostopoulou S, Maravelis AG (2016) Hydrocarbon prospectivity in the Hellenic trench system: organic geochemistry and source rock potential of upper Miocene-lower Pliocene successions in the eastern Crete Island, Greece. Int J Earth Sci 105:1859–1878

Zidianakis G, Iliopoulos G, Zelilidis A, Kovar-Eder J (2015) Myrica from the plant assemblage of Pitsidia (Crete, Late Miocene): putting the puzzle together. Palaeontogr Abt B 293:149–171

The Fossil Record of Conodonts in Greece

Olga Koukousioura and Vasiliki–Grigoria Dimou

1 Introduction

Conodonts (Conodonta Eichenberg, 1930) refer to small marine organisms of an extinct group of fossil animals that lived from the Palaeozoic to early Mesozoic eras (e.g., Spasov and Ganev 1960). Conodonts were introduced and described for the first time by Pander in 1856. His discovery of these microfossil findings of Paleozoic strata of Baltic provinces and Moscow environs was phosphatic teeth-like parts (calcium phosphate or apatitic conodont "elements") and initiated a vigorous debate regarding their biology and affinity for more than 130 years. Hence, along these years, a great number of papers dealt with their morphology, anatomy, distribution, and taxonomy (e.g., Huckriede 1958; Lindström 1964; Kozur 1975; Müller 1981; Murdock et al. 2013). Moreover, conodonts are one of the most important and widespread—though quite enigmatic as it will be explained below—microfossils, which have an excellent fossil record from the Cambrian to Triassic; thus they own biozones (e.g., Walliser 1964; Ziegler and Sandberg 1990).

The conodont fossil record from Greece includes 90 genera, 348 species and 74 subspecies from 80 localities/sections (e.g., Bender et al. 1960; Vrielynck 1978a; Kozur and Krahl 1984; Appendix). The oldest record comes from the Silurian to Carboniferous of Chios Island (e.g., Kauffmann 1965; Herget and Roth 1968) and from the Carboniferous of Attica (Caridroit et al. 2000). The rest of the localities correspond to the Triassic Period (e.g., Flament 1973; Muttoni et al. 2014).

O. Koukousioura (✉) · V. Dimou
Department of Geology, School of Geology, Aristotle University, Thessaloniki, Greece
e-mail: okoukous@geo.auth.gr; dimouvaso@geo.auth.gr

E. Vlachos (ed.), *Fossil Vertebrates of Greece Vol. 1*,
https://doi.org/10.1007/978-3-030-68398-6_3

2 Historical Overview

In 1856, Christian Heinrich Pander (one of the founders of embryology and paleontology in Russia) in his monograph introduced and described conodonts for the first time. Pander regarded these tiny specimens (minute teeth-like remains composed of pure calcium carbonate)—from which he derived the concept of "Conodonten"—as the teeth and/or jaws of an unknown group of fossil fishes. However, he reported that the structure of those teeth showed several and strong differentiations in icthyological character from all known fish, which led him to the impression that these fossils were not fish teeth. Pander also noted the lack of bone findings, indicating that conodonts did not have hard parts; thus he suggested that the fossils could be related to hagfish or lampreys. Because of this, Pander used a conservative way to describe these specimens: each species was interpreted to have teeth of a single morphology. Consequently, the systematic taxonomy of his monograph concluded the establishment of 13 genera with 55 newly named species. Later, Ulrich and Bassler (1926) of the US National Museum accepted the idea of Pander's conodonts and proposed a more extensive classification model of the group. They described many new species from Devonian and Mississippian rocks in the southeastern United States and were the first to recognize their biostratigraphic usefulness. In the early 1930s, conodonts became the main interest and paleontologic curiosities for paleontologists around the world. Among them, Branson and Mehl (1934a, b) and their students (at the University of Missouri) studied the Middle Ordovician to Lower Carboniferous of eastern and central United States under an extensive program of conodont research and produced a considerably expanded taxonomy and knowledge of their nature. Interest in conodonts grew dramatically, and the number of paleontologists interested in this enigmatic group increased greatly, due primarily to the establishment of this major research program. In 1934, Schmidt and Scott discovered groups of individual elements preserved together in the same black shale bedding plain. This important discovery led to the hypothesis that the individual elements were held in pairs (termed an apparatus) when in life, often likened to mouthparts. From the 1960s onward, conodonts have developed into one of the most important biostratigraphic tools available for the study of Palaeozoic and Triassic sequences.

3 Conodont Classification and the Great Fossil Enigma

Based on Pander's doubts about the uncertain origin of conodonts, many scenarios were suggested for their phylogenetic position. The unsolved puzzle of the true nature of conodonts remains, until nowadays, the main subject for more than a hundred papers that deal with a plethora of hypotheses on the function of the conodont and their affinity, or the identity of the organism that bore them.

Isolated conodonts are widespread and abundant. Until the 1980s, their biological affinities were still not known. Conodonts have been variously interpreted as the remains of plants (algae, vascular plants), conulariids, aschelminthes (copulatory structures of nematodes, priapulids, teeth of rotifers, gastrotrichs, kinorhynchs), and gnathostomulids; as molluscan radulae of various kinds, annelid jaws, and arthropod organs (but not since the last century); and as elements of lophophorates, chaetognaths, and chordates such as agnathans (including cyclostomes), selachians, ostracoderms, placoderms, and various kinds of primitive vertebrates (Hass 1962; Müller 1981). In the 1980s, many paleontologists (e.g., Clark 1981; Briggs et al. 1983; Gould 1983) placed conodonts in a separated phylum, Conodonta, because the morphological features of specimens do not resolve the question of conodont function and origin. According to Tillier and Cuif (1986), conodonts were invertebrates that resemble to aplacophoran molluscs, but this speculation was turned down by Briggs et al. (1987), a few years later. Janvier (1981) supported the hypotheses of Tillier and Cuif (1986) only to revise his original thoughts a few years later (e.g., Janvier 1995). Nonetheless, the great controversy lied in the phylogenetic relationship of conodonts with two other groups: chaetognaths and chordates. The chaetognaths that have only recently been the object of detailed investigation (Repetski and Szaniawski 1981; Szaniawski 1982) are small marine worms, dorsoventrally flattened, with horizontal fins, grasping spines, while they are mostly planktic (Hyman 1959). Rietschel (1973) noted the similarity between conodonts and the grasping spines of chaetognaths, suggesting that conodonts might have functioned in the same way.

As regards possibility of a vertebrate, chordate affinity for the conodonts has received more attention than any other (e.g., Schmidt in Schmidt and Müller 1964; Müller 1981). Sansom et al. (1992) and subsequent papers supported that conodont hard tissues are homologous to vertebrate tissues, since the discoveries of two findings of nearly complete animals. These results and the current knowledge of the anatomy of conodonts derived from two enlightening fossil findings, based on the form and the features of the soft-tissue remains. The first refers to a well-preserved new specimen from the Upper Ordovician Soom Shale of South Africa. This giant conodont *Promissum pulchrum* reveals details of the trunk musculature, feeding apparatus, and eyes. The preserved length is 109 mm but in its entire length reaches approximately 400 mm (Gabbott et al. 1995). The second particularly important discovery by Briggs et al. in 1983 comes from the Lower Carboniferous Granton Shrimp Bed of Edinburgh and was the species *Clydagnathus windsorensis*. This specimen refers to a wormlike, elongate (40.5 mm long and mostly less than 1.95 mm wide) organism that was most likely flattened in life. The preserved region of the head includes the conodont apparatus and the anterior one-third of the trunk. In the tail region, the presence of two distinct groups of fin rays on one side of the trunk suggests that two caudal fins may have been present (Briggs et al. 1983). Several specimens have been recovered from the same area, representing more or less complete conodont organisms, while the careful study and description of them came by Aldridge et al. (1986, 1993) and Aldridge (1987).

Although Aldridge et al. (1986) and Aldridge (1987) concluded that conodonts are related to myxinoids because of the laterally compressed body, large eyes, bilaterally feeding apparatus, myomeres, notochord, and a tail fin, there are still controversies about their phylogenetic position within the vertebrates. The presence of a notochord is a plesiomorphic chordate character shared with the protochordates (Urochordata and Cephalochordata) and the craniates, whereas the chevron-shaped muscle blocks are normally regarded as limited to the cephalochordates and craniates (e.g., Maisey 1986). Several authors related conodonts to craniates (e.g., Janvier 1983; Dzik 1986; Smith and Hall 1990). Janvier (1996) was the first who conducted computer-based parsimony analysis resulting in the affinity of conodonts with vertebrate groups, as a sister group of the lampreys. Later, Donoghue et al. (2000) reapplied a phylogenetic analysis where conodonts were placed between cyclostomes and other vertebrates, more derived than both hagfishes and lampreys, due to their calcified dermal skeleton. The position of conodonts as extinct chordates, more derived than living jawless vertebrates, but basal to those with jaws, has been established by cladistic analysis and has been summarized by Sweet and Donoghue (2001).

4 Distribution

The fossil record of conodonts is a well-studied subject in Greece already from the early 1960s. The study areas referred to localities across different regions of the central, south, as well as islands of the Hellenic territory (Fig. 1), where Paleozoic to lower Mesozoic (Triassic) sequences occur. From Central Greece, nine localities have been studied, including Pyli Trikala, Mt. East Koziakas, Mt. Othrys, Mt. Iti (Latsinies), Mt. Vardoussia, Kotsilieri, Galaxidi (Koutsouros–Kokkinovrakhos), Psilovrakhos–Karpenisi–Fragista, and Nafpaktos. Conodonts have been recorded also from Beotia (Mt. Helicon), Evia (Xirovouni–Seta), and Attica (Mt. Parnis). Three main areas of Peloponnesus were studied, Achaia (Priolithos–Drimos and Klitoria), Messinia (Mathia–Achladochori and Mt. Ithome), and Argolis (Asklepieion of Epidaurus, Theokafta, Adhami, Alogomandra, Maurovouni–Prosimi–Stefanion, Profitis Ilias, Moni Taxiarchis–Tsoukalia, Trapezona, Midhea, Karafotia–Trahla, Iliokastro–Kastro Hill). Concerning the islands, Hydra has been the subject of study of many researchers in numerous outcrops (Zogeika, Agios Nikolaos Monastery, Agia Triada, Mandraki–Hydra Chora, Malies, Klimaki, Pirghos, Agios Taxiarchis, Kaminia, Vlichos, Mt. Eros, Palamidas, Agia Marina, Episkopi, Tsigkri, Bisti, and also the adjacent small Kivotos, Petassi, and Pontikos Islands). Furthermore, one of the most studied areas in Greece is Chios Island, with more than 20 investigated localities (Kambia, Keramos, Kourounia–Nenitouria, Melanios–Agio Galas, Parparia, Potamia, Amani, Kipouries, Volisos, Katavasi, Kardamyla, Metochi–Megali Rachi, Anavatos, Vrontados, Rema Armenis, Latomi, Agios Markos, Kephalovouni, Marathovouno, Marmarotrapeza, Agia Anna,

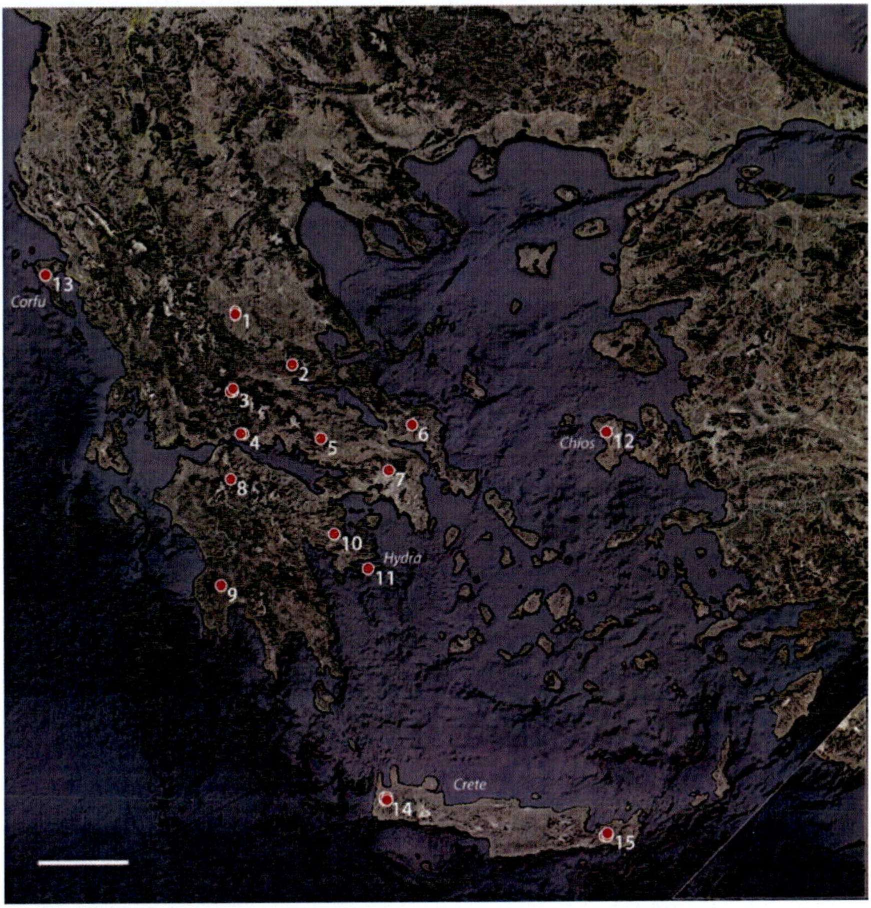

Fig. 1 Map with all the localities, where conodonts have been described: Central Greece: **1**, Pyli Trikala, Mt. Koziakas; **2**, Mt. Othrys; **3**, Mt. Iti, Mt. Vardoussia, Kotsilieri, Psilovrakhos–Karpenisi–Fragista; **4**, Galaxidi, Nafpaktos; **5**, Beotia; **6**, Evia; **7**, Attica. Peloponnesus: **8**, Achaia; **9**, Messinia; **10**, Argolis; **11**, Hydra Island; **12**, Chios Island; **13**, Corfu Island. Crete Island: **14**, Voutas, Sfinari, Kambos, Paleochora, Skafi; **15**, Myrsini, Skopi, Tripokefala, Ziros. See Appendix for more information. Image exported from Google Earth Pro © 2019, map data from US Dept. of State Geographer, SIO, NOAA, U.S. Navy, NGA, GEBCO, image from Landsat/Copernicus. Scale bar equals 80 km, North faces upward

Parthenis, Korakaris, and Agios Georgios Sykousis). Only one species is reported from Corfu Island (Foustapidima Cape), while western and east Crete Island has been investigated as well (Voutas, Sfinari, Kambos, Paleochora, Skafi, Myrsini, Skopi, Tripokefala, Ziros). Further details on all localities are provided in the Appendix.

5 Systematic Paleontology

Phylum Chordata Bateson, 1886
Subphylum Vertebrata Linnaeus, 1758
Class Conodonta Eichenberg, 1930

Remarks 348 species and 74 subspecies belonging to 90 conodont genera have been reported from Greece (Fig. 2). For space reasons, the list is given at generic

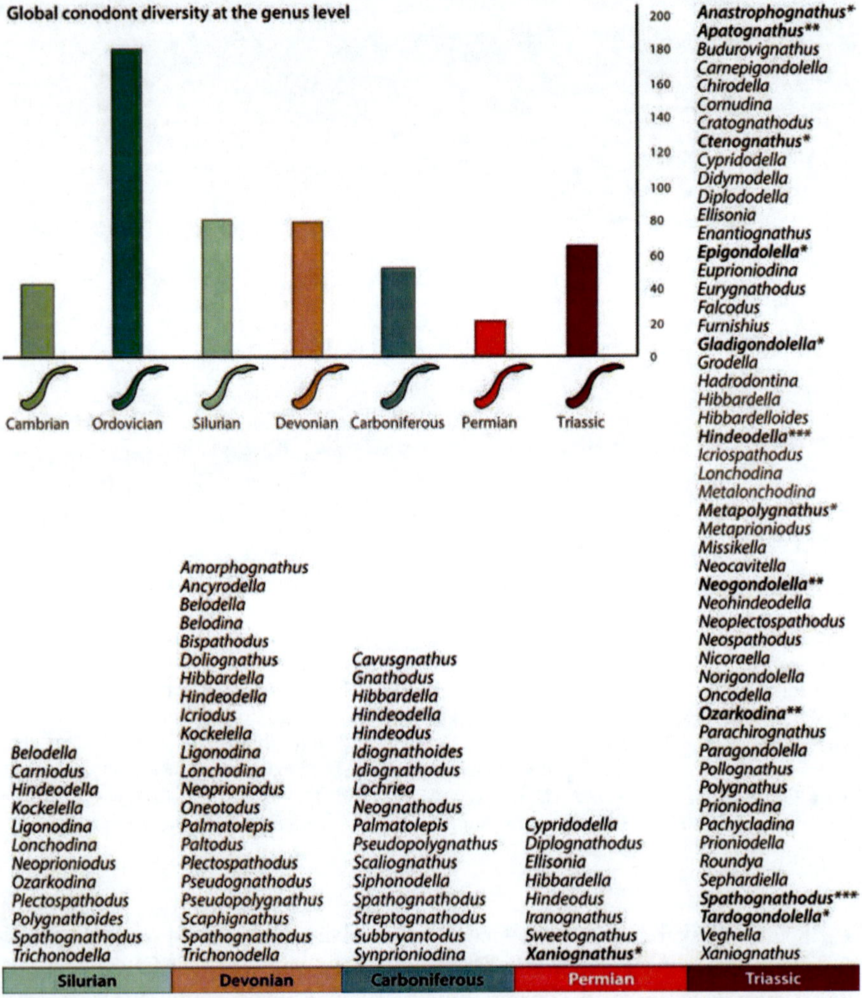

Fig. 2 The temporal distribution of conodont genera in Greece, compared to the global diversity (Data from the Paleobiology Database, accessed June 22, 2020). Genera marked with boldface contain species whose type localities are in Greece; the number of asterisks indicates the number of new taxa from Greece. Conodont silhouettes from phylopic.org and attributed to J. Headen

level and in alphabetical order. Identified species are given without authorships in the main text, with additional information for the taxa described from Greece; full authorships and their references are given in the Appendix.

Amorphognathus Branson and Mehl, 1933

Type Species *Amorphognathus ordovicica* Branson and Mehl, 1933.

Identified Species in Greece *Amorphognathus* sp.

Distribution In the Hellenic material, *Amorphognathus* sp. has been found only in the Upper Devonian of Kipouries and Volisos (Chios Island) by Herget and Roth (1968) and Roth (1968).

★*Anastrophognathus* Bender, 1968b

Type Species ★*Anastrophognathus sagittalis* Bender, 1968b.

Identified Species in Greece ★*Anastrophognathus sagittalis.*

Distribution *Anastrophognathus sagittalis* that has been described for the first time from Greece presents a wide distribution in the Hellenic peninsula, as it has been found in the late Anisian of Central Greece (Psilovrakhos; Kauffmann in Fleury 1980), the Lower to Middle Triassic of Argolis Peloponnesus (Asklipieion Epidaurus; Vrielync 1978a, b, 1980), Hydra Island (Agios Nikolaos Monastery; Dürkoop et al. 1986), Chios Island (Marathovouno; Bender 1968a, b; Tietze 1969), and Crete Island (Tripokefala and Voutas; Krahl et al. 1983, 1986).

Remarks The genus *Anastrophognathus* has been described for the first time from the Hellenic material by Bender in 1968b, with the type species *Anastrophognathus sagittalis* from the lower Anisian limestones of Marathovouno of Chios Island.

Ancoradella Walliser, 1964

Type Species *Ancoradella ploeckensis* Walliser, 1964.

Identified Species in Greece *Ancoradella ploeckensis.*

Distribution Concerning the Greek material, *Ancoradella ploeckensis* has been described only from the Silurian of Kardamyla (Chios Island) (Kauffmann 1965).

Ancyrodella Ulrich and Bassler, 1926

Type Species *Ancyrodella nodosa* Ulrich and Bassler, 1926.

Identified Species in Greece *Ancyrodella* sp.

Distribution In the Hellenic deposits, *Ancyrodella* has been found only in the Upper Devonian of Melanios–Agio Galas, Parparia, and Kipouries in Chios Island (Herget and Roth 1968; Roth 1968).

Apatognathus **Branson and Mehl, 1934a**

Type Species *Apatognathus varians* Branson and Mehl, 1934a.

Identified Species in Greece ★*Apatognathus mitzopouli*, ★*Apatognathus radiatus*, *Apatognathus ziegleri*, *Apatognathus* sp.

Distribution Representatives of this genus have been described for the first time in Greece and are distributed in the Olenekian to early Anisian of Marathovouno of Chios Island (Bender 1968b), with two species (*A. mitzopouli* and *A. radiatus*), whereas all *Apatognathus* species have been reported from other localities in Chios island as well (Parthenis, Agia Anna, Kephalovouni, Agios Markos, Rema Armenis, and Anavatos) (Bender 1968a, b; Tietze 1969; Tietze in Jacobshagen and Tietze 1974). *A. mitzopouli* has been also found in the late Anisian to Norian of Mt. Vardoussia (Ardaens 1978), while *A. ziegleri* is more widespread and has been found in the Carnian of Theokafta (Argolis) (Bender et al. 1960; Bender 1968b), the Ladinian of Asklipieion (Argolis) and the late Anisian to Carnian of Mandraki–Hydra Chora and Petassi Island (Hydra) (Huckriede 1958; Römermann 1968).

Remarks The species *Apatognathus mitzopouli* and *Apatognathus radiatus* have been identified for the first time from the Olenekian to late Anisian of Marathovouno (Chios Island), by Bender (1968b).

Belodella **Ethington, 1959**

Type Species *Belodus devonicus* Stauffer, 1940.

Identified Species in Greece *Belodella triangularis*, *Belodella* sp.

Distribution In the Greek material, *Belodella triangularis* has been found only in the Upper Devonian to Carboniferous of Kipouries while *Belodella* cf. *triangularis* in the Middle Devonian of Parthenis and *Belodella* sp. in the Silurian to Devonian of Agia Anna (Chios Island) (Roth 1968; Walisser in Tietze 1969).

Belodina **Ethington, 1959**

Type Species *Belodus compressus* Branson and Mehl, 1933.

Identified Species in Greece *Belodina triangularis.*

Distribution In the Greek material, *Belodina triangularis* has been reported only from the Devonian of Potamia and Kipouries in Chios Island (Herget and Roth 1968; Roth 1968).

Bispathodus **Müller, 1962**

Type Species *Spathodus spinulicostatus* Branson, 1934.

Identified Species in Greece *Bispathodus stabilis.*

Distribution Representatives of *Bispathodus stabilis* in the Hellenic material have been found only in the late Famennian (Upper Devonian) of Kardamyla (Chios Island) (Larghi et al. 2005).

Budurovignathus **Kozur, 1989**

Type Species *Polygnathus mungoensis* Diebel, 1956.

Identified Species in Greece *Budurovignathus hungaricus*, *Budurovignathus mungoensis.*

Distribution The two representatives of the genus *Budurovignathus* have been reported in the Hellenic peninsula, from the late Anisian to early Ladinian of Agia Triada of Hydra Island by Muttoni et al. (1994, 1997).

Carnepigondolella **Kozur, 2003**

Type Species *Metapolygnathus zoae* Orchard, 1991.

Identified Species in Greece *Carnepigondolella gulloae*, *Carnepigondolella nodosa.*

Distribution The two representatives of the genus of *Carnepigondolella* have been found in the Carnian of Agia Marina (Hydra Island) (Muttoni et al. 2014), while *C. nodosa* has been also reported from the late Carnian to early Norian of Stefanion (Argolis) (Noyan and Kozur 2007).

Carniodus **Walliser, 1964**

Type Species *Carniodus carnulus* Walliser, 1964.

Identified Species in Greece *Carniodus carinthiacus*, *Carniodus* cf. *carnulus.*

Distribution The two representatives of the genus *Carniodus* have been described from the Hellenic territory, in the Silurian of Kardamyla (Chios Island) by Kauffmann in 1965.

***Cavusgnathus* Harris and Hollingsworth, 1933**

Type Species *Cavusgnathus alta* Harris and Hollingsworth, 1933.

Identified Species in Greece *Cavusgnathus* sp.

Distribution The genus of *Cavusgnathus* in Greece has been described only from the late Carboniferous of Sfinari (West Crete Island) (Krahl et al. 1983).

***Chirodella* Hirschmann, 1959**

Type Species *Chirodella triquetra* (Tatge, 1956).

Identified Species in Greece *Chirodella dinodoides*, *Chirodella gracilis*.

Distribution *C. dinodoides* has a wide distribution in the Hellenic material characterizing the Middle to Late Triassic of Central Greece in Kotsilieri, Galaxidi, and Mt. Vardoussia, along with *C. gracilis* (Ardaens 1978), Nafpaktos and Psilovrakhos–Karpenisi–Fragista (Vrielynck in Fleury 1980), and Mt. Iti in Latsinies (Wigniole 1977), while it has also been reported from Mt. Helicon (Beotia) (Steuber 1991), Peloponnesus Achaia (Drimos) (Flament 1973), and Argolis (Asklipieion Epidaurus, Mavrovouni–Prosimi, Trapezona, Karafotia–Trahla, Profitis Ilias) (Vrielynck 1978a, 1980).

***Cornudina* Hirschmann, 1959**

Type Species *Ozarkodina breviramulis* Tatge, 1956.

Identified Species in Greece *Cornudina breviramulis*, *Cornudina breviramulis minor*, *Cornudina tortilis*, *Cornudina oezdemirae*, *Cornudina* sp.

Distribution *C. breviramulis minor* has been found in Carnian–Norian of Midhea and Karafotia–Trahla and Mavrovouni–Prosimi (Argolis) (Vrielynck 1978a, 1980), while in the Anisian–Ladinian of Mt. Helicon (Beotia) (Steuber 1991) *C. breviramulis* and *C. torilis* have been reported. Furthermore, *C. torilis* has also been found in the Anisian to Ladinian of Asklipieion (Argolis) (Vrielynck 1978a, b), while Ardaens (1978) has described *C. tortilis* from the Anisian to Norian of Mt. Vardoussia and *C. breviramulis* and *C. oezdemirae* from Galaxidi.

***Cratognathodus* Mosher, 1968**

Type Species *Prioniodina kochi* Huckriede, 1958.

Identified Species in Greece *Cratognathodus kochi*, *Cratognathodus posterognathus angulatus*, *Cratognathodus posterognathus posterognathus*.

Distribution The subspecies *C. posterognathus angulatus* and *C. posterognathus posterognathus* have been found in the Anisian to Ladinian of Mt. Helicon (Beotia) (Steuber 1991). Moreover, *C. posterognathus* and *C. kochi* have been described from the Triassic of the Adhami (Argolis) (Dürkoop et al. 1986), Karpenisi–Fragista, Mt. Vardoussia (Central Greece; Ardaens 1978; Vrielynck in Fleury 1980), and Priolithos–Drimos (Flament 1973). Additionally, *C. kochi* has been found in several late Anisian to early Ladinian localities of Hydra Island (Zogeika, Agios Nikolaos Monastery, Malies, Tsigkri, Bisti, Agia Triada, Petassi Island) and in some cases along with *C. posterognathus* (Dürkoop et al. 1986; Angiolini et al. 1992; Muttoni et al. 1994).

***Ctenognathus* Pander, 1856**

Type Species *Ctenognathus murchisoni* Pander, 1856.

Identified Species in Greece ★*Ctenognathus chionensis.*

Distribution *C. chionensis* has been described from Greece and in the Hellenic material has been found only in the Olenekian limestones of Marathovouno in Chios Island by Bender in 1968b.

***Cypridodella* Mosher, 1968**

Type Species *Cypridodella conflexa* Mosher, 1968.

Identified Species in Greece *Cypridodella conflexa, Cypridodella delicatula, Cypridodella mediocris, Cypridodella muelleri, Cypridodella* sp.

Distribution The representatives of the genus *Cypridodella* refer to the Carnian to Norian of Drimos (Achaia Peloponnesus) (Flament 1973), although Krahl et al. (1986) have reported *Cypridodella* sp. from the early to middle Permian of Myrsini, East Crete Island.

***Didymodella* Mosher, 1969**

Type Species *Dichodella alternata* (Mosher, 1968).

Identified Species in Greece *Didymodella alternata, Didymodella* sp.

Distribution *D. alternata* presents wide distribution in the Hellenic material, as it has been described from the Anisian to Ladinian of Mt. Helicon (Steuber 1991) and the Anisian to Ladinian of Asklipieion Epidaurus and Adhami (Argolis) (Dürkoop et al. 1986; Vrielynck 1978a, 1980). Also, Ardaens (1978) has reported *Didymodella* cf. *alternata* and *Didymodella* sp. from the late Anisian to Norian of Mt. Vardoussia.

***Diplododella* Ulrich and Bassler, 1926**

Type Species *Diplododella bilateralis* Ulrich and Bassler, 1926.

Identified Species in Greece *Diplododella bidentata*, *Diplododella magnidentata*, *Diplododella meissneri*, *Diplododella triassica*, *Diplododella thuringensis*.

Distribution The widely distributed species *Diplododella bidentata* has been found in the Middle to Late Triassic of Mt. Helicon (Beotia) by Steuber (1991), and Kotsilieri by Ardaens in 1978, while he has also described *D. meissneri* and *D. thuringensis* from Mt. Vardoussia. Furthermore, *D. bidentata* has been reported from the late Carnian to early Norian of Priolithos (Peloponnesus) along with *D. magnidentata* by Flament (1973), from the late Anisian to Ladinian of Asklipieion Epidaurus and Adhami (Argolis) by Dürkoop et al. (1986), and from the Carnian to Norian of Karafotia–Trahla along with *D. meissneri* (Vrielynck 1978a, 1980). Finally, Krahl et al. (1986) have found it in the Olenekian to Anisian of Skopi (East Crete Island).

***Diplognathodus* Kozur and Merrill, 1975**

Type Species *Spathognathodus coloradoensis* Murray and Chronic, 1965.

Identified Species in Greece *Diplognathodus* sp.

Distribution Concerning the Greek material, *Diplognathodus* sp. has been identified only from the early to middle Permian of Myrsini (East Crete) by Krahl et al. (1986).

***Doliognathus* Branson and Mehl, 1941**

Type Species *Doliognathus tata* Branson and Mehl, 1941.

Identified Species in Greece *Doliognathus* sp.

Distribution Herget and Roth (1968) have described the genus of *Doliognathus* in the Hellenic peninsula only from the Upper Devonian deposits of Kambia (Chios).

***Ellisonia* Müller, 1956**

Type Species *Ellisonia triassica* Müller, 1956.

Identified Species in Greece *Ellisonia* sp.

Distribution Although the genus *Ellisonia* has been identified only as *Ellisonia* sp., it has been found in several localities and detected age of the deposits of two islands, as the Wuchiapingian of Agios Taxiarchis (Hydra Island; Nestell and Wardlaw 1987) and the Lower Triassic of Kambos and Paleochora, the Induan of

Myrsini, and the Olenekian to Anisian of Skopi and Tripokefala of Crete Island by Krahl et al. (1983).

Enantiognathus Mosher and Clark, 1965

Type Species *Apatognathus inversus* Sannemann, 1955.

Identified Species in Greece *Enantiognathus bitortus*, *Enantiognathus mitzopouli*, *Enantiognathus petrae-viridis*, *Enantiognathus ziegleri*.

Distribution Dürkoop et al. (1986) reported the four representatives of the genus *Enantiognathus* from the Anisian to Ladinian of Adhami (Argolis). Furthermore, *E. bitortus* has been also described from the Olenekian to Anisian of the Skopi, Tripokefala (West Crete Island) (Krahl et al. 1986). *E. ziegleri* has been reported from Asklipieion Epidaurus, Karafotia–Trahla (Argolis) (Vrielynck 1978a, 1980), and the Carnian to Norian of Priolithos–Drimos (Achaia) (Flament 1973) in Peloponnesus, whereas it has been found in the Late Triassic of several localities in Central Greece (Ardaens 1978; Vrielynck in Fleury 1980). It has been also reported from the late Anisian to Ladinian of Agia Triada, Agia Marina, Pirghos, and Episkopi of Hydra Island (Muttoni et al. 1994; Angiolini et al. 1992) and the Olenekian to Anisian of Chios Island (Katavasi and Metochi–Megali Rachi, Roth 1968; Anavatos, Rema Armenis, Agia Anna, Parthenis and Korakaris, Tietze 1969). *E. petrae-viridis* has also a quite wide distribution as it has been found in Asklipieion Epidaurus (Argolis) by Vrielynck (1978a, 1980), the Ladinian of Voutas (Crete Island) (Kozur and Krahl 1984), and the Anisian to Norian of Central Greece (Mt. Vardoussia and Galaxidi, Ardaens 1978; Karpenisi–Fragista, Vrielynck in Fleury 1980), while *Enantiognathus* sp. has been only described from Myrsini East Crete Island (Krahl et al. 1986).

Epigondolella Mosher, 1968

Type Species *Polygnathus abneptis* Huckriede, 1958.

Identified Species in Greece *Epigondolella abneptis*, *Epigondolella bidentata*, *Epigondolella diebeli*, *Epigondolella echinata*, *Epigondolella hungarica*, *Epigondolella multidentata*, *Epigondolella mungoensis*, *Epigondolella nodosa*, *Epigondolella permica*, *Epigondolella postera*, *Epigondolella primitia*, *Epigondolella pseudodiebeli*, *Epigondolella quadrata*, ★*Epigondolella rigoi*, ★*Epigondolella stefanionensis*, *Epigondolella ziegleri*, *Epigondolella* sp.

Distribution The genus of *Epigondolella* is one of the most common genera, with 17 species, found in the Hellenic material, including two of them described from Greece. They refer mainly in the Argolis area: Upper Triassic of Stefanion (Noyan and Kozur 2007; Vrielynck 1978a, 1980), Asklipieion Epidaurus (Krystyn and Mariolakos 1975; Vrielynck 1978a, b, 1980), Moni Taxiarchis–Tsoukalia (Baumgartner 1985), Trapezona, Karafotia–Trahla, and Midhea (Vrielynck 1978a, b, 1980), whereas it has also been described from Priolithos–Drimos and Klitoria of Achaia Peloponnesus (Flament 1973; Vrielynck in Fleury 1980). The second important area is Central Greece with seven species reported from the late Triassic of Nafpaktos and Psilovrakhos–Karpenisi–Fragista (Vrielynck in Fleury 1980), Mt. Iti (Wigniole 1977), and Mt. Vardoussia, Kotsilieri, and Galaxidi (Ardaens 1978). They have been also described from Pyli Trikala (Lekkas 1986), Mt. Othrys (Ferriere 1974), and East Mt. Koziakas (Ardaens 1978), while it has been also reported from Agia Marina of Hydra Island (Muttoni et al. 2014).

Remarks Two species, *Epigondolella rigoi* and *Epigondolella stefanionensis*, have been described and named from Argolis Peloponnesus by Noyan and Kozur in 2007. The type locality of *Epigondolella rigoi* Kozur, 2007, is a cherty limestone in Stefanion section, and it corresponds to upper *E. rigoi* zone. *Epigondolella stefanionensis* Noyan 2007 has been identified in a section, consisting of a platy cherty limestone, NE of Stefanion, south of Mt. Rakhi Stefaniou, and it corresponds to the upper part of the late Carnian.

***Euprioniodina* Ulrich and Bassler, 1926**

Type Species *Euprioniodina deflecta* Ulrich and Bassler, 1926.

Identified Species in Greece *Euprioniodina mediocris*, *Euprioniodina muelleri*, *Euprioniodina multihamata*, *Euprioniodina pectiniformis*, *Euprioniodina petraeviridis*, *Euprioniodina pronoides*.

Distribution Dürkoop et al. (1986) have been the only ones who have described the genus of *Euprioniodina* in the Greek peninsula. They reported six species from the late Anisian to Ladinian of Adhami of Argolis area in Peloponnesus.

***Eurygnathodus* Staesche, 1964**

Type Species *Eurygnathodus costatus* Staesche, 1964.

Identified Species in Greece *Eurygnathodus costatus*, *Eurygnathodus paracostatus*.

Distribution Krahl et al. (1986) have found the two species of the genus *Eurygnathodus* in the Greek material in the Olenekian of Myrsini (East Crete Island), while Tietze (1969) reported *E. costatus* from the Olenekian to Ladinian of Anavatos of Chios Island.

***Falcodus* Huddle, 1934**

Type Species *Falcodus angulus* Huddle, 1934.

Identified Species in Greece ?*Falcodus* sp.

Distribution Bender et al. (1960) have described the species ?*Falcodus* sp. only from the Carnian of Theokafta (Argolis), in the Hellenic peninsula.

***Furnishius* Clark, 1959**

Type Species *Furnishius trisseratus* Clark, 1959.

Identified Species in Greece *Furnishius triserratus.*

Distribution *Furnishius triserratus* has been described in the Greek deposits only from the Lower Triassic of Kambos of West Crete Island (Krahl et al. 1983).

***Gladigondolella* Müller, 1962**

Type Species *Gladigondolella tethydis* (Huckriede, 1958).

Identified Species in Greece ★*Gladigondolella carinata*, *Gladigondolella malayensis budurovi*, *Gladigondolella malayensis malayensis*, *Gladigondolella tethydis*, *Gladigondolella triangularis*, *Gladigondolella* sp.

Distribution The genus *Gladigondolella* has been found in the Hellenic material with five species and three subspecies, including *G. carinata* described for the first time from the lower Anisian limestones of Marathovouno of Chios Island by Bender in 1968b. They refer to five main areas. In the area of Argolis Peloponnesus, three species (*G. carinata, G. malayensis*, and *G. tethydis*) have been reported from the Middle to Upper Triassic of the Theokafta, Profitis Ilias, and Asklipieion Epidaurus (Bender 1968b; Krystyn and Mariolakos 1975; Vrielynck 1978a, 1980; Baumgartner 1985; Dürkoop et al. 1986). In Crete Island the same *Gladigondolella* species are described in Skopi, Sfinari, Voutas, and Tripokefala (Krahl et al. 1983, 1986; Kozur and Krahl 1984). Four species (*G. carinata, G. malayensis budurovi, G. tethydis, G. triangularis*) have also been found in the Early to Middle Triassic of Katavasi, Metochi–Megali Rachi, Anavatos, Rema Armenis, Kephalovouni, Marathovouno, Parthenis Korakaris, and Marmarotrapeza of Chios Island (Roth 1968; Tietze 1969; Bender 1968b; Assereto et al. 1980; Tietze in Jacobshagen and Tietze 1974; Gaetani et al. 1992; Jacobshagen et al. 1993). From Hydra Island, two species (*G. malayensis malayensis* and *Gladigondolella* sp.), have been reported from the Anisian to Carnian of Agia Marina (Muttoni et al. 1994, 2014); and *G. tethydis* from Episkopi (along with *G. carinata*) and Agia Triada Agios Nikolaos Monastery, Malies, Pirghos, Vlichos, Eros, Tsigkri, Bisti, and Petassi Island (Römermann 1968; Dürkoop et al. 1986; Angiolini et al. 1992; Muttoni et al. 1994). Kauffmann in Fleury (1980), Steuber (1991), and Ardaens (1978) have found *G. tethydis* in the Olenekian to Norian of Psilovrakhos and the Anisian to Ladinian of Mt. Helicon, Mt. Vardoussia, and Galaxidi (Central Greece), while De Bono et al. (2001) described *G. tethydis* and *G. malayensis* in the Carnian of Evia.

***Gnathodus* Pander, 1856**

Type Species *Gnathodus mosquensis* Pander, 1856.

Identified Species in Greece *Gnathodus angustus*, *Gnathodus bilineatus bilineatus*, *Gnathodus commutatus commutatus*, *Gnathodus commutatus homopunctatus*, *Gnathodus commutatus nodosus*, *Gnathodus cuneiformis*, *Gnathodus delicatus*, *Gnathodus girtyi*, *Gnathodus noduliferus*, *Gnathodus* cf. *opimus*, *Gnathodus praebilineatus*, *Gnathodus punctatus*, *Gnathodus* cf. *roundyi*, *Gnathodus semiglaber*, *Gnathodus* n. sp. aff. *sicilianus*, *Gnathodus texanus*, *Gnathodus wapanuckensis*, *Gnathodus* sp.

Distribution The genus *Gnathodus* has been mainly reported from the Carboniferous of Chios Island and specifically from Keramos, Kourounia–Nenitouria, Melanios–Agio Galas, Parparia, Potamia, Kipouries, Volisos, and Agios Georgios Sykousis with 13 species and 4 subspecies (Herget and Roth 1968; Roth 1968; Tietze 1969; Groves et al. 2003; Zanchi et al. 2003). Furthermore, Krahl et al. (1983) have described one species, *G. angustus*, in the late Carboniferous of Sfinari (West Crete Island).

***Gondolella* Stauffer and Plummer, 1932**

Type Species *Gondolella elegantula* Stauffer and Plummer, 1932.

Identified Species in Greece *Gondolella aegaea*, *Gondolella arcuata*, *Gondolella asiatica*, *Gondolella bakalovi*, *Gondolella basisymmetrica*, *Gondolella b. bifurcata*, *Gondolella b. hanbulogi*, *Gondolella bulgarica*, *Gondolella constricta*, *Gondolella cornuta*, *Gondolella denuda*, *Gondolella eotrammeri*, *Gondolella excelsa*, *Gondolella excentrica*, *Gondolella f. foliata*, *Gondolella f. inclinata*, *Gondolella f. pseudobifurcata*, *Gondolella f. fueloepi*, *Gondolella gujioensis*, *Gondolella idahoensis*, *Gondolella intermedia*, *Gondolella jubata*, *Gondolella laevis*, *Gondolella liebermani*, *Gondolella mombergensis*, *Gondolella navicula*, *Gondolella navicula hallstattensis*, *Gondolella nepalensis*, *Gondolella orientalis*, *Gondolella planata*, *Gondolella polygnathiformis*, *Gondolella regale*, *Gondolella subcarinata*, *Gondolella* aff. *szaboi*, *Gondolella tadpole*, *Gondolella tethydis*, *Gondolella timorensis*, *Gondolella trammeri*, *Gondolella* sp. ex aff. *Gondolella auriformis*, *Gondolella* sp.

Distribution The genus *Gondolella* has been described with many species and subspecies from several localities of the Greek peninsula. The most complete section from Upper Carboniferous to Upper Triassic, referred to Crete Island and specifically in Sfinari and Voutas (West Crete) but also in Myrsini, Tripokefala, and Skopi (East Crete) (Krahl et al. 1983, 1986; Kozur and Krahl 1984). Most of the species were catalogued from several localities of Hydra Island (Muttoni et al. 1994, 1997; Angiolini et al. 1992; Dürkoop et al. 1986; Huckriede 1958; Bender and Kockel 1963). Argolis is the third important area where several species were found in the Triassic sediments of Asklipieion Epidaurus, Adhami, Iliokastro/Kastro Hill, Alogomandra, and Theokafta (Mauvier in Decourt 1964; Baumgartner 1985; Dürkoop et al. 1986; Krystyn and Mariolakos 1975; Bender et al. 1960; Huckriede 1958), while they have been found also in the Middle to Late Triassic of Messinia in Mathia–Achladochori and Mt. Ithome (Kozur in Thiebault 1982; Terry 1969). Representatives of this genus have also been reported from the Anisian to Ladinian of Mt. Helicon of Beotia (Steuber 1991; Clement 1977); the Olenekian to late Anisian of Mt. Parnis in Attica (Bender and Kockel 1963), Marathovouno, and Marmarotrapeza of Chios Island (Muttoni et al. 1995; Bender and Kockel 1963; Gaetani et al. 1992; Jacobshagen et al. 1993); and the Anisian to Norian of Mt. Othrys, Mt. Vardoussia, and Galaxidi (Ferriere 1974; Ardaens 1978) of Central Greece.

***Grodella* Kozur and Mostler, 1970**

Type Species *Grodella delicatula* (Mosher, 1968).

Identified Species in Greece *Grodella delicatula.*

Remarks In the Hellenic material, only the species *Grodella delicatula* has been found in the Anisian to Norian of Profitis Ilias (Peloponnesus) (Vrielynck 1978a).

***Hadrodontina* Staesche, 1964**

Type Species *Hadrodontina anceps* Staesche, 1964.

Identified Species in Greece *Hadrodontina anceps*, *Hadrodontina* sp.

Distribution Krahl et al. (1983) have found the species *Hadrodontina anceps* in the Greek material only in the Lower Triassic of Voutas (West Crete Island), while Tietze (1969) reported *Hadrodontina* sp. from the Induan to Ladinian of Marathovouno and the Carboniferous of Agios Georgios Sykousis of Chios Island.

***Hibbardella* Ulrich and Bassler, 1926**

Type Species *Prioniodus angulatus* Hinde, 1879.

Identified Species in Greece *Hibbardella lautissima*, *Hibbardella magnidentata*, *Hibbardella* cf. *nevadensis*, *Hibbardella triassica*, *Hibbardella zapfei*, *Hibbardella* sp.

Distribution In the Hellenic peninsula, the genus *Hibbardella* presents six representatives. *H. lautissima* has been reported from the Anisian to Ladinian of Mt. Helicon in Beotia (Steuber 1991) and from Adhami of Argolis area (Dürkoop et al. 1986). The most common species found is *H. magnidentata* which has been reported from the Middle to Late Triassic of Mt. Helicon of Beotia (Steuber 1991), from

Karafotia–Trahla and Stefanion of Argolis area (Vrielynck 1978a, 1980), and finally from Nafpaktos (Vrielynck in Fleury 1980). Bender (1968b) and Tietze (1969) have reported *H. triassica* from Marathovouno and Parthenis of Chios Island and *Hibardella* sp. from Metochi–Megali Rachi and Rema Armenis, the Middle Devonian of Parthenis, and the Carboniferous of Agios Georgios Sykousis (Roth 1968; Tietze 1969). Finally, Flament (1973) has found *H. zapfei* only in the Carnian to Norian of Drimos of Achaia Peloponnesus.

***Hibbardelloides* Kozur and Mostler, 1970**

Type Species *Hibbardelloides acroforme* (Mosher and Clark 1965).

Identified Species in Greece *Hibbardelloides acroforme.*

Distribution Steuber, in 1991, catalogued in the Greek material *H. acroforme* from the Anisian to Ladinian of Mt. Helicon in Beotia.

***Hindeodella* Ulrich and Bassler, 1926**

Subgenus *Hindeodella* (*Metaprioniodus*) Huddle, 1934

Type Species *Hindeodella subtilis* Ulrich and Bassler, 1926.

Identified Species in Greece *Hindeodella andrusovi*, ★*Hindeodella bitorta*, *Hindeodella bogschi*, ★*Hindeodella ceweki*, *Hindeodella equidentata*, *Hindeodella* (*Metaprioniodus*) *andrusovi*, *Hindeodella* (*Metaprioniodus*) *andrusovi andrusovi*, *Hindeodella* (*Metaprioniodus*) *andrusovi koeveskalensis*, *Hindeodella* (*Metaprioniodus*) *bicuspidata*, *Hindeodella* (*Metaprioniodus*) *koeveskalensis*, *Hindeodella* (*Metaprioniodus*) *multihamata*, *Hindeodella* (*Metaprioniodus*) *pectiniformis*, *Hindeodella* (*Metaprioniodus*) *spengleri*, *Hindeodella* (*Metaprioniodus*) *suevica*, *Hindeodella multihamata*, *Hindeodella pectiniformis*, *Hindeodella petraeviridis*, *Hindeodella priscilla*, *Hindeodella raridenticulata*, *Hindeodella spengleri sapanlii*, *Hindeodella spengleri spengleri*, ★*Hindeodella stoppeli*, *Hindeodella suevica*, *Hindeodella triassica*, *Hindeodella triassica triassica*, *Hindeodella uniforma*, *Hindeodella* sp.

Distribution The genus *Hindeodella* has been described from the Greek peninsula from several localities, including three new species. Their record started with the description of *H. equidentata*, *H. priscilla*, and *Hindeodella* sp. from the Silurian to Upper Devonian to lower Carboniferous of several localities of Chios Island (Kauffmann 1965; Herget and Roth 1968; Roth 1968) and the Permian/Triassic boundary of Episkopi Hydra Island (Römermann 1968), continued with *H. bitorta*, *H. triassica* Metochi–Megali Rachi (Roth 1968), the Lower to Middle Triassic of Chios Island (Bender 1968a, b; Tietze 1969), the Middle Triassic (Anisian to Ladinian) from Mt. Helicon of Beotia (Steuber 1991), and Alogomandra, Asklipieion Epidaurus, and Adhami of Argolis (Bender et al. 1960; Dürkoop et al. 1986;

Huckriede 1958; Vrielynck 1978a, 1980). The Middle to Late Triassic was found in Argolis (Theokafta and Profitis Ilias, Vrielynck 1978a; Mauvier in Decourt 1964; Bender et al. 1960) and in Hydra and Petassi Islands (Römermann 1968; Huckriede 1958) and in Central Greece (Ardaens 1978; Vrielynck in Fleury 1980). Finally, the Late Triassic (Carnian to Norian) has been described from Karpenisi to Fragista and Psilovrakhos (Fleury 1980) and Priolithos, Theokafta, Karafotia–Trahla, and Mavrovouni–Prosimi of Peloponnesus (Bender 1968b; Bender et al. 1960; Flament 1973; Vrielynck 1978a, 1980).

Remarks Three *Hindeodella* species have been identified for the first time from the Lower and Middle Triassic limestones of Marathovouno in Chios Island, by Bender in 1968b. *Hindeodella bitorta* and *Hindeodella ceweki* correspond to the Early Anisian, while *Hindeodella stoppeli* corresponds to the Olenekian.

Hindeodus Rexroad and Furnish, 1964

Type Species *Spathognathodus cristulus* Youngquist and Miller, 1949.

Identified Species in Greece *Hindeodus julfensis*, *Hindeodus minutus*, *Hindeodus typicalis*, *Hindeodus* sp.

Distribution The species of *H. typicalis* and *H. julfensis* have been found in the Lopingian (late Permian) of Hydra Island by Argyriou et al. (2017) and Angiolini et al. (1992) and in Wuchiapingian of Agios Taxiarchis and Episkopi by Nestell and Wardlaw (1987), while *H. minutus* (with *Hindeodus* sp.) has been described from the early to middle Permian of Myrsini (East Crete Island) and Carboniferous of Sfinari and Voutas (West Crete) by Krahl et al. in 1986 and in 1983, respectively.

Icriodus Branson and Mehl, 1938

Type Species *Icriodus expansus* Branson and Mehl, 1938.

Identified Species in Greece *Icriodus alternatus*, *Icriodus symmetricus*, *Icriodus* sp.

Distribution The two representatives of the genus *Icriodus* have been found in the late Devonian of Kambia, while *Icriodus* sp. has been reported in Katavasi, Potamia, and Kipouries of Chios Island (Herget and Roth 1968; Roth 1968).

Icriospathodus Krahl et al., 1983

Type Species *Neospathodus collinsoni* Solien, 1979.

Identified Species in Greece *Icriospathodus collinsoni*.

Distribution Krahl et al. (1983) have found the species of *Icriospathodus collinsoni* only in the Early Triassic of Voutas (West Crete Island).

***Idiognathodus* Gunnell, 1931**

Type Species *Idiognathodus claviformis* Gunnell, 1931.

Identified Species in Greece *Idiognathodus tersus*, *Idiognathodus togashii.*

Distribution Krahl et al. (1983) have reported the two representatives of *Idiognathodus* from the late Carboniferous of Sfinari from West Crete.

***Idiognathoides* Harris and Hollingsworth, 1933**

Type Species *Idiognathoides sinuata* Harris and Hollingsworth, 1933.

Identified Species in Greece *Idiognathoides attenatus*, *Idiognathoides* cf. *convexus*, *Idiognathoides ouachitensis*, *Idiognathoides sinuatus.*

Distribution Krahl et al. (1983) have reported the species of *I. sinuatus* from the late Carboniferous of Sfinari (West Crete Island), while the other representatives of this genus have been found by Tietze (1969) in the Carboniferous of Vrontados and Agios Georgios Sykousis of Chios Island.

***Iranognathus* Kozur et al., 1975**

Type Species *Iranognathus unicostatus* Kozur, Mostler and Rahimi–Yazd, 1975.

Identified Species in Greece *Iranognathus* sp.

Distribution The genus *Iranognathus* has been reported only from the early to middle Permian of Myrsini (East Crete Island) by Krahl et al. (1986).

***Kockelella* Walliser, 1957**

Type Species *Kockelella patula* Walliser, 1964.

Identified Species in Greece *Kockelella patula*, *Kockelella variabilis.*

Distribution The species of *K. patula* has been found in the Devonian of Kipouries and Volisos (Herget and Roth 1968; Roth 1968), while *K. variabilis* has been reported only from the Silurian of Kardamyla of Chios Island (Kauffmann 1965).

***Ligonodina* Ulrich and Bassler, 1926**

Type Species *Ligonodina pectinata* Ulrich and Bassler, 1926.

Identified Species in Greece *Ligonodina salopia*, *Ligonodina silurica*, *Ligonodina* sp.

Distribution Kauffmann (1965) and Herget and Roth (1968) have described the two representatives of the genus *Ligonodina* from the Silurian to Devonian of Kardamyla, Katavasi, and Kipouries, while Tietze (1969) has reported *Ligonodina* sp. from Agia Anna and Agios Georgios Sykousis of Chios Island.

Lochriea Scott, 1942

Type Species *Lochriea montanaensis* (Scott, 1942).

Identified Species in Greece *Lochriea commutata*, *Lochriea mononodosa.*

Distribution The two representative species of the genus *Lochriea* have been described from the late Visean/earlier Serpukhovian of Kourounia–Nenitouria (Groves et al. 2003) and Selino–Amani of Chios Island (Zanchi et al. 2003).

Lonchodina Ulrich and Bassler, 1926

Type Species *Lonchodina typicalis* Ulrich and Bassler, 1926.

Identified Species in Greece *Lonchodina discreta*, *Lonchodina greilingi*, *Lonchodina hungarica*, *Lonchodina latidentata*, *Lonchodina muelleri*, *Lonchodina? posterognathus*, *Lonchodina spengleri*, *Lonchodina venusta*, *Lonchodina walliseri*, *Lonchodina* sp.

Distribution In the Hellenic material, ten species have been catalogued from the genus *Lonchodina*. In Chios Island *L. greilingi* and *L. walliseri* have been reported from the Silurian to Devonian of Kardamyla, Kipouries, Volisos, Katavasi, Agia Anna Parthenis, and Agios Georgios Sykousis (Kauffmann 1965; Herget and Roth 1968; Roth 1968; Tietze 1969), while *L. spengleri*, *L latidentata*, *L. mulleri*, *L. venusta*, and *L. discrteta* from the Olenekian to Anisian of Katavasi, Metochi–Megali Rachi, Anavatos, Rema Armenis, Latomi, Marmarotapeza Kephalovouni, Agia Anna, Parthenis, Korakaris, and Marathovouno (Bender 1968a, b; Roth 1968; Tietze 1969). Furthermore, in Argolis *L. muelleri*, *L. spengleri*, *L. venusta*, *L. latidentata*, and *L. discreta* have been reported from the Middle to Upper Triassic of Adhami, Alogomandra, Theokafta, and Asklipieion Epidaurus (Bender et al. 1960; Bender 1968b; Huckriede 1958; Mauvier in Decourt 1964). *Lonchodina? posterognathus* and *L. hungarica* have been reported from the late Anisian to Norian of Mt. Vardoussia and Mt. Helicon (Ardaens 1978; Steuber 1991), and finally *Lonchodina* species have been found in the Carnian of Mandraki–Hydra Chora of Hydra Island and Petassi Island (Huckriede 1958; Römermann 1968).

***Metalonchodina* Branson and Mehl, 1941**

Type Species *Metalonchodina bidentata* (Gunnell, 1931).

Identified Species in Greece *Metalonchodina digitiformis.*

Distribution Mauvier in Decourt (1964) has catalogued *M. digitiformis* from the Anisian to Carnian of Theokafta (Argolis) in Peloponnesus.

***Metapolygnathus* Hayasi, 1968**

Type Species *Metapolygnathus communisti* Hayasi, 1968.

Identified Species in Greece *Metapolygnathus abneptis abneptis*, *Metapolygnathus bidentatus*, *Metapolygnathus communisti communisti*, *Metapolygnathus communisti parvus*, *Metapolygnathus echinatus*, *Metapolygnathus hungaricus*, *Metapolygnathus japonicus*, *Metapolygnathus mirautae*, *Metapolygnathus mostleri*, *Metapolygnathus multidentatus*, ★*Metapolygnathus multinodosus*, *Metapolygnathus mungoensis*, *Metapolygnathus nodosus*, *Metapolygnathus parvus*, *Metapolygnathus polyganthiformis*, *Metapolygnathus posterus*, *Metapolygnathus primitius*, *Metapolygnathus pseudodiebeli*, *Metapolygnathus spengleri*, *Metapolygnathus* cf. *tadpole*, *Metapolygnathus zoae*, *Metapolygnathus* sp.

Distribution The genus *Metapolygnathus* has been reported from four different Hellenic areas with 21 species and 3 subspecies. In particular, nine species have been reported from Argolis in Adhami, while Noyan and Kozur (2007) catalogued six species from the upper Carnian to lower Norian of Stefanion Section. *M. polyganthiformis* was a common species found in Karafotia–Trahla, Profitis Ilias, Asklipieion Epidaurus, and Midhea (Vrielynck 1978a, b), in Mathia Achladochori of Messinia Peloponnesus (Kozur in Thiebault 1982), but also in the Anisian to Norian of Nafpaktos, Karpenisi–Fragista, Mt. Vardoussia, and Koutsouros Kokkinovrakhos of Central Greece (Ardaens 1978; Vrielynck in Fleury 1980). *Metapolygnathus* species have been found in the Carnian–Norian of Mandraki–Hydra Chora, Kaminia, Palamidas and Agia Marina, of Hydra and Kivotos Islands (Dürkoop et al. 1986). Finally, three species have been described from Sfinari and Voutas, Crete Island (Krahl et al. 1983; Kozur and Krahl 1984).

Remarks *M. multinodosus* Noyan 2007 has been identified in Argolis in a section northeast of Stefanion, south of Mt. Rakhi Stefaniou (Noyan and Kozur 2007), and it corresponds to the latest Carnian.

***Metaprioniodus* Huddle, 1934**

Type Species *Metaprioniodus biangulatus* Huddle, 1934.

Identified Species in Greece *Metaprioniodus benderi*, *Metaprioniodus* sp.

Distribution The two representatives of *Metaprioniodus* have been found only in the Anisian to Norian of Mt. Vardoussia in Central Greece by Ardaens, in 1978.

***Misikella* Kozur and Mock, 1974**

Type Species *Misikella longidentata* Kozur and Mock, 1974.

Identified Species in Greece *Misikella hernsteini.*

Distribution Vrielynck (1978a, 1980) has reported *M. hernsteini* only from the Ladinian to Norian of Midhea and Mavrovouni–Prosimi in Argolis Peloponnesus.

***Neocavitella* Sudar and Budurov, 1979**

Type Species *Neocavitella cavitata* Sudar and Budurov, 1979.

Identified Species in Greece *Neocavitella tatrica.*

Distribution Muttoni et al. (2014) have catalogued the species of *N. tatrica* from the Ladinian of the Agia Marina (Hydra Island).

***Neognathodus* Dunn, 1970**

Type Species *Polygnathus bassleri* Harris and Hollingsworth, 1933.

Identified Species in Greece *Neognathodus bassleri, Neognathodus symmetricus, Neognathodus* sp.

Distribution Krahl et al. (1983) have found the three species of the genus *Neognathodus* only in the late Carboniferous of Sfinari (West Crete Island).

***Neogondolella* Bender and Stoppel, 1965**

Type Species *Gondolella mombergensis* Tatge, 1956.

Identified Species in Greece ★*Neogondolella aegaea, Neogondolella constricta, Neogondolella leveni, Neogondolella mombergensis, Neogondolella navicula, Neogondolella navicula hallstatensis, Neogondolella navicula navicula, Neogondolella navicula steinbergensis, Neogondolella orientalis,* ★*Neogondolella palata, Neogondolella polygnathiformis, Neogondolella regale, Neogondolella tadpole, Neogondolella timorensis timorensis, Neogondolella* sp.

Distribution In the Hellenic material, *Neogondolella* has been found mainly in Hydra Island with *N. navicula* in the Carnian to Norian of Kaminia and Petassi (Dürkoop et al. 1986; Römermann 1968) and *N. palata* in Kivotos Island (Römermann 1968), while *N. leveni* and *N. orientalis* characterized the Lopingian of Episkopi, Agios Taxiarchis, and Klimaki (Nestell and Wardlaw 1987). Five species have been catalogued from the Theokafta, Asklipieion, and Karafotia–Trahla of Argolis (Bender 1968b; Vrielynck 1978a, b, 1980). Furthermore, *N. aegaea* along with *N. mombergensis* has been found in five localities of Chios Island (Bender 1968b; Roth 1968; Besenecker et al. 1968; Tietze 1969; Assereto et al. 1980). Vrielynck in Fleury (1980) described *N. navicula steinbergensis* from the Olenekian to Norian of Psilovrakhos, and Ardaens (1978) has reported five species from the Anisian to Norian of East Mt. Koziakas, Koutsouros–Kokkinovrakhos, and Mt. Vardoussia of Central Greece.

Remarks The genus *Neogondolella* with two species have been described from Greece by Bender in 1968b, although *Neogondolella* as genus has been initially introduced by Bender and Stoppel in 1965. *Neogondolella aegaea* has been identified in the lower Anisian limestones of Marathovouno in Chios Island while *Neogondolella palata* in the early Carnian of Theokafta in Argolis.

***Neohindeodella* Kozur, 1968**

Type Species *Hindeodella triassica* Müller, 1956.

Identified Species in Greece *Neohindeodella* cf. *requiramosa*, *Neohindeodella dropla*, *Neohindeodella summesbergeri praecursor*, *Neohindeodella summesbergeri summesbergeri*, *Neohindeodella triassica*, *Neohindeodella triassica aequidentata*, *Neohindeodella triassica kobayashi*, *Neohindeodella triassica riegeli*, *Neohindeodella triassica triassica*, *Neohindeodella triassica ziegleri*.

Distribution Three species and seven subspecies have been catalogued from the genus *Neohindeodella*, in the Hellenic material. The most common found is the subspecies, *N. triassica triassica*, that characterized the Middle to Late Triassic of Mt. Helicon of Beotia (Steuber 1991), Trapezona, Asklipieion Epidaurus, Midhea, Profitis Ilias, Mavrovouni–Prosimi, and Karafotia–Trahla (along with *N. summesbergeri praecursor*) of Argolis area (Vrieynck 1978a, b, 1980) and the Olenekian to Norian of Psilovrakhos, Mt. Iti, Kotsilieri Central Greece (Wigniolle 1977; Ardaens 1978; Vrielynck in Fleury 1980). The subspecies of *N. triassica riegeli* has been found almost in the same localities as the *N. triassica triassica*, while *N. dropla* has been found in the Middle to Late Triassic of Mt. Helicon (Beotia) (Steuber 1991), Nafpaktos (Vrielynck in Fleury 1980), Mathia–Achladochori of Messinia Peloponnesus (Kozur in Thiebault 1982), Mt. Vardoussia, Mt. Koziakas, and Kokkinovrakhos along with other *Neohindeodella* species (Ardaens 1978). *N. triassica* and *Neohindeodella* sp. have been reported from the Early Triassic of Myrsini, Skopi, and Tripokefala of Crete Island (Krahl et al. 1986), and *N. triassica ziegleri* has been reported in Karpenisi–Fragista (Vrielynck in Fleury 1980).

***Neoplectospathodus* Kozur and Mostler, 1970**

Type Species *Neoplectospathodus muelleri* Kozur and Mostler, 1970.

Identified Species in Greece *Neoplectospathodus muelleri*, *Neoplectospathodus* sp.

Distribution Steuber (1991) reported the species of *N. muelleri* and *Neoplectospathodus* sp. only in the Anisian to Ladinian of Mt. Helicon of Beotia.

Neoprioniodus **Rhodes and Müller, 1956**

Type Species *Prioniodus conjuctus* Gunnell, 1931.

Identified Species in Greece *Neoprioniodus bicurvatus*, *Neoprioniodus* cf. *bicuspidatus*, *Neoprioniodus excavatus*, *Neoprioniodus latidentatus*, *Neoprioniodus multiformis*, *Neoprioniodus subcarnus*, *Neoprioniodus* sp.

Distribution The species of *N. bicurvatus* has been found only in the late Devonian of Katavasi and Agia Anna (Tietze 1969; Herget and Roth 1968) while the *N. subcarnus* only in the middle Silurian of Kardamyla (Kauffmann 1965). The rest of the species found in the Hellenic material has been reported in Potamia, Kipouries, Volisos, Anavatos, Agios Markos, and Kardamyla of Chios Island (Kauffmann 1965; Herget and Roth 1968; Roth 1968; Tietze 1969).

Neospathodus **Mosher, 1968**

Type Species *Spathognathodus cristagalli* Huckriede, 1958.

Identified Species in Greece *Neospathodus* cf. *aequiramosa*, *Neospathodus cristagalli*, *Neospathodus dieneri*, *Neospathodus germanicus*, *Neospathodus hernsteini*, *Neospathodus homeri*, *Neospathodus kockeli*, *Neospathodus longiusculus*, *Neospathodus pakistanensis*, *Neospathodus triangularis*, *Neospathodus waageni*, *Neospathodus* sp.

Distribution Almost all the representatives of the genus *Neospathodus* have been reported from the Island of Crete (Krahl et al. 1983, 1986). *N. kockeli* has only been found in the Early to Middle Triassic of Agia Triada, Vlichos, Pirghos, and Agia Marina of Hydra Island (Angiolini et al. 1992; Muttoni et al. 1997) and *N. hernsteini* in the Carnian to Norian of Mt. Iti (Wigniolle 1977), in Drimos and Klitoria of Achaia Peloponnesus (Flament 1973; Vrielynck in Fleury 1980), and in Mt Koziakas (Ardaens 1978). The most common species are *N. homeri* and *N. triangularis* that have been reported from Adhami (Argolis) (Dürkoop et al. 1986), Marathovouno and Marmarotrapeza Chios Island (Assereto et al. 1980; Muttoni et al. 1995; Gaetani et al. 1992; Jacobshagen et al. 1993), and Episkopi Hydra Island (Dürkoop et al. 1986). *Neospathodus* species have been also found in Mt. Vardoussia and Koutsouros–Kokkinovrakhos of Central Greece (Ardaens 1978).

***Nicoraella* Kozur, 1980**

Type Species *Ozarkodina kockeli* Tatge, 1956.

Identified Species in Greece *Nicoraella germanica, Nicoraella kockeli.*

Distribution *N. kockeli* has been catalogued in the Olenekian to Anisian of Adhami (Argolis Peloponnesus) while *N. germanica* in the late Anisian of Tsigkri (Hydra Island) by Dürkoop et al. (1986).

***Norigondolella* Kozur, 1990**

Type Species *Paragondolella steinbergensis* Mosher, 1968.

Identified Species in Greece *Norigondolella kozuri, Norigondolella navicula, Norigondolella steinbergensis, Norigondolella* sp.

Distribution All representatives of the genus *Norigondolella* have been catalogued from the early to middle Norian of Agia Marina, Hydra Island (Muttoni et al. 2014).

***Oncodella* Mosher, 1968**

Type Species *Oncodella idiodentica* Mosher, 1968.

Identified Species in Greece *Oncodella pausidentata.*

Distribution Vrielynck (1978a, 1980) has reported *O. pausidentata* from the late Triassic of Mavrovouni–Prosimi (Argolis), while Flament (1973) and Vrielynck in Fleury (1980) in Priolithos–Drimos and Klitoria of Achaia of Peloponnesus.

***Oneotodus* Lindström, 1954**

Type Species *Distacodus simplex* Furnish, 1938.

Identified Species in Greece *Oneotodus* sp.

Distribution Herget and Roth (1968) and Roth (1968) have found the genus *Oneotodus* only in the Devonian of Kipouries and Volisos of Chios Island.

***Ozarkodina* Branson and Mehl, 1933**

Type Species *Ozarkodina typica* Branson and Mehl, 1933

Identified Species in Greece *Ozarkodina crassa, Ozarkodina delicatula, Ozarkodina denckmanni,* ★*Ozarkodina*? *fisticulata, Ozarkodina kockeli, Ozarkodina media, Ozarkodina saginata, Ozarkodina sweeti, Ozarkodina tortilis,*

★*Ozarkodina turgida*, *Ozarkodina typica denckmanni*, *Ozarkodina* cf. *ziegleri*, *Ozarkodina ziegleri ziegleri*, *Ozarkodina* sp.

Distribution *Ozarkodina tortilis* has been the most widespread representative of the genus, in the Hellenic peninsula, as it has been reported from Mt. Helicon of Beotia (Steuber 1991), Nafpaktos, Psilovrakhos, Kotsilieri, and Mt. Vardoussia of Central Greece (Flament 1973; Ardaens 1978), Tripokefala of East Crete Island (Krahl et al. 1986), and Drimos and Priolithos of Peloponnesus (Vrielynck in Fleury; Flament 1973). It has also been found in 16 localities of Chios Island, along with other species as *Ozarkodina ?fisticulata* (Bender 1968a, b; Bender and Kockel 1963; Herget and Roth 1968; Roth 1968; Tietze 1969) and seven localities of Argolis (Bender et al. 1960, 1968b; Dürkoop et al. 1986; Vrielynck 1978a, 1980; Mauvier in Decourt 1964; Huckriede 1958), and in Petassi Island of Hydra (Römermann 1968). Furthermore, in Hydra Island *Ozarkodina* species have been reported from the Late Triassic of Mandraki–Hydra Chora (Muttoni et al. 1994; Angiolini et al. 1992; Huckriede 1958). Krahl et al. (1986) described *O. turgida* and *O. sweeti* from the Myrsini, Tripokefala, and Skopi of Crete Island. Earlier, Kauffmann (1965) catalogued four species from the Silurian of Kardamyla of Chios Island.

Remarks Two species of *Ozarkodina* have been identified and described from the Triassic limestones of Marathovouno in Chios Island by Bender in 1968b. *Ozarkodina ?fisticulata* corresponds to the Olenekian to Anisian boundary, whereas *Ozarkodina turgida* characterizes Olenekian.

Pachycladina Staesche, 1964

Type Species *Pachycladina obliqua* Staesche, 1964.

Identified Species in Greece *Pachycladina inclinata*, *Pachycladina longispinosa*, *Pachycladina symmetrica*, *Pachycladina* sp.

Distribution The representatives of the genus *Pachycladina* have been described from the Early Triassic of Sfinari and Kambos of West Crete Island (Krahl et al. 1983), while *Pachycladina* ? sp. has been found in the Olenekian to Anisian of Parthenis of Chios Island (Tietze 1969).

Palmatolepis Ulrich and Bassler, 1926

Type Species *Palmatolepis perlobata* Ulrich and Bassler, 1926.

Identified Species in Greece *Palmatolepis crepida crepida*, *Palmatolepis delicatula clarki*, *Palmatolepis delicatula delicatula*, *Palmatolepis distorta*, *Palmatolepis elongata*, *Palmatolepis gigas*, *Palmatolepis glabra elongata*, *Palmatolepis glabra glabra*, *Palmatolepis glabra pectinata*, *Palmatolepis gracilis gracilis*, *Palmatolepis gracilis sigmoidalis*, *Palmatolepis helmsi*, *Palmatolepis minuta minuta*, *Palmatolepis* *perlobata perlobata*, *Palmatolepis perlobata schindewolfi*, *Palmatolepis perlobata sigmoidea*, *Palmatolepis proversa*, *Palmatolepis punctata*, *Palmatolepis quadrantinodosa inflexa*, *Palmatolepis quadrantinodosa inflexoidea*, *Palmatolepis quadrantinodosa marginifera*, *Palmatolepis quadrantinodosa lobata*, *Palmatolepis regularis*, *Palmatolepis rhomboidea*, *Palmatolepis rugosa* cf. *ampla*, *Palmatolepis rugosa grossi*, *Palmatolepis rugosa postera*, *Palmatolepis rugosa rugosa*, *Palmatolepis rugosa trachytera*, *Palmatolepis subperlobata*, *Palmatolepis subrecta*, *Palmatolepis tenuipunctata*, *Palmatolepis termini*, *Palmatolepis triangularis*, *Palmatolepis* sp.

Distribution The genus *Palmatolepis*, from the Hellenic material, is represented by 15 species and 20 subspecies that refer only to the Chios Island. In particular, Kauffmann (1965), Herget and Roth (1968), and Roth (1968) have catalogued the largest population of species from the late Devonian to early Carboniferous of Kardamyla, Kambia, Keramos, Melanios–Agio Galas, Parparia, Potamia, Kipouries, Volisos, and Katavasi, and Larghi et al. (2005) have reported only *P. gracilis gracilis* and *P. gracilis sigmoidalis* from the late Devonian of Kardamyla.

***Paltodus* Pander, 1856**

Type Species *Paltodus subaequalis* Pander, 1856.

Identified Species in Greece *Paltodus recurvatus*, *Paltodus rotundatus*, *Paltodus* sp.

Distribution Herget and Roth (1968), Roth (1968), and Tietze (1969) have described the three representatives of the genus *Paltodus* from the Devonian of Potamia, Kipouries, Katavasi, and Agia Anna of Chios Island.

***Parachirognathus* Clark, 1959**

Type Species *Parachirognathus ethingtoni* Clark, 1959.

Identified Species in Greece *Parachirognathus petrae-viridis*.

Distribution The species *P. petrae-viridis* has been described from the Anisian of Theokafta (Argolis) and Katavasi, Anavatos, Rema Armenis, and Marathovouno in Chios Island (Roth 1968; Bender 1968b; Tietze 1969), while it has been also reported from the late Anisian to Carnian of Petassi Island (Hydra Island) (Römermann 1968).

***Paragondolella* Mosher, 1968**

Type Species *Paragondolella navicula* (Huckriede, 1958).

Identified Species in Greece *Paragondolella excelsa, Paragondolella foliata, Paragondolella inclinata, Paragondolella navicula, Paragondolella navicula navicula, Paragondolella navicula steinbergensis, Paragondolella polygnathiformis, Paragondolella praelindae, Paragondolella steinbergensis, Paragondolella tadpole.*

Distribution Five *Paragondolella* species characterize the early Carnian of Agia Marina of Hydra Island (Muttoni et al. 2014), while *P. foliata* and *P. polygnathiformis* have been also found in the Carnian of Evia (De Bono et al. 2001). Vrielynck (1978a, 1980) has reported four species from the Middle to Late Triassic of Profitis Ilias, Mavrovouni–Prosimi, Karafotia–Trahla, and Asklipieion (Argolis) that have also been found along with *P. navicula* in the Middle to Upper Triassic sediments of Priolithos (Peloponnesus) and Nafpaktos (Flament 1973, Vrielynck in Fleury 1980). *P. navicula steinbergensis* and *P. polygnathiformis* have been reported from the Carnian to Norian of Mt. Othrys (Ferriere 1974) and *P. navicula* from the Middle Triassic of Mt. Ithome (Terry 1969).

Plectospathodus Branson and Mehl, 1933

Type Species *Plectospathodus flexuosus* Branson and Mehl, 1933.

Identified Species in Greece *Plectospathodus extensus.*

Distribution *P. extensus* has been described only from the Silurian to Devonian of Kardamyla, Potamia, Katavasi, Kipouries, and Agia Anna from Chios Island (Kauffmann 1965; Herget and Roth 1968; Roth 1968).

Pollognathus Kozur and Mostler, 1970

Type Species *Pollognathus sequens* (Kozur, 1968).

Identified Species in Greece *Pollognathus germanicus.*

Distribution Vrielynck (1978a) has found the species of *P. germanicus* in the Anisian to Norian of Profitis Ilias (Argolis), while it has been found also in the same detected age of the deposits in Mt. Vardoussia (Central Greece) by Ardaens (1978).

Polygnathoides Branson and Mehl, 1933

Type Species *Polygnathoides siluricus* Branson and Mehl, 1933.

Identified Species in Greece *Polygnathoides emarginatus, Polygnathoides siluricus.*

Distribution The representatives of the genus *Polygnathoides* have been reported from the Silurian of Kardamyla and Agia Anna (Chios Island), by Kauffmann (1965) and Walliser in Tietze (1969).

***Polygnathus* Hinde, 1879**

Type Species *Polygnathus dubius* Hinde, 1879.

Identified Species in Greece *Polygnathus abneptis*, *Polygnathus carinata*, *Polygnathus* cf. *inortata*, *Polygnathus communis*, *Polygnathus linguiformis*, *Polygnathus linguiformis linguiformis*, *Polygnathus mehli*, *Polygnathus mungoensis*, *Polygnathus* cf. *pura*, *Polygnathus nodocostata*, *Polygnathus styriaca*, *Polygnathus tethydis*, *Polygnathus vogesi*, *Polygnathus* sp.

Distribution The genus *Polygnathus*, from the Hellenic material, is represented by 13 species and 1 subspecies, and they mainly refer to Chios Island. Seven species have been reported from Kambia, Melanios–Agio Galas, Potamia, Kipouries, Katavasi, Kardamyla, and Marmarotrapeza (Larghi et al. 2005; Bender and Kockel 1963; Herget and Roth 1968; Roth 1968). In spite of that, Bender et al. (1960) have catalogued *P. mehli*, *P. abneptis*, and *P. mungoensis* from the Middle to Late Triassic of the Adhami, Theokafta, and Alogomandra of Argolis, and Huckriede (1958) and Mauvier in Decourt (1964) have described *P. tethydis* from Asklipieion Epidaurus and Theokafta, respectively. Nonetheless, representatives of the genus have been found in Mt. Parnis of Attica (late Tournaisian, Caridroit et al. 2000; Olenekian to early Anisian, Bender and Kockel 1963) and Agia Triada and Mandraki–Hydra of Hydra Island (Bender and Kockel 1963; Huckriede 1958). Finally, *P. abneptis* has been also described from the Carnian of Cape Foustapidima in Corfu Island (Huckriede 1958), while Clement in 1977 has described *P. tethydis* from the Anisian of Mt. Helicon in Beotia.

***Prioniodella* Ulrich and Bassler, 1926**

Type Species *Prioniodella normalis* Ulrich and Bassler, 1926.

Identified Species in Greece *Prioniodella ctenoides*, *Prioniodella decrescens*, *Prioniodella pectiniformis*, *Prioniodella prioniodellides*.

Distribution The genus *Prioniodella* has been found in the Hellenic material in three localities with four representatives. All of them have been reported by Bender et al. (1960) and Bender (1968b), from the Carnian of the Theokafta (Argolis) and Chios Island, except for *P. decrescens*, which has been reported from the early Norian of Drimos Peloponnesus (Flament 1973) and Triassic of Mandraki–Hydra Chora and Petassi Island (Huckriede 1958; Römermann 1968). Some representatives have been reported from the Olenekian–Ladinian of Chios Island (Tietze 1969), while *P. pectiniformis* and *P. ctenoides* have been also described from the Anisian to Ladinian of Asklipieion Epidaurus (Argolis) (Huckriede 1958).

***Prioniodina* Ulrich and Bassler, 1926**
Subgenus *Prioniodina* (*Cypridodella*) Mosher, 1968

Type Species *Prioniodina subcurvata* Ulrich and Bassler, 1926.

Identified Species in Greece *Prioniodina (Cypridodella) muelleri*, *Prioniodina (Cypridodella) venusta*, *Prioniodina conflexa*, *Prioniodina ? dinodoides*, *Prioniodina excavata*, *Prioniodina kochi*, *Prioniodina latidentata*, *Prioniodina mediocris*, *Prioniodina mitzopouli*, *Prioniodina muelleri*, *Prioniodina petrae-viridis*, *Prioniodina* cf. *prona*, *Prioniodina pronoides*, *Prioniodina scolosculptura*, *Prioniodina spengleri*, *Prioniodina sweeti transita*, *Prioniodina sweeti-sweeti*, *Prioniodina venusta*, *Prioniodina* sp.

Distribution The genus *Prioniodina* is widely distributed in Hellenic peninsula, mainly in Argolis Peloponnesus within nine Middle to Upper Triassic deposits (Bender et al. 1960; Bender 1968b; Dürkoop et al. 1986; Vrielynck 1978a, 1980; Mauvier in Decourt 1964; Huckriede 1958). Other representatives have been found at the Middle to Late Triassic in Messinia (Mathia–Achladochori; Kozur in Thiebault 1982) and Achaia of Peloponnesus (Priolithos; Flament 1973). Many species have also been reported from several localities of Central Greece (Lekkas 1986; Ardaens 1978; Kauffmann in Fleury 1980; Vrielynck in Fleury 1980; Wigniole 1977). Krahl et al. (1986) have found three species in the Early Triassic of Myrsini and Tripokefala of East Crete Island, while Bender (1968b), Bender and Kockel (1963), and Tietze (1969) have described *Prioniodina* species from the Anisian of Chios Island. *P. kochi* has been also described from Mandraki (Hydra) by Huckriede (1958), while later Steuber (1991) has catalogued five species from Mt. Helicon of Beotia.

Pseudognathodus Perret, 1993

Type Species *Gnathodus homopunctatus* Ziegler, 1960.

Identified Species in Greece *Pseudognathodus homopunctatus.*

Distribution Groves et al. (2003) have described *P. homopunctatus* from the late Famennian (late Devonian) of Kourounia–Nenitouria (Chios Island).

Pseudopolygnathus Branson and Mehl, 1934b

Type Species *Pseudopolygnathus prima* Branson and Mehl, 1934b.

Included Species *Pseudopolygnathus marburgensis*, *Pseudopolygnathus micropunctata*, *Pseudopolygnathus dentilineata*, *Pseudopolygnathus triangula triangula.*

Distribution Herget and Roth (1968), Roth (1968), and Larghi et al. (2005) have reported the genus *Pseudopolygnathus* from the Hellenic material and specifically with four representatives from the late Devonian to Carboniferous of Kambia, Parparia, Kipouries, Katavasi, and Kardamyla of Chios Island.

***Roundya* Hass, 1953**

Type Species *Roundya barnettana* Hass, 1953.

Identified Species in Greece *Roundya lautissima*, *Roundya magnidentata*, *Roundya meissneri*, *Roundya* sp.

Distribution Bender et al. (1960) and Bender (1968b) have catalogued two species of the genus *Roundya* from the Carnian of Theokafta and Anisian to Ladinian of Alogomandra, while they have reported *R. magnidentata* from the Adhami of Argolis. Furthermore, *R. lautissima* has been described also from the Middle to Late Triassic of Asklipieion Epidaurus (Huckriede 1958), Theokafta Argolis (Mauvier in Decourt 1964), and Katavasi and Marathovouno of Chios Island (Roth 1968; Tietze 1969). *Roundya* species have been also reported from several localities of Chios Island (Roth 1968; Tietze 1969).

***Scaliognathus* Branson and Mehl, 1941**

Type Species *Scaliognathus anchoralis* Branson and Mehl, 1941.

Identified Species in Greece *Scaliognathus anchoralis.*

Distribution Herget and Roth (1968) have found the species *S. anchoralis* only in the early Carboniferous of Parparia and Potamia of Chios Island.

***Scaphignathus* Helms, 1959**

Type Species *Scaphignathus velifera* Ziegler, 1959.

Identified Species in Greece *Scaphignathus velifera.*

Distribution *S. velifera* has been reported only in the late Devonian of Kambia of Chios Island by Herget and Roth (1968).

***Sephardiella* March, Budurov, Hirsch and Márquez–Aliaga, 1988**

Type Species *Sephardiella mungoensis* (Diebel, 1956).

Identified Species in Greece *Sephardiella mungoensis.*

Distribution Concerning the Hellenic material, De Bono et al. (2001) have found the species of *S. mungoensis* only in the Carnian of Evia.

***Siphonodella* Branson and Mehl, 1944**

Type Species *Siphonodella duplicata* (Branson and Mehl, 1934b).

Identified Species in Greece *Siphonodella obsoleta.*

Distribution Herget and Roth (1968) have described this species from the early Carboniferous of Kambia, Melanios–Agio Galas, and Parparia of Chios Island.

***Spathognathodus* Branson and Mehl, 1941**

Type Species *Spathodus primus* Branson and Mehl, 1933.

Identified Species in Greece *Spathognathodus bidentatus*, *Spathognathodus* cf. *cristagalli*, ★*Spathognathodus gondolelloides*, ★*Spathognathodus homeri*, *Spathognathodus inclinatus*, *Spathognathodus i. inclinatus*, *Spathognathodus inortatus*, *Spathognathodus orphanus*, *Spathognathodus pennatus pennatus*, *Spathognathodus stabilis*, *Spathognathodus steinhornensis* cf. *eosteinhornensis*, *Spathognathodus steinhornensis* cf. *remscheidensis*, *Spathognathodus strigosus*, ★*Spathognathodus triangularis*, *Spathognathodus* sp.

Distribution The genus *Spathognathodus* is represented by 12 species and 4 subspecies in the Hellenic material, which are mainly referred to Chios Island. Specifically, *S. i. inclinatus* and *S. p. pennatus* have been found in the Silurian of Kardamyla (Kauffmann 1965), whereas eight species have been found only in the late Devonian to early Carboniferous in 10 different localities (Herget and Roth 1968; Roth 1968; Tietze 1969), and other species have been found in 12 localities of Olenekian to Anisian sequences (Bender and Kockel 1963; Roth 1968; Tietze 1969; Tietze in Jacobshagen and Tietze 1974). Nevertheless, *S. gondolelloides* and *S. homeri* have been catalogued by Bender and Kockel (1963), from the Olenekian to early Anisian of Mt. Parnis (Attica), while they also have been found in the Permian/Triassic boundary of Episkopi Hydra Island (Römermann 1968) and the Anisian of Beotia (Clement 1977). Finally, *S. orphanus* has been reported only from the late Carboniferous of Sfinari (West Crete Island; Krahl et al. 1983).

Remarks Bender in 1968b has described three new species of *Spathognathodus* from the Triassic limestones of Marathovouno of Chios Island. *Spathognathodus homeri* and *Spathognathodus gondolelloides* characterize the early Anisian, while *Spathognathodus triangularis* characterizes the Olenekian.

***Streptognathodus* Stauffer and Plummer, 1932**

Type Species *Streptognathodus excelsus* Stauffer and Plummer, 1932.

Identified Species in Greece *Streptognathodus elegantulus*, *Streptognathodus* ex. aff. *elongatus*, *Streptognathodus gracilis*, *Streptognathodus ruzhencevi*, *Streptognathodus* sp.

Distribution *S.* ex. aff. *elongatus* has been reported from the Carboniferous of Chios Island (Agios Georgios Sykousis; Tietze 1969), while all the other representatives of the genus *Streptognathodus* have been found in the late Carboniferous of Sfinari in West Crete Island (Krahl et al. 1983).

***Subbryantodus* Branson and Mehl, 1941**

Type Species *Subbryantodus arcuatus* Branson and Mehl, 1941.

Identified Species in Greece *Subbryantodus* sp.

Distribution *Subbryantodus* sp. has been found only in the Kasimovian of Agios Georgios Sykousis in Chios Island (Tietze 1969).

***Sweetognathus* Clark, 1972**

Type Species *Spathognathodus whitei* Rhodes, 1963.

Identified Species in Greece *Sweetognathus bogoslovskajae.*

Distribution The species *S. bogoslovskajae* has been reported only from the Permian of Myrsini (East Crete Island), by Krahl et al. (1986).

***Synprioniodina* Bassler, 1925**

Type Species *Synprioniodina alternata* Bassler, 1925.

Identified Species in Greece *Synprioniodina* sp.

Distribution In the Hellenic material *Synprioniodina* sp. has been described only from the Kasimovian of Chios (Agios Georgios Sykousis) by Tietze in 1969.

★*Tardogondolella* Bender, 1968b

Type Species ★*Tardogondolella abneptis* (Huckriede, 1958).

Identified Species in Greece ★*Tardogondolella abneptis.*

Distribution Bender (1968b) has described *T. abneptis* from the Carnian of Theokafta (Argolis Peloponnesus). Furthermore, Dürkoop et al. (1986) have described it from Hydra Island and Römermann (1969) from Petassi and Kivotos Islands, while Tietze (1969) has found it in Korakaris (Chios Island).

Remarks The genus *Tardogondolella* has been described for the first time from the Carnian of Theokafta Argolis, and it was attributed to the genus *Polygnathus.*

***Trichonodella* (*Trichognathus*) Branson and Mehl, 1933**

Type Species *Trichognathus prima* Branson and Mehl, 1933.

Identified Species in Greece *Trichonodella excavata*, *Trichonodella inconstans*, *Trichonodella symmetrica*, *Trichonodella* sp.

Distribution The representatives of the genus *Trichonodella* have been catalogued by Kauffmann (1965), Herget and Roth (1968), Roth (1968), and Tietze (1969), only from four localities of the Silurian to late Devonian of Chios Island.

Veghella **Kozur and Mostler, 1970**

Type Species *Veghella delicatula* (Budurov, 1960).

Identified Species in Greece *Veghella delicatula.*

Distribution In the Greek material, *V. delicatula* has been found only in the late Anisian to Norian of Mt. Vardoussia (Central Greece), by Ardaens (1978).

Xaniognathus **Sweet, 1970**

Type Species *Xaniognathus curvatus* Sweet, 1970.

Identified Species in Greece ★*Xaniognathus hydraensis*, *Xaniognathus turgidus*, *Xaniognathus* sp.

Distribution Krahl et al. (1983) have found *X. turgidus* and *Xaniognathus* sp. in the Early Triassic of Voutas (West Crete Island), while Nestell and Wardlaw (1987) have found *X. hydraensis* in the Wuchiapingian of Episkopi, Agios Taxiarchis, and Klimaki of Hydra Island.

Remarks *X. hydraensis* has been described for the first time by Nestel and Wardlaw in 1987, from the Wuchiapingian of Hydra Island. The type locality is approximately 115 m above Agios Taxiarchis, and 7 m below the contact with the overlying Triassic Eros limestone, and consists also a limestone of the Permian sequence of Hydra Island.

6 Concluding Remarks

According to the data presented in this review work, conodonts from the Hellenic peninsula have been studied already from the beginning of 1960 until nowadays by a great number of researchers. The existence of these assemblages has been reported from several formations/outcrops around the Greek territory, spanning the Silurian to Triassic periods. Most of the sites belong to deposits of the Sub-Pelagonian geotectonic unit, while some of them belong to Pindos and Trypali units. Finally, only one site (Corfu) is reported from the Ionian unit. The aforementioned localities belong mainly to the central and south Greece. Distinctly, 5 of them referred to the mainland, Central Greece, Beotia, Evia, Attica, and Peloponnesus, while the rest to the 4 Islands of Hydra, Chios, Corfu, and Crete, with over than 80 localities and many more outcrops. The most studied regions are Argolida, Hydra, and Chios Island; nevertheless the most stratigraphically completed ones are Chios (Silurian to Upper Triassic) and Crete (upper Carboniferous to Upper Triassic) Islands. The total number of recorded genera is 90 with 348 representative species and 74 subspecies.

From the mainland, and specifically Central Greece, outcrops have been studied from Pyli Trikala, East Mt. Koziakas, Mt. Othrys, Mt. Iti (Latsinies), Mt. Vardoussia, Kotsilieri, Galaxidi (Koutsouros–Kokkinovrakhos), Psilovrakhos–Karpenisi–Fragista, and Nafpaktos. The conodont fauna of all these localities comprises 24 genera and indicates a Middle to Late Triassic age. In Beotia in Mt. Helicon, the conodont assemblage consists of 19 Middle to Late Triassic genera, while the assemblage from Evia contains only 5 species that suggest a Carnian age.

From Attica, Mt. Parnis, two outcrops have been studied: the first one indicates a lower Carboniferous deposit in spite of the second one that implies Lower to lower Middle Triassic. The most studied region of the mainland is Peloponnesus, distinguished into three main subregions: Achaia, Messinia, and Argolis. Achaia and Messinia have been studied from three outcrops each that comprise 13 and 5 genera, respectively, indicating Middle to Late Triassic deposits. Argolis conodont fauna has been described from 11 distinguished localities with 36 genera catalogued that suggest Triassic deposits. Most importantly, from Argolis material four new species have been described, *Epigondolella rigoi* Kozur 2007, *Epigondolella stefanionensis* Noyan 2007, *Metapolygnathus multinodosus* Noyan 2007, and *Neogondolella palata* Bender 1968b and the genus *Tardogondolella* Bender 1968b.

From the Greek archipelago, the conodont research focused on four islands. The first one, Corfu Island, is less studied, with only one species reported from a Carnian formation. Secondly, from Hydra Island, 19 localities (including 3 small nearby Islands) with more than 40 outcrops have been studied that comprise a late Permian to Late Triassic conodont fauna. Moreover, from Hydra's material a new species was identified, *Xaniognathus hydraensis* Nestel and Wardlaw 1987. The conodont data, from Chios Island, that sum up into this review paper present the most stratigraphically complete record, with 17 Paleozoic, 11 Mesozoic distinguished localities, and many more outcrops, that suggest a continuous succession from Silurian to Carnian. Additionally, this material led Bender in 1968b to identify two new genera, *Anastrophognathus* and *Neogondolella*. He also described 14 new species: *Anastrophognathus sagittalis*, *Apatognathus mitzopouli*, *Apatognathus radiatus*, *Ctenognathus chionensis*, *Gladigondolella carinata*, *Hindeodella bitorta*, *Hindeodella ceweki*, *Hindeodella stoppeli*, *Neogondolella aegaea*, *Ozarkodina ? fisticulata*, *Ozarkodina turgida*, *Spathognathodus gondolelloides*, *Spathognathodus homeri*, and *Spathognathodus triangularis*. The last location, Crete Island, also presents a big stratigraphic range, with 9 localities and over than 25 genera that indicate a continuous succession from Upper Carboniferous to Upper Triassic. Our results contribute to a review of conodont assemblages and to a stratigraphic assessment of the Paleozoic and Mesozoic fossiliferous deposits from the Hellenic peninsula.

Acknowledgments The authors would like to thank Dr. Martin Langer (University of Bonn), Dr. György Less (University of Miskolc) and Dr. Viktor Karádi (Eötvös Loránd University of Budapest), for tracking down and providing fruitful bibliographic material; as well as Dr. Maria V. Triantaphyllou and Dr. Margarita D. Dimiza (National and Kapodistrian University of Athens) for their valuable scientific suggestions, advices, and discussions. The comments of two anonymous reviewers are greatly appreciated.

Appendix

List of Fossil Localities with Occurrences of Conodonts in Greece

Locality	Series/stages	Basic references
Taxa		
Central Greece		
Pyli Trikala	Upper Triassic (early to middle Norian)	1
Epigondolella echinata, Epigondolella permica, Prioniodina sweeti–sweeti		
East Mt. Koziakas	Upper Triassic (Carnian to Norian)	2
Enantiognathus ziegleri, Epigondolella bidentata, Epigondolella nodosa, Epigondolella permica, Epigondolella postera, Epigondolella pseudodiebeli, Neohindeodella sp., *Neogondolella navicula, Neogondolella navicula steinbergensis, Neogondolella polygnathiformis, Neogondolella tadpole, Neogondolella* sp., *Neospathodus hernsteini, Neospathodus* sp., *Prioniodina sweeti transita, Prioniodina (Cypridodella) muelleri, Prioniodina* sp.		
Mt. Othrys	Carnian to Norian	3
Epigondolella abneptis, Epigondolella bidentata, Epigondolella postera, Gondolella polygnathiformis, Paragondolella navicula steinbergensis, Paragondolella polygnathiformis		
Mt. Iti (Latsinies)	Upper Triassic (late Carnian to early Norian)	4
Chirodella dinodoides, Epigondolella abneptis, Epigondolella bidentata, Epigondolella permica, Epigondolella postera, Neohindeodella triassica triassica, Neospathodus hernsteini, Prioniodina excavata		
Mt. Vardoussia	Middle to Upper Triassic (late Anisian to Norian)	2, 5

Locality	Series/stages	Basic references
Conodonts, *Apatognathus mitzopouli, Chirodella dinodoides, Chirodella gracilis, Cornudina tortilis, Cratognathodus kochi, Didymodella* cf. *alternata, Didymodella* sp., *Diplododella meissneri, Diplododella thuringensis, Enantiognathus petrae-viridis, Enantiognathus ziegleri, Epigondolella abneptis, Epigondolella bidentata, Epigondolella nodosa, Epigondolella permica, Gladigondolella tethydis, Gondolella aegea, Gondolella bulgarica, Gondolella excelsa, Gondolella* sp., *Hindeodella (Metaprioniodus) andrusovi andrusovi, Hindeodella (Metaprioniodus) andrusovi koeveskalensis, Hindeodella (Metaprioniodus) suevica, Hindeodella stoppeli, Lonchodina ? posterognathus, Lonchodina hungarica, Metapolygnathus* cf. *tadpole, Metapolygnathus parvus, Metapolygnathus polygnathiformis, Metaprioniodus benderi, Metaprioniodus* sp., *Neogondolella navicula hallstattensis, Neogondolella navicula navicula, Neogondolella navicula steinbergensis, Neohindeodella* cf. *requiramosa, Neohindeodella dropla, Neohindeodella summesbergeri praecursor, Neohindeodella triassica, Neohindeodella triassica kobayashii, Neohindeodella triassica riegeli, Neohindeodella triassica triassica, Neospathodus* cf. *cristagalli, Neospathodus germanicus, Neospathodus homeri, Neospathodus triangularis, Neospathodus* sp., *Ozarkodina tortilis, Ozarkodina turgida, Pollognathus germanicus, Prioniodina (Cypridodella) muelleri, Prioniodina (Cypridodella) venusta, Prioniodina excavata, Prioniodina* sp., *Veghella delicatula*		
Kotsilieri	Middle to Upper Triassic (Anisian to Carnian)	2
Chirodella dinodoides, Diplododella bidentata, Epigondolella bidentata, Epigondolella permica, Neohindeodella triassica triassica, Ozarkodina tortilis, Prioniodina (Cypridodella) muelleri		
Galaxidi (Koutsouros–Kokkinovrakhos)	Middle to Upper Triassic (Anisian to Norian)	2, 6
Conodonts, *Chirodella dinodoides, Cornudina breviramulis minor, Cornudina oezdemirae, Enantiognathus petrae-viridis*?, *Enantiognathus ziegleri, Epigondolella permica, Gladigondolella tethydis, Gondolella* sp., *Hindeodella (Metaprioniodus) andrusovi andrusovi, Hindeodella (Metaprioniodus) splengeri, Hindeodella triassica triassica, Metapolygnathus polygnathiformis, Neogondolella navicula navicula, Neogondolella navicula steinbergensis, Neogondolella* sp., *Neohindeodella summesbergeri praecursor, Neohindeodella triassica kobayashii, Neohindeodella triassica triassica, Neospathodus* cf. *aequiramosa, Neospathodus* cf. *cristagalli, Neospathodus* aff. *germanicus, Neospathodus* cf. *kockeli, Prioniodina (Cypridodella) muelleri, Prioniodina (Cypridodella) venusta*		
Psilovrakhos–Karpenisi–Fragista	Triassic (Olenekian to Norian)	7
Anastrophognathus sagittalis, Chirodella dinodoides, Enantiognathus ziegleri, Epigondolella bidentata, Epigondolella diebeli, Epigondolella nodosa, Epigondolella permica, Epigondolella postera, Gladigondolella tethydis, Hindeodella pectiniformis, Hindeodella (Metaprioniodus) andrusovi, Hindeodella (Metaprioniodus) pectiniformis, Hindeodella (Metaprioniodus) spengleri, Neogondolella navicula steinbergensis, Neohindeodella triassica triassica, Neohindeodella triassica ziegleri, Ozarkodina tortilis, Prioniodina (Cypridodella) muelleri, Prioniodina (Cypridodella) venusta		
Nafpaktos	Middle to Upper Triassic (late Ladinian to Carnian)	8

Locality	Series/stages	Basic references
Chirodella dinodoides, Enantiognathus ziegleri, Epigondolella bidentata, Epigondolella nodosa, Epigondolella permica, Epigondolella postera, Hibbardella magnidentata, Metapolygnathus polygnathiformis, Neohindeodella dropla, Neohindeodella triassica triassica, Ozarkodina tortilis, Paragondolella excelsa, Paragondolella navicula navicula, Prioniodina (Cypridodella) muelleri		
Beotia		
Mt. Helicon	Middle Triassic (Anisian to Ladinian)	9, 10
Chirodella dinodoides, Cornudina breviramulis, Cornudina tortilis, Cratognathodus posterognathus agnulatus, Cratognathodus posterognathus posterognathus, Didymodella alternata, Diplododella bidentata, Enantiognathodus petraeviridis, Enantiognathodus ziegleri, Gladigondolella tethydis, Gondolella aegea, Gondolella bakalovi, Gondolella constricta, Gondolella eotrammeri, Gondolella excelsa, Gondolella trammeri, Hibbardella lautissima, Hibbardella magnidentata, Hibbardelloides acroforme, Hindeodella andrusovi, Hindeodella bogschi, Hindeodella multihamata, Hindeodella pectiniformis, Hindeodella spengleri sapanlii, Hindeodella spengleri spengleri, Hindeodella suevica, Lonchodina hungarica, Neohindeodella dropla, Neohindeodella triassica aequidentata, Neohindeodella triassica riegeli, Neohindeodella triassica triassica, Neoplectospathodus muelleri, Neolectospathodus sp., *Ozarkodina tortilis, Polygnathus tethydis, Prioniodina mediocris, Prioniodina muelleri, Prioniodina pronoides, Prioniodina scolosculptura, Prioniodina venusta, Spathognathodus gondolelloides, Spathognathodus homeri*		
Evia		
Xirovouni–Seta	Upper Triassic (Carnian)	11
Gladigondolella malayensis, Gladigondolella tethydis, Paragondolella foliata, Paragondolella polygnathiformis, Sephardiella mungoensis		
Attica		
Mt. Parnis	Mississippian (late Tournaisian)	12
Polygnathus mehli		
Mt. Parnis	Lower to Middle Triassic (Olenekian to early Anisian)	13
Gondolella aegea, Polygnathus tethydis, Spathognathodus gondolelloides, Spathognathodus homeri		
Peloponnesus		
Priolithos–Drimos	Upper Triassic (Carnian to Norian)	14
Chirodella dinodoides, Cratognathodus kochi, Cypridodella conflexa, Cypridodella delicatula, Cypridodella mediocris, Cypridodella muelleri, Cypridodella sp., *Diplododella bidentata, Diplododella magnidentata, Enantiognathus ziegleri, Epigondolella abneptis, Epigondolella bidentata, Epigondolella multidentata, Epigondolella primitia, Hibbardella zapfei, Hindeodella suevica, Hindeodella triassica, Hindeodella uniforma, Neospathodus hernsteini, Neospathodus* sp., *Oncodella pausidentata, Ozarkodina tortilis, Paragodolella navicula navicula, Paragondolella navicula steinbergensis, Paragondolella polygnathiformis, Prioniodella decrescens, Prioniodina excavata, Prioniodina petrae-viridis*		

Locality	Series/stages	Basic references
Klitoria	Upper Triassic (late Carnian to Norian)	8
Epigondolella abneptis, Epigondolella bidentata, Epigondolella permica, Neospathodus hernsteini, Oncodella pausidentata, Paragondolella navicula steinbergensis		
Mathia–Achladochori	Upper Triassic (late Carnian to early Norian)	15
Gondolella navicula hallstattensis, Gondolella polygnathiformis, Gondolella tadpole, Gondolella sp., *Metapolygnathus abneptis abneptis, Metapolygnathus echinatus, Metapolygnathus primitius, Metapolygnathus pseudodiebeli, Neohindeodella dropla, Neohindeodella triassica triassica, Prioniodina (Cypridodella) muelleri*		
Mt. Ithome	Middle Triassic	16
Gondolella mombergensis, Paragondolella navicula		
Asklipieion Epidaurus	Middle to Upper Triassic (Anisian to Carnian)	17–22
Anastrophognathus sagittalis, Apatognathus ziegleri, Chirodella dinodoides, Cornudina torilis, Didymodella alternata, Enantiognathus petrae-viridis, Enantiognathus ziegleri, Epigondolella abneptis, Epigondolella hungarica, Epigondolella mungoensis, Epigondolella nodosa, Epigondolella permica, Gladigondolella malayensis, Gladigondolella sp., *Gladigondolella tethydis, Gondolella excelsa, Gondolella navicula, Gondolella polygnathiformis, Hindeodella (Metaprioniodus)* cf. *koeveskalensis, Hindeodella (Metaprioniodus) multihamata, Hindeodella (Metaprioniodus) suevica, Hindeodella petrae-viridis, Hindeodella triassica, Lonchodina latidentata, Lonchodina muelleri, Lonchodina spengleri, Lonchodina venusta, Metapolygnathus polygnathiformis, Neogondolella constricta, Neogondolella mombergensis, Neogondolella palata, Neohindeodella triassica triassica, Ozarkodina saginata, Ozarkodina tortilis, Paragondolella excelsa, Paragondolella navicula, Paragondolella polygnathiformis, Paragondolella tadpole, Polygnathus tethydis, Prioniodella ctenoides, Prioniodella pectiniformis, Prioniodina (Cypridodella) muelleri, Prioniodina (Cypridodella) venusta, Prioniodina ? dinodoides, Prioniodina kochi, Roundya lautissima*		
Theokafta	Middle to Upper Triassic (Anisian to Carnian)	23–25
Apatognathus ziegleri, ? *Falcodus* sp., *Gladigondolella carinata, Gladigondolella tethydis, Gondolella navicula, Hindeodella multihamata, Hindeodella petrae-viridis, Hindeodella triassica, Lonchodina discreta, Lonchodina muelleri, Lonchodina spengleri, Lonchodina venusta, Metalonchodina digitiformis, Neogondolella navicula,* ***Neogondolella palata****, Ozarkodina saginata, Ozarkodina tortilis, Ozarkodina turgida, Parachirognathus petrae-viridis, Polygnathus abneptis, Polygnathus mungoensis, Polygnathus tethydis, Prioniodella ctenoides, Prioniodella decrescens, Prioniodella pectiniformis, Prioniodella prioniodellides, Prioniodina* cf. *prona, Prioniodina kochi, Prioniodina latidentata, Prioniodina mediocris, Roundya lautissima, Roundya magnidentata, Tardogondolella abneptis*		
Adhami	Triassic (Olenekian to middle Carnian)	19, 25, 26

Locality	Series/stages	Basic references
Cratognathodus kochi, Cratognathodus posterognathus, Didymodella alternata, Diplododella bidentata, Diplododella triassica, Enantiognathus bitortus, Enantiognathus mitzopouli, Enantiognathus petraeviridis, Enantiognathus ziegleri, Euprioniodina mediocris, Euprioniodina muelleri, Euprioniodina multihamata, Euprioniodina pectiniformis, Euprioniodina petraeviridis, Euprioniodina pronoides, Gladigondolella tethydis, Gondolella arcuata, Gondolella basisymmetrica, Gondolella bifurcata, Gondolella constricta, Gondolella cornuta, Gondolella excelsa, Gondolella excentrica, Gondolella mombergensis, Gondolella navicula, Gondolella polygnathiformis, Gondolella tadpole, Gondolella timorensis, Hibbardella lautissima, Hibbardella magnidentata, Hindeodella petrae-viridis, Hindeodella suevica, Hindeodella triassica, Hindeodella (Metaprioniodus) pectiniformis, Lonchodina muelleri, Lonchodina spengleri, Lonchodina venusta, Metapolygnathus abneptis, Metapolygnathus bidentatus, Metapolygnathus mostleri, Metapolygnathus multidentatus, Metapolygnathus mungoensis, Metapolygnathus nodosus, Metapolygnathus primitius, Metapolygnathus spengleri, Metapolygnathus hungaricus, Neospathodus homeri, Neospathodus triangularis, Nicoraella kockeli, Ozarkodina saginata, Ozarkodina tortilis, Polygnathus abneptis, Polygnathus tethydis, Prioniodina (Cypridodella) muelleri, Prioniodina conflexa, Prioniodina latidentata, Prioniodina spengleri, Prioniodina venusta, Roundya magnidentata		
Alogomandra	Middle Triassic (Anisian to Ladinian)	25
Gondolella navicula, Hindeodella multihamata, Hindeodella petrae-viridis, Hindeodella triassica, Lonchodina muelleri, Lonchodina spengleri, Lonchodina venusta, Ozarkodina tortilis, Polygnathus tethydis, Roundya lautissima, Roundya magnidentata		
Mavrovouni–Prosimi–Stefanion	Upper Triassic (late Carnian to Norian)	19, 21, 27
Carnepigondolella nodosa, Chirodella dinodoides, Cornudina breviramulis minor, Epigondolella abneptis, Epigondolella bidentata, Epigondolella echinata, Epigondolella multidentata, Epigondolella nodosa, Epigondolella permica, Epigondolella postera, Epigondolella quadrata, ***Epigondolella rigoi, Epigondolella stefanionensis****, Epigondolella ziegleri, Hibbardella magnidentata, Hindeodella (Metaprioniodus) andrusovi andrusovi, Hindeodella (Metaprioniodus) suevica, Metapolygnathus communisti communisti, Metapolygnathus communisti parvus, Metapolygnathus communisti,* ***Metapolygnathus multinodosus****, Metapolygnathus zoae, Misikella hernsteini, Neohindeodella summesbergeri praecursor, Neohindeodella summesbergeri summesbergeri, Neohindeodella triassica triassica, Oncodella pausidentata, Ozarkodina tortilis, Paragondolella navicula steinbergensis, Paragondolella navicula, Paragondolella polygnathiformis, Paragondolella steinbergensis, Paragondolella tadpole, Prioniodina (Cypridodella) muelleri, Prioniodina excavata*		
Profitis Ilias	Middle to Upper Triassic (Anisian to Norian)	19
Chirodella dinodoides, Gladigondolella tethydis, Grodella delicatula, Hindeodella (Metaprioniodus) bicuspidata, Hindeodella (Metaprioniodus) suevica, Metapolygnathus polygnathiformis, Neohindeodella dropla, Neohindeodella triassica riegeli, Neohindeodella triassica triassica, Paragondolella excelsa, Paragondolella cf. *excelsa, Pollognathus germanicus, Prioniodina (Cypridodella) muelleri*		
Moni Taxiarchis–Tsoukalia	Upper Triassic (Norian)	22
Epigondolella abneptis		

Locality	Series/stages	Basic references
Trapezona	Upper Triassic (late Carnian to Norian)	19, 20, 21
Chirodella dinodoides, Epigondolella nodosa, Epigondolella permica, Neohindeodella summesbergeri summesbergeri, Neohindeodella triassica riegeli, Neohindeodella triassica triassica, Prioniodina (Cypridodella) muelleri		
Midhea	Middle to Upper Triassic (Ladinian to Norian)	19
Cornudina breviramulis minor, Cornudina sp., *Metapolygnathus polygnathiformis, Misikella hernsteini, Neohindeodella* cf. *triassica triassica, Prioniodina (Cypridodella) muelleri, Epigondolella* sp.		
Karafotia–Trahla	Upper Triassic (Carnian to Norian)	19, 21
Chirodella dinodoides, Cornudina breviramulis minor, Diplododella bidentata, Diplododella meissneri, Enantiognathus ziegleri, Epigondolella bidentata, Epigondolella nodosa, Epigondolella permica, Epigondolella ziegleri, Hibbardella magnidentata, Hindeodella (Metaprioniodus) andrusovi andrusovi, Hindeodella (Metaprioniodus) suevica, Metapolygnathus polygnathiformis, Neogondolella navicula steinbergensis, Neohindeodella summesbergeri praecursor, Neohindeodella summesbergeri summesbergeri, Neohindeodella triassica riegeli, Neohindeodella triassica triassica, Ozarkodina tortilis, Paragondolella steinbergensis, Paragondolella tadpole, Prioniodina (Cypridodella) muelleri		
Iliokastro–Kastro Hill	Middle Triassic (Ladinian	22
Gondolella trammeri		
Hydra		
Zogeika	Middle Triassic (late Anisian)	26
Cratognathodus kochi, Gondolella excelsa, Gondolella navicula		
Agios Nikolaos monastery (4 localities)	Middle Triassic (late Anisian)	26
Anastrophognathus sagittalis, Cratognathodus kochi, Gladigondolella tethydis, Gondolella constricta, Gondolella excelsa, Gondolella mombergensis, Gondolella navicula		
Agia Triada	Middle Triassic (late Anisian/early Ladinian)	13, 17, 28, 29
Budurovignathus hungaricus, Budurovignathus mungoensis, Cratognathodus kochi, Enantiognathus ziegleri, Gladigondolella tethydis, Gondolella bifurcata bifurcata, Gondolella bifurcata hanbulogi, Gondolella bulgarica, Gondolella cornuta, Gondolella eotrammeri, Gondolella excelsa, Gondolella foliata inclinata, Gondolella fueloepi fueloepi, Gondolella fueloepi pseudobifurcata, Gondolella liebermani, Gondolella navicula, Gondolella tethydis, Gondolella trammeri, Neospathodus kockeli, Ozarkodina sp., *Polygnathus tethydis*		
Mandraki–Hydra Chora	Upper Triassic (Carnian)	17, 26

Locality	Series/stages	Basic references
Apatognathus ziegleri, *Gondolella navicula*, *Gondolella polygnathiformis*, *Hindeodella petrae-viridis*, *Lonchodina latidentata*, *Lonchodina muelleri*, *Lonchodina spengleri*, *Lonchodina venusta*, *Metapolygnathus nodosus*, *Ozarkodina saginata*, *Polygnathus tethydis*, *Prioniodella decrescens*, *Prioniodina kochi*		
Malies	Middle Triassic (late Anisian to Ladinian)	30
Cratognathodus kochi, *Gladigondolella tethydis*, *Gondolella cornuta*, *Gondolella* aff. *eotrammeri*, *Gondolella excelsa*, *Gondolella* aff. *szaboi*, *Gondolella* sp., *Gondolella trammeri*, *Ozarkodina* sp.		
Klimaki	Lopingian (Wuchiapingian)	31
Neogondolella orientalis, *Xaniognathus hydraensis*		
Pirghos	Middle Triassic (late Anisian–Ladinian)	30
Enantiognathus ziegleri, *Gladigondollela tethydis*, *Gondolella bifurcata bifurcata*, *Gondolella bifurcata hanbulogi*, *Gondolella bulgarica*, *Gondolella constricta*, *Gondolella cornuta*, *Gondolella excelsa*, *Gondolella trammeri*, *Neospathodus kockeli*, *Ozarkodina* sp.		
Agios Taxiarchis	Lopingian (Wuchiapingian)	31
Ellisonia sp., *Hindeodus julfensis*, *Neogondolella orientalis*, ***Xaniognathus hydraensis***		
Kaminia	Upper Triassic (Carnian to Norian)	26, 32
Gondolella polygnathiformis, *Gondolella tadpole*, *Metapolygnathus bidentatus*, *Metapolygnathus multidentatus*, *Metapolygnathus nodosus*, *Neogondolella navicula*, *Tardogondolella abneptis*		
Vlichos	Middle Triassic (middle to late Anisian)	26, 28
Gladigondolella tethydis, *Gondolella bifurcata bifurcata*, *Gondolella bulgarica*, *Gondolella constricta*, *Gondolella cornuta*, *Gondolella excelsa*, *Gondolella fueloepi pseudobifurcata*, *Gondolella mombergensis*, *Gondolella navicula*, *Gondolella tethydis*, *Neospathodus kockeli*		
Mt. Eros	Middle Triassic (late Anisian)	26
Gladigondolella tethydis, *Gondolella constricta*, *Gondolella excelsa*, *Gondolella mombergensis*, *Gondolella navicula*		
Palamidas	Upper Triassic (Carnian)	26
Gondolella polygnathiformis, *Metapolygnathus nodosus*, *Metapolygnathus primitius*		
Agia Marina	Middle to Upper Triassic (late Anisian to Norian)	17, 26, 28–30

Locality	Series/stages	Basic references
Carnepigondolella gulloae, *Carnepigondolella nodosa*, *Enantiognathus ziegleri*, *Epigondolella quadrata*, *Gladigondolella malayensis malayensis*, *Gladigondolella* sp., *Gladigondolella tethydis*, *Gondolella bifurcata bifurcata*, *Gondolella bifurcata hanbulogi*, *Gondolella* aff. *bifurcata hanbulogi*, *Gondolella bulgarica*, *Gondolella cornuta*, *Gondolella eotrammeri*, *Gondolella excelsa*, *Gondolella liebermani*, *Gondolella* sp., *Hindeodella petrae-viridis*, *Metapolygnathus abneptis*, *Metapolygnathus multidentatus*, *Metapolygnathus posterus*, *Neocavitella tatrica*, *Neospathodus kockeli*, *Norigondolella kozuri*, *Norigondolella navicula*, *Norigondolella steinbergensis*, *Norigondolella* sp., *Ozarkodina* sp., *Paragondolella foliata*, *Paragondolella inclinata*, *Paragondolella polygnathiformis*, *Paragondolella praelindae*, *Paragondolella tadpole*		
Episkopi	Permian to Upper Triassic (Lopingian to Carnian)	26, 30–33
Enantiognathus ziegleri, *Gladigondolella carinata*, *Gladigondolella tethydis*, *Gondolella* aff. *eotrammeri*, *Gondolella bulgarica*, *Gondolella constricta*, *Gondolella excelsa*, *Gondolella foliata foliata*, *Gondolella polygnathiformis*, *Gondolella* sp., *Gondolella tadpole*, *Hindeodella stoppeli*, *Hindeodella triassica*, *Hindeodus julfensis*, *Hindeodus typicalis*, *Neogondolella leveni*, *Neogondolella orientalis*, *Neospathodus homeri*, *Neospathodus triangularis*, *Ozarkodina* sp., *Spathognathodus gondolelloides*, *Spathognathodus homeri*, *Xaniognathus hydraensis*		
Tsigkri	Middle Triassic (late Anisian)	26
Cratognathodus kochi, *Cratognathodus posterognathus*, *Gladigondolella tethydis*, *Gondolella bulgarica*, *Gondolella constricta*, *Gondolella excelsa*, *Gondolella mombergensis*, *Gondolella navicula*, *Gondolella regale*, *Gondolella timorensis*, *Nicoraella germanica*		
Bisti	Middle to Upper Triassic (late Anisian to Carnian)	26, 30
Cratognathodus kochi, *Cratognathodus posterognathus*, *Gladigondolella tethydis*, *Gondolella* aff. *eotrammeri*, *Gondolella bifurcata bifurcata*, *Gondolella bifurcata hanbulogi*, *Gondolella bulgarica*, *Gondolella excelsa*, *Gondolella mombergensis*, *Gondolella navicula*, *Gondolella polygnathiformis*, *Gondolella tadpole*, *Gondolella trammeri*		
Kivotos Island	Upper Triassic (Carnian)	26, 32
Gondolella navicula, *Gondolella polygnathiformis*, *Metapolygnathus nodosus*, *Neogondolella palata*, *Tardogondolella abneptis*		
Petassi Island	Middle to Upper Triassic (late Anisian to Carnian)	26, 32
Apatognathus ziegleri, *Cratognathodus posterognathus*, *Gladigondolella tethydis*, *Gondolella bifurcata*, *Gondolella constricta*, *Gondolella excelsa*, *Gondolella navicula*, *Hindeodella triassica*, *Lonchodina discreta*, *Lonchodina muelleri*, *Neogondolella navicula*, *Neogondolella palata*, *Ozarkodina tortilis*, *Ozarkodina turgida*, *Parachirognathus petraeviridis*, *Prioniodella ctenoides*, *Prioniodella prioniodellides*, *Tardogondolella abneptis*		
Pontikos Island	Upper Triassic (Norian)	26
Metapolygnathus abneptis, *Metapolygnathus bidentatus*		

Locality	Series/stages	Basic references
Chios		
Kambia	Upper Devonian to lower Carboniferous	34
Doliognathus sp., *Hindeodella* sp., *Icriodus alternatus, Icriodus symmetricus, Palmatolepis delicatula clarki, Palmatolepis delicatula delicatula, Palmatolepis gracilis gracilis, Palmatolepis gracilis sigmoidalis, Palmatolepis minuta minuta, Palmatolepis perlobata perlobata, Palmatolepis quadrantinodosalobata, Palmatolepis rugosa postera, Palmatolepis rugosa rugosa, Palmatolepis rugosa trachytera, Palmatolepis tenuipunctata, Palmatolepis* sp., *Polygnathus communis, Polygnathus* cf. *inortata, Polygnathus styriaca, Polygnathus vogesi, Pseudopolygnathus marburgensis, Pseudopolygnathus micropunctata, Scaphignathus velifera, Siphonodella obsoleta, Spathognathodus inortatus, Spathognathodus stabilis*		
Keramos	Upper Devonian	34
Gnathodus bilineatus, Gnathodus commutatus commutatus, Gnathodus commutatus nodosus, Gnathodus delicatus, Gnathodus girtyi, Gnathodus semiglaber, Palmatolepis glabra glabra, Palmatolepis glabra penctinata, Palmatolepis gracilis gracilis, Palmatolepis perlobata perlobata, Palmatolepis rugosa cf. *ampla, Palmatolepis rugosa grossi, Palmatolepis minuta minuta*		
Kourounia–Nenitouria	Middle to Upper Mississippian (late Visean–Serpukhovian)	35
?*Gnathodus cuneiformis, Gnathodus bilineatus,* ?*Gnathodus praebilineatus, Pseudognathodus homopunctatus, Lochriea commutata, Lochriea mononodosa*		
Melanios–Agio Galas	Upper Devonian to lower Carboniferous	34
Ancyrodella sp., *Gnathodus bilineatus, Gnathodus bilineatus bilineatus, Gnathodus commutatus commutatus, Gnathodus commutatus homopunctatus, Gnathodus commutatus nodosus, Gnathodus delicatus, Gnathodus semiglaber, Gnathodus texanus, Palmatolepis crepida crepida, Palmatolepis distorta, Palmatolepis gigas, Palmatolepis glabra elongata, Palmatolepis glabra glabra, Palmatolepis glabra pectinata, Palmatolepis gracilis gracilis, Palmatolepis minuta minuta, Palmatolepis perlobata perlobata, Palmatolepis perlobata sigmoidea, Palmatolepis punctata, Palmatolepis quadrantinodosa marginifera, Palmatolepis quadrantinodosa* spp., *Palmatolepis* cf. *regularis, Palmatolepis rugosa grossi, Palmatolepis subperlobata, Palmatolepis subrecta, Palmatolepis tenuipunctata, Polygnathus vogesi, Polygnathus* sp., *Siphonodella obsoleta, Spathognathodus stabilis*		
Parparia	Upper Devonian to lower Carboniferous	34
Ancyrodella sp., *Gnathodus bilineatus, Gnathodus commutatus commutatus, Gnathodus commutatus homopunctatus, Gnathodus commutatus nodosus, Gnathodus delicatus, Gnathodus girtyi, Gnathodus punctatus, Gnathodus semiglaber, Gnathodus texanus, Palmatolepis crepida crepida, Palmatolepis distorta, Palmatolepis glabra elongata, Palmatolepis glabra glabra, Palmatolepis glabra penctinata, Palmatolepis gracilis gracilis, Palmatolepis minuta minuta, Palmatolepis perlobata perlobata, Palmatolepis perlobata sigmoidea, Palmatolepis proversa, Palmatolepis quadrantinodosa inflexa, Palmatolepis quadrantinodosa inflexoidea, Palmatolepis quadrantinodosa marginifera, Palmatolepis quadrantinodosa* spp., *Palmatolepis rhomboidea, Palmatolepis rugosa grossi, Palmatolepis subrecta, Palmatolepis triangularis, Pseudopolygnathus dentilineata, Pseudopolygnathus triangula triangula, Scaliognathus anchoralis, Siphonodella obsoleta, Spathognathodus strigosus*		

Locality	Series/stages	Basic references
Potamia	Upper Devonian to lower Carboniferous	34
Belodina triangularis, *Gnathodus bilineatus*, *Gnathodus commutatus commutatus*, *Gnathodus commutatus homopunctatus*, *Gnathodus delicatus*, *Gnathodus semiglaber*, *Gnathodus texanus*, *Hindeodella* aff. *equidentata*, *Hindeodella priscilla*, *Hindeodella* sp., *Icriodus* sp., *Neoprioniodus excavatus*, *Ozarkodina typica denckmanni*, *Ozarkodina* sp., *Palmatolepis crepida crepida*, *Palmatolepis delicatula clarki*, *Palmatolepis delicatula delicatula*, *Palmatolepis distorta*, *Palmatolepis* cf. *elongata*, *Palmatolepis glabra elongata*, *Palmatolepis glabra glabra*, *Palmatolepis glabra penctinata*, *Palmatolepis gracilis gracilis*, *Palmatolepis helmsi*, *Palmatolepis minuta minuta*, *Palmatolepis perlobata perlobata*, *Palmatolepis quadrantinodosa inflexoidea*, *Palmatolepis quadrantinodosa marginifera*, *Palmatolepis quadrantinodosalobata*, *Palmatolepis* cf. *regularis*, *Palmatolepis rugosa grossi*, *Palmatolepis subperlobata*, *Palmatolepis subrecta*, *Palmatolepis tenuipunctata*, *Palmatolepis triangularis*, *Paltodus* cf. *recurvatus*, *Paltodus* sp., *Plectospathodus extensus*, *Polygnathus linguiformis linguiformis*, *Polygnathus nodocostata*, *Polygnathus styriaca*, *Scaliognathus anchoralis*, *Spathognathodus steinhornensis* cf. *remscheidensis*		
Amani	Middle to Upper Mississippian (late Visean to Serpukhovian)	36
Gnathodus bilineatus, *Lochriea commutata*, *Lochriea mononodosa*		
Kipouries	Upper Devonian to Carboniferous	34, 37
Amorphognathus sp., *Ancyrodella* sp., *Belodella triangularis*, *Gnathodus bilineatus*, *Gnathodus commutatus commutatus*, *Gnathodus commutatus nodosus*, *Gnathodus* cf. *girtyi*, *Hindeodella equidentata*, *Hindeodella* aff. *equidentata*, *Hindeodella priscilla*, *Hindeodella* aff. *priscilla*, *Icriodus* sp., *Kockelella patula*, *Ligonodina* cf. *salopia*, *Ligonodina silurica*, *Lonchodina* cf. *walliseri*, *Lonchodina greilingi*, *Neoprioniodus bicurvatus*, *Neoprioniodus excavatus*, *Neoprioniodus latidentatus*, *Neoprioniodus multiformis*, *Neoprioniodus* sp., *Oneotodus* sp., *Ozarkodina* cf. *ziegleri*, *Ozarkodina denckmanni*, *Ozarkodina media*, *Ozarkodina* sp., *Palmatolepis* cf. *regularis*, *Palmatolepis distorta*, *Palmatolepis glabra elongata*, *Palmatolepis glabra glabra*, *Palmatolepis glabra pectinata*, *Palmatolepis gracilis gracilis*, *Palmatolepis gracilis sigmoidalis*, *Palmatolepis helmsi*, *Palmatolepis minuta minuta*, *Palmatolepis perlobata perlobata*, *Palmatolepis perlobata sigmoidea*, *Palmatolepis proversa*, *Palmatolepis quadrantinodosa marginifera*, *Palmatolepis quadrantinodosalobata*, *Palmatolepis rugosa postera*, *Palmatolepis rugosa rugosa*, *Palmatolepis rugosa* spp., *Palmatolepis rugosa trachytera*, *Palmatolepis subperlobata*, *Palmatolepis subrecta*, *Palmatolepis tenuipunctata*, *Palmatolepis termini*, *Palmatolepis triangularis*, *Paltodus* cf. *recurvatus*, *Paltodus rotundatus*, *Paltodus* sp., *Plectospathodus extensus*, *Polygnathus linguiformis*, *Polygnathus* cf. *pura*, *Polygnathus* sp., *Pseudopolygnathus dentilineata*, *Spathognathodus bidentatus*, *Spathognathodus inclinatus*, *Spathognathodus pennatus*, *Spathognathodus steinhornensis* cf. *eosteinhornensis*, *Spathognathodus* sp., *Trichonodella excavata*, *Trichonodella inconstans*, *Trichonodella* aff. *symmetrica*		
Volisos	middle Silurian to Devonian	34, 37

Locality	Series/stages	Basic references
Amorphognathus sp., *Gnathodus girtyi*, *Gnathodus punctatus*, *Gnathodus* sp. indet., *Gnathodus texanus*, *Hindeodella equidentata*, *Kockelella patula*, *Lonchodina greilingi*, *Neoprioniodus multiformis*, *Oneotodus* sp., *Ozarkodina media*, *Ozarkodina* cf. *ziegleri*, *Palmatolepis delicatula clarki*, *Palmatolepis delicatula delicatula*, *Palmatolepis glabra glabra*, *Palmatolepis glabra pectinata*, *Palmatolepis gracilis gracilis*, *Palmatolepis minuta minuta*, *Palmatolepis punctata*, *Palmatolepis quadrantinodosa inflexa*, *Palmatolepis quadrantinodosa* spp., *Palmatolepis quadrantinodosalobata*, *Palmatolepis* cf. *regularis*, *Palmatolepis subperlobata*, *Palmatolepis tenuipunctata*, *Palmatolepis triangularis*, *Spathognathodus inclinatus*, *Spathognathodus pennatus*, *Spathognathodus stabilis*, *Trichonodella excavata*		
Katavasi	Upper Devonian to Middle Triassic	34, 36
Enantiognathus ziegleri, *Gladigondolella tethydis*, *Hindeodella equidentata*, *Hindeodella triassica*, *Icriodus* sp., *Ligonodina* cf. *salopia*, *Ligonodina silurica*, *Lonchodina greilingi*, *Lonchodina* aff. *greilingi*, *Lonchodina latidentata*, *Lonchodina spengleri*, *Lonchodina* cf. *walliseri*, *Neogondolella mombergensis*, *Neoprioniodus bicurvatus*, *Neoprioniodus excavatus*, *Neoprioniodus latidentatus*, *Neoprioniodus multiformis*, *Ozarkodina media*, *Ozarkodina saginata*, *Ozarkodina tortilis*, *Ozarkodina turgida*, *Ozarkodina* ? *fisticulata*, *Palmatolepis distorta*, *Palmatolepis glabra elongata*, *Palmatolepis glabra glabra*, *Palmatolepis glabra pectinata*, *Palmatolepis gracilis gracilis*, *Palmatolepis gracilis sigmoidalis*, *Palmatolepis minuta minuta*, *Palmatolepis perlobata perlobata*, *Palmatolepis perlobata sigmoidea*, *Palmatolepis proversa*, *Palmatolepis punctata*, *Palmatolepis quadrantinodosa inflexoidea*, *Palmatolepis quadrantinodosa marginifera*, *Palmatolepis quadrantinodosa* spp., *Palmatolepis rugosa grossi*, *Palmatolepis rugosa postera*, *Palmatolepis rugosa rugosa*, *Palmatolepis rugosa* spp., *Palmatolepis subperlobata*, *Palmatolepis tenuipunctata*, *Paltodus* sp., *Parachirognathus petrae-viridis*, *Plectospathodus extensus*, *Polygnathus nodocostata*, *Polygnathus* sp., *Pseudopolygnathus micropunctata*, *Roundya lautissima*, *Spathognathodus bidentatus*, *Spathognathodus gondolelloides*, *Spathognathodus homeri*, *Spathognathodus inclinatus*, *Spathognathodus steinhornensis* cf. *eosteinhornensis*, *Trichonodella excavata*, *Trichonodella* aff. *excavata*, *Trichonodella inconstans*, *Trichonodella* aff. *symmetrica*		
Kardamyla	Silurian (Wenlock to Ludlow)	38
Ancoradella ploeckensis, *Carniodus carinthiacus*, *Carniodus* cf. *carnulus*, *Hindeodella equidentata*, *Kockelella variabilis*, *Ligonodina salopia*, *Ligonodina silurica*, *Lonchodina greilingi*, *Lonchodina walliseri*, *Neoprioniodus excavatus*, *Neoprioniodus latidentatus*, *Neoprioniodus multiformis*, *Neoprioniodus subcarnus*, *Ozarkodina crassa*, *Ozarkodina media*, *Ozarkodina ziegleri ziegleri*, *Ozarkodina* sp., *Plectospathodus extensus*, *Polygnathoides emarginatus*, *Polygnathoides siluricus*, *Spathognathodus inclinatus inclinatus*, *Spathognathodus pennatus pennatus*, *Trichonodella excavata*, *Trichonodella inconstans*, *Trichonodella* sp.		
Kardamyla	Upper Devonian (Famennian)	38, 39
Bispathodus stabilis, *Palmatolepis distorta*, *Palmatolepis glabra glabra*, *Palmatolepis glabra pectinata*, *Palmatolepis gracilis gracilis*, *Palmatolepis gracilis sigmoidalis*, *Palmatolepis minuta minuta*, *Palmatolepis perlobata schindewolfi*, *Palmatolepis perlobata sigmoidea*, *Palmatolepis quadrantinodosa inflexa*, *Palmatolepis quadrantinodosa marginifera*, *Polygnathus communis*, *Polygnathus vogesi*, *Polygnathus* sp., *Pseudopolygnathus marburgensis*		

Locality	Series/stages	Basic references
Metochi–Megali Rachi	Lower to Middle Triassic (Olenekian to Anisian)	37
Enantiognathus ziegleri, *Gladigondolella tethydis*, *Hibbardella* sp., *Hindeodella bitorta*, *Hindeodella triassica*, *Lonchodina latidentata*, *Lonchodina muelleri*, *Lonchodina venusta*, *Neogondolella mombergensis*, *Ozarkodina* cf. *kockeli*, *Ozarkodina* sp., *Roundya* sp., *Spathognathodus gondolelloides*, *Spathognathodus homeri*		
Anavatos	Lower to Middle Triassic (Olenekian to Ladinian)	40
Apatognathus radiatus, *Enantiognathus ziegleri*, *Eurygnathodus costatus*, *Gladigondolella carinata*, *Gladigondolella tethydis*, *Hindeodella bitorta*, *Hindeodella multihamata*, *Hindeodella raridenticulata*, *Hindeodella triassica*, *Hindeodella* sp., *Lonchodina muelleri*, *Lonchodina spengleri*, *Lonchodina venusta*, *Neogondolella aegea*, *Neogondolella mombergensis*, *Neogondolella navicula*, *Neoprioniodus* cf. *bicuspidatus*, *Ozarkodina* ? *fisticulata*, *Ozarkodina tortilis*, *Ozarkodina turgida*, *Parachirognathus petrae-viridis*, *Prioniodella pectiniformis*, *Prioniodina kochi*, *Roundya lautissima*, *Roundya* sp., *Spathognathodus gondolelloides*, *Spathognathodus homeri*, *Spathognathodus* cf. *triangularis*		
Vrontados	Carboniferous	40
Idiognathoides attenatus		
Rema Armenis	Middle Triassic (Anisian to Ladinian)	40
Apatognathus radiatus, *Enantiognathus ziegleri*, *Gladigondolella tethydis*, *Hibbardella* cf. *nevadensis*, *Hibbardella* sp., *Hindeodella multihamata*, *Hindeodella triassica*, *Lonchodina discreta*, *Lonchodina muelleri*, *Lonchodina venusta*, *Neogondolella aegaea*, *Neogondolella mombergensis*, *Neogondolella navicula*, *Ozarkodina kockeli*, *Ozarkodina saginata*, *Ozarkodina tortilis*, *Parachirognathus petrae-viridis*, *Prioniodella ctenoides*, *Prioniodella pectiniformis*, *Prioniodina kochi*, *Prioniodina* cf. *kochi*, *Prioniodina mediocris*, *Roundya magnidentata*, *Roundya* sp., *Spathognathodus gondolelloides*, *Spathognathodus homeri*		
Latomi	Lower to Middle Triassic (Induan to early Anisian)	40
Hindeodella multihamata, *Hindeodella triassica*, *Lonchodina* cf. *muelleri*, *Prioniodella prioniodellides*, *Spathognathodus homeri*		
Agios Markos	upper Carboniferous	40
Apathognathodus sp., *Hibbardella* sp., *Hindeodella* sp., *Lonchodina* sp., *Neoprioniodus* ? sp., *Ozarkodina* sp., *Roundya* sp., *Spathognathodus* sp., *Spathognathodus* sp. aff. *triangularis*		
Kephalovouni	Lower Triassic (Olenekian)	40
Apatognathus radiatus, *Apatognathus* sp., *Gladigondolella carinata*, *Hindeodella bitorta*, *Hindeodella ceweki*, *Hindeodella raridenticulata*, *Hindeodella triassica*, *Lonchodina discreta*, *Lonchodina muelleri*, *Ozarkodina* ? *fisticulata*, *Ozarkodina* cf. *delicatula*, *Ozarkodina saginata*, *Ozarkodina tortilis*, *Ozarkodina turgida*, *Roundya* sp., *Spathognathodus homeri*, *Spathognathodus triangularis*, *Spathognathodus* cf. *triangularis*		

Locality	Series/stages	Basic references
Marathovouno	Lower to Middle Triassic (Induan to Ladinian)	24, 40–44
Anastrophognathus sagittalis**,** ***Apatognathus mitzopouli*****,** ***Apatognathus radiatus*****,** *Apatognathus ziegleri*, ***Ctenognathus chionensis*****,** ***Gladigondolella carinata*****,** *Gladigondolella tethydis*, *Gladigondolella triangularis*, *Gondolella regalis*, *Gondolella timorensis*, *Hadrontina* sp., *Hibbardella triassica*, ***Hindeodella bitorta*****,** ***Hindeodella ceweki*****,** *Hindeodella multihamata*, *Hindeodella raridenticulata*, ***Hindeodella stoppeli*****,** *Hindeodella triassica*, *Hindeodella* sp., *Lonchodina* cf. *muelleri*, *Lonchodina discreta*, *Lonchodina muelleri*, *Lonchodina spengleri*, *Lonchodina venusta*, ***Neogondolella aegaea*****,** *Neogondolella mombergensis*, *Neogondolella navicula*, *Neogondolella regale*, *Neogondolella timorensis timorensis*, *Neospathodus homeri*, *Neospathodus triangularis*, ***Ozarkodina ?fisticulata*****,** *Ozarkodina saginata*, *Ozarkodina tortilis*, ***Ozarkodina turgida*****,** *Parachirognathus petrae-viridis*, *Prioniodella ctenoides*, *Prioniodella pectiniformis*, *Prioniodella prioniodellides*, *Prioniodina kochi*, *Prioniodina* cf. *prona*, *Roundya lautissima*, *Roundya* sp., *Spathognathodus* cf. *cristagalli*, ***Spathognathodus gondolelloides*****,** ***Spathognathodus homeri*****,** ***Spathognathodus triangularis***		
Marmarotrapeza	Lower to Middle Triassic (Olenekian to Anisian)	13, 40, 45, 46
Gladigondolella carinata, *Gladigondolella malayensis budurovi*, *Gladigondolella tethydis*, *Gondolella aegaea*, *Gondolella jubata*, *Gondolella mombergensis*, *Gondolella regale*, *Gondolella timorensis*, *Hindeodella* sp., *Lonchodina* cf. *muelleri*, *Neospathodus homeri*, *Neospathodus triangularis*, *Ozarkodina ?fisticulata*, *Ozarkodina saginata*, *Polygnathus carinata*, *Polygnathus tethydis*, *Prioniodina kochi*, *Spathognathodus gondolelloides*, *Spathognathodus homeri*, *Spathognathodus triangularis*, *Spathognathodus* cf. *triangularis*		
Agia Anna	middle Silurian to Upper Devonian	47
Belodella sp., *Hindeodella equidentata*, *Ligonodina* sp., *Lonchodina greilingi*, *Lonchodina walliseri*, *Neoprioniodus bicurvatus*, *Ozarkodina denckmanni*, *Paltodus recurvatus*, *Plectospathodus extensus*, *Polygnathoides siluricus*, *Spathognathodus inclinatus*, *Trichonodella symmetrica*		
Agia Anna	Lower Triassic (Olenekian)	40
Apatognathus radiatus, *Enantiognathus ziegleri*, *Hindeodella raridenticulata*, *Hindeodella triassica*, *Lonchodina venusta*, *Ozarkodina ?fisticulata*, *Ozarkodina turgida*, *Roundya* sp., *Spathognathodus homeri*, *Spathognathodus triangularis*		
Parthenis	Middle Devonian	40
Bellodella cf. *triangularis*, *Hibardella* sp., *Lonchodina* sp., *Spathognathodus* cf. *triangularis*		
Parthenis	Lower to Middle Triassic (Olenekian to early Anisian)	40, 48

Locality	Series/stages	Basic references
Apatognathus mitzopouli, Apatognathus radiatus, Apatognathus sp., *Enantiognathus ziegleri, Gladigondolella carinata, Hibbardella triassica, Hindeodella multihamata, Hindeodella* sp., *Hindeodella triassica, Lonchodina muelleri, Ozarkodina tortilis, Ozarkodina turgida, Pachycladina* ? sp., *Prioniodella prioniodellides, Prioniodina mediocris, Roundya magnidentata, Roundya meissneri, Spathognathodus gondolelloides, Spathognathodus homeri, Spathognathodus triangularis*		
Korakaris	Middle to Upper Triassic (Anisian to Carnian)	40
Enantiognathus ziegleri, Gladigondolella tethydis, Lonchodina muelleri, Neogondolella aegaea, Neogondolella navicula, Neogondolella palata, Ozarkodina saginata, Ozarkodina tortilis, Prioniodella pectiniformis, Prioniodina kochi, Roundya sp., *Spathognathodus gondolelloides, Spathognathodus* sp., *Tardogondolella abneptis*		
Agios Georgios Sykousis	Upper Pennsylvanian	40
Gnathodus cf. *noduliferus, Gnathodus* cf. *opimus, Gnathodus* cf. *roundyi, Gnathodus* n. sp., aff. *sicilianus, Gnathodus noduliferus, Gnathodus wapanuckensis, Hadrodontina ?* sp., *Hibbardella* sp., *Hindeodella* sp., *Idiognathoides* cf. *convexus, Idiognathoides ouachitensls, Idiognathoides* sp. aff. *ouachitensis, Ligonodina* ? sp., *Lonchodina* sp., *Ozarkodina delicatula, Ozarkodina* sp., *Spathognathodus* sp., *Streptognathodus* ex aff. *elongatus, Subbryantodus* sp., *Synprioniodina* sp.		
Corfu		
Foustapidima Cape	Upper Triassic (Carnian)	17
Polygnathus abneptis		
Crete		
Voutas	upper Carboniferous to Middle Triassic	49, 50
Anastrophognathus sagittalis, Ellisonia sp., *Enantiognathus petraeviridis, Gladigondolella carinata, Gladigondolella malayensis, Gondolella foliata foliata, Gondolella foliata inclinata, Gondolella jubata, Gondolella orientalis, Gondolella planata, Gondolella subcarinata, Gondolella* sp., *Hadrodontina anceps, Hindeodus minutus, Icriospathodus collinsoni, Metapolygnathus japonicus, Neospathodus cristagalli, Neospathodus dieneri, Neospathodus homeri, Neospathodus longiusculus, Neospathodus* sp., *Neospathodus triangularis, Pachycladina inclinata, Pachycladina longispinosa, Pachycladina symmetrica, Xaniognathus turgidus, Xaniognathus* sp.		
Sfinari	upper Carboniferous to Triassic	51
Cavusgnathus sp., *Gladigondolella malayensis, Gnathodus angustus, Gondolella* sp. ex aff. *G. auriformis, Gondolella denuda, Gondolella intermedia, Gondolella laevis, Gondolella nepalensis, Gondolella tadpole, Gondolella* sp., *Hindeodus minutus, Idiognathodus tersus, Idiognathodus togashii, Idiognathoides sinuatus, Metapolygnathus abneptis, Metapolygnathus mirautae, Metapolygnathus* sp., *Neognathodus bassleri, Neognathodus* sp., *Neognathodus symmetricus, Neospathodus homeri, Pachycladina* sp., *Spathognathodus orphanus, Streptognathodus elegantulus, Streptognathodus gracilis, Streptognathodus ruzhencevi, Streptognathodus* sp.		
Kambos	upper Carboniferous to Triassic	51

Locality	Series/stages	Basic references
Ellisonia sp., *Furnishius triserratus, Neospathodus dieneri, Neospathodus longiusculus, Neospathodus waageni, Pachycladina longispinosa, Pachycladina* sp., *Pachycladina symmetrica*		
Paleochora	upper Carboniferous to Triassic	51
Neospathodus homeri, Ellisonia sp.		
Skafi	upper Carboniferous to Triassic	51
Neospathodus sp., *Neospathodus homeri*		
Myrsini	lower Permian to Lower Triassic	52
Cypridodella sp., *Diplognathodus* sp., *Ellisonia* sp., *Enantiognathus* sp., *Eurygnathodus costatus, Eurygnathodus paracostatus, Gondolella asiatica, Gondolella gujioensis, Gondolella idahoensis, Gondolella intermedia, Gondolella planata, Gondolella* sp., *Hindeodus minutus, Hindeodus* sp., *Iranognathus* sp., *Neohindeodella triassica, Neospathodus cristagalli, Neospathodus dieneri, Neospathodus pakistanensis, Neospathodus waageni, Neospathodus* sp., *Ozarkodina sweeti, Ozarkodina* sp., *Prioniodina mitzopouli, Prioniodina* sp., *Sweetognathus bogolovskajae*		
Skopi	Lower to Middle Triassic (Olenekian to Anisian)	52
Diplododella bidentata, Ellisonia sp., *Enantiognathus bitortus, Gladigondolella carinata, Gondolella* ? *timorensis, Neohindeodella* sp., *Neospathodus homeri, Neospathodus* cf. *homeri, Neospathodus* cf. *triangularis, Neospathodus* sp., *Ozarkodina turgida, Ozarkodina* sp.		
Tripokefala	Lower to Middle Triassic (Olenekian to Anisian)	52
Anastrophognathus sagittalis, Enantiognathus bitortus, Ellisonia sp., *Enantiognathus* sp., *Gladigondolella tethydis, Gondolella bulgarica, Gondolella* cf. *regalis, Neohindeodella* sp., *Neospathodus homeri, Ozarkodina tortilis, Ozarkodina turgida, Prioniodina (Cypridodella) muelleri*		
Ziros	Upper Triassic	53
Conodont fragments indet.		

[1]Lekkas (1986), [2]Ardaens (1978), [3]Ferriere (1974), [4]Wigniolle (1977), [5]Celet (1977), [6]Johns (1977), [7]Vrielynck and Kauffmann in Fleury (1980), [8]Vrielynck in Fleury (1980), [9]Steuber (1991), [10]Clement (1977), [11]De Bono et al. (2001), [12]Caridroit et al. (2000), [13]Bender and Kockel (1963), [14]Flament (1973), [15]Kozur in Thiebault (1982), [16]Terry (1969), [17]Huckriede (1958), [18]Krystyn and Mariolakos (1975), [19]Vrielynck (1978a), [20]Vrielynck (1978b), [21]Vrielynck (1980), [22]Baumgartner (1985), [23]Mauvier in Decourt 1964, [24] Bender 1968b, [25]Bender et al. (1960), [26]Dürkoop et al. (1986), [27]Noyan and Kozur (2007), [28]Muttoni et al. (1994), [29]Muttoni et al. (1997), [30]Angiolini et al. (1992), [31]Nestell and Wardlaw (1987), [32]Römermann (1968), [33]Argyriou et al. (2017), [34]Herget and Roth (1968), [35]Groves et al. (2003), [36]Zanchi et al. (2003), [37]Roth (1968), [38]Kauffmann (1965), [39]Larghi et al. (2005), [40]Tietze (1969), [41]Bender (1968a), [42]Besenecker et al. (1968), [43]Assereto et al. (1980), [44]Muttoni et al. (1995), [45]Gaetani et al. (1992), [46]Jacobshagen et al. (1993), [47]Walliser in Tietze (1969), [48]Tietze in Jacobshagen and Tietze (1974), [50]Kozur and Krahl (1984), [51]Krahl et al. (1983), [52]Krahl et al. (1986), [53]Zambetakis–Lekkas (1977)

References

Aldridge RJ (1987) Conodont palaeobiology: a historical review. In: Aldridge RJ (ed) Palaeobiology of Conodonts. Ellis Horwood, Chichester, pp 11–34
Aldridge RJ, Briggs DEG, Clarkson ENK, Smith MP (1986) The affinities of conodonts new evidence from the Carboniferous of Edinburgh, Scotland. Lethaia 19:279–291
Aldridge RJ, Briggs DEG, Smith MP, Clarkson ENK, Clark NDL (1993) The anatomy of conodonts. Philos Trans R Soc Lond Ser B 340:405–421
Angiolini L, Dragonetti L, Muttoni G, Nicora A (1992) Triassic stratigraphy in the island of Hydra (Greece). Riv Ital Paleontol Stratigr 98(2):137–180
Ardaens R (1978) Géologie de la chaine du Vardoussia. Comparaison avec le massif du Koziakas (Grèce Continentale). These 3eme cycle, Univ Sci Techn Lille
Argyriou T, Romano C, Carrillo-Briceño DG, Brosse M, Hofmann R (2017) The oldest record of gnathostome fossils from Greece: Chondrichthyes from the Lopingian of Hydra Island. Palaeontol Electron 20.1(8A):1–9
Assereto R, Jacobshagen V, Kauffmann G, Nicora A (1980) The Scythian/Anisian boundary in Chios, Greece. Riv Ital Paleontol Stratigr 85:715–735
Bassler RS (1925) Classification and stratigraphic use of conodonts. Bull Geol Soc Am 36:218–220, New York
Baumgartner O (1985) Jurassic sedimentary evolution and nappe emplacement in the Argolis Peninsula (Peloponnesus, Greece). Birkhäuser Verlag, Basel
Bender H (1968a) Der Nachweis von Unter–Anis («Hydasp») auf der Insel Chios. Ann Géol Pays Hellén XIX:412–464
Bender H (1968b) Die Conodonten Chronologie der Mediterranen Trias. Ann Géol Pays Hellén XIX:465–540
Bender H, Kockel CW (1963) Die Conodonten der Griechischen Trias. Ann Géol Pays Hellén XIV:436–445
Bender H, Stoppel D (1965) Perm-Conodonten. Geol Jahrb 82:331–364
Bender H, Hirschberg KM, Leuteritz K, Mänz H (1960) Zur geologie der Olonos–Pindos– und der Pernass–Kionazone im tal des Asklepieion (Argolis). Ann Géol Pays Hellén XI:201–213
Besenecker H, Dürr S, Herget G, Jacobshagen V, Kauffmann G, Lüdtke G, Roth W, Tietze K-W (1968) Geologie von Chios (Ägäis). Geol Palaeontol 2:121–150
Branson ER (1934) Conodonts from the Hannibal Formation of Missouri. Missouri Univ Stud 8:301–343
Branson EB, Mehl MG (1933) Conodont studies. Missouri Univ Stud 8(1–4):1–343
Branson EB, Mehl MG (1934a) Conodonts from the Grassy Creek shale of Missouri. Missouri Univ Stud 8(3):171–259
Branson EB, Mehl MG (1934b) Conodonts from the Bushberg sandstone and equivalent formations of Missouri. Missouri Univ Stud 8(4):265–300
Branson EB, Mehl MG (1938) The conodont genus *Icriodus* and its stratigraphic distribution. J Paleontol 12(2):156–166
Branson EB, Mehl MG (1941) New and little known Carboniferous conodont genera. J Paleontol 15(2):97–106
Briggs DEG, Clarkson ENK, Aldridge RJ (1983) The conodont animal. Lethaia 16:1–14
Briggs DEG, Aldridge RJ, Smith MP (1987) Conodonts are not aplacophoran molluscs. Lethaia 20:381–382
Caridroit M, Ferrière J, Dégardin J-M, Vachard D, Clément B (2000) Première datation des lydiennes paléozoïques dans les Hellénides internes (mont Parnis, Grèce); implications géologiques. Acad Sci Paris, Sci Terre Planètes 331:413–418
Celet P (1977) Les bordures de la zone du Parnasse (Grèce). Evolution Paléogéographique au Mésozoique et caractères structuraux. Proceedings of the VI Colloquium on the Geology of the Aegean region, Kallergis G (ed) Inst Geol Mining Res, Athens, pp 725–740 (in French with English abstract)

Clark DL (1959) Conodonts from the Triassic of Nevada and Utah. J Paleontol 33:305–312
Clark DL (1972) Early Permian crisis and its bearing on Permo–Triassic conodont taxonomy. Geol Palaeontol 1:147–158
Clark DL (1981) Phylum Conodonta Eichenberg, 1930. In: Robison RA (ed) Treatise on invertebrate paleontology, W Supplement 2. Conodonta, W111. Geol Sac Am Univ Kansas Press, Lawrence
Clement B (1977) Relations structurales entre la Zone du Parnasse et al Zone Pélagonienne en Béotie (Grèce continentale). Proceedings of the VI Colloquium on the Geology of the Aegean region, Kallergis G (ed) Inst Geol Mining Res Athens, pp 237–251
De Bono A, Martini R, Zaninetti L, Hirsch F, Stampfli MG, Vavassis I (2001) Permo–Triassic stratigraphy of the pelagonian zone in central Evia island (Greece). Eclogae Geol Helv 94:289–311
Decourt J (1964) Contribution à l'étude géologique d'un secteur du Péloponnèse septentrional. Ann Géol Pays Hellén XV:1–418
Diebel K (1956) Conodonten in der Oberkreide von Kamerun. Andean Geol 5:424–450. (in German)
Donoghue PCJ, Forey PL, Aldridge RJ (2000) Conodont affinity and chordate phylogeny. Biol Rev 75:191–251
Dunn DL (1970) Middle Carboniferous conodonts from western United States and phylogeny of the platform group. J Paleontol 44:312–342
Dürkoop A, Richter DK, Stritzche R (1986) Fazies, Alter und Korrelarion der triadischen Rotkalke von Epidaurus, Adhami und Hydra (Griechenland). Facies 14:105–150
Dzik J (1986) Chordate affinities of the conodonts. In: Hoffman A, Nitecki MH (eds) Problematic fossil taxa, Oxford monographs on geology and geophysics, vol 5. Oxford University Press, New York, pp 240–254
Ethington RL (1959) Conodonts of the Ordovician Galena formation. J Paleontol 33:257–292
Ferriere J (1974) Étude géologique d'un secteur des zones helléniques internes subpélagonienne et pélagonienne (massif de l'Othrys–Grèce continentale). Importance et signification de la période orogénique anté–Crétacé supérieur. Bull Soc Géol Fr S7–XVI(5):543–562
Flament JM (1973) De l' Olonos au Chelmos étude géologique d' un secteur de la nappe du Pinde–Olonos. Thèse 3eme Cycle, Lille, pp 206
Fleury JJ (1980) Les zone de Gavrovo–Tripolitza et du Pinde–Olonos (Grèce continentale et Péloponnèse du Nord). Evolution d'une plate–forme et d'un basin dans leur cadre alpin. Ann Soc Géol Nord 4(1):473
Furnish WM (1938) Conodonts from the Prairie du Chien (Lower Ordovician) Beds of the Upper Mississippi Valley. J Paleontol 12:318–340
Gabbott SE, Aldridge RJ, Theron JN (1995) A giant conodont with preserved muscle tissue from the Upper Ordovician of South Africa. Nature 374:800–803
Gaetani M, Jacobschagen V, Nicora A, Kauffmann G, Tselepidis V, Fantini Sestini N, Mertmann D, Scourtsis-Coroneou V (1992) The Early–Middle Triassic boundary at Chios (Greece). Riv Ital Paleontol Stratigr 98(2):181–204
Gould SJ (1983) Nature's great era of experiments. Nat Hist 7(83):12–21
Groves RJ, Larghi C, Nicora A, Rettori R (2003) Mississippian (Lower Carboniferous) microfossils from the Chios Mélange (Chios Island, Greece). Geobios 36:379–389
Gunnell FH (1931) Conodonts from the Fort Scott limestone of Missouri. J Paleontol 5(3):244–252
Harris RW, Hollingsworth RV (1933) New Pennsylvanian conodonts from Oklahoma. Am J Sci 25:193–204
Hass WH (1953) Conodonts of the Barnett Formation of Texas. US Geol Surv Prof Pap 243-F:69–98
Hass WH (1962) Conodonts. In: Moore RC (ed) Treatise on invertebrate paleontology, part W. Miscellanea Geol Soc America Univ Kansas Press, pp 3–69
Hayasi S (1968) The Permian conodonts in chert of the Adoyama Formation, Ashio Mountains, Central Japan. Earth Sci 22(2):63–77, Tokyo
Helms H (1959) Conodonten aus dem Saalfelder Oberdevon (Thuringen). Geologie 8:634–677

Herget G, Roth W (1968) Stratigraphie des Paläozoikums im Nordwest–Teil der Insel Chios (Ägäis). Neues Jahrb Geol Palaeontol Abh 131(1):46–71

Hinde GL (1879) On conodonts from the Chazy and Cincinatti group of the Cambro-Silurian, and from the Hamilton and Genesee Shale divisions of the Devonian, in Canada and the United States. Quart J Geol Soc Lond 35:351–369

Hirschmann C (1959) Uber Conodonten aus den Oberen Muschelkalk des Thüringer Beckens. Freib Forsch H 76:33–86

Huckriede R (1958) Die Conodonten der Mediterranen Trias und ihr stratigraphischer Wert. Paläontol Z 32(3/4):141–175

Huddle JW (1934) Conodonts from the New Albany Shale of Indiana. Bull Am Paleontol 21(72):1–136

Hyman LB (1959) The Inrzertebratu. V. Smaller Coeloniate Groups. McGraw–Hill Book Company, Inc, New York/London/Toronto, 783 pp

Jacobshagen V, Tietze KW (1974) Biostratigraphische Probleme im Skyth/Anis-Grenzbereich auf der Insel Chios (Ägäis). Schrirt Erdwiss Komm 2:115–123, Wien

Jacobshagen V, Gaetani M, Nicora·A, Tselepidis·V, Kauffman·G, Mertman D, Skourtsis-Coroneou·V, Fantini Sestini·N (1993) The Early/Middle Triassic boundary on Chios island: preliminary results of a reinvestigation. Bull Geol Soc Greece XXVIII(3):25–38

Janvier P (1981) The phylogeny of the Craniata, with particular reference to the significance of fossil 'agnathans'. J Vertebr Paleontol 1:121–159

Janvier P (1983) L'animal-conodonte' enfin demasque? La Recherche 14(145):832–833

Janvier P (1995) Conodonts join the club. Nature 374:61–762

Janvier P (1996) The dawn of the vertebrates: characters versus common ascent in the rise of current vertebrate phylogenies. Palaeontology 39:407–287

Johns DR (1977) The structure and stratigraphy of the Galaxidion Region, Central Greece. Proceedings of the VI Colloquium on the Geology of the Aegean region, Kallergis G (ed) Inst Geol Mining Res, Athens, pp 715–724

Kauffmann G (1965) Fossil–belegtes Alt–Pälaozoikum im Nordostteil der lnsel Chios (Ägäis). Neues Jb Geol Paläontol Mh 7:400–404, Stuttgart

Kozur H (1968) Neue Conodonten aus dem Oberen Muschelkalk des germanischen Binnenbeckens. Mber Deutsch Akad Wiss Berlin 10(20):130–142, Berlin

Kozur H (1975) Beitrage zur Conodontefauna des Perm. Geol Paläontol Mitt Ibk 5(4):1–44

Kozur H (1980) The main events in the Upper Permian and Triassic conodont evolution and its bearing to the Upper Permian and Triassic stratigraphy. Riv Ital Paleontol Stratigr 85(3–4):741–766

Kozur H (1989) The taxonomy of the Gondolellid conodont in the Permian and Triassic. In: Ziegler W (ed) 1st International Senckenberg Conference and 5th European Conodont Symposium (ECOS V). Contribution III. Forschungsinstitut Senckenberg, Frankfurt, pp 409–469

Kozur H (1990) *Norigondolella* n. gen., eine neue obertriassische Conodontengattung. Palaontol Z 64(1/2):125–132

Kozur H (2003) Integrated ammonoid, conodont and radiolarian zonation of the Triassic. Hallesches Jahrb Geowiss 25:49–79

Kozur H, Krahl J (1984) Erster Nachweis triassischer Radiolaria in der Gruppe auf der Insel Kreta [First evidence of Triassic Radiolaria in the Phyllite Group Crete Island]. Neues Jb Geol Paläontol Mh 7:400–404

Kozur H, Mock R (1974) Zwei neue conodonten-Arten aus der Trias des Slowakischen Karstes. Cas Miner Geol 19:135–139

Kozur H, Mostler H (1970) Neue Conodonten aus der Trias. Ber Nat–Med Ver Ibk 58:429–464

Kozur H, Mostler H, Rahimi-Yazd A (1975) Beitrage zur mikrofauna permitriadischer Schichtfologen Teil II: Neue conodonten aus dem Ober Perm und der basalen Trias von Nord- und Zentraliran. Geol Paläontol Mitt Ibk 5(3):1–23

Krahl JV, Kauffmann GG, Kozur HM, Richter D, Förster OA, Heinritzi F (1983) Neue Daten zur Biostratigraphie und zur tektonischen Lagerung der Phyllit–Gruppe und der Trypali–Gruppe auf der Insel Kreta (Griechenland). Geol Rundsch 72(8):1147–1166

Krahl J, Kauffmann G, Richter D, Kozur H, Möller I, Förster O, Heinritzi F, Dornsiepen U (1986) Neue Fossilfunde in der Phyllit–Gruppe Ostkretas (Griechenland). Z Dt Geol Ges 137:523–536
Krystyn L, Mariolakos I (1975) Stratigraphie und Tektonik der Hallstätter–Kalk–Scholle von Epidauros (Griechenland). Sber Österr Akad Wiss Math–Naturwiss Kl I 184:181–195
Larghi C, Cordey F, Corradini C, Gaetani M, Nicora A (2005) Palaeozoic (Silurian and Devonian) radiolarians and conodonts from chert olistoliths of the Volissos Turbidites, Chios island, Greece. Eclogae Geol Helv 98:123–131
Lekkas E (1986) Presence of Triassic clastics at the base of Koziakas sequence, Western Thessaly. IGME, Geolog Geophys Res Spec iss, Athens, pp 235–242 (in Greek)
Lindström M (1954) Conodonts from the lowermost Ordovician strata of South-central Sweden. Geol Fören Stockh Förh 76(4):517–604
Lindström M (1964) Conodonts. Elsevier, Amsterdam, p 196
Maisey JG (1986) Head and tails: a chordate phylogeny. Cladistics 2:201–256
Mosher LC (1968) Triassic conodonts from western North America and Europe and their correlation. J Paleontol 42(4):895–946
Mosher LC (1969) Nomenclatural revisions for triassic conodonts in MOSHER 1968. J Paleonolt 43:1441, Tulsa
Mosher LC, Clark DL (1965) Middle Triassic conodonts from the Prida Formation of Northwestern Nevada. J Paleontol 39(4):351–365
Müller KJ (1956) Triassic conodonts from Nevada. J Paleontol 30(4):818–830
Müller KJ (1962) Zur systematischen Einteilung der Conodontophorida. Palaontol Z 36(1/2):109–117
Müller KJ (1981) Zoological affinities. In: Robison RA (ed) Treatise on invertebrate paleontology, part W, Miscellanea, Suppl 2, Conodonta. Geol Soc Am Univ Kansas, Lawrence, pp 78–82
Murdock D, Dong X-P, Repetski EJ, Marone F, Stampanoni M, Donoghue PH (2013) The origin of conodonts and of vertebrate mineralized skeletons. Nature 502:546–549
Murray FN, Chronic J (1965) Pennsylvanian conodonts and other fossils from insoluble residues of the Minturn Formation (Desmoinesian), Colorado. J Paleontol 39(4):594–610
Muttoni G, Channell ETJ, Nicora A, Rettori R (1994) Magnetostratigraphy and biostratigraphy of an Anisian–Ladinian (Middle Triassic) boundary section from Hydra (Greece). Palaeogeogr Palaeoclimatol Palaeoecol 111:249–262
Muttoni G, Kent DV, Gaetani V (1995) Magnetostratigraphy of a Lower–Middle Triassic boundary section from Chios (Greece). Phys Earth Planet Inter 92:245–260
Muttoni G, Kent DV, Brack P, Nicora A, Balini M (1997) Middle Triassic magnetostratigraphy and biostratigraphy from the Dolomites and Greece. Earth Planet Sci Lett 146:107–120
Muttoni G, Mazza M, Mosher D, Katz EM, Kent VD, Balini M (2014) A Middle–Late Triassic (Ladinian–Rhaetian) carbon and oxygen isotope record from the Tethyan Ocean. Palaeogeogr Palaeoclimatol Palaeoecol 399:246–259
Nestell KM, Wardlaw RB (1987) Upper Permian conodonts from Hydra, Greece. J Paleontol 61(4):758–772
Noyan FÖ, Kozur WH (2007) Revision of the late Carnian–early Norian conodonts from the Stefanion section (Argolis, Greece) and their palaeobiogeographic implications. Neues Jb Geol Paläontol Abh 245(2):159–178
Orchard MJ (1991) Late Triassic conodont biochronology and biostratigraphy of the Kunga Group, Queen Charlotte Islands, British Columbia. In: Evolution and hydrocarbon potential of the Queen Charlotte Basin, British Columbia. Geol Surv Can Pap 90(10):173–193
Pander CH (1856) Monographie der fossilen Fische des silurischen Systems der russischbaltischen Gouvernements. Akad Wiss, St Petersburg, pp 1–91
Perret M-F (1993) Recherches micropaléontologiques et biostratigraphiques (conodontes-foraminifères) dans le Carbonifère Pyrénéen. Strata 21:1–597
Repetski JE, Szaniawski H (1981) Paleobiologic interpretation of Cambrian and earliest Ordovician conodont natural assemblages. In: Taylor ME (ed) Short Pap Sec lnt Symp Cambrian Syst US Geol Surv, Open–File Report 82–743:169–172

Rexroad CB, Furnish WM (1964) Conodonts from the Pella Formation (Mississippian), south-central Iowa. J Paleontol 38:667–676
Rhodes FHT (1963) Conodonts from the topmost Tensleep sandstone of the eastern Big Horn Mountains, Wyoming. J Paleontol 37(2):401–408
Rhodes FHT, Müller KJ (1956) The conodont genus *Prioniodus* and related forms. J Paleontol 30(3):695–699
Rietschel S (1973) Zur Deutung der Conodonten. Nat Mus 103:409–418
Römermann H (1968) Geologie von Hydra (Griechenland). Geol Palaeontol Marburg 2:163–171
Roth G (1968) Geologie von NW-Chios (Ägäis). PhD thesis, University of Marburg
Sannemann D (1955) Beitrag zur untergliederung des Oberdevons nach Conodonten. Neues Jb Geol Paläontol Abh 100:324–331
Sansom IJ, Smith MP, Armstrong HA, Smith MM (1992) Presence of the earliest vertebrate hard tissues in conodonts. Science 256:1308–1311
Schmidt H, Müller KJ (1964) Weitere Funde von Conodonten–Gruppen aus dem oberen Karbon des Sauerlandes. Paläontol Z 38:105–135
Scott HW (1942) Conodont assemblage from the Heath Formation, Montana. J Paleontol 16:293–300
Smith MM, Hall BK (1990) Development and evolutionary origins of vertebrate skeletogenic and odontogenic tissues. Biol Rev 65:277–373
Solien MA (1979) Conodont biostratigraphy of the Lower Triassic Thaynes Formation, Utah. J Paleontol 53(2):276–306
Spasov C, Ganev M (1960) Karnische Conodonten ans dem Luda–Kamcia–Teil des Ostbalkans. Bulgaria Trav Géol Ser Paliontol 2:77–99
Staesche U (1964) Conodonten aus dem Skyth von Südtirol. Neues Jb Geol Paläontol Abh 119:247–306, Stuttgart
Stauffer CR (1940) Conodonts from the Devonian and associated clays of Minnesota. J Paleontol 14:417–435
Stauffer RC, Plummer JH (1932) Texas Pennsylvanian conodonts and their stratigraphic relations. Univ Texas Bull 3201:13–50
Steuber T (1991) Conodont stratigraphy, depositional environments and stable isotope composition of the Triassic in the Helicon Mountains (Beotia, Greece). Proceedings of the 5th Congress. Thessaloniki, May 1990. Bull Geol Soc Greece XXV:415–528
Sweet WC (1970) Permian and Triassic conodonts from a section at Guryul Ravine, Vihi district, Kashmir. Univ Kansas Paleontol Contr 49:1–10
Sweet WC, Donoghue PCJ (2001) Conodonts: past, present, future. J Paleontol 75:1174–1184
Szaniawski H (1982) Chaetognath grasping spines recognised among Cambrian protoconodonts. J Paleontol 56:806–810
Tatge U (1956) Conodonten aus dem germanischen Muschelkalk. Paläontol Z 30(1/2):108–127
Terry J (1969) Etude géologique d'un secteur de la Messénie septentrionale, Grèce. DEA Lille, p 112
Thiebault F (1982) Evolution géodynamique des Héllénides externs en Péloponnèse méridional (Grèce). Soc Géol Nord S6–2:574
Tietze KW (1969) Geologie von Mittel-Chios (Agais). PhD thesis, Diss Univ Marburg
Tillier S, Cuif J-P (1986) L'animal–conodonte est–il un Mollusque Aplacophore? C R Acad Sci, Paris 303(II):627–632
Ulrich EO, Bassler RS (1926) A classification of the toothlike fossils, conodonts, with descriptions of American Devonian and Mississippian species. US Nat Mus Proc 68(2613):1–63
Vrielynck B (1978a) Données nouvelles sur les zones internes du Péloponnèse: Les massifs á l' est de la Plaine d'Argos (Grèce). Doctoral thesis, Univ Sci Et Techn Lille
Vrielynck B (1978b) Données nouvelles sur les zones internes du Péloponnèse: Les massifs á l' est de la Plaine d'Argos (Grèce). Ann Géol Pays Hellén XXIX/2:440–462
Vrielynck B (1980) Precisions sur la stratigraphie du Trias d'Argolide, (Péloponnèse, Grèce) et consequences structurales. Bull Soc Geol Fr 7(XXII(3)):345–352

Walliser OH (1957) Conodonten aus dem Oberen Gotlandicum Deutschland und der Karnischen Alpen. Notizbl Hess Landesamt Bodenforschung Wiesbaden 85:28–52

Walliser OH (1964) Conodonten des Silurs. Abh Hess Land Bodenf Wiesbaden 41:106

Wigniole E (1977) Données nouvelles sur la géologie du massif de l'Iti (Grèce continentale). Ann Soc Géol Nord XCVII(3):239–251

Zanchi A, Garzanti E, Larghi C, Angiolini L, Gaetani M (2003) The Variscan orogeny in Chios (Greece): carboniferous accretion along a Palaeotethyan active margin. Terra Nova 15:213–223

Ziegler W (1959) Conodonten aus Devon und Karbon Süd-westeuropas und Bemerkungen zur bretonischen Faltung (Montagne Noire, Massiv Mouthoumet, Span. Pyrenäen). N J Geol Paläeontol Mh 7:289–309

Ziegler W (1960) Paläontologischer Anhang. In: Kronberg P, Pilger A, Scherp A, Ziegler W (eds) Spuren Altvariscischer Bewegungen im nordöstlichen Teil des Rheinischen Schiefergebirges, Fortschr Geolog Rheinland Westfalen 3:1–46, pp 35–43

Ziegler W, Sandberg CA (1990) The Late Devonian standard conodont zonation. Cour Forschungsinst Senck 121:1–115

Zambetakis-Lekkas A (1977) La série de Mangassa. Stratigraphie, Paléogéographie, Tectonique. VI Colloquium on the geology of the Aegean region, Athens 1977, Proc I:103–109

The Fossil Record of Ray-Finned Fishes (Actinopterygii) in Greece

Thodoris Argyriou

1 Introduction

A continuously thickening count of over 33,000 living species, already surpassing that of all other vertebrate species combined, renders Actinopterygii (ray-finned bony fishes) the most diverse clade of vertebrate animals (Nelson et al. 2016; Eschmeyer and Fong 2019). This vast taxonomic richness is coupled with an astounding array of anatomical, feeding, reproductive, locomotory, and other biological specializations, which enabled ray-fins to conquer most aquatic niches on the planet (from abyssal plains to hypersaline volcanic lakes), and to even glide, or crawl, outside the aquatic medium (Helfman et al. 2009; Nelson et al. 2016; Froese and Pauly 2019).

Actinopterygian origins date back to at least the late Silurian (~425 Ma), as inferred by the age of the first fossils of their sarcopterygian sister group (Zhu et al. 2009). The oldest definitive actinopterygian fossils are dated in the Early and Middle Devonian (Giles et al. 2015; Lu et al. 2016). Similarly, all major living actinopterygian lineages (Cladistia; Chondrostei; Neopterygii) are widely thought to have diverged in the Paleozoic (Near et al. 2012; Broughton et al. 2013; Sallan 2014; Betancur-R et al. 2015; Friedman 2015), but their first accepted occurrences are in early and middle Mesozoic rocks (Olsen 1984; Hilton and Forey 2009; Giles et al. 2017). The current taxonomic dominance of actinopterygians relative to other aquatic vertebrates was foreshadowed by their impressive Triassic radiation(s), in the

T. Argyriou (✉)
UMR 7207 (MNHN – Sorbonne Université – CNRS) Centre de Recherche en Paléontologie – Paris, Muséum national d'Histoire naturelle, Paris cedex 05, France

Department of Earth and Environmental Sciences, Paleontology & Geobiology, Ludwig-Maximilians-Universität München, Richard-Wagner-Str. 10, München, Germany

GeoBio-Center, Ludwig-Maximilians-Universität München, Richard-Wagner-Str. 10, München, Germany
e-mail: t.argyriou@lrz.uni-muenchen.de

E. Vlachos (ed.), *Fossil Vertebrates of Greece Vol. 1*,
https://doi.org/10.1007/978-3-030-68398-6_4

aftermath of the catastrophic Permian-Triassic extinction (~251 Ma; Friedman and Sallan 2012; Romano et al. 2016), and the appearance of Telostei in the Middle Triassic (total group: Tintori et al. 2015; Arratia 2017). The Cretaceous–Paleogene Extinction (K/Pg) led to the complete eradication or decimation of several long-lasting Mesozoic actinopterygian lineages, especially in marine waters (Friedman 2009; Friedman and Sallan 2012; Guinot and Cavin 2016). Paleogene and younger radiations in the newly emptied ecological spaces marked the beginning of a new era in the evolutionary history of the clade, in which Teleostei, largely represented by members of the ostariophysan (mostly in freshwaters) and acanthomorph (spiny-rayed fishes) super-radiations, are established as dominant vertebrate components of global aquatic niches (Alfaro et al. 2009, 2018; Sibert and Norris 2015).

The long evolutionary history of ray-finned fishes is evidenced by a particularly rich fossil record. Stratified aquatic deposits can yield large numbers of articulated, yet flattened individuals (for some examples from Greece, see Gaudant 2001, 2004). Laminated rocks, corresponding to rapid deposition and low energy and low bottom oxygenation environments, can sometimes allow the preservation of very delicate structures, such as gut contents, embryos, or even soft tissues and pigments (e.g., Viohl 1990; Vinther 2015; Argyriou et al. 2016; Maxwell et al. 2017), allowing for painting a more vivid picture of past ecosystems and biotic interactions within them, while providing important insights into the evolution of key anatomical structures and biological specializations. Three-dimensional anatomical structure is only rarely preserved in fossil actinopterygians, but such fossils can allow the study of structures that are lost in flattened individuals (e.g., braincases, sensory endocasts, gill skeletons), and are of tremendous value to paleontologists, especially in light of modern fossil imaging and investigation techniques (Friedman et al. 2015; Argyriou et al. 2018). Equally important biological and geological information often derives from studies of identifiable ichthyoliths (isolated bones, scales and teeth, Bellas et al. 1998; Argyriou et al. 2015; Vasileiadou et al. 2017) and otoliths, (e.g., Agiadi et al. 2013, 2019; Agiadi this volume). Due to their exquisite fossil record, actinopterygians constitute an excellent means for studying the evolution of aquatic life on the planet, and its response to major biotic crises and extinctions (Friedman and Sallan 2012; Guinot and Cavin 2016; Romano et al. 2016). At the same time, the largely ectothermic traits, in conjunction with the strict environmental and habitat preferences of most ray-fins (e.g., Helfman et al. 2009), establish their fossil record as a powerful tool for making paleoenvironmental and paleobiogeographical inferences (Gaudant 2002; Landini and Sorbini 2005a; Argyriou et al. 2015; Carnevale et al. 2018), often through the scope of isotope geochemistry (e.g., Otero et al. 2011).

Historically, the study of fossil actinopterygians has been associated with several challenges, chief of which is the sheer richness of actinopterygian biodiversity and fossil record, which encompasses numerous groups of tens or hundreds of millions of years of divergent evolutionary histories (Friedman and Sallan 2012; Near et al. 2012; Broughton et al. 2013; Betancur-R et al. 2015; Friedman 2015; Nelson et al. 2016). As a result, commonly used synapomorphies for many clades of ray-fins—mostly teleosts—are established on modern taxa alone, and their validity has not been questioned by large-scale phylogenetic analyses (see, e.g., Wiley and Johnson

2010). Placing early diverging fossil taxa without closely related modern representatives on the tree of life is an extremely difficult task and constitutes a frontier in current paleoichthyology (Sallan 2014; Friedman 2015; Giles et al. 2017; Argyriou et al. 2018). Even when modern-looking forms are examined, their study usually requires consultation of extensive comparative materials in collections. Often limited anatomical information coming from two-dimensionally flattened fossils, as well as prevalent issues related to structural homology—mostly encountered when examining deeply nested forms (Schultze 2008)—pose additional obstacles for deciphering the fossil record and history of the clade.

Fossil actinopterygians form an extremely important, but historically overlooked, component of Greek vertebrate paleontological research. This synthesis aims at bringing together available information on the rich record of body fossils of the clade in Greece and critically assessing historical attributions in terms of validity and reliability—at least according to the opinion of the author, and based on a critical review of the literature. By placing all relevant occurrences in an up-to-date systematic, geographic, and stratigraphic context, it is anticipated that this chapter will fuel future paleo-ichthyological research in Greece.

2 Historical Overview and Collection Information

Actinopterygians from the Miocene of Kymi, Euboea, were the first to attract scientific attention on fishes from Greece, during the 1800s (de Waldheim 1838, 1850; Valenciennes 1861; Gaudry 1862; Unger 1867), but these early authoritative—rather than synapomorphy-based—taxonomic assessments are in need of revision, especially since they are not accompanied by adequate figures of the material. Fossil cyprinodontids were discovered in the Messinian of Crete by Raulin (1861) and were later revisited by Woodward (1901). In the same work, Woodward updated the taxonomic status of fossil percomorphs from Kymi, first described by de Waldheim (1838, 1850), on the basis of material in the collections of what is nowadays the Natural History Museum of London (but see Centropomidae section below). After a four-decade hiatus, Weiler described a fossil clupeid from the Miocene of Serres Basin (Weiler 1943).

The second half of the twentieth century saw an increase in interest in the study of Greek ray-fin fossil record. Kottek described some fragmentary leptolepid remains from the Toarcian (Early Jurassic) of Leukada, which derive from a since unexplored, and allegedly rich fossil fish bonebed near the village of Anavryssada (Kottek 1964). This represents the stratigraphically oldest occurrence of fossil ray-fins in Greece known to date. Melentis was the first Greek paleontologist to publish on fossil actinopterygians (Melentis 1967a, b). Unfortunately, his accounts of a †pycnodontid from the Cenomanian (Late Cretaceous) of Lindos, Rhodos, should be treated with extreme caution (see †Pycnodontidae section below). Important, but largely overlooked and preliminary taxonomic lists and figures of Late

Cretaceous–Paleogene actinopterygians from Eurytania, have been provided in a stratigraphic monograph by Koch and Nikolaus (1969).

Symeonidis and Bachmayer set the stage for later work on the diverse Neogene marine ichthyofaunas of Crete (Symeonidis 1969; Bachmayer and Symeonidis 1978; Bachmayer et al. 1984). Their taxonomic results in the case of the so-called †*Mene psarianosi* (= †*Alectis psarianosi*) and their newly erected syngnathiform species have been since revised (Bannikov 1987; Bürgin 1994; Gaudant 2001, 2004). Müller and Strauch (1994) also provided a short account of a fossil assemblage from the Pliocene of Prassas, Crete, which was subsequently revised (Gaudant 2001). Another interesting assemblage, containing both articulated skeletons, ichthyoliths and otoliths from the Messinian of Karpathos, was presented by Bellas et al. (Bellas et al. 1998). The bulk of research on Neogene marine actinopterygians was led by Gaudant between the late 1970s and the early 2010s. These works expanded on the alpha-taxonomic status and paleoenvironmental implications of fossils from Aegina (Gaudant et al. 2010b), Attiki (Gaudant and Symeonidis 1995), Crete (Gaudant 1979, 1980, 2001, 2004, 2014b; Gaudant et al. 1994, 1997a) and Gavdos Island (Gaudant et al. 2005, 2006), highlighted the richness and potential of the Miocene–Pliocene actinopterygian archive of Greece, and also contributed data for downstream analyses of Mediterranean actinopterygian diversity changes during the latter part of the Cenozoic (Gaudant 2002; Landini and Sorbini 2005a). Yet, detailed descriptions of Neogene actinopterygians from Greece and their systematic treatment are still lacking for the most part.

Few treatments of the Greek record of ray-fins have been published in the past two decades. Careful anatomical and systematic work on Tortonian (Přikryl and Carnevale 2017) and Pliocene (Sorbini and Tyler 2004) actinopterygians from Crete revealed the presence of new fossil species, indicating that actinopterygian diversity and taxonomy in the Neogene of Crete, and conceivably elsewhere in Greece, remain underappreciated. The study of Maastrichtian actinopterygians from Gavdos Island represents the first recent attempt to investigate the pre-Cenozoic fossil record of actinopterygians of the country (Cavin et al. 2012). The first investigation of microscopic ichthyoliths from Greek freshwater deposits (early Miocene, Lesvos Island) was only published a few years ago (Vasileiadou et al. 2017). Furthermore, a recently completed MSc. thesis has shed light on the late Miocene freshwater–brackish clupeids of Aidonochori, Serres Basin, Northern Greece (Kevrekidis 2016). Finally, a new study dealing with the Maastrichtian–Paleogene actinopterygian assemblages from Eurytania has been made available while this work was in press (Argyriou and Davesne 2021).

Over 170 years of research have contributed significantly toward creating a framework for the temporal, stratigraphic, and geographical distribution of fossil ray-fins in Greece, while also providing taxonomic assessments of relevant assemblages. At the same time, past research efforts have been sporadic and/or have largely concluded at a preliminary level, without tapping the potential informativeness of this rich record for a variety of subjects. These include the evolutionary histories and paleobiogeography of Eastern Mediterranean marine and freshwater faunas, or the study of major events, such as the K/Pg Extinction and the Messinian Salinity

Crisis. Unlike late Cenozoic mammals (see relevant chapters in this volume), research on the actinopterygian record of Greece has not received enough scientific attention and is still in its infancy (but see otoliths in Agiadi this volume). Detailed descriptions of Greek fossils are largely wanting, and older taxonomic assessments are in dire need of revision, posing obstacles for further analytical work. This is further exacerbated by the lack of organized collections of fossil fishes and recent comparative material in Greek institutions. A rudimentary-only collection of Cretaceous–Pliocene fossil fishes is housed at the AMPG (organized by the author and S. Roussiakis). Additionally, numerous teleostean skeletons from several Neogene localities of Crete (mainly *Bregmaceros* and *Spratelloides* from the Heraklion Basin) form part of the collections of the Natural History Museum of Crete. By contrast, rich collections from the Cretan Neogene are housed in the Natural History Museum of Vienna (Gaudant 2001, 2004). Some few findings from the Miocene of Kymi, Euboea (Gaudry 1862; T.A. personal observations and communication with Giorgio Carnevale) and the Zanclean of Aegina Island (Gaudant et al. 2010b) are housed in the MNHN, Paris, while additional fossils from the Miocene of Crete and possibly Kymi are housed in the Natural History Museum of London (Woodward 1901).

3 Phylogenetic Relationships

Actinopterygians are widely accepted as a monophyletic group, despite owing their name to their rayed fins, which are also present in sarcopterygians (Friedman 2015). Combinations of hard-tissue synapomorphies [(i) absence of jugal canal; (ii) presence of median aortic canal in the occiput; (iii) absence of endoskeletal basal plates of dorsal and anal fins; (iv) narrow interorbital septum; (v) lateral cranial canal; (vi) posteriorly expanded supratemporals; (vii) acrodin caps on teeth; (viii) fenestrated metapterygoidean portion of palatoquadrate; (ix) perforate propterygium embraced by fin rays; (x) presence of ganoine] are employed to distinguish Paleozoic actinopterygians from other osteichthyes, though, naturally, distinguishing lines dim when examining deeply nested taxa (Friedman 2015; Giles et al. 2015; Lu et al. 2016).

The relationships among actinopterygians—neopterygians and teleosts in particular—had been a matter of controversy (e.g., see Patterson 1973 vs. Grande 2010, for the status of Holostei; Patterson and Rosen 1977 vs. Arratia 1997, for identifying the deepest branch of crown group Teleostei; or Davesne et al. 2016 for a summary of the complicated history of acanthomorph interrelationships) before the establishment of reliable molecular systematic methodologies (Near et al. 2012, 2013; Betancur-R et al. 2015; Hughes et al. 2018). More recent morphology-based investigations are slowly bridging the gap between molecular and morphological phylogenetic hypotheses (e.g., Grande 2010; Davesne et al. 2016). It is now accepted that extant Actinopterygii are distributed across three major lineages of inferred Paleozoic origins: the basally diverging Cladistia (ropefish and bichirs); Chondrostei

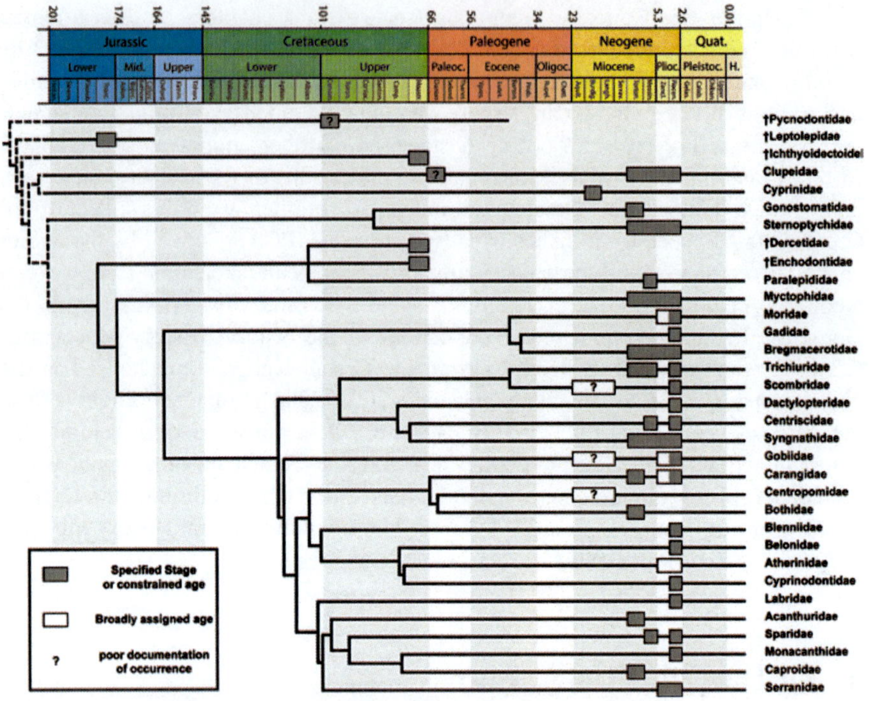

Fig. 1 Fossil record and time-calibrated phylogenetic relationships of major groups of Greek fossil fishes (skeletal record only). Only occurrences considered as valid herein are taken into account. Occurrences marked in darker gray boxes correspond to better constrained ages (usually biostratigraphically). White occurrence boxes correspond to broadly or tentatively assigned ages. Phylogenetic placement and time calibrations for extant groups follow Betancur-R et al. 2017. Placement of fossil taxa is as follows: †Pycnodontidae are established as non-teleosteomorphs; †Leptolepididae and †Ichthyodectoidei (†Ichthyodectiformes) sensu Arratia 2017; †Dercetidae and †Enchodontidae as closely related to modern Aulopifomes sensu Davis and Fielitz 2010, although an agnostic placement of fossil taxa relative to extant ones is preferred here. Results from Argyriou and Davesne (2021) were not incorporated in the figures, or the text, as this work became available after the preparation of this chapter

(sturgeons and paddlefish); and Neopterygii, with the latter including the nowadays depauperate Holostei (gars and bowfin) and the overabundant Teleostei (Near et al. 2012; Betancur-R et al. 2015, 2017; Nelson et al. 2016; Hughes et al. 2018). The relationships of the major groups of ray-finned fishes found in Greece are shown in Fig. 1.

Teleostei account for ~99% of living actinopterygian species and are strongly supported as monophyletic by both morphological and molecular data (Near et al. 2012; Betancur-R et al. 2013, 2015; Broughton et al. 2013; Arratia 2017; López-Arbarello and Sferco 2018). Numerous hard-tissue synapomorphies support teleost monophyly, including: (i) orbital region of skull narrower than postorbital; (ii) presence of one suborbital bone between the posterodorsal margin of the infraorbital series and the opercular apparatus; (iii) presence of two supramaxillae; (iv) lower jaw articulation located ventral to the posterior margin of the orbit; (v) articular fused with angular and retroarticular; (vi) notch on ascending margin of dentary; (vii) coronoid process formed by angular and dentary; (viii) lateral ridge on dentary separating dental and splenial portions of lower jaw; (ix) long epineural processes; (x) presence of 20–26 principal rays; (xi) presence of an enlarged dorsal scute preceding caudal fin (sensu Arratia 2017).

Apart from one problematic occurrence of a Cenomanian †pycnodontid in Rhodos Island (Melentis 1967a, b), the Greek actinopterygian archive is exclusively populated by teleosts (Appendix). †Leptolepid remains from the Toarcian of Leukada (Kottek 1964) represent the only ascertained stem teleost occurrence in Greece while all remaining occurrences correspond to members of crown group lineages. Reliably documented Mesozoic actinopterygians from Greece have been ascribed to extinct families (e.g., †Dercetidae, †Enchodontidae; see Koch and Nikolaus 1969; Cavin et al. 2012), whereas Cenozoic fossils have been predictably included in extant taxa (families, genera, and in some cases to the species-level; see Appendix) on the basis of broad morphological and meristic similarities. The phylogenetic affinities of Greek actinopterygian fossils have not yet been explored, to the exclusion of *Enchodus* cf. *dirus* from the Maastrichtian of Gavdos Island (Cavin et al. 2012).

4 Distribution in Greece and Nature of Deposits

As noted above, ray-finned fishes, and teleosts in particular, are expected to have gradually expanded their occupation of most aquatic ecological niches on the planet, especially after the Paleozoic (Friedman and Sallan 2012). They are thus expected to be present in sedimentary rocks that were deposited in a large spectrum of paleoenvironments, ranging from offshore and deep marine waters, to fluvial, or lacustrine freshwaters. However, freshwater deposits (and included fossils) are more difficult to be preserved in the geological record, due to their relative ephemerality and patchiness. It is thus unsurprising that the vast majority of fossil ray-fins in Greece come from marine deposits. To paint a more complete picture of the Greek fossil record, locality information mined from gray literature (e.g., conference, abstracts field guides, and unpublished theses), stratigraphic and sedimentological works from the 1970s to the 2010s, or deriving from personal observations of the author and communication with colleagues is also taken into account. A brief summary of fossiliferous localities, ages, and paleoenvironments is given below, and more extensively in the Appendix; these lists might not be exhaustive.

Marine Occurrences At least 33 indisputably marine, or likely marine localities have been recognized in Greece, although not all have been subjected to proper collecting and study (Fig. 2). The marine actinopterygian record of Greece begins with

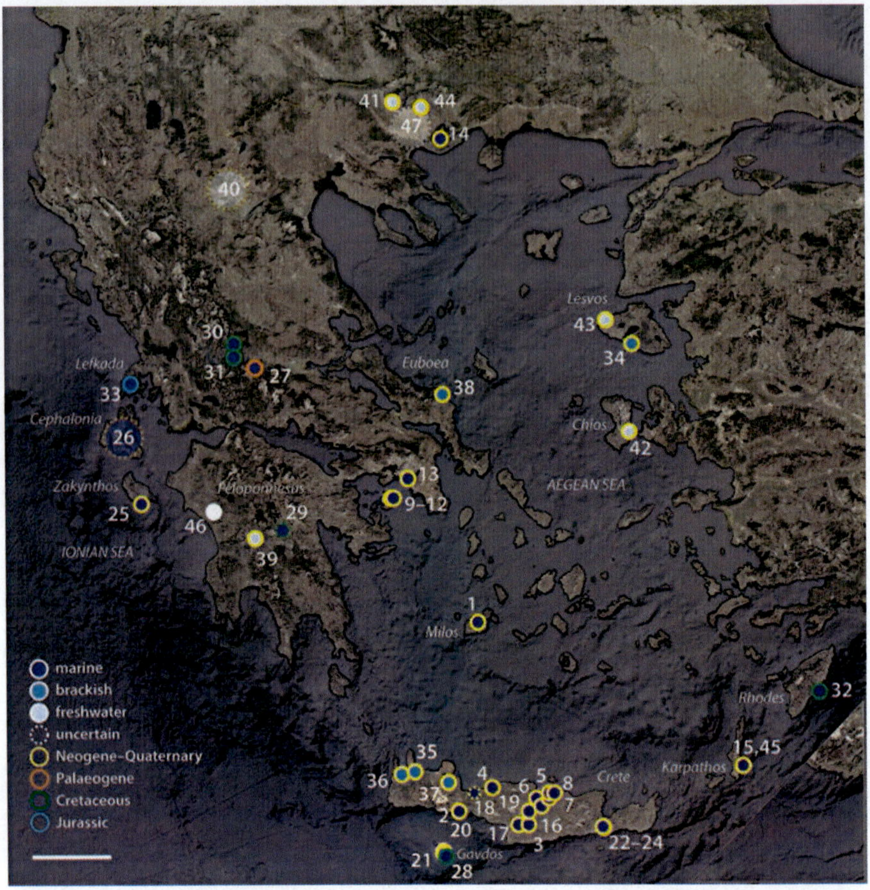

Fig. 2 Map of the occurrences of fossil ray-finned fishes from Greece. See Appendix for further information. **Marine occurrences: 1**, Northern Milos (Pliocene–Pleistocene); **2**, near Frangokastello (middle–late Pliocene); **3**, Rouphas (Piacenzian, late Pliocene); **4**, Stavromenos (Piacenzian); **5**, Amnissos (Piacenzian); **6**, Voutes (Piacenzian); **7**, Prassas (Piacenzian); **8**, Gournes (Piacenzian); **9**, Aghia Marina (Piacenzian?,); **10**, Aghios Thomas (Zanclean); **11**, Faros, Aegina (Zanclean?, early Pliocene?); **12**, Temple of Aphrodite, Aegina (Zanclean?, early Pliocene?); **13**, Kalamaki (Pliocene); **14**, Akropotamos (Messinian); **15**, 'Unit b', Pigadia Basin (Messinian); **16**, Tsangaraki (late Messinian); **17**, Mires (late Messinian); **18**, unknown locality near Rethymnon (Messinian?); **19**, Psalida (Tortonian-early Messinian); **20**, Metochia (Tortonian–Messinian); **21**, Karave (Tortonian); **22**, Aghia Paraskevi Chapel (Tortonian); **23**, 400 m from Aghia Paraskevi Chapel (Tortonian); **24**, type strata of Makrylia Fm. (Tortonian), **25**, Keri (Tortonian); **26**, unknown locality in Kefallonia (late Miocene?); **27**, Aghioi Theodoroi (Paleocene); **28**, Gavdos (Pindos unit; late Maastrichtian); **29**, near Megalopoli (Maastrichtian); **30**, Aghia Trias-Domnianoi Road (Maastrichtian); **31**, near Karpenisi (late Maastrichtian); **32**, Lindos (Cenomanian?); **33**, Anavryssada (Toarcian). **Brackish and potentially brackish occurrences: 34**, Vatera (Pliocene–Pleistocene); **35**, Voukolies (Messinian); **36**, Kalergiana (Messinian); **37**, Neo Khorio (Messinian); **38**, Kymi (early Miocene?). **Freshwater occurrences: 39**, Marathousa 1 (middle Pleistocene); **40**, Florina–Ptolemais–Servia Basin (Messinian–Zanclean); **41**, Ano Metochi (Messinian); **42**, Thymiana (Burdigalian-Langhian); **43**, Lapsarna (Burdigalian); **44**, Aidonochori. **Uncertain environment: 45**, 'Unit e', Pigadia Basin (middle Pleistocene), **46**, Gryllos-Mesovouni (Pliocene-Pleistocene); **47**, lignite mine in Strymonas valley (Pliocene). See Appendix for more information. Image exported from Google Earth Pro © 2019, map data from US Dept. of State Geographer, SIO, NOAA, U.S. Navy, NGA, GEBCO, image from Landsat/Copernicus. Scale bar equals 80 km, North faces upward

some poorly preserved †leptolepids from the Toarcian of the Ionian Unit, Leukada Island (Kottek 1964). Other published Mesozoic occurrences are restricted to the Cenomanian of Lindos, Rhodos Island (Melentis 1967a, b), and the terminal Maastrichtian of Pindos Unit, in Eurytania (Koch and Nikolaus 1969; Argyriou and Davesne 2021) and Gavdos Island (Cavin et al. 2012). Unpublished teleost remains are also present in Late Cretaceous exposures of Pindos Unit in Northern and central Peloponnese (T.A. personal observations and personal communication with E. Koskeridou, G. Theodorou and K. Trikolas).

The Cenozoic record of Greece is very rich, but is largely skewed toward the late Miocene and Pliocene. The oldest reported Cenozoic occurrences come from flysch deposits of the Pindos Unit in Eurytania and are dated as Paleocene–early Eocene (Koch and Nikolaus 1969; Argyriou and Davesne 2021). Rich Tortonian marine assemblages can be found in the vicinity of Ierapetra and possibly Heraklion, Crete (Symeonidis 1969; Gaudant et al. 1997a; Gaudant 2004), and Gavdos Island (Gaudant et al. 2005), while a fish-bearing, bituminous horizon in Keri, Zakynthos has been identified, but not yet studied (Dermitzakis 1979; T.A. personal observations). Messinian fossiliferous deposits predating the Messinian Salinity Crisis are widespread in Crete, Gavdos (Gaudant 1980; Gaudant et al. 1997b, 2006) and Karpathos (Bellas et al. 1998), but only one poorly studied horizon is known in continental Greece, in Acropotamos, Kavala (Dermitzakis et al. 1986). Teleost teeth broadly ascribed to the late Miocene are also present in Kefalonia (pers. obs. on AMPG material; S. Roussiakis pers. comm.). The most notable early Pliocene occurrences come from the diatomitic horizons of Aegina Island, but despite their potential informativeness regarding the state of the Mediterranean in the immediate aftermath of the MSC, they remain understudied (Gaudant et al. 2010b; Argyriou and Theodorou 2011). The late Pliocene of Heraklion Basin, Crete, hosts exceptionally preserved and diverse ichthyofaunas (Gaudant et al. 1994; Gaudant 2001). Additional Pliocene occurrences have been reported in Rethymno, Chania, and Attiki regions (Gaudant and Symeonidis 1995; Gaudant 2001; Sorbini and Tyler 2004). Late Pliocene and Pleistocene fossiliferous horizons situated in Milos and Rhodos wait to be sampled and studied (Cornée et al. 2006; Calvo et al. 2012). Finally, wooden crates containing what appear to be unstudied marine teleost fossils from the Neogene are hosted in the collections of the AMPG (T.A. personal observations). Some of these fossils are wrapped in newspapers that treat the Russo-Japanese war of 1904–1905 as current news, but no locality or collection information has been retrieved so far.

Freshwater and Brackish Occurrences Approximately six freshwater, and five brackish or potentially brackish occurrences have been reported. Burdigalian ichthyoliths from Lapsarna, Lesvos (Vasileiadou et al. 2017), and the early Miocene of SE Chios (Reichenbacher and Böhme 2004) constitute the oldest known freshwater fish assemblages in Greece. Ray-fin remains have been reported from the Messinian of Metochi, Serres (Georgalis et al. 2017); the Messinian–Zanclean of Florina-Ptolemais-Servia (Hordijk and de Bruijn 2009); and the Pleistocene of Marathousa, Megalopolis (Panagopoulou et al. 2018), but have not yet been studied.

The understudied assemblages from Kymi, Euboea, are tentatively dated as early Miocene, and tentatively interpreted as brackish, based on the supposed euryhaline affinities of recognized taxa (de Waldheim 1838, 1850; Gaudry 1862; Unger 1867; Woodward 1901; Kottis et al. 2002). Occurrences of †*Aphanius crassicaudus* in gypsiferous evaporitic deposits in Neo Khorio, Voukolies, and Kallergiana, Crete, have been identified as brackish and concurrent with the peak of the Messinian Salinity Crisis and the supposed desiccation of the Mediterranean in the late Messinian (Freuenthal 1969; Gaudant 1979, 2014b). However, there is evidence of primarily marine clupeids in the same beds, complicating this narrative (Gaudant 1980). The Pliocene clupeids from Serres Basin (Weiler 1943), and unidentified Plio-Pleistocene ray-fins from Vatera, Lesbos (Drinia et al. 2002), might also correspond to brackish, or freshwater deposits. New information on fresh or brackish waters of Greece is coming from the ongoing study of fossil clupeids from Aidonochori, Serres Basin (Kevrekidis 2016). Another possible addition to this list is the middle Pleistocene of Pigadia Basin in Karpathos Island. There, a vertebra possibly belonging to an ostariophysan—a group commonly found in freshwaters—has been found (Bellas et al. 1998).

5 Systematic Paleontology

Due to the taxonomic and phylogenetic breadth covered in this chapter, this section is organized in monophyletic (when possible) families. Synapomorphies for each family are not given, as this lies beyond the scope of this work. Instead, the reader is referred to Wiley and Johnson (2010), Nelson et al. (2016), and Betancur-R et al. (2017) for basic information on morphological synapomorphies and redirection to specialized literature. Taxonomic ranks superior to the genus level are unnamed, since they are obsolete in a modern phylogenetic context. Information on naming authorities and type genera and species is largely based on Wiley and Johnson (2010) and Eschmeyer and Fong (2019). The order of families and classification are based on the scheme proposed by Betancur-R et al. (2017), with fossil families being included according to their widely accepted position (see caption of Fig. 1). References regarding naming authorities, which are not also cited in the running text are not given in the reference list due to space constrains. Throughout the text and figures, extinct taxa are preceded by the dagger symbol '†', following Patterson and Rosen (1977). Fossils that were not properly described and/or figured, or the geological context of which is problematic are only listed in the Appendix.

Actinopterygii sensu Goodrich, 1930

Neopterygii Regan, 1923

†Pycnodontiformes Berg, 1937

†Pycnodontidae Agassiz, 1833

Type Genus †*Pycnodus* Agassiz, 1833.

†*Coelodus* Heckel 1854

Type Species †*Coelodus saturnus* Heckel, 1854.

Nomenclatural and Taxonomical History Initially described as †*Coelodus muensteri* (Melentis 1967a, b; Fig. 3a), but the specific attribution was subsequently challenged and removed (Schultz et al. 1997).

Distribution In Greece, this taxon is known only from the Cenomanian of Lindos, Rhodos (Melentis 1967a, b). Outside Greece, this taxon has been reported from Europe and the Americas, but only its European occurrences have been revised. The latter include the Upper Jurassic–Cenomanian of UK and continental Europe, including the Balkans as well (Schultz et al. 1997).

Comments This is a problematic occurrence, as the original locality and lithological context are conjecturally established. Prior to its study (Melentis 1967a, b), the †pycnodontid in question was part of the private collection of a cleric, who had in turn bought it from a fossil vendor, as evidenced by the transaction receipt that accompanies the specimen in the collections of the AMPG (S. Roussiakis pers. comm.; Fig. 3b). The fossiliferous facies that might have produced this fossil have not been since identified and explored for additional vertebrate remains. Therefore, the possibility that this specimen was not originally found in Rhodos, or even Greece, cannot be fully ruled out.

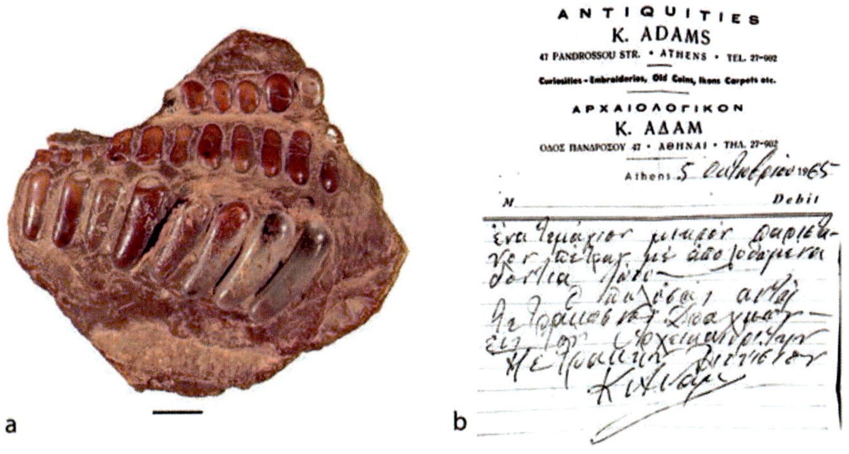

Fig. 3 (**a**) Lingual view of the prearticular toothplate of †*Coelodus* sp. (†Pycnodontidae; formerly assigned to †*Coelodus muensteri* by Melentis 1967a, b), from the Cenomanian of Lindos, Rhodos Island; (**b**) Scan of the sale receipt accompanying the original fossil (See text for details). Scale bar equals 1 cm

Teleostei Müller, 1846 (sensu Arratia 2017)

†Leptolepididae Pictet, 1851

Type Genus †*Leptolepis* Agassiz, 1832.

Nomenclatural and Taxonomical History †Leptolepididae has a complicated taxonomic history and has been previously treated as a grade-like assemblage of primitive teleosts (Nybelin 1974; Arratia 1999, 2017; Arratia and Hikuroa 2010; Konwert and Stumpf 2017).

Distribution In Greece, this taxon is known from the Toarcian of the Ionian Unit, and particular from Aghios Ilias, Anavryssada, Lefkada Island (Kottek 1964). Outside Greece, †Leptolepididae s.l. appeared in the Early Jurassic of Western and Central Europe and seem to have had achieved a cosmopolitan distribution by the Middle Jurassic (Arratia and Hikuroa 2010; Konwert and Stumpf 2017).

Comments A partial maxilla and associated dentalosplenial of a †leptolepidid-like actinopterygian have been found in the Toarcian of Aghios Ilias (Anavryssada exposure), Lefkada, and have been identified as †*Leptolepis* cf. *coryphaenoides* (Kottek 1964). The material resembles †*Leptolepis*, but lacks any synapomorphies of the genus (Arratia 2017; Konwert and Stumpf 2017). This occurrence corresponds to the oldest record of actinopterygians in Greece, but should be treated as cf. †Leptolepididae indet. until more complete material becomes available.

†Ichthyodectiformes Bardack and Springle, 1969

†Ichthyodectoidei Romer, 1966

Distribution In Greece, †Ichthyodectoidei have only been reported from the Maastrichtian (Late Cretaceous) of the Pindos Unit, Gavdos Island (Cavin et al. 2012). Outside Greece, †Ichthyodectiformes are found in Bathonian (Middle Jurassic)–Maastrichtian (Late Cretaceous) marine, and, rarely, in brackish and freshwater deposits of all continents (Arratia et al. 2004; Cavin et al. 2013).

Comments This occurrence is based on a partially preserved postcranium from the Maastrichtian of Gavdos, which exhibits aspects of the caudal part of the vertebral column and associated dorsal, anal and caudal fins (Cavin et al. 2012). Similarities of the Gavdos specimen with †*Saurodon elongatus* from the Maastrichtian of Nardó, Italy, have been noted, but the specimen was left in open nomenclature due to its incomplete preservation (Cavin et al. 2012).

Clupeiformes Bleeker, 1859

Clupeidae Cuvier, 1817

***Alosa* Linck, 790**

Type Species *Alosa alosa* (Linnaeus, 1758).

Included Taxa †*Alosa crassa* (Sauvage, 1873); *Alosa tanaica* (Grim, 1901); †*Alosa elongata* Agassiz, 1843.

Nomenclatural and Taxonomical History †*Alosa crassa* is often included in the extant genus *Sardina* Antipa 1904, as †*Sardina crassa*, or †*Sardina* cf. *crassa* (Dermitzakis et al. 1986; Gaudant 2004, 2014b; Gaudant et al. 2005). The original generic attribution *Alosa* is herein maintained, following Grande (1985) and Carnevale (2007).

Distribution In Greece, marine occurrences of †*Alosa* cf. *crassa*, or *Alossa* sp. are known from the Tortonian of the Makrilia Formation, in the vicinity of Ierapetra, Crete, the Messinian and Pliocene of Crete and the Tortonian–Messinian of Gavdos Island. *Alosa elongata* has been signaled in the Pliocene of Heraklion basin, in Crete. *A. tanaica* (first described as *Caspialoassa nordmanni* is known from the Pliocene of Strymonas valley in Serres (see Appendix for detailed information). Outside Greece, fossils affiliated with †*Alosa crassa* are common in the Mediterranean Neogene, with records in Algeria, Spain, Italy, and Cyprus (Sauvage 1873; Arambourg 1925, 1927; Gaudant et al. 2000, 2015; Landini and Sorbini 2005a; Carnevale 2007; Gaudant 2008a, b). Also, †*Alosa elongata* is known from the late Miocene of Algeria, Spain, Italy (Arambourg 1927; Landini and Sorbini 2005a; Gaudant 2008b; Gaudant et al. 2008, 2010a). Fossils attributed to *A. tanaica* have been reported from the Paratethyan Miocene (Paucă 1931; Rückert-Ülkümen 1994; Schwarzhans et al. 2017.

Comments Most of the attributions to *Alosa* sp. refer to isolated scales, or bones of larger sized clupeids. The alosin from the Pliocene of Serres Basin, was originally attributed to *Caspialosa nordmanni* (Antipa 1904; Weiler 1943), which is considered a junior synonym of the Black Sea species *A. tanaica* (Grim 1901). This species is often mentioned as *A. nordmanni* (Schwarzhans et al. 2017) and, adding to the confusion, some of the contained fossils have been moved to †*Alosa pinarhisariensis* (Rückert-Ülkümen 1994). The specific attribution of the Serres alosin is ambiguous at best. However, future studies on newly discovered fossils from the late Miocene of Aidonochori, Serres Basin, might help shed some light on the complicated systematics of freshwater–brackish clupeids from the area (Kevrekidis 2016).

Sardina Antipa, 1904

Type Species *Sardina pilchardus* (Walbaum, 1792). Only included species.

Distribution In Greece, this taxon is known from the Pliocene of Heraklion Basin and Rethymno, Crete (see Appendix). Outside Greece, *Sardina* fossils are known from the Pliocene and Pleistocene of Italy (Landini and Sorbini 2005a). *S. pilchardus* is present in the Western Atlantic, the Mediterranean, and the Black and Azov seas (Froese and Pauly 2019).

Comments Many of the *Sardina* sp., or *Sardina*? sp. occurrences in Greece correspond to poorly preserved, and/or disarticulated material, and might show taxonomic overlap with *Alosa* spp.

***Spratelloides* Bleeker, 1851**

Type Species *Clupea argyrotaeniata* (Bleeker, 1849) [= junior synonym of *Spratelloides gracilis* (Temminck and Schlegel, 1846)].

Included Taxa *S. gracilis* (Temminck and Schlegel, 1846).

Distribution In Greece, this taxon is known from the Tortonian of the Makrilia Formation, vicinity of Ierapetra, Crete, the Zanclean of Aegina Island, the Pliocene of Attiki, the Pliocene of Heraklion Basin and Rethymno, in Crete (Appendix and Fig. 4). Outside Greece, fossils affiliated with *S. gracilis* are known from the late Miocene, Pliocene and early Pleistocene of Algeria, Spain, and Italy (Arambourg 1927; Gaudant 2002, 2008b; Landini and Sorbini 2005a; Carnevale et al. 2006; Gaudant et al. 2015). Nowadays, *S. gracilis* is present in tropical waters of the Indo-Pacific Realm (Froese and Pauly 2019).

Cypriniformes Bleeker, 1859
Cyprinidae Cuvier, 1817

Type Genus *Cyprinus* Linnaeus, 1758.

†*Palaeoleuciscus* Obrhelová, 1967

Type Species †*Palaeoleuciscus chartaceus* (Laube, 1905).

Distribution In Greece, this taxon is known only from the early Miocene of Lapsarna Locality, Lesvos Island, Greece (Vasileiadou et al. 2017; Appendix).

Fig. 4 *Spratelloides* cf. *gracilis* (Clupeidae) from the Zanclean diatomitic marls near Mesagros, Aegina Island (AMPG collection). Scale bar equals 1 cm

Outside Greece, its oldest fossils come from the Oligocene of Germany and Switzerland (Cavender 1991; Böhme and Ilg 2003). There numerous records of tooth and skeletal remains of the genus in the Miocene of Europe and Anatolia as well (e.g., Cavender 1991; Böhme and Ilg 2003; Rückert-Ülkümen and Yiğitbaş 2007; Böhme 2010; Gaudant 2014a, 2015).

***Barbus* Cuvier and Cloquet, 1816**

Type Species *Barbus barbus* (Linnaeus, 1758).

Distribution In Greece, this taxon is only known from the early Miocene of Lapsarna Locality, Lesvos Island, Greece (Vasileiadou et al. 2017; Appendix). Outside Greece, *Barbus*-like (?*Barbus* sp.) pharyngeal teeth first appear in the Oligocene of Afroarabia (Otero and Gayet 2001). *Barbus* remains are quite common in Neogene and younger freshwater deposits in Europe, Africa, the Arabian subcontinent, and Anatolia (Cavender 1991; Stewart 2001, 2003; Böhme and Ilg 2003; Přikryl et al. 2016b; Stewart and Murray 2017). The distribution of modern *Barbus* s.s. species ranges from Europe to the Caspian Sea, Anatolia, and Tigris and Euphrates drainages, with approx. 13 species found in Greek freshwaters (Froese and Pauly 2019).

Comments It is widely accepted that modern and fossil species lumped under the genus *Barbus* correspond to different lineages, rendering the genus paraphyletic (Berrebi and Tsigenopoulos 2003; Kottelat and Freyhof 2007). *Barbus* s.l. is used to indicate uncertainty relative to the placement of fossils relative to modern lineages, whereas *Barbus* s.s. indicates affinities with the European and Anatolian subgenus *Barbus*. Isolated pharyngeal teeth or bones are difficult to attribute to specific lineages, yet similarities of the Lapsarna *Barbus* with *Barbus* s.s. have been noted (Vasileiadou et al. 2017).

***Cyprinion* Heckel, 1843**

Type Species *Cyprinion macrostomus* Heckel, 1843.

Distribution In Greece, pharyngeal bones and teeth of *Cyprinion* have been found in the early Miocene of Lapsarna Locality, Lesvos Island, Greece (Vasileiadou et al. 2017; Appendix). Outside Greece, pharyngeal teeth putatively attributed to the genus have also been found in the early Miocene of Turkey (Böhme and Ilg 2003). Modern representatives of *Cyprinion* have a distribution that ranges from the Arabian subcontinent to Myanmar (Froese and Pauly 2019).

Comments According to Vasileiadou et al. (2017), the Lapsarna material likely represents a new species, which has not yet been described formally.

Aulopiformes Rosen, 1973
†Dercetidae Woodward, 1901

Type Species †*Dercetis scutatus* (Agassiz, 1834).

Distribution In Greece, fossils of †Dercetidae are so far known from the Late Maastrichtian of Pindos Unit, in the vicinity of Karpenisi, Eurytania (Koch and Nikolaus 1969). Outside Greece, †dercetids are known from the Late Cretaceous–Paleocene of Europe, the Middle East, North and Central Africa, and South America (summarized in da Silva and Gallo 2011, 2016; Vernygora et al. 2018).

Comments The †dercetids found in the vicinity of Karpenisi have been tentatively ascribed to either the aulopiform †*Rhynchodercetis* or, erroneously, the stem teleost †*Belonostomus* (Koch and Nikolaus 1969). Only one †dercetid-like elongate lower jaw, bearing numerous minute teeth, has been figured, but this material has not been described or treated systematically (Koch and Nikolaus 1969, Table 33:3), and bears no apomorphies to support any attribution below the Family-level. No repository information has been provided for these fossils.

†Enchodontidae Woodward, 1901

Type Genus †*Enchodus* Agassiz, 1835.

†*Enchodus* Agassiz, 1835

Type Species †*Enchodus lewesiensis* (Mantell, 1822).

Distribution In Greece, fossils of †*Enchodus* cf. *dirus* are known from the late Maastrichtian, Pindos Unit, Gavdos Island (Cavin et al. 2012). Additional †enchodontid remains, possibly referable to †*Enchodus*, have been collected from Maastrichtian deposits of the Pindos Unit, exposed on the road from Aghia Triada to Domniani, near Karpenisi, Eurytania (Koch and Nikolaus 1969). Outside Greece, the oldest disarticulated remains referred to †*Enchodus* sp. have been collected in the Early Cretaceous of Spain (Kriwet 2003), while articulated remains are known from the Albian–Cenomanian of Mexico (Fielitz and González-Rodríguez 2010). This genus achieved a cosmopolitan distribution during the Late Cretaceous, with occurrences in North and South America, England and western continental Europe, Africa, India, and the Middle East (summarized in Goody 1969; Chalifa 1996; Cavin et al. 2012; da Silva and Gallo 2016; Holloway et al. 2017).

Comments A relatively well-preserved skull of †*Enchodus*, also exhibiting features of the palate and jaws, was found in the late Maastrichtian deposits of Gavdos Island and was attributed to a juvenile individual (Cavin et al. 2012). The specimen was included in a phylogenetic analysis alongside other †enchodontids, and was resolved as sister taxon to †*Enchodus dirus* (Cavin et al. 2012), which, in turn, is found in several Late Cretaceous localities of the Western Interior Seaway, North America (Goody 1969). The Gavdos and North American specimens seem to only differ in size and ornamentation, and their seemingly close relationship supports complex biogeographical scenarios, involving the ability of, at least some, Late Cretaceous †enchodontids to cross oceanic basins (Cavin et al. 2012).

Paralepididae Gill, 1872

Type Genus *Paralepis* Cuvier, 1817.

Paralepis Cuvier, 1817

Type Species *Paralepis coregonoides* Risso, 1820.

Included Taxa †*Paralepis albyi* (Sauvage, 1870).

Distribution In Greece, this taxon is known from the Messinian of Gavdos (Metochia: †*P. albyi*; Gaudant et al. 2006). Outside Greece, †*P. albyi* is present in the late Miocene of Spain and Italy (Arambourg 1925; Gaudant 2002; Carnevale 2007). Two species of the deep-water genus *Paralepis* are nowadays present in the Mediterranean (Froese and Pauly 2019).

Stomiatiformes sensu Rosen, 1973
Gonostomatidae Gill, 1893

Type Genus *Cyclothone* Goode and Bean, 1883.

Cyclothone Goode and Bean, 1883

Type Species *Cyclothone acclinidens* Garman, 1899.

Included Taxa †*Cyclothone gaudanti* Přikryl and Carnevale, 2017.

Remarks The modern species *Cyclothone braueri*, *Cyclothone microdon*, and *Cyclothone pygmaea* have been reported from the Italian Pliocene and Pleistocene (Landini and Sorbini 2005a; Přikryl and Carnevale 2017). The genus has a cosmopolitan distribution in modern deep marine waters (Froese and Pauly 2019).

★†*Cyclothone gaudanti* Přikryl and Carnevale, 2017

Nomenclatural and Taxonomical History *Cyclothone* sp. in Gaudant 2004 (initial identification); *Cyclothone gaudanti* Přikryl and Carnevale, 2017 (new species).

Type Material NHMW 1999z0042/0020, nearly complete, articulated individual.

Type Locality Locality 400 m to the North of Aghia Paraskevi Chapel, vicinity of Ierapetra, Crete, Greece, Makrilia Formation.

Distribution This species is only known from the Tortonian of Makrilia Formation, in the vicinity of Ierapetra, Crete (Aghia Paraskevi Chapel Locality, and from a locality 400 m to the North of Aghia Paraskevi Chapel; Gaudant 2004, Přikryl and Carnevale 2017). These localities constitute the whole range of this taxon.

Remarks *Cyclothone gaudanti* demonstrates affinities with the small-sized and lightly pigmented species of the genus, such as *Cyclothone alba*, *C. braueri*, and *Cyclothone signata* (Přikryl and Carnevale 2017). Based on its affinities and the depositional environment of Makrilia Formation, it is estimated that this species possibly occupied the upper mesopelagic layer (Přikryl and Carnevale 2017).

Sternoptychidae Duméril, 1806

Type Genus *Sternoptyx* Hermann, 1781.

***Maurolicus* Cocco, 1838**

Type Species *Maurolicus muelleri* (Gmelin, 1789). Only included species.

Distribution In Greece, this taxon is known with the species *M. muelleri* from the Tortonian of the Makrilia Formation, in the vicinity of Ierapetra, Crete (Gaudant 2004), the Messinian of Gavdos (Gaudant et al. 2006), and the Messinian of Mires (Gaudant et al. 1997a). Furthermore, *Maurolicus* cf. *M. muelleri* has been reported from the Pliocene of Heraklion as well (Gaudant 2001; Appendix). Outside Greece, *M. muelleri* has a continuous presence in the Mediterranean Realm (including Algeria, Italy and Spain) since the late Miocene (Gaudant 2002, 2008b; Landini and Sorbini 2005a; Gaudant et al. 2008, 2010a), and is characterized by a cosmopolitan distribution in modern bathypelagic waters (Froese and Pauly 2019).

***Argyropelecus* Cocco, 1829**

Type Species *Argyropelecus hemigymnus* Cocco, 1829.

Distribution In Greece, the only record of *Argyropelecus* sp. comes from the Pliocene (Zanclean) of Aegina Island (Argyriou and Theodorou 2011), and is figured here for the first time (Fig. 5). Outside Greece, the fossil record of the genus dates back to the Oligocene of Romania, Poland, and the Caucasus (summarized in Carnevale 2003).

Remarks So far, only two species have been recognized on the basis of body fossils in the Mediterranean region: the late Miocene †*A. logearti* in Algeria (Arambourg 1927; Carnevale 2003), and the extant *A. hemigymnus* in the Pliocene and Pleistocene of Italy (Carnevale 2003, 2007; Landini and Sorbini 2005a).

Myctophiformes Regan, 1911
Myctophidae Gill, 1893

Type Genus *Myctophum* Rafinesque, 1810a.

***Myctophum* Rafinesque, 1810a**

Type Species *Myctophum punctatum* Rafinesque, 1810b.

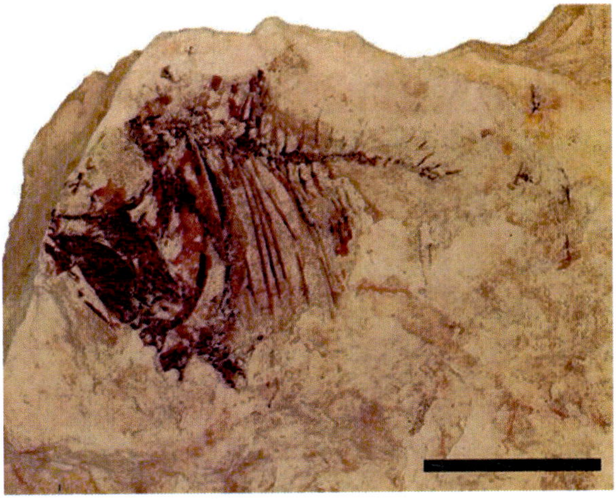

Fig. 5 *Argyropelecus* sp. (Sternoptychidae), from the Zanclean diatomitic marls near Mesagros, Aegina Island (AMPG collection). This specimen is figured here for the first time. Scale bar equals 1 cm

Included Taxa †*"Myctophum" licatae* Sauvage, 1870; †*"Myctophum" dorsale* Sauvage, 1870.

Distribution In Greece, myctophid body fossils are known from the Tortonian of Makrilia Formation, vicinity of Ierapetra (Gaudant 2004), whereas occurrences identified to the species level are known from the Tortonian–Messinian of Messara, Heraklion (Psalidha, Mires: †*"Myctophum" licatae*, †*"Myctophum" dorsale*; Tsangaraki: †*"Myctophum" licatae*; Gaudant et al. 1997a), and the Tortonian (Metochia, Karave: †*"Myctophum" licatae*) and Messinian (Metochia: †*M. licatae*, †*M. dorsale*; Fig. 5) of Gavdos (Gaudant et al. 2005, 2006). A single myctophid tooth was also recovered from the Messinian strata of 'Unit b' of Pigadia Basin in Karpathos (Bellas et al. 1998). Outside Greece, these two Miocene species have been recognized in several Mediterranean localities in Algeria, Spain, Italy, and Cyprus (Arambourg 1925, 1927; Gaudant 2002, 2008a, b; Merle et al. 2002; Gaudant et al. 2010a). Several extant species of *Myctophum* have been reported from Pliocene and Pleistocene localities of the Mediterranean (Landini and Sorbini 2005a). Nowadays, *Myctophum* species have a cosmopolitan bathypelagic distribution (Froese and Pauly 2019).

Remarks Greek body fossils of Myctophidae were lumped by Gaudant in the extant genus *Myctophum*, following his taxonomic reappraisal of material previously studied by Arambourg (1925), from the late Miocene of Licata, Sicily (Gaudant and Ambroise 1999). However, Gaudant's taxonomical scheme relied heavily on putatively overlapping ratios of body measurements and the overall presence or absence of photophores, and did not take important anatomical characters of synapomorphical value into proper account (see e.g., Paxton 1972; Carnevale 2007;

Fig. 6 Myctophidae indet. From the Messinian of Metochia C, Gavdos Island (AMPG collection; photograph provided by JD Carrillo). Scale bar equals 1 cm

Denton 2013). Detailed considerations of fossil myctophids from e.g., the Serravalian of Italy revealed the simultaneous presence of multiple genera—but not *Myctophum*—in the Mediterranean Neogene (Carnevale 2007), signaling the necessity for future revision of fossil myctophid diversity in the entirety of the Mediterranean Basin, including Greece. Given the scope of the present work, I refrain from reappraising Greek myctophid fossils and retain Gaudant's classification in inverted commas (also employed by Gaudant in his later works). New material from Metochia C, Gavdos, collected by the author (e.g., Fig. 6), preserves delicate anatomical structures that can facilitate relevant undertakings in the future.

Gadiformes sensu Endo, 2002
Gadidae Rafinesque, 1810b

Type Genus *Gadus* Linnaeus, 1758.

Micromesistius Gill, 1863

Type Species *Micromesistius poutassou* (Risso, 1827).

Distribution In Greece, the only fossil occurrence of *Micromesistius* sp. comes from the Pliocene Amnissos, Heraklion Basin (Gaudant 2001). Outside Greece, this taxon is known from the Pliocene to Pleistocene of Italy (Landini and Sorbini 2005a). The genus is nowadays present in temperate Atlantic and Pacific waters, including the Mediterranean (Froese and Pauly 2019).

Moridae Goode and Bean, 1896

Type Genus *Mora* Risso, 1827.

Gadella **Lowe, 1843**

Type Species *Gadella gracilis* Lowe, 1843.

Distribution In Greece, occurrences of *Gadella* are known from the Pliocene of Attiki (Gaudant and Symeonidis 1995) and Heraklion Basin (Gaudant 2001). Outside Greece, this taxon is known from the Pliocene of Italy (Landini and Sorbini 2005a). *Gadella* is still present in tropical to subtropical deep waters around the world, including the Mediterranean (Froese and Pauly 2019).

Bregmacerotidae Gill, 1872

Type Genus *Bregmaceros* Thompson, 1840.

Bregmaceros Thompson, 1840

Type Species *Bregmaceros mcclellandi* Thompson, 1840.

†*Bregmaceros albyi* (Sauvage, 1880)

Distribution This is one of the most widely distributed taxa in Greece, with occurrences from the Tortonian of Makrilia Formation, vicinity of Ierapetra, Crete (Symeonidis 1969; Bürgin 1994; Gaudant 2004), the Tortonian and Messinian of Metochia, Gavdos (Gaudant et al. 2005, 2006), the Tortonian–early Messinian of Messara, Heraklion, Crete (Gaudant et al. 1997a), the Messinian of Karpathos (Bellas et al. 1998), the Pliocene (cf. Zanclean) of Aegina Island (Argyriou and Theodorou 2011; Fig. 7), the Pliocene of Heraklion (Gaudant et al. 1994; Müller and Strauch 1994; Gaudant 2001) and the Piacenzian of Rethymno (Symeonidis and Schultz 1973; see Appendix for more information). Outside Greece, body fossils of †*Bregmaceros albyi* are known from the Miocene of Algeria, Italy, Malta, and Spain, and from the Pliocene of Italy and, possibly (as †*B.* cf. *albyi*), Cyprus (Gaudant 2002; Landini and Sorbini 2005a; Přikryl et al. 2016a; and references therein). The species is also present in the early Pleistocene of Italy (Landini and Sorbini 2005a).

Acanthomorpha Rosen, 1973
Atheriniformes sensu Dyer and Chernoff, 1996
Atherinidae Risso, 1827

Type Genus *Atherina* Linnaeus, 1758.

Distribution In Greece, atherinids are known only from the Pliocene of Attiki (Gaudant and Symeonidis 1995). Outside Greece, atherinids are known to have been present in Tethys since at least the Eocene of Monte Bolca, Italy (Carnevale et al. 2014), but their more recent fossil record in the area is largely composed of otoliths (Landini and Sorbini 2005a). Body fossils of atherinids are known from the Miocene of the Paratethys (Schwarzhans et al. 2017) and California (Jordan and Gilbert 1919).

Fig. 7 *Bregmaceros* cf. *albyi* (Bregmacerotidae), from the Zanclean diatomitic marls near Mesagros, Aegina Island. This specimen is figured here for the first time (AMPG collection). Scale bars equal 1 cm

Beloniformes Berg, 1937
Belonidae Bonaparte, 1832

Type Genus *Belone* Cuvier, 1817.

Belone **Cuvier, 1817**

Type Species *Belone belone* (Linnaeus, 1760).

Distribution In Greece, this taxon is only known from the Pliocene of Heraklion Basin (Gaudant et al. 1994; Gaudant 2001). Outside Greece, fossils of this taxon are more common, coming from the Oligocene of the Caucasus (Smirnov 1936), the Miocene of the USA (de Sant'Anna et al. 2013), and the Miocene to Pleistocene of Italy (Arambourg 1925; Landini and Sorbini 2005a). The genus is nowadays present in the Atlantic Ocean, the Mediterranean, and the Black Sea (Froese and Pauly 2019).

Scomberesox Lacepède, 1803

Type Species *Scomberesox saurus* (Walbaum, 1792).

Distribution In Greece, the only possible fossils of this genus come from the Pliocene of Heraklion Basin (Gaudant 2001). Outside Greece, fossils of *Scomberesox* are known from the Miocene of USA, Algeria, and Italy, and the Pleistocene of Italy (Jordan and Gilbert 1919; Gaudant 2002, 2008b; Landini and Sorbini 2005a). Nowadays, the genus has a circumglobal distribution, which also includes the Mediterranean (Froese and Pauly 2019).

Blenniiformes sensu Springer, 1993
Blenniidae Rafinesque, 1810a

Type Genus *Blennius* Linnaeus, 1758.

***Salaria* Forskål, 1775**

Type Species Not designated.

Included Taxa *Salaria pavo* (Risso, 1810).

Nomenclatural and Taxonomical History Initially ascribed by Gaudant et al. (1994) to the genus *Blennius*, as *Blennius* cf. *B. pavo* (Risso 1810). This species is now included in the genus *Salaria*, as *S. pavo* (see Eschmeyer and Fong 2019).

Distribution In Greece, this taxon is known only from the Pliocene of Heraklion Basin (Gaudant et al. 1994). So far no additional body-fossil records exist for this taxon. Nowadays, *S. pavo* is present in shallow marine to brackish waters along the Western Atlantic, Mediterranean, and Black Sea coasts (Froese and Pauly 2019).

Perciformes sensu Betancur-R et al. 2017
Serranidae Swainson, 1839

Type Genus *Serranus* Cuvier, 1817.

***Serranus* Cuvier, 1817**

Type Species *Serranus cabrilla* (Linnaeus, 1758).

Distribution In Greece, this taxon is certainly known from the Pliocene of Heraklion Basin (Gaudant 2001), while the presence of the family has been signaled in the Pliocene of Aegina Island as well (Gaudant et al. 2010b). Outside Greece, *Serranus* fossils are known from the Miocene of Algeria, and the Plio-Pleistocene of Sardinia (Abbazzi et al. 2004; Gaudant 2008b). *Serranus* has a modern cosmopolitan, Tropical–Temperate, marine distribution (Froese and Pauly 2019).

Cyprinodontiformes sensu Parenti, 1981
Cyprinodontidae Gill, 1865

Type Genus *Cyprinodon* Lacepède, 1803.

***Aphanius* Nardo, 1827**

Type Species *Aphanius nanus* Nardo, 1827.

Included Taxa †*A. crassicaudus* Agassiz, 1839.

Nomenclatural and Taxonomical History Initially included in the genus †*Lebias* (Agassiz, 1833–1843), as †*L. crassicaudus* (Raulin 1861). Woodward erected the genus *Pachylebias* to accommodate pachyostotic cyprinodontiform material from the Messinian of Italy and Crete (Woodward 1901). †*Pachylebias* was later found to be a junior synonym of the extant genus *Aphanius* (Gaudant 1979).

Distribution In Greece, this taxon is known from the Messinian of Chania, Crete (Raulin 1861; Woodward 1901; Freuenthal 1969; Gaudant 1979, 2014b). Outside Greece, *Aphanius* is an essential component of Messinian oligohaline fish assemblages of the Mediterranean, including occurrences in Algeria, Spain, Morocco, and Italy (Gaudant 2002; Carnevale et al. 2018).

Scombriformes sensu Betancur-R et al. 2017
Scombridae Rafinesque, 1815

Type Genus *Scomber* Linnaeus, 1758.

Scomber Linnaeus, 1758

Type Species *Scomber scombrus* Linnaeus, 1758 (= *S. scomber* L., 1758).

Distribution In Greece, this genus is only known from the Pliocene of Heraklion Basin (Gaudant 2001). Outside Greece, the genus is known from the Oligocene of central Europe and the Caucasus, and the Neogene of Russia, the United States, Japan, Algeria, Italy, Croatia, and Serbia (Arambourg 1927; Yabumoto and Uyeno 1994; Landini and Sorbini 2005a; Micklich and Hildenbrandt 2010; Gregorová 2011; Nazarkin and Bannikov 2014). Modern representatives of the genus are widespread in tropical to temperate waters (Froese and Pauly 2019).

Scomberomorus Lacepède, 1801

Type Species *Scomberomorus regalis* (Bloch, 1793) (= *S. plumierii* Lacepède, 1801).

Distribution In Greece, this genus is only known from the Pliocene of Heraklion Basin (Gaudant 2001). Outside Greece, the genus appears in the Paleocene of Turkmenistan (Monsch and Bannikov 2012). It is also known the Paleogene of Kazakhstan, Georgia, the United States, Germany, and the Neogene of Japan, Poland, and the United States (Schultz 1979; Yabumoto and Uyeno 1994; Weems 1999; Micklich and Hildenbrandt 2010; Monsch and Bannikov 2012; Weems et al. 2017). The genus is found today in tropical-subtropical waters of the Atlantic, Indian, and Pacific oceans, and the Western Mediterranean (Froese and Pauly 2019).

Trichiuridae Rafinesque, 1810a

Type Genus *Trichiurus* Linnaeus, 1758.

Lepidopus Goüan, 1770

Type Species *Lepidopus caudatus* Euphrasen, 1788.

Included Taxa †*L. albyi* Sauvage, 1870.

Distribution In Greece, this genus is known from the Tortonian of Makrilia Formation, in the vicinity of Ierapetra, Crete (Symeonidis 1969; Gaudant 2004), the Tortonian and Messinian of Gavdos (Gaudant et al. 2005), the Messinian of Messara (as †*Lepidopus albyi* in the text, but *Lepidopus* sp. in the respective figure caption; Gaudant et al. 1997a), and the Pliocene of Heraklion Basin and Rethymno (Gaudant et al. 1994; Gaudant 2001; see Appendix for further information). Outside Greece, †*Lepidopus albyi* is known from the late Miocene of Algeria, Spain, Italy, Cyprus, and Pliocene of Italy (Arambourg 1925, 1927; Gaudant 2002; Merle et al. 2002; Landini and Sorbini 2005a). A second fossil species, †*Lepidopus proargenteus*, is known from the late Miocene of Algeria and Italy (Arambourg 1927; Gaudant 2008b; Gaudant et al. 2010a). Remains affiliated with the living *L. caudatus* have been reported from the Pliocene of Italy (Landini and Sorbini 2005a).

Remarks Symeonidis tentatively ascribed fragmentary remains of a juvenile *Lepidopus* from the Tortonian of Crete to the Oligocene Paratethyan species †*Anenchelum glaricianum* (his †*Lepidopus* cf. *glaricianus*, Symeonidis 1969). Gaudant revised this attribution to *Lepidopus* sp. (Gaudant 2004).

Gobiiformes Günther, 1880
Gobiidae Cuvier, 1817

Type Genus *Gobius* Linnaeus, 1758.

***Gobius* Linnaeus, 1758**

Type Species *Gobius niger* Linnaeus, 1758.

Distribution In Greece, unidentified records of this genus are reported from the Pliocene of Heraklion Basin (Gaudant et al. 1994) and Attiki (Gaudant and Symeonidis 1995). Outside Greece, *Gobius* skeletons first appear in the early Miocene of the Czech Republic (Reichenbacher et al. 2018), but are found in a wide array of Neogene fossil sites in North Africa, Europe, and Asia Minor (Arambourg 1927; Gaudant 2002; Böhme and Ilg 2003; Reichenbacher et al. 2018). *Gobius* has a circumglobal modern distribution (Froese and Pauly 2019).

Comments Gobioids are among the largest clades of vertebrates (Nelson et al. 2016). In many cases, included genera and species are diagnosed on the basis of very delicate anatomical features and soft tissue structures, which are not often visible/preserved in fossils (Reichenbacher et al. 2018). Thus, Greek gobiid fossils only loosely affiliate with *Gobius* (Gaudant et al. 1994; Gaudant and Symeonidis 1995), and might actually encompass other genera as well. It should be noted here that poorly documented occurrences of Gobiidae in the Miocene of Kymi have been ascribed to the genus *Periophthalmus* (Unger 1867; Appendix), but, due to lack of descriptive or pictorial documentation, these are not further treated here.

Carangiformes sensu Betancur-R et al. 2017
Carangidae Rafinesque, 1815

Type Genus *Caranx* Lacepède, 1801.

Caranx Lacepède, 1801

Type Species *Caranx hippos* (Linnaeus, 1766).

Distribution In Greece, this genus is only known from the Pliocene of Heraklion Basin (Gaudant 2001). Outside Greece, this taxon is known from the Miocene of Algeria, Italy (Arambourg 1925, 1927), but also from the Paratethys (Schwarzhans et al. 2017).

Alectis **Rafinesque, 1815**

Type Species *Alectis ciliaris* (Bloch, 1787).

★†*Alectis psarianosi* (Symeonidis, 1969)

Nomenclatural and Taxonomical History This species, based on a single specimen from the Tortonian of Crete, was formerly included in the genus *Mene* (Menidae), as †*Mene psarianosi* (Symeonidis 1969). It was later reattributed to the carangid genus *Alectis*, but the species name was retained (Bannikov 1984, 1987; Gaudant 2004).

Type Material AMPG 185/1969, complete, laterally compressed individual (Fig. 8).

Type Locality Tortonian of Makrilia Formation, vicinity of Ierapetra, Crete (Aghia Paraskevi Chapel Locality; Symeonidis 1969; Gaudant 2004).

Distribution Known only from the type locality.

Remarks Bannikov (1987) dispelled the original, poorly supported attribution of this unique fossil to Menidae by Symeonidis (1969), after recognizing typical carangid features, including a high rhomboidal body shape; a well-developed fronto-occipital crest; and the relative length of fin rays in its median fins. Gaudant (2004) re-examined the specimen, providing its first detailed anatomical description and confirming Bannikov's assessment, but without attempting additional comparisons. According to Bannikov (1987), this fossil species differs from modern species in having "somewhat shorter anterior rays of the soft unpaired fins." Whether this difference is sufficient to maintain †*Alectis psarianosi* as a valid fossil species remains to be investigated in the future, given the fact that the fossil corresponds to a somewhat small, possibly immature, individual.

Fig. 8 The type specimen (AMPG 185/1969) of †*Alectis psarianosi* (Carangidae) from the Tortonian of Aghia Paraskevi Chapel, vicinity of Ierapetra, Crete. Scale bars equal 1 cm

***Trachurus* Rafinesque, 1810a**

Type Species *Trachurus trachurus* (Linnaeus, 1758) (= *T. saurus* Rafinesque, 1810b).

Distribution In Greece, occurrence of this genus comes from the Tortonian of Makrilia Formation, in the vicinity of Ierapetra, Crete (Gaudant 2004), the Pliocene of Attiki (Gaudant and Symeonidis 1995), and the Pliocene of Heraklion (Gaudant 2001; see Appendix for further information). Outside Greece, fossils more or less securely attributed to *Trachurus* are known from the Miocene of Algeria and Spain, the Miocene and Pliocene of Italy and the Pliocene of Cyprus (Arambourg 1927; Landini and Sorbini 2005a; Gaudant 2008a, b; Gaudant et al. 2010a; Gaudant and Courme 2014). Three species of *Trachurus* are native to modern Mediterranean pelagic habitats (Froese and Pauly 2019).

Spariformes sensu Akazaki, 1962
Sparidae Rafinesque, 1810a

Type Genus *Sparus* Linnaeus, 1758.

***Boops* Cuvier, 1814**

Type Species *Boops boops* (Linnaeus, 1758).

Distribution In Greece, this genus is known from the Pliocene of Heraklion Basin with various occurrences (Gaudant et al. 1994; Gaudant 2001). Outside Greece, *Boops* fossils have been found in the Miocene of Algeria and the Miocene and Pliocene of Italy (Arambourg 1927; Gaudant 2002). Nowadays, *Boops* is represented by two species, one of which, *B. boops*, occurs in the Mediterranean (Froese and Pauly 2019).

Comments Gaudant et al. (1994) initially ascribed the Amnissos boopsins to *Boops* cf. *B. boops*. In a subsequent work, Gaudant (2001) noted differences in meristic features of the dorsal and caudal fins of the *Boops* from Amnissos, from both the Miocene species †*B. roulei* (Arambourg 1927), and the extant *B. boops*, leaving the Cretan fossils in open nomenclature, and implying that they might correspond to a new species. This might also be true for the remaining Heraklion Basin boopsins. As evidenced by isolated teeth, additional sparids were likely present in the Messinian of Karpathos (Bellas et al. 1998).

Centriscidae Bonaparte, 1831

Type Genus *Centriscus* Linnaeus, 1758.

Aeoliscus Jordan and Starks, 1902

Type Species *Aeoliscus strigatus* (Günther, 1861).

Nomenclatural and Taxonomical History Centriscids from the Pliocene of Amnissos were initially thought to correspond to a new fossil species (†*Amphisile cretensis* Bachmayer et al. 1984). Gaudant re-examined these specimens and tentatively ascribed these fossils to the extant genus *Amphisile* (Cuvier 1817), as *Amphisile* cf. *strigata* (Gaudant 2001). *Amphihisile strigata* is, however, a synonym of *Aeoliscus. strigatus* (Günther 1861. In a subsequent work, Gaudant corrected the taxonomic designation of the Cretan material to *Aeoliscus* cf. *strigatus* (Gaudant and Courme 2014).

Distribution In Greece, this taxon is known from the Pliocene of Heraklion Basin, (Bachmayer et al. 1984; Gaudant 2001). Outside Greece, *Aeoliscus. strigatus* is also known from the late Miocene and Pliocene of Italy and the Pliocene of Cyprus (Gaudant 2002; Landini and Sorbini 2005a; Gaudant and Courme 2014). This species is nowadays encountered in tropical Indo-Pacific waters (Froese and Pauly 2019).

Remarks Given the invalidity of the genus *Amphisile* (see above), the occurrence of *Amphisile* sp. in the Messinian of Gavdos (Gaudant et al. 2006) is hereby treated conservatively as Centriscidae indet.

Syngnathidae Rafinesque, 1810a

Type Genus *Syngnathus* Linnaeus, 1758.

***Syngnathus* Linnaeus, 1758**

Type Species *Syngnathus acus* Linnaeus, 1758.

Included Taxa †*Syngnathus albyi* Sauvage, 1870.

Nomenclatural and Taxonomical History Two fossil species of *Syngnathus* have been erected on the basis of material from the Neogene of Crete: the late Miocene †*S. kaehsbaueri* Bachmayer and Symeonidis (1978) from Aghia Paraskevi Chapel, Makrilia Formation, in the vicinity of Ierapetra, and the Pliocene ★†*S. heraklionis* Bachmayer et al., 1984 from Amnissos, near Heraklion. Reinvestigations of the type material of both species found no anatomical differences from the Miocene species †*S. albyi* (Sauvage 1870), establishing the Cretan species as invalid and referred their type material to †*Syngnathus* cf. †*S. albyi* (Gaudant 2001, 2004).

Distribution In Greece, occurrences of this taxon are known from the Tortonian of Makrilia Formation, near Ierapetra, Crete (Bachmayer and Symeonidis 1978; Gaudant 2004), the Tortonian and Messinian of Metochia, Gavdos (Gaudant et al. 2005, 2006), the Messinian of Messara, Crete (Gaudant et al. 1997a), and the Pliocene of Heraklion, Crete (Bachmayer et al. 1984; Gaudant et al. 1994; Gaudant 2001). Remains of short-snouted *Syngnathus* cf. *S. acus* were also found in the Pliocene (cf. Zanclean) of Aegina (Argyriou and Theodorou 2011, Fig. 9). Outside Greece, the fossil species †*S. albyi* is common in "middle" and late Miocene and possibly Pliocene faunas of Algeria, Spain, Italy, and Cyprus (Arambourg 1925, 1927; Gaudant et al. 2000, 2010a; Gaudant 2002, 2008b; Merle et al. 2002; Landini and Sorbini 2005a; Carnevale 2007), whereas records of this species in the Miocene of Paratethys are considered dubious (Schwarzhans et al. 2017). The modern species *S. acus* and *S. phlegon* have been reported from the late Pliocene and Pleistocene of Italy (Landini and Sorbini 2005a).

Dactylopteridae Gill, 1861

Type Genus *Dactylopterus* Lacepède, 1801.

***Dactylopterus* Lacepède, 1801**

Type Species *Dactylopterus volitans* (Linnaeus, 1758).

Distribution In Greece, this taxon is only known from the Pliocene of Heraklion Basin. Outside Greece, it is known from the Pliocene of Italy and Cyprus (Landini and Sorbini 2005a; Gaudant and Courme 2014). One species survives in the Atlantic and the Mediterranean (Froese and Pauly 2019).

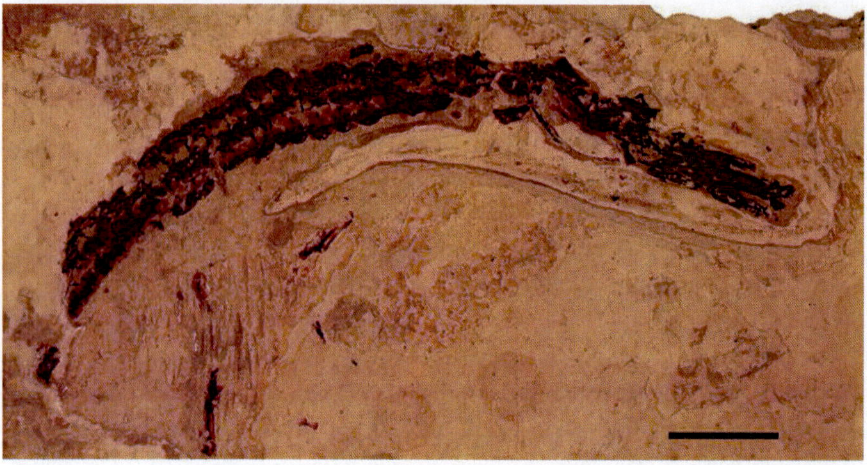

Fig. 9 *Syngnathus* cf. *acus* (Syngnathidae), from the Zanclean diatomitic marls near Mesagros, Aegina Island (AMPG collection). This specimen is figured here for the first time. Scale bar equals 1 cm

Pleuronectiformes Bleeker, 1859
Bothidae Smitt, 1892

Type Genus *Bothus* Rafinesque, 1810a.

***Arnoglossus* Bleeker, 1862**

Type Species *Pleuronectes arnoglossus* (Bloch and Schneider, 1801) (junior synonym of *Arnoglossus laterna* Walbaum, 1792).

Distribution In Greece, this taxon is only known from the Tortonian of Makrilia Formation, in the vicinity of Ierapetra, Crete (Gaudant 2004). Outside Greece, *Arnoglossus* is reported in the Miocene of Algeria and Spain, and from the Miocene, Pliocene, and Pleistocene of Italy (Landini and Sorbini 2005a; Gaudant 2008a, b; Gaudant et al. 2010a). *Arnoglossus* has a cosmopolitan modern distribution (Froese and Pauly 2019).

Centropomidae sensu Greenwood, 1976

Type Genus Not established.

Latinae indet.

Nomenclatural and Taxonomical History This fossil was initially employed to establish the genus †*Allocotus* by de Waldheim (1838). In a following publication, de Waldheim, erected a new genus and species, †*Platacanthus ubinoi*, to accommodate the same individual (de Waldheim 1850). The Musée de la Société Impériale des Naturalistes, Moscow, was listed as repository of this specimen (Fig. 9a). Woodward reassigned †*P. ubinoi* to the modern genus *Morone* (his *Labrax*), utilizing additional material from the Tertiary of Kymi, Greece, deposited in what is nowadays the Natural History Museum, UK (Woodward 1901), which, however, might correspond to a different taxon. Given the unknown location and status of the type material for these taxa, and their almost complete disappearance from the literature (but see Bannikov 1993), I suggest that †*Allocotus* should be treated as invalid and †*Morone ubinoi* as *nomen dubium*. Adding to the confusion, additional percomorph taxa have been signaled in the Miocene of Kymi, but all these old attributions are doubtful and poorly documented (Valenciennes 1861; Gaudry 1862; Unger 1867; Appendix). The holotype of one of these, known as †*Smerdis isabellae* Gaudry 1862, has been synonymized by MNHN, Paris, staff with Woodward's †*Morone* (= *Labrax*) *ubinoi* in the early 1900s (T.A. personal observations).

Distribution In Greece, this taxon is known (possibly) from the early Miocene of an unknown locality near Kymi, Euboea. Outside Greece, *Lates*-like Centropomidae were present in Eocene–Neogene freshwater, brackish and sometimes marine deposits in Europe and Afroarabia (Murray and Attia 2004; Otero 2004; Argyriou et al. 2015). A total of fourteen species (Eschmeyer and Fong 2019) survive in African fresh–estuarine waters, and Indo-Pacific coastal environments (Froese and Pauly 2019).

Comments The original description of this taxon does not contain clear information about its geographical or lithological provenance, other than that the fossil comes from some calcareous, oolithic horizon in Greece, which was thought to resemble the Solnhofen limestone (de Waldheim 1850). This specimen was not found by the author himself, but was sent to him by a certain Dr. Ubino who found, or purchased it during his travels through Greece (de Waldheim 1838). Woodward traces the provenance of the material in British collections to the freshwater deposits of Kymi, Greece. It is thus logical to hypothesize that this specimen derives from the classical Miocene plant and fish-bearing horizons of Kymi (e.g., Unger 1867; Kottis et al. 2002), yet this needs to be corroborated by future sampling. Although the drawings of the specimen are rather uninformative (de Waldheim 1838, 1850), the figured individual shows a preopercle bearing a horizontal anteroventral arm, which, in turn, forms a robust posterior spinous process, and three robust ventral processes (Fig. 10a). The possible resemblance of this animal to the percichthyid †*Cyclopoma* (= *Cyclopome*), or the centropomid *Lates* was noted by de Waldheim (1838). The imaged preopercle differs clearly from that of the preopercle of the Eocene genus †*Cyclopoma* (Agassiz 1833–1843). The shape and orientation of these processes resemble instead the preopercles of Latinae (Latidae in Otero 2004; now included within Centropomidae; see Greenwood 1976, Betancur-R et al. 2017; Fig. 10b), leaving little room for doubt that the illustrated skull is a *Lates*-like centropomid. Despite this profound resemblance, and unlike Latinae (Otero 2004),

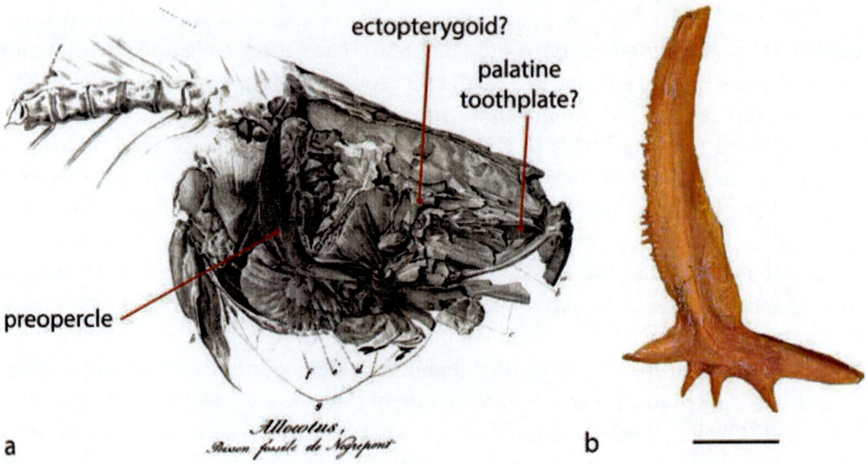

Fig. 10 (**a**) Specimen used to establish †*Allocotus* sp. (de Waldheim 1838) and subsequently †*Platacanthus ubinoi* (de Waldheim 1850), but hereby reassigned to Centropomidae indet.; modified from de Waldheim 1838, scale not provided in original work. (**b**) Preopercle of *Lates* cf. *niloticus* from the Mio-Pliocene of Sahabi, Libya (collected and attributed by the author; the specimen belongs to the collections of the University of Benghazi, Libya), for comparison with the illustrated preopercle of (**a**). Scale bar equals 1 cm

the posterior margin of the dorsal limb of the Kymi specimen is figured as non-serrate, but this feature might have been poorly preserved, or illustrated. Additional similarities between the illustrated Kymi specimen and *Lates*-like animals (Otero 2004) can be found in the acute ventral and anterior processes of the ectopterygoid, or what appears to be a broad palatine toothplate, in the anterior portion of the skull, and in what appears to be a single, posterodorsal process of the opercle. Woodward, possibly based on material from Kymi housed in British collections, mentioned that †*Morone* (= *Labrax*) *ubinoi* lacks a posterior spinous process of the preopercle, and also exhibits two feeble spines on the posterior margin of the opercle (Woodward 1901). It is thus possible that Woodward lumped two different taxa, one being a moronid and the other being de Waldheim's centropomid, into one. This highlights the need to restudy Woodward's material. This notion is reinforced by personal observations on Gaudry's material, which was used to establish the species †*Smerdis isabellae* and is now housed at the MNHN, Paris. The latter would require a detailed study to clarify its taxonomic status, which lies beyond the context of this work.

Labridae Cuvier, 1817

Type Genus *Labrus* Linnaeus, 1758.

Distribution In Greece, this group is known from the Pliocene of Heraklion Basin (Gaudant 2001). Ichthyoliths from the Pliocene of Aghia Marina, Aegina Island, have been tentatively attributed to the sparid †*Sargus* cf. *oweni* (Leivaditis 1974), which was in turn found to be a junior synonym of the labrid †*Trigonodon jugleri* (Schultz and Bellwood 2004). This might correspond to another occurrence of Labridae in Greece, but it is so far poorly substantiated. Labrids appear in the Eocene of Italy (Bannikov and Carnevale 2010). Mostly disarticulated (e.g., oral or pharyngeal teeth), but also articulated remains are common in Neogene deposits around the world (e.g., Algeria, Italy, France, Portugal, Poland, Austria, Nigeria, the United States, and New Zealand; summarized in Bellwood et al. 2019).

Acanthuriformes sensu Betancur-R et al. 2017
Acanthuridae Bleeker, 1859

Type Genus *Acanthurus* Forskål, 1775.

Distribution In Greece, this taxon is known from the Tortonian of Makrilia Formation, in the vicinity of Ierapetra, Crete (Gaudant 2004), and the Tortonian of Gavdos (Gaudant et al. 2005). Outside Greece, articulated acanthurid fossils are known from the Paleogene of Europe (e.g., Italy, Switzerland, Russia) (Tyler and Bannikov 2000; Tyler and Micklich 2011), while their Miocene occurrences (e.g., Austria and Croatia) usually correspond to isolated teeth (Schultz 2003; Tripalo et al. 2016). Nowadays, the family is absent from the Mediterranean, but is otherwise characterized by circumtropical distribution (Froese and Pauly 2019).

Comments The Gavdos specimen is probably a juvenile (Gaudant et al. 2005).

Tetraodontiformes Berg, 1937
Monacanthidae Nardo, 1842

Type Genus *Monacanthus* Oken, 1817.

†*Frigocanthus* Sorbini and Tyler, 2004

Type Species †*Frigocanthus stroppanobili* Sorbini and Tyler, 2004.

Included Taxa †*Frigocanthus margaritatus* Sorbini and Tyler, 2004.

Distribution In Greece, †*F. margaritatus* is known from the Pliocene of Heraklion Basin (Gaudant 2001; Sorbini and Tyler 2004), whereas †*F. stroppanobili* from the middle-late Pliocene of Chania, Crete (Sorbini and Tyler 2004). Outside Greece, this taxon is known from the Pliocene and Pleistocene of Italy (Sorbini and Tyler 2004).

Caproiformes sensu Rosen, 1984
Caproidae Lowe, 1843

Type Genus *Capros* Lacepède, 1802.

***Capros* Lacepède, 1802**

Type Species *Capros* (= *Zeus*) *aper* (Linnaeus, 1758).

Distribution In Greece, this genus is known only from the Tortonian of Makrilia Formation, in the vicinity of Ierapetra (Gaudant 2004). Outside Greece, fossil species of the genus have been recognized in the Oligocene of the Carpathians, the Caucasus, Germany, and France (Baciu et al. 2005b; Přikryl et al. 2016c), and the late Miocene of Italy and Algeria (Baciu et al. 2005a). The living species *C. aper* is known from the Pliocene–Pleistocene of Italy (Landini and Sorbini 2005a).

6 Conclusions and Avenues for Future Research

Over a century and a half of exploration and sporadic scientific research on the actinopterygian fossil record of Greece has led to the discovery of numerous fossil sites, and laid the foundations for further work on the regional fossil record of the clade. Actinopterygian fossils in Greece have a patchy record that spans over 180 million years, from the Toarcian (Lower Jurassic) to the Quaternary, with the vast majority of known occurrences concentrated in the late Neogene (late Miocene–Pliocene) and being of clear marine affinities (summarized in the Appendix). Little information is available for older, or freshwater and brackish localities. The pre-Neogene fossil record of Greece merits further scientific attention and exploration, especially given the huge record of sedimentary rocks and complex geological history of Greece. Deposits predating the Jurassic have not yet, to the author's knowledge, been investigated extensively for potential paleoichthyological content (excluding conodonts), and like in the case of chondrichthyans (Argyriou et al. 2017), actinopterygian fossils are anticipated in Paleozoic–early Mesozoic deposits.

Historically, alpine sedimentary formations of Greece (e.g., the "External Hellenides"), which record a major part of the Mesozoic and early Cenozoic, have been considered to be of principally tectonostratigraphic importance to the geological and paleontological community, and have remained outside the interest of vertebrate paleontologists. Yet, preliminary investigations have indicated the yet untapped paleoichthyological potential of such deposits. The presumably rich Toarcian bonebeds of Ionian Unit exposures in Lefkada have been totally neglected since their first discovery (Kottek 1964), despite their potential informativeness on early teleost diversification. Moreover, understudied fish assemblages from the late Cretaceous–Paleogene of Pindos Unit can shed light on the evolution and biogeography of Tethyan faunas (Koch and Nikolaus 1969; Cavin et al. 2012; Argyriou and Davesne 2021; pers. obs. on the K/Pg of Eurytania, Northern–Central Peloponnese and Gavdos) during this critical time interval of global marine restructuring, which is otherwise poorly documented in the global fossil record. Further research should be directed toward freshwater and brackish faunas, which are so far known from Miocene or younger deposits (e.g., Miocene of Kymi Euboea: de Waldheim 1838,

1850; Unger 1867; Miocene of Chios: Reichenbacher and Böhme 2004), and their insights into the evolution of modern European freshwater ichthyofaunas and their biogeographical and paleohydrological inferences.

Late Miocene–Pliocene fossiliferous localities and horizons, especially the ones in Crete and Gavdos, host diverse actinopterygian assemblages of often exquisite preservation quality (Gaudant et al. 1994, 1997a, 2005, 2006; Gaudant 2001, 2004). Numerous families and genera have already been recognized in these assemblages, largely on the basis of broad morphological and meristic similarities with late Miocene taxa from Italy and Algeria (e.g., Arambourg 1925, 1927; Gaudant 2008b). Previous works, however, have treated these assemblages from an alpha-taxonomic point of view, and provided some limited information on their paleoenvironmental significance. Additional taxonomic diversity, including new species, remains to be identified in these sites, pending detailed anatomical descriptions (Přikryl and Carnevale 2017), and is a prerequisite for further systematic and biogeographic work. The importance of following this line of research cannot be overstressed, especially when taking into account the scarcity of information for this part of (Neo) Tethys, in comparison to what is now the Western and Central Mediterranean (e.g., Arambourg 1925, 1927; Landini and Sorbini 2005a; Carnevale 2007; Gaudant 2008a, b). Research on Greek fossils is anticipated to unveil further clues for tracing the gradual emergence of the modern endemic and Atlantic hallmark of Mediterranean marine faunas from the tropical, Indopacific-like faunas of the Miocene, and the role of major events, such as the Messinian Salinity Crisis and Pleistocene cooling(s) in this process (Gaudant et al. 1994, 1997a, 2005, 2006; Gaudant 2001, 2002, 2004; Landini and Sorbini 2005a, b).

This work collates available information on actinopterygian fossils and their spatiotemporal distribution in Greece, as well as their potential significance. It is hoped to form a stimulating basis for further paleoichthyological research in the country. As a concluding remark, properly curated paleoichthyological collections of Greek fish fossils are crucial for invigorating paleoichthyology in the country, and, given funding and staff limitations, Greek institutions and museums are encouraged to work toward this goal.

Acknowledgments I am indebted to my former advisors and mentors G. Theodorou and P. Pavlakis (both NKUA) for supporting and encouraging me early to work on the fossil fishes of Greece, and beyond. S. Roussiakis (NKUA) is thanked for providing collection assistance and scientific discussion, and K. Agiadi (NKUA) is thanked for useful and stimulating discussions on the particulars of this chapter. Several colleagues are thanked for helping the author (re)assess some of the information presented here on the field. Listed in alphabetic order are: K. Agiadi (NKUA); A. Alexopoulos (NKUA); J. Carrillo-Briceño (Univ. Zurich); L. Cavin (MNHN, Geneva); D. Davesne (Oxford Univ.); I. Dimitriou; C. Fassoulas (NHM, Crete); A. Filipidi (Univ. Utrecht); C. Giamali (NKUA); G. Iliopoulos (Univ. Patras); E. Koskeridou (NKUA); M. Louvari (NKUA); S. Roussiakis (NKUA); G. Theodorou (NKUA); K Trikolas. Giorgio Carnevale (Univ. Torino), Charalampos Kevrekidis (LMU Munich) and C. Romano (Univ. Zurich) are thanked for their insightful reviews, which improved the quality of this work. This research was funded by a Swiss National Science Foundation early postdoctoral mobility grant (P2ZHP3_184216) to the author.

Appendix

Tables with a detail list of the occurrences of ray-finned fishes in Greece.

Table 1 List of Greek localities with known occurrences of ray-finned fishes have not yet been studied and/or identified to lower taxonomic levels

Localities[PBDB No]	Age (MN)	Taxon
Marathousa 1[199937–40]	Middle Pleistocene	Teleostei indet.[1]
'Unit e', Pigadia Basin	Middle Pleistocene	Teleostei indet. (cf. Ostariophysi)[2]
Gryllos-Mesovouni	Plio–Pleistocene	Teleostei indet.[3]
Vatera[207301]	Plio–Pleistocene	Teleostei indet.[4]
Northern Milos	Plio–Pleistocene	Teleostei indet.[5]
Faros, Aegina	? Zanclean	Teleostei indet.[6]
Temple of Aphrodite, Aegina	? Zanclean	Teleostei indet.[6]
Florina/Ptolemais	Messinian–Zanclean	Teleostei indet.[7]
Neo Khorio[207611]	Messinian	Teleostei indet.[8]
Keri, Zakynthos	Tortonian	Teleostei indet.[9]
Kefalonia	? Late Miocene	Teleostei indet.*
Nenita beds, Chios[207609]	Early Miocene (MN5)	Teleostei indet.[10, i]
Near Megalopoli	Late Maastrichtian	Teleostei indet.**

MN Mammal Neogene Zone
*Pers. obs. and comm. with S. Roussiakis
**Pers.obs. and comm. with E. Koskeridou, G. Theodorou, and K. Trikolas
[i]Families recognized on the basis of otoliths and/or ichthyoliths, treated by Agiadi (this volume)
[1]Panagopoulou et al. (2018), [2]Bellas et al. (1998), [3]Kamperis (1987), [4]Drinia et al. (2002), [5]Calvo et al. (2012), [6]Leivaditis (1974), [7]Hordijk and de Bruijn (2009), [8]Freuenthal (1969), [9]Dermitzakis (1979), [10]Reichenbacher and Böhme (2004)

Table 2 List of Greek localities with known occurrences of non-teleostean actinopterygians (†Pycnodontidae), and stem (†Ichthyodectoidei, †Leptolepididae) and early diverging crown teleosts (†Osmeroididae) in chronological order

Localities[PBDB No]	Age	Taxon
Gavdos[207367]	Late Maastrichtian	†Ichthyodectoidei indet.[1]
Unspecified localities near Karpenisi	Late Maastrichtian	†<u>Osmeroididae</u> †*Osmeroides* sp.2,**
Lindos, Rhodes[207610]	?Cenomanian	†<u>Pycnodontidae</u> †*Coelodus* sp.[3–5]
Anavryssada	Toarcian	†Leptolepididae indet.[6, *]

*Taxonomy revised, or updated in this study
**Poorly documented occurrence, not treated further in this chapter
[1]Cavin et al. (2012), [2]Koch and Nikolaus (1969), [3]Melentis (1967a), [4]Melentis (1967b), [5]Schultz et al. (1997), [6]Kottek (1964)

Table 3 List of Greek localities with known occurrences of otocephalan ray-finned fishes: families Clupeidae and Cyprinidae

Localities[PBDB No]	Age	Taxon
Stavromenos[162721]	Piacenzian	Clupeidae *Alosa* sp.[1] *Sardina* sp.[1], *Spratelloides* sp.[1],
Roufas[207304]	Piacenzian	Clupeidae *Alosa* sp.[1] *Sardina* sp.[1],
Armissos[207303]	Piacenzian	Clupeidae *Alosa* cf. *A. elongata*[1,2] *Sardina* cf. *S. pilchardus*[1,2], *Spratelloides* cf. *S. gracilis*[1,2],
Voutes[207357]	Piacenzian	Clupeidae *Alosa* sp.[2], ? *Sardina* sp.[2], *Spratelloides gracilis*[2],
Gournes[207359]	Piacenzian	Clupeidae *Alosa* sp.[2], ? *Sardina* sp.[2], *Spratelloides gracilis*[2]
Aghios Thomas, Aegina[207360]	Zanclean	Upper part of section: Clupeidae *Spratelloides gracilis*[3]
		Middle part of section: Clupeidae indet.[*]
		Mesagros – Alones road: Clupeidae *Spratelloides* cf. *S. gracilis*[4]
Kalamaki	Pliocene	Clupeidae *Spratelloides gracilis*[5]
Strymonas	Pliocene	Clupeidae ? *Alosa tanaica*[6,*]
Aidonochori	Late Miocene	Clupeidae indet.7
Metochia C[207361]	Messinian	Clupeidae *Alosa* sp.[8]
Akropotamos	Messinian	Clupeidae †*Alosa crassa*[9,*]
Ano Metochi 3[31928]	Messinian	Cyprinidae indet.[10]
Mires[207364]	Messinian	Clupeidae †*Alosa crassa*[11,*]

Localities[PBDB No]	Age	Taxon
Unknown Cretan locality, possibly near Rethymnon	?Messinian	Clupeidae †*Alosa crassa*[12,13]
Karave[60775]	Tortonian	Clupeidae †*Alosa crassa*[14,*]
Chapel Aghia Paraskevi[207356]	Tortonian	Clupeidae *Alosa* sp.[15], †*Alosa* cf. *A. crassa*[15,*], *Spratelloides* sp.[15]
Type strata of Makrilia Fm., 4 km to the North of Chapel Aghia Paraskevi	Tortonian	Clupeidae indet.[16]
Lapsarna[186558]	Burdigalian	Cyprinidae *Barbus* sp.[17], *Cyprinion* sp.[17], †*Palaeoleuciscus* sp.[17]
Aghioi Theodori or Myriki[207612]	Danian	Clupeidae indet.[18,**]
Unspecified localities near Karpenisi	Late Maastrichtian	Clupeomorpha incertae sedis †*Scombroclupea* sp.[18,**]

*Taxonomy revised, or updated in this study

**Poorly documented occurrence, not treated further in this chapter

[1]Gaudant (2001), [2]Gaudant et al. (1994), [3]Gaudant et al. (2010b), [4]Argyriou and Theodorou (2011), [5]Gaudant and Symeonidis (1995), [6]Weiler (1943), [7]Kevrekidis (2016), [8]Gaudant et al. (2006), [9]Dermitzakis et al. (1986), [10]Georgalis et al. (2017), [11]Gaudant et al. (1997a), [12]Gaudant (1980), [13]Gaudant (2014b), [14]Gaudant et al. (2005), [15]Gaudant (2004), [16]Bürgin (1994), [17]Vasileiadou et al. (2017), [18]Koch and Nikolaus (1969)

Table 4 List of Greek localities with known occurrences of aulopiform ray-finned fishes: families †Dercetidae, †Enchodontidae, and Paralepididae

Localities[PBDB No]	Age	Taxon
Metochia C[207361]	Messinian	Paralepididae †*Paralepis albyi*[1]
Gavdos[207367]	Late Maastrichtian	†Enchodontidae †*Enchodus* cf. †*E. dirus*[2],
Aghia Trias–Domiani Road[207613]	Late Maastrichtian	†Dercetidae †Dercetidae indet.[3,*] †Enchodontidae ? †*Enchodus* indet.[3], †Ichthyotringidae[**] †*Ichthyotringa* sp. [3,*,**]
Unspecified localities near Karpenissi[207615]	Late Maastrichtian	†Enchodontidae †*Eurypholis* indet.[3,**]

*Taxonomy revised, or updated in this study

**Poorly documented occurrence, not treated further in this chapter

[1]Gaudant et al. (2006), [2]Cavin et al. (2012), [3]Koch and Nikolaus (1969)

Table 5 List of Greek localities with known occurrences of stomiatiform ray-finned fishes: families Gonostomatidae and Sternoptychidae

Localities[PBDB No]	Age	Taxon
Armissos[207303]	Piacenzian	Sternoptychidae *Maurolicus* cf. *M. muelleri*[1]
Aegina Island	Zanclean	Sternoptychidae *Argyropelecus* sp.[2]
Metochia C[207361]	Messinian	Sternoptychidae *Maurolicus muelleri*[3]
Tsangaraki[207364]	Messinian	Sternoptychidae *Maurolicus muelleri*[4]
Mires[207364]	Messinian	*Maurolicus muelleri*[4]
Chapel Aghia Paraskevi[207356]	Tortonian	Gonostomatidae **†*Cyclothone gaudanti***[5–6],
		Sternoptychidae *Maurolicus muelleri*[5]
400 m from Chapel Aghia Paraskevi[207355]	Tortonian	Gonostomatidae **†*Cyclothone gaudanti***[5–6],
		Sternoptychidae *Maurolicus muelleri*[5]

[1]Gaudant (2001), [2]Argyriou and Theodorou (2011), [3]Gaudant et al. (2006), [4]Gaudant et al. (1997a), [5]Gaudant (2004), [6]Přikryl and Carnevale (2017)

Table 6 List of Greek localities with known occurrences of myctophiform ray-finned fishes: family Myctophidae

Localities[PBDB No]	Age	Taxon
Armissos[207303]	Piacenzian	Myctophidae indet.[1]
Metochia C[207361]	Messinian	†"*Myctophum*" *dorsale*[2], †"*Myctophum*" *licatae*[2]
Tsangaraki[207364]	Messinian	†"*Myctophum*" *licatae*[3]
Mires[207364]	Messinian	†"*Myctophum*" *licatae*[3], †"*Myctophum*" *dorsale*[3]
Psalida[207365]	Messinian	†"*Myctophum*" *licatae*[3], †"*Myctophum*" *dorsale*[3]
'Unit b', Pigadia Basin	Messinian	Myctophidae indet.[4]
Metochia B[185752]	Tortonian	†"*Myctophum*" *licatae*[5]
Karave[60775]	Tortonian	†"*Myctophum*" *licatae*[5]
Chapel Aghia Paraskevi[207356]	Tortonian	†"*Myctophum*" *licatae*[6]
400 m from Chapel Aghia Paraskevi[207355]	Tortonian	†"*Myctophum*" sp.[6], †"*Myctophum*" *licatae*[6]
Type strata of Makrilia Fm., 4 km to the North of Chapel Aghia Paraskevi	Tortonian	Myctophidae indet.[7]

[1]Gaudant (2001), [2]Gaudant et al. (2006), [3]Gaudant et al. (1997a), [4]Bellas et al. (1998) [5]Gaudant et al. (2005), [6]Gaudant (2004), [7]Bürgin (1994)

Table 7 List of Greek localities with known occurrences of gadiform ray-finned fishes: families Gadidae, Moridae, and Bregmacerotidae

Localities[PBDB No]	Age	Taxon
Stavromenos[162721]	Piacenzian	Bregmacerotidae *Bregmaceros* sp.[1,2]
Roufas[207304]	Piacenzian	Bregmacerotidae *Bregmaceros* cf. †*B. albyi*[1]
Armissos[207303]	Piacenzian	Bregmacerotidae *Bregmaceros* cf. †*B. albyi*[1,3]
		Gadidae *Micromesistius* sp.[1],
		Moridae *Gadella* sp.[1],
Voutes[207357]	Piacenzian	Bregmacerotidae †*Bregmaceros albyi*[3]
Prassas[207358]	Piacenzian	Bregmacerotidae †*Bregmaceros albyi*[3]
Gournes[207359]	Piacenzian	Bregmacerotidae †*Bregmaceros albyi*[3]
Aghios Thomas, Aegina[207360]	Zanclean	Middle part of section: Bregmacerotidae *Bregmaceros* cf. †*B. albyi*[4]
		Mesagros – Alones Road:Bregmacerotidae *Bregmaceros* cf. †*B. albyi*[4]
Kalamaki	Pliocene	Moridae ? *Gadella* sp.[6]
Metochia C[207361]	Messinian	Bregmacerotidae †*Bregmaceros albyi*[7]
Mires[207364]	Messinian	Bregmacerotidae †*Bregmaceros albyi*[8]
Psalida[207365]	Messinian	Bregmacerotidae †*Bregmaceros albyi*[8]
'Unit b', Pigadia Basin	Messinian	†*Bregmaceros albyi*[9]
Metochia B[185752]	Tortonian	Bregmacerotidae †*Bregmaceros albyi*[10],
Chapel Aghia Paraskevi[207356]	Tortonian	Bregmacerotidae †*Bregmaceros albyi*[11]
400 m from Chapel Aghia Paraskevi[207355]	Tortonian	Bregmacerotidae †*Bregmaceros albyi*[11]
Type strata of Makrilia Fm., 4 km to the North of Chapel Aghia Paraskevi	Tortonian	†*Bregmaceros albyi*.[12]

[1]Gaudant (2001), [2]Symeonidis and Schultz (1973), [3]Gaudant et al. (1994), [4]Argyriou and Theodorou (2011), [5]Weiler (1943), [6]Gaudant and Symeonidis (1995), [7]Gaudant et al. (2006), [8]Gaudant et al. (1997a), [9]Bellas et al. (1998), [10]Gaudant et al. (2005), [11]Gaudant (2004), [12]Bürgin (1994)

Table 8 List of Greek localities with known occurrences of acanthomorph ray-finned fishes: families listed alphabetically Acanthuridae, Atherinidae, Belonidae, Blenniidae, Bothidae, Caproidae, Carangidae, Centriscidae, Centropotamidae, Cyprinodontidae, Dactylopteridae, Gobiidae, Labridae, Monacanthidae, Scombridae, Serranidae, Sparidae, Syngnathidae, and Trichiuridae. Additional families (Ambassidae, Moronidae, Moronidae, and Percichthyidae) are indicatively listed as problematic occurrences in the Miocene of Kymi, but are not further treated

Localities[PBDB No]	Age	Taxon
Frangokastello[207302]	middle–late Pliocene	Monacanthidae †*Frigocanthus stroppanobili*[1]
Stavromenos[162721]	late Pliocene	Trichiuridae *Lepidopus* sp.[2]
Roufas[207304]	Piacenzian	Trichiuridae *Lepidopus* cf. *caudatus*[2]
Armissos[207303]	Piacenzian	Belonidae *Belone sp.*[2], ? *Scomberesox* sp.[2],
		Carangidae *Caranx* sp.[2], *Trachurus* sp.[2],
		Centriscidae *Aeoliscus* cf. *A. strigatus*[2–3],
		Dactylopteridae *Dactylopterus* sp.[2],
		Labridae indet.[2]
		Monacanthidae †*Frigocanthus margaritatus*[1,2]
		Percomorpha indet.[2,*],
		Scombridae
		Scomberomorus sp.[2], *Scomber* sp.[2],
		Serranidae *Serranus* sp.[2],
		Sparidae *Boops* sp.[2,4],
		Syngnathidae *Syngnathus* cf. †*S. albyi*[2–3],
		Trichiuridae *Lepidopus* sp.[2,4],
Voutes[207357]	Piacenzian	Belonidae *Belone* sp.[4],
		Syngnathidae *Syngnathus* sp.[4],
		Trichiuridae *Lepidopus* sp.[4]
Prassas[207358]	Piacenzian	Blenniidae *Salaria* cf. *S. pavo*[4,*],
		Percomorpha indet.[4,*]
		Sparidae *Boops* cf. *B. boops*[4],

Localities[PBDB No]	Age	Taxon
Gournes[207359]	Piacenzian	Gobiidae *Gobius* s.l. sp.[4]
		Sparidae,
		Boops cf. *B. boops*[4],
Aghia Marina Aegina	Piacenzian?	?Labridae[*,**] ?†*Trigonodon* cf. *jugleri*[5,*,**]
Aghios Thomas, Aegina[207360]	Zanclean	Upper part of section: Serranidae indet.[6]
		Mesagros – Alones Road: Percomorpha indet.[*]
		Syngnathidae *Syngnathus* cf. *S. acus*[7],
		Percomorpha indet[*].
Kalamaki	Pliocene	Atherinidae Atherinidae indet.[8]
		Carangidae *Trachurus* sp.[8],
		Gobiidae *Gobius* s.l. sp.[8],
		Percomorpha indet.[8,*]
Metochia C[207361]	Messinian	Centriscidae ? *Aeoliscus* sp.[9,*]
		Syngnathidae *Syngnathus* cf. †*S. albyi*[9]
		Trichiuridae *Lepidopus* sp.[9],
Voukolies[207362]	Messinian	Cyprinodontidae †*A. crassicaudus*[10,*]
Kaleryiana[207363]	Messinian	Cyprinodontidae †*A. crassicaudus*[10]
Tsangaraki[207364]	Messinian	Syngnathidae †*Syngnathus albyi*[11]
Mires[207364]	Messinian	Pleuronectiformes indet.[11,*]
		Syngnathidae †*Syngnathus albyi*[11],
		Trichiuridae *Lepidopus* sp.[11],
Psalida[207365]	Messinian	Percomorpha indet.[11,*]
'Unit b', Pigadia Basin	Messinian	*Dentex* sp.[12] *Pagellus* sp.[12]
Metochia B[185752]	Tortonian	Syngnathidae *Syngnathus* cf. †*S. albyi*[13]
		Trichiuridae *Lepidopus* sp. [13]
Karave[60775]	Tortonian	Acanthuridae indet.[13]

Localities[PBDB No]	Age	Taxon
Chapel Aghia Paraskevi[207356]	Tortonian	Acanthuridae indet.[14],
		Bothidae *Arnoglossus* sp.[14],
		Caproidae *Capros* sp.[14],
		Carangidae †***Alectis psarianosi*[14,15]**, *Trachurus* sp.[13],
		Percomorpha indet.[14,*],
		Syngnathidae *Syngnathus* cf. †*S. albyi*[14,16]
		Trichiuridae *Lepidopus* sp.[14,15]
Type strata of Makrilia Fm., 4 km to the North of Chapel Aghia Paraskevi	Tortonian	*Lepidopus* sp.[17]
Unknown localities near Kymi	? early Miocene	? Ambassidae indet.[18,19,*,**],
		Centropomidae cf. Latinae indet.[20,21,*]
		? Gobiidae ? *Periophtalmus* sp.[22,**],
		? Moronidae indet.[23,*,**]
		? Scombridae indet.[,18,19**],
		? Percichthyidae indet.[22,**]

[*]Taxonomy revised, or updated in this study

[**]Poorly documented occurrence, not treated further in this chapter

[1]Sorbini and Tyler (2004), [2]Gaudant (2001), [3]Bachmayer et al. (1984), [4]Gaudant et al. (1994), [5]Leivaditis (1974), [6]Gaudant et al. (2010b), [7]Argyriou and Theodorou (2011), [8]Gaudant and Symeonidis (1995), [9]Gaudant et al. (2006), [10]Gaudant (2014b), [11]Gaudant et al. (1997a), [12]Bellas et al. (1998), [13]Gaudant et al. (2005), [14]Gaudant (2004), [15]Symeonidis (1969), [16]Bachmayer and Symeonidis (1978), [17]Bürgin (1994), [18]Valenciennes (1861), [19]Gaudry (1862), [20]de Waldheim (1838), [21]de Waldheim (1850), [22]Unger (1867), [23]Woodward (1901)

References

Abbazzi L, Angelone C, Arca M, Barisone G, Bedetti C, Delfino M, Kotsakis T, Marcolini F, Palombo MR, Pavia M, Piras P, Rook L, Torre D, Tuveri C, Valli AMF, Wilkens B (2004) Plio-Pleistocene fossil vertebrates of Monte Tuttavista (Orosei, Eastern Sardinia, Italy), an overview. Riv Ital Paleontol Stratigr 110:681–706

Agiadi K (this volume) The fossil otolith record of fishes (Vertebrata: Teleostei) in Greece. In: Vlachos E (ed) The fossil vertebrates of Greece vol. 1 – Basal vertebrates, amphibians, reptiles, afrotherians, glires, and primates. Springer – Nature Publishing Group, Cham

Agiadi K, Koskeridou E, Triantaphyllou M, Girone A, Karakitsios V (2013) Fish otoliths from the Pliocene Heraklion Basin (Crete Island, Eastern Mediterranean). Geobios 46:461–472

Agiadi K, Vasileiou G, Koskeridou E, Moissette P, Cornée J-J (2019) Coastal fish otoliths from the early Pleistocene of Rhodes (eastern Mediterranean). Geobios 55:1–15

Alfaro ME, Santini F, Brock C, Alamillo H, Dornburg A, Rabosky DL, Carnevale G, Harmon LJ (2009) Nine exceptional radiations plus high turnover explain species diversity in jawed vertebrates. Proc Natl Acad Sci 106:13410–13414

Alfaro ME, Faircloth BC, Harrington RC, Sorenson L, Friedman M, Thacker CE, Oliveros CH, Černý D, Near TJ (2018) Explosive diversification of marine fishes at the Cretaceous–Palaeogene boundary. Nat Ecol Evol 2:688–696

Antipa G (1904) Die Clupeinen des westlichen Teiles des Schwarzen Meeres und der Donaumündungen. Anz Akad Wiss Wien 41:299–303

Arambourg C (1925) Révision des poissons fossiles de Licata (Sicile). Masson, Paris

Arambourg C (1927) Les poissons fossiles d'Oran. Matériaux pour la Carte géologique d'Algérie. Jules Careonel, Alger

Argyriou T, Davesne D (2021) Offshore marine actinopterygian assemblages from the Maastrichtian–Paleogene of the Pindos Unit in Eurytania, Greece. PeerJ 9:e10676

Argyriou T, Theodorou G (2011) New findings from the Pliocene (Zanclean) ichthyofauna of Aegina island, Greece. Paper presented at the European Association of Vertebrate Paleontologists 9th annual meeting, Heraklion, 2011

Argyriou T, Cook TD, Muftah AM, Pavlakis P, Boaz NT, Murray AM (2015) A fish assemblage from an early Miocene horizon from Jabal Zaltan, Libya. J Afr Earth Sci 102:86–101

Argyriou T, Clauss M, Maxwell EE, Furrer H, Sánchez-Villagra MR (2016) Exceptional preservation reveals gastrointestinal anatomy and evolution in early actinopterygian fishes. Sci Rep 6:18758

Argyriou T, Romano C, Carrillo-Briceño JD, Brosse M, Hofmann R (2017) The oldest record of gnathostome fossils from Greece: Chondrichthyes from the Lopingian of Hydra Island. Palaeontol Electron 20.1.8A:1–9

Argyriou T, Giles S, Friedman M, Romano C, Kogan I, Sánchez-Villagra MR (2018) Internal cranial anatomy of early Triassic species of †*Saurichthys* (Actinopterygii: †Saurichthyiformes): implications for the phylogenetic placement of †saurichthyiforms. BMC Evol Biol 18:161

Arratia G (1999) The monophyly of Teleostei and stem-group teleosts. Consensus and disagreements. In: Arratia G, Schultze HP (eds) Mesozoic fishes 2 – systematics and fossil record. Dr. Friedrich Pfeil, Munich, pp 265–334

Arratia G (1997) Basal teleosts and teleostean phylogeny. Palaeo Ichthyol 7:5–168

Arratia G (2017) New Triassic teleosts (Actinopterygii, Teleosteomorpha) from northern Italy and their phylogenetic relationships among the most basal teleosts. J Vertebr Paleontol 37:e1312690

Arratia G, Hikuroa DCH (2010) Jurassic fishes from the Latady Group, Antarctic Peninsula, and the oldest teleosts from Antarctica. J Vertebr Paleontol 30:1331–1342

Arratia G, Scasso RA, Kiessling W (2004) Late Jurassic fishes from Longing Gap, Antarctic Peninsula. J Vertebr Paleontol 24:41–55

Bachmayer F, Symeonidis N (1978) Eine fossile Seenadel (Syngnathidae) aus dem Obermiozän von Ierapetra, Insel Kreta (Griechenland). Ann Naturhist Mus Wien 81:121–127

Bachmayer F, Kähsbauer P, Symeonidis N (1984) Weitere fossile Seenadelreste aus dem Obermiozän der Insel Kreta (Griechenland). Ann Naturhist Mus Wien 86A:1–6

Baciu D-S, Bannikov AF, Santini F (2005a) A new species of Caproidae (Acanthomorpha, Teleostei) from the Messinian (upper Miocene) of Oran (Algeria). Geodiversitas 27:381–390

Baciu D-S, Bannikov AF, Tyler JC (2005b) Revision of the fossil fishes of the family Caproidae (Acanthomorpha) Miscelanea Paleontologica nr. 8. Studi e ricerche sui giacimenti terziari di Bolca XI, Verona, pp 7–74

Bannikov AF (1984) The new subfamily of carangid fishes. Studi e ricerche sui giacimenti terziari di Bolca 4:319–321

Bannikov AF (1987) The phylogenetic relationships of the carangid fishes of the subfamily Caranginae. Paleontol J 21:44–54

Bannikov AF (1993) A new species of bass (Teleostie, Moronidae) from the Sarmatian of Moldova. Paleontol J 27:54–62

Bannikov AF, Carnevale G (2010) *Bellwoodilabrus landinii* n. gen., n. sp., a new genus and species of labrid fish (Teleostei, Perciformes) from the Eocene of Monte Bolca. Geodiversitas 32:201–220

Bellas SM, Kussius K, Kommerell JK, Kriwet J (1998) Integrated biostratigraphical approach of the Neogene Pigadia basin of Karpathos Island (Dodecanes Group, Greece) - Implications on the depositional palaeoenvironment based on calcareous nannofossils, ostracodes and facies development data with special references to Messinian fish assemblages. Berliner geowissenschaftliche Abhandlungen Reihe E 28:129–156

Bellwood DR, Schultz O, Siqueira AC, Cowman PF (2019) A review of the fossil record of the Labridae. Ann Naturhist Mus Wien 121:125–194

Berrebi P, Tsigenopoulos CS (2003) Phylogenetic organization of the genus Barbus sensu stricto: a review based on data obtained using molecular markers. In: The freshwater fishes of Europe 5. AULA, Wiebelsheim, pp 11–22

Betancur-R R, Broughton RE, Wiley EO, Carpenter K, López JA, Li C, Holcroft NI, Arcila D, Sanciangco M, Cureton Ii JC, Zhang F, Buser T, Campbell MA, Ballesteros JA, Roa-Varon A, Willis S, Borden WC, Rowley T, Reneau PC, Hough DJ, Lu G, Grande T, Arratia G, Ortí G (2013) The tree of life and a new classification of bony fishes. PLoS Curr 5:ecurrents.tol.53ba26640df0ccaee75bb165c8c26288

Betancur-R R, Ortí G, Pyron RA (2015) Fossil-based comparative analyses reveal ancient marine ancestry erased by extinction in ray-finned fishes. Ecol Lett 18:441–450

Betancur-R R, Wiley EO, Arratia G, Acero A, Bailly N, Miya M, Lecointre G, Ortí G (2017) Phylogenetic classification of bony fishes. BMC Evol Biol 17:162

Böhme M (2010) Ectothermic vertebrates (Actinopterygii, Allocaudata, Urodela, Anura, Crocodylia, Squamata) from the Miocene of Sandelzhausen (Germany, Bavaria) and their implications for environment reconstruction and palaeoclimate. Paläontol Z 84:3–41

Böhme M, Ilg A (2003) fosFARbase, Vol. 2019

Broughton RE, Betancur-R R, Li C, Arratia G, Ortí G (2013) Multi-locus phylogenetic analysis reveals the pattern and tempo of bony fish evolution. PLoS Curr 5:ecurrents.tol.2ca8041495ffafd0c92756e75247483e

Bürgin T (1994) Echte Knochenfische (Actinopterygii; Teleostei) aus dem Oberen Miozän der Insel Kreta (Griechenland): Vorläufige Resultate. Eclog Geol Helv 87:1007–1022

Calvo JP, Triantaphyllou MV, Regueiro M, Stamatakis MG (2012) Alternating diatomaceous and volcaniclastic deposits in Milos Island, Greece. A contribution to the upper Pliocene–lower Pleistocene stratigraphy of the Aegean Sea. Palaeogeogr Palaeoclimatol Palaeoecol 321–322:24–40

Carnevale G (2003) Redescription and phylogenetic relationships of *Argyropelecus logearti* (Teleostei: Stomiiformes: Sternoptychidae), with a brief review of fossil *Argyropelecus*. Riv Ital Paleontol Stratigr 109:63–76

Carnevale G (2007) Fossil fishes from the Serravallian (Middle Miocene) of Torricella Peligna, Italy. Paleontographica Ital 91:1–67

Carnevale G, Landini W, Sarti G (2006) Mare versus Lago-mare: marine fishes and the Mediterranean environment at the end of the Messinian Salinity Crisis. J Geol Soc 163:75–80

Carnevale G, Bannikov AF, Marramà G, Tyler JC, Zorzin R (2014) The Pesciara-Monte Postale Fossil-Lagerstätte: 2. Fishes and other vertebrates. Rend Soc Paleontol Ital 4:37–63

Carnevale G, Dela Pierre F, Natalicchio M, Landini W (2018) Fossil marine fishes and the 'Lago Mare'; event: Has the Mediterranean ever transformed into a brackish lake? Newsl Stratigr 51:57–72

Cavender TM (1991) The fossil record of the Cyprinidae. In: Winfield IJ, Nelson JS (eds) Cyprinid Fishes: systematics, biology and exploitation. Springer Netherlands, Dordrecht, pp 34–54

Cavin L, Alexopoulos A, Piuz A (2012) Late Cretaceous (Maastrichtian) ray-finned fishes from the island of Gavdos, southern Greece, with comments on the evolutionary history of the aulopiform teleost *Enchodus*. Bull Soc Géol Fr 183:561–572

Cavin L, Forey PL, Giersch S (2013) Osteology of Eubiodectes libanicus (Pictet & Humbert, 1866) and some other ichthyodectiformes (Teleostei): phylogenetic implications. J Syst Palaeontol 11:115–177
Chalifa Y (1996) A new species of *Enchodus* (Aulopiformes: Enchodontidae) from the Northern Negev, Israel, with comments on the evolutionary trends in the Enchodontoidei. In: Arratia G, Viohl G (eds) Mesozoic fishes - systematics and paleoecology. Dr. Friedrich Pfeil, München, pp 349–367
Cornée J-J, Moissette P, Joannin S, Suc J-P, Quillévéré F, Krijgsman W, Hilgen F, Koskeridou E, Münch P, Lécuyer C, Desvignes P (2006) Tectonic and climatic controls on coastal sedimentation: the Late Pliocene–Middle Pleistocene of northeastern Rhodes, Greece. Sediment Geol 187:159–181
da Silva HMA, Gallo V (2011) Taxonomic review and phylogenetic analysis of Enchodontoidei (Teleostei: Aulopiformes). An Acad Bras Cienc 83:483–511
da Silva HMA, Gallo V (2016) Distributional patterns of enchodontoid fishes in the Late Cretaceous. Cretac Res 65:223–231
Davesne D, Gallut C, Barriel V, Janvier P, Lecointre G, Otero O (2016) The phylogenetic intrarelationships of spiny-rayed fishes (Acanthomorpha, Teleostei, Actinopterygii): Fossil Taxa increase the congruence of morphology with molecular data. Front Ecol Evol 4:129
Davis MP, Fielitz C (2010) Estimating divergence times of lizardfishes and their allies (Euteleostei: Aulopiformes) and the timing of deep-sea adaptations. Mol Phylogenet Evol 57:1194–1208
de Sant'Anna VB, Collette BB, Godfrey SJ (2013) †Belone countermani, a new Miocene needlefish (Belonidae) from the St. Marys Formation of Calvert Cliffs, Maryland. Proc Biol Soc Wash 126:137–150
de Waldheim GF (1838) Recherches sur les ossemens fossiles de la Russie II. Semen, Moscow
de Waldheim GF (1850) Sur une poisson fossile de la Grèce. Bull Socimperiale des naturalistes Moscou 23:285–288
Denton JSS (2013) Lanternfish (Teleostei, Myctophiformes, Myctophidae) body fossils from the Modelo Formation (upper Miocene) of Los Angeles County. Calif J Vertebr Paleontol 33:786–793
Dermitzakis MD (1979) The Neogene of Zakynthos. In: Symeonidis N, Papanikolaou D, Dermitzakis MD (eds) Field guide to the Neogene of Megara – Pelopponisos – Zakynthos. Department Geology & Paleontology, University of Athens, Athens, pp 45–67
Dermitzakis MD, Georgiades-Dikeoulia E, Velitzelos E (1986) Ecostratigraphic observations on the Messinian deposits of Akropotamos area (Kavala, N. Greece). Ann Géol Pays Héllen 33:367–376
Drinia H, Dermitzakis MD, Kouli K, Tsourou T (2002) Sedimentary facies analysis and paleoenvironmental interpretation of Vatera Formation, Lesvos Island, Greece. Ann Géol Pays Héllen 39:15–35
Eschmeyer WN, Fong JD (2019) Species by family/subfamily in the Catalog of Fishes, vol 2016. California Academy of Sciences, San Francisco
Fielitz C, González-Rodríguez KA (2010) A new species of *Enchodus* (Aulopiformes: Enchodontidae) from the Cretaceous (Albian to Cenomanian) of Zimapán, Hidalgo, México. J Vertebr Paleontol 30:1343–1351
Freuenthal T (1969) Stratigraphy of Neogene deposits in the Khania Province, Crete, with special reference to Foraminifera of the family Planorbulinidae and the genus *Heterostegina*. Utrecht Mircropaleontol Bull 1:1–208
Friedman M (2009) Ecomorphological selectivity among marine teleost fishes during the end-Cretaceous extinction. Proc Natl Acad Sci 106:5218–5223
Friedman M (2015) The early evolution of ray-finned fishes. Palaeontology 58:213–228
Friedman M, Sallan LC (2012) Five hundred million years of extinction and recovery: a phanerozoic survey of large-scale diversity patterns in fishes. Palaeontology 55:707–742
Friedman M, Beckett HT, Close RA, Johanson Z (2015) The English chalk and London clay: two remarkable British bony fish Lagerstätten. Geol Soc Lond, Spec Publ 430:165

Froese R, Pauly D (2019) FishBase, Vol. 08/2019. World Wide Web electronic publication

Gaudant J (1979) «Pachylebias crassicaudus» (Agassiz) (Poisson téléostéen, cyprinodontiforme), un constituant majeur de l'ichthyofaune du Messinien continental du bassin méditerranéen. Geobios 12:47–73

Gaudant J (1980) Sur la présence d'Alosa crassa SAUVAGE (Poissons téléostéens, Clupeidae) dans les gypses messiniens de Crète occidentale. Proc K Ned Akad Wet B 83:263–268

Gaudant J (2001) Amnissos: un gisement clé pour la connaissance de l'ichthyofaune du Pliocène supérieur de Crète. Ann Naturhist Mus Wien 102A:131–187

Gaudant J (2002) La crise messinienne et ses effets sur l'ichthyofaune néogène de la Méditerranée: le témoignage des squelettes en connexion de poissons téléostéens. Geodiversitas 24:691–710

Gaudant J (2004) Additions à l'ichthyofaune tortonienne du bassin de Ierapetra (Crète orientale, Grèce). Ann Naturhist Mus Wien 105:257–285

Gaudant J (2008a) L'ichthyofaune messinienne du basin de Sorbas (Almería, Espagne) et ses rapports avec l'environment sédimentaire. Rev Esp Paleont 23:211–223

Gaudant J (2008b) Paléobiodiversité et paléoenvironnements: l'exemple des gisements de poissons téléostéens du Messinien préévaporitique d'Oran et du bassin du Chélif (Algérie). Geodiversitas 30:141–163

Gaudant J (2014a) Articulated skeletons of *Palaeoleuciscus* aff. *oeningensis* (Agassiz) (Teleostean fish, Cyprinidae) in the lower Miocene of Frankfurt on the Main (Germany), with a note on cyprinid pharyngeal teeth found in the lower Miocene of the Mainz basin. Palaeobiodiv Palaeoenviron 94:481–486

Gaudant J (2014b) Fish mass mortality in the Upper Miocene laminated gypsum of western Crete (Hania Province, Greece). Palaeodiversity 7:39–45

Gaudant J (2015) Re-examination of the upper Miocene freshwater fish fauna from Höwenegg (Hegau, Germany). Swiss J Palaeontol 134:117–127

Gaudant J, Ambroise D (1999) Réexamen critique des Myctophidae Messiniens de Licata (Sicile, Italie): conséquences taxonomiques. Cybium 23:131–145

Gaudant J, Courme M-D (2014) A new Pliocene marine fish fauna from north-western Cyprus, with a micropalaeontological account. Neues Jahrb Geol Palaontol Abh 272:205–212

Gaudant J, Symeonidis N (1995) Découverte d'une frayère à Spratelloides (poissons téléostéens) dans le Pliocène moyen de Kalamaki (Attique, Grèce). Ann Géol Pays Héllen 36:331–349

Gaudant J, Delrieau B, Dermitzakis MD, Symeonidis NK (1994) Découverte d'une ichthyofaune marine dans les diatomites du Pliocène supérieur (Plaisancien) des environs d'Héraklion (Crète centrale, Grèce). C R Acad Sci Paris 319:589–596

Gaudant J, Fourtanier E, Lauriat-Rage A, Tsagaris S, Vénec-Peyré M-T, Zorn I (1997a) Découverte d'une ichthyofaune marine dans le Messinien préévaporitique de la Messara (Crète centrale, Grèce): interprétation paléoécologique. Géol Méditerr 24:175–195

Gaudant J, Saint-Martin J-P, Bessedik M, Mansour B, Moissette P, Rouchy J-M (1997b) Découverte d'une frayère de poissons téléosteens dans des diatomites messiniennes du Djebel Murdjadjo (environs d'Oran, Algérie). J Afr Earth Sci 24:511–529

Gaudant J, Barker MJ, Courme M-D, di Stephano A, Martill DM, Vénec-Peyré M-T, Zorn I, Panayides I (2000) Alassa: a new fossil fish fauna from the middle Miocene Serravallian of Cyprus. Paper presented at the 3rd, International conference on the geology of the eastern Mediterranean, Nicosia, Cyprus, 2000

Gaudant J, Tsaparas N, Antonarakou A, Drinia H, Dermitzakis MD (2005) The Tortonian fish fauna of Gavdos Island (Greece). C R Palevol 4:687–695

Gaudant J, Tsaparas N, Antonarakou A, Drinia H, Saint-Martin S, Dermitzakis MD (2006) A new marine fish fauna from the pre-evaporitic Messinian of Gavdos Island (Greece). C R Palevol 5:795–802

Gaudant J, Cavallo O, Courme-Rault M-D, Fornaciari E, Lauriat-Rage A (2008) Paléontologie du gisement de poissons fossiles du Messinien préévaporitique de Roddi, près d'Alba (Piémont, Italie). Riv Piemont Storia Nat 29:3–60

Gaudant J, Courme-Rault M-D, Fornaciari E, Fourtanier E (2010a) The Upper Miocene fossil fish locality of Pecetto di Valenza (Piedmont, Italy): a multidisciplinary approach. Boll Soc Paleontol Ital 49:203–225

Gaudant J, Courme-Rault M-D, Saint-Martin S (2010b) On the fossil fishes, diatoms, and foraminifera from Zanclean (Lower Pliocene) diatomitic sediments of Aegina Island (Greece): a stratigraphical and palaeoenvironmental study. Palaeodiversity 3:141–149

Gaudant J, Soria MJ, Fierro I, Saint Martin S (2015) The Upper Miocene fish fauna of the Fortuna Basin, as observed in the surroundings of Albanilla (Murcia Province, Spain). Span J Palaeontol 30:67–78

Gaudry A (1862) Animaux fossiles et géologie de l'Attique d'apres les recherches faites en 1855–56 et en 1860 sous les auspices de l'Academie des sciences par Albert Gaudry. Savy, F, Paris

Georgalis GL, Villa A, Delfino M (2017) Fossil lizards and snakes from Ano Metochi – a diverse squamate fauna from the latest Miocene of northern Greece. Hist Biol 29:730–742

Giles S, Coates MI, Garwood RJ, Brazeau MD, Atwood R, Johanson Z, Friedman M (2015) Endoskeletal structure in *Cheirolepis* (Osteichthyes, Actinopterygii), an early ray-finned fish. Palaeontology 58:849–870

Giles S, Xu G-H, Near TJ, Friedman M (2017) Early members of 'living fossil' lineage imply later origin of modern ray-finned fishes. Nature 549:265–268

Gmelin JF (1789) Pisces; Caroli a Linné. Systema naturae per regna tria naturae: secundum classes, ordines, genera, species, cum characteribus, differentiis, synonymis, locis, Lipsiae

Goody PC (1969) The relationships of certain Upper Cretaceous teleosts with special reference to myctophoids. Bull Br Mus (Nat Hist) Geol Ser Suppl 7:1–255

Grande L (1985) Recent and fossil clupeomorph fishes with materials for revision of the subgroups of clupeoids. Bull Am Mus Nat Hist 181:231–372

Grande L (2010) An empirical synthetic pattern study of gars (Lepisosteiformes) and closely related species, based mostly on skeletal anatomy: the resurrection of Holostei. American Society of Ichthyologists and Herpetologists Special Publication 6. Copeia Suppl 10:871

Greenwood PH (1976) A review of the family Centropomidae (Pisces, Perciformes). Bull Br Mus (Nat Hist) Zool 29:1–81

Gregorová R (2011) Fossil fish fauna (Teleostei, Selachii) from the Dynów marlstone (Rupelian, NP 23) of the Menilitic Formation at the locality of Litenčice (Czech Republic). Acta Mus Moraviae – Sci Geol 2:3–33

Grim O (1901) The herrings of the Sea of Azov. Vestnik Rybopromysi St. Petersburg 16

Guinot G, Cavin L (2016) 'Fish' (Actinopterygii and Elasmobranchii) diversification patterns through deep time. Biol Rev 91:950–981

Helfman GS, Collette B, Facey DE, Bowen BW (2009) The diversity of fishes. Wiley-Blackwell, Chichester

Hilton EJ, Forey PL (2009) Redescription of †*Chondrosteus acipenseroides* Egerton, 1858 (Acipenseriformes, †Chondrosteidae) from the Lower Lias of Lyme Regis (Dorset, England), with comments on the early evolution of sturgeons and paddlefishes. J Syst Palaeontol 7:427–453

Holloway WL, Claeson KM, Sallam HM, El-Sayed S, Kora M, Sertich JJW, O'Connor PM (2017) A new species of the neopterygian fish *Enchodus* from the Duwi Formation, Campanian, Late Cretaceous, Western Desert, central Egypt. Acta Palaeontol Pol 62:603–611

Hordijk K, de Bruijn H (2009) The succession of rodent faunas from the Mio/Pliocene lacustrine deposits of the Florina-Ptolemais-Servia Basin (Greece). Hell J Geosci 44:21–103

Hughes LC, Ortí G, Huang Y, Sun Y, Baldwin CC, Thompson AW, Arcila D, Betancur-R R, Li C, Becker L, Bellora N, Zhao X, Li X, Wang M, Fang C, Xie B, Zhou Z, Huang H, Chen S, Venkatesh B, Shi Q (2018) Comprehensive phylogeny of ray-finned fishes (Actinopterygii) based on transcriptomic and genomic data. Proc Natl Acad Sci 115:6249–6254

Jordan DS, Gilbert JZ (1919) Fossil fishes of the (Miocene) Monterey Formations of Southern Californi; pp. 13–60, Fossil fishes of Southern California. Stanford University Press

Kamperis EG (1987) Geological and petrogeological study of Northwest Peloponnese. School of mining & metallurgical engineering. National Technical University of Athens, Athens

Kevrekidis C (2016) Taxonomic study of selected Clupeidae (herrings) from the Miocene of the Mediterranean and Paratethys: comparative meristics and morphometry: In Faculty of geosciences, M.Sc. Thesis, Ludwig-Maximilians-Universität München, Munich

Koch KE, Nikolaus HJ (1969) Zur Geologie des Ostpindos – Flyschbeckens und seiner Umrandung. Institute for geology and subsurface research, Athens

Konwert M, Stumpf S (2017) Exceptionally preserved Leptolepidae (Actinopterygii, Teleostei) from the late early Jurassic Fossil-Lagerstätten of Grimmen and Dobbertin (Mecklenburg-Western Pomerania, Germany). Zootaxa 4243:249–296

Kottek A (1964) Fischreste aus dem griechischen Lias. Ann Géol Pays Héllen 39:175–181

Kottelat M, Freyhof J (2007) Handbook of European freshwater fishes. Publications Kottelat, Cornol and Berlin

Kottis T, Ioakim C, Velitzelos E, Kvaček Z, Denk T, Velitzelos D (2002) Field trip guidebook Evia Island. Neogene sedimantary basin of Kymi/Aliveri, Athens

Kriwet J (2003) Lancetfish teeth (Neoteleostei, Alepisauroidei) from the Early Cretaceous of Alcaine, NE Spain. Lethaia 36:323–331

Landini W, Sorbini C (2005a) Evolutionary dynamics in the fish faunas of the Mediterranean basin during the Plio-Pleistocene. Quat Int 140–141:64–89

Landini W, Sorbini C (2005b) Evolutionary trends in the Plio-Pleistocene ichthyofauna of the Mediterranean Basin: nature, timing and magnitude of the extinction events. Quat Int 131:101–107

Leivaditis GI (1974) Geological and geomorphological observations on Aegina Island: In Department of Geology, PhD thesis, National and Kapodistrian University of Athens

López-Arbarello A, Sferco E (2018) Neopterygian phylogeny: the merger assay. R Soc Open Sci 5:172337

Lu J, Giles S, Friedman M, den Blaauwen JL, Zhu M (2016) The oldest actinopterygian highlights the cryptic early history of the hyperdiverse ray-finned fishes. Curr Biol 26:1602–1608

Maxwell EE, Argyriou T, Stockar R, Furrer H (2017) Evolutionary implications of the ontogeny and reproductive biology of Saurichthyidae (Actinopterygii). Zitteliana 91:62

Melentis JK (1967a) Studien über fossile Vertebraten Griechenlands. 21. Über Coelodus münsteri AG. (Pisces) aus dem Cenoman von Griechenland. Ann Géol Pays Héllen 18:456–462

Melentis JK (1967b) Über *Coelodus münsteri* AG. (Pisces) aus dem Cenoman von Griechenland. Praktika Akadimias Athinon 42:115–123

Merle D, Lauriat-Rage A, Gaudant J, Pestrea S, Courme-Rault M-D, Zorn I, Blanc-Valleron M-M, Rouchy J-M, Orszag-Sperber F, Krijgsman W (2002) Les paléopeuplements marins du Messinien pré-évaporitique de Pissouri (Chypre, Méditerranée orientale): aspects paléoécologiques précédant la crise de salinité messinienne. Geodiversitas 24:669–689

Micklich N, Hildenbrandt HL (2010) Emergency excavation in the Grube (Unterfeld) Frauenweiler clay pit (Oligocene Rupelian; Baden-Wurttemberg, S Germany): new records and palaeoenvironmental information. Kaupia 17:3–21

Monsch K, Bannikov AF (2012) New taxonomic synopses and revision of the scombroid fishes (Scombroidei, Perciformes), including billfishes, from the Cenozoic of territories of the former USSR. Earth Environ Sci Trans R Soc Edinb 102:253–300

Müller AH, Strauch F (1994) Eine pliozäne Fischfauna von Prassas bei Heraklion, Kreta. Münster Forsch Geol Palaeontol 76:243–253

Murray AM, Attia YS (2004) A new species of lates (Teleostei: Perciformes) from the Lower Oligocene of Egypt. J Vertebr Paleontol 24:299–308

Nazarkin MV, Bannikov AF (2014) Fossil mackerel (Actinopterygii: Scombridae: *Scomber*) from the Neogene of South-Western Sakhalin, Russia. Zoosyst Rossica 23:158–163

Near TJ, Eytan RI, Dornburg A, Kuhn KL, Moore JA, Davis MP, Wainwright PC, Friedman M, Smith WL (2012) Resolution of ray-finned fish phylogeny and timing of diversification. Proc Natl Acad Sci 109:13698–13703

Near TJ, Dornburg A, Eytan RI, Keck BP, Smith WL, Kuhn KL, Moore JA, Price SA, Burbrink FT, Friedman M, Wainwright PC (2013) Phylogeny and tempo of diversification in the superradiation of spiny-rayed fishes. Proc Natl Acad Sci 110:12738–12743

Nelson JS, Grande TC, Wilson MVH (2016) Fishes of the world. Wiley, Hoboken, p i-xli + 707

Nybelin O (1974) A Revision of the Leptolepid fishes. Acta Regiae Societatis Scientiarum et Litterarum Gothoburgensis Zoologica 9:1–202

Offshore marine actinopterygian assemblages from the Maastrichtian–Paleogene of the Pindos Unit in Eurytania, Greece

Olsen PE (1984) The skull and pectoral girdle of the parasemionotid fish *Watsonulus eugnathoides* from the Early Triassic Sakamena Group of Madagascar, with comments on the relationships of the holostean fishes. J Vertebr Paleontol 4:481–499

Otero O (2004) Anatomy, systematics and phylogeny of both recent and fossil latid fishes (Teleostei, Perciformes, Latidae). Zool J Linnean Soc 141:81–133

Otero O, Gayet M (2001) Palaeoichthyofaunas from the Lower Oligocene and Miocene of the Arabian Plate: palaeoecological and palaeobiogeographical implications. Palaeogeogr Palaeoclimatol Palaeoecol 165:141–169

Otero O, Lécuyer C, Fourel F, Martineau F, Mackaye HT, Vignaud P, Brunet A (2011) $\delta^{18}O$ of fishes indicates a Messinian increasing of the aridity in Central Africa. Geology 39:435–438

Panagopoulou E, Tourloukis V, Thompson N, Konidaris G, Athanassiou A, Giusti D, Tsartsidou G, Karkanas P, Harvati K (2018) The lower palaeolithic site of Marathousa 1, Megalopolis, Greece: overview of the evidence. Quat Int 497:33–46

Patterson C (1973) Interrelationships of holosteans. In: Greenwood PH, Miles RS, Patterson C (eds) Interrelationships of fishes. Academic Press, London, pp 233–305

Patterson C, Rosen DE (1977) Review of ichthyodectiform and other Mesozoic teleost fishes and the theory and practice of classifying fossils. Bull Am Mus Nat Hist 158:81–172

Paucă M (1931) *Alosa nordmanni* Antipa aus dem Pontien von Targu-Jiu. Bull Sec Sci Acad Roum 16:106–107

Paxton JR (1972) Osteology and relationships of the lanternfishes (Family Myctophidae). Bull Nat Hist Mus Los Angeles Cty, Sci 13:1–81

Přikryl T, Carnevale G (2017) Miocene bristlemouths (Teleostei: Stomiiformes: Gonostomatidae) from the Makrilia Formation, Ierapetra, Crete. C R Palevol 16:266–277

Přikryl T, Brzobohatý R, Gregorová R (2016a) Diversity and distribution of fossil codlets (Teleostei, Gadiformes, Bregmacerotidae): review and commentary. Palaeobiodiv Palaeoenviron 96:13–39

Přikryl T, Gómez de Soler B, Campeny G, Oms O, Roubach S, Blain H-A, Agustí J (2016b) Fish fauna of the Camp dels Ninots locality (Pliocene; Caldes de Malavella, province of Girona, Spain) – first results with notes on palaeoecology and taphonomy. Hist Biol 28:347–357

Přikryl T, Kania I, Krzemiński W (2016c) Synopsis of fossil fish fauna from the Hermanowa locality (Rupelian; Central Paratethys; Poland): current state of knowledge. Swiss J Geosci 109:429–443

Raulin V (1861) Description physique et naturelle de l'isle de Crète. Actes Soc Linnéenne Bord 23:321–383

Reichenbacher B, Böhme M (2004) Fish remains (otoliths, teeth) from the Miocene of Chios (Greece) – new palaeobiogeographic and palaeoclimatic results. Paper presented at the 5th International Symposium on Eastern Mediterranean Geology, Thessaloniki, Greece, 2004

Reichenbacher B, Gregorová R, Holcová K, Šanda R, Vukić J, Přikryl T (2018) Discovery of the oldest Gobius (Teleostei, Gobiiformes) from a marine ecosystem of early Miocene age. J Syst Palaeontol 16:493–513

Risso A (1810) Ichthyologie de Nice, ou histoire naturelle des poissons du Département des Alpes Maritimes. F. Schoell, Paris

Romano C, Koot MB, Kogan I, Brayard A, Minikh AV, Brinkmann W, Bucher H, Kriwet J (2016) Permian–Triassic Osteichthyes (bony fishes): diversity dynamics and body size evolution. Biol Rev 91:106–147

Rückert-Ülkümen N (1994) Zur systematischen Stellung einiger Clupeidae aus Thrakien, Türkei. Mitt Bayer Staatssammlung Paläontol Hist Geol 34

Rückert-Ülkümen N, Yiğitbaş E (2007) Pharyngeal teeth, lateral ethmoids, and jaw teeth of fishes and additional fossils from the late Miocene (late Khersonian/early Maeotian) of Eastern Paratethys (Yalova, near İstanbul, Turkey). Turk J Earth Sci 16:211–224

Sallan LC (2014) Major issues in the origins of ray-finned fish (Actinopterygii) biodiversity. Biol Rev 89:950–971

Sauvage MH-E (1873) Memoire sur la faune ichthyologigue de la periode Tertiare et plus speciale-ment sur les poissons fossiles d'Oran et de Licata. Ann Sci Geol 4:1–272

Schultz O (1979) Supplementary notes on elasmobranch and teleost fish remains from the Korytnica Clays (Middle Miocene; Holy Cross Mountains, Central Poland). Acta Palaeontol Pol 29:287–293

Schultz O (2003) The Middle Miocene Fish Fauna (excl. otolithes) from Mühlbach am Manhartsberg and Grund near Hollabrunn, Lower Austria. Ann Naturhist Mus Wien 104A:185–193

Schultz O, Bellwood DR (2004) *Trigonodon oweni* and *Asima jugleri* are different parts of the same species *Trigonodon jugleri*, a Chiseltooth Wrasse from the lower and middle Miocene in Central Europe (Osteichthyes, Labridae, Trigonodontinae). Ann Naturhist Mus Wien 105A:287–305

Schultz O, Paunović M, von Summesberger H (1997) Der Nachweis von Coelodus (Osteichthyes, Pycnodontidae) im Turonien (Oberkreide) von Gams bei Hieflau, Steiermark, Österreich, und aus der Oberkreide von Kroatien und Italien. Ann Naturhist Mus Wien 98:73–141

Schultze HP (2008) Nomenclature and homologization of cranial bones in actinopterygians. In: Arratia G, Schultze HP, Wilson MVH (eds) Mesozoic fishes 4. Dr. Friedrich Pfeil, München, pp 23–48

Schwarzhans W, Carnevale G, Bannikov AF, Japundžić S, Bradić K (2017) Otoliths in situ from Sarmatian (Middle Miocene) fishes of the Paratethys. Part I: Atherina suchovi Switchenska, 1973. Swiss J Palaeontol 136:7–17

Sibert EC, Norris RD (2015) New age of fishes initiated by the Cretaceous–Paleogene mass extinction. Proc Natl Acad Sci 112:8537–8542

Smirnov VP (1936) Fishes of the North Caucasian Oligocene. Chernorechenskii District (in Russian). Travaux scientifiques de l'Université de l'Usbekistan 1:1–96

Sorbini C, Tyler JC (2004) Review of the fossil file fishes of the family Monacanthidae (Tetraodontiformes), Pliocene and Pleistocene of Europe, with a new genus, *Frigocanthus*, and two new species related to the recent *Aluterus*. Boll Mus Civico Storia Nat Verona 28:41–76

Stewart KM (2001) The freshwater fish of Neogene Africa (Miocene-Pleistocene): systematics and biogeography. Fish Fish 2:177–230

Stewart KM (2003) Fossil fish remains from the Pliocene Kanapoi Site, Kenya. In: Harris JM, Leakey MG (eds) Geology and vertebrate paleontology of the Early Pliocene site of Kanapoi, Northern Kenya. Natural History Museum of Los Angeles County, Los Angeles, pp 21–38

Stewart KM, Murray AM (2017) Biogeographic implications of fossil fishes from the Awash River, Ethiopia. J Vertebr Paleontol 37:e1269115

Symeonidis N (1969) Fossile Fische aus der Gegend von Ierapetra (Kreta). Ann Géol Pays Héllén 21:501–530

Symeonidis N, Schultz O (1973) Bemerkungen zur neogenen Fischfauna Kretas und Beschreibung zweier Fundstellen mit miozänen Mollusken, Echiniiden etc., Insel Kreta, Griechenland. Ann Naturhist Mus Wien 77:141–147

Tintori A, Sun Z, Ni P, Lombardo C, Jiang D, Motani R (2015) Oldest stem Teleostei from the late Ladinian (Middle Triassic) of Southern China. Riv Ital Paleontol Stratigr 121:285–296

Tripalo K, Japundžić S, Sremac J, Bošnjak M (2016) First record of Acanthuridae (surgeonfish) from the Miocene deposits of the Medvednica Mt. Geol Croat 69:201–204

Tyler JC, Bannikov AF (2000) A new species of the surgeon fish genus Tauichthys from the Eocene of Monte Bolca, Italy (Perciformes, Acanthuridae). Boll Mus Civico Storia Nat Verona 24:29–36

Tyler JC, Micklich NR (2011) A new genus and species of surgeon fish (Perciformes, Acanthuridae) from the Oligocene of Kanton Glarus, Switzerland. Swiss J Palaeontol 130:203–216

Unger F (1867) Die fossile Flora von Kumi auf der Insel Euboea. Denkschriften K Akad Wiss Math Naturwiss Kl 27:27–87

Valenciennes A (1861) Rapport sur les collections des espèces de Mammifères déterminées par leurs nombreux ossements fossiles rectieillis par M. Albert Gaudry, à Pikermi, près d'Athènes, pendant son voyage en Attique. C R Hebd Séances Acad Sci 52:1295–1300

Vasileiadou K, Böhme M, Neubauer TA, Georgalis GL, Syrides GE, Papadopoulou L, Zouros N (2017) Early Miocene gastropod and ectothermic vertebrate remains from the Lesvos Petrified Forest (Greece). Palaeontol Z 91:541–564

Vernygora O, Murray AM, Luque J, Ruge MLP, Fonseca MEP (2018) A new Cretaceous dercetid fish (Neoteleostei: Aulopiformes) from the Turonian of Colombia. J Syst Palaeontol 16:1057–1071

Vinther J (2015) A guide to the field of palaeo colour. BioEssays 37:643–656

Viohl G (1990) Piscivorous fishes of the Solnhofen lithographic limestone. In: Boucot AJ (ed) Evolutionary paleobiology of behavior and coevolution. Elsevier, Amsterdam, pp 287–303

Weems RE (1999) Part 4. Actinopterygian fishes from the Fisher/Sullivan site. In: Weems RE, Grimsley GJ (eds) Early Eocene vertebrates and plants from the Fisher/Sullivan site (Nanjemoy Formation) Stafford County, Virginia. Commonwealth of Virginia Department of Mines, Minerals and Energy Division of Mineral Resources, Charlottesville, pp 53–99

Weems RE, Edwards LE, Landacre B (2017) Geology and biostratigraphy of the Potomac River cliffs at Stratford Hall, Westmoreland County, Virginia. Field Guid 47:125–152

Weiler W (1943) Über einen pliozänen Fischrest aus Mazedonien. Z Deut Geol Ges 95:210–213

Wiley EO, Johnson GD (2010) A teleost classification based on monophyletic groups. In: Nelson JS, Schultze HP, Wilson MVH (eds) Origin and phylogenetic interrelationships of Teleosts. Dr. Friedrich Pfeil, Munich, pp 123–182

Woodward AS (1901) Catalogue of the fossil fishes in the British Museum (National History). Part IV. Containing the actinopterygian Teleostomi of the suborders Isospondyli (in part), Ostariophysi, Apodes, Percesoces, Hemibranchii, Acanthopterygii, and Anacanthini. British Museum (Natural History), London

Yabumoto Y, Uyeno T (1994) Late Mesozoic and Cenozoic fish faunas of Japan. Island Arc 3:255–269

Zhu M, Zhao W, Jia L, Lu J, Qiao T, Qu Q (2009) The oldest articulated osteichthyan reveals mosaic gnathostome characters. Nature 458:469–474

The Fossil Otolith Record of Fishes (Vertebrata: Teleostei) in Greece

Konstantina Agiadi

1 Introduction

The study of otoliths has contributed significantly to the knowledge of the Cenozoic fossil fish record of Greece. Otoliths are the aragonitic, incrementally growing structures found in the inner ear of teleost fishes, which facilitate sound and balance perception (Schulz-Mirbach et al. 2018). There are three sets of otoliths in each fish: the sagittae, lapilli, and asterisci. The sagitta is the otolith used in most applications. Henceforth, the term "otolith" will refer only to sagittal otoliths. The main diagnostic features of otoliths are the overall shape, the thickness, and the morphology of the inner face (Fig. 1).

The species-specific morphology and high frequency of otoliths in marine and lake sediments renders them invaluable to the identification of past fish faunas (Nolf 1985). Due to their small size and robust structure, as well as the fact that their preservation seems to be independent of that of the fish skeletons, they can be obtained through sampling of sedimentary sequences in regular intervals (Agiadi et al. 2018). Therefore, they are useful for investigating the evolution of fish faunas through time. Moreover, they are reliable paleobathymetric and paleoecologic indicators (Nolf and Brzobohatý 1994a; Girone 2000; Agiadi et al. 2010, 2011), as well as good recorders of paleoceanographic and paleoclimatic parameters (Ivany et al. 2000; Andrus et al. 2002). Since fish inhabit the entire water column, the otoliths found within marine sediments reflect the paleoceanographic conditions in the surface, intermediate, as well as the deep water levels and the bottom of the sea. Thus, otoliths provide information on the temperature, salinity, bathymetry, substratum,

K. Agiadi (✉)
Department of Historical Geology and Paleontology, Faculty of Geology and Geoenvironment, National and Kapodistrian University of Athens, Athens, Greece

Department of Paleontology, University of Vienna, Vienna, Austria
e-mail: kagiadi@geol.uoa.gr

E. Vlachos (ed.), *Fossil Vertebrates of Greece Vol. 1*,
https://doi.org/10.1007/978-3-030-68398-6_5

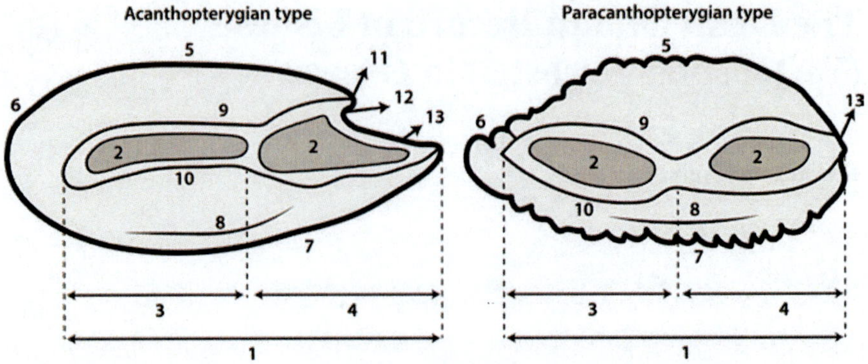

Fig. 1 Morphological nomenclature applied for the description and systematic taxonomy of fossil fish otoliths (modified from Nolf 1985). On the left is a schematic representation of the Acanthopterygian type of otolith, while on the right the Paracanthopterygian type. **1** Sulcus, **2** Colliculum, **3** Cauda, **4** Ostium, **5** Dorsal rim, **6** Posterior rim, **7** Ventral rim, **8** Ventral fissure/furrow, **9** Crista superior, **10** Crista inferior, **11** Antirostrum, **12** Excissura, **13** Rostrum. Both sketches correspond to a left otolith

water oxygenation, productivity, temperature gradient, oceanic circulation, terrigenous input, and other parameters. Moreover, otoliths are formed by the accretion of concentric layers of proteins and aragonite crystallites throughout the fish's life duration (Degens et al. 1969; Campana 1999). Taking advantage of that, numerous biogeochemical proxies based on fish otoliths have been developed and applied to great success (Thorrold et al. 1997; Ghosh et al. 2007).

The studies of fish otoliths and skeletal parts complement each other. Indeed, there are significant differences between the two records. Comparing the otolith-based with the skeletal-based records of the Messinian of Italy, Girone et al. (2010) observed that the skeletal-based record was much richer than the otolith-based record, mainly due to the greater number of studies on the subject, but there were notable taxonomic discrepancies between the two records, in both the species and genus level. Similarly, Agiadi et al. (2018) reviewed the Pliocene–Pleistocene fish records in Greece obtained through both approaches and made two surprising observations. First, the otolith-based record is much richer than the skeletal one. At least to some extent, this is due to lack of research effort targeting fish skeletal remains. Nevertheless, individual studies on fish otoliths yielded greater numbers of species than studies on skeletal material. Moreover, only three fish species had been identified through both otoliths and skeletal parts, suggesting that the two approaches are suitable for identifying different fish species. For example, the skeleton-based record is rich in Clupeidae remains, which are not as often found in otolith assemblages, possibly because their otoliths are small and fragile. It may be that the different taphonomic and preservation requirements favor different taxonomic groups. Therefore, in order to study the paleobiogeographic evolution of the fish fauna in a basin, it is important to combine results from both approaches. Whereas preserved skeletons provide data on the morphology and ecology of fish, otoliths give a more complete picture of the assemblage.

2 Historical Overview

The landmark handbooks by Nolf (1985, 2013) provide overviews of the fish otolith record worldwide, and they are not repeated here. Notably, the systematic study of fish otoliths is a recent development, and indeed the fossil otolith record is far from complete. The first reference of fossil otoliths from the Greek realm is that of Marcopoulou-Diacantoni and Kagiouzis (2001), reporting on few otoliths in the marine Pliocene deposits south of Rethymnon (Crete Island). Keupp and Bellas (2002) also reported some findings from the Tortonian of Chania Prefecture. Reichenbacher and Boehme (2004) presented some otoliths of lake fish from the southeastern part of Chios Island. Next, Tsaparas and Marcopoulou-Diacantoni (2005) reported on otoliths from the Tortonian–Messinian marine sediments on Gavdos Island. The first extensive studies on fossil fish otoliths from Greece were those on the marine Pleistocene of Zakynthos and Kephallonia Islands (Agiadi et al. 2010, 2011), including the first paleoecologic and paleobathymetric analysis of otolith assemblages in Greece. Since then, many other major studies have followed. The Zanclean marine fish otoliths of Crete were reported on by Agiadi et al. (2013a) and used for estimating the paleodepth in the same area (Agiadi et al. 2013b). The Messinian marine fish otolith assemblages of the Ionian realm were presented in the study by Karakitsios et al. (2017). Lake fish otoliths were reported by Vasileiadou et al. (2017) from the Burdigalian of Lesvos Island. Next, the Tortonian–Messinian marine fish record from Crete Island was presented (Agiadi et al. 2017; Moissette et al. 2018). After that, the Pleistocene marine fish assemblages of the southeastern Aegean Sea were presented and analyzed in the recent studies on Rhodes Island (Agiadi et al. 2018, 2019). Finally, the early Zanclean shallow-water marine fish assemblages of southwest Peloponnese were presented by Agiadi et al. (2020).

3 Methodological Particularities in the Study of Fossil Fish Otoliths

A main issue in the study of the evolution of fish faunas is stratigraphic continuity and control. As a standard, fish otolith studies require large bulk sediment samples. In older studies (Anfossi and Mosna 1972; Aruta and Greco 1980), researchers approached the sampling procedure with the notion of horizon sampling. They generally took bulk sediment samples from horizons or beds that provided macroscopic evidence of fish otolith presence. As a result, these studies produced sporadic fish distributions and stratigraphic continuity was lost. In more recent studies (Nolf and

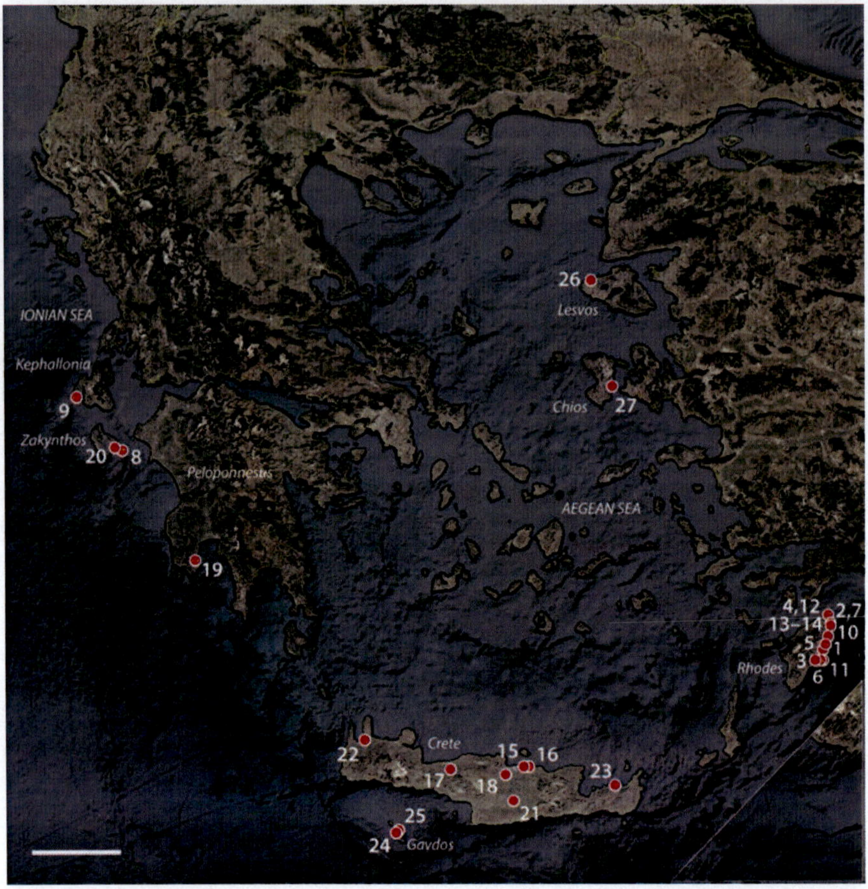

Fig. 2 Map of the occurrences of fossil otoliths of fishes from Greece. See Appendix for further information and text for details. **Pleistocene: 1**, Tsampika, Rhodes Island; **2**, Kallithea beach, Rhodes Island; **3**, Lardos, Rhodes Island; **4**, Sgourou, Rhodes Island; **5**, Haraki, Rhodes Island; **6**, Pefka, Rhodes Island; **7**, Kallithea road, Rhodes Island; **8**, Gerakas, Zakynthos Island; **9**, Akrotiri, Kephallonia Island; **10**, Cape Vagia, Rhodes Island; **11**, Lindos, Rhodes Island; **12**, Kritika North, Rhodes Island; **13**, Faliraki South, Rhodes Island; **14**, Faliraki North, Rhodes Island; **Pliocene: 15**, Amnissos, Heraklion, Crete; **16**, Gournes, Heraklion, Crete; **17**, Prassies, Rethymnon, Crete; **18**, Voutes, Heraklion, Crete; **19,** Agia Triada, SW Peloponnese; **Miocene: 20**, Kalamaki, Zakynthos Island; **21**, Kapariana, Heraklion, Crete; **22**, Potamida, Chania, Crete; **23**, Faneromeni, Heraklion, Crete; **24**, Agios Ioannis, Gavdos Island; **25**, Potamos, Gavdos Island; **26**, Lapsarna, Lesvos Island; **27**, Nenita, Chios Island. See Appendix for more information. Image exported from Google Earth Pro © 2019, map data from US Dept. of State Geographer, SIO, NOAA, U.S. Navy, NGA, GEBCO, image from Landsat/Copernicus. Scale bar equals 80 km, North faces upward

Cappetta 1988; Girone and Varola 2001; Girone 2003), researchers have realized this drawback, and sampling is now made in a more organized manner, taking multiple large bulk sediment samples along biostratigraphically dated outcrops at regular intervals. This new approach enables direct correlation of the otolith study results with those from other studies on the same outcrops and allows reconstructing the evolution of the fish fauna through time in the specific study area.

4 Distribution

Teleost fish today occupy almost all water bodies on Earth and express a wide and diverse variety of adaptations, even to extreme environments. Therefore, it is expected that fish remains, and more specifically otolith remains, may be found in marine sediments from very different marine and freshwater paleoenvironmental settings. The Appendix contains the current state of knowledge regarding the distribution of fossil fish otoliths in the deposits of the Greek realm (Fig. 2). Thus far, the marine fish otolith record includes findings only from the late Cenozoic, namely the Tortonian–Messinian of Crete (Agiadi et al. 2017; Moissette et al. 2018) and Gavdos (Tsaparas and Marcopoulou-Diacantoni 2005), the Messinian of Zakynthos (Karakitsios et al. 2017), the Pliocene of Crete (Agiadi et al. 2013a, b; Marcopoulou-Diacantoni and Kagiouzis 2001) and Peloponnese (Agiadi et al. 2020), and the Pleistocene of Rhodes (Agiadi et al. 2018, 2019; Cornée et al. 2019), Zakynthos, and Kephallonia (Agiadi et al. 2010, 2011). Freshwater fish have been reported based on their otoliths from the Burdigalian of Lesvos (Vasileiadou et al. 2017) and the Burdigalian–Langhian of Chios (Reichenbacher and Boehme 2004).

5 Systematic Paleontology

Despite the recent development of the systematic study of fossil otoliths, at least 122 species, 95 genera, and 49 teleost fish families have been recognized in Greece. In this chapter, all valid taxa reported in Greece are mentioned, along with their distribution in the fossil record focusing on the Mediterranean and the Paratethys. Moreover, species of particular paleontological or paleobiogeographic interest are given special attention. For specimens that are unidentifiable to the species and/or genus level, only their record in Greece is mentioned. Type material and locality are mentioned only when they refer to otoliths. The classification scheme adopted is that for fish by Nelson et al. (2016). The names of fossil taxa are preceded by † (Patterson and Rosen 1977).

Teleostei Müller, 1845
Albuliformes Greenwood, 1977
Albulidae Bleeker, 1859

Type Genus *Albula* Scopoli, 1777.

Included Taxa The genus *Albula*, which has ten species, and the genus *Pterothrissus* with two species are the only living representatives of this family (Nelson et al. 2016).

***Pterothrissus* Hilgendorf, 1877**

Type Species *Pterothrissus gissu* Hilgendorf, 1877.

†*Pterothrissus compactus* Schwarzhans, 1981

Type Material Schwarzhans 1981, left otolith SMF P. 6239 (holotype) (Fig. 9 in Schwarzhans 1981) and four paratypes SMF P. 6240–6244 (only three figured in Figs. 10, 11, and 12 by Schwarzhans 1981).

Type Locality Le Puget sur Argens (South France), Zanclean.

Distribution In Greece, this extinct otolith-based species is only known from the Zanclean of Peloponnese (Fig. 3; Agiadi et al. 2020). Moreover, it has been reported from the Zanclean of southeast France (Schwarzhans 1981, 1986), northern Italy (Nolf and Cavallo 1995), and Portugal (Nolf 2013).

Remarks The otoliths of this species have medium thickness and a rectangular inner face. The sulcus is clearly divided into an oval-shaped ostium placed in the anterodorsal region of the inner face, and a long a rather thin cauda.

Congidae Kaup, 1856

Type Genus *Conger* Bosc, 1817.

Included Taxa The family Congidae contains 194 extant species (Nelson et al. 2016). In the Greek fossil otolith record, Congidae is represented by: *Conger conger* (Linnaeus, 1758); *Gnathophis mystax* (Delaroche, 1809); †*Pseudophichthys escavaratierensis* (Nolf and Cappetta, 1988); *Pseudophichthys splendens* (Lea, 1913); †*Rhynchoconger pantanellii* (Bassoli, 1906).

Fig. 3 Scanning electron microscopic image of a typical otolith of *Pterotrissus compactus*, from the Zanclean of Agia Triada (Agiadi et al. 2020). Scale bar equals 1 mm

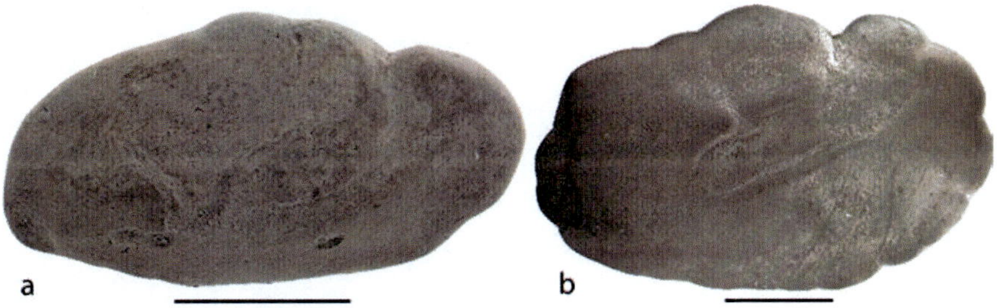

Fig. 4 Scanning electron microscopic images of typical otoliths of Congidae from Greece. **a** *Conger conger* (scale bar equals 200 μm), and **b** *Gnathophis mystax* (scale bar equals 500 μm). Both come from the Middle Pleistocene of Rhodes Island (Agiadi et al. 2018)

Distribution The extant species *Conger conger, Gnathophis mystax*, and *Pseudophichthys splendens* have all been reported from the Pleistocene of Rhodes (Agiadi et al. 2018; Fig. 4). In addition, *Conger conger* and *Gnathophis mystax* are known from the Zanclean of Peloponnese (Agiadi et al. 2020), whereas *Pseudophichthys splendens* was found in the Zanclean of Crete, along with †*Pseudophichthys escavaratierensis* (Agiadi et al. 2013a). Finally, the Mediterranean endemic species †*Rhynchoconger pantanellii* has been reported from the Tortonian–Zanclean of Crete (Agiadi et al. 2013a, 2017; Moissette et al. 2018). Outside Greece, *C. conger* has been reported from the Pleistocene of Italy; *G. mystax* from the Zanclean of Italy; *P. splendens* from the Messinian and Piacenzian of Italy; †*P. escavaratierensis* from the Zanclean of Italy and France (Nolf and Cappetta 1988; Nolf and Cavallo 1995; Nolf and Girone 2006); and †*Rhynchoconger pantanellii* from the Oligocene of Hungary (D. Nolf and Brzobohatý 1994b), the late Miocene of Austria (Nolf and Brzobohatý 2009), the Tortonian and the Zanclean of Italy (Anfossi et al. 1982; Schwarzhans 1979; Lin et al. 2015), and the Zanclean of Spain (Nolf et al. 1998).

Remarks Congidae otoliths are convex in both their inner and outer face, and their sulcus is elongated, undivided, and opens toward the dorsal rim.

Engraulidae Gill, 1861

Type Genus *Engraulis* Cuvier, 1816.

Included Taxa Overall, there are 146 living species belonging to this family (Nelson et al. 2016). Only the European anchovy *Engraulis encrasicolus* (Linnaeus, 1758) has been found in the Greek fossil otolith record.

Distribution *Engraulis encrasicolus* has been recorded in the Middle–Upper Pleistocene of Rhodes (Agiadi et al. 2018; Fig. 5) and in the Pliocene and Pleistocene of Italy (Landini and Sorbini 2005; Girone et al. 2006; Girone 2007).

Fig. 5 Scanning electron microscopic image of a typical otolith of *Engraulis encrasicolus*, from the Middle–Upper Pleistocene of Rhodes (Agiadi et al. 2018). Scale bar equals 500 μm

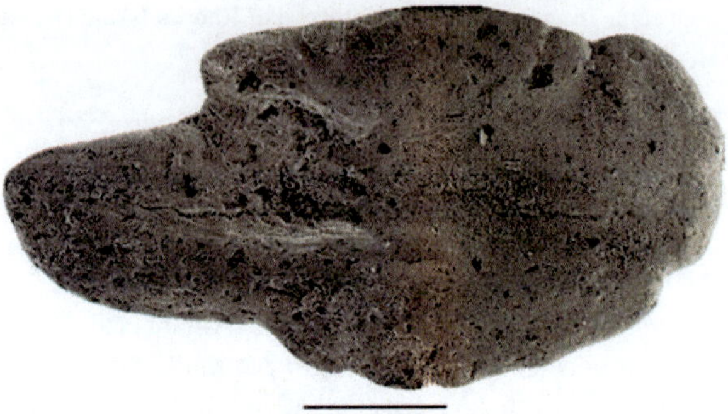

Fig. 6 Scanning electron microscopic image of a typical otolith of *Sardina pilchardus*, from the Zanclean of Agia Triada, Peloponnesus (Agiadi et al. 2020). Scale bar equals 200 μm

Remarks The otoliths of *E. encrasicolus* are elongated, short, thin, and rather fragile. The sulcus is thin and elongated, and their rostrum is very long.

Clupeidae Cuvier, 1816

Type Genus *Clupea* Linnaeus, 1758.

Included Taxa There are 218 extant species in this family (Nelson et al. 2016). In the Greek fossil otolith record, there are only *Sardina pilchardus* (Walbaum, 1792), *Sardinella maderensis* (Lowe, 1838), and *Spratelloides* Bleeker, 1851.

Distribution *Sardina pilchardus* (Fig. 6) has been reported from the Gelasian of Rhodes (Agiadi et al. 2019); the Messinian and the Piacenzian of Crete (Gaudant et al. 1994; Gaudant 2001, 2002); and the Messinian, the Piacenzian, and the Early-Middle Pleistocene of northern Italy (Girone et al. 2006; Girone 2007; Girone et al. 2010). *Sardinella maderensis* has been reported from the Zanclean of northeast Spain (Nolf et al. 1998) and northern Italy (Nolf and Girone 2006), and the Calabrian of Rhodes (Agiadi et al. 2018). *Spratelloides* species are known from the Messinian and the Pliocene of northern Italy (Landini and Sorbini 1993; Carnevale et al. 2006); the Tortonian–Piacenzian of Crete (Gaudant 1993; Gaudant et al. 1994; Gaudant 2001, 2004); and the Zanclean of Peloponnese (Agiadi et al. 2020).

Remarks Clupeidae otoliths are also elongated, short and thin, like those of Engraulidae. However, their main body (excluding the long rostrum) is more rectangular, and they are generally a bit higher.

Argentinidae Bonaparte, 1846

Type Genus *Argentina* Linnaeus, 1758.

Included Taxa Argentinidae include 27 extant species (Nelson et al. 2016). Only *Argentina sphyraena* (Linnaeus, 1758) has been found in the Greek fossil otolith record.

Distribution This species has been found in the Middle Pleistocene of Rhodes (Agiadi et al. 2018; Fig. 7). Previously, it was reported from the Messinian and Early Pleistocene of northern Italy (Girone et al. 2006, 2010).

Remarks The otoliths of this family are thin, with an approximately triangular inner face and a long rostrum. The posterior side is very high, and the sulcus is elongated and divided.

Microstomatidae Bleeker, 1859

Type Genus *Microstoma* Cuvier, 1816.

Included Taxa Twenty modern species are placed in this family (Nelson et al. 2016). Only *Nansenia groenlandica* (Reinhardt, 1840) appears in the fossil otolith record of Greece.

Fig. 7 Scanning electron microscopic image of a typical otolith of *Argentina sphyraena*, from the Middle Pleistocene of Rhodes Island (Agiadi et al. 2018). Scale bar equals 1 mm

Distribution This species has been found in the Gelasian–Calabrian of Kephallonia and Zakynthos (Agiadi et al. 2010, 2011), and the Early–Middle Pleistocene of Italy (Girone 2003; Girone et al. 2006), denoting cold intervals.

Remarks The otoliths of this family are elongated with a very long rostrum and a rather rounded posterior side.

Sternoptychidae Duméril, 1805

Type Genus *Sternoptyx* Hermann, 1781.

Included Taxa This family includes 73 modern species (Nelson et al. 2016). The Greek fossil otolith record includes *Maurolicus muelleri* (Gmelin, 1789) and *Polyipnus polli* Schultz, 1961.

Distribution *Maurolicus muelleri* has been reported from the pre-evaporitic Messinian of Zakynthos (Karakitsios et al. 2017), the Pliocene of Crete (Gaudant 2001; Agiadi et al. 2013a), and the Gelasian–Middle Pleistocene of Rhodes (Agiadi et al. 2018; Fig. 8a), Kephallonia, and Zakynthos (Agiadi et al. 2010, 2011). Outside of Greece, this species is common in the Messinian–Pleistocene of Italy (Nolf and Cavallo 1995; Girone and Varola 2001; Girone 2003; Nolf and Girone 2006; Girone 2007; Girone et al. 2010) and south France (Nolf and Cappetta 1988), and the Messinian of Crete and Cyprus (Gaudant 2002). *Polyipnus*

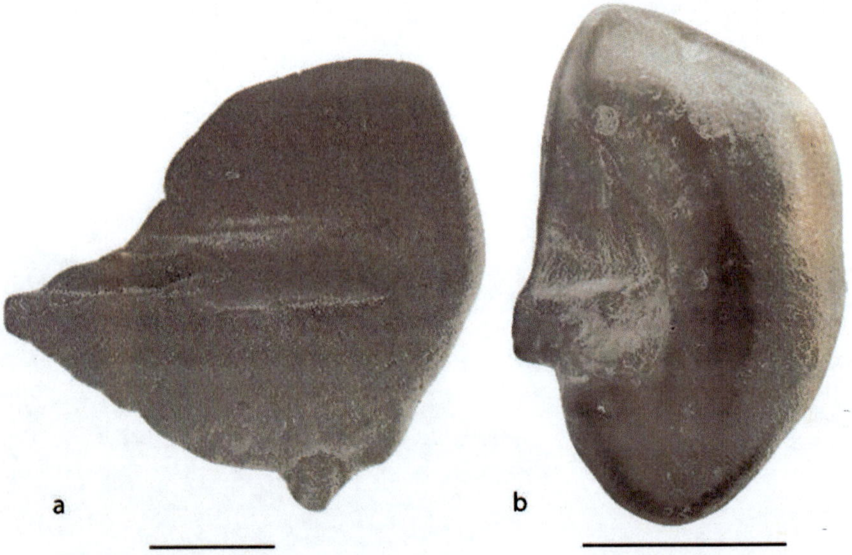

Fig. 8 Scanning electron microscopic images of typical otoliths of Sternoptychidae from Greece. **a** *Maurolicus muelleri*, and **b** *Polyipnus polli*. Both come from the Middle Pleistocene of Rhodes Island (Agiadi et al. 2018). Scale bars equal 200 μm

polli is known from the Middle Pleistocene of Rhodes (Agiadi et al. 2018; Fig. 8b) and Italy (Girone and Varola 2001).

Remarks The otoliths of the species belonging to this family are triangular or more circular and quite high.

Phosichthyidae Weitzman, 1974

Type Genus *Phosichthys* Hutton, 1872.

Included Taxa There are 24 modern species in this family (Nelson et al. 2016). *Vinciguerria attenuata* (Cocco, 1838), *Vinciguerria lucetia* (Garman, 1899), and *Vinciguerria poweriae* (Cocco, 1838) are the species found in the fossil otolith record of Greece.

Distribution *Vinciguerria* cf. *V. attenuata* was identified in the Middle Pleistocene sediments of Rhodes (Agiadi et al. 2018), whereas *V. attenuata* was previously known from the Pleistocene of Italy (Girone et al. 2006). *Vinciguerria* cf. *V. lucetia* was reported from the Pleistocene of Kephallonia (Agiadi et al. 2010). Previously, *V. lucetia* was known from the Miocene of south France (Steurbaut 1979). *Vin. poweriae* is a common species also in the Pliocene of Crete (Agiadi et al. 2013a) and the Pleistocene of Rhodes (Agiadi et al. 2018), but also found in the Pliocene and Pleistocene of Italy (Girone 2000, 2003; Girone et al. 2006; Nolf and Girone 2006; Girone 2007) and south France (Nolf and Cappetta 1988).

Remarks The otoliths of Phosichthyidae are very small, thin, and fragile. They have a long and thin rostrum and an elongated, very thin sulcus. The posterior side is high.

Stomiidae Bleeker, 1859

Type Genus *Stomias* Cuvier, 1816.

Included Taxa There are 286 modern species in this family (Nelson et al. 2016). Only *Chauliodus sloani* Bloch and Schneider, 1801 is found in the Greek fossil otolith record.

Distribution This species has been reported from the Gelasian–Calabrian of Zakynthos (Agiadi et al. 2010, 2011) and the Early Pleistocene of Italy (Girone 2003).

Remarks Stomiidae otoliths are very small and round in shape with a very thin, but deep sulcus.

Chlorophthalmidae Garman, 1899

Type Genus *Chlorophthalmus* Bonaparte, 1840.

Included Taxa This family includes 17 modern species (Nelson et al., 2016).

Chlorophthalmus Bonaparte, 1840

Type Species *Chlorophthalmus agassizi* Bonaparte, 1840.

Chlorophthalmus agassizi Bonaparte, 1840

Distribution This species is abundant in the deep eastern Ionian Sea today, and it is only known as a fossil from Greek localities. *Chlorophthalmus* cf. *C. agassizi* was first found in the Middle Pleistocene of Kephallonia (Agiadi et al. 2010). Later, *Chlorophthalmus agassizi* was discovered in the Zanclean sediments of Crete (Agiadi et al. 2013a) and the Gelasian–Calabrian of Rhodes (Agiadi et al. 2018).

Remarks The otoliths of this species are elongate with an equally elongated and wide sulcus.

Scopelarchidae Alcock, 1896

Type Genus *Scopelarchus* Alcock, 1896.

Included Taxa This family includes 18 modern species (Froese and Pauly, 2019).

Distribution *Scopelarchus analis* has been found in the Calabrian of Kephallonia (Agiadi et al. 2010) and Rhodes (Agiadi et al. 2018; Fig. 9). In addition, it is known from the pre-evaporitic Messinian and the Pleistocene of Italy (Girone et al. 2006, 2010).

Remarks The otoliths of Scopelarchidae are rectangular in shape, with a central and divided sulcus with two rounded colliculi, and an anteroventral extension.

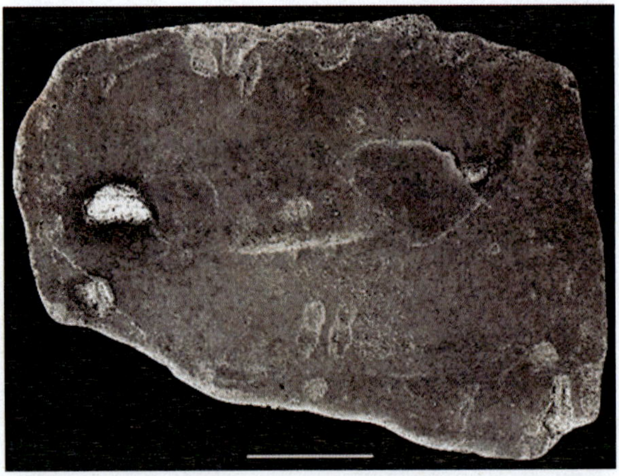

Fig. 9 Scanning electron microscopic image of a typical otolith of *Scopelarchus analis*, from the Middle Pleistocene of Rhodes Island (Agiadi et al. 2018). Scale bar equals 1 mm

Myctophidae Gill, 1893

Type Genus *Myctophum* Rafinesque, 1810.

Included Taxa *Benthosema glaciale* (Reinhardt, 1837); *Benthosema suborbitale* (Gilbert, 1913); †*Bolinichthys italicus* (Anfossi and Mosna, 1971); *Ceratoscopelus maderensis* (Lowe, 1839); †*Diaphus acutirostrum* (Holec, 1975); *Diaphus adenomus* Gilbert, 1905; †*Diaphus befralai* Brzobohatý and Nolf, 2000; †*Diaphus cavallonis* Brzobohatý and Nolf, 2000; *Diaphus holti* Tåning, 1918; †*Diaphus pedemontanus* (Robba, 1970); *Diaphus rafinesquii* (Cocco, 1838); †*Diaphus rubus* (Girone et al., 2010); *Diaphus splendidus* (Brauer, 1904); *Diaphus taaningi* Norman, 1930; *Electrona risso* (Cocco, 1829); *Hygophum benoiti* (Cocco, 1838); *Hygophum hygomii* (Lütken, 1892); *Lampadena dea* Fraser-Brunner, 1949; *Lampanyctus crocodilus* (Risso, 1810); †*Lampanyctus latesulcatus* (Nolf and Steurbaut, 1983); *Lobianchia dofleini* (Zugmayer, 1911); †*Myctophum coppa* (Girone et al., 2010); †*Myctophum fitchi* (Schwarzhans, 1979); *Myctophum punctatum* Rafinesque, 1810; *Notoscopelus elongatus* (Costa, 1844); *Notoscopelus resplendens* (Richardson, 1845); *Protomyctophum arcticum* (Lütken, 1892); †*Scopelopsis pliocenicus* (Anfossi and Mosna, 1976); *Symbolophorus veranyi* (Moreau, 1888). Overall, today, this family has 248 species (Froese and Pauly, 2019).

Distribution Fish otoliths from species belonging to this family are, by far, the most common in the Cenozoic marine record of the Mediterranean and worldwide (Schwarzhans and Aguilera 2013). Several authors (Brzobohatý and Nolf 1996, 2000; Girone and Nolf 2002; Girone et al. 2010) have presented thorough revisions of the European record of Myctophidae. These mesopelagic fish are the dominant family in deep pelagic environments today, and have been reported in Greece from the Tortonian until the Middle Pleistocene (Fig. 10). Particularly interesting is the record of some of these species in response to major paleoclimatic shifts. Such is the case for *Benthosema glaciale* and *Protomyctophum arcticum*, whose abundance and presence, respectively, seem to reflect the glacial intervals in the Pleistocene (Agiadi et al. 2018). Moreover, there is a notable change from the large, robust otoliths of the Tortonian and the Zanclean, such as those of *Diaphus adenomus*, *Diaphus splendidus*, and †*Diaphus cavallonis*, to the smaller and more refined forms of the Pleistocene, including *Diaphus taaningi*, *Diaphus rafinesquii*, and *Diaphus holti*. In addition, the endemic species †*Diaphus rubus* and †*Myctophum coppa* evolve during the Messinian in the Mediterranean (Girone et al. 2010) and are also recorded in the pre-evaporitic Messinian of Zakynthos (Karakitsios et al. 2017) and Crete (Moissette et al. 2018).

Remarks Myctophid otoliths are oval or square in shape, with a wide, divided sulcus, and clearly defined rostrum, antirostrum, and excissura. The otoliths of several species in this family, when not eroded, show denticles on their ventral rim. In addition, a unique characteristic of this family is the collicular crest underlying the caudal colliculum.

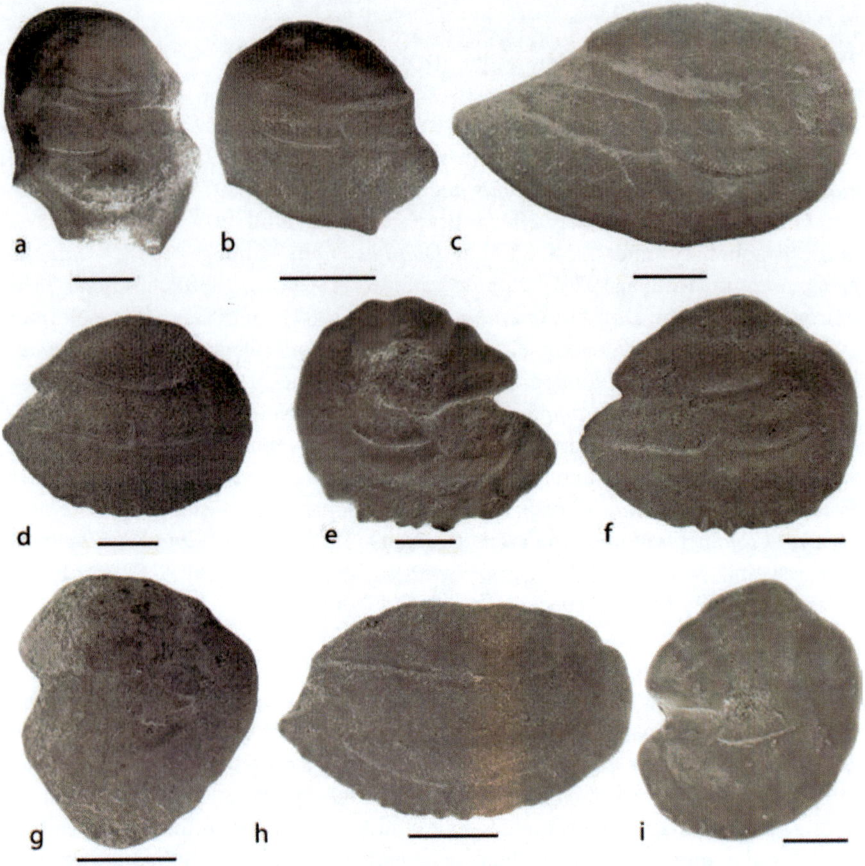

Fig. 10 Scanning electron microscopic images of typical otoliths of Myctophidae from Greece. **a** *Benthosema glaciale*, **b** *Benthosema suborbitale*, **c** *Ceratoscopelus maderensis*, **d** *Diaphus cavallonis*, **e** *Diaphus holti*, **f** *Diaphus rafinesquii*, **g** *Hygophum benoiti*, **h** *Notoscopelus elongatus*, and **i** *Protomyctophum arcticum*. **a**, **c**, and **e–i** are from the Middle Pleistocene of Rhodes Island (Agiadi et al. 2018), **b** and **d** are from the Tortonian of Potamida, Crete Island (Agiadi et al. 2017). Scale bars: 500 μm for **a–e** and **h–i**, 1 mm for **f**, 200 μm for **g**

Macrouridae Bonaparte, 1831

Type Genus *Macrourus* Bloch, 1786.

Included Taxa †*Bathygadus novus* (Bassoli, 1906); *Coelorinchus* Giorna, 1809; *Trachyrincus scabrus* (Rafinesque, 1810). This family contains 406 modern species (Froese and Pauly 2019).

Distribution The deepwater species †*B. novus* has been identified in the late Eocene, the Tortonian, and the Zanclean of Italy (Nolf and Steurbaut 1983, 2004; Nolf and Girone 2006; Lin et al. 2015), in the Eocene of south France (Lin et al.

2017c), and in the Oligocene and Miocene of central Paratethys (Brzobohatý 1995). The only record of this species in Greece is from the Tortonian–Messinian of Crete (Agiadi et al. 2017). *Coelorinchus* spp. are common in the deepwater Mediterranean record, and a thorough review was provided by Nolf and Girone (2000). In Greece, specimens unidentifiable to the species level have only been reported from the Middle Pleistocene of Rhodes (Agiadi et al. 2018). The modern species *Trachyrincus scabrus* was identified in the Tortonian–Messinian of Crete (Agiadi et al. 2017), as well as in the Tortonian, the Piacenzian, and the Pleistocene of Italy (Girone et al. 2006; Girone 2007; Lin et al. 2015).

Remarks Macrourid otoliths are usually large and have a rather irregular shape, where the anterior side is higher. The sulcus is central, extending posterodorsally, and divided.

Moridae Moreau, 1881

Type Genus *Mora* Risso, 1827.

Included Taxa *Laemonema* Günther, 1862; *Physiculus huloti* Poll, 1953. There are 108 species in this family today (Froese and Pauly 2019).

Distribution Specimens attributed to *Laemonema*, but unidentifiable to the species level, have been found in the Zanclean of Crete (Agiadi et al. 2013a) and the Early-Middle Pleistocene of Kephallonia (Agiadi et al. 2010) and Rhodes (Agiadi et al. 2018). Outside Greece, *Laemonema* species have been reported from the Messinian, the Piacenzian, and the Pleistocene of Italy (Girone et al. 2006; Girone 2007; Girone et al. 2010). *Physiculus* aff. *P. huloti* was found in the Messinian of Zakynthos (Karakitsios et al. 2017), in the Tortonian, Messinian and Zanclean of Italy (Steurbaut 1979; Nolf and Steurbaut 1983; Nolf and Girone 2006; Girone et al. 2010), in the Pliocene of Spain (Nolf and Martinell 1980), in the Chattian–Langhian of France (Nolf and Brzobohatý 2002), and in the middle Miocene of Czech Republic (Brzobohatý et al. 2007).

Remarks The otoliths of Moridae are very different from those of other families. They are elongate and very thick. They do not have a rostrum. The cauda opens posteriorly, and there is a ridge-like, sharp caudal colliculum (Schwarzhans 2019).

Bregmacerotidae Gill, 1872

Type Genus *Bregmaceros* Thompson, 1840.

Included Taxa This family contains 14 modern species (Froese and Pauly 2019).

***Bregmaceros* Thompson, 1840**

Type Species *Bregmaceros mcclellandi* Thompson, 1840.

†*Bregmaceros albyi* Sauvage, 1880

Distribution †*Bregmaceros albyi* is a very common species in the Neogene fish skeletal record across the Mediterranean and one of the few fossil species that has been also found as skeletons with otoliths in situ. The otoliths of this species are also rather common, often forming monospecific or oligospecific assemblages in the Miocene. Otoliths of †*Bregmaceros albyi* have been found in the Messinian–Piacenzian of Crete (Marcopoulou-Diacantoni and Kagiouzis 2001; Agiadi et al. 2013a; Moissette et al. 2018) and the Zanclean of Peloponnese (Agiadi et al. 2020). Otoliths of *Bregmaceros*, unidentifiable to the species level, have also been found in the Tortonian of Crete (Moissette et al. 2018), in the Gelasian–Calabrian of Zakynthos (Agiadi et al. 2010), and in the Upper Pleistocene of Rhodes (Cornée et al. 2019).

Remarks The otoliths of Bregmacerotidae are star-like in shape and very thin with a central and divided sulcus (Fig. 11).

Gadidae Rafinesque, 1810

Type Genus *Gadus* Linnaeus, 1758.

Included Taxa *Gadiculus argenteus* Guichenot, 1850; †*Paratrisopterus labiatus* (Schubert, 1905); *Gadiculus thori* Schmidt, 1913; *Micromesistius poutassou* (Risso, 1827); *Gaidropsarus* Rafinesque, 1810; *Phycis blennoides* (Brünnich, 1768). Gadidae contain 23 modern species (Froese and Pauly, 2019).

Fig. 11 Scanning electron microscopic image of a typical otolith of *Bregmaceros albyi*, from the Messinian of Faneromeni, Crete Island (Moissette et al. 2018). Scale bar equals 200 μm

Distribution *Gadiculus argenteus* is an extant species that is known from the Mediterranean area already from the middle Miocene (Steurbaut 1979; Nolf and Brzobohatý 2002; Girone et al. 2006, 2010). In Greece, *Gadiculus argenteus* has been reported from the Messinian of Zakynthos (Karakitsios et al. 2017) and the Gelasian–Upper Pleistocene of Kephallonia, Zakynthos (Agiadi et al. 2010, 2011), and Rhodes (Agiadi et al. 2018). † *Paratrisopterus labiatus* was a very common species in the Pliocene Mediterranean. Outside Greece, it is known from the Messinian–Gelasian of Italy, the Pliocene of South France, the Zanclean of Spain, and already from the middle Miocene of the Paratethys, the North Sea Basin, and the Portuguese coast (Nolf and Martinell 1980; Nolf and Cappetta 1988; Nolf and Cavallo 1995; Nolf et al. 1998; Nolf and Girone 2006; Girone et al. 2006; Girone 2007; Girone et al. 2010; Schwarzhans et al. 2017). However, its eastern Mediterranean distribution was more expanded, with known records in the Tortonian–Piacenzian of Crete (Gaudant 2001; Agiadi et al. 2013a; Moissette et al. 2018), the Messinian and the Gelasian of Zakynthos (Agiadi et al. 2010, 2011; Karakitsios et al. 2017), the Gelasian of Kephallonia (Agiadi et al. 2010, 2011), and the Gelasian–Calabrian of Rhodes (Agiadi et al. 2018). *Gadiculus thori* is an extant species inhabiting the North Atlantic today, which has only been found only in the Upper Pleistocene of Rhodes (Agiadi et al. 2018) and in the Gelasian–Middle Pleistocene Italy (Girone et al. 2006). The extant species *Micromesistius poutassou* is found today in the Mediterranean and has been reported as a fossil from the Zanclean–Middle Pleistocene of Italy (Girone et al. 2006). Its otoliths were found in the Gelasian–Upper Pleistocene of Rhodes (Agiadi et al. 2018). Specimens attributed to *Gaidropsarus* have been reported from the Gelasian–Calabrian of Rhodes and Kephallonia (Agiadi et al. 2010, 2011, 2018). *Gaidropsarus* otoliths have also been found in the Pleistocene of Italy (Girone et al. 2006). Finally, *Phycis blennoides* has been reported from the Gelasian and the Upper Pleistocene of Rhodes (Agiadi et al. 2018). Previously, it was reported from the Tortonian, and the Pliocene–Pleistocene of Italy (Nolf and Steurbaut 1983; Girone 2003; Girone et al. 2006; Girone 2007), and already from the middle Miocene of the Paratethys (Brzobohatý et al. 2007).

Remarks The otoliths of Gadidae are rectangular-to-irregular in shape, some more elongated than others, with a central, wide, and divided sulcus (Fig. 12).

Carapidae Poey, 1867

Type Genus *Carapus* Rafinesque, 1810.

Included Taxa This family contains 36 modern species (Froese and Pauly 2019), of which only *Carapus acus* (Brünnich, 1768) and *Echiodon dentatus* (Cuvier, 1829) have been found in the Greek fossil record.

Distribution *Carapus acus* has been reported in Greece from the Zanclean of Peloponnese (Agiadi et al. 2020), and the Gelasian–Calabrian of Rhodes (Agiadi

Fig. 12 Scanning electron microscopic images of typical otoliths of Gadidae from Greece. **a** *Gadiculus argenteus*, **b** *Gadiculus thori*, **c** *Gaidropsarus* sp., **d** *Phycis blennoides*, all from the Middle Pleistocene of Rhodes Island (Agiadi et al. 2018). Scale bars: 1 mm for **a–b**, 200 μm for **c**, 500 μm for **d**

et al. 2018). It is also known from the Pliocene–Pleistocene of Italy (Girone et al. 2006). *Echiodon dentatus* is known from the Pliocene of Crete (Agiadi et al. 2013a) and the Gelasian–Calabrian of Rhodes (Agiadi et al. 2018). In addition, it has been found in the Early–Middle Pleistocene of Italy (Nolf and Girone 2000; Girone and Varola 2001).

Remarks The otoliths of Carapidae are thick, oval in shape with a central, closed, and undivided sulcus.

Ophidiidae Rafinesque, 1810

Type Genus *Ophidion* Linnaeus, 1758.

Included Taxa *Hoplobrotula* Gill, 1863; *Monomitopus magnus* Carter and Cohen, 1985; *Monomitopus metriostoma* (Vaillant, 1888); *Ophidion barbatum* Linnaeus, 1758. This family includes 258 modern species (Froese and Pauly 2019).

Distribution Otoliths unidentifiable to the species level that were attributed to *Hoplobrotula* as well as specimens identified as *Monomitopus* aff. *M. magnus* were found in the Tortonian–Messinian sediments of Crete (Agiadi et al. 2017). *Monomitopus metriostoma* has only been reported from the Calabrian of Rhodes (Agiadi et al. 2018). As for *Hoplobrotula*, several different species assigned to this genus have been found already from the early Oligocene, the Burdigalian, and the Tortonian–Zanclean of Italy (Robba 1970; Nolf and Cavallo 1995; Nolf and Brzobohatý 2004; Nolf and Steurbaut 2004; Nolf and Girone 2006; Girone et al. 2010; Lin et al. 2015), as well as from the Priabonian of northern Italy and South France (Girone and Nolf 2009). *Ophidion barbatum* has been reported from the Zanclean of Peloponnese (Agiadi et al. 2020) and the Gelasian–Calabrian of Rhodes (Agiadi et al. 2018). Elsewhere in the Mediterranean, it has been found in the Pleistocene of Italy (Girone and Varola 2001; Girone et al. 2006).

Remarks Detailed descriptions of Ophidiiforms' peculiar otoliths may be found in Schwarzhans and Aguilera (2016). These otoliths are thick and massive with an oval or rectangular, elongate shape. Their sulcus is elongate, wide, and divided.

Bythitidae Gill, 1861

Type Genus *Bythites* Reinhardt, 1835.

Included Taxa *Bellotia apoda* Giglioli, 1883; *Grammonus ater* (Risso, 1810); †*Grammonus bassolii* (Nolf, 1980). This family includes 198 modern species (Froese and Pauly 2019).

Nomenclatural and Taxonomical History *Grammonus ater* (Risso, 1810) was originally assigned as *Oligopus ater* Risso, 1810.

Distribution *Bellotia* cf. *B. apoda* has been found in the Gelasian–Calabrian of Zakynthos (Agiadi et al. 2010, 2011) and the Pleistocene of Italy (Girone et al. 2006). *Grammonus ater* is an extant species that has been reported from the Zanclean–Piacenzian of Crete (Agiadi et al. 2013a; Fig. 13a) and the Gelasian–Upper Pleistocene of Rhodes (Agiadi et al. 2018). Outside Greece, it has also been found in the Pleistocene of Italy (Girone and Varola 2001; Girone et al. 2006). †*Grammonus bassolii* was found in the Tortonian–Messinian of Crete (Agiadi et al. 2017) and the Zanclean of Peloponnese (Agiadi et al. 2020; Fig. 13b). It was also found in the Tortonian of Italy (Lin et al. 2015; Lin et al. 2017a).

Fig. 13 Scanning electron microscopic images of typical otoliths of Bythitidae from Greece. **a** *Grammonus ater* from the Zanclean of Voutes, Crete Island (Agiadi et al. 2013b), and **b** *Grammonus bassolii* from the Zanclean of Agia Triada, Peloponnesus (Agiadi et al. 2020). Scale bars equal 500 μm

Remarks These otoliths are oval, elongate, and thick. Their inner face is convex as is their outer face. The sulcus is off-centered and may or may not be divided.

Gobiidae Cuvier, 1816

Type Genus *Gobius* Linnaeus, 1758.

Included Taxa *Aphia minuta* (Risso, 1810); *Callogobius* Bleeker, 1874; *Chromogobius zebratus* (Kolombatovic, 1891); *Deltentosteus quadrimaculatus* (Valenciennes, 1837); †*Gobius bicornutus* (Lin et al., 2015); †*Gobius brioche* (Lin et al., 2015); *Gobius bucchichi* Steindachner, 1870; *Gobius cobitis* Pallas, 1814; *Gobius couchi* Miller and El-Tawil, 1974; *Gobius geniporus* Valenciennes, 1837; *Gobius niger* Linnaeus, 1758; *Gobius paganellus* Linnaeus, 1758; *Gobius vittatus* Vinciguerra, 1883; *Lesueurigobius friesii* (Malm, 1874); *Lesueurigobius sanzi* (de Buen, 1918); *Lesueurigobius suerii* (Risso, 1810); *Mesogobius* Bleeker, 1874; *Pomatoschistus marmoratus* (Risso, 1810); *Thorogobius* Miller, 1969. Overall, this family today includes 1850 species (Froese and Pauly 2019).

Distribution Gobiidae is the most abundant and diverse family in sediments deposited in continental shelfs worldwide today, but also in the Neogene (Appendix, Fig. 14). In Greece, the Tortonian record of Crete includes *Deltentosteus* aff. *D. quadrimaculatus*, *Lesueurigobius* aff. *L. friesii*, and *Pomatoschistus* sp. (Agiadi et al. 2017; Moissette et al. 2018). In the Messinian, some new endemic species appear (Moissette et al. 2018), specifically †*Gobius bicornutus* and †*Gobius brioche* from Crete, which have also been reported from the Tortonian of Italy (Lin et al. 2015; Lin et al. 2017a), as well as *Mesogobius* sp., which was previously found in the pre-evaporitic Messinian of Italy (Girone et al. 2010). *Aphia minuta*, *Callogobius* sp., *Chromogobius zebratus*, *Deltentosteus quadrimaculatus*, *Gobius bucchichi*, *Gobius cobitis*, *Gobius* cf. *G. couchi*, *Gobius* cf. *G. geniporus*, *Gobius niger*, *Gobius paganellus*, *Lesueurigobius friesii*, *Lesueurigobius sanzi*, *Lesueurigobius suerii*, *Pomatoschistus marmoratus*, and *Thorogobius* sp. have been found in the Pliocene (Agiadi et al. 2013a, 2020). The Pleistocene record includes *Aphia minuta*, *Chromogobius zebratus*, *Deltentosteus quadrimaculatus*, *Gobius cobitis*, *Gobius geniporus*, *Gobius niger*, *Gobius paganellus*, *Gobius vittatus*, *Lesueurigobius friesii*, *Lesueurigobius sanzi*, *Lesueurigobius suerii*, and *Pomatoschistus* sp. (Agiadi et al. 2018, 2019).

Remarks Gobies have very particular otoliths. They are square, rectangular, or rounded, and very robust. The sulcus is central or off-central, usually well divided and closed.

Pomacentridae Bonaparte, 1832

Type Genus *Pomacentrus* Lacepède, 1802.

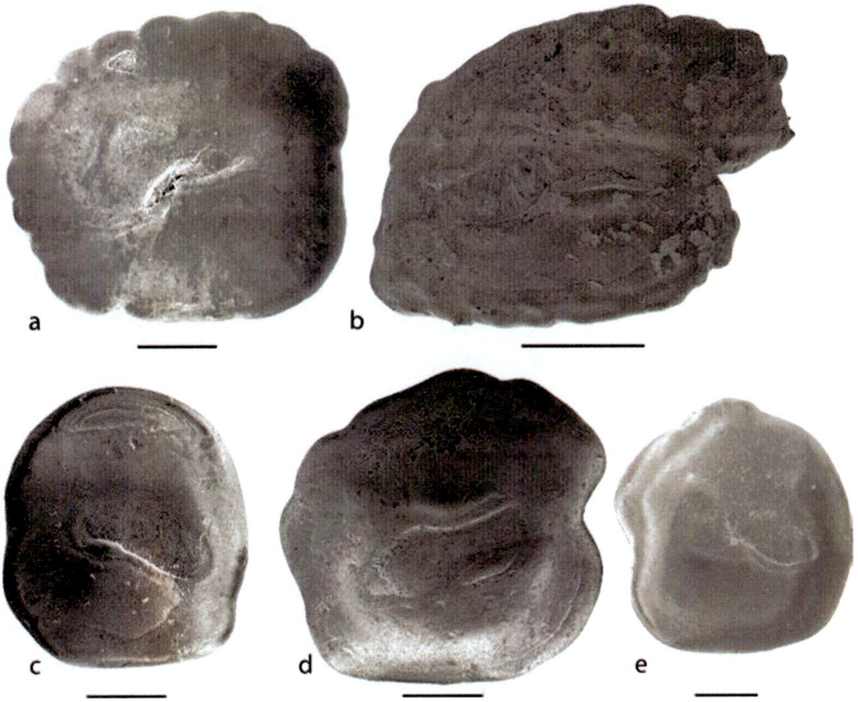

Fig. 14 Scanning electron microscopic images of typical otoliths of Gobiidae from Greece. **a** *Deltentosteus quadrimaculatus*, **b** *Gobius paganellus*, **c** *Aphia minuta*, **d** *Lesueurigobius friesii*, **e** *Pomatoschistus marmoratus*. **a** and **c** are from the Middle Pleistocene of Rhodes Island (Agiadi et al. 2018), **b** is from the Gelasian of Rhodes Island (Agiadi et al. 2019), **d** is from the Zanclean of Voutes, Crete Island (Agiadi et al. 2013b), and **e** is from the Zanclean of Agia Triada, Peloponnesus (Agiadi et al. 2020). Scale bars equal 500 μm for **a**, **b**, and **d**, and 200 μm for **c** and **e**

Included Taxa From the 405 species recorded in this family, which are all extant (Froese and Pauly 2019), only *Chromis chromis* (Linnaeus, 1758) has been found in Greece as a fossil.

Distribution *Chromis chromis* has only been found in the Gelasian–Upper Pleistocene of Rhodes (Agiadi et al. 2018, 2019).

Remarks These otoliths are fusiform in shape with a clear rostrum, and a sulcus divided in a triangular ostium and a long cauda that is slightly bent toward the posterior end.

Blenniidae Rafinesque, 1810

Type Genus *Blennius* Linnaeus, 1758.

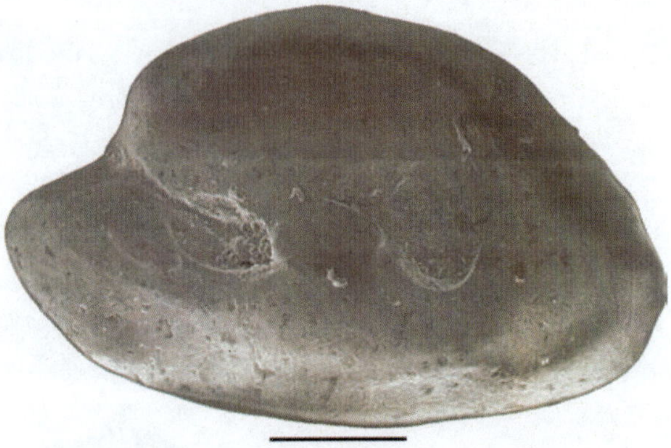

Fig. 15 Scanning electron microscopic image of a typical otolith of *Blennius ocellaris*, from the Zanclean of Agia Triada, Peloponnesus (Agiadi et al. 2020). Scale bar equals 100 μm

Included Taxa This family has 401 modern species (Froese and Pauly 2019). Only *Blennius ocellaris* Linnaeus, 1758 has been reported in Greece as a fossil.

Distribution *Blennius ocellaris* has been reported from the Tortonian–Messinian of Crete (Agiadi et al. 2017), the Zanclean of Peloponnese (Agiadi et al. 2020; Fig. 15), the Upper Pleistocene of Rhodes (Agiadi et al. 2018), but also from the Early–Middle Pleistocene of Italy (Girone and Varola 2001; Girone et al. 2006).

Remarks Blenniid otoliths are somewhat triangular in shape with the larger being the ventral side and a clear sulcus and excissura.

Atherinidae Risso, 1827

Type Genus *Atherina* Linnaeus, 1758.

Included Taxa Only *Atherina boyeri* Risso, 1810 has been found as a fossil in Greece among the 71 modern species (Froese and Pauly 2019).

Distribution This extant species is known from the Zanclean of Crete (Agiadi et al. 2013a, Fig. 16) and the Gelasian–Upper Pleistocene of Rhodes (Agiadi et al. 2018). It has also been reported from the Messinian as well as the Pleistocene of Italy (Gaudant 2002; Girone et al. 2006, 2010).

Remarks The otoliths of Atherinidae are oval in shape with a clear small, round ostium, opening on the anterior side, and a long but thin cauda.

Carangidae Rafinesque, 1815

Type Genus *Caranx* Lacepède, 1801.

Included Taxa Among the 146 modern species (Froese and Pauly 2019) known in this family, only *Trachurus mediterraneus* (Steindachner, 1868) and *Trachurus trachurus* (Linnaeus, 1758) have been found in the fossil record of Greece.

Fig. 16 Scanning electron microscopic image of a typical otolith of *Atherina boyeri*, from the Zanclean of Voutes, Crete Island (Agiadi et al. 2013b). Scale bar equals 500 μm

Distribution *Trachurus mediterraneus* and *T. trachurus* have both been reported from the Gelasian of Rhodes (Agiadi et al. 2019). *Trachurus trachurus* is also known from the Messinian of Italy (Gaudant 2002; Girone et al. 2010). Both species are known from the Pleistocene of Italy (Girone 2005; Girone et al. 2006).

Remarks These otoliths are rectangular elongated with a long protruding rostrum. The ventral side is almost flat. The sulcus is long and wide and opens with a triangular ostium.

Citharidae Günther, 1864

Type Genus *Citharus* Artedi, 1793.

Included Taxa There are six modern species in this family (Froese and Pauly 2019), but only *Citharus linguatula* (Linnaeus, 1758) has been found in Greece as a fossil.

Distribution In Greece, this species is known only from the Zanclean of Crete (Agiadi et al. 2013a). However, it has also been reported from the Tortonian and the Zanclean of Italy (Lin et al. 2015; Nolf and Girone 2006), and the middle–late Miocene of South France (Nolf and Břzobohatý 2002) and the Paratethys (Nolf and Brzobohatý 2009).

Remarks These otoliths are oval or fusiform in shape. The sulcus is open and divided.

Scophthalmidae Chabanaud, 1933

Type Genus *Scophthalmus* Rafinesque, 1810.

Included Taxa This family contains nine modern species (Froese and Pauly 2019).

***Scophthalmus* Rafinesque, 1810**

Type Species *Scophthalmus rhombus* (Linnaeus, 1758).

***Scophthalmus rhombus* (Linnaeus, 1758)**

Distribution The only known record of this species as a fossil is from the Zanclean of Peloponnese (Agiadi et al. 2020, Fig. 17).

Remarks These otoliths are round approximately with a rather irregular rim and an open, divided, and shallow sulcus.

Bothidae Smitt, 1892

Type Genus *Bothus* Rafinesque, 1810.

Included Taxa This family includes 164 modern species (Froese and Pauly 2019). Only †*Arnoglossus kokeni* (Bassoli, 1906) and *Arnoglossus laterna* (Walbaum, 1792) have been found as fossils in Greece.

Distribution †*Arnoglossus kokeni* is an extinct Mediterranean species that has been reported from the Zanclean of Crete (Agiadi et al. 2013a, Fig. 18a) and Peloponnese (Agiadi et al. 2020), as well as from the Gelasian–Calabrian of Rhodes (Agiadi et al. 2018). It has also been found in the Tortonian–Zanclean and the Early–Middle Pleistocene of Italy (Nolf and Cavallo 1995; Schwarzhans 1999; Girone 2000; Nolf and Girone 2000; Girone and Varola 2001; Nolf and Girone 2006; Girone et al. 2010; Lin et al. 2015; Lin et al. 2017a), and the Pliocene of South France (Nolf and Cappetta 1988) and Spain (Nolf and Martinell 1980). *Arnoglossus laterna* is known from the Zanclean of Peloponnese (Agiadi et al.

Fig. 17 Scanning electron microscopic image of a typical otolith of *Scophthalmus rhombus*, from the Zanclean of Agia Triada, Peloponnesus (Agiadi et al. 2020). Scale bar equals 200 μm

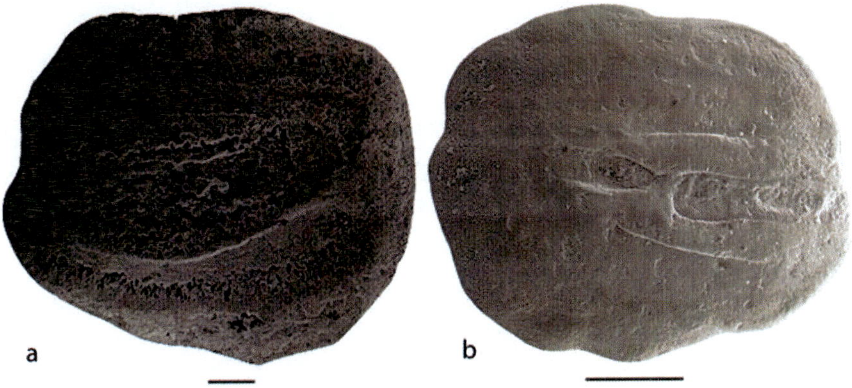

Fig. 18 Scanning electron microscopic images of typical otoliths of Bothidae from Greece. **a** *Arnoglossus kokeni* and **b** *Arnoglossus laterna*, both from the Zanclean of Agia Triada, Peloponnesus (Agiadi et al. 2020). Scale bars equal 500 μm for **a** and 200 μm for **b**

2020; Fig. 18b) and the Gelasian of Rhodes (Agiadi et al. 2019). This species has previously been reported from the Messinian–Pleistocene of Italy (Landini and Varola 1983; Landini and Menesini 1986; Schwarzhans 1979; Gaudant 2002; Girone et al. 2010).

Remarks These otoliths are small, mostly round or oval, but sometimes more square, with a small, thin, open, rectangular, and divided sulcus.

Soleidae Bonaparte, 1832

Type Genus *Solea* Quensel, 1806.

Included Taxa Soleidae contain 180 modern species (Froese and Pauly 2019), from which only *Buglossidium luteum* (Risso, 1810) and *Microchirus variegatus* (Donovan, 1808) are found in the Greek fossil record.

Distribution Both these species have only been reported from the Zanclean of Peloponnese (Agiadi et al. 2020), and previously reported from the Zanclean of Italy (Nolf and Girone 2006). Moreover, *Microchirus variegatus* is known from the Tortonian of Italy as well (Lin et al. 2015; Lin et al. 2017a).

Remarks The otoliths of soleids are small and round with a small, thin sulcus.

Cynoglossidae Jordan, 1888

Type Genus *Cynoglossus* Hamilton, 1822.

Included Taxa This family contains 145 modern species (Froese and Pauly 2019).

***Cynoglossus* Hamilton, 1822**

Type Species *Cynoglossus lingua* Hamilton, 1822.

†*Cynoglossus obliqueventralis* Schwarzhans, 1999

Type Material SMF P 9327 (holotype) Fig. 905 in Schwarzhans (1999).

Type Locality Oued Beth, NW Morocco; Lower Pliocene.

Distribution This species has been found in the Zanclean of Peloponnese (Agiadi et al. 2020) and Italy (Nolf and Girone 2006).

Remarks Cynoglossidae have otoliths that are irregular in shape and the sulcus somewhat follows the general shape of the otolith. Although the sulcus is open, the caudal part is much larger than the ostium part, but a clear division is not visible usually.

Trichiuridae Rafinesque, 1810

Type Genus *Trichiurus* Linnaeus, 1758.

Included Taxa This family includes 45 modern species (Froese and Pauly 2019). *Aphanopus carbo* Lowe, 1839 and *Lepidopus caudatus* (Euphrasen, 1788) have been found as fossils in Greece.

Distribution *Aphanopus carbo* is an extant species that is also known from the Tortonian of Italy (Lin et al. 2017a). *Aphanopus* aff. *A. carbo* has been reported from the Gelasian–Middle Pleistocene of Zakynthos, Kephallonia (Agiadi et al. 2010, 2011), and Italy (Girone et al. 2006). Otoliths of *Lepidopus caudatus* have been reported from the pre-evaporitic Messinian of Zakynthos (Karakitsios et al. 2017) and Italy (Girone et al. 2010). *Lepidopus* cf. *L. caudatus* was mentioned in the Tortonian of Italy (Lin et al. 2015).

Remarks Trichiuridae otoliths are elongated, clearly rounded with a long sulcus.

Trachinidae Rafinesque, 1815

Type Genus *Trachinus* Linnaeus, 1758.

Included Taxa Only *Trachinus draco* Linnaeus, 1758 has been found as fossil in Greece, among the nine modern species forming this family (Froese and Pauly 2019).

Distribution In Greece, this species has only been reported from the Gelasian of Rhodes (Agiadi et al. 2019). Elsewhere, it is known from the Tortonian–Zanclean of Italy (Nolf and Girone 2006; Girone et al. 2010; Lin et al. 2015; Lin et al. 2017a).

Remarks These otoliths are fusiform with a long but very thin sulcus opening in a pointy small rostrum. The excissura is visible.

Acropomatidae Gill, 1893

Type Genus *Acropoma* Temminck and Schlegel, 1843.

Included Taxa Acropomatidae include 37 modern species (Froese and Pauly 2019).

***Verilus* Poey, 1860**

Type Species *Verilus sordidus* Poey, 1860.

†*Verilus mutinensis* (Bassoli, 1906)

Nomenclatural and Taxonomical History †*Parascombrops mutinensis* in Bassoli 1906 (new species); *Verilus mutinensis* in Schwarzhans and Prokofiev 2017 (new combination).

Distribution This is a very common species in the Neogene Mediterranean otolith record. In Greece, it has been reported from the Pliocene of Crete (Agiadi et al. 2013a) and Peloponnese (Agiadi et al. 2020), and the Pleistocene of Rhodes (Agiadi et al. 2018) and Kephallonia (Agiadi et al. 2010, 2011). Previously, it has been found in the Pliocene of Italy, South France, and Spain (Anfossi and Mosna 1979; Schwarzhans 1979; Nolf and Martinell 1980; Anfossi et al. 1982; Nolf and Cappetta 1988; Nolf and Cavallo 1995; Nolf et al. 1998; Nolf and Girone 2006; Girone 2007). In Italy, it has also been reported from the pre-evaporitic Messinian and the Pleistocene (Girone et al. 2006, 2010).

Remarks These otoliths are elongated oval with a wide sulcus that is clearly divided into a long, thick cauda and an oval ostium that opens toward the anterodorsal side.

Epigonidae Poey, 1861

Type Genus *Epigonus* Rafinesque, 1810.

Included Taxa Epigonidae includes 45 modern species (Froese and Pauly 2019). Only *Epigonus denticulatus* Dieuzeide, 1950 has been found as a fossil in Greece.

Distribution *Epigonus* aff. *E. denticulatus* has been found in the Zanclean of Crete (Agiadi et al. 2013a) and in the Tortonian and Zanclean of Italy (Nolf and Girone 2006; Lin et al. 2015).

Remarks These otoliths are oval to fusiform in shape, rather thick and robust. Their sulcus is open, divided into an elongated thick cauda and a large oval ostium. The characteristic of this family is that there is a clear break at the point where the ostium is divided from the cauda, and the ostium extends further toward the ventral margin.

Mullidae Rafinesque, 1815

Type Genus *Mullus* Linnaeus, 1758.

Included Taxa This family includes 88 modern species (Froese and Pauly 2019). Only *Mullus barbatus* Linnaeus, 1758 and *Mullus surmuletus* Linnaeus, 1758 have been found in Greece as fossils.

Distribution Both these species have only been found in the Gelasian of Rhodes (Agiadi et al. 2019). Outside Greece, *Mullus barbatus* has been reported from the Pleistocene of Italy (Girone et al. 2006), whereas *M. surmuletus* is known from the early–middle Miocene of Spain (Hoedemakers and Batllori 2004).

Remarks These otoliths are irregular in shape, but the ventral rim is characteristically flat. There is a clear large opening in the anterior rim, where the sulcus begins.

Haemulidae Gill, 1885

Type Genus *Haemulon* Cuvier, 1829.

Included Taxa This family includes 134 modern species (Froese and Pauly 2019). Only *Pomadasys incisus* (Bowdich, 1825) has been found so far as a fossil in Greece.

Distribution This species has been found in the Zanclean of Peloponnese (Agiadi et al. 2020). It has also been reported from the Tortonian and the Zanclean of Italy (Nolf and Girone 2006; Lin et al. 2015; Lin et al. 2017a).

Remarks These otoliths are fusiform in shape. The sulcus is open, divided into a long cauda that bends ventrally at the posterior end, and an oval ostium.

Cepolidae Rafinesque, 1810

Type Genus *Cepola* Linnaeus, 1764.

Included Taxa This family includes 45 modern species (Froese and Pauly 2019). Only *Cepola macrophthalma* (Linnaeus, 1758) has been found in Greece as a fossil.

Distribution This species has been reported from the Zanclean of Crete (Agiadi et al. 2013a, Fig. 19) and Peloponnese (Agiadi et al. 2020), as well as from the Pleistocene of Rhodes (Agiadi et al. 2018, 2019). Elsewhere, it is known from the Tortonian of Italy (Lin et al. 2015; Lin et al. 2017a).

Fig. 19 Scanning electron microscopic image of a typical otolith of *Cepola macrophthalma*, from the Zanclean of Voutes, Crete Island (Agiadi et al. 2013b). Scale bar equals 500 μm

Remarks The otoliths of this family are fusiform with a sulcus that is divided into two almost equal parts.

Sparidae Rafinesque, 1810

Type Genus *Sparus* Linnaeus, 1758.

Included Taxa *Boops boops* (Linnaeus, 1758); *Dentex dentex* (Linnaeus, 1758); *Dentex gibbosus* (Rafinesque, 1810); *Dentex macrophthalmus* (Bloch, 1791); *Dentex maroccanus* Valenciennes, 1830; *Diplodus annularis* (Linnaeus, 1758); *Oblada melanura* (Linnaeus, 1758); *Pagellus acarne* (Risso, 1827); *Pagellus bogaraveo* (Brünnich, 1768); *Pagellus erythrinus* (Linnaeus, 1758); *Pagrus pagrus* (Linnaeus, 1758); *Spicara smaris* (Linnaeus, 1758); *Spicara maena* (Linnaeus, 1758). Overall, this family includes 159 modern species (Froese and Pauly 2019).

Distribution These are typical fish found in the continental shelf throughout the Quaternary (Fig. 20). Only *Pagellus* cf. *P. acarne* has been reported from the Tortonian–Messinian of Crete (Agiadi et al. 2017). The Zanclean of Greece contains *Boops boops, Dentex gibbosus, Dentex macrophthalmus, Dentex maroccanus, Diplodus annularis, Oblada melanura, Pagellus acarne, Pagellus bogaraveo, Pagellus erythrinus*, and *Spicara maena* (Agiadi et al. 2013a, 2020). In the Pleistocene, the record includes *Boops boops, Dentex dentex, Dentex gibbosus, Dentex macrophthalmus, Dentex maroccanus, Diplodus* cf. *D. annularis, Oblada melanura, Pagellus bogaraveo, Pagellus erythrinus, Pagrus pagrus*, and *Spicara smaris* (Agiadi et al. 2018, 2019). *Dentex macrophthalmus, Dentex* aff. *D. maroccanus*, and *Pagellus acarne* are also known from the Tortonian of Italy (Lin et al. 2015). *Dentex* aff. *D. macrophthalmus* and *Dentex* aff. *D. maroccanus* have been reported from the pre-evaporitic Messinian of Italy (Girone et al. 2010). The Zanclean of Italy also includes *Boops boops, Dentex macrophthalmus, Dentex* aff. *D. maroccanus, Diplodus annularis, Oblada melanura, Pagellus acarne, Pagellus* aff. *P. bogaraveo*, and *Pagellus erythrinus* (Nolf and Girone 2006), whereas only *Dentex maroccanus, Pagellus erythrinus,* and *Spicara smaris* are mentioned in the Pleistocene (Girone et al. 2006).

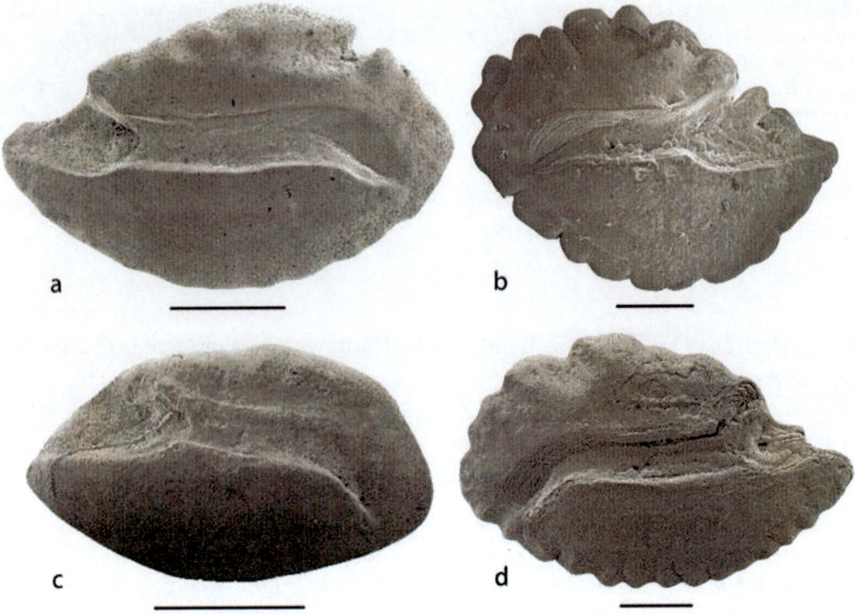

Fig. 20 Scanning electron microscopic images of typical otoliths of Bothidae from Greece. **a** *Boops boops*, **b** *Dentex maroccanus*, **c** *Oblada melanura*, and **d** *Pagellus acarne*. **a**, **b**, and **d** are from the Zanclean of Agia Triada, Peloponnesus (Agiadi et al. 2020), whereas **c** is from the Zanclean of Voutes, Crete Island (Agiadi et al. 2013b). Scale bars equal 500 μm for **a**, **b**, and **d**, and 1 mm for **c**

Remarks Sparidae otoliths are fusiform with a divided sulcus into an oval or triangular ostium and an elongate cauda that is bent ventrally at the posterior end.

6 Concluding Remarks

Otoliths present an invaluable tool for reconstructing Cenozoic fish faunas. Undoubtedly, the fossil otolith record of Greece is far from complete, and many areas remain untouched by researchers. First, research on the freshwater fish faunas of the past has been extremely limited, and, in fact, there are no significant studies published yet from any locality in Greece. On the other hand, regarding the marine record, this so far holds no data for the earlier part of the Cenozoic, specifically before the Tortonian. In addition, there is a gap of knowledge regarding the Piacenzian record, which may be due to lack of studies on deposits covering this stage, as well as taphonomic biases. In fact, the Piacenzian marine fish faunas of Crete have been studied based on skeletal remains recovered from several localities (Gaudant et al. 1994; Gaudant 2001), which, however, did not yield many otoliths. Therefore, it is necessary to stress here that it is necessary to study both otoliths and skeletons, if we are to obtain a full picture and study the evolution of the teleost fish fauna. In addition, it is vital that we aim to recover fish skeletons with otoliths in situ (e.g., Gierl et al. 2013; Schwarzhans and Carnevale 2017), because this is the only way to reconcile otolith-based with skeleton-based identifications of extinct species.

Moreover, bridging the Middle Pleistocene fossil record with the biological data on the modern marine fish fauna of the Mediterranean Sea remains problematic. Indeed, with few exceptions (Cornée et al. 2019), it is difficult to obtain information on the Upper Pleistocene, since such deposits are scarce on land. On the other hand, the archaeological record may hold material and data on the Holocene record (Vika and Theodoropoulou 2012), but otoliths from archaeological excavations, although they should exist, have not been reported yet. An alternative approach would be to recover otoliths from surficial sea bottom sediments (Schwarzhans 2013; Lin et al. 2017b; Lin et al. 2018; Agiadi and Albano 2020; Albano et al. 2020), but this has not been attempted in Greece yet. Future works should focus on such approaches in order to gain a more complete record of the teleost fish fauna of the Greek seas.

Acknowledgments The author would like to thank Dr. Werner Schwarzhans and Prof. Gary Stringer for carefully reviewing the manuscript for this chapter. This research has been co-funded by the European Social Fund and Greek national funds through the action "Postdoctoral Research Fellowships" of the program "Human Resource Development, Education and Lifelong Learning" 2014–2020, which is implemented by the State Scholarships Foundation (I.K.Y.).

Appendix

List of the Greek localities containing otolith fossils of fishes (Teleostei). Locality numbers refer to the collection numbers of the PaleoBiology Database (PBDB)

Locality[PBDB]	Age (MN; GPTS in Ma)	Taxon
Tsampika, Rhodes Island[207572]	Late Pleistocene, Ionian (0.256–0.126)	*Bregmaceros* sp.[1,*]
Kallithea beach, Rhodes Island[207570]	Late Pleistocene, Ionian (0.8–0.7)	*Engraulis encrasicolus*[2], *Vinciguerria poweriae*[2], *Ceratoscopelus maderensis*[2], *Hygophum benoiti*[2], *Myctophum punctatum*[2], *Aphia minuta*[2], *Gadiculus argenteus*[2], *Atherina boyeri*[2], *Cepola macrophthalma*[2], *Deltentosteus quadrimaculatus*[2], *Gobius* cf. *paganellus*[2], *Lesueurigobius suerii*[2]

Locality[PBDB]	Age (MN; GPTS in Ma)	Taxon
Lardos, Rhodes Island[207568]	Late Pleistocene, Ionian (0.8–0.7)	*Engraulis encrasicolus*[2], Clupeidae[2], *Gonostoma* sp.[2], *Maurolicus muelleri*[2], *Vinciguerria* cf. *attenuata*[2], *Benthosema glaciale*[2], *Ceratoscopelus maderensis*[2], *Diaphus holti*[2], *Diaphus rafinesquii*[2], *Diaphus taaningi*[2], *Electrona risso*[2], *Hygophum benoiti*[2], *Hygophum hygomii*[2], *Lampanyctus crocodilus*[2], *Lobianchia dofleini*[2], *Myctophum punctatum*[2], *Notoscopelus elongatus*[2], *Protomyctophum arcticum*[2], *Scopelopsis pliocenicus*[2], *Aphia minuta*[2], *Conger conger*[2], *Gnathophis mystax*[2], *Argentina sphyraena*[2], *Polyipnus polli*[2], *Trachurus* sp.[2], *Coelorinchus* sp.[2], *Spicara smaris*[2], *Gaidropsarus* sp.[2], *Phycis blennoides*[2], *Gadiculus argenteus*[2], *Gadiculus thori*[2], *Micromesistius poutassou*[2], *Grammonus ater*[2], *Atherina boyeri*[2], *Scorpaena* sp.[2], *Dentex* sp.[2], *Pagellus erythrinus*[2], *Oblada melanura*[2], *Cepola macrophthalma*[2], *Chromis chromis*[2], *Blennius ocellaris*[2], Cottidae indet.[2], *Deltentosteus quadrimaculatus*[2], *Gobius niger*[2], *Lesueurigobius friesii*[2], *Lesueurigobius suerii*[2], *Pomatoschistus* sp.[2]
Sgourou, Rhodes Island[207573]	Middle Pleistocene, Calabrian (1.8–1.0)	*Hygophum benoiti*[*], *Notoscopelus elongatus*[*]
Haraki, Rhodes Island[207574]	Middle Pleistocene, Calabrian (1.8–1.0)	*Notoscopelus elongatus*[*]
Pefka, Rhodes Island[207575]	Middle Pleistocene, Calabrian (1.8–1.0)	*Benthosema glaciale*[*], *Diaphus* sp.[*], *Hygophum benoiti*[*], *Myctophum punctatum*[*], *Notoscopelus elongatus*[*]
Kallithea road, Rhodes Island[207571]	Middle Pleistocene, Gelasian–Calabrian (1.8–1.0)	Clupeidae indet.[2], *Maurolicus muelleri*[2], *Benthosema glaciale*[2], *Benthosema suborbitale*[2], *Ceratoscopelus maderensis*[2], *Diaphus holti*[2], *Diaphus rafinesquii*[2], *Diaphus taaningi*[2], *Electrona risso*[2], *Hygophum benoiti*[2], *Hygophum hygomii*[2], *Lampadena* sp.[2], *Lampanyctus crocodilus*[2], *Lobianchia dofleini*[2], *Myctophum punctatum*[2], *Notoscopelus elongatus*[2], *Scopelopsis pliocenicus*[2], *Aphia minuta*[2], *Conger conger*[2], *Pseudophichthys splendens*[2], *Polyipnus polli*[2], *Chlorophthalmus agassizi*[2], *Laemonema* sp.[2], *Trachurus* sp.[2], *Spicara smaris*[2], *Gaidropsarus* sp.[2], *Gadiculus argenteus*[2], *Micromesistius poutassou*[2], *Ophidion barbatum*[2], *Monomitopus metriostoma*[2], *Grammonus ater*[2], *Echiodon dentatus*[2], *Dentex dentex*[2], *Dentex gibbosus*[2], *Diplodus* cf. *annularis*[2], *Oblada melanura*[2], *Pagellus bogaraveo*[2], *Pagellus erythrinus*[2], *Pagrus pagrus*[2], *Cepola macrophthalma*[2], *Chromis chromis*[2], *Deltentosteus quadrimaculatus*[2], *Gobius niger*[2], *Gobius paganellus*[2], *Lesueurigobius friesii*[2], *Lesueurigobius sanzi*[2], *Lesueurigobius suerii*[2], *Pomatoschistus* sp.[2], *Arnoglossus* sp.[2]
Gerakas, Zakynthos Island[207576]	Middle Pleistocene, Gelasian–Calabrian (1.95–0.97)	*Nansenia groenlandica*[3–4], *Maurolicus muelleri*[3–4], *Vinciguerria poweriae*[3–4], *Chauliodus* aff. *sloani*[3–4],

Locality[PBDB]	Age (MN; GPTS in Ma)	Taxon
		Benthosema glaciale[3–4], Electrona risso[3–4], Hygophum benoiti[3–4], Hygophum hygomii[3–4], Myctophum punctatum[3–4], Symbolophorus aff. veranyi[3–4], Ceratoscopelus maderensis[3–4], Ceratoscopelus sp.1[3–4], Diaphus holti[3–4], Diaphus rafinesquii[3–4], Diaphus taaningi[3–4], Diaphus sp.1[3–4], Diaphus sp.2[3–4], Lobianchia dofleini[3–4], Notoscopelus elongatus[3–4], Scopelopsis pliocenicus[3–4], Bregmaceros sp.[3–4], Lampanyctus crocodilus[3–4], Gadiculus argenteus[3–4], Trachurus sp.[3–4], Bellotia cf. apoda[3–4], Deltentosteus aff. quadrimaculatus[3–4], Lesueurigobius sp.[3–4], Aphanopus aff. carbo[3–4]
Akrotiri, Kephallonia Island[207577]	Early–Middle Pleistocene, Gelasian–Ionian (1.96–0.61)	Nettastomatidae[3–4], Nansenia groenlandica[3–4], Alepocephalus sp.[3–4], Gonostoma sp.[3–4], Maurolicus muelleri[3–4], Vinciguerria cf. lucetia[3–4], Vinciguerria poweriae[3–4], Chlorophthalmus cf. agassizi[3–4], Scopelarchus analis[3–4], Benthosema glaciale[3–4], Ceratoscopelus maderensis[3–4], Ceratoscopelus sp.1[3–4], Diaphus holti[3–4], Diaphus rafinesquii[3–4], Diaphus taaningi[3–4], Diaphus sp.1[3–4], Diaphus sp.3[3–4], Electrona risso[3–4], Hygophum benoiti[3–4], Hygophum hygomii[3–4], Lampanyctus crocodilus[3–4], Lobianchia dofleini[3–4], Myctophum punctatum[3–4], Notoscopelus elongatus[3–4], Scopelopsis pliocenicus[3–4], Gaidropsarus sp.[3–4], Gadiculus argenteus[3–4], Paratrisopterus labiatus[3–4], Trachurus sp.[3–4], Sparidae indet.[3–4], Laemonema sp.[3–4], Verilus mutinensis = Parascombrops mutinensis[3–4], Deltentosteus quadrimaculatus[3–4], Lesueurigobius sp.[3–4], Aphanopus aff. carbo[3–4]
Cape Vagia, Rhodes Island[207569]	Early–Middle Pleistocene, Gelasian–Calabrian (2.0–1.0)	Engraulis encrasicolus[2], Sardinella maderensis[2], Maurolicus muelleri[2], Nettastomatidae indet.[2], Scopelarchus analis[2], Benthosema glaciale[2], Ceratoscopelus maderensis[2], Diaphus holti[2], Diaphus rafinesquii[2], Diaphus taaningi[2], Hygophum benoiti[2], Lampanyctus crocodilus[2], Lobianchia dofleini[2], Notoscopelus elongatus[2], Aphia minuta[2], Pseudophichthys splendens[2], Gaidropsarus sp.[2], Gadiculus argenteus[2], Paratrisopterus labiatus[2], Micromesistius poutassou[2], Ophidion barbatum[2], Monomitopus metriostoma[2], Grammonus ater[2], Carapus acus[2], Atherina boyeri[2], Verilus mutinensis[2], Dentex sp.[2], Oblada melanura[2], Cepola macrophthalma[2], Deltentosteus quadrimaculatus[2], Gobius cf. paganellus[2], Lesueurigobius friesii[2], Lesueurigobius suerii[2], Pomatoschistus sp.[2], Arnoglossus kokeni[2]
Lindos, Rhodes Island[2075778]	Early–Middle Pleistocene, Gelasian–Ionian (2.09–0.467)	Vinciguerria poweriae*, Benthosema glaciale*, Ceratoscopelus maderensis*, Ceratoscopelus sp.1*, Diaphus rafinesquii*, Diaphus taanigni*, Electrona risso*, Hygophum benoiti*, Hygophym hygomii*, Lampadena dea*, Lampanyctus crocodilus*, Lobianchia dofleini*, Myctophum punctatum*, Notoscopelus resplendens*, Notoscopelus elongatus*, Scopelopsis pliocenicus*
Kritika North, Rhodes Island[207579]	Early Pleistocene, Gelasian (2.58–1.8)	Sardina pilchardus[5], Conger conger[5], Gnathophis mystax[5], Hygophum hygomii[5],

<table>
<tr><th>Locality[PBDB]</th><th>Age (MN; GPTS in Ma)</th><th>Taxon</th></tr>
<tr><td colspan="3">Diaphus sp.[5], Ophidion barbatum[5], Carapus acus[5], Trachurus trachurus[5], Trachurus mediterraneus[5], Aphanius sp.[5], Haemulidae indet.[5], Boops boops[5], Dentex macrophthalmus[5], Spicara smaris[5], Sparidae indet.[5], Chromis chromis[5], Cepola macrophthalma[5], Trachinus draco[5], Aphia minuta[5], Deltentosteus quadrimaculatus[5], Gobius paganellus[5], Gobius vittatus[5], Gobius sp.[5], Lesueurigobius friesii[5], Lesueurigobius sanzi[5], Gobiidae indet.[5], Arnoglossus laterna[5]</td></tr>
<tr><td>Faliraki South, Rhodes Island[207580]</td><td>Early Pleistocene, Gelasian (2.58–1.8)</td><td>Engraulis encrasicolus[5], Conger conger[5], Ceratoscopelus maderensis[5], Diaphus sp.[5],</td></tr>
<tr><td colspan="3">Ophidion barbatum[5], Carapus acus[5], Grammonus ater[5], Trachurus mediterraneus[5], Atherina sp.[5], Phycis blennoides[5], Haemulidae indet.[5], Apogon sp.[5], Dentex maroccanus[5], Pagellus bogaraveo[5], Spicara smaris[5], Sparidae indet.[5], Mullus barbatus[5], Mullus surmuletus[5], Chromis chromis[5], Cepola macrophthalma[5], Chromogobius zebratus[5], Deltentosteus quadrimaculatus[5], Gobius cobitis[5], Gobius geniporus[5], Gobius paganellus[5], Gobius vittatus[5], Gobius sp.[5], Lesueurigobius friesii[5], Lesueurigobius suerii[5], Gobiidae indet.[5]</td></tr>
<tr><td>Faliraki North, Rhodes Island[207581]</td><td>Early Pleistocene, Gelasian (2.58–1.8)</td><td>Ceratoscopelus maderensis[5], Trachurus mediterraneus[5], Atherina sp.[5], Phycidae indet.[5], Spicara smaris[5], Aphia minuta[5], Gobius paganellus[5], Gobius vittatus[5], Gobius sp.[5], Lesueurigobius friesii[5], Gobiidae indet.[5],</td></tr>
<tr><td>Amnissos, Heraklion, Crete[207303]</td><td>Late Pliocene, Piacenzian (3.6–3.0)</td><td>Notoscopelus resplendens[*], Bregmaceros albyi[*], Lesueurigobius suerii[*]</td></tr>
<tr><td>Gournes, Heraklion, Crete[207359]</td><td>Late Pliocene, Piacenzian (3.6–3.0)</td><td>Diaphus cavallonis[*], Notoscopelus resplendens[*], Bregmaceros albyi[*], Grammonus ater[*], Paratrisopterus labiatus[*], Verilus mutinensis[*], Callogobius sp.[*], Deltentosteus quadrimaculatus[*], Gobius paganellus[*], Lesueurigobius friesii[*], Lesueurigobius suerii[*]</td></tr>
</table>

Locality[PBDB]	Age (MN; GPTS in Ma)	Taxon
Prassies, Rethymnon, Crete[207582]	Early Pliocene, Zanclean (3.8–3.6)	*Ceratoscopelus maderensis*[6,*], *Bregmaceros albyi*[6,*]
Voutes, Heraklion, Crete[207357]	Early Pliocene, Zanclean (3.8–3.6)	*Maurolicus muelleri*[7,*], *Vinciguerria poweriae*[7,*], *Ceratoscopelus maderensis*[7,*], *Diaphus* aff. *adenomus*[7,*], *Diaphus cavallonis*[7,*], *Diaphus holti*[7,*], *Diaphus rafinesquii*[7,*], *Diaphus* aff. *splendidus*[7,*], *Diaphus taaningi*[7,*], *Hygophum benoiti*[7,*], *Hygophum hygomii*[7,*], *Lobianchia dofleini*[7,*], *Myctophum fitchi*[7,*], *Notoscopelus resplendens*[7,*], *Scopelopsis pliocenicus*[7,*], *Bregmaceros albyi*[7,*], *Pseudophichthys escavaratierensis*[7,*], *Pseudophichthys splendens*[7,*], *Rhynchoconger pantanellii*[7,*], *Chlorophthalmus agassizi*[7,*], *Laemonema* sp.[7,*], *Paratrisopterus labiatus*[7,*], *Grammonus ater*[7,*], *Echiodon dentatus*[7,*], *Atherina boyeri*[7,*], *Apogon* sp.[7,*], *Epigonus* aff. *denticulatus*[7,*], *Verilus mutinensis*[7,*], *Oblada melanura*[7,*], *Pagellus bogaraveo*[7,*], *Pagellus erythrinus*[7,*], *Cepola macrophthalma*[7,*], *Aphia minuta*[7,*], *Deltentosteus* aff. *quadrimaculatus*[7,*], *Gobius niger*[7,*], *Gobius* cf. *paganellus* [7,*], *Gobius* sp.1[7,*], *Lesueurigobius friesii*[7,*], *Lesueurigobius sanzi*[7,*], *Lesueurigobius suerii*[7,*], "*Gobidarum*" sp.1[7,*], *Citharus linguatula*[7,*], *Arnoglossus kokeni*[7,*]
Agia Triada, SW Peloponnese[207769]	Early Pliocene, Zanclean (5.08–3.81; MPL2 and MPL3)	*Pterothrissus compactus*[8], *Conger conger*[8], *Gnathophis mystax*[8], *Spratelloides* sp.[8], *Ceratoscopelus maderensis*[8], *Diaphus rafinesqui*[8], *Diaphus splendidus*[8], *Diaphus* sp.[8], *Electrona risso*[8], *Hygophum hygomii*[8], *Notoscopelus resplendens*[8], *Bregmaceros albyi*[8], *Carapus acus*[8], *Ophidion barbatum*[8], *Grammonus bassoli*[8], *Apogon* sp.[8], *Aphia minuta*[8], *Chromogobius zebratus*[8], *Deltentosteus quadrimaculatus*[8], *Gobius bucchichi*[8], *Gobius cobitis*[8], *Gobius* cf. *couchi*[8], *Gobius* cf. *geniporus*[8], *Gobius* cf. *paganellus*[8], *Gobius*? sp. 1[8], *Gobius*? sp. 2[8], *Gobius*? sp. 3[8], *Gobius* sp.[8], *Lesueurigobius friesii*[8], *Lesueurigobius sanzi*[8], *Lesueurigobius suerii*[8], *Lesueurigobius* sp.[8], *Pomatoschistus marmoratus*[8], *Thorogobius* sp.[8], Ambassidae indet.[8], *Blennius ocellaris*[8], Citharidae indet.[8], *Scophthalmus rhombus*[8], *Arnoglossus kokeni*[8], *Arnoglossus laterna*[8], *Arnoglossus* cf. *laterna*[8], *Arnoglossus* sp.[8], *Achirus* sp.[8], *Buglossidium luteum*[8], *Microchirus variegatus*[8], *Cynoglossus obliqueventralis*[8], *Verilus mutinensis*[8], Mullidae indet.[8], Percidae indet.[8], *Pomadasys incisus*[8], *Cepola macrophthalma*[8], *Boops boops*[8], *Dentex gibbosus*[8], *Dentex macrophthalmus*[8], *Dentex maroccanus*[8], *Diplodus annularis*[8], *Oblada melanura*[8], *Pagellus acarne*[8], *Pagellus bogaraveo*[8], *Pagellus erythrinus*[8], *Spicara maena*[8]
Kalamaki, Zakynthos Island[207583]	Late Miocene, Messinian (6.5–5.97)	*Maurolicus muelleri*[9], *Vinciguerria poweriae*[9], *Ceratoscopelus maderensis*[9], *Diaphus cavallonis*[9], *Diaphus* cf. *pedemontanus*[9], *Diaphus rafinesquii*[9], *Diaphus rubus*[9], *Diaphus taaningi*[9], *Myctophum coppa*[9], *Physiculus* aff. *huloti*[9], *Gadiculus argenteus*[9], *Paratrisopterus labiatus*[9], *Lepidopus caudatus*[9], *Buglossidium* sp.[9]

Locality[PBDB]	Age (MN; GPTS in Ma)	Taxon
Kapariana, Heraklion, Crete[207592]	Late Miocene, Messinian (6.83–6.5)	*Bolinichthys italicus*[10,*], *Ceratoscopelus maderensis*[10,*], *Diaphus acutirostrum*[10,*], *Diaphus rubus*[10,*], *Lampanyctus latesulcatus*[10,*], *Lobianchia dofleini*[10,*], *Myctophum coppa*[10,*], *Bregmaceros albyi*[10,*]
Potamida, Chania, Crete[207584–91]	Late Miocene, Tortonian–Messinian (7.51–6.66)	*Pseudophichthys* sp.[11], *Rhynchoconger pantanellii*[11], *Benthosema suborbitale*[11], *Ceratoscopelus maderensis*[11], *Diaphus* aff. *acutirostrum*[11], *Diaphus befralai*[11], *Diaphus cavallonis*[11], *Diaphus rafinesquii*[11], *Diaphus rubus*[11], *Diaphus* aff. *splendidus*[11], *Hygophum hygomii*[11], *Lampadena dea*[11], *Lampanyctus latesulcatus*[11], *Lobianchia dofleini*[11], *Myctophum coppa*[11], *Myctophum fitchi*[11], *Myctophum punctatum*[11], *Notoscopelus elongatus*[11], *Scopelopsis pliocenicus*[11], *Trachyrinchus scabrus*[11], *Bathygadus novus*[11], *Paratrisopterus labiatus*[11], *Echiodon* sp.[11], *Hoplobrotula* sp.[11], *Monomitopus* aff. *magnus*[11], *Grammonus bassolii*[11], *Atherina* sp.[11], *Scopelogadus* sp.[11], *Hoplostethus* sp.[11], *Pagellus* cf. *acarne*[11], *Blennius ocellaris*[11], *Deltentosteus* aff. *quadrimaculatus*[11], *Lesueurigobius* aff. *friesii*[11], *Pomatoschistus* sp.[11], *Gobius brioche*[11], *Mesogobius* sp.[11]
Faneromeni, Heraklion, Crete[207593]	Late Miocene, Tortonian–Messinian (7.58–6.83)	*Rhynchoconger pantanellii*[10,*], *Ceratoscopelus maderensis*[10,*], *Diaphus acutirostrum*[10,*], *Diaphus cavallonis*[10,*], *Diaphus rafinesquii*[10,*], *Diaphus splendidus*[10,*], *Hygophum hygomii*[10,*], *Lobianchia dofleini*[10,*], *Myctophum fitchi*[10,*], *Bregmaceros* sp.[10,*], *Gobius bicornuta*[10,*], *Lesueurigobius* sp.[10,*]
Agios Ioannis, Gavdos Island[60746]	Late Miocene, Tortonian (11.6–9.77)	*Diaphus cavallonis*[12,*], (formerly misidentified as *Diaphus dumerili* (Bleeker, 1856) *Diaphus holti*[12,*], *Notoscopelus resplendens*[12,*]

Locality[PBDB]	Age (MN; GPTS in Ma)	Taxon
Potamos, Gavdos Island[207594]	Late Miocene, Tortonian (11.6–9.77)	*Diaphus cavallonis*[12,*], *Diaphus holti*[12,*], *Notoscopelus resplendens*[12,*]
Lapsarna, Lesvos Island[186558]	Early Miocene, Burdigalian (18.4–15.67)	*Barbus* sp.[13], *Palaeoleuciscus* sp.[13]
Nenita, Chios Island[207609]	Early Miocene, Burdigalian–Langhian (20.44–13.82)	*Aphanius* sp.[14], *Prolebias* sp.[14], *Barbus* sp.[14], *Palaeocarassius* sp.[14], *Phoxinellus* sp.[14], *Esox* sp.[14], *Morone* sp.[14]

GPTS Geomagnetic Polarity Time Scale
[a]This study
[1]Cornée et al. 2019, [2]Agiadi et al. 2018, [3]Agiadi et al. 2010, [4]Agiadi et al. 2011, [5]Agiadi et al. 2020, [6]Marcopoulou–Diacantoni and Kagiouzis 2001, [7]Agiadi et al. 2013b, [8]Agiadi et al. 2019, [9]Karakitsios et al. 2017, [10]Moissette et al. 2018, [11]Agiadi et al. 2017, [12]Tsaparas and Marcopoulou–Diacantoni 2005, [13]Vasileiadou et al. 2017, [14]Reichenbacher & Boehme 2004

References

Agiadi K, Albano PG (2020) Holocene fish assemblages provide baseline for the rapidly changing eastern Mediterranean. The Holocene. https://doi.org/10.1177/0959683620932969

Agiadi K, Triantaphyllou M, Girone A, Karakitsios V, Dermitzakis M (2010) Paleobathymetric interpretation of the fish otoliths from the lower - middle quaternary deposits of Kephallonia and Zakynthos islands (Ionian Sea, Western Greece). Riv Ital Paleontol Stratigr 116:63–78

Agiadi K, Triantaphyllou M, Girone A, Karakitsios V (2011) The early quaternary palaeobiogeography of the eastern Ionian deep-sea teleost fauna: a novel palaeocirculation approach. Palaeogeogr Palaeoclimatol Palaeoecol 306:228–242

Agiadi K, Koskeridou E, Triantaphyllou M, Girone A, Karakitsios V (2013a) Fish otoliths from the Pliocene Heraklion Basin (Crete Island, Eastern Mediterranean). Geobios 46:461–472

Agiadi K, Koskeridou E, Triantaphyllou M, Karakitsios V (2013b) Paleobathymetry of a Late Pliocene Voutes coast (Heraklion basin, Crete). Bull Geol Soc Greece 47:52–61

Agiadi K, Antonarakou A, Kontakiotis G, Kafousia N, Moissette P, Cornée J-J, Manoutsoglou E, Karakitsios V (2017) Connectivity controls on the late Miocene eastern Mediterranean fish fauna. Int J Earth Sci 106:1147–1159

Agiadi K, Girone A, Koskeridou E, Moissette P, Cornée J-J, Quillévéré F (2018) Pleistocene marine fish invasions and paleoenvironmental reconstructions in the eastern Mediterranean. Quat Sci Rev 196:80–99

Agiadi K, Vasileiou G, Koskeridou E, Moissette P, Cornée JJ (2019) Coastal fish otoliths from the early Pleistocene of Rhodes (eastern Mediterranean). Geobios 55

Agiadi K, Giamali C, Girone A, Moissette P, Koskeridou E, Karakitsios V (2020) The Zanclean marine fish fauna and palaeoenvironmental reconstruction of a coastal marine setting in the eastern Mediterranean. Palaeobiodiv Palaeoenviron

Andrus CFT, Crowe DE, Sandweiss DH, Reitz EJ, Romanek CS (2002) Otolith δ18O record of mid-Holocene Sea surface temperatures in Peru. Science 295:1508–1511
Anfossi G, Mosna S (1972) Otoliti del Pliocene inferior di Lugagnano (Piacenza). Atti Inst Geol Univ Pavia 23:90–118
Anfossi G, Mosna S (1979) La fauna ittiologica di Monteu Roero (Alba, Italia NW). Otoliti. Atti Inst Geol Univ Pavia 27:111–132
Anfossi G, Brambilla G, Mosna S (1982) La fauna del Pliocene di Taino (Varese). Atti Inst Geol Univ Pavia 30:83–102
Aruta L, Greco A (1980) Otoliti dell' Emiliano di localita Olivella (PA) e del Pliocene superior di contrada Pipitone (AG) (Sicilia occidentale). Nat Sicil 4:101–117
Brzobohatý R (1995) Macrouridae (Teleostei, Otolithen) im Oligozän und Miozän der zentralen Paratethys und ihre paläogeographische. Bedeutung Jahrb Geol Bundesanst Wien 138:615–637
Brzobohatý R, Nolf D (1996) Myctophid otoliths (teleostean fish) from the European tertiary: revision of the genera Benthosema, Hygophum, Lampadena, Notoscopelus and Symbolophorus. Bull Inst R Sci Nat Belgique Sci Terre 66:151–176
Brzobohatý R, Nolf D (2000) Diaphus otoliths from the European Neogene (Myctophidae, Teleostei). Bull Inst R Sci Nat Belgique Sci Terre 70:185–206
Brzobohatý R, Nolf D, Kroupa O (2007) Fish Otoliths from the Middle Miocene of Kienberg at Mikulov, Czech Republic, Vienna Basin: Their paleoenvironmental and paleogeographic significance. Bull Inst R Sci Nat Belgique Sci Terre 77:167–196
Campana SE (1999) Chemistry and composition of fish otoliths: pathways, mechanisms and applications. Mar Ecol Prog Ser 188:263–297
Carnevale G, Landini W, Sarti G (2006) Mare versus Lago-mare: marine fishes and the Mediterranean environment at the end of the Messinian Salinity Crisis. J Geol Soc 163:75–80
Cornée J-J, Quillévéré F, Moissette P, Fietzke J, López-Otálvaro GE, Melinte-Dobrinescu M, Philippon M, van Hinsbergen DJJ, Agiadi K, Koskeridou E, Münch P (2019) Tectonic motion in oblique subduction forearcs: insights from the revisited Middle and Upper Pleistocene deposits of Rhodes, Greece. J Geol Soc 176:78–96
Degens ET, Deuser WG, Haedrich RL (1969) Molecular structure and composition of fish otoliths. Mar Biol 2:105–113
Froese R, Pauly P (2019) FishBase. World Wide Web electronic publication. www.fishbase.org, version (08/2019)
Gaudant J (1993) Découverte de Poissons fossiles dans le Tortonien diatomitique du bassin de Hellín (Province d'Albacete, Espagne). Acta Geol Hispánica 28:47–54
Gaudant J (2001) Amnissos: un gisement cle pour la connaissance de l' ichthyofauna du Pliocene superieur de Crete. Ann Naturhist Mus Wien 102:131–187
Gaudant J (2002) La crise messinienne et ses effets sur l'ichthyofaune néogène de la Méditerranée: le témoignage des squelettes en connexion de poissons téléostéens
Gaudant J (2004) Additions à l'ichthyofaune tortonienne du bassin de Ierapetra (Crète orientale, Grèce). Ann Naturhist Mus Wien 105:257–285
Gaudant J, Delrieu B, Dermitzakis M, Symeonidis N (1994) Decouverte d' une ichthyofauna marine dans les diatomites du Pliocene superieur (Piacenzian) des environs d' Heraklion (Crete centrale, Grece). C R Acad Sci Paris 319:589–596
Ghosh P, Eiler J, Campana SE, Feeney RF (2007) Calibration of the carbonate 'clumped isotope' paleothermometer for otoliths. Geochim Cosmochim Acta 71:2736–2744
Gierl C, Reichenbacher B, Gaudant J, Erpenbeck D, Pharisat A (2013) An extraordinary gobioid fish fossil from Southern France. PLoS One 8(5):e64117
Girone A (2000) The use of fish otoliths for paleobathymetric evaluation of the lower to middle Pleistocene deposits in Southern Italy. Boll Soc Paleontol Ital 39:235–242
Girone A (2003) The Pleistocene bathyal teleostean fauna of Archi (southern Italy): Palaeoecological and palaeobiogeographic implications. Riv Ital Paleontol Stratigr 109:99–110

Girone A (2005) Response of otolith assemblages to sea-level fluctuations at the Lower Pleistocene Montalbano Jonico section (southern Italy). Boll Soc Paleontol Ital 44:35–45

Girone A (2007) Piacenzian otolith assemblages from northern Italy (Rio Merli section, Emilia Romagna). Boll Soc Paleontol Ital 45:159–170

Girone A, Nolf D (2002) Lampadena ionica: a new teleost from the Mediterranean Pleistocene. Riv Ital Paleontol Stratigr 108:493–500

Girone A, Nolf D (2009) Fish otoliths from the Priabonian (Late Eocene) of North Italy and South-East France – their paleobiogeographical significance. Rev Micropaleontol 52:195–218

Girone A, Varola A (2001) Fish otoliths from the middle Pleistocene deposits of Montalbano Jonico (Southern Italy). Boll Soc Paleontol Ital 40:431–443

Girone A, Nolf D, Cappetta H (2006) Pleistocene fish otoliths from the Mediterranean Basin: a synthesis. Geobios 39:651–671

Girone A, Nolf D, Cavallo O (2010) Fish otoliths from the pre-evaporitic (Early Messinian) sediments of northern Italy: their stratigraphic and palaeobiogeographic significance. Facies 56:399–432

Hoedemakers K, Batllori J (2004) Fish otoliths from the early and middle Miocene of the Penedès (Catalunya, Spain). Batalleria:105–134

Ivany LC, Patterson WP, Lohmann KC (2000) Cooler winters as a possible cause of mass extinctions at the Eocene/Oligocene boundary. Nature 407:887

Karakitsios V, Roveri M, Lugli S, Manzi V, Gennari R, Antonarakou A, Triantaphyllou M, Agiadi K, Kontakiotis G, Kafousia N, de Rafelis M (2017) A record of the Messinian salinity crisis in the eastern Ionian tectonically active domain (Greece, eastern Mediterranean). Basin Res 29:203–233

Keupp H, Bellas S (2002) Miozan-Fossilien aus NW-Kreta III Die Beckenfazies. Fossilien 19:34–40

Landini W, Menesini E (1986) L' ittiofauna pliocenica della Sez, di Stuni e suoi rapport con l' ittiofauna plio-pleistocenica della Vrica (Crotone, Calabria). Boll Soc Paleontol Ital 25:41–63

Landini W, Sorbini C (1993) Biogeographic and palaeoclimatic relationships of the Middle Pliocene ichthyofauna of the Samoggia Torrent (Bologna, Italy), in: Ciencias Da Terra. Presented at the 1st R.C.A.N.S. Congress, pp 83–89

Landini W, Sorbini C (2005) Evolutionary trends in the Plio-Pleistocene ichthyofauna of the Mediterranean Basin: nature, timing and magnitude of the extinction events. Quat Int 131:101–107

Landini W, Varola A (1983) L 'ittiofauna del Pleistocene inferior di Matera. Thalass Salentina 13:16–45

Lin C-H, Girone A, Nolf D (2015) Tortonian fish otoliths from turbiditic deposits in Northern Italy: taxonomic and stratigraphic significance. Geobios 48:249–261

Lin C-H, Brzobohatý R, Nolf D, Girone A (2017a) Tortonian teleost otoliths from northern Italy: taxonomic synthesis and stratigraphic significance. Eur J Taxon 2017:1–44

Lin C-H, Taviani M, Angeletti L, Girone A, Nolf D (2017b) Fish otoliths in superficial sediments of the Mediterranean Sea. Palaeogeogr Palaeoclimatol Palaeoecol 471:134–143

Lin C-H, Nolf D, Steurbaut E, Girone A (2017c) Fish otoliths from the Lutetian of the Aquitaine Basin (SW France), a breakthrough in the knowledge of the European Eocene ichthyofauna. J Syst Palaeontol 15:879–907

Lin C-H, Chiang Y-P, Tuset VM, Lombarte A, Girone A (2018) Late quaternary to recent diversity of fish otoliths from the Red Sea, Central Mediterranean, and NE Atlantic sea bottoms. Geobios

Marcopoulou-Diacantoni A, Kagiouzis G (2001) Otoliths from the lower Pliocene of the section Prassies (Rethymnon, NW-Crete), systematics – paleoecology. Bull Geol Soc Greece 34:577–584

Moissette P, Cornée J-J, Antonarakou A, Kontakiotis G, Drinia H, Koskeridou E, Tsourou T, Agiadi K, Karakitsios V (2018) Palaeoenvironmental changes at the Tortonian/Messinian boundary: a deep-sea sedimentary record of the eastern Mediterranean Sea. Palaeogeogr Palaeoclimatol Palaeoecol 505:217–233

Nelson JS, Grande TC, Wilson MVH (2016) Fishes of the world, 5th edn. Wiley, Hoboken
Nolf D (1985) Otolithi Piscium, handbook of paleoichthyology. G. Fischer, Stuttgart
Nolf D (2013) The diversity of fish otoliths, past and present. Royal Belgian Institute of Natural Sciences, Brussels
Nolf D, Brzobohatý R (1994a) Fish otoliths as paleobathymetric indicators. Paleontol Evol 24–25:255–264
Nolf D, Brzobohatý R (1994b) Fish otoliths from the late Oligocene (Eger and Kiscell Formations) in the Eger area (northeastern Hungary). Bull Inst R Sci Nat Belg Sci Terre 64:225–252
Nolf D, Brzobohatý R (2002) Otolithes de poissons du paléocanyon de saubrigues (chattien à langhien), aquitaine méridionale, France. Rev Micropaleontol 45:261–296
Nolf D, Brzobohatý R (2004) Otolithes de poissons du Miocène inférieur Piémontais. Riv Piemontese Storia Nat 25:69–118
Nolf D, Brzobohatý R (2009) Lower Badenian fish otoliths of the Styrian and Lavanttal basins, with a revision of Weinfurter's type material. Ann Naturhist Mus Wien 111:323–355
Nolf D, Cappetta H (1988) Otolithes de poissons pliocenes du Sud-Est de la France. Bull Inst R Sci Nat Belg Sci Terre 58:209–271
Nolf D, Cavallo O (1995) Otolithes de poisons du Pliocene inferieur de Monticello d' Alba (Piemont, Italie). Riv Piemontese Stor Nat 15:11–40
Nolf D, Girone A (2000) Otolithes de Poissons du Pleistocene Inferieur (Santernien) de Morrona (Sud Est de Pisa). Riv Piemontese Storia Nat 21:3–18
Nolf D, Girone A (2006) Otolithes de poissons du Pliocene Inferieur (Zancleen) des environes d' Alba (Piemont) et de la cote Ligure. Riv Piemontese Stor Nat 27:77–114
Nolf D, Martinell J (1980) Otolithes de Teleosteens du Pliocene des environs de Figueras (Catalogne). Geol Palaentologica 14:209–234
Nolf D, Steurbaut E (1983) Révision des otolithes de téléostéens du Tortonien stratotypique et de Montegibbio (Miocène Supérieur de l'Italie septentrionale). Meded Van Werkgr Voor Tert Kwartaire Geol 20:143–197
Nolf D, Steurbaut E (2004) Otolithes de Poissons de l'Oligocene inferieur du Bassin Liguro-Piemontais oriental, Italie. Riv Piemontese Storia Nat 25:21–68
Nolf D, Mane R, Lopez A (1998) Otolithes de poissoins du Pliocene inferieur de Papiol, pres de Barcelone. Palaeovertebrata Montp 27:1–17
Patterson C, Rosen DE (1977) Review of ichthyodectiform and other Mesozoic teleost fishes, and the theory and practice of classifying fossils. Bull Am Nat Hist Mus 158:81–172
Reichenbacher B, Boehme M (2004) Fish remains (otoliths, teeth) from the Miocene of Chios (Greece) – new palaeobiogeographic and palaeoclimatic results. 5th interbational symposium on Eastern Mediterranean geology, Thessaloniki, pp 338–339
Robba E (1970) Otoliti del Tortoniano-tipo (Piemonte). Riv It Paleontol Stratigr 76:89–172
Schulz-Mirbach T, Olbinado M, Rack A, Mittone A, Bravin A, Melzer RR, Ladich F, Heß M (2018) In-situ visualization of sound-induced otolith motion using hard X-ray phase contrast imaging. Sci Rep 8:3121
Schwarzhans W (1979) Otolithen aus dem Unter-Pliozan von Sud-Sizilien und aus der Toscana. Berl Geowiss Abh 8:1–52
Schwarzhans W (1981) Die Entwicklung der Familie Pterothrissidae (Elopomorpha: Pisces), rekonstruiert nack Otolithen. Senckenb Lethaea 62:77–91
Schwarzhans W (1986) Die Otolithen des Unter Pliozän von Le Puget, S Frankreich. Senckenberg Lethaea 67:219–273
Schwarzhans W (1999) A comparative morphological treatise of recent and fossil otoliths of the order Pleuronectiformes , Piscium Catalogus: Otolithi Piscium. Verlag Dr. Friedrich Pfeil, Munchen
Schwarzhans W (2019) A comparative morphological study of recent otoliths of the Moridae (Gadiformes). In: Schwarzhans W. Reconstruction of the fossil marine bony fish fauna (Teleostei) from the Eocene to Pleistocene of New Zealand by means of otoliths. Mem Soc Italian Sci Natur Mus Storia Natur Milan 46:355–370

Schwarzhans W, Aguilera O (2013) Otoliths of the Myctophidae from the Neogene of tropical America. Palaeo Ichthyol 13:83–150

Schwarzhans W, Aguilera O (2016) Otoliths of the Ophidiiformes from the Neogene of tropical America. Palaeo Ichthyol 14:91–124

Schwarzhans W, Carnevale G (2017) Otoliths in situ from Sarmatian (Middle Miocene) fishes of the Paratethys. Preface: a first attempt to fill the gap between the otolith and skeletal records of teleost fishes. Swiss J Palaeont 136(1):1–6

Schwarzhans W, Carnevale G, Bratishko A, Japundzic S, Bradic K (2017) Otoliths in situ from Sarmatian (Middle Miocene) fishes of the Paratethys. Part II: Gadidae and Lotidae. Swiss J Palaeont 136:19–43

Schwarzhans W (2013) Otoliths from dredges in the Gulf of Guinea and off the Azores – an actuo-paleontological case study. Palaeo Ichthyol 13:7–40

Steurbaut E (1979) Les otolithes de Téléostéens des Marnes de Saubrigues (Miocène d'Aquitane méridionale, France). Palaeontogr Beitr Natur Vorz Paläozoology Stratigr 166:50–91

Thorrold SR, Campana SE, Jones CM, Swart PK (1997) Factors determining δ 13C and δ 18O fractionation in aragonitic otoliths of marine fish. Geochim Cosmochim Acta 61:2909–2919

Tsaparas N, Marcopoulou-Diacantoni A (2005) Otoliths from the Middle to Upper Miocene of the Gavdos Island (South Greece). Systematics - Paleoecology. Rev Paleobiologie 24:617–628

Vasileiadou K, Böhme M, Neubauer TA, Georgalis GL, Syrides GE, Papadopoulou L, Zouros N (2017) Early Miocene gastropod and ectothermic vertebrate remains from the Lesvos Petrified Forest (Greece). Pal Z 91:541–564

Vika E, Theodoropoulou T (2012) Re-investigating fish consumption in Greek antiquity: results from δ13C and δ15N analysis from fish bone collagen. J Archaeol Sci 39:1618–1627

The Fossil Record of Amphibians (Amphibia: Urodela and Anura) in Greece

Massimo Delfino and Georgios L. Georgalis

1 Introduction

Extant amphibians form an extremely diverse group that is currently represented by more than 8000 living species of gymnophionans or caecilians (214 species), urodelans or caudates (742 species of newts and salamanders), and anurans or salientians (7164 species of frogs and toads) (Frost 2020). The separation of these groups goes back at least to the earliest Mesozoic since a stem frog, *Triadobatrachus massinoti* (Piveteau, 1936), has been found in Early Triassic rocks of Madagascar (Ascarrunz et al. 2016).

Europe hosts only about 30 urodelans and 40 anurans (as well as two invasive alien species) but not gymnophionans (Sillero et al. 2014). Of these, 8 urodelans and 18 anurans currently inhabit Greece (Frost 2020). Even if apparently gymnophionans never occurred in Greece (and the whole Europe; see Estes 1981), the Greek fossil record testifies for a past diversity higher than the current as shown in the following sections.

M. Delfino (✉)
Dipartimento di Scienze della Terra, Università di Torino, Torino, Italy

Institut Català de Paleontologia Miquel Crusafont, Universitat Autònoma de Barcelona, Barcelona, Spain
e-mail: massimo.delfino@unito.it

G. L. Georgalis
Dipartimento di Scienze della Terra, Università di Torino, Torino, Italy

Palaeontological Institute and Museum, University of Zurich, Zurich, Switzerland
e-mail: georgios.georgalis@pim.uzh.ch

E. Vlachos (ed.), *Fossil Vertebrates of Greece Vol. 1*,
https://doi.org/10.1007/978-3-030-68398-6_6

2 Historical Overview

The study of Greek fossil amphibians is a rather recently developed field. The oldest paper dealing with this subject seems to be the one published in 1975 by Schneider, who studied a Pleistocene herpetofauna from the island of Chios. In the following years, Kotsakis dedicated a paper to the amphibians and reptiles of the Bate Cave in Crete (1977) and another one to the presence of *Discoglossus* in Crete (1982). Other papers by the same author (Kotsakis et al. 1979, Kotsakis 1986, 1991) included the Greek fossil amphibians in the broader context of vertebrate palaeobiogeography in the eastern Mediterranean area. Remarkable is the contribution by Sanchiz (1984), who provided a careful description and some figures of the anurans from three Greek sites. In the following decades, the fossil amphibians from Greece did not became the subject of papers dedicated to them, but were only cursorily mentioned in faunal lists, even without precise information on the referred material (Vugt et al. 2000), or cited in a book on the Pleistocene herpetofaunas of Europe (Holman 1998), or listed in the volume dedicated to the fossil frogs and toads of the world, the Handbuch der Palaoherpetologie, a work of reference published by Sanchiz in 1998. In the last few years, a few papers have been published on the herpetofaunas, and therefore also on the amphibians, of several Greek localities (Georgalis et al. 2016, 2019a, b; Vasileiadou et al. 2017; Loréal et al. 2020); generally thanks to abundant, mostly unstudied remains stored in the collections of Utrecht University (the Netherlands).

3 Phylogenetic Relationships

Despite some past issues related to the monophyly of the extant clades of amphibians (Cannatella 2007), their phylogenetic relationships were recently confirmed thanks to the genetic analysis of a broad sample of species (Pyron and Wiens 2011): gymnophionans have a basal position in the tree of life of extant amphibians, and both urodelans and anurans are derived and sister taxa. Because none of the very primitive (stem) amphibian taxa have been described on the basis of Greek remains, and none of the materials found in Greece has been analysed in a phylogenetic context, the Greek fossil record does not provide significant information to the knowledge of the phylogenetic relationships of the amphibians, but, conversely is quite interesting in terms of palaeobiogeography.

4 Distribution

The knowledge of the Greek amphibians is based on materials coming from very few localities whose age does not extend in the deep past beyond the early Miocene (see Fig. 1 and Appendix). Remarkable is the presence of several taxa of urodelans

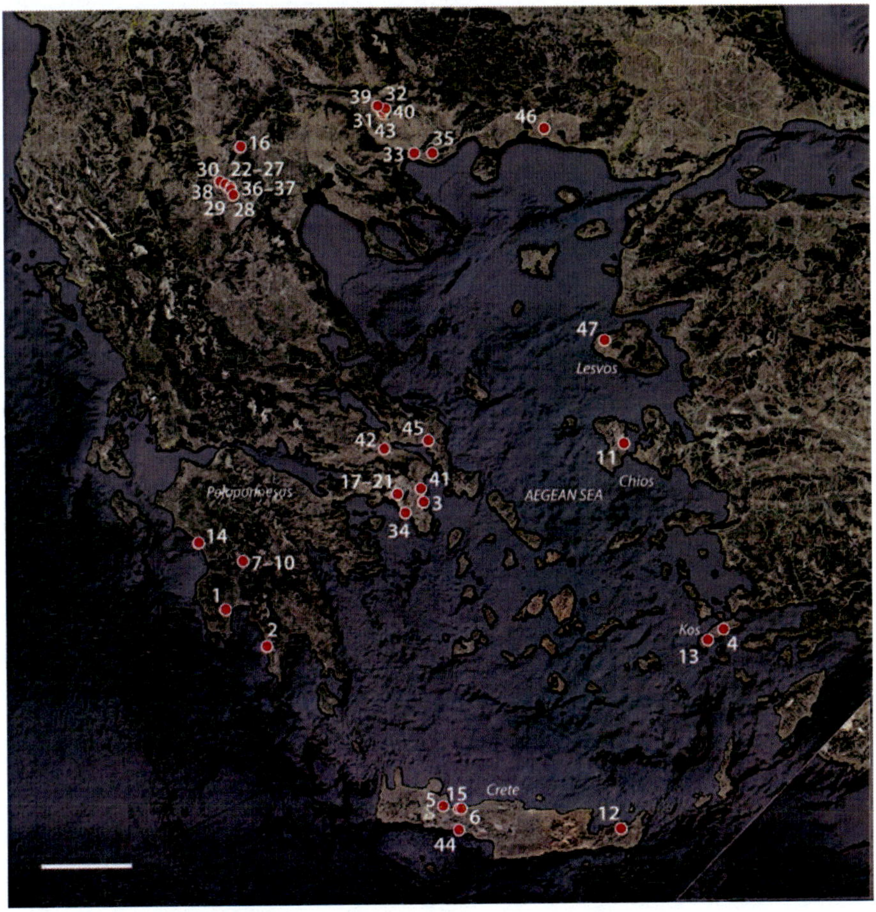

Fig. 1 Map of the occurrences of fossil amphibians from Greece. See Tables 1 and 2 for further information and text for details. **Holocene: 1**, Nichoria; **Holocene/Pleistocene**: **2**, Kalamakia, **3**, Vraona Cave; **Pleistocene: 4**, Pili B, **5**, Liko Cave, **6**, Bate Cave, **7–10**, Choremi localities, **11**, Latomi 1, **12**, Sitia 1, **13**, Laghada, **14**, Kaiafas; **15**, Mavro Mouri IV Cave, **16**, Arnissa, **17–21**, Tourkovounia localities; **Pliocene: 22–27**, Vorio localities, **28**, Notio 1, **29**, Tomea Eksi 3, **30**, Kardia, **31**, Spilia; **Pliocene/Miocene: 32**, Maramena; **Pliocene–Miocene: 33**, Rema Aslan 1, **34**, Varkiza localities; **Miocene: 35**, Rema Marmara 2, **36–37**, Tomea Eksi localities, **38**, Komanos localities, **39**, Monasteri, **40**, Ano Metochi, **41**, Pikermi 4, **42**, Biodrak C, **43**, Lefkon, **44**, Plakias, **45**, Aliveri, **46**, Karydia, **47**, Lapsarna. See Appendix for more information. Image exported from Google Earth Pro © 2019, map data from US Dept. of State Geographer, SIO, NOAA, US Navy, NGA, GEBCO, image from Landsat/Copernicus. Scale bar equals 80 km. North faces upward

in the single locality of Maramena, which, as far amphibians and reptiles are concerned, is by far the most informative locality of Greece and the richest herpetofauna of its age in the whole Europe (Georgalis et al. 2019b). The cryptobranchid remains from Maramena represent the first finding in the Balkans and significantly enlarge the range of this group of giant salamanders that are now extirpated in

Europe. Another extinct genus of urodelans found in Greece is *Mioproteus*; however, this identification should only be considered as tentative (Vasileiadou et al. 2017). The presence of *Salamandrina* and *Ommatotriton* at Maramena significantly extends the range of these two genera that are currently extirpated from Greece and the Balkans (Valakos et al. 2008; Sillero et al. 2014; Georgalis et al. 2019b; Macaluso et al. 2021). Newts of the genus *Lissotriton* have been also identified in Maramena; interestingly, in this latter locality, there have been described two species of this genus. The extant *Salamandra*, a common and widespread faunal element of the Greek current herpetofauna, has also been identified in Maramena. The most widespread, abundant and common fossil frog in Greek localities is by far *Latonia*, being recorded already in early Miocene localities (Georgalis et al. 2019a), with the genus persisting also across the latest Miocene/earliest Pliocene boundary, as it is testified by its presence in Maramena and the Ptolemais area (Georgalis et al. 2019b; Loréal et al. 2020). The extant genera *Hyla* and *Pelophylax* are known also from Maramena, whereas *Rana* is known from the Ptolemais area, attesting the presence of these three, now widespread in Greece, lineages already since the late Neogene of the region. If confirmed, the putative presence of *Discoglossus* in Crete (Kotsakis 1982) could be particularly interesting because this genus is currently absent on the island (but see below for an alternative identification).

5 Systematic Paleontology

Amphibia Linnaeus, 1758
Urodela Duméril, 1806

Remarks The earliest records of urodelans in Greece are of early Miocene age (see Appendix, Table S1). This material originates from the localities of Lapsarna (Lesvos) and Karydia (Eastern Macedonia and Thrace). Among them, the Lapsarna record has been identified to the genus level (*Mioproteus*), whereas the Karydia one cannot be more precisely determined (Vasileiadou et al. 2017; Georgalis et al. 2019a). Nevertheless, even if represented solely by an isolated tibia (Fig. 2a), the Karydia record reveals a wider distribution for Greek urodelans during the early Neogene.

Proteidae Gray, 1825
***Mioproteus* Estes and Darevsky, 1977**

Type Species *Mioproteus caucasicus* Estes and Darevsky, 1977.

?*Mioproteus* sp.

Remarks The tentative identification of *Mioproteus* in the early Miocene of the Lesvos Petrified forest is based on eight incomplete trunk vertebrae whose elongation recalls the general shape of *Mioproteus* vertebrae (Vasileiadou et al. 2017). However, the vertebrae from Lesvos do not preserve the robust posterior basapoph-

Fig. 2 Selected fossils of Urodela from Greece. **a** Left tibia (AMPG KR3 037) of Urodela indet. From the early Miocene of Karydia, in dorsal view. **b** Atlas of Cryptobranchidae (UU MAA 7441) from the latest Miocene/earliest Pliocene of Maramena, in anterior view. Scale bar equals 2 mm

yses that characterize *Mioproteus*, and therefore, further material is needed to support this identification. Proteids currently inhabit a small area of southern Europe (not in Greece) with a single species, *Proteus anguinus* Laurenti, 1768, that lives in subterranean waters (Sillero et al. 2014; Speybroeck et al. 2016). The extinct genus *Mioproteus* had a much broader range including northern Caucasus and Kazakhstan (in Europe: Austria, Czech Republic, France, Germany, Hungary, Moldova, Poland, Romania and Ukraine) and probably a different ecology as shown by other extant proteids (Venczel and Codrea 2019).

Cryptobranchidae Fitzinger, 1826
Cryptobranchidae indet.

Remarks One incomplete atlas (Fig. 2b; and possibly a second, fragmentary one) from the late Miocene/early Pliocene of Maramena has been referred to an undetermined cryptobranchid because of the relatively large size and its general morphology (Georgalis et al. 2019b). Cryptobranchid salamanders are not present anymore in Europe, but from the late Oligocene to the Miocene, they were present in the central parts of the continent (Böhme et al. 2012). Renown is the famous case of the 1 meter long fossil from southern Germany, called *Homo diluvii testis* and considered to be a man witness of the Deluge, but that later turned out to be the skeleton of an extinct giant cryptobranchid salamander, named *Andrias scheuchzeri* (Holl, 1831) (Scheuchzer 1726; Holl 1831). The finding from Maramena represents the sole known fossil record of this clade in the Mediterranean region and its southernmost distribution. Whether the Maramena material pertains to the extant genus *Andrias* Tschudi, 1837, which was abundant in the Neogene of Central Europe and has currently modern representatives living in Eastern Asia, can only be verified by the discovery of more complete material.

Salamandridae Goldfuss, 1820
***Salamandrina* Fitzinger, 1826**

Type Species *Salamandra perspicillata* Savi, 1821.

***Salamandrina* sp.**

Remarks Two tiny urodelan vertebrae characterized, among others, by a sort of zygosphene and by a forked neural spine, testify for the presence of a spectacled salamander *Salamandrina* in the late Miocene/early Pliocene of Maramena (Georgalis et al. 2019b; Fig. 3a). This genus is currently endemic of the Italian Peninsula (Sillero et al. 2014), but the fossil records testify for a former broader range extending at least from Spain to Hungary, and including also Sardinia (Italy) and Germany (Macaluso et al. 2021). The record from Maramena represents the easternmost occurrence of the genus.

***Lissotriton* Bell, 1839**

Type Species *Salamandra punctata* Latreille, 1800.

***Lissotriton vulgaris* (Linnaeus, 1758)**
***Lissotriton* sp. (*Lissotriton vulgaris* group) and *Lissotriton* sp.**

Remarks The identification of this genus at Maramena is based on very few vertebrae some of which could belong to the *Lissotriton vulgaris* group because of their relatively large size, the presence of small subcentral foramina, and the high neural spine and neural arch (Georgalis et al. 2019b; Fig. 3b). Besides this form, in Maramena, there was identified also a second species of the genus that had some overall resemblance with the extinct *Lissotriton rohrsi* (Herre, 1955) from the Miocene of Central Europe (Herre 1955; Ivanov 2008). The newts of the *Lissotriton vulgaris* group have a broad range in Europe that includes also Greece (Valakos et al. 2008; Sillero et al. 2014; Speybroeck et al. 2016).

***Ommatotriton* Gray, 1850**

Type Species *Triton vittatus* Gray in Jenyns, 1835.

***Ommatotriton* sp.**

Remarks The genus *Ommatotriton* is represented at Maramena by vertebral and appendicular remains (Georgalis et al. 2019b; Fig. 3c). The vertebrae are characterized by a triangular ventral lamina, two medium-sized subcentral foramina, a low and bifurcated neural arch, and a low neural spine which is slightly higher than the neural arch. The femur morphology resembles that of *Ommatotriton*, in terms of general shape and size, as well as the size and dimensions of the trochanteric crest and intertrochanteric groove and therefore tentatively referred to this taxon. The presence of *Ommatotriton* in Greece is quite relevant due to the fact that this newt is currently distributed in the Middle East, eastern Anatolia and the Caucasus (van Riemsdijk et al. 2017). At the same time, being the first reliable find

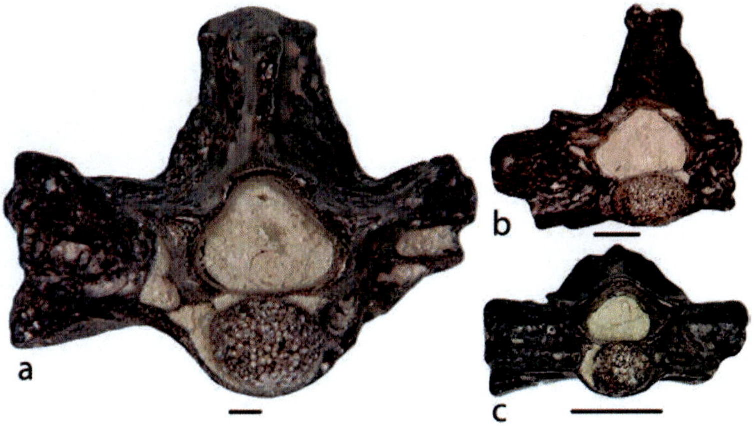

Fig. 3 Selected specimens of trunk vertebrae of salamanders from the latest Miocene/earliest Pliocene of Maramena, all in anterior view. **a** *Salamandrina* sp. (UU MAA 7483), **b** *Lissotriton* sp. (UU MAA 7517), and **c** *Ommatotriton* sp. (UU MAA 7489). Scale bars equal 1 mm

of *Ommatotriton* in the fossil record, it extends this lineage back to the latest Miocene/earliest Pliocene (Georgalis et al. 2019b).

***Salamandra* Garsault, 1764**

Type Species *Salamandra terrestris* Bonnaterre, 1789.

***Salamandra* sp.**

Remarks Twelve trunk and six caudal vertebrae, once again from Maramena, testify for the presence of an unknown member of the genus *Salamandra* (Georgalis et al. 2019b; Fig. 4). The referral of the material to this genus is based on the dorsoventrally flattened centrum, and low neural spine and neural arch. It is not possible to distinguish this material from both extant and fossil specimens of the recent species *Salamandra salamandra* (Linnaeus, 1758), as well as from the extinct taxon *Salamandra sansaniensis* Lartet, 1851. Currently, six extant species of the genus *Salamandra* are known in Europe, Northwest Africa and Middle East (Sparreboom 2014), and their comparative osteology is nearly totally unexplored (see Villa et al. 2014), a fact that hinders the identification of isolated vertebra fossils at specific rank.

Anura Fischer, 1813
Alytidae Fitzinger, 1843
***Latonia* Meyer, 1843**

Type Species *Latonia seyfriedi* Meyer, 1843.

***Latonia gigantea* (Lartet, 1851)**
***Latonia* cf. *gigantea* and *Latonia* sp.**

Fig. 4 A trunk vertebra of *Salamandra* sp. (UU MAA 7495) from the latest Miocene/earliest Pliocene of Maramena, in lateral view. Scale bar equals 2 mm

Remarks Anuran remains referred to *Latonia* have been described or at least listed for a number of Greek localities whose age ranges from the early Miocene to the early Pliocene (Fig. 5a–d). They have been formally described from the early Miocene of Aliveri (Georgalis et al. 2019a) and Karydia (Georgalis et al. 2019a), as well as the latest Miocene/earliest Pliocene of Maramena (Georgalis et al. 2019b) and the Ptolemais area (Loréal et al. 2020). Nevertheless, a number of as yet unpublished reports of the presence of this genus in various Greek localities was listed by Sanchiz (1998): Biodrak C, Lefkon, Ano Metochi, Monasteri, Pikermi 4, Rema Marmara 2, Rema Aslan 1, and Spilia. A fragmentary vertebra and a partial tibiofibula from the earliest late Miocene (early Tortonian, MN9) of Plakias in Crete have been tentatively referred to alytids (as cf. Alytidae indet.; Georgalis et al. 2016) and could well belong to *Latonia*, pending the retrieval of more informative remains in the same locality.

The identification of fossils of this frog is relatively easy, being their morphology very similar to that of the only extant species, *Latonia nigriventer* (Mendelssohn and Steinitz 1943), that now lives only in a small area in Israel (Biton et al. 2013). Fossil specimens of extinct species of *Latonia* are, however, much larger in size than those of the extant *L. nigriventer* and can have a much more marked cranial ornamentation. The material from Maramena is particularly abundant, but because of the absence of dermal ornamentation on the maxillae and since the comparative diagnosis of the several species of *Latonia* is hindered by the fact that some of them are currently represented only by specimens preserved on slabs, whereas others solely by disarticulated elements, the fossil material has been referred only at genus level.

To the contrary, the remains, likewise abundant, from Karydia have been tentatively referred to non-adult specimens of *Latonia gigantea* (as *Latonia* cf. *gigantea*) because the general morphology of the frontoparietal is comparable with early ontogenetic stages of that species (Georgalis et al. 2019a; Fig. 5e). The Karydia specimens remain so far the only *Latonia* material from Greece that can be (at least tentatively) referred to the species level (Georgalis et al. 2019b).

Latonia is known to have achieved a rather broad geographic distribution across various late Palaeogene, Neogene and early Quaternary localities of Europe and Anatolia (Roček 1994, 2013; Vasilyan et al. 2014; Villa et al. 2019; Syromyatnikova et al. 2019a, 2019b) being its last occurrence datum in Europe from the Early

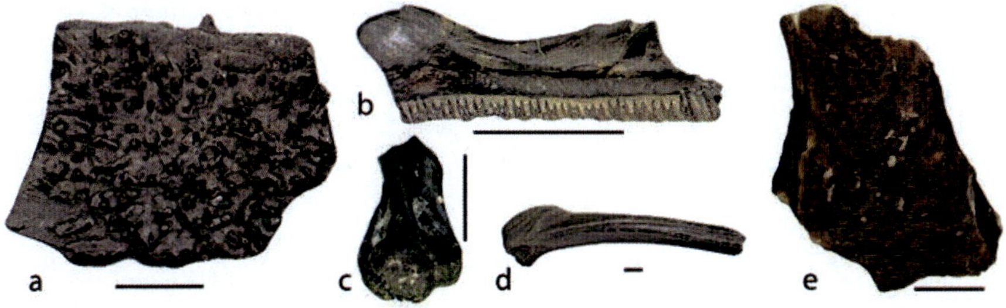

Fig. 5 Selected specimens of *Latonia* from Greece. **a–d** *Latonia* sp. from various localities: **a** a frontoparietal fragment (UU VOR3A 1518), in dorsal view, from the early Pliocene of Vorio 3A (scale bar equals 1 mm); **b** a left maxilla (UU MAA 7227) in medial view from the latest Miocene/earliest Pliocene of Maramena (scale bar equals 10 mm); **c** a right humerus (UU MAA 7225) in ventral view from the latest Miocene/earliest Pliocene of Maramena (scale bar equals 5 mm); and **d** a right ilium (UU VOR3 1518) in lateral view from the early Pliocene of Vorio 3 (scale bar equals 1 mm). **e** A frontoparietal of *Latonia* cf. *gigantea* (AMPG KR3 038) from the early Miocene of Karydia, in dorsal view (scale bar equals 1 mm)

Pleistocene of Pietrafitta in central Italy (Sorbelli et al. in press). Noteworthy is that according to Syromyatnikova et al. (2019b), *L. gigantea* could be a junior synonym of *L. seyfriedi*.

***Discoglossus* Otth, 1837**
***Discoglossus* sp.**

Type Species *Discoglossus pictus* Otth, 1837.

Remarks Kotsakis (1982) reported the presence of *Discoglossus* in the Early Pleistocene of Sitia (a site located on the road from Sitia to Psilokefalo; Aghios Nikolaos, Crete) on the basis of an isolated right humerus characterized, among others, by a moderate length (about 15 mm) and a paraventral crest. The record has been later cited by several authors (see e.g., Holman 1998; Sanchiz 1998), but since *Discoglossus* does not belong to the extant herpetofauna of the Balkans and Crete (Valakos et al. 2008; Sillero et al. 2014) and since another alytid, *Latonia* (rather common in the Greek fossil record, see above), has a humeral morphology that is very close to that of *Discoglossus*, the proper identification of this material should be re-evaluated. Worth mentioning is that the moderate size of the humerus is not an argument for its taxonomic identification, because, even if some extinct species of *Latonia* were very large in size, the extant *L. nigriventer* is actually rather small and its humerus has a size comparable to that of the humerus from Crete (see Biton et al. 2013, 2016).

Hylidae Rafinesque, 1815
***Hyla* Laurenti, 1768**

Type Species *Hyla viridis* Laurenti, 1768.

***Hyla* sp.**

Remarks Several localities (Ano Metochi, Kaiafas, Maramena, Pikermi, Spilia), whose age ranges from late Miocene to Pleistocene, yielded anuran fossils that have been referred to tree frogs of the genus *Hyla* (Sanchiz 1998; Georgalis et al. 2019b; Villa et al. 2020). However, actually, only those from Maramena have been fully described and figured (Georgalis et al. 2019b), with all the rest of occurrences having been simply listed by Sanchiz (1998) in the reference work 'Handbuch der Palaoherpetologie'. The fossils from Maramena comprise an ilium (Fig. 6a) and 10 humeri that show the typical morphology of the tree frogs. In particular, the ilium has a broad preacetabular zone and the humeri have a displaced distal articular surface. Since it is not possible to identify at species level the numerous (> 35; Frost 2020) extant *Hyla* species on the basis of their osteology, the fossil remains are referred only at genus level, even if it is clear that the morphology of the fossils is congruent with the European (and perhaps Asian) species. That is why some remains have been originally referred to the group named after *Hyla arborea* (Linnaeus, 1758), the species that formerly included several lineages that now have their own specific status. *Hyla arborea* and *Hyla orientalis* Bedriaga, 1890 are now present in Greece (Frost 2020).

Ranidae Batsch, 1796
Pelophylax Fitzinger, 1843

Type Species *Rana esculenta* Linnaeus, 1758.

***Pelophylax bedriagae* (Camerano, 1881), *Pelophylax cretensis* (Beerli, Hotz, Tunner, Heppich and Uzzell, 1994) and *Pelophylax ridibundus* (Pallas, 1771)**

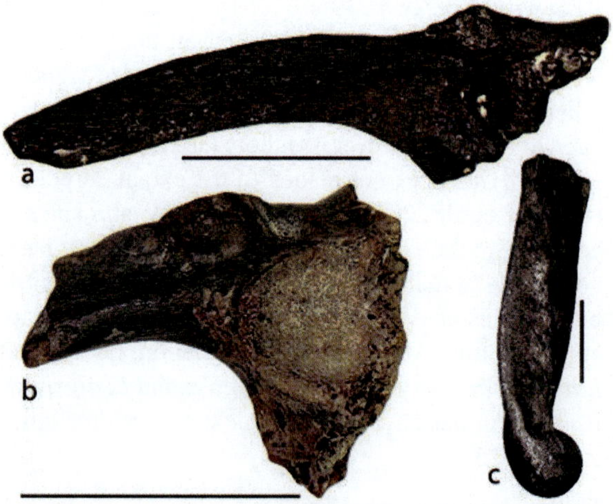

Fig. 6 Select specimens of Hylidae and Ranidae from Greece. **a** A left ilium of *Hyla* sp. (UU MAA 7118) in lateral view from the latest Miocene/earliest Pliocene of Maramena (scale bar equals 2 mm), **b** a left ilium of *Pelophylax* sp. (UU MAA 7232) in lateral view from the latest Miocene/earliest Pliocene of Maramena (scale bar equals 5 mm), and **c** a left humerus of *Rana* sp. (UU NO1 1003) in medial view from the early Pliocene of Notio 1 (scale bar equals 1 mm)

Remarks Green frogs of the genus *Pelophylax* have a rather uniform skeletal morphology that hinders the identification at species level of their fossils. Their osteology, with few exceptions (Böhme 1977; Böhme & Günther 1979; Bailon 1999; Blain et al. 2015), is also poorly known and generally understudied. Therefore, some specific identification of green frogs remains coming from Greek sites (see Appendix, Table S2) was likely based more (or only) on geographic provenance than on morphology. It is the case, for example, of *Pelophylax* cf. *cretensis* from Sitia 1 and 2 (Crete; Sanchiz 1984, 1998; Holman 1998), as that is the only native green frog species of Crete (Sillero et al. 2014). Other remains were referred to both *Pelophylax ridibundus* (in most of the cases as *Pelophylax* cf. *ridibundus*) and *Pelophylax* cf. *bedriagae* (both currently present in Greece; Sillero et al. 2014), but all these remains should be re-described and their identification re-evaluated after a thorough study of the osteology of all extant species. Continental green frogs reported in Table S2 are therefore referred to *Pelophylax* gr. *ridibundus*. Remains of *Pelophylax* sp. have also been found in the diverse herpetofauna of Maramena (Fig. 6b).

***Rana* Linnaeus, 1758**

Type Species *Rana temporaria* Linnaeus, 1758.

***Rana graeca* Boulenger, 1891**

Remarks This species, which is currently present in Greece (Sillero et al. 2014), has been cited from two localities, the Holocene archaeological site of Nichoria (*R. graeca*; Sloan and Duncan 1978; Sanchiz 1998) and the Pleistocene locality of Kaiafas, where the material has been tentatively identified as *Rana* cf. *graeca* (Holman 1998; Sanchiz 1998). Since the osteology of this species is nearly unknown (Picariello et al. 1996) and the identifications of the fossils are not accompanied by descriptions and figures, these data should be considered with caution pending further confirmation.

***Rana dalmatina* Fitzinger, 1838**

Remarks Rauscher (1995) listed the presence of *Rana dalmatina* in the late Quaternary of Vraona Cave on the basis of two ilia and few other less diagnostic elements; however, he provided no figure of this material. Even if the ilium can be a diagnostic element for brown frogs (Bailon 1999), this specific identification should be confirmed by full description (and desirably a figure). The agile frog is currently rare but present in small areas of continental Greece (Sillero et al. 2014).

***Rana* sp.**

Remarks Material that was recently described by Loréal et al. (2020) from the early Pliocene of the area of Ptolemais represents the oldest occurrence of the genus *Rana* in the Greek area (Fig. 6c). Nevertheless, this material is not diagnostic to the species level, thus not affording any further insight about its precise affinities. Worth mentioning is that Roger & Darlas (2008) listed *Rana* sp. from the Late Pleistocene layers of the Palaeolithic site of Kalamakia, but, in the total absence of a formal description and any figure, it is not clear if they referred the material (not reported in detail) to *Rana* s.s. or to *Pelophylax* sp.

Pelobatidae Bonaparte, 1850
***Pelobates* Wagler, 1830**

Type Species *Bufo fuscus* Laurenti, 1768.

***Pelobates syriacus* Boettger, 1889**
***Pelobates* cf. *syriacus* and *Pelobates* sp.**

Remarks *Pelobates syriacus* is the only spadefoot species currently inhabiting Greece (Valakos et al. 2008; Sillero et al. 2014). Spadefoot fossil remains have been reported for eight sites whose age ranges from the late Miocene to the Late Pleistocene (see Appendix, Table S2), and most of them have been tentatively referred to this species as *Pelobates* cf. *syriacus* (see Schneider 1975; Sanchiz 1984, 1998; Holman 1998). The descriptions and drawings of *Pelobates* cf. *syriacus* from Pili 2 (Kos Island) provided by Sanchiz (1984) are the only chance to check the correctness of the identification of this species based on Greek materials—we agree with such taxonomic arrangement, and hence, Kos remains so far the only area in Greece that yielded fossil remains of *Pelobates* with certainty.

Bufonidae Gray, 1825
***Bufo* Garsault, 1764**

Type Species *Rana bufo* Linnaeus, 1758.

***Bufo* sp.**

Remarks Fossil remains of this species have been so far reported from three Late Pleistocene–Holocene localities (Kalamakia, Nichoria, Vraona Cave; Sloan & Duncan 1978; Rauscher 1995; Sanchiz 1998). The identification of *Bufo* at Vraona Cave (based on a single tibiofibula) is not supported by a description or a figure and therefore should be considered with caution (Rauscher 1995). This species is currently found throughout the mainland but only on few of the main islands (Valakos et al. 2008).

***Bufotes* Rafinesque, 1815**

Type Species *Bufo vulgaris* Laurenti, 1768.

***Bufotes viridis* (Laurenti, 1768)**

Remarks Green toads are a speciose group of anurans (Frost 2020) whose external morphology is rather uniform, and the skeletal morphology, never studied in detail, is likely very similar if not identical. Fossil remains of this group of toads are generally very common in Mediterranean palaeontological and archaeological sites, and they are usually referred to *Bufotes viridis* (formerly *Bufo viridis*), even if they do not necessarily belong to this species. At the moment, a proper identification of osteological (fossil) material should be therefore *Bufotes* gr. *viridis* (and this name has been applied in Table S2). Greece, which is currently inhabited by *Bufotes viridis* only (Frost 2020), is not an exception, and green toads have been reported from about 10 fossil sites (see Table S2) whose age ranges from the late Miocene to the Holocene. The fossil material has been referred to *Bufotes viridis* (in some cases as *Bufotes* cf. *viridis*), but it is not unlikely that other cryptic species, extinct or extant, could be represented considering that at present at least four species of green toads inhabit the Mediterranean area (see Dufresnes et al. 2019) and they are not identifiable on the basis of their skeleton.

Acknowledgments MD thanks C Doukas and the Department of Historical Geology and Paleontology at the National and Kapodistrian University of Athens for having hosted him for a research stay in 2003–2004 (Italian CNR grant no. 203.22-26.02.02) that later on, with the precious collaboration of many students and colleagues, gave origin to a series of papers that contributed to the knowledge of the evolutionary history of the Greek herpetofaunas. GLG acknowledges support from Forschungskredit of the University of Zurich, grant no. [FK-20-110]. Elena Syromyatnikova and Andrea Villa critically commented and improved this chapter.

Appendix

List of the Greek localities containing amphibian fossils: urodelans in Table S1 and anurans in Table S2. Locality numbers refer to the collection numbers of the PaleoBiology Database (PBDB).

Table S1 List of the Greek localities containing urodelan fossils

Locality[PBDB No]	Age (MN)	Taxa
Kaiafas[34766]	Early Pleistocene	Salamandridae indet. [1]
Maramena[32189]	Late Miocene/early Pliocene (MN13/14)	Cryptobranchidae indet.[2] *Salamandrina* sp.[2] *Lissotriton* sp. [*Lissotriton vulgaris* group][2] *Lissotriton* sp.[2] *Ommatotriton* sp.[2] *Salamandra* sp.[2]
Karydia[69435]	Early Miocene (MN4a)	Urodela indet.[3]
Lapsarna[186558]	Early Miocene (?MN3)	?*Mioproteus* sp.[4]

MN Mammal Neogene zone
[1]Villa et al. 2020, [2]Georgalis et al. 2019b, [3]Georgalis et al. 2019a, [4]Vasileiadou et al. 2017

Table S2 List of the Greek localities containing anuran fossils

Locality[PBDB No]	Age (MN)	Taxa
Nichoria[207978]	Holocene	*Rana graeca*[1–2] *Bufo bufo*[1–2]
Kalamakia[184245]	Late Pleistocene–Holocene	*Rana* sp.[3] *Bufo bufo*[3]
Vraona Cave[183130]	Late Pleistocene–Holocene	*Rana dalmatina*[4] *Bufo bufo*[4] *Bufotes* gr. *viridis*[4]
Pili B[204953]	Late Quaternary (probably Late Pleistocene)	*Pelobates syriacus*[2,5–6] *Pelophylax* cf. *bedriagae*[2,5–6] *Bufotes* gr. *viridis*[2,5–6]
Liko Cave B[192951]	Late Pleistocene	*Pelophylax* gr. *P. ridibundus*[2,7] *Bufo* sp. or *Bufotes* sp.[2,7]
Bate Cave[183122]	Late Pleistocene	*Bufotes* gr. *viridis*[2,7–11]
Choremi 1–4[200078–82]	Middle Pleistocene	*Pelophylax* gr. *ridibundus*[12]
Latomi 1[204077]	Middle Pleistocene	*Pelobates* sp.[2,6–7,10,12–13] *Pelophylax* cf. *bedriagae*[2,6–7,10,12–13] *Bufotes* gr. *viridis*[2,6–7,10,12–13]
Sitia 1[202328]	Early–Middle Pleistocene	*Discoglossus* sp. (or *Latonia* sp.?) [2,7,9] *Pelophylax* gr. *ridibundus*[2,7,9]
Laghada B[205041]	Early Pleistocene	*Rana* sp.[2,5]
Kaiafas[34766]	Early Pleistocene	*Hyla* gr. *arborea*[2,7,14] *Rana* cf. *graeca*[2,7] *Bufo* sp. or *Bufotes* sp.[2,7]
Mavro Mouri IV[192442]	Pleistocene	*Bufotes* gr. *viridis*[2,7,15]
Arnissa[204957]	Pleistocene	*Pelobates* sp.[2,7] *Pelophylax* gr. *ridibundus*[2,7] *Bufotes* gr. *viridis*[2,7]
Tourkovounia 1, 2, 3, 4, 5[34589, 34592, 34767]	Late Pliocene to Middle Pleistocene	*Pelobates* cf. *syriacus*[2,7] *Pelophylax* gr. *ridibundus*[2,7] *Bufotes* gr. *viridis*[2,7]
Vorio 8[207946]	Early Pliocene (MN15)	*Latonia* sp.[16]
Vorio 4[207947]	Early Pliocene (MN15)	*Latonia* sp.[16] Ranidae indet.[16] Anura indet.[16]
Vorio 3A[205086]	Early Pliocene (MN15)	*Latonia* sp.[16] *Rana* sp.[16] Ranidae indet.[16] Anura indet.[16]
Vorio 3[205086]	Early Pliocene (MN15)	*Latonia* sp.[16] Anura indet.[16]
Spilia-4[31907]	Early Pliocene (MN15)	*Hyla* sp.[2] *Pelobates* cf. *syriacus*[2] *Pelophylax* gr. *ridibundus*[2] *Bufotes* gr. *viridis*[2]

(continued)

Table S2 (continued)

Locality[PBDB No]	Age (MN)	Taxa
Tomea Eksi 3 (= Ptolemais 3)[205088]	Early Pliocene (MN15)	*Latonia* sp.[16] Ranidae indet.[16] Anura indet.[16]
Spilia[31229]	Early Pliocene (MN15)	*Latonia* sp.[2]
Notio 1[205081]	Early Pliocene (MN15)	*Latonia* sp.[16] *Rana* sp.[15] Ranidae indet.[16] Anura indet.[16]
Vorio 2[207948]	Early Pliocene (MN14)	*Latonia* sp.[16] Ranidae indet.[16] Anura indet.[16]
Vorio 1[205087]	Early Pliocene (MN14)	*Latonia* sp.[16] Ranidae indet.[16]
Kardia[35087]	Early Pliocene (MN14)	*Pelophylax* gr. *ridibundus*[2]
Trilofon	Miocene–Pliocene	cf. *Rana* sp.[2]
Varkiza 1[204166] and 2[204167]	Miocene–Pliocene	*Bufotes* gr. *viridis*[2]
Rema Aslan 1[205075]	Miocene–Pliocene	cf. *Latonia* sp.[2]
Maramena[32189]	Late Miocene/early Pliocene (MN13/14)	*Latonia* sp.[17] *Hyla* sp.[17] *Pelophylax* sp.[17]
Rema Marmara 2[32190]	Late Miocene (MN13)	*Pelobates* sp.[2] *Latonia* sp.[2]
Tomea Eksi 2[205089]	Late Miocene (MN13)	*Latonia* sp.[16] *Rana* sp.[16] Anura indet.[16]
Tomea Eksi 1 (= Ptolemais 1)[205090]	Late Miocene (MN13)	*Latonia* sp.[6] Ranidae indet.[16] Anura indet.[16]
Komanos 1[205083–4]	Late Miocene (MN13)	*Latonia* sp.[16]
Monasteri[31940]	Late Miocene (MN13)	*Latonia* sp.[2] *Pelobates* cf. *syriacus*[2]
Ano Metochi[31924, 31928]	Late Miocene (MN13)	*Latonia* sp.[2] *Hyla* sp.[2] *Pelobates* sp.[2] *Pelophylax* gr. *ridibundus*[2] *Bufotes* gr. *viridis*[2]
Pikermi 4[207979]	Late Miocene (MN12/13)	*Latonia* sp.[2] *Hyla* sp.[2]
Biodrak C[205342]	Late Miocene (MN10)	*Latonia* sp.[2] *Pelophylax* sp. or *Rana* sp.[2]
Lefkon[31223]	Late Miocene (MN10)	*Latonia* sp.[2] *Pelophylax* gr. *P. ridibundus*[2]
Plakias[183151]	Late Miocene (MN9)	cf. Alytidae indet.[18]
Aliveri[68032]	Early Miocene (MN4a)	cf. *Latonia* sp.[19]
Karydia[69435]	Early Miocene (MN4a)	*Latonia* cf. *gigantea*[19]

MN Mammal Neogene zone

[1]Sloan and Duncan 1978, [2]Sanchiz 1998, [3]Roger and Darlas 2008, [4]Rauscher 1995, [5]Sanchiz 1984, [6]Kotsakis 1991, [7]Holman 1998, [8]Kotsakis 1977, [9]Kotsakis 1982, [10]Kotsakis 1991, [11]Kotsakis et al. 1979, [12]Vugt et al. 2000, [13]Schneider 1975, [14]Villa et al. 2020, [15]Caloi et al. 1988, [16]Loréal et al. 2020, [17]Georgalis et al. 2019b, [18]Georgalis et al. 2016, [19]Georgalis et al. 2019a

References

Ascarrunz E, Rage J-C, Legreneur P, Laurin M (2016) *Triadobatrachus massinoti*, the earliest known lissamphibian (Vertebrata: Tetrapoda) re-examined by µCT scan, and the evolution of trunk length in batrachians. Contrib Zool 85:201–234

Bailon S (1999) Différenciation ostéologique des anoures (Amphibia, Anura) de France. In: DesseJ, Desse-Berse tN (eds) Fiches d'ostéologie animale pour l'Archéologie, Série C: Varia. Centre de Recherches Archéologiques-CNRS, Valbonne, 41 pp

Bedriaga von J (1890) Die Lurchfauna Europa's. I. Anura. Froschlurche. Bull Soc Impériale Naturalistes Moscou, Nouv sér 3:466–622

Beerli P, Hotz H, Tunner HG, Heppich S, Uzzell T (1994) Two new water frog species from the Aegean Islands Crete and Karpathos (Amphibia, Salientia, Ranidae). Notulae Naturae Acad Nat Sci Philadelphia 470:1–9

Bell T (1839) A history of British reptiles. John van Voorst, London

Biton R, Geffen E, Vences M, Cohen O, Bailon S, Rabinovich R, Malka Y, Oron T, Boistel R, Brumfeld V, Gafny S (2013) The rediscovered *Hula* painted frog is a living fossil. Nat Commun 2959:1–6

Biton R, Boistel R, Rabinovich R, Gafny S, Brumfeld V, Bailon S (2016) Osteological observations on the alytid anura *Latonia nigriventer* with comments on functional morphology, biogeography, and evolutionary history. J Morphol 277:1131–1145

Blain H-A, Lózano-Fernández I, Böhme G (2015) Variation in the ilium of central European water frogs *Pelophylax* (Amphibia, Ranidae) and its implications for species-level identification of fragmentary anuran fossils. Zool Stud 54:5

Boettger O (1889) Ein neuer *Pelobates* aus Syrien. Zool Anz 12:144–147

Böhme G (1977) Zur Bestimmung quartärer Anuren Europas an Hand von Skelettementen. Wiss Z Humboldt Univ Berlin, Math-Nat 3:283–300

Böhme G, Günther R (1979) Osteological studies in the European water frogs *Rana ridibunda*, *Rana lessonae* and *Rana "esculenta"* (Anura, Ranidae). Mitt Zool Mus Berlin 55:203–215

Böhme M, Vasilyan D, Winklhofer M (2012) Habitat tracking, range dynamics and palaeoclimatic significance of Eurasian giant salamanders (Cryptobranchidae) – indications for elevated Central Asian humidity during Cenozoic global warm periods. Palaeogeogr Palaeoclimatol Palaeoecol 342–343:64–72

Boulenger GA (1891) Description of a new European frog. Ann Mag Nat Hist 8:344–353

Caloi L, Kotsakis T, Palombo MR (1988) La fauna a vertebrati terrestri del Pleistocene delle isole del Mediterraneo. Bull Écol 19:131–151

Camerano L (1881) Recherches sur les variations de la *Rana esculenta* et du *Bufo viridis* dans le Bassin de la Méditerranée. C R Assoc Fr Avancement Sci 10:680–690

Cannatella DC (2007) An integrative phylogeny of amphibia. In: Narins PM, Feng AS, Fay RR, Popper AN (eds) Hearing and sound communication in amphibians. Springer, New York, pp 13–43

Dufresnes C, Mazepa G, Jablonski D, Oliveira RC, Wenseleers T, Shabanov DT, Auer M, Ernst R, Koch C, Ramírez-Chaves HE, Mulder KP, Simonov E, Tiutenko A, Kryvokhyzha D, Wennekes PL, Zinenko OI, Korshunov OV, Al-Johany AM, Peregontsev EA, Masroor R, Betto-Colliard C, Denoël M, Borkin LJ, Skorinov DV, Pasynkova RA, Mazanaeva LF, Rosanov JM, Dubey S, Litvinchuk S (2019) Fifteen shades of green: the evolution of *Bufotes* toads revisited. Mol Phylogenet Evol 141:106615

Estes R, Darevsky I (1977) Fossil amphibians from the Miocene of the North Caucasus, USSR. J Paleontol Soc India 20:164–169

Estes R (1981) Gymnophiona, Caudata. Handbuch der Paläoherpetologie, part 2. Gustav Fischer Verlag, Stuttgart/New York

Fitzinger LJFJ (1826) Neue Classification der Reptilien nach ihren natürlichen Verwandtschaften nebsteiner Verwandschafts-Tafel und einemVerzeichnisse der Reptilien-Sammlung des K. K. Zoologischen Museums zu Wien. JG Heubner, Wien

Fitzinger LJFJ (1843) Systema Reptilium. Fasciculus Primus. Braumüller et Seidel, Wien

Frost DR (2020) Amphibian Species of the World: an online reference. Version 6.0 http://research.amnh.org/herpetology/amphibia/index.html.doi.org/10.5531/db.vz.0001. Accessed 30 Jan 2020

Garsault de FA (1764) Les Figures des Plantes et Animaux d'Usage en Medecine, Décrits dans la Matiere Medicale de Mr. Geoffroy Medecin, Dessinés d'après Nature. Chez l'auteur, Paris

Georgalis GL, Villa A, Vlachos E, Delfino M (2016) Fossil amphibians and reptiles from Plakias, Crete: a glimpse into the earliest late Miocene herpetofaunas of southeastern Europe. Geobios 49:433–444

Georgalis GL, Villa A, Ivanov M, Roussiakis S, Skandalos P, Delfino M (2019a) Early Miocene herpetofaunas from the Greek localities of Aliveri and Karydia – bridging a gap in the knowledge of amphibians and reptiles from the early Neogene of southeastern Europe. Hist Biol 31:1045–1064

Georgalis GL, Villa A, Ivanov M, Vasilyan D, Delfino M (2019b) Fossil amphibians and reptiles from the Neogene locality of Maramena (Greece), the most diverse European herpetofauna at the Miocene/Pliocene transition boundary. Palaeontol Electronica 22.3.68:1–99

Gray JE (1850) Catalogue of the specimens of Amphibia in the collection of the British Museum. Part II. Batrachia Gradientia, etc. British Museum (Natural History), London

Herre W (1955) Die fauna der Miozänens palten füllung von Neudorf a.D. March (Črs): Amphibia (Urodela). Österreichische Akad Wiss, Math-Naturwiss Kl Abt I, Sitzungsber 164:783–803

Holl F (1831) Handbuch der Petrefaktenkunde. Dresden

Holman JA (1998) Pleistocene amphibians and reptiles in Britain and Europe. Oxford monographs on geology and geophysics no. 38. Oxford University Press, New York

Ivanov M (2008) Early Miocene amphibians (Caudata, Salientia) from the Mokrá-Western Quarry (Czech Republic) with comments on the evolution of early Miocene amphibian assemblages in Central Europe. Geobios 41:465–492

Jenyns L (1835) A manual of British vertebrate animals. Pitt Press, Cambridge

Kotsakis T (1977) I resti di Anfibi e Rettili pleistocenici della grotta << Bate >> (Rethymnon, Creta). Rend Accad Naz Lincei 63:571–582

Kotsakis T (1982) Presence du genre *Discoglossus* Otth (Discoglossidae, Anura, Amphibia) dans le Villafranchien de l'ile de Crete. Geol Romana 21:185–189

Kotsakis T (1986) Vertebrati insulari e paleogeografia: alcuni e sempi. Boll Soc Paleontol It 24:225–244

Kotsakis T (1991) Insular and non insular vertebrate fossil faunas in the eastern Mediterranean islands. Accad Naz Lincei, Atti Convegni Lincei 85:289–334

Kotsakis T, Petronio C, Sirna G (1979) The quaternary vertebrates of the Aegean Islands: palaeogeographical implications. Ann Géol Pays Helléniques 30:31–64

Lartet E (1851) Notice sur la Colline de Sansan, Suivie d'une Récapitulation des Diverses Espèces d'Animaux Vertébrés Fossiles, Trouvéssoit à Sansan, soit dans d'autres Gisements du Terrain TertiaireMiocène dans le Bassin Sous-Pyrénéen. JA Portes, Auch

Latreille PA (1800) "An. VIII". Histoire naturelle des salamandres de France, Précédée d'un Tableau Méthodique des Autres Reptiles Indigènes. Imprimerie de Crapelet, Paris

Laurenti JN (1768) Austriaci Viennensis specimen medicum, exhibens synops in reptilium emendatam cum experimentis circa Venena et antidota reptilium Austriacorum quod authoritate et consensu. Joannis Thomae de Trattnern, Caes. Reg. Maj. Aulae Typographi et Bibliopolae, Wien

Linnaeus C (1758) Systema Naturae per regna trianaturae, secundum classes, ordines, genera, species, cum characteribus, differentiis, synonymis, locis. Laurentii Salvii, Stockholm

Loréal E, Villa A, Georgalis GL, Delfino M (2020) Amphibians and reptiles from the late Miocene and early Pliocene of the Ptolemais area (Western Macedonia, Greece). Ann Paléontol 106:102407

Macaluso L, Villa A, Pitruzzella G, Rook L, Carnevale G, Delfino M (2021) A progressive extirpation: an overview of the fossil record of Salamandrina (Salamandridae, Urodela). Hist Biol https://doi.org/10.1080/08912963.2021.1888946

Mendelssohn H, Steinitz H (1943) A new frog from Palestine. Copeia 1943:231–233

Meyer von H (1843) Mittheilungen an professor Bronn gerichtet. N Jahrb Geol Paläontol, Geognas, Geol, Petrefactenkunde 1843:579–590

Otth A (1837) Beschreibun gainer neuen europäischen Froschgattung, *Discoglossus*. N Denkschriften Allgemeinen Schweizerischen Ges Gesammten Naturwiss 1:1–8

Pallas PS (1771) Reisedurchverschiedene Provinzen des Russischen Reichs. Theil 1. Gedruckt bey der Kayserlichen Academie der Wissenschaften, St. Pétersbourg

Picariello O, Scillitani G, Usai I (1996) Skeletal morphology of *Rana italica* Dubois, 1987 and *Rana graeca* Boulenger, 1891 (Amphibia, Ranidae) and its systematic implications. Preliminary results. Stud Trentini Sci Nat Acta Biol 71:223–231

Piveteau J (1936) Une forme ancestrale des amphibiens anoures dans le Trias inférieur de Madagascar. C R Hebd Séances Acad Sci 202:1607–1608

Pyron RA, Wiens JJ (2011) A large-scale phylogeny of Amphibia including over 2800 species, and a revised classification of extant frogs, salamanders, and caecilians. Mol Phylogenet Evol 61(2):543–583

Rafinesque CS (1815) Analyse de la Nature ou Tableau de l'Univers et des Corps Organisés. Jean Barravecchia, Palermo

Rauscher KL (1995) Die herpetofauna der Vraona-Höhle (Attika) in Griechenland. In: Symeonidis N, Rabeder G (eds) Das Jungpleistozan in der <<Höhle>> von Vraona auf Attica in Griechenland. Ann Géol Pays Helléniques 36:39–41

Roček Z (1994) Taxonomy and distribution of Tertiary discoglossids (Anura) of the genus *Latonia* v. Meyer, 1843. Geobios 27:717–751

Roček Z (2013) Mesozoic and Tertiary Anura of Laurasia. Palaeobiodiv Palaeoenviron 93:397–439

Roger T, Darlas A (2008) Microvertébrés, paléo-environnement et paléoclimat de la grotte de Kalamakia (Péloponnèse, Grèce). In: Darlas A, Mihailović D (eds) The Palaeolithic of the Balkans. Proceedings of the XV World Congress of the International Union for Prehistoric and Protohistoric Sciences, BAR International Series, pp 77–84

Sanchiz FB (1984) Algunas batracofaunas pleistocénicas de islas del Mediterráneo oriental. Actas II reunión Iberoamericana sobre conservación y Zoologia de Vertebrados. Caceres, 21–30 Junio 1984. Ed. Estación Biológica de Doñana C.S.IC., Sevilla, pp 59–69

Sanchiz FB (1998) Salientia. In: Handbuch der Palaoherpetologie, part 4. Gustav Fischer Verlag, Stuttgart/New York

Savi P (1821) Descrizione (inedita) di una nuova specie di Salamandra terrestre, *Salamandra perspicillata* Nob. Biblioteca Italiana, Ossia Giornale di Lett, Sci Arti, Milano 22:228–230

Scheuchzer JJ (1726) Homo diluvii testis, Züric

Schneider B (1975) Eine mittelpleistozaene Herpetofauna von der Insel Chios, Aegaeis. Senckenberg Biol 56:191–198

Sillero N, Campos J, Bonardi A, Corti C, Creemers R, Crochet PA, Isailović JC, Denoël M, Ficetola GF, Gonçalves J, Kuzmin S, Lymberakis P, de Pous P, Rodríguez A, Sindaco R, Speybroeck J, Toxopeus B, Vieites DR, Vences M (2014) Updated distribution and biogeography of amphibians and reptiles of Europe based on a compilation of countrywide mapping studies. Amphibia-Reptilia 35:1–31

Sloan RE, Duncan MA (1978) Zooarchaeology of Nichoria. In: Rapp G, Aschenbrenner SE (eds) Excavations at Nichoria in Southwest Greece, Site, environs and techniques, vol I. University of Minnessota Press, Minneapolis, pp 60–77

Sorbelli L, Villa A, Gentili S, Cherin M, Carnevale G, Tschopp E, Delfino M (in press) The early Pleistocene ectothermic vertebrates of Pietrafitta (Italy) and the last European occurrence of *Latonia*. C R Palevol

Sparreboom M (2014) Salamanders of the Old World: the Salamanders of Europe, Asia and Northern Africa, 1st edn. KNNV Publishing, Zeist

Speybroeck J, Beukema W, Bok B, Van Der Voort J, Velikov I (2016) Field guide to the amphibians and reptiles of Britain and Europe. Bloomsbury, London/New York

Syromyatnikova E, Georgalis GL, Mayda S, Kaya T, Saraç G (2019a) A new early Miocene herpetofauna from Kilçak, Turkey. Rus J Herpetol 26:205–224

Syromyatnikova E, Roček Z, van de Velde S (2019b) New discoveries in the frog *Latonia seyfriedi* (Anura: Alytidae) and their impact on taxonomy of the genus *Latonia*. PalZ 93:669–677

Tschudi von JJ (1837) Über den *Homo diluvii testis*, *Andrias Scheuchzeri*. N Jahrb Mineral, Geognos, Geol Petrefakten-Kunde 5:545–547

Valakos ED, Pafilis P, Sotiropoulos K, Lymberakis P, Maragou P, Foufopoulos J (2008) Amphibians and Reptiles of Greece. Chimaira, Frankfurt

van Riemsdijk I, Arntzen JW, Bogaerts S, Franzen M, Litvinchuk SN, Olgun K, Wielstra B (2017) The Near East as a cradle of biodiversity: a phylogeography of banded newts (genus *Ommatotriton*) reveals extensive inter- and intraspecific genetic differentiation. Mol Phylogenet Evol 114:73–81

Vasileiadou K, Böhme M, Neubauer TA, Georgalis GL, Syrides GE, Papadopoulou L, Zouros N (2017) Early Miocene gastropod and ectothermic vertebrate remains from the Lesvos Petrified Forest (Greece). Paläontol Z 91:541–564

Vasilyan D, Schneider S, Bayraktutan MS, Şen Ş (2014) Early Pleistocene freshwater communities and rodents from the Pasinler Basin (Erzurum Province, north-eastern Turkey). Turk J Earth Sci 23:293–307

Venczel M, Codrea VA (2019) A new proteid salamander from the lower Oligocene of Romania, with notes on the paleobiogeography of Eurasian proteids. J Vertebr Paleontol 38:e150802

Villa A, Andreone F, Boistel R, Delfino M (2014) Skull and lower jaw osteology of the Lanza's salamander, *Salamandra lanzai* (Amphibia, Caudata). In: Capula M, Corti C (eds) Scripta Herpetologica. Studies on amphibians and reptiles in honour of Benedetto Lanza. Monografie della Societas Herpetologica Italica, vol III, Edizioni Belvedere, Latina, pp 171–200

Villa A, Delfino M, Luján ÀH, Alméčija S, Alba DM (2019) First record of *Latonia gigantea* (Anura, Alytidae) from the Iberian Peninsula. Hist Biol 31:371–382

Villa A, Georgalis GL, Delfino M (2020) The latest early Pleistocene amphibians and reptiles from Kaiafas (Greece) and the first record of fossil *Ophiomorus* (Squamata, Scincidae). Geobios 62:79–90

Vugt van N, Bruijn de H, Kolfschoten van M, Langereis GG (2000) Magneto-and cyclostratigraphy and mammal-fauna's of the Pleistocene lacustrine Megalopolis Basin, Peloponnesos, Greece. In: Vugt van N (ed) Orbital forcing in late Neogene lacustrine basins from the Mediterranean. A magnetostratigraphic and cyclostratigraphic study. Geologica Ultraiectina, vol 189, pp 69–92

Wagler J (1830) Natürliches System der Amphibien, mit vorangehender Classification der Säugthiere und Vogel. Ein Beitrag zur vergleichenden Zoologie. JG Cotta, München/Stuttgart/Tübingen

The Fossil Record of Lizards and Snakes (Reptilia: Squamata) in Greece

Georgios L. Georgalis and Massimo Delfino

1 Introduction

Lizards and snakes—together comprising the clade of squamates—represent the most diverse clade of reptiles with more than 10,000 extant species (e.g., Uetz and Stylianou 2018). They are widely distributed across all tropical and temperate regions of the planet, having colonized even deserts and the most remote oceanic islands (Pianka and Vitt 2003; Wallach et al. 2014). Squamate fossils are already known since the Mesozoic, being diverse throughout the Cenozoic, especially during the Palaeogene (Rage 2013). During their long evolutionary history, they have achieved a considerably large number of shapes, bauplans, sizes, and functional morphologies (Rage 2013). In Greece, the fossil record of squamates is relatively scarce, though important taxa have been established from the country as well as lineages now extirpated from Europe have been identified in the area.

G. L. Georgalis (✉)
Palaeontological Institute and Museum, University of Zurich, Zurich, Switzerland

Dipartimento di Scienze della Terra, Università di Torino, Torino, Italy
e-mail: georgios.georgalis@pim.uzh.ch

M. Delfino
Dipartimento di Scienze della Terra, Università di Torino, Torino, Italy

Institut Català de Paleontologia Miquel Crusafont, Universitat Autònoma de Barcelona, Barcelona, Spain
e-mail: massimo.delfino@unito.it

E. Vlachos (ed.), *Fossil Vertebrates of Greece Vol. 1*,
https://doi.org/10.1007/978-3-030-68398-6_7

2 Historical Overview

Squamates are known from the Greek fossil record already since the second half of the nineteenth century. The first scientific publication dealing with fossil squamates (and in fact with fossil reptiles in general) from the Greek area was made by Owen (1857) who described the gigantic viperid snake *Laophis crotaloides* Owen, 1857 from the Neogene of the Thessaloniki area. Other important discoveries subsequently took place in the next decades of the nineteenth century. Interestingly, all of them consisted of large-sized forms, i.e., varanids and pythonids (Gaudry 1862a, b, 1862–1867; Römer 1870; Weithofer 1888). Since then, however, and despite the importance of these early finds, squamate fossils received only minor attention in palaeontological studies from Greece, with only a few published works during the twentieth century (Schneider 1975; Kotsakis 1977; Szyndlar and Zerova 1990; Szyndlar 1991a, b, 1995; Richter 1995).

The twenty-first century, on the other hand, witnesses a massive wave of descriptions of squamates from many different Neogene and Quaternary localities across the country, including new taxa as well as novel occurrences of clades previously unrecognized in the area (Conrad et al. 2012; Georgalis et al. 2016a, b, c, 2017a, b, 2018b, c, 2019a, b; Vasileiadou et al. 2017; Villa et al. 2018, 2020; Georgalis 2019; Loréal et al. 2020). Many of these finds were collected by the University of Utrecht during the last quarter of the twentieth century, being the by-product of search of fossil mammals. In many cases, these new occurrences expand the geographical or temporal distribution of several clades, or improve our knowledge on the taxonomy and morphology of this diverse group of reptiles.

3 Phylogenetic Relationships

The phylogenetic relationships of many squamate lineages are still at a state of flux, mainly due to inconsistent results arising from molecular and morphological studies, or combination of both (e.g., Vidal and Hedges 2005, 2009; Wiens et al. 2010; Gauthier et al. 2012; Pyron et al. 2013). As such, the traditional division of squamates into three suborders (Lacertilia, Serpentes, and Amphisbaenia) (e.g., Estes 1983) is no longer acceptable. In fact, Lacertilia is currently ubiquitously considered paraphyletic, while most plylogenetic analyses place snakes (Serpentes) somewhere within anguimorphs (i.e., relatively close to anguids and varanids) and amphisbaenians close to lacertids (e.g., Vidal et al. 2005, 2009; Müller et al. 2011; Wiens et al. 2012; Pyron et al. 2013).

4 Distribution

Squamate fossil remains from Greece are known across most of the Neogene and Quaternary, spanning several localities in the mainland of the country and the islands. The oldest records are from the early Miocene of Thrace, Euboea, and Lesvos (Römer 1870; Georgalis et al. 2016b, 2019a; Vasileiadou et al. 2017). After that, there is a relative hiatus in the squamate fossil record, as the immediately younger remains are of late Miocene age. In fact, the majority of the so far described lizard and snake fossils from Greece are of late Miocene age, originating from Crete, Central and Western Macedonia, Samos, and Central Greece (Gaudry 1862a, b, 1862–1867; Weithofer 1888; Nopcsa 1908; Richter 1995; Szyndlar 1995; Conrad et al. 2012; Georgalis et al. 2016c, 2017a, 2018b, 2019b; Villa et al. 2018; Loréal et al. 2020). Pliocene records are relatively few, originating from Rhodes, Central and Western Macedonia, and Central Greece (Owen 1857; Szyndlar and Zerova 1990; Szyndlar 1991a, b; Georgalis et al. 2016a, 2018c; Loréal et al. 2020). Finally, Quaternary records are also few with a so far relatively limited number of remains originating from Chios, Central Greece, Kos, and Crete (Schneider 1975; Kotsakis 1977; Szyndlar 1991a, b; Georgalis et al. 2017b). All squamate lineages that are present in the extant herpetofauna have so far been identified in the fossil record of the country, with the exception of gekkotans; these occurrences are summarized in Fig. 1 and the Appendix.

5 Systematic Paleontology

Squamata Oppel, 1811b
Iguania Cuvier, 1817
Acrodonta Cope, 1864
Chamaeleonidae Rafinesque, 1825
***Chamaeleo* Laurenti, 1768**

Type Species *Chamaeleo parisiensium* Laurenti, 1768 (junior synonym of *Chamaeleo chamaeleon* [Linnaeus, 1758]).

***Chamaeleo andrusovi* Čerňanský, 2010**
Chamaeleo* cf. *andrusovi

Remarks Chamaeleonids are present in the extant herpetofauna of Greece (Valakos et al. 2008); however, the single known fossil occurrence of the group from the country has been described from the early Miocene of Aliveri, Euboea (Georgalis et al. 2016b). It consists of cranial material with a complex ornamentation and well-spaced and complex clusters of tubercles, features which evoked strong resemblance with a slightly coeval taxon from Central Europe, that is, *Chamaeleo andrusovi* from the early Miocene (MN4) of Dolnice, Czech Republic, and was accordingly tentatively referred to the same species (Georgalis et al. 2016b).

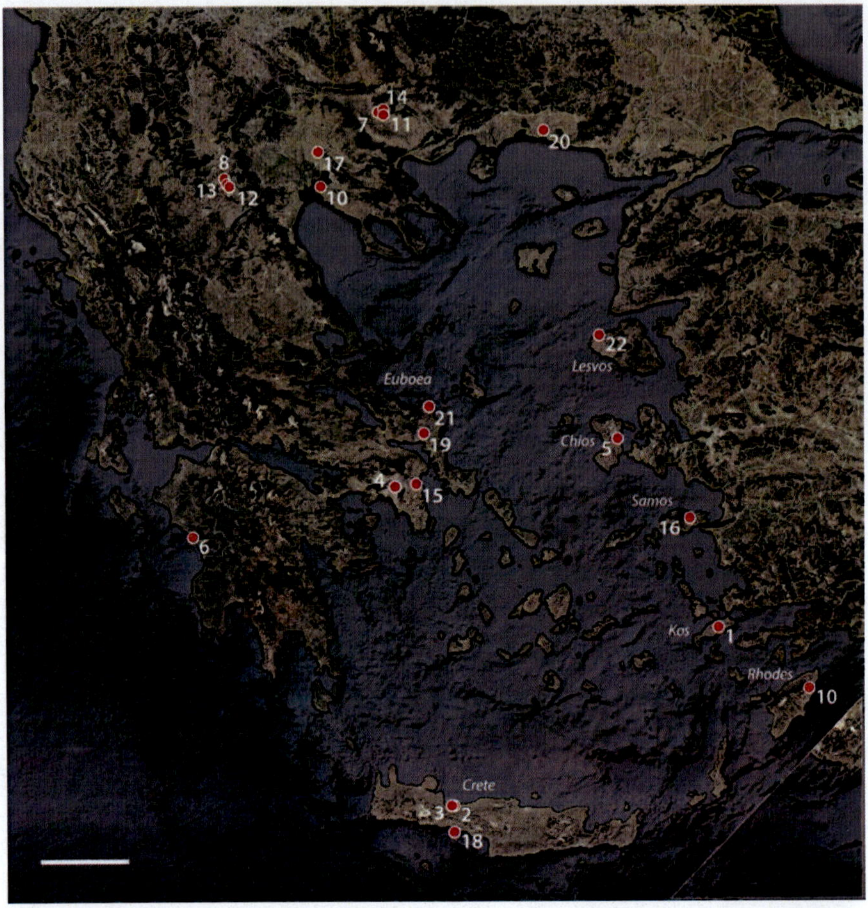

Fig. 1 Map of the occurrences of fossil squamates from Greece. See Appendix for further information. **1**, Pili B; **2**, Gerani Cave; **3**, Gerani Cave; **4**, Tourkobounia localities; **5**, Latomi; **6**, Kaiafas; **7**, Spilia-4; **8**, Vorio localities; **9**, Megalo Emvolon; **10**, Maritsa; **11**, Maramena; **12**, Tomea Eksi localities; **13**, Komanos; **14**, Ano Metochi localities; **15**, Pikermi localities; **16**, Mytilinii; **17**, Ravin de la Pluie; **18**, Plakias; **19**, Aliveri; **20**, Karydia; **21**, Kymi; **22**, Lapsarna. See Appendix for more information. Image exported from Google Earth Pro © 2019, map data from the US Department of State Geographer, SIO, NOAA, US Navy, NGA, GEBCO, image from Landsat/Copernicus. Scale bar equals 80 km, North faces upward

According to such taxonomic affinities, it is evident that the Aliveri chameleons represented a totally distinct lineage than the two species of *Chamaeleo* that currently occur in the country (Georgalis et al. 2016b). On the basis of its stratigraphic age and its geographic position, it was suggested that the Aliveri finds indicated that chameleons probably used the "*Gomphotherium*-Landbridge" that united Afro-Arabia with Eurasia during the early Miocene in order to disperse from Africa to Europe through the Balkan area (Georgalis et al. 2016b). It is worth noting that in Aliveri, there were found also chamaeleonid tooth-bearing bones, which possessed a characteristic acrodont dentition—though it is probable that they also pertain to the same species—they bear no diagnostic feature for species level determination and were accordingly treated as indeterminate chamaeleonids (Georgalis et al. 2016b).

Agamidae Spix, 1825
Agaminae Spix, 1825

Remarks Agamines are present in the extant herpetofauna of Greece, represented solely by *Stellagama stellio* (Linnaeus, 1758), which currently achieves a rather disjunct distribution in continental and insular parts of the country (Valakos et al. 2008). It is as yet unclear if this disjunct and peculiar distribution of this species in the extant herpetofauna of the country is indeed natural or simply the product of anthropogenic transportation during Antiquity (Delfino et al. 2008). Nevertheless, their single known occurrence in the fossil record of Greece is documented in the latest Miocene/earliest Pliocene (MN13/14) of Maramena (Richter 1995; Georgalis et al. 2019b). Agamids are characterized by an acrodont dentition on their maxillae and dentaries, which however becomes pleurodont in the respective anteriormost portions (Delfino et al. 2008). Although represented by relatively abundant cranial and postcranial remains, due to our current insufficient knowledge of the skeletal anatomy of extant agamids, the material was only identified as indeterminate agamines (Georgalis et al. 2019b). The purported presence of an uromastycine agamid, based on non-figured appendicular material from the Pleistocene of Crete (Mangili 1980) has been rejected as probably instead pertaining to a large lacertid (Georgalis et al. 2016b).

Laterata Vidal and Hedges, 2005
Lacertiformata Vidal and Hedges, 2005
Lacertidae Oppel, 1811b

Remarks Lacertids constitute the most diverse and speciose reptile group in Europe (Arnold et al. 2007). The group is particularly diverse in Greece with multiple species distributed in both mainland and the islands including a large number of endemic taxa (Arnold et al. 2007; Bellati et al. 2014). Lacertids are characterized by heterodont and pleurodont dentition composed of mono-, bi- and tricuspid teeth (Augé 2005; Černanský and Syromyatnikova 2019; Villa and Delfino 2019). Their fossil record in Greece is also relatively abundant, at least by fossil squamate standards; indeed, lacertids are already known since the early Miocene across different parts of the country, such as Aliveri (Euboea), Karydia (Thrace), and Lapsarna (Lesvos) (Vasileiadou et al. 2017; Georgalis et al. 2019a). They are also found in several late Miocene and Pliocene localities across Greece (Georgalis et al. 2019b; Georgalis 2019). Their post-Neogene record though is relatively scarce with Pleistocene occurrences known from the island of Chios (Schneider 1975), Kaiafas in western Peloponnese (at least two morphotypes; Villa et al. 2020), and possibly also from Crete (Mangili 1980). Nevertheless and despite this abundance of remains and the many distinct genera currently inhabiting the area, all such Greek fossil

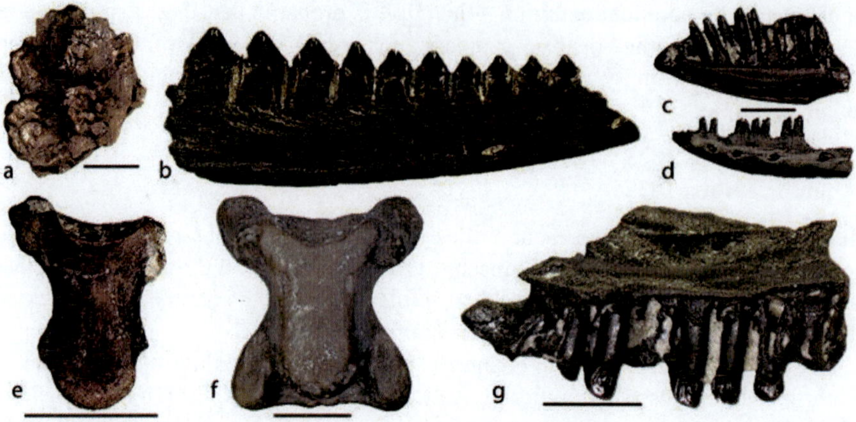

Fig. 2 **a** *Chamaeleo* cf. *andrusovi* from Aliveri, UU AL 3501, skull roofing bone in dorsal view, **b** Agaminae indet. from Maramena, UU MAA 7041, right dentary in lateral view, **c** Lacertidae indet. from Maramena, UU MAA 7050, right dentary in medial view, **d** Lacertidae indet. from Notio 1, UU NO1 1037, left dentary in lateral view, **e** Amphisbaenia indet. from Plakias, UU PL 733, presacral vertebra in ventral view, **f** Amphisbaenia indet. from Spilia-4, UU SP4 501, presacral vertebra in ventral view, **g** aff. *Palaeocordylus* sp. from Maramena, UU MAA 7047, right maxilla in medial view, Scales equal 1 mm

remains represent indeterminate lacertids and cannot be identified beyond the family level. The only potential informative occurrence is a single dentary fragment from the latest Miocene/earliest Pliocene of Maramena, whose characteristic symmetrical tricuspid tooth morphology is kind of reminiscent of the Central European genus *Miolacerta* Roček, 1984 (Georgalis et al. 2019b). However, still, in that case, the material is too limited in order to decipher such generic placement with confidence (Fig. 2).

Podarcis Wagler, 1830

Type Species *Seps muralis* Laurenti, 1768.

Podarcis erhardii (Bedriaga, 1882)
Podarcis cf. *erhardii*

Remarks This record was based on material from the Late Pleistocene cave of Bate, Crete, identified as such by Kotsakis (1977). However, the anatomy of most extant small lacertid species of the Mediterranean is not adequately known (e.g., Villa et al. 2017; however, see Villa and Delfino 2019) and therefore, this identification likely based more on biogeography than on anatomical features, and should be confirmed by further fossil material and a better knowledge of the osteology of Cretan *Podarcis* spp.

Amphisbaenia Gray, 1844

Remarks Amphisbaenians are legless squamates that were for long time considered a distinct suborder within Squamata, besides lizards and snakes (e.g., Estes 1983). However, recent advances in molecular phylogenies, coupled with advances in morphology and the fossil record, denote that they are lizards more closely related to lacertids (Müller et al. 2011; Čerňanský et al. 2015). Amphisbaenians have a relatively rich fossil record in Europe, achieving also a high diversity during their Cenozoic past (Augé 2012; Folie et al. 2013; Čerňanský et al. 2015), however, their extant representatives in the continent are members of a single genus, *Blanus* Wagler, 1830, confined in the two margins of the Mediterranean Sea (Sindaco et al. 2014). In Greece, amphisbaenians are represented solely by the species *Blanus strauchi* (Bedriaga 1884), distributed in few islands in southeastern Aegean Sea, very close to the adjacent Anatolian coast (Valakos et al. 2008; Sindaco et al. 2014). Their Greek fossil record is scarce, but nevertheless, signals a much broader geographic distribution, as amphisbaenian remains are known from the late Miocene (MN9) of Plakias, Crete (Georgalis et al. 2016c), and the early Pliocene (MN15) of Spilia-4, Central Macedonia (Georgalis et al. 2018c). Especially for the latter record, it is of importance because it represents the youngest record of amphisbaenians in continental Eastern Europe (Georgalis et al. 2018c). Their remains in Greece consist so far only of isolated vertebrae. Amphisbaenian vertebrae are characterized by a dorsoventrally compressed centrum with a flat ventral surface and roughly parallel lateral margins, massive and rounded synapophyses, absence of zygosphene, and a dorsally weakly convex neural arch lacking a neural spine (Estes 1983; Georgalis et al. 2018c). Although this combination of features allows referral of these Greek fossils to this group, amphisbaenian vertebrae are not further informative for species or even genus level identification (Estes 1983). Nevertheless, it seems most probable that these remains pertain to the genus *Blanus*, especially when taking into consideration that the latter genus is known from nearby Anatolia since the middle Miocene (Georgalis et al. 2018a).

Scinciformata Vidal and Hedges, 2005
Scincidae Oppel, 1811

Remarks Scincids are currently components of the extant herpetofauna of Greece, being represented by four different genera distributed in both continental and insular environments (Valakos et al. 2008). Nevertheless, their fossil record in the area is extremely poor, being so far represented solely by limited cranial material from the latest Miocene/earliest Pliocene of Maramena (that was tentatively referred to this clade; Georgalis et al. 2019b), the early Pliocene of the Ptolemais area (Loréal et al. 2020), and the Early Pleistocene of Kaiafas (Villa et al. 2020). Among these three occurrences, only the latter can afford a more precise taxonomic determination, allowing its attribution to *Ophiomorus* Duméril and Bibron, 1839; this was the first time that this genus was identified in the fossil record globally (Villa et al. 2020). The Greek scincids are characterized by pleurodont teeth bearing two parallel cusps and absent or light lingual striae (Villa and Delfino 2019).

Cordyliformes Fitzinger, 1826
Cordylidae Fitzinger, 1826
***Palaeocordylus* Roček, 1984**

Type Species *Palaeocordylus bohemicus* Roček, 1984.

aff. *Palaeocordylus* sp.

Remarks Cordylids are currently distributed solely in sub-Saharan Africa, however, their Neogene fossil record testifies a much broader geographic range encompassing also Central Europe (Roček 1984; Čerňanský 2012). Recently, this European distribution was significantly expanded by remains from northern Greece, more specifically the nearby localities in Central Macedonia, Ano Metochi-2 and Ano Metochi-3 (Georgalis et al. 2017a), and Maramena (Richter 1995; Georgalis et al. 2019b). This Greek material was initially identified as an indeterminate "scincoid" (Richter 1995), later as indeterminate cordylids (Georgalis et al. 2017a), whereas more recently, the abundant and better preserved cranial material from Maramena afforded a more accurate taxonomic identification (Georgalis et al. 2019b) possessing certain resemblance with the genus *Palaeocordylus* from the early Miocene of the Czech Republic (Roček 1984; Čerňanský 2012). Whether it represents a congeneric or distinct form from *Palaeocordylus* could not be verified on the basis of the available Greek material. Interestingly, the Maramena material comprises by far the youngest record of cordylids in Europe, denoting their presence in the latest Miocene/earliest Pliocene boundary (Georgalis et al. 2019b). The Greek aff. *Palaeocordylus* sp. is characterized by pleurodont dentition with teeth bearing two parallel cusps and a prominent and dense striation on their lingual side (Georgalis et al. 2019b).

Anguimorpha Fürbringer, 1900
Anguidae Gray, 1825
Anguinae Gray, 1825

Remarks Anguines are represented in the extant herpetofauna of Greece by two genera of legless lizards, that is, *Anguis* and *Pseudopus* (Valakos et al. 2008). Nevertheless, their fossil record attests a larger diversity, encompassing, besides these two genera, also the currently extinct in Europe, *Ophisaurus*. In fact, anguines are among the most common squamates in the Greek fossil record; besides the three above-mentioned genera that are discussed separately below, several other indeterminate occurrences based on osteoderms and caudal vertebrae are known from the early Miocene of Aliveri and Karydia (Georgalis et al. 2019a), the late Miocene of Ravin de la Pluie (Georgalis et al. 2018b), the late Miocene of Pikermi (Georgalis 2019), the late Miocene of Ano Metochi (Georgalis et al. 2017a), the latest Miocene/earliest Pliocene of Maramena (Georgalis et al. 2019b), the late Miocene and early Pliocene of the Ptolemais area (Loréal et al. 2020), and the Late Pleistocene of Kalamakia Cave (Kolendrianou et al. 2020).

Anguis Linnaeus, 1758

Type Species *Anguis fragilis* Linnaeus, 1758.

Remarks *Anguis* is represented in the extant Greek herpetofauna by three species, that is, *Anguis cephallonica* Werner, 1894, *Anguis fragilis* Linnaeus, 1758, and *Anguis graeca* Bedriaga, 1881 (Gvoždík et al. 2010). Its sole known fossil records from Greece are documented in the latest Miocene/earliest Pliocene of Maramena and the early Pliocene of the Ptolemais area, thus extending the presence of the genus in the area already by the Neogene (Georgalis et al. 2019b; Loréal et al. 2020). Both these occurrences are known exclusively by vertebral material; vertebrae of this genus are characterized by a dorsoventrally compressed centrum with straight and parallel lateral margins, flat ventral surface, and the absence of precondylar constriction (Klembara 1981; Čerňanský et al. 2019). Therefore, although assignment to the genus level is clear, no further precise referral to the species level could be made.

Ophisaurus Daudin, 1803b

Type Species *Anguis ventralis* Linnaeus, 1766.

Remarks *Ophisaurus* is a genus of legless lizards, currently distributed in North America, Asia, and northern Africa (Čerňanský et al. 2019). In Europe, the genus is currently extinct but has an extensive fossil record in the late Palaeogene and especially the early Neogene (Georgalis et al. 2017a; Klembara and Rummel 2018), surviving in the western parts of the continent even up to the Quaternary (Blain and Bailon 2019). *Ophisaurus* is characterized by conical and slightly posteriorly curved teeth and a surangular spine on the dentary (Klembara et al. 2014), while its presacral vertebrae can be distinguished on the basis of its dorsoventrally compressed and subtriangular vertebral centrum with concave lateral margins, a flat ventral surface, and the absence of a precondylar constriction (Klembara 1981; Čerňanský et al. 2019). In the fossil record of Greece, *Ophisaurus* represents the most abundant anguine, known by both cranial and vertebral material. Its remains having been described from the early Miocene (MN4) of Karydia (Georgalis et al. 2019a), the late Miocene (MN13) of Ano Metochi (Georgalis et al. 2017a), and the latest Miocene/earliest Pliocene (MN13/14) of Maramena (Georgalis et al. 2019b). Notably, the Maramena material denotes the presence of as many as five different morphotypes of *Ophisaurus* in a single locality (Georgalis et al. 2019b). Whether these morphotypes pertain indeed to distinct species or simply represent some kind of unknown variation, cannot be assessed with certainty (Georgalis et al. 2019b). In any case, the Maramena material represents the youngest occurrence of *Ophisaurus* in Eastern Europe (Georgalis et al. 2019b) (Fig. 3).

Pseudopus Merrem, 1820

Type Species *Lacerta apoda* Pallas, 1775.

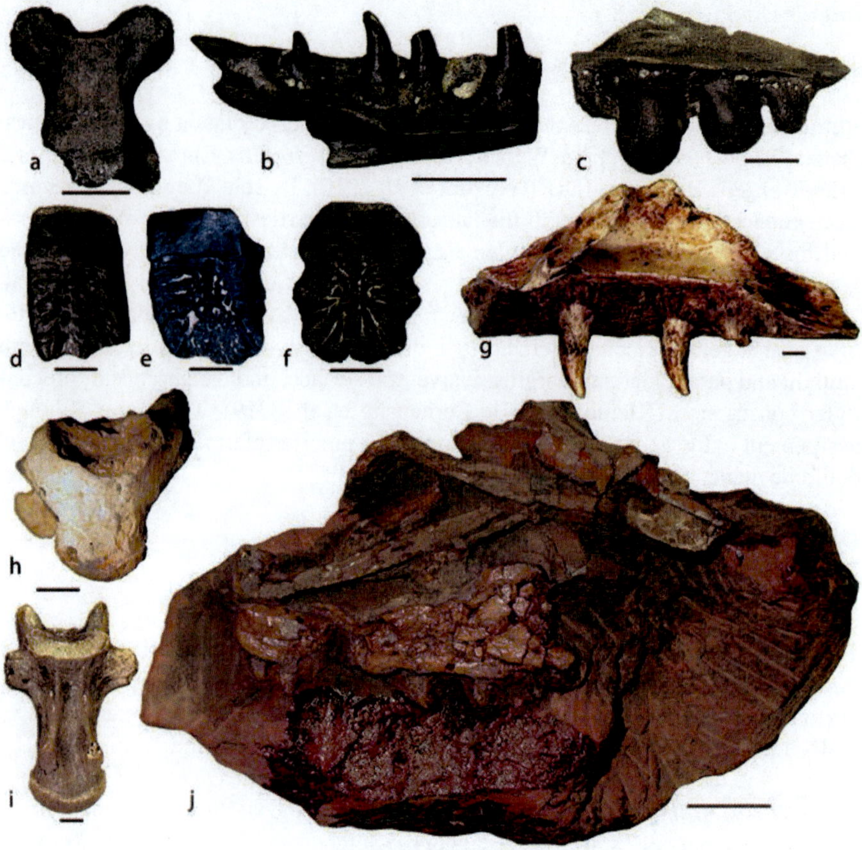

Fig. 3 **a** *Anguis* sp. from Notio 1, UU NO1 1026, presacral vertebra in ventral view, **b** *Ophisaurus* sp. from Ano Metochi, UU AM3 510, left dentary in medial view, **c** *Pseudopus* sp. from Maramena, UU MAA 7171, right maxilla in medial view, **d** Anguinae indet. from Aliveri, UU AL 3545, osteoderm in dorsal view, **e** Anguinae indet. from Ano Metochi, UU AM3 516, osteoderm in dorsal view, **f** Anguinae indet. from Maramena, UU MAA 7168, osteoderm in dorsal view, **g** *Varanus* sp. from Tourkobounia 5, UU TB5 1001, right maxilla in medial view, **h** *Varanus* sp. from Ravin de la Pluie, LGPUT RPl 297, presacral vertebra in ventral view, **i** *Varanus* sp. from Maramena, UU MAA 7198, caudal vertebra in ventral view, **j** holotype of *Varanus marathonensis* from Pikermi, IPUW 1888–001-001. All scale bars equal 1 mm, exc. **h** which equals 5 mm and **j** which equals 10 mm

Remarks *Pseudopus apodus* is the largest lizard occurring in the extant herpetofauna of Europe (up to 1.4 m according to Valakos et al. 2008). This genus of legless lizards is known in the fossil record of Europe already since the late Palaeogene, achieving its maximum diversity during the early Neogene, when several distinct species are known from different parts of the continent (Klembara et al. 2010; Klembara 2012; Jandzik et al. 2018; Klembara and Rummel 2018). *Pseudopus* is characterized by robust, bulbous, cylindrical, and striated sub-pleurodont teeth on its maxillae and dentaries (Klembara et al. 2014), while its presacral vertebrae can be distinguished by the dorsoventrally compressed and triangular vertebral centrum, straight lateral margins, a convex ventral surface, and the absence of precondylar constriction (Čerňanský et al. 2019). In Greece, the genus is known so far exclusively from cranial and vertebral material from the latest Miocene/earliest Pliocene of Maramena (Georgalis et al. 2019b) plus remains tooth-bearing, vertebral, and osteoderm material from the Early Pleistocene of Kaiafas (Villa et al. 2020). The Maramena material bears resemblance in certain features with the Neogene–Quaternary *Pseudopus pannonicus* (Kormos, 1911), the largest known anguine (Georgalis et al. 2019b). Osteoderm material from the Pleistocene Cave of Kalamakia in Peloponnese was referred to *Pseudopus apodus* by Kolendrianou et al. (2020); however, this material does not bear diagnostic features for such species or even genus level determination and should be better treated as an indeterminate anguine.

Platynota Duméril and Bibron, 1835
Varanidae Gray, 1827 (sensu Estes et al. 1988)
***Varanus* Merrem, 1820**

Type Species *Lacerta varia* White, 1790.

Remarks Varanids, commonly known as monitor lizards, are the largest lizards inhabiting the planet today (Pianka et al. 2004). Teeth of *Varanus* are characterized by the presence of plicidentine (Kearney and Rieppel 2006). Their vertebrae are distinguished by the presence of precondylar constriction, cotyle and condyle markedly depressed dorsoventrally, cotyle facing anteroventrally, and neural arch clearly inclined anteriorly with a distinct anterior part (pars tectiformis) (Hoffstetter 1969; Bailon and Rage 1994; Delfino et al. 2013).

The earliest occurrence of *Varanus* in Greece is documented in the late Miocene (MN10) locality of Ravin de la Pluie, near Thessaloniki, Central Macedonia (Georgalis et al. 2018b). This material was based on isolated vertebrae and appendicular elements and therefore could not be identified to the species level; nevertheless though, the appendicular remains from Ravin de la Pluie lead to the assumption that Miocene varanids from Europe possessed comparatively short and robustly built limbs (Georgalis et al. 2018b). Other Neogene indeterminate remains of *Varanus* have been described from the late Miocene of Pikermi (Gaudry 1862a, b, 1862–1867; Georgalis 2019) and the latest Miocene/earliest Pliocene of Maramena (Georgalis et al. 2019b). Interestingly, the isolated varanid teeth from Maramena are characterized by four different types of serration—this limited material is, however, not sufficient for assuming if these different types of serration correspond indeed to different species of *Varanus* or simply to some kind of intraspecific variation or even taphonomic biases (Georgalis et al. 2019b). The youngest occurrence of varanids from Greece is documented in the Middle Pleistocene of Tourkobounia 5, near Athens, based on cranial material (Georgalis et al. 2017b). This occurrence is also the youngest record of varanids in the whole Europe, marking thus that monitor lizards survived in the continent also in the Quaternary only to become ultimately extinct rather recently (Georgalis et al. 2017b). In total, three species of *Varanus* have been named from Greece, but only one of them is currently considered valid (see the next section).

★*Varanus marathonensis* Weithofer, 1888
(= ★*Varanus amnhophilis* Conrad, Balcarcel and Mehling, 2012)

Nomenclatural and Taxonomic History *Varanus marathonensis* in Weithofer 1888 (new species); *Varanus marathonis* in Sickenberg 1971 (incorrect subsequent spelling). The validity of *Varanus marathonensis* has never been questioned and it has been considered the senior synonym of *Varanus atticus* (by Fejervary 1918, Kuhn 1939, Villa et al. 2018), *Varanus amnhophilis* (by Villa et al. 2018), and (partially) of *Varanus deserticolus* (by Fejervary 1918).

Type Material IPUW 1888-001-001(lectotype by the subsequent designation of Villa et al. 2018), left portion of a skull, including the maxilla, the premaxilla, and the prefrontal. An additional specimen exists in the type series (thus being a paralectotype), that is, a left supraorbital, which is now probably lost.

Type Locality Pikermi, Attica, Central Greece; late Miocene (MN12–13).

Distribution Besides its type locality, this species is known from the late Miocene of Mytilinii, Samos, Greece (type material of *Varanus amnhophilis* Conrad et al. 2012, and referred material of Villa et al. 2018). Outside Greece, it has also been reported from the middle Miocene (MN7/8), Abocador de Can Mata, Catalonia, Spain (referred material of Villa et al. 2018), and the late Miocene (MN10) of Cerro de los Batallones, Madrid, Spain (referred material of Villa et al. 2018). This species could be also potentially present in the early Pliocene (MN15) of Csarnóta 1and the late Pliocene (MN16) of Beremend 1, both in Southern Transdanubia, Hungary. These potential occurrences are based on the referred material of Fejérváry 1918, but the conspecificity of both records with *V. marathonensis* has been recently rejected by Villa et al. (2018).

Remarks The first fossil lizard species named from Greece, *Varanus marathonensis*, was established by Weithofer (1888) on the basis of cranial material from the classical locality of Pikermi, near Athens. After more than one century, the species remains still a pivotal taxon for our understanding of the evolution and palaeodiversity of monitor lizards. *Varanus marathonensis* was recently redescribed in much detail by Villa et al. (2018) who also expanded its geographic and stratigraphic distribution by referring to the same species material from the middle and late Miocene of Spain and the late Miocene of the type locality, Pikermi, as well as Samos. The same authors also synonymized with *V. marathonensis* the species *Varanus amnhophilis* from the late Miocene of Samos (Villa et al. 2018). The latter taxon was established by Conrad et al. (2012) on the basis of remains of skull, braincase, mandible, clavicle, and few vertebrae from the Mytillinii Formation in Samos. Conrad et al. (2012) considered that *V. amnhophilis* was distinct enough from other varanids and so they erected a new monotypic genus to accommodate it *Varaneades*. On the other hand, Villa et al. (2018), on the basis also of new material from Samos, suggested that *V. amnhophilis* fell within the variability range of the widespread *V. marathonensis* and as such they synonymized both.

★*Varanus atticus* Nopcsa, 1908 (*nomen dubium*)
Besides *V. marathonensis*, another congeneric species has been established from the classical locality of Pikermi. This is *Varanus atticus* Nopcsa, 1908, based on a single large vertebra (MNHN.F.PIK 3715; holotype) that was originally described several decades earlier by Gaudry (1862a, b, 1862–1867). This specimen is of historical importance as it represents the first description of fossils pertaining to monitor lizards in Europe, and the second in the world (see Georgalis 2019). In any case, vertebrae of varanids do not possess enough diagnostic features for species level identification (Georgalis et al. 2018b; Villa et al. 2018); thus, *Varanus atticus* is most probably conspecific of *V. marathonensis* (see Villa et al. 2018, for example), but on the current total absence of articulated cranial and vertebral material from Pikermi that could eventually confirm or dispute such synonymy, it is best considered as a nomen dubium, with the material pertaining to an indeterminate species of *Varanus* (Georgalis 2019).

Serpentes Linnaeus, 1758
Scolecophidia Duméril and Bibron, 1844

Remarks Scolecophidians, commonly known as worm snakes or blind snakes, are the tiniest among snakes. Their vertebrae are characterized by a rather small size, absence of neural spine, absence of hypapophysis and haemal keel, synapophyses not divided into diapophyses and parapophyses, absence of paracotylar foramina, and cotyle and condyle being flattened dorsoventrally (Rage 1984; Szyndlar 1991a). They are currently present in Greece, being represented by the typhlopid *Xerotyphlops vermicularis* (Merrem, 1820), this being also the sole extant scolecophidian in Europe. Due to their extremely small vertebrae and fragile skulls, their fossils are rather rare (Mead 2013; Georgalis et al. 2017a). Nevertheless, scolecophidians are relatively well recorded as fossils in the Greek area: their earliest occurrence is documented in the late Miocene (MN13) of Ano Metochi (Georgalis et al. 2017a). Other scolecophidian remains are also known from the latest Miocene/earliest Pliocene (MN13/14) of Maramena (Georgalis et al. 2019b), the early Pliocene (MN14) of Maritsa, Rhodes (Szyndlar 1991a), the late Pliocene (MN16) of Tourkobounia 1, near Athens (Szyndlar 1991a), and the Middle Pleistocene of Tourkobounia 2 and 5 (Szyndlar 1991a).

Alethinophidia Nopcsa, 1923
Constrictores Oppel, 1811a (sensu Georgalis and Smith 2020)
Pythonoidea Fitzinger, 1826 (sensu Wallach et al. 2014)
Pythonidae Fitzinger, 1826
***Python* Daudin, 1803a**

Type Species *Coluber molurus* Linnaeus, 1758.

★*Python euboicus* Römer, 1870

Nomenclatural and Taxonomic History *Python euboicus* in Römer 1870 (new species); *Heteropython euboicus* in Rochebrune 1880 (new genus and new combination); *Python euboeicus* in Hoernes 1884 (incorrect subsequent spelling); *Heteropython enboicus* in Palacký 1898 (incorrect subsequent spelling). Taking into consideration that the holotype of *P. euboicus* is lost, Rage (1984) considered that the large taxon from Euboea should better be considered to be a nomen dubium, an opinion shared also later by others (Szyndlar 1991a; Szyndlar and Rage 2003; Wallach et al. 2014).

Type Material University of Wrocław, Poland (holotype), an articulated skeleton with portions of mandible, currently lost.

Type Locality Kymi, Euboea Island, Central Greece; MN3/4, Burdigalian, early Miocene.

Distribution Known only from the type locality.

Remarks Already since its original description, the holotype and only known specimen of *Python euboicus* was considered as one of the most beautifully preserved snake fossils (Römer 1870). Indeed, based on an articulated portion of the vertebral column, including several vertebrae and ribs along with a portion of the mandible, *Python euboicus* was astonishingly preserved. Rochebrune (1880) erected the monotypic genus *Heteropython* in order to accommodate this species, as he felt that it was distinct enough from *Python*. Taking into consideration that the holotype of *P. euboicus* is lost, Rage (1984) considered that the large taxon from Euboea should better be considered to be a nomen dubium, an opinion shared also later by others (Szyndlar 1991a; Szyndlar and Rage 2003; Wallach et al. 2014). Our recent investigation at the collections of the University of Wrocław found no trace of this specimen (GLG). Therefore, unfortunately, what only remains of this enigmatic species is the original lithograph and the description in Römer's (1870) paper. Nevertheless, even from only the lithograph it is clear that the holotype bears the typical characteristic of the genus *Python* such as the large vertebral size, the wider than long centrum, and the shallow interzygapophyseal constriction (Szyndlar and Rage 2003). Vertebrae that approach the morphology of *P. euboicus* have been recently described from the middle Miocene of Anatolia (Georgalis et al. 2020).

Booidea Gray, 1825 (sensu Pyron et al. 2014)
Erycidae Bonaparte, 1831 (sensu Pyron et al. 2014)

Remarks Besides, the Pliocene and Quaternary remains that can be confidently assigned to the genus *Eryx* (see the next section), Georgalis (2019) described a tiny trunk vertebra from the late Miocene of Pikermi, Attica, which he tentatively referred to erycids. Nevertheless, unlike most other snakes, the most definite diagnostic part in the erycid skeleton lies in the caudal vertebrae, which are characterized by the presence of distinctive and complex processes (Hoffstetter and Rage 1972; Rage 1984; Szyndlar 1991a). As such, on the absence of any caudal vertebrae from Pikermi, no certainty about the erycid affinities of this specimen can be provided. Nevertheless, if indeed an erycid, the Pikermi vertebra would represent the oldest record of that lineage not only from Greece but also from the whole Eastern Mediterranean (Georgalis 2019).

***Eryx* Daudin, 1803b**

Type Species *Boa turcica* Olivier, 1801, junior synonym of *Eryx jaculus* (Linnaeus, 1758).

***Eryx jaculus* (Linnaeus, 1758)**

Remarks Commonly called the Sand Boa, *Eryx jaculus* occurs also today in the extant herpetofauna of Greece (Valakos et al. 2008). The species is also known in Quaternary deposits from two Aegean islands, that is, Latomi, Chios (Schneider 1975), and Pili B, Kos (Szyndlar 1991a).

cf. *Eryx* sp.

Remarks An early Pliocene record from Maritsa, Rhodes was described and referred to cf. *Eryx* by Szyndlar (1991a). It is probable that this material represents *Eryx jaculus* or some closely related form—if so, it would extend the fossil record of this species back to the Neogene.

Caenophidia Hoffstetter, 1939
Colubroidea Oppel, 1811b (sensu Zaher et al. 2009)
Colubridae Oppel, 1811b

Remarks The division of colubrids into "colubrines" and natricines on the basis of their absence or presence respectively of hypapophyses throughout the vertebral column is somewhat arbitrary and certainly does not reflect modern approaches in snake systematics using molecular data (e.g., Lawson et al. 2005; Vidal et al. 2007). Nevertheless, its prevailing usage in palaeontological studies (e.g., Rage 1984; Szyndlar 1991a, b, 1995, 2005, 2012), which anyway cannot afford such molecular data, incline us to follow this taxonomic scheme. We made nevertheless an exception for *Malpolon*, which is placed in psammophiids, a lineage that current molecular studies place very close to elapids instead of colubrids (e.g., Kelly et al. 2008).

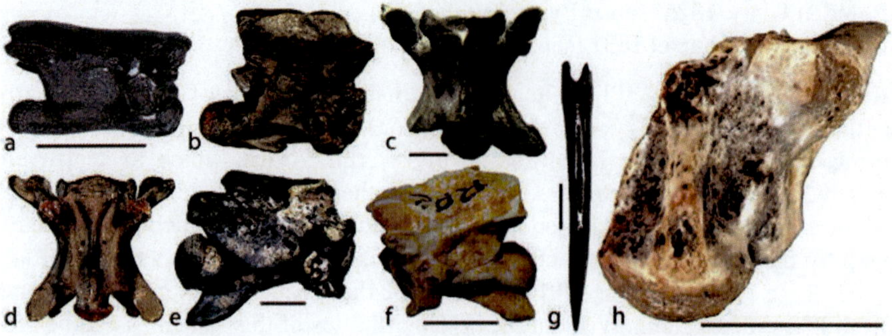

Fig. 4 **a** Scolecophidia indet. from Maramena, UU MAA 7374, trunk vertebra in right lateral view, **b** *Periergophis micros* from Maramena, UU MAA 7615, holotype posterior trunk vertebra in right lateral view, **c** "Colubrinae" indet. from Ano Metochi, UU AM3 529, trunk vertebra in dorsal view, **d** *Natrix* sp. from Maramena, UU MAA 7699, trunk vertebra in ventral view, **e** *Natrix* sp. from Ano Metochi, UU AM2 515, trunk vertebra in right lateral view, **f** *Naja* cf. *romani* from Ravin de la Pluie, LGPUT RPl 301, trunk vertebra in left lateral view, **g** "Oriental viper" from Maramena, UU MAA 7281, fang in anterior view, **h** *Laophis crotaloides* from Megalo Emvolon, UU KB 3, trunk vertebra in ventral view. Scale bars equal 1 mm, exc. **g** which equals 5 mm and **h** which equals 10 mm

Whatever their exact interrelationships may have been, colubrids are documented in the Greek area already since the early Miocene (MN4), known almost simultaneously from Aliveri, Euboea, and Karydia, Thrace (Georgalis et al. 2019a). Notably, the colubrids from Karydia and Aliveri are characterized by a peculiar structure that was termed as paracentral ridge (Georgalis et al. 2019a). It was suggested that this early Miocene form that was present in both Karydia and Aliveri probably represented a new taxon, however, on the basis of such fragmentary material, it is not further possible to assess its exact taxonomic affinities (Fig. 4).

"Colubrinae" Oppel, 1811b (sensu Szyndlar 1991a)
***Dolichophis* Gistel, 1868**

Type Species *Coluber caspius* Gmelin, 1789.

Remarks *Dolichophis* is represented in the extant herpetofauna of Greece by two species, *Dolichophis caspius* and *Dolichophis jugularis* (Linnaeus, 1758). Vertebral material from the late Miocene (MN13) of Ano Metochi referred to cf. *Dolichophis*, tentatively suggests that this lineage existed in the area already by that time (Georgalis et al. 2017a).

***Elaphe* Fitzinger in Wagler, 1833**

Type Species *Elaphe parreyssii* Fitzinger in Wagler, 1833 (= *Elaphe sauromates* [Pallas, 1814]).

Remarks *Elaphe quatuorlineata* (Lacépède, 1789) is distributed throughout most of continental Greece as well as numerous Aegean and Ionian islands, whereas *E. sauromates* occurs only in the northeastern portion of the country (Wallach et al. 2014). Material from the Middle Pleistocene of Tourkobounia 2 has been referred to *E. quatuorlineata* (Szyndlar 1991a).

***Hierophis* Fitzinger in Bonaparte, 183**

Type Species *Coluber viridiflavus* Lacépède, 1789.

***Hierophis gemonensis* (Laurenti, 1768)**
Hierophis* cf. *gemonensis

Remarks *Hierophis* is present in the extant herpetofauna of Greece, being represented by the species *Hierophis gemonensis* distributed across much of the continental area plus several islands, and *Hierophis viridiflavus* confined to Gyaros island, probably the product of human transportation during Antiquity (Valakos et al. 2008). Material from the Late Pleistocene of Bate Cave, Crete has been tentatively referred to *Hierophis* cf. *gemonensis* by Kotsakis (1977).

***Hierophis hungaricus* (Bolkay, 1913)**
Hierophis* cf. *hungaricus

Remarks Vertebral material from the latest Miocene/earliest Pliocene (MN13/14) of Maramena that was originally described as an indeterminate "colubrine" by Szyndlar (1995) was subsequently suggested to bear resemblance with the late Miocene Hungarian taxon *H. hungaricus* (Venczel 1998; Szyndlar 2005) and it was recently formally reassigned to *Hierophis* cf. *hungaricus* by Georgalis et al. (2019b).

***Paraxenophis* Georgalis, Villa, Ivanov, Vasilyan, and Delfino, 2019b**

Type Species *Paraxenophis spanios* Georgalis et al., 2019b.

★*Paraxenophis spanios* Georgalis, Villa, Ivanov, Vasilyan, and Delfino, 2019b

Type Material UU MAA 7645 (holotype), a mid-trunk vertebra; UU MAA 7536 (paratype), a posterior trunk vertebra; UU MAA 7533 (paratype), a posterior trunk vertebra.

Type Locality M1 site, Maramena, Central Macedonia; MN13/14, latest Miocene/earliest Pliocene.

Distribution Known only from the type locality.

Remarks This recently established species is based on several isolated vertebrae pertaining to different portions of the skeleton (Georgalis et al. 2019b). *Paraxenophis spanios* was differentiated from other snakes by a series of features, the most peculiar and characteristic of which was the presence of distinct lateral notches on either side of its haemal keel (Georgalis et al. 2019b). Its exact affinities with other colubroid taxa cannot be resolved with certainty. *Paraxenophis spanios* is so far known exclusively from its type locality, Maramena (Georgalis et al. 2019b).

***Periergophis* Georgalis, Villa, Ivanov, Vasilyan, and Delfino, 2019b**

Type Species *Periergophis micros* Georgalis et al. 2019b.

★*Periergophis micros* Georgalis, Villa, Ivanov, Vasilyan, and Delfino, 2019b

Type Material UU MAA 7615 (holotype), a posterior trunk vertebra; UU MAA 7551 (paratype), an anterior trunk vertebra; UU MAA 7614 (paratype), a mid-trunk vertebra; UU MAA 7618 (paratype), a cloacal vertebra; UU MAA 7636 (paratype), a caudal vertebra.

Type Locality M1 site, Maramena, Central Macedonia, Greece; MN13/14, latest Miocene/earliest Pliocene.

Distribution Known exclusively from M1 and M3 sites, Maramena (Georgalis et al. 2019b).

Remarks *Periergophis micros* is a tiny and rather bizarre snake that was recently established on abundant isolated vertebrae from two different sites within Maramena (Georgalis et al. 2019b). The large number of available vertebrae from Maramena along with the fact that they are known from practically every portion of the column, renders its intracolumnar variation as relatively well understood (Georgalis et al. 2019b). *Periergophis micros* was diagnosed on the basis of a number of characters, among which a unique autapomorphy: the presence of distinct tubercles in the haemal keel of posterior mid- and especially posterior trunk vertebrae (haemal keel tubercles) and in the haemapophyses of caudal vertebrae (haemapophyseal tubercles) (Georgalis et al. 2019b). These bizarre structures (haemal keel tubercles and haemapophyseal tubercles) were newly introduced terms for accommodating these anatomical novelties, and they are unknown in any other snake, extinct or extant (Georgalis et al. 2019b). The exact phylogenetic affinities of *Periergophis micros* with other colubroid taxa remain unresolved. Taking into consideration its peculiar vertebral anatomy along with its age, it could be probable that *P. micros* was perhaps the product of a potential dispersal from Africa to southeastern Europe that was facilitated by the Messinian Salinity Crisis.

Natricinae Bonaparte, 1838

Remarks Apart from finds that are referred to the genus *Natrix* (see the next section), natricines are represented in several fossil localities in Greece (Appendix), but many of these records cannot be assigned to the genus level. However, these records help establishing the temporal distribution of this clade in Greece, known probably already since the early Miocene of Lapsarna, Lesvos (Vasileiadou et al. 2017) and continuing to the late Miocene of Plakias, Crete (Georgalis et al. 2016c), the latest Miocene/earliest Pliocene of Maramena (Georgalis et al. 2019b), the late Miocene and early Pliocene of the Ptolemais area (Loréal et al. 2020), and the Middle Pleistocene of Latomi, Chios (Schneider 1975).

Natrix Laurenti, 1768

Type Species *Coluber natrix* Linnaeus, 1758.

Remarks The genus *Natrix* is an abundant and ubiquitous element of the extant Greek herpetofauna, being represented by two species, that is, *Natrix natrix* and *Natrix tessellata* (Laurenti, 1768) (Valakos et al. 2008). The earliest occurrence of *Natrix* in the Greek fossil record is documented in the late Miocene (MN13) of Ano Metochi (Georgalis et al. 2017a). Other remains of the genus are known from the latest Miocene/earliest Pliocene of Maramena (Georgalis et al. 2019b), the early Pliocene of the Ptolemais area (Loréal et al. 2020), and the Early Pleistocene of Gerani cave in Crete (Szyndlar 1991b).

Natrix tessellata (Laurenti, 1768)
Natrix cf. *tessellata*

Remarks Natricine remains from the Early Pleistocene of Gerani cave in Crete have been tentatively referred to the extant species *Natrix tessellata* (as *N.* cf. *tessellata* by Szyndlar 1991b), which still occurs in the island. Such taxonomic referral demonstrates that this form was already present in the area already since the beginning of the Quaternary.

Natrix rudabanyaensis Szyndlar, 2005
Natrix cf. *rudabanyaensis*

Remarks The abundant Maramena material denoted the presence of two distinct congeneric forms of *Natrix*, one of which shows resemblance with the Central European species *Natrix rudabanyaensis* (Georgalis et al. 2019b).

Elapoidea Vidal, Delmas, David, Cruaud, Couloux, and Hedges, 2007
Elapidae Boié, 1827
Micrurus Wagler, 1824

Type Species *Micrurus spixii* Wagler, 1824.

Remarks Coral snakes pertaining to the genus *Micrurus* have currently a distribution confined to the Neotropics (Wallach et al. 2014). Nevertheless, their fossil record denotes that during the Neogene, the genus was present also in Europe (Rage and Holman 1984). Their vertebrae are characterized by relatively small dimensions, elongated centrum, short hypapophysis directed more posteriorly than ventrally, hypapophyses present throughout the trunk region, and low neural spine

(Rage and Holman 1984; Ivanov and Böhme 2011). The single probable known occurrence of *Micrurus* in Greece is based on an incomplete trunk vertebra from Maramena (Georgalis et al. 2019b), identified as cf. *Micrurus* sp. If correct, this would represent both the easternmost European record of the genus but also its youngest occurrence from the whole continent (Georgalis et al. 2019b).

***Naja* Laurenti, 1768**

Type Species *Coluber naja* Linnaeus, 1758.

Remarks Cobras of the extant genus *Naja* were widespread in Neogene ecosystems across many parts of Europe (Szyndlar and Rage 1990; Szyndlar and Zerova 1990; Georgalis et al. 2018b). Vertebrae of *Naja* are characterized by being relatively heavily built, the continuous presence of hypapophyses throughout the trunk region, a depressed neural arch, well-developed subcentral ridges, and short condylar neck (Rage 1984; Szyndlar 1991b). In Greece, except for *Naja* cf. *romani* from Ravin de la Pluie (see the next section), all other occurrences of *Naja* cannot be identified beyond the generic level. These include remains of the latest Miocene/ earliest Pliocene (MN13/14) of Maramena (Georgalis et al. 2019b) and the late Pliocene (MN16) of Tourkobounia 1, Attica (Szyndlar and Zerova 1990). An even younger, purported occurrence of a cobra from the Middle Pleistocene of Latomi, Chios (Schneider 1975) was subsequently rejected (Szyndlar 1991b; Georgalis et al. 2018b). It seems thus that cobras became extinct from the area sometime prior to the onset of the Quaternary.

***Naja romani* (Hoffstetter, 1939)**
Naja* cf. *romani

Remarks The earliest occurrence of *Naja* in the Greek fossil record is documented in the late Miocene (MN10) of Ravin de la Pluie, near Thessaloniki, which was referred to *Naja* cf. *romani* (Georgalis et al. 2018b). This is the sole Greek record that could be assigned to the species level, as all other sporadic occurrences of *Naja* in the area could be identified solely to the genus level (see the earlier section).

Psammophiidae Bourgeois, 1968
***Malpolon* Fitzinger, 1826**

Type Species *Coluber monspessulanus* Hermann, 1804.

Remarks Unlike its related elapids, trunk vertebrae of *Malpolon* do not possess hypapophyses throughout the column, but instead, hypapophyses are restricted solely in the anterior trunk region (Szyndlar 1991a). *Malpolon* is present in the extant herpetofauna of Greece, being represented by *Malpolon insignitus* (Geoffroy Saint-Hilaire, 1827) (Wallach et al. 2014). Nevertheless, the fossil record of the genus in the area is rather patchy, being so far known exclusively by vertebral material from the late Pliocene of Tourkobounia 1, Attica, referred by Szyndlar (1991a) to *Malpolon* sp. Whether they pertain to the extant species *M. insignitus* or some ancestral form cannot be confirmed, however, that material clearly denotes the presence of this lineage in the Greek region already by the Neogene.

Viperoidea Oppel, 1811b
Viperidae Oppel, 1811b

Remarks Indeterminate viperids are known already since the early Miocene of Aliveri (Georgalis et al. 2019a), these representing the oldest occurrence of this clade in the country. More indeterminate viperids are known from the late Miocene of Ano Metochi 3 (Georgalis et al. 2017a, 2019b). Besides these indeterminate finds, more precise determinations have been made with Greek remains pertaining to *Vipera* and the extinct genus *Laophis* (see the next section).

***Laophis* Owen, 1857**

Type Species *Laophis crotaloides* Owen, 1857.

★*Laophis crotaloides* Owen, 1857

Nomenclatural and Taxonomic History *Laophis crotaloïdes* in Owen 1857 (new species); *Laophis crotaloides* in Rochebrune 1880 (justified emendation). Rage (1984) regarded the taxon to be a nomen dubium, a taxonomic opinion that was subsequently followed by other workers as well (Szyndlar 1991b; Szyndlar and Rage 1999, 2002; Wallach et al. 2014); the validity of this taxon has been reinstated by Georgalis et al. (2016a).

Type Material Unknown collection (holotype), 13 trunk vertebrae, now lost (Georgalis et al. 2016a).

Type Locality Megalo Emvolon (formerly Karaburun or Karabournou), near Thessaloniki, Central Macedonia, Greece; MN15, early Pliocene.

Distribution Known only from the type locality (Georgalis et al. 2016a).

Remarks This is the first fossil reptile described from Greece. Established by the prominent British palaeontologist Richard Owen (1857) on the basis of 13 large vertebrae in the vicinity of Thessaloniki. Owen (1857) already highlighted its large size and recognized *Laophis crotaloides* as the largest viperid. Nevertheless, eventually the material became lost and this hindered significantly the understanding of the affinities of this taxon, especially when taking into consideration that the only remaining source were the original Owen's (1857) figures and descriptions. Recently, Georgalis et al. (2016a) described a single, incomplete large trunk vertebra originating from the type locality Megalo Emvolon. These authors referred this vertebra to *Laophis crotaloides*, and revalidated the species providing an emended diagnosis (Georgalis et al. 2016a). Futhermore, Georgalis et al. (2016a) confirmed the gigantic nature of this viperid snake, as the vertebra from Megalo Emvolon has a centrum length surpassing the 16 mm.

***Vipera* Laurenti, 1768**

Type Species *Coluber aspis* Linnaeus, 1758.

***Vipera* sp.**
"Oriental Vipers" Complex

Remarks Primarily on the basis of vertebral features, the small-sized European vipers (comprising the "*Vipera berus*" and "*Vipera aspis*" Complexes) have been differentiated from the larger ones, which are included in the "Oriental Vipers" Complex (Szyndlar and Rage 1999, 2002). The latter group comprises the extant genera *Macrovipera* Reuss, 1927, and *Montivipera* Nilson et al., 1999. Both groups are currently represented in the extant herpetofauna of the country, but only members of the latter are known in the fossil record (Wallach et al. 2014). Large-sized forms attributed to the "Oriental vipers" have been described from the latest Miocene/earliest Pliocene (MN13/14) of Maramena (Szyndlar 1995; Georgalis et al. 2019b), the late Pliocene (MN16) of Tourkobounia 1 (Szyndlar 1991b), and the Middle Pleistocene of Latomi, Chios (Schneider 1975).

Acknowledgments We are grateful to Hans de Bruijn and Wilma Wessels (University of Utrecht) for the loan of material—without the study of the material collected by the University of Utrecht that has resulted in a number of different publications, this current review paper would have been definitely much shorter. GLG acknowledges support from Forschungskredit of the University of Zurich, grant no. [FK-20-110]. The quality of the manuscript was enhanced by useful comments and suggestions provided by the two reviewers, Andrea Villa and Elena Syromyatnikova.

Appendix

List of the Greek localities containing squamate fossils. Locality numbers refer to the collection numbers of the PalaeoBiology Database (PBDB)

Locality	Age (MN)	Taxa
Pili B[204953]	Late Quaternary	*Eryx jaculus*[1]
Gerani Cave[195229]	Late Pleistocene	*Natrix* cf. *tessellata*[2]
Bate Cave[183122]	Late Pleistocene	*Podarcis* cf. *erhardii*[3], *Hierophis* cf. *gemonensis*[3]
Tourkobounia 5[34592]	Middle Pleistocene	*Varanus* sp.[1,4], Scolecophidia indet.[1,4]
Tourkobounia 2[34767]	Middle Pleistocene	*Elaphe quatuorlineata*[1], *Zamenis situla*[1] Scolecophidia indet.[1]

Locality	Age (MN)	Taxa
Latomi[204077]	Middle Pleistocene	Lacertidae indet.[5], "Colubrinae" indet.[5], *Eryx jaculus*[5], Natricinae indet. [5], "Oriental vipers complex" indet.[5]
Kaiafas[34766]	Early Pleistocene	Lacertidae indet. (A)[6], Lacertidae indet. (B)[6], Lacertidae indet.[6], Laterata indet. or Scinciformata indet.[6], *Ophiomorus* sp.[6], *Pseudopus* sp.[6], Squamata indet.[6], "Colubrinae" indet.[6]
Tourkobounia 1[34589]	late Pliocene (MN16)	*Naja* sp.[1,2,7], *Malpolon* sp.[1,2,7], Scolecophidia indet.[1,2,7], "Oriental vipers complex" indet.[1,2,7]
Spilia-4[31229]	early Pliocene (MN15)	Amphisbaenia indet.[8]
Vorio 8[207964]	early Pliocene (MN15)	Caenophidia indet.[9]
Vorio 3A[205086]	early Pliocene (MN15)	*Natrix* sp.[9], Caenophidia indet.[9]
Vorio 3[205086]	early Pliocene (MN15)	*Natrix* sp.[9], Caenophidia indet.[9]
Megalo Emvolon[36579]	early Pliocene (MN15)	***Laophis crotaloides***[10,11]
Vorio 2[207948]	early Pliocene (MN14)	*Natrix* sp.[9]
Vorio 1[205087]	early Pliocene (MN14)	*Natrix* sp.[9]
Maritsa[68046]	early Pliocene (MN14)	cf. *Eryx* sp.[1], Scolecophidia indet.[1]

Locality	Age (MN)	Taxa
Maramena[32189]	late Miocene/early Pliocene (MN13/14)	Agaminae indet.[12,14], Lacertidae indet.[14], ?Lacertidae indet.[14], aff. *Palaeocordylus* sp.[12,14], ?Scincidae indet.[14], *Anguis* sp.[14], *Ophisaurus* sp. (morph. 1–5)[12,14], *Pseudopus* sp.[14], Anguinae indet.[14], *Varanus* sp.[14], *Hierophis* cf. *hungaricus*[14], ***Paraxenophis spanios***[14], ***Periergophis micros***[14], "Colubrinae" indet. (morph. 1)[14], *Natrix* aff. *rudabanyaensis*[14], *Natrix* sp.[14], cf. *Micrurus* sp.[14], *Naja* sp.[14], Scolecophidia indet.[14], "Oriental vipers complex" indet.[13,14]
Tomea Eksi 2[205088]	late Miocene (MN13)	Lacertidae indet.[9], Laterata or Scinciformata indet.[9], Squamata indet.[9], "Colubrinae" indet.[9], Natricinae indet.[9], Colubridae indet.[9], Caenophidia indet.[9]
Tomea Eksi 1[205089]	late Miocene (MN13)	Lacertidae indet.[9], Squamata indet.[9], Natricinae indet.[9], Caenophidia indet.[9]
Komanos 1[205083–84]	late Miocene (MN13)	Anguinae indet.[9]
Ano Metochi 2[31924]	late Miocene (MN13)	Lacertidae indet.[15], aff. *Palaeocordylus* sp.[15], *Ophisaurus* sp.[15], Anguinae indet.[15], *Natrix* sp.[15], Serpentes indet.[15]
Ano Metochi 3[31928]	late Miocene (MN13)	Lacertidae indet.[15], aff. *Palaeocordylus* sp.[15], *Ophisaurus* sp.[15], Anguinae indet.[15], cf. *Dolichophis* sp.[15], "Colubrinae" indet.[15], *Natrix* sp.[15], Scolecophidia indet.[15], Viperidae indet.[14,15]

Locality	Age (MN)	Taxa
Pikermi (including both classical locality and Chomateri and PV1 sites) **182754**, [195562, 202630]	late Miocene (MN11–13)	Lacertidae indet.[16], ?Anguinae indet.[16], ***Varanus marathonensis***[20–22], *Varanus* sp.[16–19], ?Erycidae indet.[16], Colubridae indet.[16], Squamata indet.[16]
Mytilinii[95691]	late Miocene (late MN11 – early MN13)	*Varanus marathonensis*[22,23]
Ravin de la Pluie[191070]	late Miocene (MN10)	?Anguinae indet.[24], *Varanus* sp.[24], "Colubrinae" indet.[24], *Naja* cf. *romani*[24]
Plakias[183151]	late Miocene (MN9)	Amphisbaenia indet.[25], ?Natricinae indet.[25]
Aliveri[68032]	early Miocene (MN4a)	*Chamaeleo* cf. *andrusovi*[27], Chamaeleonidae indet.[26], Lacertidae indet.[27], Laterata indet. Or Scinciformata indet.[27], Anguinae indet.[27], Colubridae indet.[27], Viperidae indet.[27], Serpentes indet.[27]
Karydia[69435]	early Miocene (MN4a)	Lacertidae indet.[27], Laterata indet. or Scinciformata indet.[27], cf. *Ophisaurus* sp.[27], Anguinae indet.[27], Colubridae indet.[27], Serpentes indet.[27]
Kymi[206392]	early Miocene (MN3/4)	***Python euboicus***[28]
Lapsarna[186558]	early Miocene (?MN3)	Lacertidae indet.[29], ?Natricinae indet.[29], Serpentes indet.[29], Squamata indet.[29]

MN Mammal Neogene Zone

[1]Szyndlar 1991a, [2]Szyndlar 1991b, [3]Kotsakis 1977, [4]Georgalis et al. 2017b, [5]Schneider 1975, [6]Villa et al. 2020, [7]Szyndlar and Zerova 1990, [8]Georgalis et al. 2018c, [9]Loréal et al. 2020, [10]Owen 1857, [11]Georgalis et al. 2016a, [12]Richter 1995, [13]Szyndlar 1995, [14]Georgalis et al. 2019b, [15]Georgalis et al. 2017a, [17]Gaudry 1862a, [18]Gaudry 1862b, [19]Gaudry 1862–1867, [20]Weithofer 1888, [21]Nopcsa 1908, [22]Villa et al. 2018, [23]Conrad et al. 2012, [24]Georgalis et al. 2018b, [25]Georgalis et al. 2016c, [26]Georgalis et al. 2016b, [27]Georgalis et al. 2019a, [28]Römer 1870, [29]Vasileiadou et al. 2017

References

Arnold EN, Arribas O, Carranza S (2007) Systematics of the Palaearctic and Oriental lizard tribe Lacertini (Squamata: Lacertidae: Lacertinae), with descriptions of eight new genera. Zootaxa 1430:1–86

Augé M (2005) Evolution des lézards du Paléogène en Europe. Mém Mus National Hist Nat Paris 192:1–369

Augé M (2012) Amphisbaenians from the European Eocene: a biogeographical review. Palaeobiodivers Palaeoenviron 92:425–443

Bailon S, Rage J-C (1994) Squamates Néogènes et Pléistocènes du Rift occidental, Ouganda. In: B. Senut and M. Pickford (eds) Geology and Palaeobiology of the Albertine Rift Valley, Uganda–Zaire. Vol. 2: Palaeobiology. CIFEG Occasional Publications 1994/29, pp 129–135

Bedriaga von J (1881) Die Amphibien und Reptilien Griechenlands. Bull Soc Impériale Nat Moscou 56:242–310

Bedriaga von J (1882) Die Amphibien und Reptilien Griechenlands. Bull Soc Impériale Nat Moscou 56:43–103

Bedriaga von J (1884) Nachträgliche Bemerkungüber *Amphisbaena strauchi* v. *Bedr*. Zool Anz 7:346

Bellati A, Carranza S, Garcia-Porta J, Fasola M, Sindaco R (2014) Cryptic diversity within the *Anatololacerta* species complex (Squamata: Lacertidae) in the Anatolian peninsula: evidence from a multi-locus approach. Mol Phylogenet Evol 82:219–233

Blain H-A, Bailon S (2019) Extirpation of *Ophisaurus* (Anguimorpha, Anguidae) in Western Europe in the context of the disappearance of subtropical ecosystems at the Early-Middle Pleistocene transition. Palaeogeogr Palaeoclimatol Palaeoecol 520:96–113

Bolkay J (1913) Additions to the fossil herpetology of Hungary from the Pannonian and Praeglacial periode. Mitt Jahrb K Ungarischen Geol Reichsanstalt 21:217–230

Bonaparte CL (1832–1841) Iconographia della Fauna Italica per le Quattro Classi degli Animali Vertebrati. Tomo II. Anfibi. Dalla Tipografia Salviucci, Roma

Čerňanský A (2010) A revision of chamaeleonids from the Lower Miocene of the Czech Republic with description of a new species of *Chamaeleo* (Squamata, Chamaeleonidae). Geobios 43:605–613

Čerňanský A (2012) The oldest known European Neogene girdled lizard fauna (Squamata, Cordylidae), with comments on Early Miocene immigration of African taxa. Geodiversitas 34:837–848

Černanský A, Syromyatnikova EV (2019) The first Miocene fossils of *Lacerta* cf. *trilineata* (Squamata, Lacertidae) with a comparative study of the main cranial osteological differences in green lizards and their relatives. PLoS One 14:e0216191

Čerňanský A, Augé M, Rage J-C (2015) A complete mandible of a new amphisbaenian reptile (Squamata, Amphisbaenia) from the late Middle Eocene (Bartonian, MP 16) of France. J Vertebr Paleontol 35:e902379

Čerňanský A, Yaryhin O, Cicekovά J, Werneburg I, Hain M, Klembara J (2019) Vertebral comparative anatomy and morphological differences in anguine lizards with a special reference to *Pseudopusapodus*. Anat Rec 302:232–257

Conrad JL, Balcarcel AM, Mehling CM (2012) Earliest example of a Giant monitor lizard (*Varanus*, Varanidae, Squamata). PLoS One 7(8):e41767

Daudin FM (1803a) Histoire naturelle, génerale et particulière des reptiles; Ouvragefaisant suite aux Oeuvres de Leclerc de Buffon, et partie du courscompletd'histoire naturelle rédigé par C.S. Sonnini, membre de plusieurssociétéssavantes. Tome cinquième. F. Dufart, Paris

Daudin FM (1803b) Histoire naturelle, générale et particulière des reptiles: ouvragefaisant suite à l'Histoire Naturelle générale et particulière, composée par Leclerc de Buffon, et rédigée par C.S. Sonnini. Tome septième. F. Dufart, Paris

Delfino M, Kotsakis T, Arca M, Tuveri C, Pitruzella G, Rook L (2008) Agamid lizards from the Plio-Pleistocene of Sardinia (Italy) and an overview of the European fossil record of the family. Geodiversitas 30:641–656

Delfino M, Rage J-C, Bolet A, Alba DM (2013) Early Miocene dispersal of the lizard *Varanus* into Europe: reassessment of vertebral material from Spain. Acta Palaeontol Pol 58:731–735
Estes R (1983) Sauria Terrestria, Amphisbaenia. In: Wellnhofer P (ed) Encyclopedia of Paleoherpetology, part 10a. Gustav Fischer Verlag, Stuttgart/New York
Estes R, de Queiroz K, Gauthier JA (1988) Phylogenetic relationships within Squamata. In: Estes R, Pregill GK (eds) Phylogenetic relationships of the lizard families: essays commemorating C.L. Camp. Stanford University Press, Stanford, pp 119–281
Fejérváry de GJ (1918) Contributions to a monography on fossil Varanidae and on Megalanidae. Ann Hist-Nat Mus Nat hungarici 16:341–467
Fitzinger LJFJ (1826) Neue Classification der Reptiliennachihren Natürlichen Verwandtschaften. Nebsteiner Verwandtschafts-Tafel und einemVerzeichnisse der Reptilien-Sammlung des k. k. zoologischen Museums zu Wien. J. G. Huebner, Wien
Folie A, Smith R, Smith T (2013) New amphisbaenian lizards from the early Paleogene of Europe and their implications for the early evolution of modern amphisbaenians. Geol Belgica 16:227–235
Fürbringer M (1900) Zurvergleichen den anatomie des Brustschulter apparates und der Schultermuskeln. Jenaische Zeit 34:215–718
Gaudry A (1862–1867) Animaux fossiles et géologie de l'Attique. Savy, Paris
Gaudry A (1862a) Note sur les débris d'Oiseaux et de Reptiles trouvés à Pikermi, Grèce, suivie de quelques remarques de paléontologie générale. Bull Soc Géol Fr 19:629–640
Gaudry A (1862b) Résultats des fouilles exécutées en Grèce sous les auspices de l'Academie. C R Hebd Séances Acad Sci 54:502–505
Gauthier JA, Kearney M, Maisano JA, Rieppel O, Behike ADB (2012) Assembling the squamate tree of life: perspectives from the phenotype and the fossil record. Bull Peabody Mus Nat Hist 53:3–308
Geoffroy-Saint-Hilaire I (1827) Description des reptiles qui se trouventen Égypte. In: de Savigny MJ-CL (ed) Description de l'Égypte, ou recueil des observations et des recherches qui ontété faitesen Égypte pendant l'expédition de l'armée Française. Histoire naturelle. Tome premier. Imprimerie Impériale, Paris, pp 121–160
Georgalis GL (2019) Poor but classic: the squamate fauna from the late Miocene of Pikermi, near Athens, Greece. C R Palevol 18:801–815
Georgalis GL, Smith KT (2020) Constrictores Oppel, 1811 – the available name for the taxonomic group uniting boas and pythons. Vertebrate Zool 70:291–304
Georgalis GL, Szyndlar Z, Kear BP, Delfino M (2016a) New material of *Laophis crotaloides*, an enigmatic giant snake from Greece, with an overview of the largest fossil European vipers. Swiss J Geosci 109:103–116
Georgalis GL, Villa A, Delfino M (2016b) First description of a fossil chamaeleonid from Greece and its relevance for the European biogeographic history of the group. Sci Nat 103:12
Georgalis GL, Villa A, Vlachos E, Delfino M (2016c) Fossil amphibians and reptiles from Plakias, Crete: a glimpse into the earliest late Miocene herpetofaunas of southeastern Europe. Geobios 49:433–444
Georgalis GL, Villa A, Delfino M (2017a) Fossil lizards and snakes from Ano Metochi – a diverse squamate fauna from the latest Miocene of northern Greece. Hist Biol 29:730–742
Georgalis GL, Villa A, Delfino M (2017b) The last European varanid: demise and extinction of monitor lizards (Squamata, Varanidae) from Europe. J Vertebr Paleontol 37:e1301946
Georgalis GL, Halaçlar K, Mayda S, Kaya T, Ayaz D (2018a) First fossil find of the *Blanus strauchi* complex (Amphisbaenia, Blanidae) from the Miocene of Anatolia. J Vertebr Paleontol 38:e1437044
Georgalis GL, Rage J-C, de Bonis L, Koufos GD (2018b) Lizards and snakes from the late Miocene hominoid locality of Ravin de la Pluie (Axios Valley, Greece). Swiss J Geosci 111:169–181
Georgalis GL, Villa A, Delfino M (2018c) The last amphisbaenian (Squamata) from continental Eastern Europe. Ann Paléontol 104:155–159
Georgalis GL, Villa A, Ivanov M, Roussiakis S, Skandalos P, Delfino M (2019a) Early Miocene herpetofaunas from the Greek localities of Aliveri and Karydia – bridging a gap in the knowl-

edge of amphibians and reptiles from the early Neogene of southeastern Europe. Hist Biol 31:1045–1064

Georgalis GL, Villa A, Ivanov M, Vasilyan D, Delfino M (2019b) Fossil amphibians and reptiles from the Neogene locality of Maramena (Greece), the most diverse European herpetofauna at the Miocene/Pliocene transition boundary. Palaeontol Electronica 22.3.68:1–99

Georgalis GL, Mayda S, Alpagut B, Şarbak A, Güler G (2020) The westernmost Asian record of pythonids (Serpentes): the presence of *Python* in a Miocene hominoid locality of Anatolia. J Vertebr Paleontol 40:e1781144

Gistel JNFX (1868) Die lurche Europas. Ein Beitragzur Lehre von der geographischen Verbreitung der selben. In: Gistel J (ed) Blicke in das Leben der Natur und des Menschen. Verlag von Ed. Wartig, Leipzig, pp 144–167

Gmelin JF (1789) Caroli a Linne...Systema naturae per regna tria natural, secundum classes, ordines, genera, species, cum characteribus differentilis, synonymis, locis. Tomus I, Editio decima tertia, aucta, reformata. Pars III. Amphibia et Pisces. Georg. Emanuel Beer, Lipsiae, p 1038–1516

Gvoždík V, Jandzik D, Lymberakis P, Jablonski D, Moravec J (2010) Slow worm, *Anguis fragilis* (Reptilia: Anguidae) as a species complex: genetic structure reveals deep divergences. Mol Phylogenet Evol 55:460–472

Hermann J (1804) Observationes zoologicaequibus novae complures, aliaeque animalium species describuntur et illustrantur, opus posthumus edidit Fridericus Ludovicus Hammer. Pars prior, observationum quatuor centurias continens. Amandum Koenig, Parisiis

Hoernes R (1884) Elemente der Palaeontologie (Palaeozoologie). Verlag von Veit & Co, Leipzig

Hoffstetter R (1939) Contribution à l'étude des Elapidæ actuels et fossiles et de l'ostéologie des Ophidiens. Archives du Muséum d'Histoire Naturelle de Lyon 15:1–78

Hoffstetter R (1969) Présence de Varanidae (Reptilia, Sauria) dans le Miocène de Catalogne. Considérations sur l'histoire de la famille. Bull Mus Nat Hist Nat 40:1051–1064

Hoffstetter R, Rage J-C (1972) Les Erycinæ fossiles de France (Serpentes, Boidæ). Compréhension et histoire de la sous-famille. Annales de Paléontologie 58:81–124

Ivanov M, Böhme M (2011) Snakes from Griesbeckerzell (Langhian, Early Badenian), North Alpine Foreland Basin (Germany), with comments on the evolution of snake faunas in Central Europe during the Miocene Climatic Optimum. Geodiversitas 33:411–449

Jandzik D, Jablonski D, Zinenko O, Kukushkin OV, Moravec J, Gvoždík V (2018) Pleistocene extinctions and recent expansions in an anguid lizard of the genus *Pseudopus*. Zool Scripta 47:21–32

Kearney M, Rieppel O (2006) An investigation into the occurrence of plicidentine in the teeth of squamate reptiles. Copeia 2006:337–350

Kelly CMR, Barker NP, Villet MH, Broadley DG, Branch WR (2008) The snake family Psammophiidae (Reptilia: Serpentes): Phylogenetics and species delimitation in the African sand snakes (*Psammophis* Boie, 1825) and allied genera. Mol PhylogenetEvol 47:1045–1060

Klembara J (1981) Beitrag zur Kenntnis der Subfamilie Anguinae (Reptilia, Anguidae). Acta Univ Carolinae Geol 2:121–168

Klembara J (2012) A new species of *Pseudopus* (Squamata, Anguidae) from the early Miocene of Northwest Bohemia (Czech Republic). J Vertebr Paleontol 32:854–866

Klembara J, Rummel M (2018) New material of *Ophisaurus*, *Anguis* and *Pseudopus* (Squamata, Anguidae, Anguinae) from the Miocene of the Czech Republic and Germany and systematic revision and palaeobiogeography of the Cenozoic Anguinae. Geol Mag 155:20–44

Klembara J, Böhme M, Rummel M (2010) Revision of the Anguine Lizard *Pseudopus laurillardi* (Squamata, Anguidae) from the Miocene of Europe, with Comments on Paleoecology. J Paleontol 84(2):159–196

Klembara J, Hain M, Dobiašová K (2014) Comparative anatomy of the lower jaw and dentition of *Pseudopus apodus* and the interrelationships of species of subfamily Anguinae (Anguimorpha, Anguidae). Anat Rec 297:516–544

Kolendrianou M, Ligkovanlis S, Maniakas I, Tzortzi M, Iliopoulos G (2020) The Palaeolithic cave of Kalamakia (Mani Peninsula), Greece: new insights on the palaeoenvironment using microvertebrates and mesowear analysis of ruminant teeth. Heliyon 6:e03958

Kormos T (1911) A Polgárdi pliocén csontlelet. Földtani Közlöny 41:48–64
Kotsakis T (1977) I resti di anfibi e rettili pleistocenici della grotta "Bate" (Rethymnon, Creta). Atti Accad Naz Lincei 63:571–582
Kuhn O (1939) Squamata: Lacertilia et Ophidia. Fossilium Catalogus. I: Animalia. Pars 86. Verlag Gustav Feller, Neubrandenburg
Lacépède BGE (1789) Histoire naturelle des serpens. Tome second. Hotel de Thon, Paris
Laurenti JN (1768) Specimen medicum, exhibens synopsin reptiliumemendatam cum experimentis circa venena et antidota reptiliumaustracorum, quod authoritate et consensu. Joan. Thomae nob. De Trattnern, Vienna
Lawson R, Slowinski JB, Crother BI, Burbrink FT (2005) Phylogeny of the Colubroidea (Serpentes): new evidence from mitochondrial and nuclear genes. Mol Phylogenet Evol 37:581–601
Linnaeus C (1758) Systema Naturae per regna trianaturae, secundum classes, ordines, genera, species, cum characteribus, differentiis, synonymis, locis. Laurentii Salvii, Stockholm, Holmiae
Linnaeus C (1766) Systema naturæ per regna trianaturæ, secundum classes, ordines, genera, species, cum characteribus, differentiis, synonymis, locis. Tomus I. Editioduodecima, reformata. Laurentii Salvii, Stockholm, Holmiae
Loréal E,Villa A, Georgalis GL, Delfino M(2020) Amphibians and reptiles from the late Miocene and early Pliocene of the Ptolemais area (Western Macedonia, Greece). Ann Paléontol
Mangili G (1980) Fossils reptiles of Simonelli cave. Accad Naz Lincei Probl Attu Sc Cult Quaderno 249:121–122
Mead JI (2013) Scolecophidia (Serpentes) of the Late Oligocene and Early Miocene, North America, and a fossil history overview. Geobios 46:225–231
Merrem B (1820) Versucheines systems der Amphibien, vol 8. J. C. Krieger, Marburg
Müller J, Hipsley CA, Head JJ, Kardjilov N, Hilger A, Wuttke M, Reisz R (2011) Eocene lizard from Germany reveals amphisbaenian origins. Nature 473:364–367
Nopcsa F (1908) Zur Kenntnis der fossilen Eidechsen. Beitr Paläontol Geol Osterreich-ungarns Orients 21:33–62
Olivier GA (1801) Voyage dans l'Empire Othoman, l'Egypte et de la Perse, fait par ordre de Gouvernement, pendant les six premières années de la République. Avec Atlas. Tome premier. H. Agasse, Paris
Oppel M (1811a) Suite du 1er. memoire sur la classification des reptiles. Ord. II. Squammata mihi. Sect. II. Ophidii. Ord. III. Ophidii, Brongniart. Annales du Museum d'histoire Naturelle, Paris, 16:376–393
Oppel M (1811b) Die Ordnungen, Familien und Gattungen der Reptilien als Prodrom einer Naturgeschichte derselben. Joseph Lindauer, Munich, pp 87
Owen R (1857) On the fossil vertebræ of a Serpent (*Laophis crotaloïdes*, Ow.) discovered by Capt. Spratt, R.N., in a tertiary formation at Salonica. Quart J Geol Soc London 13:196–199
Palacký J (1898) La distribution des ophidiens sur le globe. Mém Soc Zool Fr 11:88–125
Pallas PS (1775) *Lacerta apoda* descripta. Nov Comm Acad Sci Imperialis Petropolitanae 19:435–454
Pallas PS (1814) Zoographia Rosso-Asiatica, sistens omnium animalium in extenso Imperio Rossico et adjacentibus maribus observatorum recensionem, domicilia, mores et descriptiones, anatomen atque icones plurimorum. Volumen tertium. Animalia monocardiaseufrigidi sanguinis imperii Rosso-Asiatici. Officina Caesarea Academiae Scientiarum, Petropoli
Pianka ER, Vitt LJ (2003) Lizards: windows to the evolution of diversity. University of California Press, Berkeley
Pianka ER, King D, King RA (2004) Varanoid Lizards of the World. Indiana University Press, Bloomington. https://doi.org/10.2307/j.ctt2005wjp
Pyron RA, Burbrink FT, Wiens JJ (2013) A phylogeny and revised classification of Squamata, including 4161 species of lizards and snakes. BMC Evol Biol 13:93
Pyron RA, Reynolds GA, Burbrink FT (2014) A taxonomic revision of boas (Serpentes: Boidae). Zootaxa 3846:249–260

Rafinesque CS (1815) Analyse de la nature ou tableau de l'univers et des corps organisés. Constantine Samuel Rafinesque-Schmaltz, Palermo

Rage JC (1984) Serpentes. In: Wellnhofer P (ed) Encyclopedia of paleoherpetology. Gustav Fischer, Stuttgart/New York

Rage J-C (2013) Mesozoic and Cenozoic squamates of Europe. Palaeobiodiv Palaeoenviron 93:517–534

Rage J-C, Holman JA (1984) Des serpents (Reptilia, Squamata) de type nord-américain dans le Miocène français. Evolution parallèle ou dispersion? Geobios 17:89–104. https://doi.org/10.1016/s0016-6995(84)80007-8

Richter A (1995) The vertebrate locality Maramena (Macedonia, Greece) at the Turolian-Ruscinian Boundary (Neogene). 3. Lacertilia (Squamata, Reptilia). Münchner Geowiss Abh A 28:35–39

Roček Z (1984) Lizards (Reptilia, Sauria) from the lower Miocene locality Dolnice (Bohemia, Czechoslovakia). Rozpravy Ceskoslovenské Akad Ved, Rada Mat a prírodních Ved 94:3–69

Rochebrune de AT (1880) Revision des ophidiens fossiles du Museum d'Histoire Naturelle. Nouv Arch Mus Hist Nat 3:271–296

Römer F (1870) Über *Python Euboïcus*, eine fossile Riesenschlange aus tertiärem Kalkschiefer von Kumi auf der Insel Euboea. Z Deut Geol Ges 22:582–590

Schneider B (1975) Ein mittelpleistozäne Herpetofauna von der Insel Chios, Ägäis. Senckenberg Biol 56:191–198

Sickenberg O (1971) Revision der Wirbeltierfauna der HohlePetralona. AnnGeol Pays Helleniques 23:230–264

Sindaco R, Kornilios P, Sacchi R, Lymberakis P (2014) Taxonomic reassessment of *Blanus strauchi* (Bedriaga, 1884) (Squamata: Amphisbaenia: Blanidae), with the description of a new species from south-east Anatolia (Turkey). Zootaxa 3795:311–326

Szyndlar Z (1991a) A review of Neogene and Quaternary snakes of Central and Eastern Europe. Part I: Scolecophidia, Boidae, Colubrinae. Estudios Geol 47:103–126

Szyndlar Z (1991b) A review of Neogene and Quaternary snakes of Central and Eastern Europe. Part II: Natricinae, Elapidae, Viperidae. Estudios Geol 47:237–266

Szyndlar Z (1995) The vertebrate locality Maramena (Macedonia, Greece) at the Turolian-Ruscinian Boundary (Neogene). 4. Serpentes (Squamata, Reptilia). Münchner Geowiss Abh A 28:35–39

Szyndlar Z (2005) Snake fauna from the Late Miocene of Rudabánya. In: Bernor RL, Kordos L, Rook L (eds) Multidisciplinary research at Rudabánya. Palaeontographia Ital 90:31–52

Szyndlar Z (2012) Early Oligocene to Pliocene Colubridae of Europe: a review. Bull Soc Géol Fr 183:661–681

Szyndlar Z, Rage J-C (1990) West Palearctic cobras of the genus *Naja* (Serpentes: Elapidae): interrelationships among extinct and extant species. Amphibia-Reptilia 11:385–400

Szyndlar Z, Rage J-C (1999) Oldest fossil vipers (Serpentes: Viperidae) from the Old World. Kaupia 8:9–20

Szyndlar Z, Rage J-C (2002) Fossil record of the true vipers. In: Schuett G, Hoggren M, Douglas M, Greene H (eds) Biology of the vipers. Eagle Mountain Publishing, Eagle Mountain, pp 419–441

Szyndlar Z, Rage J-C (2003) Non-erycine Booidea from the Oligocene and Miocene of Europe. Institute of Systematics and Evolution of Animals. Polish Academy of Sciences, Kraków

Szyndlar Z, Zerova G (1990) Neogene cobras of the genus *Naja* (Serpentes: Elapidae) of East Europe. Ann Naturhist Mus Wien 91:53–61

Uetz P, Stylianou A (2018) The original descriptions of reptiles and their subspecies. Zootaxa 4375:257–264

Valakos ED, Pafilis P, Sotiropulos K, Lymberakis P, Maragou P, Foufopulos J (2008) The amphibians and reptiles of Greece. Frankfurt contribution to natural history, vol 32. Edition Chimaira, Frankfurt am Main

Vasileiadou K, Böhme M, Neubauer TA, Georgalis GL, Syrides GE, Papadopoulou L, Zouros N (2017) Early Miocene gastropod and ectothermic vertebrate remains from the Lesvos Petrified Forest (Greece). Paläontol Z 91:541–564

Venczel M (1998) Late Miocene snakes (Reptilia: Serpentes) from Polgárdi (Hungary): a second contribution. Acta Zool Cracoviensia 4:1–22

Vidal N, Hedges BS (2005) The phylogeny of squamate reptiles (lizards, snakes, and amphisbaenians) inferred from nine nuclear protein-coding genes. C R Biol 328:1000–1008

Vidal N, Hedges BS (2009) The molecular evolutionary tree of lizards, snakes, and amphisbaenians. C R Biol 332:129–139

Vidal N, Delmas A-S, David P, Cruaud C, Couloux A, Hedges SB (2007) The phylogeny and classification of caenophidian snakes inferred from seven nuclear protein-coding genes. C R Biol 330:182–187

Villa A, Delfino M (2019) A comparative atlas of the skull osteology of European lizards (Reptilia: Squamata). Zool J Linnean Soc 187:829–928

Villa A, Tschopp E, Georgalis GL, Delfino M (2017) Osteology, fossil record and palaeodiversity of the European lizards. Amphibia-Reptilia 38:79–88

Villa A, Abella J, Alba DM, Almécija S, Bolet A, Koufos GD, Knoll F, Luján ÁH, Morales J, Robles JM, Sánchez IM, Delfino M (2018) Revision of *Varanus marathonensis* (Squamata, Varanidae) based on historical and new material: morphology, systematics, and paleobiogeography of the European monitor lizards. PLoS One 13:e0207719

Villa A, Georgalis GL, Delfino M (2020) The latest early Pleistocene amphibians and reptiles from Kaiafas (Greece) and the first record of fossil *Ophiomorus* (Squamata, Scincidae). Geobios

Wagler JG (1824) Serpentum brasiliensium species novae ou Histoire naturelle des espèces nouvelles de serpens, recueillies et observées pendant le voyage dans l'intérieur du Brésil dans les années 1817, 1818, 1819, 1820, exécuté par ordre de Sa Majesté le Roi de Bavière, publiée par Jean de Spix,...décrited' après les notes du voyageur par Jean Wagler. Franc. Seraph. Hübschmann, Monachii

Wagler JG (1830) Natürliches System der Amphibien, mit vorangehender Classification der Saugthiere und Vogel. Ein Beitrag zur vergleichenden Zoologie. J.G. Cotta schenBuchhandlung, Munchen/Stuttgart/Tubingen

Wagler JG (1833) Descriptiones et Icones Amphibiorum. Sumtibus J.G. Cottae, Monachii, Stuttgartiae, Tubingae

Wallach V, Williams KL, Boundy J (2014) Snakes of the world: a catalogue of living and extinct species. CRC Press, Boca Raton/London/New York

Weithofer A (1888) Beiträge zur Kenntniss der Fauna von Pikermi bei Athen. Beit Paläontol Österreich-Ungarns 6:225–292

Werner F (1894) Die Reptilien- und Batrachierfauna der jonischen Inseln. Verh Zool Bot Ges Wien 44:225–237

White J (1790) Journal of a voyage to New South Wales, with sixty–five plates of non descript animals, birds, lizards, serpents, curious cones of trees and other natural productions. Debrett, London

Wiens JJ, Kuczynski CA, Townsend T, Reeder TW, Mulcahy DG, Sites JW Jr (2010) Combining phylogenomics and fossils in higher-level squamate reptile phylogeny: molecular data change the placement of fossil taxa. Syst Biol 59:674–688

Wiens JJ, Hutter CR, Mulcahy DG, Noonan BP, Townsend TM, Sites JW, Reeder TW (2012) Resolving the phylogeny of lizards and snakes (Squamata) with extensive sampling of genes and species. Biol Lett 8:1043

Zaher H, Grazziotin FG, Cadle JE, Murphy RW, Cesar de Moura-Leite J, Bonatto SL (2009) Molecular phylogeny of advanced snakes (Serpentes, Caenophidia) with an emphasis on South American xenodontines: a revised classification and descriptions of new taxa. Pap Avulsos Zool 49(11):115–153

Zittel KA (1887–1890) Handbuch der Paläontologie. Palaeozoologie. III. Pisces, Amphibia, Reptilia, Aves. Druck und Verlag von R. Oldenbourg, Munchen/Leipzig

The Fossil Record of Crocodylians (Reptilia: Crocodylia) in Greece

Georgios L. Georgalis and Massimo Delfino

1 Introduction

Crocodylians are large, carnivorous, and semiaquatic reptiles, currently distributed across Africa, southern Asia, Australia, and southern North, Central, and South America (Trutnau and Sommerlad 2006). However, their fossil record denotes that they had a much wider geographic distribution, with an array of currently extinct taxa that achieved several different bauplans, ecologic adaptations, and size ranges, conquering many different palaeoenvironments during the Late Mesozoic and throughout the Cenozoic (Brochu 2001a, b, 2003; Scheyer et al. 2013; Mannion et al. 2015, 2019; Godoy et al. 2019).

Despite the astonishing Cenozoic diversity that encompassed also Europe (among the recent publications see: Brochu 2006, 2007; Delfino et al. 2007, 2019; Piras et al. 2010; Delfino and Rook 2008; Delfino and Smith 2009, 2012; Martin 2010; Martin and Gross 2011; Delfino and Rossi, 2013; Martin et al. 2014; Díaz Aráez et al. 2017; Macaluso et al. 2019; Rio et al. 2019; Zoboli et al. 2019), crocodylians are only scarcely known in the southeastern margins of the continent, particularly in Greece. We here review this so far poor, fossil record that nevertheless confirms the presence of these fascinating reptiles in the area.

G. L. Georgalis (✉)
Palaeontological Institute and Museum, University of Zurich, Zurich, Switzerland

Dipartimento di Scienze della Terra, Università di Torino, Torino, Italy
e-mail: georgios.georgalis@pim.uzh.ch

M. Delfino
Dipartimento di Scienze della Terra, Università di Torino, Torino, Italy

Institut Català de Paleontologia Miquel Crusafont, Universitat Autònoma de Barcelona, Barcelona, Spain
e-mail: massimo.delfino@unito.it

E. Vlachos (ed.), *Fossil Vertebrates of Greece Vol. 1*,
https://doi.org/10.1007/978-3-030-68398-6_8

2 Historical Overview

Although crocodylian fossil remains have been known in the Balkans since several decades (e.g., Huene and Nikoloff 1963; Schleich 1994; Jovanović 1995;Vasilyan 2019; Georgalis et al. 2021), they instead have only recently been described from the Greek area (Georgalis et al. 2016, 2019; Vasileiadou et al. 2017).

3 Phylogenetic Relationships

Extant crocodylians are ranked into three distinct families, that is, Crocodylidae, Gavialidae, and Alligatoridae (Trutnau and Sommerlad 2006). As for the exact interrelationships between true and false gharial families (lineages of the extant genera *Gavialis* Oppel, 1811 and *Tomistoma* Müller, 1846, respectively), there is not as yet a full consensus, as topologies resulting from phylogenetic analyses based on morphology (also including the fossil record) contradict those based on molecular data (see Gold et al. 2014 and references therein).

4 Distribution

Crocodylian fossils from the Greek area are rather scarcely known (Fig. 1 and Table 1), a fact that also resides to their finds being only much recently described and, above all, that they are represented only by little informative skeletal elements, such as isolated teeth and one osteoderm. Nevertheless, even this poor fossil record attests a relatively wide geographic distribution spanning from the early Miocene and ultimately terminating during the early late Miocene (Georgalis et al. 2016, 2019; Vasileiadou et al. 2017). The earliest crocodylian remains are known from the early Miocene of Lapsarna in Lesvos Island (Vasileiadou et al. 2017) and Aliveri in Euboea Island (Georgalis et al. 2019). Crocodylian remains from the early Miocene of nearby Anatolia (Syromyatnikova et al. 2019) evince that this group was widespread in the Eastern Mediterranean during the early Neogene. The only other record of crocodylians in Greece originates from the late Miocene (MN9) of Plakias in Crete (Georgalis et al. 2016). The latter find is of special interest as it represents not only the youngest record of the group from Greece but from the whole northeastern Mediterranean (Georgalis et al. 2016). In fact, the Plakias crocodylian represents one of the youngest crocodiles from the whole Europe, being surpassed only by the slightly younger remains from the centralnorth and northwestern Mediterranean (see Georgalis et al. 2016 for details). It is nevertheless unclear whether these late Miocene European–Mediterranean crocodylian records (including the Plakias one) represent relics of the once widespread European lineages or are simply the product of dispersal(s) from northern Africa, taking into

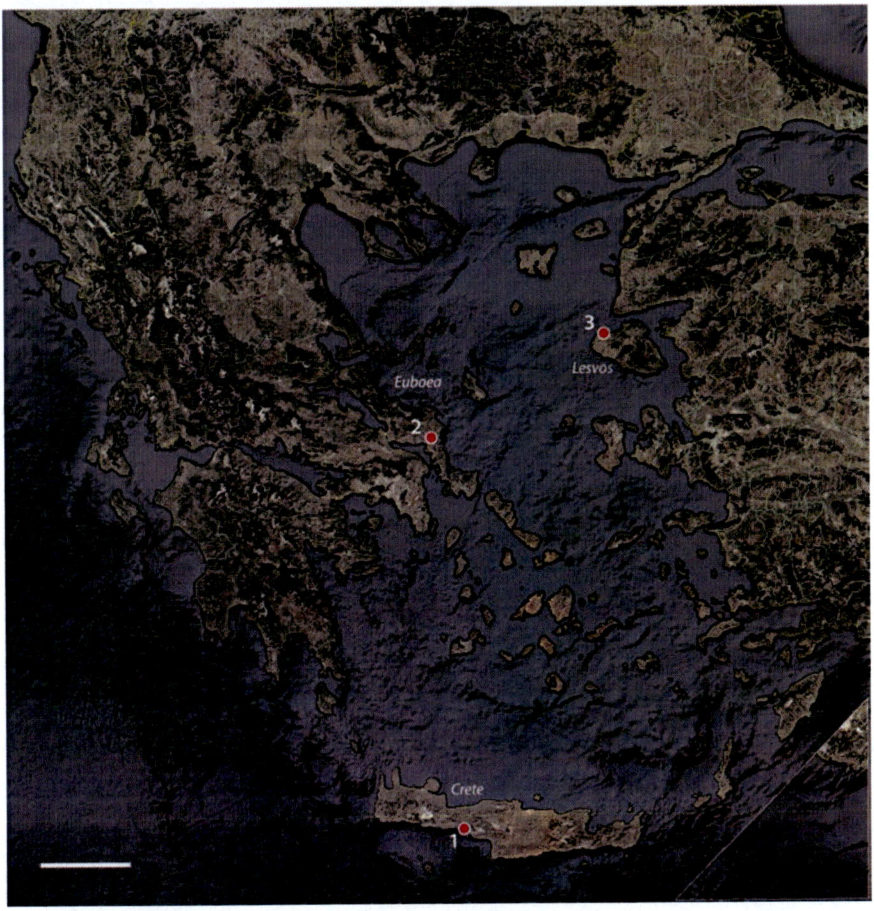

Fig. 1 Map of the occurrences of fossil crocodylians from Greece. See Table 1 for further information. **1**, Plakias; **2**, Aliveri; **3**, Lapsarna. See Appendix for more information. Image exported from Google Earth Pro © 2019, map data from US Dept. of State Geographer, SIO, NOAA, U.S. Navy, NGA, GEBCO, image from Landsat/Copernicus. Scale bar equals 80 km, North faces upward

Table 1 List of the Greek localities containing crocodylian fossils. Locality numbers refer to the collection numbers of the PaleoBiology Database (PBDB)

Locality[PBDB No]	Age (MN)	Taxon
Plakias[183151]	late Miocene (MN9)	Crocodylia indet.[1]
Aliveri[68032]	early Miocene (MN4a)	Crocodylia indet.[2]
Lapsarna[186558]	early Miocene(?MN3)	Crocodylia indet.[3]

MN Mammal Neogene Zone
[1]Georgalis et al. 2016, [2]Georgalis et al. 2019, [3]Vasileiadou et al. 2017

consideration also the marine dispersal capabilities of extant (Campbell et al. 2010; Nicolaï and Matzke 2019) and probably extinct taxa (Delfino et al. 2020a).

5 Systematic Paleontology

Crocodylia Gmelin, 1789

Remarks The so far existing fossil crocodylian material from Greece consists exclusively of isolated teeth (Fig. 2; Georgalis et al. 2016, 2019; Vasileiadou et al. 2017), with the single exception of an osteoderm (Vasileiadou et al. 2017). Crocodylian teeth are in most cases not diagnostic for identification at the genus or family level. The same applies for the known osteoderm, which was never figured anyway. Crocodylian teeth have a conical crown and—if preserved—a variably long, hollow, tubular root. When the root is broken off or reabsorbed, the base of the crown is clearly concave. The crown is characterized by a mesio-distal carina separating the lingual and the labial surface. It is unclear whether the isolated teeth from Greece belong to one or more distinct taxa because they show a generalized morphology—accordingly, size differences cannot be interpreted as indicative of taxonomic diversity or simply due to ontogenetic status. However, at least the late Miocene remains could belong to the genus *Crocodylus* that seems to be present in

Fig. 2 Selected specimens of fossil crocodylians from Greece. **a–b** Isolated teeth from Aliveri (UU AL 3538 in mesial and 3536 in lingual views, respectively; see also Georgalis et al. 2019), and **c–d** isolated teeth from the late Miocene of Plakias (UU PL 706 in labial and 708 in mesial views, respectively; see also Georgalis et al. 2016). Collection of the Faculty of Geosciences, University of Utrecht (The Netherlands). Scale bar equals 1 mm

several southern European late Miocene localities (Montoya et al. 2006; Delfino et al. 2007; Delfino and Rook 2008; Delfino and Rossi 2013; Delfino et al. 2020b). Pending the recovery and description of more complete material, at the moment we can only say that Greece was inhabited during part of the Miocene by indeterminate crocodylians. This is, nevertheless, an important conclusion that clearly confirms the presence of these reptiles in the country. These established crocodylian occurrences in Greece also highlight the recent advances on our understanding of Greek palaeoherpetofauna.

Acknowledgments We wish to thank the two reviewers, Elena Syromyatnikova and Andrea Villa, for their useful comments, which enhanced the quality of the manuscript. GLG acknowledges support from Forschungskredit of the University of Zurich, grant no. [FK-20-110].

References

Brochu CA (2001a) Crocodylian snouts in space and time: phylogenetic approaches toward adaptive radiation. Am Zool 41:564–585

Brochu CA (2001b) Congruence between physiology, phylogenetics and the fossil record on crocodylian historical biogeography, pp 9–28. In: Grigg GC, Seebacher F, Franklin CE (eds) Crocodilian biology and evolution. Surrey Betty & Sons, Chipping Norton, pp 1–446

Brochu CA (2003) Phylogenetic approaches toward crocodylian history. Ann Rev Earth Planet Sc 31:357–397

Brochu CA (2006) *Eosuchus* (Crocodylia, Gavialoidea) from the Lower Eocene of the Isle of Sheppey, England. J Vert Paleontol 26(2):466–470

Brochu CA (2007) Systematics and taxonomy of Eocene Tomistomine crocodylians from Britain and Northern Europe. J Palaeontol 50(4):1–13

Campbell HA, Watts ME, Sullivan S, Read AM, Choukroun S, Irwin SR, Franklin CE (2010) Estuarine crocodiles ride surface currents to facilitate long-distance travel. J Anim Ecol 79:955–964

Delfino M, Rook L (2008) African crocodylians in the Late Neogene of Europe. A revision of *Crocodylus bambolii* Ristori, 1890. J Paleontol 82(2):336–343

Delfino M, Rossi MA (2013) Fossil crocodylid remains from Scontrone (Tortonian, Southern Italy) and the Late Neogene Mediterranean biogeography of crocodylians. Geobios 46:25–31

Delfino M, Smith T (2009) A reassessment of the morphology and taxonomic status of *'Crocodylus' depressifrons* Blainville, 1855 (Crocodylia, Crocodyloidea) based on the early Eocene remains from Belgium. Zool J Linnean Soc 156:140–167

Delfino M, Smith T (2012) Reappraisal of the morphology and phylogenetic relationships of the middle Eocene alligatoroid *Diplocynodon deponiae* (Frey, Laemmert, and Riess, 1987) based on a three-dimensional specimen. J Vert Paleontol 32:1358–1369

Delfino M, Böhme M, Rook L (2007) First European evidence for transcontinental dispersal of *Crocodylus* (late Neogene of southern Italy). Zool J Linnean Soc London 149:293–307

Delfino M, Martin JE, de Lapparent de Broin F, Smith T (2019) Evidence for a pre-PETM dispersal of the earliest European crocodyloids. Hist Biol 31:845–852

Delfino M, Iurino D, Mercurio B, Piras P, Rook L, Sardella R (2020a) Old African fossils provide new evidence for the origin of the American crocodiles. Sci Rep 10:11127

Delfino M, Luján ÀH, Abella J, Alba DM, Böhme M, Pérez-Ramos A, Tschopp E, Morales J, Montoya P (2020b) Late Miocene crocodylian remains from Venta del Moro (Iberian

Peninsula) provide further insights on the dispersal of *Crocodylus* across the late Miocene Tethys. J Paleontol 95:184. https://doi.org/10.1017/jpa.2020.62
Díaz Aráez JL, Delfino M, Luján ÀH, Fortuny J, Bernardini F, Alba DM (2017) New remains of *Diplocynodon* (Crocodylia: Diplocynodontidae) from the early Miocene of the Iberian Peninsula. C R Palevol 16:12–26
Georgalis GL, Čerňanský A, Mayda S (2021) Late Paleogene herpetofaunas from the crossroads between two continents - new amphibian and reptile remains from the Oligocene of southern Balkans and Anatolia. C R Palevol
Georgalis GL, Villa A, Vlachos E, Delfino M (2016) Fossil amphibians and reptiles from Plakias, Crete: a glimpse into the earliest late Miocene herpetofaunas of southeastern Europe. Geobios 49:433–444
Georgalis GL, Villa A, Ivanov M, Roussiakis S, Skandalos P, Delfino M (2019) Early Miocene herpetofaunas from the Greek localities of Aliveri and Karydia – bridging a gap in the knowledge of amphibians and reptiles from the early Neogene of southeastern Europe. Hist Biol 31:1045–1064
Godoy PL, Benson RBJ, Bronzati M, Butler RJ (2019) The multi-peak adaptive landscape of crocodylomorph body size evolution. BMC Evol Biol 19(1):167
Gold MEL, Brochu CA, Norell MA (2014) An expanded combined evidence approach to the *Gavialis* problem using geometric morphometric data from crocodylian braincases and eustachian systems. PLoS One 9(9):e105793
Huene von F, Nikoloff I (1963) Ein pliozänes Krokodil in Bulgarien. N Jahr Geol Paläontol Abh 118:266–271
Jovanović M (1995) *Crocodylus moraviensis* – a new species of tertiary crocodile in the marls in Popovac near Paraćin (Serbia). C R Séances Soc Serbe Géol:35–38
Macaluso L, Martin JE, Del Favero L, Delfino M (2019) Revision of the crocodilians from the Oligocene of Monteviale (NE Italy) and the diversity of European eusuchians across the Eocene-Oligocene boundary. J Vert Paleontol:e1601098
Mannion PD, Benson RBJ, Carrano MT, Tennant JP, Judd J, Butler RJ (2015) Climate constrains the evolutionary history and biodiversity of crocodylians. Nat Commun 6:8438
Mannion PD, Chiarenza AA, Godoy PL, Cheah YN (2019) Spatiotemporal sampling patterns in the 230 million year fossil record of terrestrial crocodylomorphs and their impact on diversity. Palaeontology 62(4):615–637
Martin JE (2010) A new species of *Diplocynodon* (Crocodylia, Alligatoroidea) from the Late Eocene of the Massif Central, France, and the evolution of the genus in the climatic context of the Late Palaeogene. Geol Mag 147(4):596–610
Martin JE, Gross M (2011) Taxonomic clarification of *Diplocynodon* Pomel, 1847 (Crocodilia) from the Miocene of Styria, Austria. N Jahrb Geol PaläontolAbh 261:177–193
Martin JE, Smith T, Lapparent de Broin F d, Escuillie F, Delfino M (2014) Late Paleocene eusuchian remains from Mont de Berru, France and the origin of the alligatoroid *Diplocynodon*. Zool J Linnean Soc 172:867–891
Montoya P, Morales J, Robles F, Abella J, Benavent JV, Sánchez FR (2006) Las nuevas excavaciones (1995–2006) en el yacimiento del Mioceno final de Venta del Moro, Valencia. Estudios Geol 62:313–326
Nicolaï MPJ, Matzke NJ (2019) Trait-based range expansion aided in the global radiation of Crocodylidae. Glob Ecol Biogeogr 28:1244. https://doi.org/10.1111/geb.12929
Piras P, Colangelo P, Adams DC, Buscalioni AD, Cubo J, Kotsakis A, Meloro C, Raia P (2010) The *Gavialis-Tomistoma* debate: the contribution of skull ontogenetic allometry and growth trajectories to the study of crocodylian relationships. Evol Dev 12:568–579
Rio JP, Mannion PD, Tschopp E, Martin JE, Delfino M (2019) Reappraisal of the morphology and phylogenetic relationships of the alligatoroid crocodilian *Diplocynodon hantoniensis* from the late Eocene of the United Kingdom. Zool J Linnean Soc. https://doi.org/10.1093/zoolinnean/zlz034

Scheyer TM, Aguilera OA, Delfino M, Fortier DC, Carlini AA, Sánchez R, Carrillo-Briceño JD, Quiroz L, Sánchez-Villagra MR (2013) Crocodylian diversity peak and extinction in the late Cenozoic of the northern Neotropics. Nat Commun 4:1907

Schleich HH (1994) Fossil Schildkröten- und Krokodilreste aus dem Tertiär Thrakiens (W-Türkei). Cour Forschung Senckenberg 173:137–151

Syromyatnikova E, Georgalis GL, Mayda S, Kaya T, Saraç G (2019) A new early Miocene herpetofauna from Kilçak, Turkey. Rus J Herpetol 26:205–224

Trutnau L, Sommerlad R (2006) Crocodilians. Their natural history & captive husbandry. Chimaira, Frankfurt am Main

Vasileiadou K, Böhme M, Neubauer TA, Georgalis GL, Syrides GE, Papadopoulou L, Zouros N (2017) Early Miocene gastropod and ectothermic vertebrate remains from the Lesvos Petrified Forest (Greece). Palaeontol Z 91:541–564

Vasilyan D (2019) Fish, amphibian and reptilian assemblage from the middle Miocene locality Gračanica – Bugojno palaeolake. Bosnia Herzegovina Palaeobiodiv Palaeoenviron 100:437. https://doi.org/10.1007/s12549-019-00381-8

Zoboli D, Sanciu L, Pillola GL, Delfino M (2019) An overview of the crocodylian fossil record from Sardinia (Italy). Ann Paléontol 105(2):123–137

The Fossil Record of Turtles and Tortoises (Reptilia: Testudines) in Greece

Evangelos Vlachos

1 Introduction

Turtles are amniotes with a fully developed turtle shell, included in the clade Testudinata, whereas crown turtles are included in the more derived clade Testudines (see Joyce et al. 2004). The extant diversity of turtles comprises at least 356 species distributed globally (except the Arctic and Antarctica), adapted to the majority of continental and marine environments (TTWG 2017 and references therein). The basic distinction of extant crown turtles (order Testudines) is in two suborders, Pleurodira and Cryptodira, depending on the kind of neck retraction. The extant turtle fauna of Greece includes nine species, all within Cryptodira: the marine turtles of the superfamily Chelonioidea (three species, but only *Caretta caretta* nests on Greek coasts) and the continental turtles of the superfamily Testudinoidea (nine species) (see TTWG 2017 and references therein for further information). Up to now, there is no evidence of chelonioid turtles in the fossil record of Greece. All remaining continental turtles that currently inhabit Greece (exc. the introduced species) have been positively identified in the fossil record of the country (Vlachos 2015). For some clades (e.g., Geoemydidae and Testudinidae), additional species have been found in Greece, including at least two species of giant tortoises (Testudinidae). Finally, one pleurodire taxon is represented as well in the fossil record of Greece (Georgalis et al. 2013).

Because of the presence of a solid shell, turtles have great fossilization potential and represent a relatively common and easily identifiable part of the vertebrate fossil record, as even the smallest shell fragments can be confidently identified at least

E. Vlachos (✉)
Museo Paleontológico Egidio Feruglio, CONICET, Trelew, Chubut, Argentina
e-mail: evlacho@mef.org.ar; evlacho@geo.auth.gr

E. Vlachos (ed.), *Fossil Vertebrates of Greece Vol. 1*,
https://doi.org/10.1007/978-3-030-68398-6_9

at the family level. Fossil turtles from Greece have received focused research interest during the last decade.

2 Historical Overview

As fossil turtles have been found in some of the earliest discovered localities in the country (e.g., Pikermi and Samos), the published record of fossil turtles from Greece covers a long time period, from 1862 to the present. Roughly, the history of studies on fossil turtles from Greece can be divided into three stages, as the history of vertebrate fossils of Greece (Introduction, this book).

Early Stage This stage includes all the works that have been done on fossil turtles in Greece until the end of the World War II. During the Early Stage, several prominent foreign workers contributed with publications on fossil turtles, naming several of the Greek taxa as well. The earliest reference of a fossil turtle from Greece was made by Gaudry (1862) as part of his classical works about the fossil fauna from the Late Miocene of Pikermi, near Athens, including the type specimens of *Testudo marmorum*—this is the first fossil turtle taxon named from Greece. Shortly after, Forsyth Mayor (1891) mentioned fossil turtle occurrences in the Late Miocene of Samos. At the beginning of the twentieth century, del Campana (1917, 1919) described a shell from the species *Testudo amiatae* and eggs of a giant tortoise, from the locality Capudjlar (= Pylea), near Thessaloniki. Few years later, Arambourg and Piveteau (1929) published material from several sites near Thessaloniki, including several specimens of small- and large-sized testudinids. Around the same time, Barnum Brown collected specimens of a giant tortoise from various quarries in Samos Island, which remained unpublished until recently (Vlachos et al. 2019a). The Early Stage culminates with Szalai's (1931) description of the first giant tortoise taxon from Greece, "*Testudo*" *schafferi* from the Late Miocene of Samos Island. All fossil specimens collected during the Early Stage are housed in foreign institutions outside Greece. The only exception to this rule is the collection of fossil tortoises in Samos Island by Th. Skouphos (AMPG), specimens that, nevertheless, remained unknown until recently (Vlachos et al. 2019c).

Intermediate Stage This stage includes all post-World War II works of the twentieth century. The main difference of this stage is that it includes mostly works made by Greek researchers, who enriched significantly the collections of this country. Most of these works described turtles in the context of their accompanied fauna, and certainly turtles were far from a solid research interest at that time. Paraskevaidis (1955) described two endocasts (steinkerns) of two testudinids from the Lower to Middle Miocene of Chios Island, which unfortunately have been lost during the World War II. Later, Melentis (1966) described turtle fossils from the Pleistocene of Megalopolis basin, which are the first freshwater turtle remains from Greece. Bachmayer and Symeonidis published a series of papers describing new turtle material from Greece, mainly from the Late Miocene of Pikermi and the Pliocene of

Northern Greece (Bachmayer 1967; Bachmayer and Symeonidis 1970, 1977; Bachmayer et al., 1979), as well as from the Pleistocene of Crete Island (Bachmayer et al. 1975). Additional materials from caves in Crete (Kotsakis 1977; Brinkerink 1996) and Tilos (Bachmayer and Symeonidis 1975) islands were added during the last part of the century. The last works of the Intermediate Stage include isolated occurrences of turtles from new localities in the country (Schleich 1982; Rauscher 1995; Schmidt-Kittler 1995; van Vugt et al. 2000), but without detailed information. During that stage, only a new subspecies has been erected (*Testudo marginata cretensis* Bachmayer et al. 1975) from Crete, whereas the remaining occurrences were attributed to the previously named species (e.g., *Testudo marmorum*) or were not identified to the specific level.

Modern Stage This stage includes all works of the twenty-first century and is when fossil turtles begin to be considered as the topic of specific research and fieldwork. The first work that truly tried to place the turtle fossil record from Greece in a global context is done by Lapparent de Broin (2002), where she described new specimens of a giant tortoise from Lesvos Island, together with specimens of a small testudinid. Although these materials were quite fragmentary, Lapparent de Broin (2002) discussed extensively the evolution of giant testudinids in the area. Recently, researchers working exclusively with turtles (E. Vlachos) or with other reptiles as well (G. Georgalis) have presented a wealth of new information on the fossil record of the country. Some highlights include the first and only pleurodire in Greece (Georgalis et al. 2013), first occurrences of soft-shelled turtles (Vlachos et al. 2015b; Georgalis et al. 2016; Vlachos and Bacharidis 2018), new species of small-sized (Vlachos and Tsoukala 2016) and giant testudinids (Vlachos et al. 2014), a new species of a geoemydid (Vlachos et al. 2019a), and numerous works on new occurrences (Tsoukala et al. 2011; Vlachos and Tsoukala 2014; Garcia et al. 2016; Vlachos and Delfino 2016; Georgalis et al. 2019a, b; Athanassiou et al. 2018; Loreál et al. 2020) and revision of previously published material (Vlachos et al. 2015a) or both (Vlachos 2015). The most recent effort to provide an extended summary of the published specimens was made by Georgalis and Kear (2013), which mostly summarizes our knowledge of the Early and Intermediate Stages under modern concepts of turtle taxonomy. However, as the turtle record of the country has been greatly enriched and critically reevaluated since then, it seems timely and necessary to provide an updated review for the purposes of this volume—with the rate that turtles are being described from Greece lately, this will certainly not be the final word.

3 Phylogenetic Relationships

The phylogenetic relationships of the fossil turtles from Greece are currently well-known in most cases, as a result of the growth of our knowledge during the modern times, as explained above. The majority of the named and valid species have been included in various cladistic analyses during the last 5 years. As most of the species belong to clades with extant representatives—and thus with well-known phylogenetic relationships—their position is generally well-supported. As such, the fossil turtles from Greece can be placed within a solid taxonomic and evolutionary framework.

The phylogenetic relationships of the only pleurodire turtle from Greece, *Nostimochelone lampra*, remain unclear. In its original description (Georgalis et al. 2013) and later (Georgalis and Kear 2013), it has been included within Podocnemidoidea. Based on the preserved information, it is even possible to further refine its placement within this superfamily, within Bothremydidae (Vlachos 2015)—it would be necessary, however, to complete the observations on the dorsal side of the carapace. The phylogenetic relationships of this diverse clade are now well-known (Ferreira et al. 2018 and references therein). Although the main temporal distribution of the family is in the Cretaceous, some bothremydids could extend to the Paleogene (Gaffney et al. 2006). It is therefore necessary to fully corroborate in the future the bothremydid affinities of *Nostimochelone*, as it could represent the youngest stratigraphically member of this clade.

With the exception of some more exclusive clades, the overall relationships of Trionychidae are obscure and largely unresolved (Vitek and Joyce 2015, and references therein). The fossil soft-shelled turtles from Greece are few and known only by isolated shell fragments that cannot be identified at the species level. It is therefore impossible to state anything on their phylogenetic relationships.

An interesting topic, from a phylogenetic point of view, is the position of the newly described *Mauremys aristotelica* by Vlachos et al. (2019a). This taxon exhibits conspicuously wide vertebral scutes, which are similar to some degree to other taxa from the Old World like *Clemmydopsis*, *Sakya*, *Sarmatemys*, and *Shansiemys* (Danilov 2005, Danilov et al. 2017, Vlachos et al. 2019a; and references therein). These taxa—which actually exhibit an enormous variety of scute morphology associated with the wide vertebral scutes—have been considered as representing separate clades of unclear placement within the total clade of Geoemydidae. This hypothesis would make the appearance of the wide vertebral scutes homoplastic within the total clade of Geoemydidae. The recent analysis of Vlachos et al. (2019a), however, demonstrates that it is more parsimonious to consider that all these taxa form a clade characterized by wide vertebral scutes, retrieved deeply nested within *Mauremys*.

On the other hand, the diverse testudinid records from Greece also bring important information on the phylogenetic relationships of the family Testudinidae. Although the taxonomy of many extinct members of this family remained problematic, the renewed interest on tortoise fossils from Europe and Greece over the last years has offered plenty of opportunities to investigate further the phylogeny of the tortoises (Pérez-García and Vlachos 2014; Vlachos 2015; Vlachos and Tsoukala 2016; Luján et al. 2016; Pérez-García et al. 2017a; Vlachos and Rabi 2018). Now it becomes clear that derived tortoises from Europe and Greece (i.e., Testudininae) belong mainly to two clades: Geochelonini and Testudinini.

The giant tortoises from Greece, *Titanochelon bacharidisi* and *Titanochelon schafferi*, are placed within the genus/clade *Titanochelon* together with numerous other species from the Neogene of Europe (see Pérez-García and Vlachos 2014). Both species have been previously included in the waste-basket genus *Cheirogaster*, which at a certain point including all European Cenozoic large-sized tortoises without a cervical scute (Lapparent de Broin 2001, 2002). Recent works suggest that *Cheirogaster* is placed basally within Testudinidae (Pérez-García and Vlachos 2014; Vlachos and Rabi 2018), and derived giant tortoises like *Titanochelon* are sister to the extant African *Stigmochelys*.

On the other hand, the small-sized tortoises are placed within Testudinini. It is now unclear whether all small-sized tortoises from Greece are placed within crown *Testudo*, as *Testudo marmorum* from Pikermi (Vlachos et al. 2019b) or those from Samos and Chios could be located along the stem that leads to modern *Testudo* (Vlachos et al. 2019c).

Besides the specimens attributed to the *Testudo graeca* complex, *Testudo hellenica* and *Testudo brevitesta* are probably placed in successive, more derived positions leading to the extant *Testudo marginata*. Depending on the situation of *Testudo marmorum* and its hypo-xiphiplastral hinge (see below), the marble tortoise could also be placed within this clade, in between *Te. hellenica* and *Te. brevitesta*. This entire clade has been named *Chersus* by Parham in Parham et al. (2006) and includes also the extant *Testudo kleinmanni* from Africa. This is a clade that probably has a South Balkan origin, and the fossil record from Greece helps in demonstrating the appearance and gradual evolution of the development of the elongated and posteriorly flared posterior carapacial border of the extant marginated tortoise (Vlachos and Tsoukala 2016).

4 Distribution

The geographic distribution of the testudine fossils from Greece largely follows the number and distribution of the fossil localities of the country. There is a handful of Early and Middle Miocene in Greece, but by the Late Miocene they already show a more extensive distribution in both continental and insular Greece (see Fig. 1 and Appendix). In the Pliocene fossil turtles are mostly found in Northern Greece, whereas in the Pleistocene they occur in the Peloponnesus peninsula as well (Appendix). Geoemydids are present in Greece at least since the early Late Miocene (MN9), whereas emydids occur only since the Late Pleistocene. It is worthy to note that geoemydids are able to tolerate salty water compared to emydids, which could lead to different geographical distributions as geoemydids are able to surpass barriers that emydids cannot (see Vamberger et al. 2014 and references therein). The temporal distribution of small-sized testudinids (MN10–present) in Greece is more extensive than that of the giant tortoises (MN12–Early Pleistocene), which disappear in the fossil record of the country few thousand years before their extinction from continental Europe (Pérez-García et al. 2017b). Up to now, no giant tortoises have been found in Crete, which otherwise contains occurrences of Trionychidae, Geoemydidae, and small-sized Testudinidae.

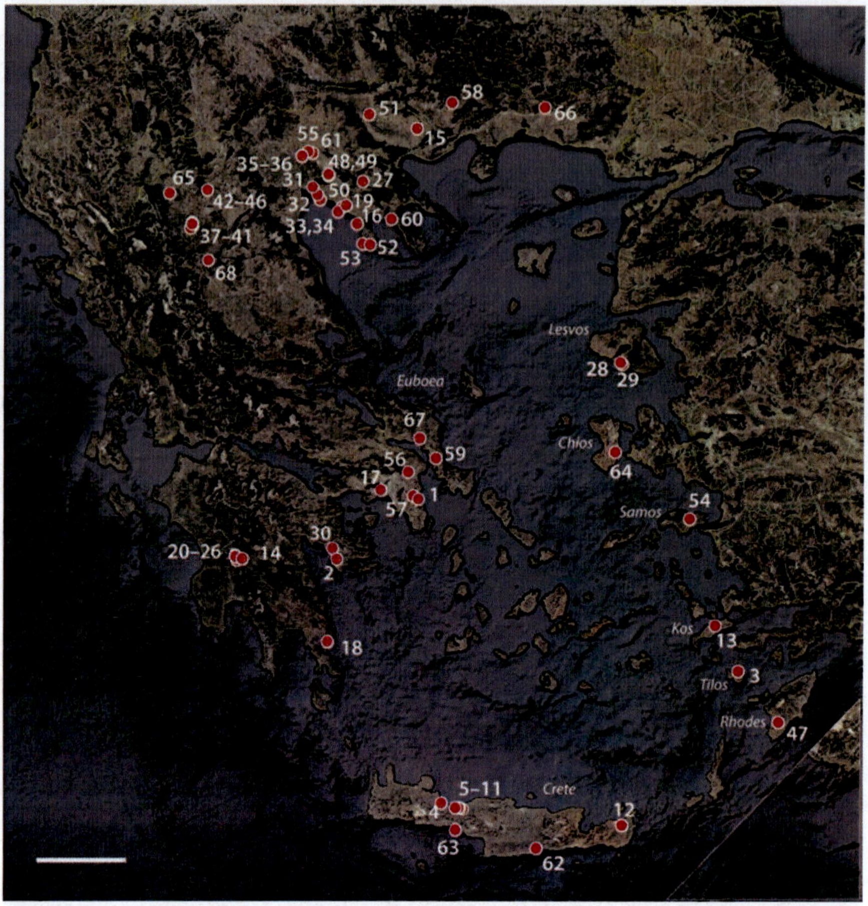

Fig. 1 Map of Greece showing the various occurrences of turtle and tortoise fossils. **Holocene–Pleistocene**: **1**, Vravrona; **2**, Franchthi Cave. **Pleistocene**: **3**, Charkadio Cave; **4**, Liko Cave; **5**, Gerani IV Cave; **6**, Zourida Cave; **7**, Bate Cave; **8**, Simonelli Cave; **9**, Rethymnon fissure; **10**, Koumpes I Cave; **11**, Mavro Mouri IV Cave; **12**, Siteia I; **13**, Kos; **14**, Megalopolis; **15**, Xerias; **16**, Potidea; **17**, Korydallos; **18**, Lakonia; **19**, Petralona Cave; **20–23**, Kyparissia localities; **24–26**, Choremi localities; **27**, Kalamoto-2; **28**, Vatera-F; **29**, Vatera-T; **30**, Karnezeika. **Pliocene**: **31**, Megalo Emvolon; **32**, Nea Michaniona; **33**, Epanomi; **34**, Nea Kallikrateia; **35**, Gefira-1; **36**, Gefira-2; **37–41**, Milia localities; **42–46**, Ptolemais localities; **47**, Apolakkia; **48**, Pylea; **49**, Allatini. **Pliocene/Miocene boundary**: **50**, Nea Silata; **51**, Maramena. **Miocene**: **52**, Kryopigi; **53**, Fourka; **54**, Samos localities; **55**, Vathylakkos-3; **56**, Liossati; **57**, Pikermi; **58**, Platania; **59**, Halmyropotamos; **60**, Nikiti-2; **61**, Ravin de la Pluie; **62**, Kastellios localities; **63**, Plakias; **64**, Michalos; **65**, Nostimo; **66**, Karydia; **67**, Aliveri; **68**, Megalos Lakkos. See Appendix for more information. (Image exported from Google Earth Pro © 2019, map data from US Dept. of State Geographer, SIO, NOAA, US Navy, NGA, GEBCO, image from Landsat/Copernicus. Scale bar equals 80 km, North faces upward)

Certainly, there are many more latest Pleistocene–Holocene occurrences of turtles from Greece coming from numerous zooarchaeological sites. As in the majority of these sites turtle specimens have received little or no attention at all, these occurrences are not included herein.

5 Systematic Paleontology

All turtles from Greece belong to the crown clade Testudines. There is a single representative of Pleurodira, and all other turtle taxa from Greece are members of Cryptodira representing at least four different families. At least 14 different taxa can be identified in the Greek fossil record. The most diverse clade is Testudinidae, comprising more than half of the total diversity of Greek fossil turtles with at least two giant and seven small-sized tortoise species. Freshwater turtles of the clade Geoemydidae follow with at least two species of the genus *Mauremys*, whereas the remaining three families/clades are represented by a single taxon each. Herein, the previously used phylogenetic nomenclature is abandoned.

Testudines Batsch, 1788

Comments Occurrences of Testudines indet. belong to two categories: studied material that cannot be further diagnosed (e.g., the material from Karydia and Aliveri in Georgalis et al. 2019a) and material that has never been studied and is only referred to as "turtles" (e.g., the Kastellios material in de Bruijn et al. 1971).

Pleurodira Cope, 1864
Podocnemidoidea Cope, 1868 (sensu Gaffney et al. 2011)

Type Genus *Podocnemis* Wagler, 1830.

★*Nostimochelone lampra* Georgalis, Velitzelos, Velitzelos, Kear, 2013

Nomenclatural and Taxonomical History *Labrochelone nostimiensis* and *Lamprochelone nostimiensis* in Mueller-Töwe et al. 2011 (nomina nuda); *Nostimochelone lampra* in Georgalis et al. 2013 (new species)

Type Material NMP V1 (holotype), partial carapace and plastron (Figs. 17.2–3 in Georgalis et al. 2013 and Fig. 3 in Georgalis and Kear 2013; Fig. 2).

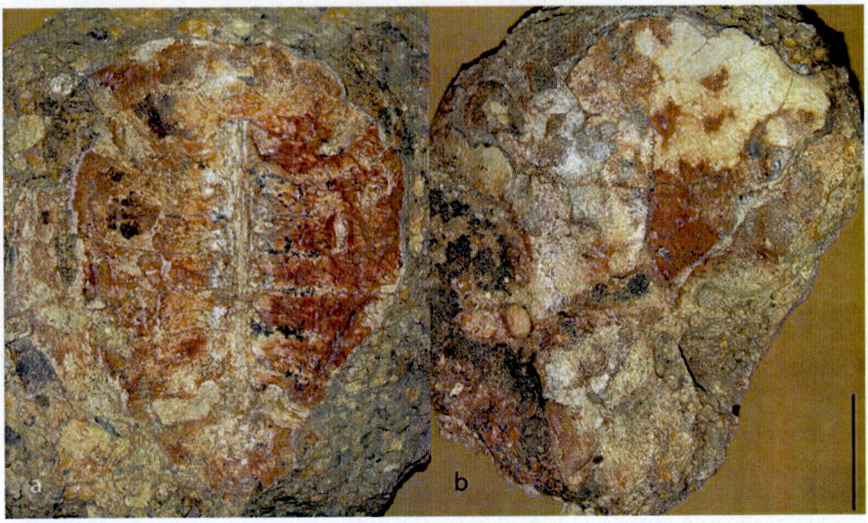

Fig. 2 The holotype (NMP V1) of *Nostimochelone lampra*, the only pleurodire turtle known in the fossil record of Greece. **a** Visceral view of the carapace, **b** ventral view of the plastron. Scale bar equals 5 cm

Type Locality Outside Nostimo village, Kastoria; Zeugostasion Formation, Early Miocene (Late Burdigalian).

Distribution Known only from its type locality.

Taxonomic Remarks The pleurodiran affinities of this taxon are quite clear, as already pointed out by Georgalis et al. (2013), including the sutural articulation between the pelvis and the shell and the presence of mesoplastra. Although the dorsal surface of the carapace is not available for observation, a combination of characters allows diagnosis within Podocnemidoidea. Georgalis et al. (2013) suggested that some character states are similar to the derived condition seen in Podocnemididae. Inclusion of this taxon in the matrices of Gaffney et al. (2006, 2011) revealed that *Nostimochelone* is placed within Bothremydidae, either within Kurmademydini or within Bothremydini (Vlachos 2015). However, observations on the dorsal side of the carapace would be necessary for a definite conclusion regarding the affinities of this taxon. As such, *Nostimochelone* is placed herein only within Podocnemidoidea as originally described by Georgalis et al. (2013).

Comments *Nostimochelone lampra* is the only pleurodire known from Greece and one of the oldest occurrences of turtles in the country. I believe that it is absolutely necessary to fully prepare the type specimen, to allow observations on the dorsal side of the carapace and the visceral side of the plastron. Although the phylogenetic relationships of Pleurodira are heavily based on cranial characters (see Gaffney et al. 2006, Ferreira et al. 2018, and references therein), the complete information of the shell morphology of *Nostimochelone* could provide important clues on its affinities. Especially if indeed this taxon is placed within Bothremydidae, this could have important implications on the temporal distribution of this clade. In any case, *Nostimochelone* is one of the youngest pleurodire occurrences from Europe (Georgalis et al. 2013).

Cryptodira Cope, 1868
Trionychidae Gray, 1825

Type Genus *Trionyx* Geoffroy Saint-Hilaire, 1809.

Distribution Early Miocene (Late Aquitanian) of Northwestern Greece, early Late Miocene (MN9) of Crete, to the Pliocene of Northern Greece (Vlachos et al. 2015b; Georgalis et al. 2016; Vlachos and Bacharidis 2018).

Comments This clade has not been documented in Greece until recently (Georgalis and Kear 2013), when new fieldwork efforts and investigations were able to find new soft-shelled turtles in the fossil record of Greece (Fig. 3; Vlachos et al. 2015b; Georgalis et al. 2016; Vlachos and Bacharidis 2018). Even with these few findings,

Fig. 3 Fossils of soft-shelled turtles from Greece. **a–e** LGPUT GAS 8a–e, partial carapace plates of the Trionychidae indet. from the Pliocene of Gefira (Northern Greece), **f–g** LGPUT GNE 001, partial costal plate of the Trionychidae indet. from the Early Miocene of Megalos Lakkos (Neochori, Grevena, Northern Greece). Scale bar equals 5 cm

we can reconstruct that this clade was present for a long period of time in Greece, spanning almost the entire Neogene—this is the turtle clade with the longest temporal distribution in the country. The few recovered plates have been clearly identified as belonging to soft-shelled turtles both on morphological (Vlachos et al. 2015b; Georgalis et al. 2016; Vlachos and Bacharidis 2018) and histological (Vlachos et al. 2015b) evidence. It is, however, impossible to further refine this identification based on the available information and especially on the absence of cranial or plastral elements (see Vitek and Joyce 2015; Georgalis and Joyce 2017 and references therein). It is worthy to note that, based on the currently available fossil record, the soft-shelled turtle specimen from Megalos Lakkos (near Neochori, Grevena, Northwestern Greece; Vlachos and Bacharidis 2018) is the oldest turtle occurrence from Greece.

Emydidae Bell, 1825

Type Genus *Emys* Duméril, 1805.

Comments Emydidae is a diverse family of testudinoids, known in the Old World with a single extant species, the European pond turtle *Emys orbicularis*—a species that forms part of the extant herpetofauna of Greece. The fossil material from Greece that can be attributed to this clade is limited and belongs to the extant species, *Emys orbicularis*. All remaining extant emydids are distributed to the New World, with the exception of *Emys trinacris* in Sicily (see TTWG 2017 for discussion). Fossil emydids known from shell remains can be readily distinguished based on the presence of a wide costovertebral tunnel and wide rib heads, the absence of musk ducts and their foramina, and the expansion of the vertebral V onto the pygal bone. There are few occurrences of this clade in the fossil record of Greece, all of which are identified as *Emys orbicularis*.

◉*Emys orbicularis* (Linnaeus, 1758)—the European pond turtle

Nomenclatural and Taxonomical History The original name and combination is *Testudo orbicularis* Linnaeus, 1758. This name has a long history of combinations and synonymy, summarized in TTWG 2017 (and references therein), and is beyond the scope of this chapter.

Type Specimen Museum für Naturkunde, Magdeburg (MNM) Z-4180 (neotype), adult male specimen (Fritz 1994; not figured).

Type Locality Mecklenburg Lake Plateau, Germany (based on the neotype).

Distribution Extant *Emys orbicularis* and its subspecies show, mainly, a European distribution from Portugal in the west to Turkmenistan in the east and Tunisia in the south to Estonia in the north (see TTWG 2017 for further details). In the Greek fossil record, this species is mainly known from the middle Pleistocene to Early Holocene of Peloponnesus peninsula (see Appendix).

Taxonomic Remarks The main character of diagnostic value for this species is the presence of a hinge between the hyoplastron and the hypoplastron. This character allows an immediate distinction from the sympatric geoemydids (which are more extensive temporally and geographically; see below). More characters from the shell facilitate further the identification of even isolated plates, as the fossils from Greece that can be attributed to the European pond turtle are few and quite fragmented. The most reliable characters to diagnose *Emys orbicularis* in the Greek fossil record are the presence of a hyo-hypoplastral hinge seen as a complete coincidence of the hyo-hypoplastral suture and the pectoro-abdominal sulcus; a nuchal plate overlapped laterally by the pleural I; a lyre-shaped vertebral I; and small pygal that is covered by the vertebral V and the marginals XII.

Comments An excellent example to illustrate the diagnostic potential of the abovementioned characters in isolated fragments is from the turtle specimens from Megalopolis. Originally, these specimens were attributed to the geoemydid "*Clemmys*" *caspica* by Melentis (1966). Even though the original material cannot be located, the published information suffices to identify that some of these specimens actually belonged to *Emys orbicularis* (see Vlachos and Delfino 2016). The earliest published records of *Emys orbicularis* come from the Middle Pleistocene of various levels in Choremi section (van Vugt et al. 2000; not described) and various localities in Kyparissia (Athanassiou et al. 2018). Recently, during a reexamination of the fossil record of Kalamoto-2 (KAL) locality in Mygdonia basin, E. Tsoukala brought to my attention a single chelonian specimen which is briefly described in the box below. Recent studies and discoveries (Vlachos and Delfino 2016; Athanassiou et al. 2018) have helped to establish the presence of *Emys orbicularis* in the Middle to Late Pleistocene of Peloponnesus. This region has been suggested as one of the pleniglacial refugia of this species, prior to its northward expansion during the warmer parts of the Holocene (see Vlachos and Delfino 2016 and references therein).

The Kalamoto Pond Turtle

LGPUT KAL 180 is a partial left xiphiplastron, measuring approximately 26 mm in length and 38 mm in width (scale bar is 10 mm). The expanded covering of the scutes on the visceral side excludes its identification as a testudinid. Based on the preserved sutures, the position of this fragment can be reconstructed as depicted, showing a posterior plastral lobe with gently curved and converging borders. The femoro-anal sulcus appears to be diagonally placed in respect to the midline, being straight medially and S-curved laterally. Based on the preserved part, I can safely estimate that the anal scutes had an extensive covering on the ventral side of the xiphiplastron. The documented morphology is consistent with *Emys orbicularis*. As Kalamoto-2 is dated to the early Middle Pleistocene (MNQ20), this occurrence would predate the occurrences of this taxon from Choremi and Kyparissia, making it the

(continued)

oldest confirmed occurrence of *Emys orbicularis* in Greece. Note that previous reports on the presence of emydids in the Pliocene of Ptolemais remain unconfirmed based on new sampling efforts (Georgalis and Kear 2013, Loreál et al. 2020).

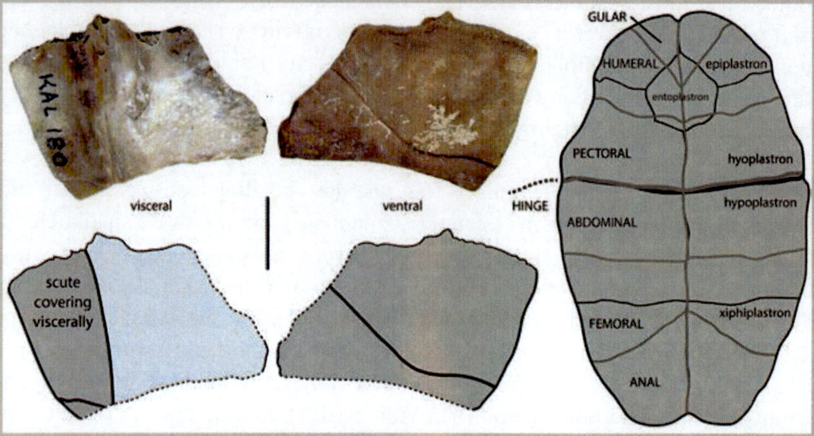

Geoemydidae Theobald, 1868

Type Genus *Geoemyda* Gray, 1834.

Comments Geoemydidae is a diverse family of testudinoids, which shows the exact opposite geographical distribution compared to Emydidae (see above). A single genus is distributed in the New World (*Rhinoclemmys*), whereas the rest of the diversity of this clade is found in the Old World. Fossil geoemydids known from shell remains can be readily distinguished based on the presence of a narrow costovertebral tunnel with wide rib heads, the presence of musk ducts and their foramina, and the expansion of the marginals XII onto the suprapygal II.

Mauremys Gray, 1869

Type Species *Emys fuliginosus* Gray, 1860, currently considered a subjective synonym of *Emys leprosa* Schoepff in Schweigger, 1812 (see TTWG 2017).

Comments This genus contains at least nine extant species and several subspecies, distributed in south Europe, northern Africa, Middle East, and Asia. Both extant and extinct geoemydids of Greece are placed within the genus *Mauremys*, represented in the extant herpetofauna with the species *Mauremys rivulata*. The fossil record of this clade is extensive in the Old World, and the members of *Mauremys* sensu lato

Fig. 4 **a** LGPUT GAS 34 (holotype) partial carapace of *Mauremys aristotelica* in dorsal view (scale bar equals 5 cm), **b** reconstruction of *Mauremys aristotelica*, showing the hypothesized shell coloration that could have served as a camouflage for this species (drawing: Jorge González, Argentina)

can be distinguished by a vertebral I that contacts the marginal II, an entoplastron that is crossed by the humero-pectoral sulcus on its posterior part, short anal scutes, and a deep rounded anal notch. There are several occurrences of geoemydids in the fossil record of Greece. Those from the Pleistocene are attributed to the extant *Mauremys rivulata*. The occurrences from several collections in the Late Pliocene of Milia (Northern Greece) and from the Late Miocene of Plakias (Crete) cannot be identified beyond the genus level. The occurrence from Plakias is the oldest occurrence of the family in Greece (see Georgalis et al. 2016).

★*Mauremys aristotelica* Vlachos, Sterli, Syrides, Vasileiadou, 2019a

Type Material LGPUT GAS 34 (holotype), partial carapace and plastron (Figs. 2 and 3 in Vlachos et al. 2019a; Fig. 4a).

Type Locality Gefira-2 (GAS-2) sandpit, near Gefira village, lower Axios valley, northern Greece, Pliocene.

Distribution Besides its type locality, this species is found in several localities in the latest Miocene/earliest Pliocene of Northern Greece (Appendix).

Taxonomic Remarks This turtle is clearly a member of Geoemydidae because of the presence of anterior and posterior musk duct foramina, a neural series of a quadrangular neural I followed by anterior short-sided hexagonal neurals, wide and flat rib heads near the neural/costal suture, and divided last marginals whose sulcus expands onto the suprapygal II. The combination of characters like the contact between vertebral I and marginal II, an entoplastron crossed by the humero-pectoral sulcus on its posterior part, short anal scutes, and a deep and rounded anal notch, allows its attribution to the genus *Mauremys*. This species is characterized by four extremely wide vertebral scutes which almost reach the marginals; this morphology distinguishes this taxon from other *Mauremys* and geoemydid taxa with wide vertebral scutes (see Vlachos et al. 2019a for detailed comparisons).

Comments *Mauremys aristotelica* comprises the only turtle taxon from Greece with such wide vertebral scutes, and its fossils can be easily distinguished even with isolated plates (e.g., Maramena material in Georgalis et al. 2019b). In the past, numerous geoemydids from the Neogene of Eurasia have been named and described with such scutes, but mostly attributed to other genera like *Sarmatemys*, *Sakya*, *Clemmydopsis*, *Melanochelys*, and *Shansiemys*. In some cases, the morphology radically differs from "regular" geoemydids, not only because of the great width of the scutes but also because some anterior pleural scutes might be missing (e.g., *Clemmydopsis*) or supernumerary pleural scutes were present (e.g., *Sakya*). It is not strange, therefore, why the dominant idea was that these taxa represented distinct lineages that developed wide vertebrals separately. However, the recent phylogenetic analysis by Vlachos et al. (2019a) demonstrated that it is more parsimonious to consider that the wide vertebrals occurred once and all these peculiar morphologies are autapomorphies of the various species. We also argued and speculated that this morphology could be a result of heterochronic changes to retain longer the juvenile pattern of wider vertebrals with darker coloration, as a kind of camouflage (Fig. 4b). It is, therefore, quite intriguing to investigate further the anatomy and evolutionary history of these geoemydids with wide vertebrals.

◉*Mauremys rivulata* (Valenciennes in Bory de Saint-Vincent, 1833)—western Caspian turtle

Nomenclatural and Taxonomical History The original name and combination is *Emys rivulata*, published by Valenciennes in 1833 in loose leaf with sets of plates (see TTWG 2017, and references therein). Since then, a long nomenclatural and taxonomical history is recorded, summarized in Mantziou and Rifai (2014) and TTWG (2017), and is beyond the scope of this chapter.

Type Material Unspecified.

Type Locality Originally not designated, but subsequently referred, as near Methoni (Modona), Peloponnesus (see TTWG 2017 and references therein).

Distribution Extant *Mauremys rivulata* shows a South Balkan-southeastern Europe-eastern Mediterranean distribution (see Mantziou and Rifai 2014). In the Greek fossil record, this species is known from the Pleistocene of Crete (various localities) and Kos, as well as the Peloponnesus peninsula; it is further reported in the recent deposits of Vraona Cave in Attica (Appendix).

Taxonomic Remarks The fossils of *Mauremys* are relatively easily distinguished from those belonging to Emydidae (see above). The fossils from Greece attributed to *M. rivulata* present all the morphological characters of this species, including the hexagonal neurals with short anterolateral sides; the entoplastron being covered by the gular, humeral, and pectoral scutes; the absence of hyo-hypoplastral hinge; and the deep and rounded anal notch. The presence of narrow vertebral scutes allows a clear distinction from *M. aristotelica*. The distinction between the extant *M. rivulata* and *M. caspica* is not possible to be made with the information preserved in the fossils, as the extant species are morphologically distinguished mainly by their coloration pattern (see Fritz et al. 2008)—the rest of the anatomy of plates and scutes is virtually identical. As such, the attribution of the Greek Pleistocene–Holocene material to *M. rivulata* is made based on biogeographic considerations as well.

Comments Although there is some hybridization between the two eastern Mediterranean *Mauremys*, namely, *Mauremys rivulata* and *Mauremys caspica*, it is now clear the area of South Balkans is inhabited by *M. rivulata*, whereas *M. caspica* is distributed in Asia Minor, Middle East, and further to the east (Fritz et al. 2008, Vamberger et al. 2017, and references therein). A notable finding is that hybridization between the two extant species is limited and localized, and the distribution of *M. rivulata* is modeled to be quite similar to its actual one at least since the Last Glacial Maximum (Vamberger et al. 2017). Based on all the above, I think that it is safe to assume at least the Pleistocene fossil record of *Mauremys* in Greece can be attributed to the extant *M. rivulata*.

Testudinidae Batsch, 1788
Testudininae Batsch, 1788

Type Genus *Testudo* Linnaeus, 1758.

Comments Derived tortoises of the clade Testudininae comprise various extant and extinct species that were formerly included in the so-called *Testudo* and *Geochelone* complexes (Fritz and Bininda-Emonds 2007). During the recent years, based both on molecular and morphological evidence, the composition of both complexes has been clarified and the inclusion of extinct taxa has been clearer: tortoises formerly included in *Geochelone* are members of Geochelonini (i.e., the unranked clade Geochelona sensu Vlachos and Rabi 2018), and those formerly included in *Testudo* are members of Testudinini (i.e., the unranked clade Testudona Parham in Parham et al. 2006). The use of both *Geochelone* and *Testudo* has been restricted to much more exclusive clades, although still some points are up to debate (e.g., the extent and definition of *Testudo* in respect to *Chersine hermanni*; Fritz and Bininda-Emonds 2007). The Greek fossil record is composed of testudininan species from both main clades—geocheloninans do not, however, contribute to the extant tortoise diversity of the country. Members of Testudininae are easily distinguished in the fossil record based on several synapomorphies from the shell, including a well-developed epiplastral lip; well-developed anterior and posterior buttresses; coincidence between the costo-peripheral suture and pleuro-marginal sulci; presence of a supracaudal scute; and short pectoral scutes, shorter than half of the medial length of hyoplastra.

Geochelonini Crumly in Swingland and Klemmens, 1989

Comments The members of Geochelonini represented in the Greek fossil record can be easily distinguished from those belonging to Testudinini (see below) based on several characters from the shell: large size, with individuals that reach or exceed one meter in carapace length; absence of cervical scute; contact between marginal 6 and pleural 3; and absence of a plastral hinge.

***Titanochelon* Pérez-García & Vlachos,** 2014

Type Species *Testudo bolivari* Hernández-Pacheco, 1917.

Included Taxa This genus includes all post-Paleogene large-sized tortoises from the European continent without a cervical scute (Pérez-García and Vlachos 2014), formerly included in the wastebasket extinct genus *Cheirogaster* Bergounioux, 1935. The Greek fossil record contains at least two different species of titanochelons, the Pliocene *Titanochelon bacharidisi* and the Late Miocene *Titanochelon schafferi*.

Comments TEWG (2015) suggested that the gender of *Titanochelon* is feminine, thus suggesting some mandatory changes in the terminations of the combined specific names. Actually, the name falls within the category of a genus group name formed by a Greek word Latinized with a change of ending (*-chelon* instead of the feminine *-chelone*) and should take the gender appropriate to the changed suffix (ICZN, Art. 30.1.3). Thus, the name *Titanochelon* should be considered of neuter gender; the names of the Greek species are not affected because they are based on personal names. Besides the occurrences in Northern Greece and Samos Island that can be identified to a species-level, several other localities contain occurrences of titanochelons that can be only identified to a genus-level (Appendix): the famous Late Miocene site of Pikermi and its neighboring Liossati; the Late Miocene sites Kryopigi and Fourka in Chalkidiki, Northern Greece; various sublocalities in the Pliocene site of Milia (Grevena); and the Vatera F-site in Lesvos Island. These fragmentary occurrences are important to be further investigated in the future, to provide further support on the biogeographic distinction of the two giant tortoise taxa from Greece, as suggested below. Selected *Titanochelon* specimens are shown in in Fig. 5.

★*Titanochelon bacharidisi* (Vlachos, Tsoukala, Corsini, 2014**)**

Nomenclatural and Taxonomical History *Cheirogaster bacharidisi* in Vlachos et al. 2014 (new species); *Titanochelon bacharidisi* in Pérez-García and Vlachos 2014 (new combination).

Type Material LGPUT EPN 100–199 (holotype), an almost complete skeleton of a female giant tortoise (Figs. 3 and 4 in Vlachos et al. 2014; Fig. 5a–c).

Type Locality Epanomi, Western Chalkidiki, Greece; Gonia Formation, Pliocene.

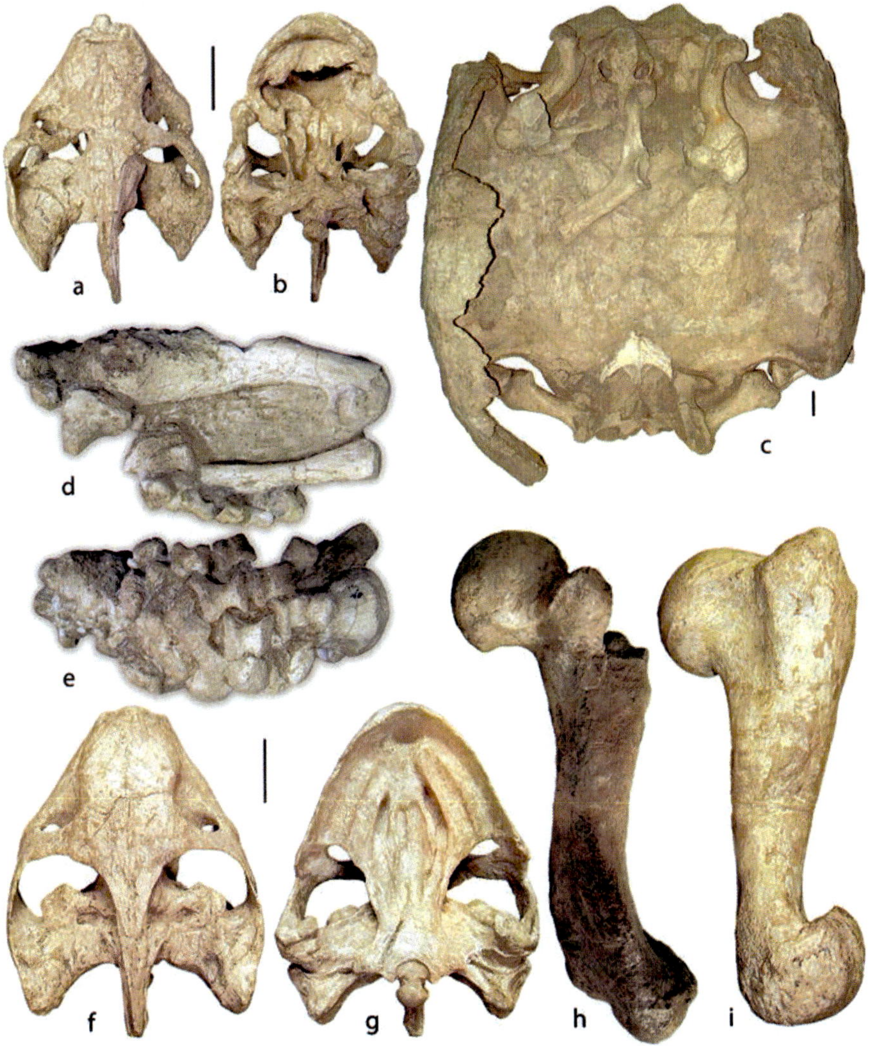

Fig. 5 Selected specimens of giant tortoises from Greece. **a–e** *Titanochelon bacharidisi* from the Pliocene of Northern Greece (**a–b** LGPUT EPN 102, holotype skull in dorsal and ventral view, **c** LGPUT EPN 101, holotype plastron in visceral view, **d–e** LGPUT KLK 504–528, articulated right posterior limb in dorsal and posterior view), **f–i** *Titanochelon schafferi* from the Late Miocene of Samos (**f–g** NHMW 2009z0103/0001, lectotype skull in dorsal and ventral view, **h** AMNH 1878, left humerus in posterior view, **i** NHMW 1911/0005/0275, paralectotype left femur in anterior view). Scale bar equal 5 cm

Distribution Latest Miocene to Pliocene of circum-Thermaikos Gulf area (N. Greece) and Late Pliocene of Bulgaria (Stojanov 2009; Vlachos et al. 2014).

Taxonomic Remarks This species is a member of the derived geocheloninan clade *Titanochelon*, sharing several characters with other titanochelons, including presence of frontals narrower than prefrontals, a long crista supraoccipitalis, the presence of an elliptical to circular premaxillary pit, two suprapygals, the first surrounding the lenticular second one, absence of cervical scute, a convex to flat dorsal lip of the epiplastra, a humero-pectoral sulcus that is perpendicular to the axial plane in the medial region, but with a well-developed lateral change of curvature, very short pectoral scutes in the sagittal plane, but markedly expanded, toward the anterior region, laterally. Among other titanochelons, *Titanochelon bacharidisi* is diagnosed because of the position of the fossa orbitalis in the midway between the apertura narium externa and the posterior tip of the squamosal; a zygomatic arch that is thick and short; squamosal processes converging posteriorly toward the midline; extended ventral emargination, short gulars contacting the anterior part of the entoplastron, which is covered medially only by humeral scutes; and its posterior border coinciding with the humero-pectoral sulcus. This species was also characterized by the presence of at least one conical, bony, thigh tubercle on each side of the tail (found in anatomical position in the Nea Kallikratia material shown in Fig. 5d and isolated in the Epanomi holotype), similar to the morphology of the extant African spurred tortoise (*Centrochelys sulcata*).

Comments This species is based on several complete skeletons from Western Chalkidiki, on the eastern side of the Thermaikos Gulf, in Thessaloniki (Vlachos et al. 2014; Vlachos 2015). Besides these materials, the giant tortoise fossils from the historical locality "Falaise de Karabouroun"[1] (in Arambourg and Piveteau 1929)

[1] Arambourg and Pivetaeu (1929) mention only three occurrences of turtles in their work, which have created some confusion in the literature (Lapparent de Broin 2002; Georgalis & Kear 2013, Vlachos et al. 2014). From the locality "Ravin de Vatilük" (which corresponds to Vathylakkos-3), they mention a "*Testudo* of large size," whereas from "Falaise de Karabouroun" (which corresponds to a locality around Megalo Emvolon), they mention a "*Testudo* of medium size" and a "*Testudo* sp."—these two localities are located in different basins and are of Late Miocene and Pliocene age, respectively. This material is now stored in MNHN. If we are to make sense of these occurrences, I trust two lines of evidence: the exact wording of the occurrences in the original text and the accompanying label information of these fossils in MNHN. First, plate I in Arambourg and Piveteau (1929) depicts a block of bones from Vathylakkos. The caption states that a *Testudo* is present among these bones, referring to a leg osteoderm and a pelvis of giant tortoise; this specimen forms part of the exhibition of the MNHN. On the other hand, plate XII contains a partial skull, a radius, and ulna, referred to *Testudo* sp. (note: as the occurrence from Falaise de Karabouroun); but the plate is cited in the text as "*Testudo* of large size" (note: as the occurrence from Vathylakkos). These bones, and other giant tortoise bones as well of similar morphology, form part of the MNHN collections and are accompanied by labels (old and new) that only state "Karabouroun" as the locality—conveniently, Lapparent de Broin (2002) only states Thessaloniki as the locality, with a Late Miocene age. These bones are of similar size, color, and morphology as all other giant tortoise fossils from Western Chalkidiki. Based on all the above, my interpretation

can be conferred to this species (Vlachos 2015). Outside Greece, this species can be identified in the Pliocene of Bulgaria (Stojanov 2009, re-identified in Vlachos et al. 2014 and Vlachos 2015). This is a giant tortoise that was smaller in size than its Late Miocene predecessors in Greece and a species that shows numerous differences in skull and shell morphology. It is certainly the most complete giant tortoise taxon from Greece and one of the most complete in Europe. Its distinction from the Late Miocene *Titanochelon schafferi* from Samos (see below) agrees with the current views regarding the differences of the South Balkan and Asia Minor faunas (Kostopoulos 2009). Thus, considering that Lesvos Island still formed part of the Asia Minor in the Early Pleistocene, the giant tortoises from Thermaikos Gulf probably represent the last giant tortoises in the South Balkans.

★*Titanochelon schafferi* (Szalai, 1931)

Nomenclatural and Taxonomical History *Testudo schafferi* in Szalai, 1931 (new species); ?*Geochelone schafferi* in Auffenberg, 1974 (new combination); ?*Cheirogaster schafferi* in Lapparent de Broin, 2001 (new combination); *Titanochelon schafferi* in Pérez-García and Vlachos, 2014 (new combination). In numerous works, Chkhikvadze (2001, 2006, 2007) identified the Samos giant tortoise material in AMNH as part of ?*Centrochelys natadzei* from the Late Miocene of Georgia, but we have recently showed that the Georgian species is quite different from the Samos one (Vlachos et al. 2019c).

Type Material NHMW 2009z0103/0001 (lectotype), a complete skull (Fig. 5f–g).

Type Locality Samos, precise locality unknown; Mytilini Formation, Late Miocene, late MN11 to early MN13.

Distribution Besides its type locality and other coeval localities in Samos, this species is also identified by Vlachos et al. (2019c) in the Late Miocene of Turkey.

Taxonomic Remarks This large giant tortoise that could reach 180 cm in length is a member of *Titanochelon* clade based on a series of characters: frontals narrower than prefrontals; long crista supraoccipitalis; circular pit in the premaxillae in ventral view; angle of the gularo-humeral sulcus and the sagittal axis higher than 45°; humero-pectoral sulcus perpendicular to the axial plane with a well-developed lateral change in curvature; short pectoral scutes; and curved diaphysis of the humerus. It differs from other members of this clade based on the following combination of characters: short anterior part of the cranium relative to the posterior part, with the

(stated already in Vlachos et al. 2014, Vlachos 2015) is that the giant tortoise fossils in MNHN come from the Pliocene of Karabouroun (= Megalo Emvolon). The occurrence of a giant tortoise in Vathylakkos is confirmed but corresponds to the two specimens from the bone conglomerate of plate I in Arambourg and Piveteau (1929). This occurrence is marked in Appendix as *Titanochelon* sp. The "*Testudo* of medium size" is not controversial and refers to the small testudinid from the locality (Appendix).

fossa orbitalis as a reference; reduced lower temporal emargination; straight crista supraoccipitalis; entoplastron not covered by gular scutes; humero-pectoral sulcus posterior to the entoplastron; humerals expanded on the posterior epiplastra; asymmetrical distal parts of the humerus; and broad fused medial and lateral centrale. See Vlachos et al. (2019c) for further information.

Comments Our knowledge on the anatomy and phylogenetic relationships of the Samos giant tortoise has improved enormously recently with the description of, probably, all the known material coming from the so-called old collections in the island (Vlachos et al. 2019c). We are, therefore, much more confident to claim the distinction of this species from *Titanochelon bacharidisi* from the Pliocene of Greece (see above). Also, *Ti. schafferi* can be confidently identified in the Miocene of mainland Turkey as well (Late Miocene of Konya-Hatunsaray-Kayadibi). Still, we do not know if *Ti. schafferi* was the sister species of *Ti. bacharidisi*; the missing carapace in the Samos taxon and the conflicting skull characters hinder phylogenetic resolution. However, both taxa are of crucial importance for understanding the evolution, and ultimately the extinction, of this clade in eastern Mediterranean.

Fossilized Eggshells of Giant Tortoises

Besides the occurrences of titanochelons mentioned above, the Greek fossil record contains at least two examples of fossilized eggs and eggshell specimens that could be attributed to giant tortoises. A clutch of at least five chelonian eggs is known from limit Miocene/Pliocene of Pylea (Northern Greece; del Campana 1919; Vlachos et al. 2015a), followed by a recent discovery of another clutch from the Early Pliocene of Apolakkia in Rhodes Island (Mueller-Töwe et al. 2011) (Appendix). They have been attributed to the ootaxon Testudoolithidae, although only the Apolakkia specimens have been properly analyzed with X-ray diffraction specters that confirmed the typical aragonite content of chelonian eggshells.

Testudinini Batsch, 1788

Comments The members of Testudinini represented in the Greek fossil record can be easily distinguished by those belonging to Geochelonini (see above) based on several characters from the shell: small size, with individuals do not exceed 50 cm in carapace length; presence of cervical scute; no contact between marginal 6 and pleural 3. Selected testudininan specimens from Greece are shown in Fig. 6.

"*Testudo*" aff. *antiqua* Bronn, 1831

Referred Material AMPG 1999, an almost complete carapace, and GNHM 22, the anterior half of a carapace and plastron (see Vlachos et al. 2019c:fig. 2).

Distribution This taxon is known certainly from the Late Miocene (late MN11–early MN13) of Samos Island.

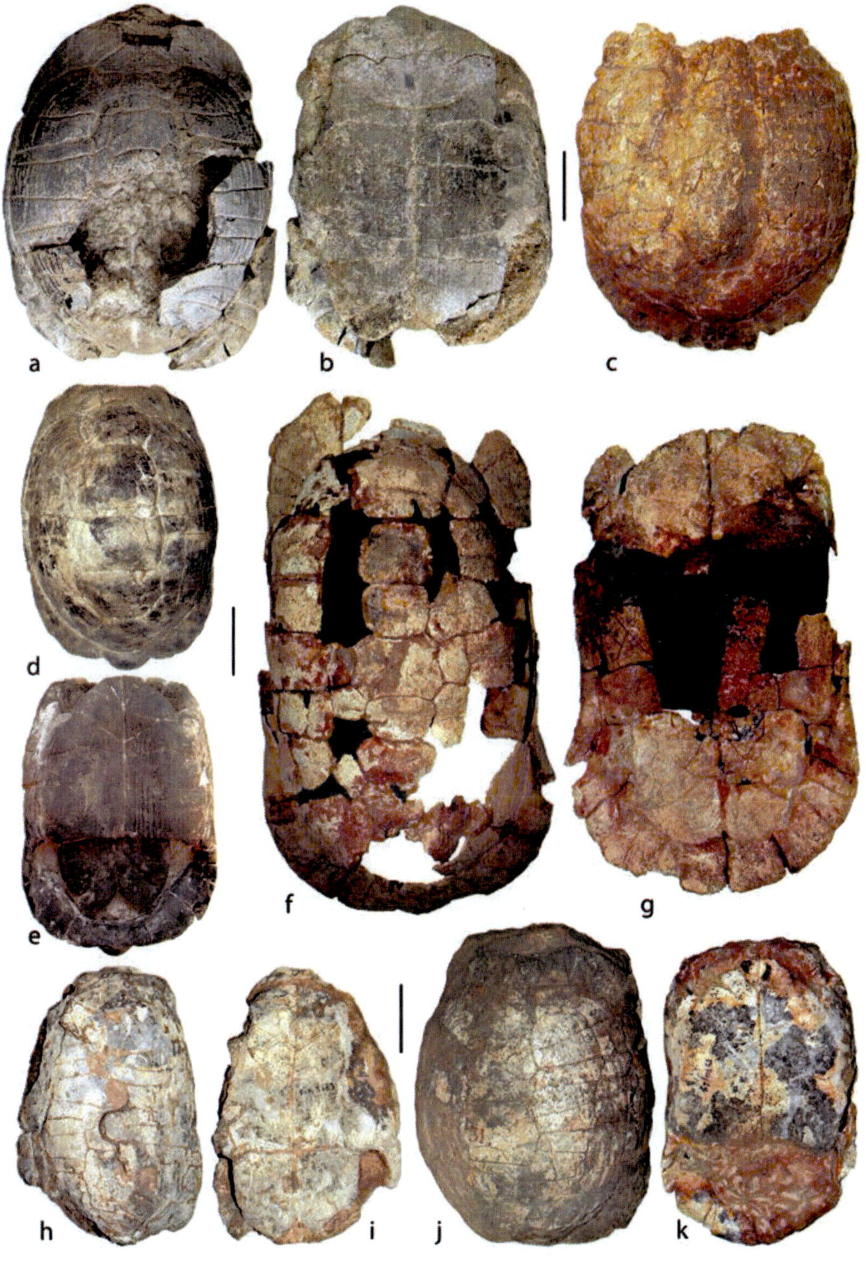

Fig. 6 Selected small-sized testudinids from Greece. **a–b** LGPUT MP 40, partial shell of *Chersine hermanni* from the ?Pleistocene of an unknown precise locality in Chalkidiki in dorsal and ventral view; **c** LGPUT MIL 495 (holotype), partial carapace of *Testudo brevitesta* from the Late Pliocene of Milia in dorsal view; **d–e** NHMW 1980z0069/0000, shell of *Testudo graeca* from the Pliocene of Megalo Emvolon in dorsal and ventral view; **f–g** AMPG 3/1974 (holotype of *Te. m. cretensis*), partial shell of *Testudo marginata* from the Pleistocene of Gerani Cave (Crete) in dorsal and ventral view; **h–k** selected shell specimens of *Testudo marmorum* from the Late Miocene of Pikermi (**h–i** MNHN PIK 3683a, lectotype in dorsal and ventral view, **j** MNHN PIK 3863b, paralectotype in dorsal view, **k** AMPG 1970d referred specimen in ventral view). Scale bars equal 5 cm

Taxonomic Remarks The combined morphology of these two shell remains from Samos presents an interesting combination of characters: there is a symmetrical contact between costal VII and suprapygal I, enclosing costal VIII (in AMPG 1999), and long medially pectoral scutes that are just over half that of the medial hyoplastral suture (in GMNH 22). This combination of characters could allow the distinction of a new species, which could represent a basal testudininan or a stem *Testudo* (see Vlachos et al. 2019c). But the fact that the two characters are observed in two different specimens of unclear association and unprecise stratigraphic position, this taxon remains with open nomenclature. As the costal configuration is observed in at least one specimen of "*Testudo*" *antiqua* from Germany as a teratomorphy (see Corsini et al. 2014:952, fig. 2.1), the open nomenclature specific prefix "aff." is used to denote taxonomic novelty, with reference to "*Te.*" *antiqua* for the abovementioned morphological similarity—this should not be confused, however, as an argument of close relationship between the Samos small testudinid and "*Te.*" *antiqua.*

Comments It is possible that the material reported by Paraskevaidis (1940) from the Middle Miocene of Chios Island belongs to this taxon as well. This material comprises two endocasts (steinkerns) of turtle shells and currently is lost, but the reconstructed morphology is similar to the Samos taxon (see Vlachos et al. 2019c:fig. 9). But as we are unable to study the original specimens—which also did not preserve any bone rather than the imprints of plates—this material is currently identified as cf. "*Testudo*" sp., pending future discoveries that might clarify the situation. Overall, none of the character states that characterize this taxon have been observed in the small testudinid occurrences of the continental Greece, which all appear to represent more derived *Testudo* species.

◉*Chersine hermanni* Gmelin, 1789—Hermann's tortoise

Nomenclatural and Taxonomical History Originally, this species was combined with *Testudo*. Mainly because of the absence of a hypo-xiphiplastral hinge in the plastron—a character that defines a "true" *Testudo* or *Testudo* s.s.—Hermann's tortoise has suffered numerous generic combinations. The latest attempt involved the erection of a new genus, *Eurotestudo* Lapparent de Broin et al., 2006, for this species, but Fritz and Bininda-Emonds (2007) and Bour and Ohler (2008) argued that the name *Chersine* Merrem, 1820, was available for this species.

Type Material Strasbourg University Zoological Museum 121 (Bour 1987).

Type Locality Collobrières, Var, France (fixed by Bour 1987).

Distribution This species has an extended extant distribution along the coastlines of north Mediterranean. The subspecies *Ch. h. boettgeri* is mainly found in the Balkan peninsula, whereas the subspecies *Ch. h. hermanni* is mainly distributed in Italy (including Sicily and Sardinia) and the coasts of France (including Corsica) and Spain (including Baleares) (see TTWG 2017 for details). In the past, the distribution of this species was much more extensive in continental Europe (Lapparent de

Broin 2001), reaching even as north as Scandinavia during the warmer parts of the Holocene (see Sommer et al. 2007 and references therein). In the fossil record of Greece, it is known only from the latest Pleistocene to Early Holocene of Franchthi Cave and from an unknown locality from the ?Pleistocene of Chalkidiki.

Taxonomic Remarks Within the more inclusive *Testudo* clade (or *Testudo* s.l.), fossils of the Hermann's tortoises are easily distinguished by those of the derived *Testudo*, because of the absence of a hinge between the hypoplastra and the xiphiplastra. Another character that is quite useful, especially when working with isolated fragments, is the presence of marginals XII crossing the pygal plate.

Comments The fossils of this taxon are quite scarce in the fossil record of Greece. There is an almost complete specimen from the ?Pleistocene of an unknown precise locality in Chalkidiki (Fig. 6a–b), but as we do not have detailed information on its geographical and stratigraphical position, this specimen is not so useful to draw broader conclusions. The other occurrence of this taxon is from Franchthi Cave. This material is placed in a well-defined stratigraphical and chronological context and is thought to represent remnants of the food remains from small game prey of the humans that inhabited the cave. Although *Chersine hermanni* has an extensive temporal presence in the Franchthi Cave (throughout the dated deposits), it is usually represented by a single individual in each level. The absence of other occurrences of this species in the pre-Late Pleistocene fossil record (which is dominated by hinged *Testudo* spp.) could provide a constraint for the timing of the colonization of South Balkans by Hermann's tortoise (Lenk et al. 1999; Fritz et al. 2007a).

Testudo **Linnaeus, 1758**

Type Species *Testudo graeca* Linnaeus, 1758.

Included Taxa Extant members of *Testudo* include the *Testudo graeca* complex (with at least five subspecies), *Testudo marginata*, and *Testudo kleinmanni*. Several extinct species have been named or considered as members of *Testudo*, but few of them can actually be demonstrated to be members of this clade (see Luján et al. 2016; Vlachos and Rabi 2018). It is however clear that this clade has a fossil record that extends at least till the Late Miocene, as the Late Miocene *Testudo marmorum* and *Testudo hellenica* from Greece are certainly the oldest members of this clade.

Comments The concept of *Testudo* herein is kept in its strict sense, i.e., including only these testudininans with a hypo-xiphiplastral hinge that can be clearly grouped with *Testudo graeca* under a cladistic context. Other extant small testudinids that have been traditionally under *Testudo* (*hermanni*, *horsfieldii*) are considered as part of distinct genera, *Chersine* and *Agrionemys*, respectively. These genera are considered as subgenera of *Testudo* by Luján et al. (2016).

★***Testudo brevitesta*** **Vlachos and Tsoukala,** 2016

Nomenclatural and Taxonomical History *Testudo brevitesta* (new species in Vlachos and Tsoukala 2016); *Testudo* (*Testudo*) *brevitesta* (referral to the subgenus level in Luján et al. 2016).

Type Material LGPUT MIL 495, partial carapace (Fig. 2 in Vlachos and Tsoukala 2016; Fig. 6c).

Type Locality Milia 4, Municipality of Grevena, Greece, Late Pliocene, MN16a.

Taxonomic Remarks This is a small-sized testudinid that is closer morphological to the extant *Testudo marginata* in having a posteriorly flared carapacial border, but without the elongation of the posterior carapace seen in the marginated tortoise.

Comments This species was recovered in conflicting phylogenetic positions in Vlachos and Tsoukala (2016), either as the sister of *Testudo marginata* or in a more derived position just basal to *Testudo marmorum* and *Testudo kleinmanni*. Shortly after, Luján et al. (2016) included this taxon in a new matrix that was more focused in resolving the interrelationships within Testudinini and recovered *Testudo brevitesta* as the sister of *Testudo marginata*, forming a clade that is sister to *Testudo marmorum*. This topology makes sense from a temporal point of view.

◉*Testudo graeca* Linnaeus, 1758—the Mediterranean spur-thighed tortoise

Type Material The holotype of *Testudo graeca* Linnaeus, 1758, is the specimen figured by Edwards and reproduced also by Bour (1987:fig. 1). The lectotype of *Testudo ibera* Pallas, 1814, is the specimen figured by Pallas (1814:pl. II, fig. 22–3), reproduced also by Bour (1987:fig. 3).

Type Locality The locality of the type of *Te. graeca* of Edwards is Santa Cruz near Oran, Algeria. The type locality of *Testudo ibera* was first reported as the area of southern Caucasus and the southern coast of Crimea but later restricted to Tbilisi, Georgia, by Bour (1987; see also TTWG 2017 and references therein).

Distribution This species, and according to the various concepts of authors, is divided in several subspecies occupying collectively the circum-Mediterranean area (Fritz et al. 2007b, 2009; TTWG 2017; and references therein). At present, the subspecies *Testudo graeca ibera* is currently distributed in Central Macedonia and Thrace, with the present-day Axios River as a boundary. The rest of the continental Greece is occupied by *Testudo marginata*, although there is a sympatric area in Central Macedonia (TTWG 2017). The earliest fossils of this species can be traced in the Late Miocene of Platania (Drama). During the Pliocene the species is found in several localities in Northern Greece (Pylea, Allatini, Megalo Emvolon), whereas in the Pleistocene it is also identified in Lesvos Island and in Petralona Cave (Appendix). It is worthy to note that the currently known distribution of this species in the fossil record of Greece is in good accordance with the extant distribution of

Te. g. ibera, i.e., not found in continental Greece beyond Axios River in Thermaikos basin (Central Macedonia).

Taxonomic Remarks The Greek fossils are herein attributed simply to the *Testudo graeca* complex, without any attempt to define it to a subspecific level in the lack of clear morphological characters. Coloration and the presence of keratinous thigh spurs are used to distinguish extant specimens, but these characters cannot be used in the fossil record. The presence of a cervical scute, the pattern of the neural plates being alternatively rectangular/octagonal, the coincidence of the costo-peripheral suture and the pleuro-marginal sulci, and the gulars that cover the anterior part of the entoplastron are characters diagnostic to the generic level, whereas the wide vertebrals (as opposed to the narrower vertebrals of *Chersine*), the presence of a hypo-xiphiplastral hinge, the posterior lobe being formed entirely by the xiphiplastra, and the absence of elongated posterior carapace (as opposed to the elongated carapace of *Te. marginata*) and posteriorly flared posterior peripherals (as opposed to *Te. marginata* and *Te. brevitesta*) allow a clear attribution of several fossil shells from Greece to the *Te. graeca* complex (e.g., Fig. 6.d–e).

Comments Two specimens from Pylea and Allatini, Northern Greece, have been previously attributed to the Italian taxon *Testudo amiatae* by del Campana (1917) and Bachmayer and Symeonidis (1970), respectively. The revised information on the morphology of *Te. amiatae* showed that there is no hinge between the hypoplastra and xiphiplastra. Similarly, the revision of the Greek specimens attributed to *Te. amiatae* by Vlachos et al. (2015a) indicated the clear presence of a hypoplastral hinge in the Pylea specimen, and the morphology of both specimens is consistent with the anatomy of the *Testudo graeca* complex. As such, the occurrence of the Italian *Te. amiatae* in Greece is refuted, and the presence of the aforementioned species is restricted to Tuscany (Italy; see Vlachos et al. 2015a for further information). Based on molecular clock analyses, the existence of the total clade of *Testudo graeca* complex clade ranges from the Late Oligocene (Pereira et al. 2017; global phylogeny of turtles) to the Late Miocene (Fritz et al. 2009; phylogeny restricted to Testudinidae); this is of course because of the taxon sampling and the selected calibration points. The oldest possible *Testudo graeca* from Greece is traced at the Late Miocene of Platania (Drama, Vlachos and Tsoukala 2014; as *Testudo* cf. *graeca*), although there are older mentioned occurrences in the Middle Miocene of Turkey (Staesche et al. 2007), which are nevertheless based on quite fragmentary material and their identification is only tentative.

★*Testudo hellenica* Garcia, Pinton, Valentin, Kostopoulos, Merceron, de Bonis, Koufos, 2020

Nomenclatural and Taxonomical History *Testudo hellenica* (nomen nudum in García et al. 2017); *Testudo hellenica* (new species in García et al. 2020).

Type Material LGPUT RPL 216 (holotype), an almost complete carapace (García et al. 2020:fig. 3).

Type Locality Ravin de la Pluie, lower Axios valley; Nea Messimvria Formation, Late Miocene, Vallesian, MN10.

Taxonomic Remarks The presence of the hypo-xiphiplastral hinge clearly places this taxon within crown *Testudo*. A combination of characters, including the tectiform shape of the carapace, the quite long and rounded epiplastral lip, and the suprapygal configuration with a straight suture between the two suprapygals (see Garcia et al. 2020), supports its distinction within known members of *Testudo*.

Comments The first mention of this species was made in an abstract (García et al. 2017), which fails to conform with the requirements of the ICZN; as such *Testudo hellenica* García et al., 2017, is a nomen nudum. However, the name is available and was used subsequently by the same authors and for the same concept (García et al. 2020). The phylogenetic analysis of Garcia et al. (2020) places *Te. hellenica* as the sister to the so-called *Chersus* clade, which comprises {*Te. marmorum* {*Te. brevitesta* + *Te. marginata*}}. Thus, it is placed deep within crown *Testudo*, which in turn would place the origin of *Te. kleinmanni* and *Te. graeca* lineages prior to the Vallesian. This is the oldest occurrence of a crown *Testudo* known to date.

◉*Testudo marginata* Schoepff, 1793—the marginated tortoise

Type Material Strasbourg University Zoological Museum (MZS) 143 (lectotype designated by Bour 1987).

Type Locality Central Greece (restricted by Bour 1987).

Distribution Extant *Testudo marginata* is mainly distributed in continental Greece and some Aegean Islands, as well as in Albania (TTWG 2017). In the Greek fossil record, it is found within its extant range but also in various localities of Crete and Tilos Islands (Appendix).

Taxonomic Remarks This is a species that is easily distinguished from other hinged *Testudo*, based on its larger size, the elongated posterior part of the carapace, and the posteriorly flared and elongated posterior peripheral plates.

Comments Although found in several localities in continental Greece, the most important fossils of the marginated tortoise are known from Crete Island (Fig. 6f–g). A series of characters led Bachmayer et al. (1975) to propose a new subspecies for the Cretan populations of marginated tortoises, but I believe that they are not enough to support this distinction (see inset below).

★***Testudo marginata cretensis*** **Bachmayer et al., 1975**
[subjective junior synonym of *Testudo marginata*]

Type Material AMPG 3/1974, an almost complete shell (Bachmayer et al. 1975, Figs. 2, 4–6, Pls. 19–23; Fig. 6f–g)

Type Locality Gerani IV Cave, Crete Island; Pleistocene

Distribution This subspecies has been identified in several other caves in Crete, but all these fossils are considered herein as *Testudo marginata* (Appendix).

Comments All the diagnostic characters given by Bachmayer et al. (1975) to support the validity and distinction of this subspecies are shown as incorrect or part of the variation of non-Cretan marginated tortoises as well (Vlachos 2015). The supposed quadrangular neural 1 is actually hexagonal with short postero-lateral sides (as in many other marginated tortoises and including the one shown by Bachmayer et al. 1975:fig. 3). The octagonal neural 2 is of no diagnostic value, as most testudinids show this condition. The third (pygal of equal dimensions) and fourth (narrow gulars) characters are quite variable in all *Testudo marginata* material. A recent attempt to define another species within *T. marginata* was by Bour (1996) with the erection of a "dwarf" species of *T. marginata*, named *Testudo weissingeri*. But mitochondrial studies (Fritz et al. 2005) do not support this distinctiveness. Based on all the available evidence, it seems that the small differences observed in the Gerani and Zourida specimens are not adequate enough in order to successfully distinguish the Crete tortoises from *T. marginata*.

★***Testudo marmorum*** **Gaudry, 1862**

Type Material MNHN PIK 3683a (lectotype by subsequent designation of Vlachos et al. 2019b), a partial shell (Fig. 3.A–G in Vlachos et al. 2019b; Fig. 6h–i).

Type Locality Pikermi, Attica, Greece; classical layers, 7.33–7.29 Ma, Late Miocene, Late Tortonian (MN12).

Distribution Besides its type locality, this species is further referred from the Late Miocene of Nikiti-2 in Chalkidiki and possibly in Halmyropotamos, as well as in Azmaka (Bulgaria).

Taxonomic Remarks *Testudo marmorum* is clearly distinguished by other *Testudo* based on the following combination of characters: cervical scute is short and narrow, vertebral 1 is lyre-shaped, pleural 1 covers the lateral parts of the nuchal, there is no coincidence between the anterior pleuro-marginal sulci and the costo-peripheral suture and the first two marginals expand on the costal 1, suprapygal 1 is boomerang-shaped embracing the smaller suprapygal 2, and gular scutes are short and cover the anterior part of the entoplastron (Vlachos et al. 2019b).

Comments Although *Testudo marmorum* is the first fossil turtle species to be named from Greece, its anatomy and phylogenetic relationships have only recently been reviewed (Vlachos et al. 2019b). The revision of previously published (MNHN type specimens, AMPG referred specimens; e.g., Fig. 6h–k) and the description of new specimens (NHMUK) from Pikermi revealed the possibility of variability in the presence and morphology of the plastral hinge. Some specimens from Pikermi (AMPG collection) do not show the typical mobile posterior lobe. The two possible explanations are either the presence of a different, hingeless species, or these distinct morphologies are included within the variability of the species (Vlachos et al. 2019b). If the latter scenario is true, this is probably an expression of the sexual dimorphism in *Te. marmorum*, where female individuals are hingeless—this is the opposite in extant *Chersine hermanni*. However, having the hypoplastral hinge variable in *Te. marmorum* challenges its traditional placement within crown *Testudo*, as its plesiomorphic (within *Testudo*) characters would favor the placement of this taxon along the stem of *Testudo*. With at least 17 individuals referred to this species—ranging from complete to partial shells, without any other skeletal remains—*Te. marmorum* is the fossil turtle taxon with the most abundant material in Greece.

6 Concluding Remarks

Specific interest on the turtle fossils from Greece during the last decade has improved significantly our knowledge of the diversity, alpha taxonomy, and distribution of this important vertebrate group in the country. With at least 68 occurrences, turtles are among the richest vertebrate clades (at order level) in Greece, albeit less diverse than others (e.g., some carnivoran orders, fishes, and squamates). Nevertheless, at least 14 species have been identified, more than half of them belonging to terrestrial turtles or tortoises. This might be, of course, the result of bias toward terrestrial fossil localities, but Testudinidae is the most diverse extant family of turtles in Greece as well. Besides the marine turtles, all native extant species of Greece have been identified in the fossil record. Both testudinids and geoemydids have been identified with confidence at least since the Late Miocene, whereas emydids only since the

Pleistocene. Furthermore, several records of soft-shelled turtles have been published recently, ranging from the Early Miocene to the Pliocene; this is a clade that is currently absent from Greece, although some individuals from Turkey might wander in Greek waters occasionally. Finally, at least one pleurodire species has been also reported from Greece. Overall, it seems that both at a family level and at the specific level (for some families at least), the diversity was higher in the past. In several cases, the Greek fossil record of turtles has made significant impact regionally and continentally and has contributed significantly in the advancement of our knowledge of the anatomy and evolution of several clades. Hopefully, this increased interest in the fossil record of turtles in the country will continue. With this solid background on the fossil record of turtles in Greece, I also hope that it will be possible a complete reevaluation of the turtle occurrences that are being recovered in several zooarchaelogical sites.

Acknowledgments I would like to deeply thank E. Tsoukala for the continuous guidance, inspiration, and support over the years. Many thanks to J. González for the drawing of *Mauremys aristotelica*, to G. Georgalis for discussions and for sharing information on the bone conglomerate from Vathylakkos (MNHN), and to the late R. Bour for sharing literature on the extant species of Greece. Most of the information summarized herein is the result of my doctoral research, and therefore I would like to thank E. Tsoukala, G. Theodorou, and A. Kotsakis (members of my Committee); G. Koufos, G. Syrides, M. Delfino, and M. de la Fuente (reviewers of the thesis); and the School of Geology (AUTH) for all their support. Last but not least, I would like to deeply thank M. Delfino and G. Georgalis for comments that improved this manuscript.

Appendix List of the Greek localities containing turtle fossils. Type localities are marked with bold. Locality numbers refer to the collection numbers of the PaleoBiology Database (PBDB)

Locality[PBDB No]	Age (MN; GPTS in Ma)	Taxa
Vraona Cave[183130]	Recent	*Emys orbicularis*[1] *Mauremys rivulata*[1]
Franchthi Cave[182709]	latest Pleistocene/ Holocene	*Chersine hermanni*[2] *Emys orbicularis*[2]
Charkadio Cave[182809]	latest Pleistocene	*Testudo marginata*[3]
Gerani IV Cave[182810]	**Pleistocene**	*Testudo marginata* (former ***T. m. cretensis***)[4]
Zourida Cave[182811]	Pleistocene	*Testudo marginata*[4]
Liko Cave[92902]	Pleistocene	*Testudo marginata*[5,6] *Mauremys rivulata*[6]
Simonelli Cave[183124]	Pleistocene	*Testudo marginata*[5,6]
Koumpes I Cave[192441]	Pleistocene	*Testudo marginata*[5,6] *Mauremys rivulata*[6]
Mavro Mouri IV[192442]	Pleistocene	*Testudo marginata*[5,6] *Mauremys rivulata*[6]
Siteia I[202328]	Pleistocene	*Mauremys rivulata*[5,6]
Rethymnon fissure[192443]	Pleistocene	*Mauremys rivulata*[5,6]
Bate Cave[183122]	Pleistocene	*Testudo marginata*[7]

(continued)

Appendix (continued)

Locality[PBDB No]	Age (MN; GPTS in Ma)	Taxa
Kos Island[192444]	Pleistocene	*Testudo marginata*[6] *Mauremys rivulata*[6]
Megalopolis[182721]	Pleistocene	*Emys orbicularis*[2] *Mauremys rivulata*[2]
Xerias[182679]	Pleistocene	*Testudo* cf. *marginata*[9]
Potidea[192445]	Pleistocene	*Testudo graeca*[6]
Chalkidiki[202329]	Pleistocene	*Chersine hermanni*[6]
Korydallos[182753]	Pleistocene	*Testudo* sp.[10]
Lakonia[52486]	Pleistocene	*Testudo marginata*[11]
Petralona Cave[183123]	late M. Pleistocene	*Testudo graeca*[6]
Kyparissia 1[194472]	Middle Pleistocene	*Testudo marginata*[12] *Emys orbicularis*[12]
Kyparissia 3[194474]	Middle Pleistocene	*Emys orbicularis*[12]
Kyparissia T[194476]	Middle Pleistocene	*Emys orbicularis*[12]
Kyparissia[194477]	Middle Pleistocene	*Testudo marginata*[12] *Emys orbicularis*[12]
Choremi 2[200080]	Middle Pleistocene	*Emys orbicularis*[13]
Choremi 3[200081]	Middle Pleistocene	*Emys orbicularis*[13]
Choremi 4[200082]	middle Pleistocene	*Emys orbicularis*[13]
Kalamoto 2[200084]	Middle Middle Pleistocene	*Emys orbicularis*[*]
Vatera-F site[183341]	Early Pleistocene	*Titanochelon* sp.[14]
Vatera-T site[183345]	Early Pleistocene	*Testudo graeca*[14]
Karnezeika[202122]	Villafranchian	Testudines indet.[15]
Aggelochori/ Megalo Emvolon/ Falaise de Karabouroun[36579]	Pliocene	*Testudo graeca*[6] *Titanochelon* cf. *bacharidisi*[6,16]
Epanomi[182672]	**Pliocene**	***Titanochelon bacharidisi*[16]**
Nea Michaniona[182674]	Pliocene	*Titanochelon bacharidisi*[16]
Nea Kallikratia[182673]	Pliocene	*Titanochelon bacharidisi*[16]
Gefira-1[182685]	Pliocene	Trionychidae indet.[17]
Gefira-2[196230]	**Pliocene**	***Mauremys aristotelica*[18]**
Milia-1a[185861]	late Pliocene	*Testudo brevitesta*[19]
Milia-2[185862]	late Pliocene	*Testudo brevitesta*[19]
Milia-4[182686]	late Pliocene	***Testudo brevitesta*[19]**
Milia-5[185859]	late Pliocene	*Testudo brevitesta*[19] *Mauremys* sp.[19] *Titanochelon* sp.[19]
Milia-10[185860]	late Pliocene	*Testudo brevitesta*[19]

(continued)

Appendix (continued)

Locality[PBDB No]	Age (MN; GPTS in Ma)	Taxa
Notio 1[205081]	late Pliocene (MN15)	Testudinoidea indet.[20], cf. *Testudo* sp.[20]
Vorio 3&3A[205086]	late Pliocene (MN15)	Testudinoidea indet.[20]
Vorio 2[207948]	late Pliocene (MN14)	Testudinoidea indet.[20]
Tomea Eksi 3[205088]	late Pliocene (MN14)	Testudinoidea indet.[20]
Vorio 1[205087]	late Pliocene (MN14)	Testudinoidea indet.[20]
Apolakkia[184242]	early Pliocene	*Testudoolithus* sp.[21] cf. *Testudo* sp.[21]
Allatini[182683]	latest Miocene/ earliest Pliocene	*Mauremys aristotelica*[18, 22] *Testudo graeca*[22]
Pylea[182684]	latest Miocene/ earliest Pliocene	Testudoolithidae indet.[22] *Testudo graeca*[22]
Nea Silata[191612]	latest Miocene/ earliest Pliocene	*Testudo* sp.[18] *Mauremys aristotelica*[18]
Maramena[32189]	latest Miocene/ earliest Pliocene (MN13/MN14)	*Testudo* sp.[6, 23] *Mauremys aristotelica*[24] ?Testudinidae indet.[24] Testudines indet.[24]
Platania[182682]	late Miocene	*Testudo* cf. *graeca*[25]
Kryopigi[157582]	late Miocene (MN12–13; 7.3–6.4)	*Titanochelon* sp.[6]
Fourka[202330]	late Miocene	*Titanochelon* sp.[6]
Vathylakkos 3[182750]	Middle Turolian (MN12; ~7.3)	*Titanochelon* sp.[26]
Pikermi[182754]	**late Miocene (MN12; 7.3–7.1)**	*Titanochelon* sp.[27, 28] ***Testudo marmorum*[27, 28]**
Pikermi Valley-1[202630]	middle Turolian (MN12)	*Testudo* cf. *marmorum*[28]
Liossati[183121]	late Miocene	*Titanochelon* sp.[28, 29]
Halmyropotamos[202213]	late Miocene (?MN12)	*Testudo marmorum*[30]
Samos[182751/192446]	late Miocene	*Titanochelon schafferi*[6, 31] *'Testudo'* aff. *antiqua*[32]
Nikiti-2[73869]	late Miocene (MN 11)	*Testudo marmorum*[33]
Ravin de la Pluie[191070]	**late Miocene (MN10; ~9.3)**	***Testudo hellenica*[34]**
Kastellios[205360–205365]	late Miocene, late Vallesian, MN 10 (9.3–9.1 Ma)	Testudines indet.[35]
Plakias[183151]	late Miocene	*Mauremys* sp.[36] Trionychidae indet.[36]
Michalos[182752]	middle Miocene	cf. *'Testudo'* sp.[37]

Locality[PBDB No]	Age (MN; GPTS in Ma)	Taxa
Nostimo[183348]	**early Miocene**	***Nostimochelone lampra*[38]**
Karydia[69435]	early Miocene	Testudines indet.[39]
Aliveri[68032]	early Miocene	Testudines indet.[39]
Megalos Lakkos[197718]	early Miocene	Trionychidae indet.[40]

GPTS Geomagnetic Polarity Time Scale
[*]this study
[1]Rauscher (1995), [2]Vlachos and Delfino (2016), [3]Bachmayer and Symeonidis (1975), [4]Bachmayer et al. (1975),[5]Brinkerink (1996), [6]Vlachos (2015), [7]Malatesta et al. (1980), [8]Kotsakis (1977), [9]Tsoukala et al. (2011), [10]Paraskevaidis (1961), [11]Schleich (1982), [12]Athanassiou et al. (2018), [13]van Vugt et al. (2000), [14]Lapparent de Broin (2002), [15]Kokotini et al. (2019), [16]Vlachos et al. (2014), [17]Vlachos et al. (2015b), [18]Vlachos et al. (2019a), [19]Vlachos and Tsoukala (2016), [20]Loreál et al. (2020), [21]Mueller-Töwe et al. (2011), [22]Vlachos et al. (2015a), [23]Schmidt-Kittler (1995), [24]Georgalis et al. (2019b), [25]Vlachos and Tsoukala (2014), [26]Arambourg and Piveteau (1929), [27]Bachmayer (1967), [28]Vlachos et al. (2019b), [29]Bachmayer and Symeonidis (1977), [30]Melentis (1969), [31]Szalai (1931), [32]Vlachos et al. (2019c), [33]Garcia et al. (2016), [34]García et al. (2020), [35]de Bruijn et al. (1971), [36]Georgalis et al. (2016), [37]Paraskevaidis (1955), [38]Georgalis et al. (2013), [39]Georgalis et al. (2019a), [40]Vlachos and Bacharidis (2018)

References

Arambourg C, Piveteau J (1929) Les vertébrés du Pontien de Salonique. Ann Paleontol 18:57–139

Athanassiou A, Michailidis D, Vlachos E, Tourloukis V, Thompson N, Harvati K (2018) Pleistocene vertebrates from the Kyparíssia lignite mine, Megalopolis Basin, S. Greece: Testudines, Aves, Suiformes. Quat Int 497:178–197

Auffenberg W (1974) Checklist of fossil land tortoises (Testudinidae). Bull Florida State Mus 18:121–253

Bachmayer F (1967) Eine Riesenschildkröte aus den altpliozänen Shichten von Pikermi (Griechenland). Praktika Akad Athenae 42:303–317

Bachmayer F, Symeonidis N (1970) Die fossilen Schildkrötenreste des geologisch-paläontologischen Institutes der Universität von Athen. Ann Geol Pays Helléniques 22:227–246

Bachmayer F, Symeonidis N (1975) Schildkrötenreste (*Testudo marginata* SCHOEPFF) aus der Höhle „Charkadio“ auf der Insel Tilos (Dodekanes, Griechenland). Ann Geol Pays Helléniques 26:324–327

Bachmayer F, Symeonidis N (1977) Eine neue „Pikermi“ Fundstelle im Gebiet von Liossati (Kiourka) nördlich von Athen (Griechenland) (Beschreibung einer Riesenschildkröte). Ann Geol Pays Helléniques 28:8–16

Bachmayer F, Brinkerink JP, Symeonidis N (1975) Pleistozäne Schildkröten aus Höhlen der Insel Kreta. Ann Geol Pays Helleniques 27:110–122

Bachmayer F, Mlynarski M, Symeonidis N (1979) Fossile Schildkröten aus dem Pliozän von Megalo Emvolo (Karaburun) bei Saloniki (Griechenland). Ann Geol Pays Helléniques 29:267–276

Bour R (1987) L'identité des tortues terrestres européennes: spécimens-types et localités-types. Rev Fr Aquariol 13:111–122

Bour R (1996) Une nouvelle espèce de tortue terrestre dans le Péloponnèse (Grèce). Dumerilia 2:23–54

Bour R, Ohler A (2008) *Chersine* Merrem, 1820 and *Chersina* Gray, 1831: a nomenclatural survey. Zootaxa 1752:66–68

Brinkerink HP (1996) Tortoises and turtles from the Pleistocene of Crete. In: Reese D (ed) Pleistocene and Holocene Fauna of Crete and its first settlers, Monographs in world archaeology vol 28, pp 207–209

Bronn HG (1831) *Testudo antiqua*, eine im Süsswasser-Gypse von Hohenhöwen untergegangen Art. Nova Acta Leopold 15:202–251

Chkhikvadze VM (2001) (The systematic position of some Asian fossil turtles). Works of the Tbilisi state Pedagogical University named after S.-S.Orbeliani 10:235–240

Chkhikvadze VM (2006) A brief catalog of modern and native land turtles of Northern Eurasia. Prometheus, Tbilisi 7:276–283

Chkhikvadze VM (2007) (Short catalog of fossil turtles of northern Eurasia) problems in Paleobiology. Tbilisi 2:126–137

Corsini JA, Böhme M, Joyce WG (2014) Reappraisal of *'Testudo' antiqua* (Testudines, Testudinidae) from the Miocene of Hohenhöwen, Germany. J Paleontol 88:948–966

Danilov I (2005) Die fossilen Schildkröten Europas. In: Fritz U (ed) Handbuch der Reptilien und Amphibien Europas. Schildkröten (Testudines) II, Cheloniidae, Dermochelyidae, Fossile Schildkröten Europas. Aula, Wiebelsheim, pp 329–448

Danilov I, Syromyatnikova E, Sukhanov V (2017) Subclass Testudinata. In: Lopatin AV, Zelenkov NV (eds) Fossil vertebrates of Russia and adjacent countries, fossil reptiles and birds, part 4. GEOS, Moscow, pp 27–395

de Bruijn H, Sondaar P, Zachariasse WJ (1971) Mammalia and foraminifera from the Neogene of Kastellios hill (Crete): correlations of continental and marine biozones. Proc K Ned Akad Wet 74:1–22

del Campana D (1917) Resti di "*Testudo*" nel Miocene Superiore di Capudjlar presso Salonicco. Boll Soc Geol Italiana 36:1–6

del Campana D (1919) Uova fossili di chelonio nel Miocene Superiore di Capudjlar presso Salonicco. Boll Soc Geol Italiana 38:69–78

Ferreira GS, Bronzati M, Langer MC, Sterli J (2018) Phylogeny, biogeography and diversification patterns of side-necked turtles (Testudines: Pleurodira). R Soc Open Sci 5:171773

Forsyth Mayor CI (1891) Considérations nouvelles sur la faune des vertébrés du Miocène supérieur dans l'île de Samos. C R Hebd. Seances Acad Sci 113:608–610

Fritz U (1994) Zur innerartlichen Variabilität von *Emys orbicularis* (Linnaeus, 1758). 4. Variabilität und Zoogeographie im pontokaspischen Gebiet mit Beschreibung von drei neuen Unterarten. Zool Abh Staat Mus Tier Dresden 48:53–93

Fritz U, Bininda-Emonds OR (2007) When genes meet nomenclature: tortoise phylogeny and the shifting generic concepts of *Testudo* and *Geochelone*. Zoology 110:298–307

Fritz U, Široký P, Kami H, Wink M (2005) Environmentally caused dwarfism or a valid species—is *Testudo weissingeri* Bour, 1996 a distinct evolutionary lineage? New evidence from mitochondrial and nuclear genomic markers. Mol Phylogenet Evol 37:389–401

Fritz U, Guicking D, Kami H, Arakelyan M, Auer M, Ayaz D, Fernández C, Bakiev A, Celani A, Džukic G, Fahd S, Havaš P, Joger U, Khabibullin V, Mazanaeva L, Široký P, Tripepi S, Vélez A, Antón G, Wink M (2007a) Mitochondrial phylogeography of European pond turtles (*Emys orbicularis*, *Emys trinacris*)–an update. Amphibia-Reptilia 28:418–426

Fritz U, Hundsdörfer AK, Široký P, Auer M, Kami H, Lehmann J, Mazanaeva LF, Türkozan O, Wink M (2007b) Phenotypic plasticity leads to incongruence between morphology–based taxonomy and genetic differentiation in western Palaearctic tortoises (*Testudo graeca* complex; Testudines, Testudinidae). Amphibia-Reptilia 28:97–121

Fritz U, Ayaz D, Buschbom J, Kami HG, Mazanaeva LF, Aloufi AA, Auer M, Rifai L, Šilić T, Hundsdörfer AK (2008) Go east: phylogeographies of *Mauremys caspica* and *M. rivulata*–discordance of morphology, mitochondrial and nuclear genomic markers and rare hybridization. J Evol Biol 21:527–540

Fritz U, Harris JD, Fahd S, Rouag R, Gracià Martinez E, Giménez Casalduero A, Široký P, Kalboussi M, Jdeidi T, Hundsdörfer AK (2009) Mitochondrial phylogeography of *Testudo graeca* in the Western Mediterranean: old complex divergence in North Africa and recent arrival in Europe. Amphibia-Reptilia 30:63–80

Gaffney ES, Tong H, Meylan PA (2006) Evolution of the side–necked turtles: the families Bothremydidae, Euraxemydidae, and Araripemydidae. Bull Am Mus Nat Hist 300:1–698

Gaffney ES, Meylan PA, Wood RC, Simons E, Campos DA (2011) Evolution of the side–necked turtles: the family Podocnemididae. Bull Am Mus Nat Hist 350:1–236

Garcia G, Kostopoulos DS, Valentin X, Koufos GD (2016) Testudines. Geobios 49:23–27

Garcia G, de Lapparent de Broin F, Pinton A, Kostopoulos DS, Merceron G, Valentin X, de Bonis L, Koufos GD (2017) The earliest modern Testudinidae: a new *Testudo sensu stricto* from the late Miocene (Vallesian) of Greece. In: Abstracts of 15th congress of the regional committee on Mediterranean Neogene Straigraphy, National and Kapodistrian University of Athens, Athens, 3–6 September 2017

Garcia G, Pinton A, Valentin X, Kostopoulos DS, Merceron G, de Bonis L, Koufos GD (2020) The earliest known crown-*Testudo* tortoise from the late Miocene (Vallesian, 9 Ma) of Greece. PLoS One. https://doi.org/10.1371/journal.pone.0224783

Gaudry A (1862) Résultats des fouilles exécutées en Grèce sous les auspices de l'Académie. C R Acad Sci Paris 52:297–300

Geoffroy Saint–Hilaire E (1809) Mémoire sur les tortues molles, nouveau genre sous le nom de *Trionyx*, et sur la formation des carapaces. Ann Mus Hist Naturelle 14:1–20

Georgalis GL, Joyce WG (2017) A review of the fossil record of Old World turtles of the clade *Pan-Trionychidae*. Bull Peabody Mus Nat Hist 58(1):115–208

Georgalis GL, Kear BP (2013) The fossil turtles of Greece: an overview of taxonomy and distribution. Geobios 46:299–311

Georgalis GL, Velitzelos E, Velitzelos DE, Kear BP (2013) *Nostimochelone lampra* gen. et sp. nov., an enigmatic new Podocnemidoidean turtle from the early Miocene of northern Greece. In: Brinkman D, Holroyd P, Gardner J. (eds) Morphology and evolution of turtles, proceedings of the Gaffney turtle symposium, Springer, Dordrecht, pp 277–287

Georgalis GL, Villa A, Vlachos E, Delfino M (2016) Fossil amphibians and reptiles from Plakias, Crete: a glimpse into the earliest late Miocene herpetofaunas of southeastern Europe. Geobios 49:433–444

Georgalis GL, Villa A, Ivanov M, Roussiakis S, Skandalos P, Delfino M (2019a) Early Miocene herpetofaunas from the Greek localities of Aliveri and Karydia–bridging a gap in the knowledge of amphibians and reptiles from the early Neogene of southeastern Europe. Hist Biol 31(8):1045–1064

Georgalis GL, Villa A, Ivanov M, Vasilyan D, Delfino M (2019b) Fossil amphibians and reptiles from the Neogene locality of Maramena (Greece), the most diverse European herpetofauna at the Miocene/Pliocene transition boundary. Palaeontol Electronica 22.3(68):1–99

Gmelin JF (1789) Caroli a Linné, Systema Naturae per regna tria naturae secundum classes, ordines, genera, species, cum characteribus, differentiis, synonymis, locis. Ed. 13. Tom. I. Pars III. GE Beer, Leipzig

Gray JE (1834) Characters of two new genera of reptiles (*Geoemyda* and *Gehyra*). Proc Zool Soc London 1834:99–101

Gray JE (1860) Description of a new species of *Emys* lately living in the gardens of the zoological society. Proc Zool Soc London 1860(28):232–233

Gray JE (1869) Description of *Mauremys laniaria*, a new freshwater tortoise. Proc Zool Soc London 1869:499–500

Hernández–Pacheco E (1917) El problema de la investigación científica en España. Asoc Española Progreso Ciencias 2:63–93

Joyce WG, Parham JF, Gauthier JA (2004) Developing a protocol for the conversion of rank-based taxon names to phylogenetically defined clade names, as exemplified by turtles. J Paleontol 78:989–1013

Kokotini M, Kargopoulos N, Iliopoulos G, Roussiakis S, Skandalos P, Michailidis D, Svorligkou G, Kampouridis P, Theodorou G (2019) Karnezeika (Argolis, Peloponnese): preliminary data concerning a new Villafranchian locality of Southern Greece. Bull Geol Soc Greece, Spl Publ 7:GSG2019–GSG2104

Kostopoulos DS (2009) The Pikermian event: temporal and spatial resolution of the Turolian large mammal fauna in SE Europe. Palaeogeogr Palaeoclimatol Palaeoecol 274:82–95

Kotsakis T (1977) I resti di anfibi e rettili pleistocenici della grotta "Bate" (Rethymnon, Creta). Rendivoti Accad Naz Lincei 8(63):571–582

Lapparent de Broin F d (2001) The European turtle fauna from the Triassic to the present. Dumerilia 4:155–217

Lapparent de Broin F d (2002) A giant tortoise from the late Pliocene of Lesvos Island (Greece) and its possible relationships. Ann Geol Pays Helléniques 34:99–130

Lapparent de Broin F d, Bour R, Perälä J (2006) Morphological definition of *Eurotestudo* (Testudinidae, Chelonii): first part. Ann Paleontol 92:255–304

Lenk P, Fritz U, Joger U, Wink M (1999) Mitochondrial phylogeography of the European pond turtle, *Emys orbicularis* (Linnaeus 1758). Mol Ecol 8:1911–1922

Luján ÀH, Delfino M, Robles JM, Alba DM (2016) The Miocene tortoise *Testudo catalaunica* Bataller, 1926, and a revised phylogeny of extinct species of genus *Testudo* (Testudines: Testudinidae). Zool J Linnean Soc 178:312–342

Malatesta A, Angelelli F, Caloi L, Kotsakis T (1980) Dwarf deer and other late Pleistocene fauna of the Simonelli cave in Crete. Quaderni Accad Naz Lincei 209:1–122

Mantziou G, Rifai L (2014) *Mauremys rivulata* (Valenciennes in Bory de saint-Vincent 1833)—Western Caspian turtle, Balkan terrapin. Conservation Biology of Freshwater Turtles and Tortoises: A compilation Project of the IUCN/SSC Tortoise and Freshwater Turtle Specialist Group, Chelonian Research Monographs

Melentis JK (1966) *Clemmys caspica* aus dem Pleistozän des Beckens von Megalopolis im Peloponnes (Griechenland). Ann Geol Pays Helléniques 17:169–181

Merrem B (1820) Versuch eines Systems der Amphibien. Tentamen Systematis Amphibiorum. J.C. Krieger, Marburg

Mueller–Töwe IJ, Kjeldahl–Vallon TA, Milàn J, Vallon LH, Theodorou G, Lindgren J, Roussiakis S, Bromley RG (2011) First chelonian eggs and carapace fragments from the Pliocene of Rhodes, Greece. N Jahr Geol Palaeontol Abh 262:309–322

Pallas PS (1814) Zoographia Rosso-Asiatica. III. Animalia Monocardia seu Frigidi Sanguinis Imperii Rosso-Asiatici. Officina Caes. Academiae Scientiarum, Petropolis

Paraskevaidis I (1955) Zwei schildkrötenreste aus dem obermiozän von Chios. Ann Geol Pays Helléniques 6:133–141

Paraskevaidis I (1961) Neue Funde Quartärer Wirbeltierreste in Attika. Ann Geol Pays Helléniques 12:149–152

Parham JF, Macey JR, Papenfuss TJ, Feldman CR, Türkozan O, Polymeni R, Boore J (2006) The phylogeny of Mediterranean tortoises and their close relatives based on complete mitochondrial genome sequences from museum specimens. Mol Phylogenet Evol 38:50–64

Pereira AG, Sterli J, Moreira FR, Schrago CG (2017) Multilocus phylogeny and statistical biogeography clarify the evolutionary history of major lineages of turtles. Mol Phylogenet Evol 113:59–66

Pérez-García A, Vlachos E (2014) New generic proposal for the European Neogene large testudinids (Cryptodira) and the first phylogenetic hypothesis for the medium and large representatives of the European Cenozoic record. Zool J Linnean Soc 172:653–719

Pérez-García A, Ortega F, Jiménez Fuentes E (2017a) Taxonomy, systematics, and diversity of the European oldest testudinids. Zool J Linnean Soc 177:648–675

Pérez-García A, Vlachos E, Arribas A (2017b) The last giant continental tortoise of Europe: a survivor in the Spanish Pleistocene site of Fonelas P-1. Palaeogeogr Palaeoclimatol Palaeoecol 470:30–39

Rauscher KL (1995) Die herpetofauna der Vraona–Hoehle (Attika) in Griechenland. Ann Geol Pays Helléniques 36:39–41
Schleich H (1982) *Testudo marginata* Schoepff aus plio/pleistozänen ablagerungen SE–Lakoniens (Peloponnes, Griechenland). Palaeontol Z 56:259–264
Schmidt–Kittler N (ed) (1995) The vertebrate locality Maramena (Macedonia, Greece) at the Turolian–Ruscinian boundary (Neogene). Pfeil, München
Schoepff JD (1793) Historia Testudinum Iconibus Illustrata. Ioannis Iacobi Palm, Erlangae
Schweigger AF (1812) Prodromus monographiae Cheloniorum. Königsberger Arch Naturwissen Math 1(271–368):406–462
Sommer RS, Persson A, Wieseke N, Fritz U (2007) Holocene recolonization and extinction of the pond turtle, Emys orbicularis (L., 1758), in Europe. Quat Sci Rev 26:3099–3107
Staesche K, Karl HV, Staesche U (2007) Fossile Schildkröten aus der Türkei. In: Staesche U (ed) Fossile Schildkröten aus drei Kontinenten, Geol Jahrb, vol 98, pp 91–149
Stojanov A (2009) Erster Nachweis einer Riesenlandschildkr ¨ ote (*Geochelone* s.l. Gray, 1872) aus Bulgarien. Rev Paléobiol 28:457–470
Szalai T (1931) *Testudo schafferi* nov. sp. eine Riesenschildkröte aus dem Pliozän von Samos. Ann Naturhist Mus Wien 46:153–163
Tsoukala E, Mol D, Pappa S, Vlachos E, van Logchem W, Vaxevanopoulos M, Reumer J (2011) *Elephas antiquus* in Greece: new finds and a reappraisal of older material (Mammalia, Proboscidea, Elephantidae). Quat Int 245:339–349
Turtle Taxonomy Extinction Group (2015) Turtles and Tortoises of the world During the Rise and Global Spread of Humanity: First Checklist and Review of Extinct Pleistocene and Holocene Chelonians. In: Rhodin AGJ, Thomson S, Georgalis GL, Karl H-V, Danilov IG, Takahashi A, de la Fuente MS, Bourque JR, Delfino M, Bour R, Iverson JB, Shaffer HB, van Dijk PP (eds) Conservation biology of freshwater turtles and tortoises: a compilation project of the IUCN/SSC tortoise and freshwater turtle specialist group. Chelonian Res Monographs 5:1–66
Turtle Taxonomy Working Group (2017) Turtles of the world: annotated checklist and atlas of taxonomy, synonymy, distribution, and conservation status (8th Ed.). In: Rhodin AGJ, Iverson JB, van Dijk PP, Saumure RA, Buhlmann KA, Pritchard PCH, Mittermeier RA (eds) Conservation biology of freshwater turtles and tortoises: a compilation project of the IUCN/SSC tortoise and freshwater turtle specialist group. Chelonian Res Monographs 7:1–292
Vamberger M, Stuckas H, Ayaz D, Lymberakis P, Široký P, Fritz U (2014) Massive transoceanic gene flow in a freshwater turtle (Testudines: Geoemydidae: *Mauremys rivulata*). Zool Scripta 43:313–322
Vamberger M, Stuckas H, Vargas-Ramírez M, Kehlmaier C, Ayaz D, Aloufi AA, Lymberakis P, Široký P, Fritz U (2017) Unexpected hybridization patterns in near eastern terrapins (*Mauremys caspica, M. rivulata*) indicate ancient gene flow across the Fertile Crescent. Zool Scripta 46:401–413
Van Vugt N, de Bruijn H, van Kolfschoten M, Langereis CG, Okuda M (2000) Magneto- and cyclostratigraphy and mammal-faunas of the Pleistocene lacustrine Megalopolis Basin, Peloponnesos, Greece. Geol Ultraiect 189:1–167
Vitek NS, Joyce WG (2015) A review of the fossil record of New World turtles of the clade Pan-Trionychidae. Bull Peabody Mus Nat Hist 56:185–245
Vlachos E (2015) The fossil chelonians of Greece. Systematics – evolution – stratigraphy – palaeoecology. Ph.D. dissertation, Aristotle University of Thessaloniki, Greece
Vlachos E, Bacharidis N (2018) A trionychid turtle from the late Aquitanian (early Miocene) of Neochori (Grevena, Western Macedonia, Greece). Boll Soc Paleont Italiana 57:255–257
Vlachos E, Delfino M (2016) Food for thought: sub-fossil and fossil chelonian remains from Franchthi Cave and Megalopolis confirm a glacial refuge for *Emys orbicularis* in Peloponnesus (S. Greece). Quat Sci Rev 150:158–171
Vlachos E, Rabi M (2018) Total evidence analysis and body size evolution of extant and extinct tortoises (Testudines: Cryptodira: Pan-Testudinidae). Cladistics 34:652–683

Vlachos E, Tsoukala E (2014) *Testudo* cf. *graeca* from the new late Miocene locality of Platania (Drama basin, N. Greece) and a reappraisal of previously published specimens. Bull Geol Soc Greece 48:27–40

Vlachos E, Tsoukala E (2016) The diverse fossil chelonians from Milia (late Pliocene, Grevena, Greece) with a new species of *Testudo* Linnaeus, 1758 (Testudines: Testudinidae). Pap Palaeontol 2:72–86

Vlachos E, Tsoukala E, Corsini J (2014) *Cheirogaster bacharidisi* sp. nov., a new species of a giant tortoise from the Pliocene of Thessaloniki (Macedonia, Greece). J Vertebr Paleontol 34:560–575

Vlachos E, Cerda I, Tsoukala E (2015a) The first record of a soft-shelled turtle (Testudines: *Pan-Trionychidae*) from southern Balkans (Pliocene, Gefira, N. Greece) and new information from bone histology. Sci Nat 102:45. https://doi.org/10.1007/s00114-015-1295-2

Vlachos E, Delfino M, Kotsakis T (2015b) The chelonians from the latest Miocene-earliest Pliocene locality of Allatini and Pylea (East Thessaloniki, Macedonia, Greece). C R Palevol 14:187–205

Vlachos E, Sterli J, Vasileiadou K, Syrides G (2019a) A new species of *Mauremys* (Testudines, Geoemydidae) from the late Miocene – Pliocene of Central Macedonia (northern Greece) with exceptionally wide vertebral scutes. Pap Palaeontol 5:177–195

Vlachos E, Georgalis GL, Roussiakis S, Böhme M, Theodorou G (2019b) The Pikermian tortoises (Testudines, Testudinidae) from the late Miocene of South Balkans. J Vertebr Paleontol 39(5):e1711520. https://doi.org/10.1080/02724634.2019.1711520

Vlachos E, Pérez-García A, Roussiakis S, Georgalis GL, Kear B (2019c) Late Miocene tortoises from Samos island in Greece: implications for European Neogene testudinid systematics and distributions. J Vertebr Paleontol 39(6):e1722950. https://doi.org/10.1080/02724634.2019.1722950

Wagler JG (1830) Natürliches System der Amphibien, mit vorangehender Classification der Säugthiere und Vögel. J.G. Cotta'schen Buchhandlung, München

The Fossil Record of Aardvarks (Mammalia: Tubulidentata: Orycteropodidae) in Greece

George D. Koufos

1 Introduction

The Tubulidentata or aardvarks (= earth pigs in Afrikaans) are a peculiar group of mammals from which only one species, *Orycteropus afer* (Pallas 1766), is still living in Sub-Saharan Africa, from Senegal to Ethiopia, and down to South Africa. One of the most distinctive characters of the aardvarks is the structure of their teeth. The teeth lack enamel and consist of dentine tubes, connected by cement; this tubular structure gave the name Tubulidentata to this mammalian order. The premolars are peg-like with angular apex (when worn) and the molars are eight-shaped in occlusal view. The living *O. afer* presents a singular body plan with a pig-like body, rabbit-like ears, kangaroo-like head, and giant rat-like tail. It is an exceptionally fast digger and can open rapidly ant and termite nests to feed on them. The aardvark is a nocturnal, solitary, and myrmecophagous animal; with its long and sticky tongue, it can collect ten thousand of insects during a night. The oldest-known representative of the order Tubulidentata comes from the early Miocene of East Africa and belongs to the genus *Myorycteropus*, but its origin, probably in the early Paleogene of Africa, remains obscure (Lehmann 2009).

2 Historical Overview

The first collected material of Tubulidentata in Greece originates from Samos and has been described as a new species under the name *Orycteropus gaudryi* by Forsyth Major (1888). Colbert (1941) studied the rich collection of cranial and postcranial

G. D. Koufos (✉)
Department of Geology, Laboratory of Geology and Paleontology,
Aristotle University of Thessaloniki, Thessaloniki, Greece
e-mail: koufos@geo.auth.gr

E. Vlachos (ed.), *Fossil Vertebrates of Greece Vol. 1*,
https://doi.org/10.1007/978-3-030-68398-6_10

remains of the Samos aardvark, housed at the American Museum of Natural History in New York (Brown collection). New orycteropodid material, collected during the new campaigns on the island of Samos from 1994 to 2007, was recently described by Koufos (2009). The taxon is also recognized in two localities of the Evia Island, but the material is still undescribed (Woodward 1901; Theodorou et al. 2003; Kampouridis et al. 2019). Finally, in three Late Miocene localities from Axios Valley (Macedonia, Greece), more fossil orycteropodids have been discovered (Bonis et al. 1994).

3 Phylogenetic Relationships

Although the aardvarks are studied since the eighteenth century, the order Tubulidentata was erected by Huxley (1872). It is not before the middle of the 1970s that the studies of the aardvarks have gone from simple descriptions in studies of their phylogenetic relationships (e.g., Patterson 1975, 1978; MacPhee 1994; van der Made 2003; Pickford 2004, 2005; Lehmann et al. 2005). The most recent and complete phylogeny of the Orycteropodidae gives also a new classification of the family, based on the study of the material and a parsimony analysis (Lehmann 2009). The author separates four genera, three African *Myorycteropus, Leptorycteropus*, *Orycteropus*, and the African-European one *Amphiorycteropus*. All have African origin, and *Amphiorycteropus* migrated to Eurasia in the middle Miocene making its first appearance in the Turkish locality of Paşalar at ~16.0 Ma with the species aff. *Amphiorycteropus* cf. *seni*. The migration age fits quite well with the final connection of Africa with Eurasia at ~17.0–15.0 Ma (Rögl 1999), as well as with the arrival of several African elements recognized in the neighboring middle Miocene locality Thymiana B in Chios Island (Aegean Sea, Greece) dated at ~15.5 Ma (Koufos 2013 and references therein). The genus *Amphiorycteropus* dispersed in Eurasia and divided into two main clades: the clade of *Amphiorycteropus* (*browni-gaudryi-depereti*) and the clade of aff. *Amphiorycteropus* (*seni-pottieri*). The first clade survived until the early Pliocene with *A. depereti* and the second until the early Vallesian with aff. *A. pottieri* (Fig. 1). Both clades are represented in the Greek fossil record.

4 Distribution

Fossil orycteropodids are known from Africa, Europe, and Western Asia, but their fossil remains are scarce. The most fossiliferous sites with orycteropodids are known from Kenya in Africa as well as from Greece and Turkey. Details on Greek occurrences of orycteropodids are given in Fig. 2 and Appendix.

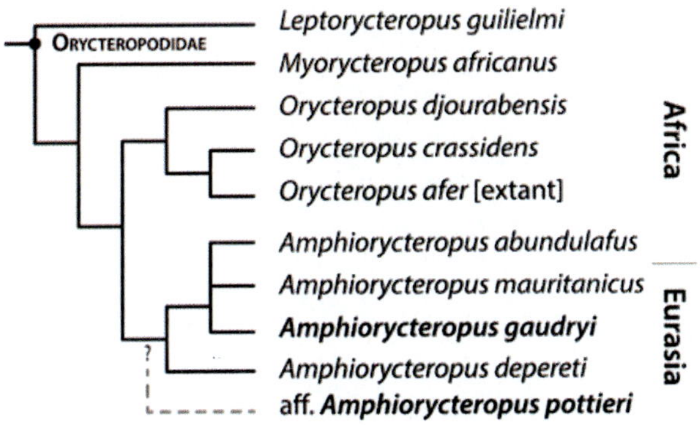

Fig. 1 Phylogeny of the Orycteropodidae, highlighting the position of the Greek orycteropodid taxa (with boldface; modified from Lehmann 2009)

5 Systematic Paleontology

Orycteropodidae Gray, 1821
***Amphiorycteropus* Lehmann, 2009**

Type Species *Orycteropus gaudryi* Forsyth Major, 1888.

Remarks Besides the certain occurrences of *Amphiorycteropus* in Greece, the genus may be present in Prochoma 1, in the Axios Valley (Macedonia, Greece) where it is represented by two isolated phalanges (Bonis et al. 1994). The genus is also traced from the locality Kerassia-1, (Theodorou et al. 2003; Kampouridis et al. 2019), and Prokopi in Evia Island (Woodward 1901).

★*Amphiorycteropus gaudryi* (Forsyth Major, 1888)

Nomenclatural and Taxonomical History *Orycteropus gaudryi* in Forsyth Major 1888 (new species); *Amphiorycteropus gaudryi* in Lehmann 2009 (new genus).

Type Material There is no defined holotype. The material described by Forsyth Major (1888) and housed in the Musée Géologique Cantonal de Lausanne can be considered as its syntypes (Lehmann 2007).

Type Locality Unknown locality of Samos Island, late Miocene, MN11–13.

Distribution The species is known from the late Miocene of Iran, Italy, Turkey, Ukraine, and Moldova. In Greece, it is known from the localities Mytilini 1A and 1B, Quarries 1, 4, and 5 of Samos Island, Dytiko 1 in the Axios Valley, as well as Prokopion (old name Achmet Aga) and Kerassia in the Evia Island.

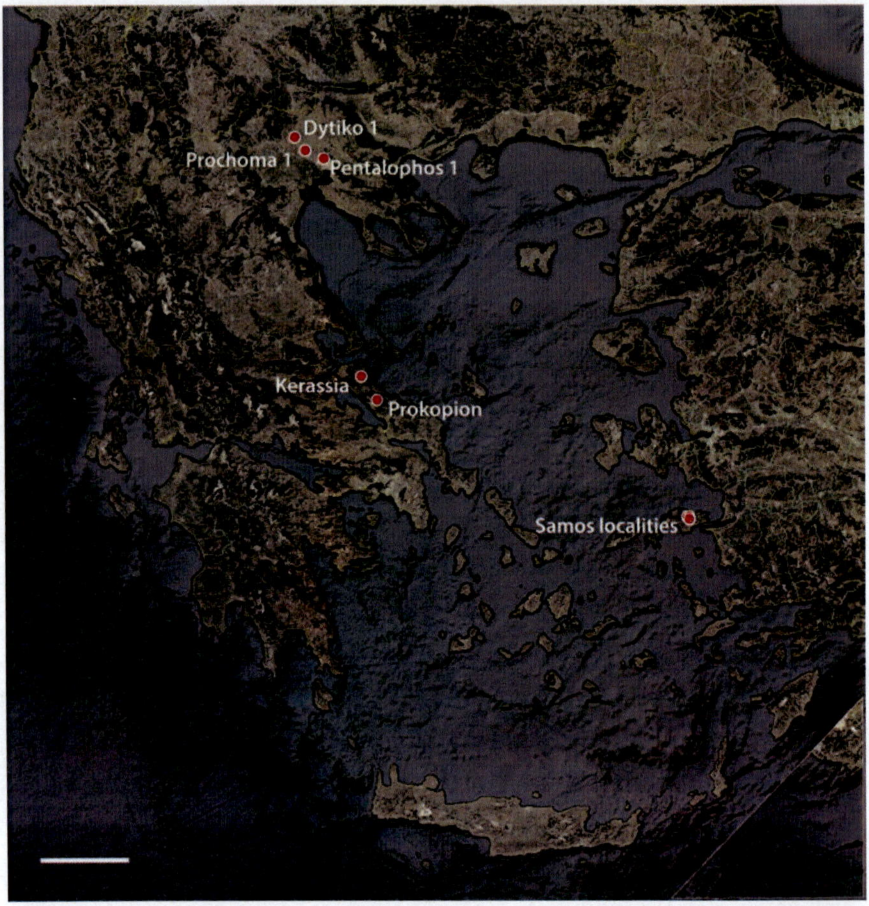

Fig. 2 Map of Greece showing the geographic distribution of the most important localities with aardvark fossils. See Appendix for more information. Image exported from Google Earth Pro © 2019, map data from US Dept. of State Geographer, SIO, NOAA, US Navy, NGA, GEBCO, image from Landsat/Copernicus. Scale bar equals 80 km, North faces upward

Remarks The taxon is mainly known from Samos by a rich collection housed in various museums (Fig. 3). Some characters of the cranial material are the elongated and relatively narrow palate; the elongated and wide nasals, separated by a shallow groove between them; the bulging frontal region; the small diastemas between the premolars; the single-lobed P1 and P2 with one occlusal wearing facet inclined distally; the P3 and P4 with one lobe and two occlusal wearing facets forming a sharp angle; the bilobed molars; the triangular-shaped M3; the elongated mandibular corpus with slightly concave inferior border; the small diastemas between the lower premolars; the large mental foramen situated almost in front of the m1; the position of the anterior margin of the masseteric fossa below the distal margin of the m3; the single-lobed premolars with two occlusal wearing facets forming an angle;

Fig. 3 Selected specimens of *Amphiorycteropus gaudryi* from the late Miocene (MN11–13) of Samos Island. (**a**) cranium and mandible from an unknown precise locality in lateral view (NHML-SAM-M.5690), (**b**) cranium from an unknown precise locality in occlusal view (NHMW-SAM-A.4760), (**c–d**) right mandibular fragment with p2–m3 from Mytilini 1 in **c** lateral and **d** occlusal views (MTLB-49). Scale bars equal 1 cm

the 8-shaped molars, the largest than all molars m2; the broader distal lobe than the mesial one in the m1 and m2; the small, triangular-shaped m3 with narrow distal lobe; and the well-developed lingual and buccal grooves that allow attribution to *A. gaudryi* (Koufos 2009 and references therein). The diagnosis of the genus *Amphiorycteropus* (= *A. gaudryi*) is given by Lehmann (2009).

aff. *Amphiorycteropus pottieri* (Ozansoy, 1965)

Nomenclatural and Taxonomical History *Orycteropus pottieri* in Ozansoy 1965 (new species); aff. *Amphiorycteropus pottieri* in Lehmann 2009 (new combination). The orycteropodids from Pentalophos 1 (Greece) and Sinap (Turkey), first described as *"Orycteropus" pottieri*, share some apomorphies with species of the genus *Amphiorycteropus*, but more material is necessary for a definite decision; thus, it is recommended to place this material in *affinis* with the genus *Amphiorycteropus* (Lehmann 2009).

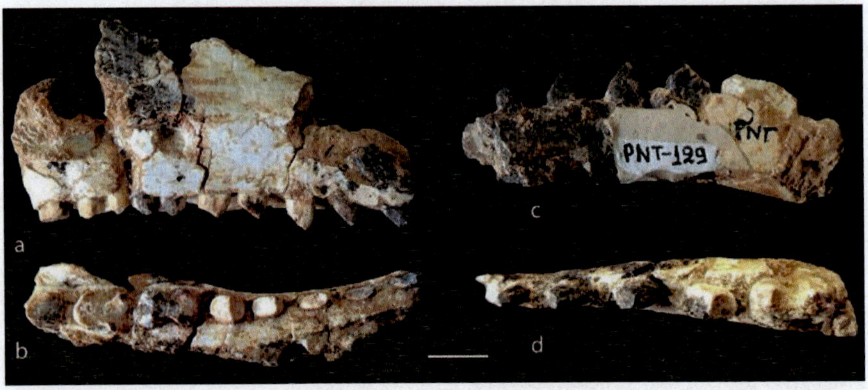

Fig. 4 Selected specimens of aff. *Amphiorycteropus pottieri* from Pentalophos 1 (Axios Valley). (**a–b**) right maxillary fragment (PNT 127) with C–M3 in **a** lateral and **b** occlusal views, (**c–d**) right mandibular fragment (PNT 129) with p1–m1 in **c** buccal and **d** occlusal views. Scale bar equals 1 cm

Type Material MNHN-TRQ-1003 (holotype), right hemimandible, described and figured by Ozansoy (1965) and housed at the Muséum national d'Histoire naturelle Paris (Lehmann 2007).

Type Locality Middle Sinap formation (late Miocene, Vallesian), near Yassiören (Ozansoy 1965).

Distribution Besides the type locality, the species has been discovered during the new campaigns of excavations in Sinap area (localities 12, 72, and 108) and from the Greek late Miocene locality Pentalophos 1 of the Axios Valley.

Remarks The material from Pentalophos 1 includes the posterior part of a cranium, maxillary, mandibular, and some postcranial remains (Fig. 4). The Pentalophos 1 material is characterized by large infraorbital foramen situated above M1; V-shaped nuchal line on the cranium; temporal lines high on the cranium; small, elliptical-shaped, and more oblique than that of *A. gaudryi* occipital condyle; large and oval foramen magnum vs rounded in *A. gaudryi*; well-developed paroccipital processes; mastoid process with ridges and grooves in their buccal surface; straight and low mandibular corpus; concave upper and straight lower tooth row; presence of a canine; remarkable diastemas between the premolars and between the canine and first premolar; large diastema between p4 and m1; uniform premolars with one lobe and angular occlusal outline; mesial wear facet smaller and more inclined than the distal one in the upper premolars while in the lower one it is the opposite; 8-shaped first and second molars in both jaws with deep lingual and buccal grooves between the lobes; rectangular outline of the lower and trapezoid outline of the upper molars; lobes with rounded mesial and distal margins; M1 and m1 with distal lobe broader than the mesial one (the opposite in the M2 and m2); small and triangular-shaped M3, and m3 with short and rounded distal lobe; the short cumulated length of the upper molars; and proximodistally elongated astragalus, which allow its attribution to this taxon (Bonis et al. 1994).

Acknowledgments I would like to thank T. Lehmann for comments that improved this manuscript.

Appendix

List of the Greek localities containing aardvark fossils. Type localities are marked with bold. Locality numbers refer to the collection numbers of the Paleobiology Database (PBDB)

Localities[PBDB No]	Age (MN; GPTS in Ma)	Taxon
Samos[202120*]	**Turolian (MN11–13)**	***Amphiorycteropus gaudryi*[1]**
Samos Q5[95690]	Late Turolian (MN13; 6.9–6.7)	*Amphiorycteropus gaudryi*[2]
Dytiko 1[32374]	Late Turolian (MN13)	*Amphiorycteropus gaudryi*[3]
Mytilini 1A[202215]	Middle Turolian (MN12; 7.3–7.1)	*Amphiorycteropus gaudryi*[4]
Mytilini 1B[202216]	Middle Turolian (MN12; 7.3–7.1)	*Amphiorycteropus gaudryi*[4]
Samos Q1[95691]	Middle Turolian (MN12; 7.3–7.1)	*Amphiorycteropus gaudryi*[2]
Prochoma 1[202222]	Middle Turolian (MN12; ~7.4 Ma)	*Amphiorycteropus* sp.[3]
Samos Q4[95689]	Middle Turolian (MN12; 7.65–7.45)	*Amphiorycteropus gaudryi*[2]
Kerassia 1[195432]	Middle Turolian (MN12)	*Amphiorycteropus gaudryi*[5]
Prokopion[214079]	?Middle Turolian (?MN12)	*Amphiorycteropus gaudryi*[6]
Pentalophos 1[202119]	Early Vallesian (MN9)	aff. *Amphiorycteropus pottieri*[3]

GPTS Geomagnetic Polarity Time Scale
*Precise locality unknown
[1]Forsyth Major (1888), [2]Colbert (1941), [3]Bonis et al. (1994), [4]Koufos (2009), [5]Kampouridis et al. (2019), [6]Woodward (1901)

References

Colbert EH (1941) A study of *Orycteropus gaudryi* from the island of Samos. Bull Am Mus Nat Hist 78:305–351

de Bonis L, Bouvrain G, Geraads D, Koufos GD, Sen S (1994) The first aardvarks from the Miocene of Macedonia (Greece). Neues Jahrb Geol Palaeontol Monatsh 194:343–360

Forsyth Major CJ (1888) Sur un gisement d'ossements fossiles dans l'île de Samos, contemporains l'âge de Pikermi. C R Acad Sci Paris 118:608–610

Huxley T (1872) A manual of the anatomy of vertebrated animals. Appleton, New York, pp 288–289

Kampouridis P, Roussiakis S, Kargopoulos N, Giaourtsakis I, Dimakopoulos G, Iliopoulos G, Svorligkou G, Theodorou G (2019) Faunal diversity at the Turolian locality of Kerassia (Northern Euboea, Greece). Bull Geol Soc Greece, sp publ 7:52–53

Koufos GD (2009) Tubulidentata. In: Koufos GD, Nagel D (eds) The late Miocene mammal faunas of the Mytilinii Basin, Samos Island, Greece: New Collection. Beitr Palaeont 31:107–125

Koufos GD (2013) Neogene mammal biostratigraphy and chronology of Greece. In: Wang X, Flynn LJ, Fortelius M (eds) Fossil mammals of Asia. Neogene biostratigraphy and chronology. Columbia University Press, New York, pp 595–621

Lehmann T (2007) Amended taxonomy of the order Tubulidentata (Mammalia, Eutheria). Ann Transvaal Mus 44:179–196

Lehmann T (2009) Phylogeny and systematics of the Orycteropodidae (Mammalia, Tubulidentata). Zool J Linnean Soc 155:649–702

Lehmann T, Vignaud P, Likius A, Brunet M (2005) A new species of Orycteropodidae (Mammalia, Tubulidentata) in the Mio-Pliocene of northern Chad. Zool J Linnean Soc 143:109–131

MacPhee RDE (1994) Morphology, adaptations, and relationships of *Plesiorycteropus*, and a diagnosis of a new order of eutherian mammals. Bull Am Mus Nat Hist 220:1–214

Ozansoy F (1965) Étude des gisements continentaux et des Mammifères du Cénozoique de Turquie. Mem Soc Geol Fr 102:1–92

Patterson B (1975) The fossil aardvarks (Mammalia, Tubulidentata). Bull Mus Comp Zool Cambridge 147:185–237

Patterson B (1978) Pholidota and Tubulidentata. In: Maglio VJ, Cooke HBS (eds) Evolution of African mammals. Harvard University Press, Cambridge, pp 268–278

Pickford M (2004) Southern Africa: a cradle of evolution. South Afr J Sci 100:205–214

Pickford M (2005) *Orycteropus* (Tubulidentata, Mammalia) from Langebaanweg and Baard's Quarry, early Pliocene of South Africa. C R Palevol 4:715–726

Rögl F (1999) Circum-Mediterranean Miocene paleogeography. In: Rössner G, Heissig K (eds) The Miocene Land Mammals of Europe. Verlag Dr Friedrich Pfeil, München, pp 39–48

Theodorou G, Athanassiou A, Iliopoulos G, Roussiakis S (2003) Preliminary remarks on the late Miocene herbivores of Kerassia (Northern Euboea, Greece). Deinsea 10:519–530

van der Made J (2003) The aardvark from the Miocene hominoid locality Çandir, Turkey. Courier Forschung Inst Senckenberg 240:133–147

Woodward AS (1901) On the bone-beds of Pikermi, Attica and on similar deposits in Northern Euboea. Geol Mag 4:481–486

The Fossil Record of Hyraxes (Mammalia: Hyracoidea: Pliohyracidae) in Greece

George D. Koufos

1 Introduction

The hyracoids (hyraxes or dassies) are a peculiar group of mammals, preserving some primitive characters and being related to the elephants and sirenians, although they are small and their appearance is more rodent like. Their similar dental pattern with horses, rhinoceroses and extinct palaeotheres and the presence of five toes in the forefoot (Myers 2000) support their similarities with the perissodactyls. The modern hyracoids belong to the family Procaviidae and have a short skull with short muzzle, prominent postorbital processes and small auditory bullae. Their dental formula is 1, 0, 4, 3/2, 0, 4, 3 = 34. There is only one upper incisor and two lower ones, which are specialized; the upper incisor is pointed, has a continuous growth like that of rodents and lacks enamel in its posterior surface; the lower ones have styloid form. The incisors are followed by a large diastema as in the rodents. The molars are selenolophodonts and they resemble those of rhinocerotids. The hyraxes are living on rocky areas or on the trees (arboreal forms) feeding on leaves, barks, grasses and some insects. The modern forms are living in the sub-Saharan Africa and the Middle East (Vaughan et al. 2015; Myers 2000).

2 Historical Overview

The hyracoids are recognized early in the Greek fossil record in Pikermi, described as *Leptodon graecus* by Gaudry (1860). Later, they were found and described in Samos Island with two species (Osborn 1898; Forsyth Major 1899; Schlosser

G. D. Koufos (✉)
Department of Geology, Laboratory of Geology and Paleontology, Aristotle University of Thessaloniki, Thessaloniki, Greece
e-mail: koufos@geo.auth.gr

E. Vlachos (ed.), *Fossil Vertebrates of Greece Vol. 1*,
https://doi.org/10.1007/978-3-030-68398-6_11

1899a, b). Among the Halmyropotamos material, stored in the Palaeontological and Geological Museum of Athens University, there are some hyracoid remains, described by Melentis (1967). Kuss (1976) found and described some remains in the locality Melambes (Crete Island). Recently, some hyaracoid fossils were discovered and described from Samos (Koufos 2009). All fossil hyracoid remains belong to the family Pliohyracidae, characterized by high-crowned cheek teeth, spatulated and triangular incisors and reduction in the i3 (Meyer 1978). The genus *Pliohyrax* with two species dominated in the Greek fossil record, whereas *Prohyrax* is probably present.

3 Phylogenetic Relationships

Hyraxes share many morphological features with the elephants and sirenians, suggesting a common ancestor; recent molecular data also support close relationships among these three mammalian groups. The first presence of hyraxes has been traced in the basal Eocene of Algeria and Morocco (Benoit et al. 2016; Gheerbrant et al. 2003). Fossil evidence from the Eocene–Oligocene of Fayum (Egypt) indicates that the hyraxes were the dominant browsers and grazers in Africa during this time span; they were common, represented by seven different genera (Myers 2000; Hoek 2018). During the beginning of the Miocene, the arrival of the artiodactyls restricted hyraxes to the tree and rocky habitats, which are not favourable for these mammals (Rasmussen 1989; Hoek 2018). Simultaneously, this faunal change pushed them to the north in Asia, as indicated by their presence in Saudi Arabia and Pakistan (Thomas et al. 1982; Pickford 1986). During the late Oligocene to late Miocene, a second radiation of hyraxes is recognized in Africa with the presence of the most primitive representative of the family Pliohyracidae. The latest representatives of this family are traced in the Plio-Pleistocene of China; the main representative of that period is the genus *Pliohyrax*. The Pliohyracidae disappeared at the Early Pleistocene after a presence of about 50 Ma. However, a hyracoid clade survived and gave the family Procaviidae, which survives until now (Rasmussen 1989).

4 Distribution

In Greece, the hyracoids are scarce and they are known from Pikermi, Samos Island (Mytilinii 1C, unknown locality/ies for the old material) and Halmyropotamos in Evia Island (Fig. 1 and Appendix). The majority of the material is of late Miocene age and belongs to the genus *Pliohyrax*. Several hyracoid remains are known from the late Miocene of Turkey, where the genus *Pliohyrax* is recognized in several localities (Baudry 1994; Kaya et al. 2012). It is also known from the wider Mediterranean region, as from Spain, France, Iran and Afghanistan (Dubrovo 1978; Sen 1998; Pickford 2009).

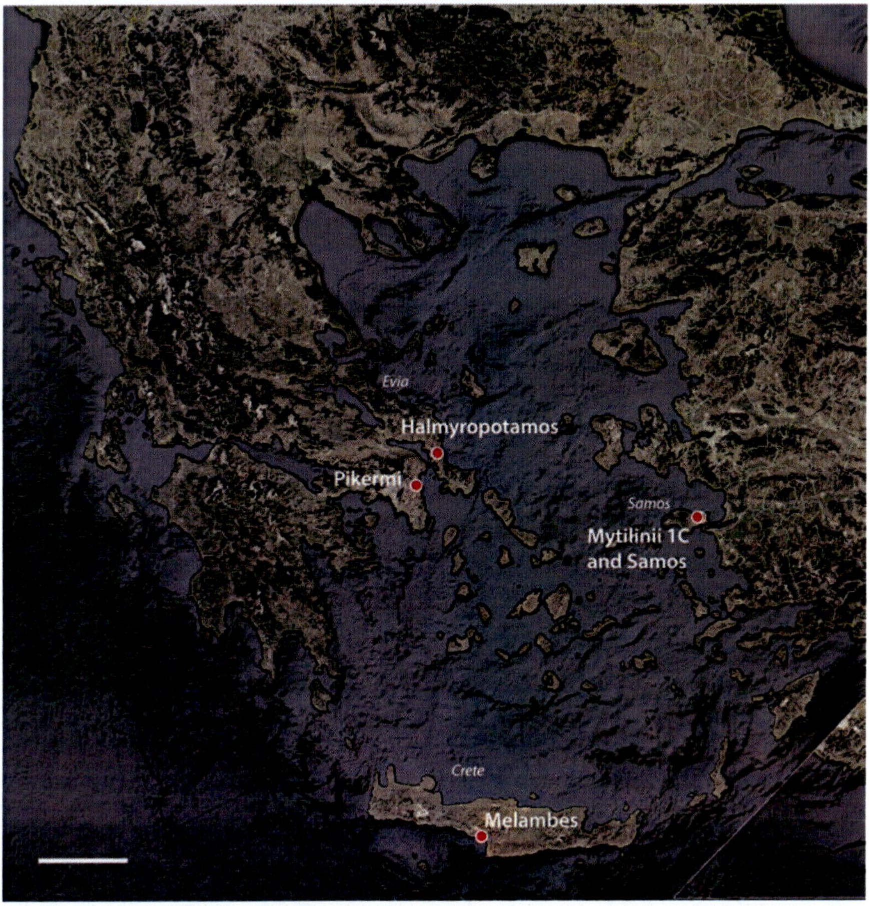

Fig. 1 Map of Greece showing the geographic distribution of the most important localities with hyracoid fossils. See Appendix for more information. Image exported from Google Earth Pro © 2019, map data from US Dept. of State Geographer, SIO, NOAA, U.S. Navy, NGA, GEBCO, image from Landsat/Copernicus. Scale bar equals 80 km, North faces upward

5 Systematic Paleontology

Pliohyracidae Osborn, 1899
***Pliohyrax* Osborn, 1899**

Type Species *Pliohyrax kruppii* Osborn, 1899.

★*Pliohyrax graecus* (Gaudry, 1860)

Nomenclatural and Taxonomical history *Leptodon graecus* in Gaudry 1860 (new species); *Pliohyrax graecus* in Forsyth Major 1899 (new combination).

Type Specimen MNHN-PIK-3389 (holotype) mandible with p4–m2 (right) and p3–m3 (left) (Gaudry 1862–1867: p. 215; pl. 34, figs 1–2), housed in the Muséum nationale d'Histoire naturelle, Paris (Fig. 2a–b).

Type Locality Pikermi, Attica, Greece, late Miocene, MN12.

Distribution Besides the type locality, *P. graecus* is known from the Greek late Miocene localities Halmyropotamos, Mytilinii-1C (MTLC) and unknown locality/ies of Samos (see Fig. 1 and Appendix). The species is also referred from the Miocene of Spain, Turkey, Iran and Afghanistan.

Remarks The known material of *P. graecus* from Greece is poor, including a few cranial and mandibular remains (Fig. 2a–c). The species is characterized by a larger size than *P. kruppii*, flattened premaxilla between the first incisors, presence of buccal cingulum in the I3, metacone like a bulge in the external wall of the upper molars, thick and deep mandibular corpus, premolariform canine with triangular occlusal outline, increase in the molar size posteriorly and elongated m3 with a third lobe (Fischer and Heizmann 1992; Koufos 2009).

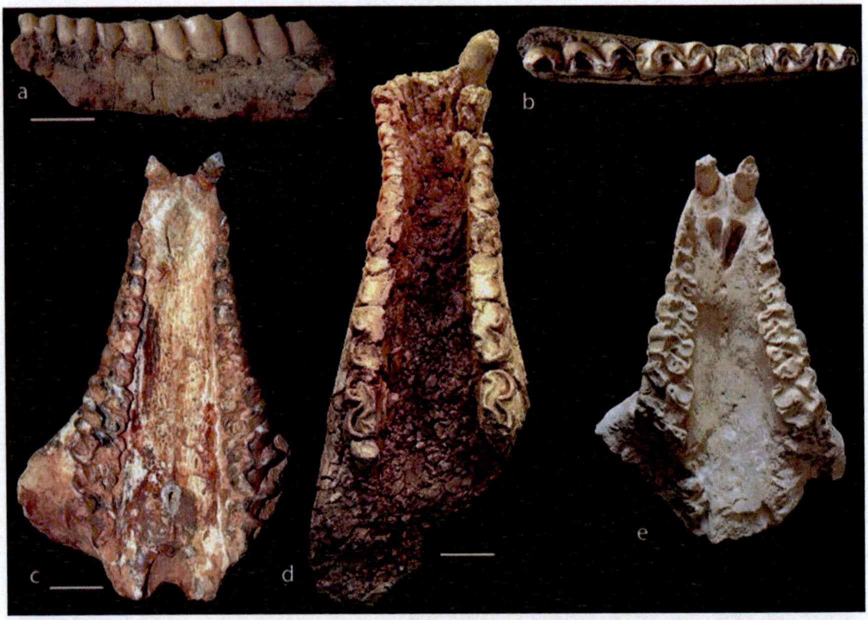

Fig. 2 Selected specimens of *Pliohyrax* from Greece. (**a–d**), *Pliohyrax graecus*: **a–b,** mandibular fragment with p3–m3 (holotype) from the late Miocene of Pikermi in **a** buccal and **b** occlusal views, (**c**) maxilla with both tooth rows from the late Miocene of Pikermi in occlusal view and (**d**) mandible with both tooth rows from the late Miocene of Mytilinii 1C (Samos Island) in occlusal view, (**e**) *Pliohyrax kruppii,* maxilla with both tooth rows (cast), from an unknown Late Miocene locality of Samos Island in occlusal view. Scale bars equal 3 cm

★*Pliohyrax kruppii* Osborn, 1899

Type Specimen SAM-44032 (holotype) cranial fragment with the dentition, described by Osborn (1899: taf. 2, figs 1–4) and housed in the Natural History Museum of Stuttgart, Germany (Fig. 1e).

Type Locality Precise locality unknown, Samos Island, late Miocene, MN11–13.

Distribution It is only known from Samos. It is also referred from Turkey.

Remarks Besides the holotype, there is a mandible in the Bayerische Staatssammlung für Paläontologie und Historische Geologie, Munich, originated from an unknown locality of Samos and described by Schlosser (1899a, b) as *Leptodon graecus*, which could belong to *P. kruppii* (Fischer and Heizmann 1992). The main characters of the species are the small size (smaller than *P. graecus*), the long distance between infraorbital foramen and maxillary groove, the deep cutting of the premaxilla between the two I1; the elevated position of the orbits, the strongly invested with enamel mesial face, the absence of a buccal cingulum in the I1 and I2, the very small but functional I2 and I3 separated from the median incisors by a long diastema, the premolariform canine, the submolariform P1–P3, the completely molariform P4 and the increase in the molar size posteriorly (Osborn 1899; Fischer and Heizmann 1992).

Prohyrax Stromer, 1922

Type Species *Prohyrax tertiarius* Stromer, 1922.

Remarks A mandibular fragment of a hyracoid is known from the locality Melambes, Crete; it was originally described as *?Pliohyrax* sp. by Kuss (1976). Later, the Melambes hyracoid was attributed as cf. *Prohyrax hendeyi* (van der Made 1996). The single specimen, combined with a poorly known associated fauna and a doubtful age of the locality, does not allow a precise determination of this material.

Acknowledgments I would like to thank R. Tabuce for comments that improved this manuscript.

Appendix

List of the Greek localities containing hyrax fossils

Localities[PBDB No]	Age (MN; GPTS in Ma)	Taxon
Pikermi[182754]	**Middle Turolian (MN12; 7.3–7.1)**	***Pliohyrax graecus***[1]
Mytilinii 1C[202217]	Middle Turolian (MN12; 7.3–7.1)	*Pliohyrax graecus*[2]
Halmyropotamos [202,213]	?Middle Turolian (MN12)	*Pliohyrax graecus*[3]
Samos[202120*]	**Turolian (MN11–13)**	*Pliohyrax graecus*[4] ***Pliohyrax kruppii***[5]
Melambes[208183]	Middle–late Miocene (MN6–10)	?*Prohyrax hendeyi*[6, 7]

Type localities are marked with bold. Locality numbers refer to the collection numbers of the PaleoBiology Database (PBDB)

GPTS Geomagnetic Polarity Time Scale

[1]Gaudry 1860, [2]Koufos 2009, [3]Melentis 1967, [4]Osborn 1889, [5]Schlosser 1899, [6]Fischer and Heizmann 1992, [7]Kuss 1976

*Precise locality unknown

References

Baudry M (1994) Les gisements de mammifères du Miocène supérieur de Kemiklitepe, Turquie: 6. Hyracoidea. Bull Mus National Hist Nat 16:113–141

Benoit J, Crochet J-Y, Mahboubi M, Jaeger J-J, Bensalah M, Adaci M, Tabuce R (2016) New material of *Seggeurius amourensis* (Paenungulata, Hyracoidea), including a partial skull with intact basicranium. J Vertebr Paleontol 36:e1034352–e1034358

Dubrovo IA (1978) New data on fossil hyracoidea. Palaeontol J 12:375–383

Fischer MS, Heizmann EPJ (1992) Über neogene Hyracoiden (Mammalia). Die Gattung *Pliohyrax*. N Jahrb Geol Palaeontol Abh 186:321–344

Forsyth Major CJ (1899) *Pliohyrax graecus* from Samos. Geol Mag, Decade 4(6):547–553

Gaudry A (1860) Résultats des fouilles entreprises en Grèce sous les auspices de l'Académie. C R Acad Sci Paris 51:926–929

Gaudry A (1862–1967) Animaux fossils et géologie de l'Attique. Editions Savy, Paris

Gheerbrant EM, Sudre JE, Cappetta HE, Mourer-Chauviré CÉ, Bourdon E, Iarochene M, Amaghzaz M, Bouya B (2003) Les localités à mammifères des carrières de Grand Daoui, bassin des Ouled Abdoun, Maroc, Yprésien: premier état des lieux. Bull Soc Geol Fr 174:279–293

Hoek H (2018) Hyracoidea. IUCN Afrotheria Specialist Group. www.afrotheria.net/hyracoidea.php

Kaya TT, Mayda S, Kostopoulos DS, Alcicek MC, Merceron G, Tan A, Karakutuk S, Giesler AK, Scott RS (2012) Şerefköy-2, a new late Miocene mammal locality from the Yatağan formation, Muğla, SW Turkey. C R Palevol 11:5–12

Koufos GD (2009) The late Miocene mammal faunas of the Mytilinii Basin, Samos Island, Greece: new collection. 7. Hyracoidea. Beit Palaeontol 31:127–137

Kuss E (1976) Ein erster Fund aus dem Vallesian von Kreta/Griechenland. N Jahrb Geol Palaeontol Monatch 3:157–162

Melentis JK (1967) Studien über fossile Vertebraten Griechenlands. 19. Die Pikermifauna von Halmyropotamos (Euböa, Griechenlands). I Teil: Odontologie und Kraniologie. Ann Geol Pays Helleniques 18:283–411

Meyer GE (1978) Hyracoidea. In: Maglio VJ, Cooke HBS (eds) Evolution of African mammals, vol 1. Harvard University Press, Cambridge, pp 284–314

Myers P (2000) Hyracoidea (On-line), Animal Diversity Web. http://animaldiversityorg/accounts/Hyracoidea/. Accessed 27 Mar 2018

Osborn HF (1898) On *Pliohyrax kruppii* Osborn, a fossil hyracoid from Samos, Lower Pliocene, in the Stuttgart collection. A new type and the first known Tertiary hyracoid. In: Proceedings of the 4th International Congress of Zoology, Cambridge

Pickford M (1986) Première découverte d'une hyracoide Paléogène en Eurasie. C R Acad Sci Paris 303:1251–1254

Pickford M (2009) New Neogene hyracoid specimens from the Peri-Tethys region and East Africa. Palaeontol Res 13:265–278

Rasmunssen DT (1989) The evolution of the Hyracoidea: a review of the fossil evidence. In: Prothero DR, Snoch RM (eds) The evolution of Perissodactyls. Oxford University Press, Oxford, pp 57–78

Schlosser M (1899a) Über neue Funde von *Leptodon graecus* Gaudry und die systematische Stellung dieses Säugetieres. Zool Anz 597:378–380

Schlosser M (1899b) Über neue Funde von *Leptodon graecus* Gaudry und die systematische Stellung dieses Säugetieres. Zool Anz 598:385–387

Sen S (1998) The age of the Molayan mammal locality, Afghanistan. Geobios 31:385–391

Thomas H, Sen S, Khan M, Battail B, Ligabue G (1982) The lower Miocene fauna of Al-Sarrar (eastern province, Saudi Arabia). Atlal J Saudi Arabia Archaeol 5:109–136

Van der Made J (1996) Pre-Pleistocene land mammals from Crete. In: Reese DS (ed) The Pleistocene and Holocene Fauna of Crete and its First Settlers, Monogr World Archaeol 28:69–79

Vaughan TA, Ryan JM, Czaplewski NJ (2015) Mammalogy, 6th edn. Jones & Bartlett Learning, Burlington

The Fossil Record of the Neogene Proboscidea (Mammalia) in Greece

George E. Konidaris and Evangelia Tsoukala

1 Introduction

The order Proboscidea is represented today by the largest living terrestrial mammals, the African savanna and forest elephants *Loxodonta africana* and *Loxodonta cyclotis*, respectively, and the Asian elephant *Elephas maximus*, which are all locally restricted and considered as threatened to extinction. The extant elephants are relics of a group that was once extremely diversified and widely distributed across Africa, Europe, Asia, and the Americas, especially during the Miocene (Shoshani and Tassy 1996a). Most proboscideans are characterized by the presence of tusks and molars of specialized morphology, making them easily recognizable in the fossil record and placing them among the most iconic vertebrates that have ever lived on this planet.

The earliest known proboscideans are *Phosphatherium* and *Daouitherium* from the Early Eocene of Morocco, at ~55.0 Ma (Gheerbrant et al. 1996, 2002). The cladistic position of *Eritherium* from the Late Paleocene (~60.0 Ma) of Morocco (Gheerbrant 2009) remains unresolved; it may be a stem proboscidean, or otherwise a sister group either to both Proboscidea and Sirenia or to all tethytherians (Gheerbrant et al. 2018). If *Eritherium* belongs to Proboscidea, then this species will represent the earliest proboscidean and will indicate a Paleocene emergence for the order. Since their appearance, the evolutionary history of proboscideans is marked by three major radiation events: (1) the Eocene radiation of primitive lophodont taxa, endemic in the, at that time, Afro-Arabian continent; (2) the Miocene

G. E. Konidaris (✉)
Palaeoanthropology, Senckenberg Centre for Human Evolution and Palaeoenvironment, Eberhard Karls University of Tübingen, Tübingen, Germany
e-mail: georgios.konidaris@uni-tuebingen.de

E. Tsoukala
Laboratory of Geology and Paleontology, Department of Geology, School of Geology, Aristotle University, Thessaloniki, Greece
e-mail: lilits@geo.auth.gr

E. Vlachos (ed.), *Fossil Vertebrates of Greece Vol. 1*,
https://doi.org/10.1007/978-3-030-68398-6_12

radiation of gomphotheres sensu lato and stegodonts; (3) the Mio–Pliocene emergence of the family Elephantidae (Shoshani and Tassy 1996b). Not only the emergence and the subsequent first radiation of proboscideans took place in Africa, but also the other two radiations appear to have been triggered also in this continent (Sanders et al. 2010).

Proboscidean remains are common in the Neogene faunas of Greece, and especially during the late Miocene, including several taxa. They are present in almost all the large mammal fossiliferous sites of the country, with relatively abundant material. The Greek fossil record includes proboscideans from all three radiations. The first two radiations are presented in this chapter and comprise members of the families Deinotheriidae, Mammutidae, Amebelodontidae, Choerolophodontidae, Gomphotheriidae, and Stegodontidae; the third radiation of Elephantidae is the subject of the next chapter (Athanassiou this volume). For the sake of simplification, we use the informal name "Neogene proboscideans of Greece" for the taxa described in the present study, acknowledging however that the elephantid *Mammuthus* (presented in the following chapter) appeared in Greece during the late Pliocene and the fact that *Anancus*' biostratigraphic range extends to the Early Pleistocene.

2 Historical Overview

The presence of proboscideans in the Greek fossil record was recognized already from the first excavations that were carried out in the nineteenth century in the rich late Miocene vertebrate localities of Pikermi (Attica, discovered in 1836) and Samos [Aegean Sea; Forsyth-Major (1888, 1894)] and the beginning of the twentieth century in Axios valley (Central Macedonia, discovered in 1915–1916; Arambourg and Piveteau (1929)]. The first systematic studies on Miocene proboscideans included important material from these localities and were conducted, among others, by Wagner (1848, 1857), Gaudry (1862–1867), and Vacek (1877) for Pikermi; Forsyth-Major (1894), Schlesinger (1917, 1922), and Lehmann (1950) for Samos; and Arambourg and Piveteau (1929) for Axios valley ("Salonique"). These publications served (and still do) as a reference point for future research on Neogene proboscideans. In all of the abovementioned localities, the fossils occur in different stratigraphic levels from multiple quarries and unfortunately—as is the case with several historical collections—most of the fossils lack precise stratigraphic information. Subsequent important Miocene proboscidean specimens were studied in Paraskevaidis (1940) and Tobien (1980) from the middle Miocene of Chios Island, Melentis (1967, 1969) from the late Miocene of Halmyropotamos, Koufos (1980) from the late Miocene of Axios valley, Tassy (1985) from the late Miocene of Pikermi, Tsoukala and Melentis (1994) from Kassandra Peninsula in Chalkidiki, Koufos et al. (2003) from the early Miocene of Gavathas on Lesvos Island, Theodorou et al. (2003) from the late Miocene of Kerassia, Athanassiou (2004), Poulakakis et al. (2005) and Iliopoulos et al. (2014) from the late Miocene of Crete Island, and Lazaridis and Tsoukala (2014a) from the late Miocene of Kryopigi. Furthermore, the new series of excavations in Axios valley (started in 1972; Koufos

2013), Samos (started in 1993; Koufos and Nagel 2009), and Pikermi (started in 2009; Theodorou et al. 2010), as well as in Nikiti (started in 1990; Koufos and Kostopoulos 2016), which aimed in the discovery of new stratified and dated material, provided a wealth of specimens, including several proboscideans. Therefore, in the recent years the need has arisen for a complete revision and study of the taxonomy, biostratigraphy, and paleoecology of the Miocene proboscideans. This was part of the investigations of Konidaris (2013) and of subsequent studies (Konidaris and Koufos 2009, 2013a, 2016, 2019; Konidaris et al. 2014, 2016, 2017; Konidaris and Roussiakis 2019).

The Pliocene–Early Pleistocene sediments of Greece have been also intensively explored during the last decades. From this period, mammutids and gomphotheres are well-known from the recent excavations in Milia (started in 1996), which provided a wealth of proboscidean specimens (Tsoukala 2000; Tsoukala and Mol 2016) and in Sesklo (Symeonidis and Tataris 1983; Athanassiou 2016), but also from Apolakkia (Theodorou et al. 2000), Vatera (de Vos et al. 2002), and Gephyra (Crégut-Bonnoure and Tsoukala 2017).

The extensive and systematic fieldwork during the last decades in Greece has greatly increased the number of proboscidean specimens, which together with the specimens from the old collections, constitute a rich proboscidean sample. Recent advances in proboscidean taxonomy, along with biostratigraphic correlations and magnetostratigraphic calibrations, allow a more comprehensive classification and biostratigraphy of the Neogene proboscideans from Greece, which are presented in this chapter.

3 Phylogenetic Relationships

The order Proboscidea consists predominantly of fossil taxa, which complicates their classification and phylogeny, in particular of the early representatives. Based on recent findings and new cladistic analyses, Proboscidea together with the orders Sirenia (sea cows) and the Hyracoidea (hyraxes), as well as with the extinct Embrithopoda (e.g., *Arsinoitherium zitelli* from the Early Oligocene of Fayum in Egypt), belong to the clade of Paenungulata; in turn, proboscideans, sirenians, and Embrithopoda constitute the clade of Tethytheria (Gheerbrant et al. 2018) (Fig. 1). Anthracobunidae and Desmostylia, which were usually classified within tethytheres, are in fact related to Laurasian Euungulata and Perissodactyla (Cooper et al. 2014). Proboscidean synapomorphies include the well-developed zygomatic process of the maxillary, which contributes significantly to the ventral border of the orbit and to the zygomatic arch, the relatively large size of the *pars mastoidea* of the periotic, the hypoconulid in a labial position, and possibly also the loss of i3 and (d) p1 (Gheerbrant et al. 2005). Well-recognized proboscidean features, such as the large size, the presence of a trunk (proboscis), the huge-sized and pneumatized crania with retracted nasal apertures, the hypertrophy of the second incisors to form upper tusks, the graviportal stance, the shortening of the tooth row, the horizontal

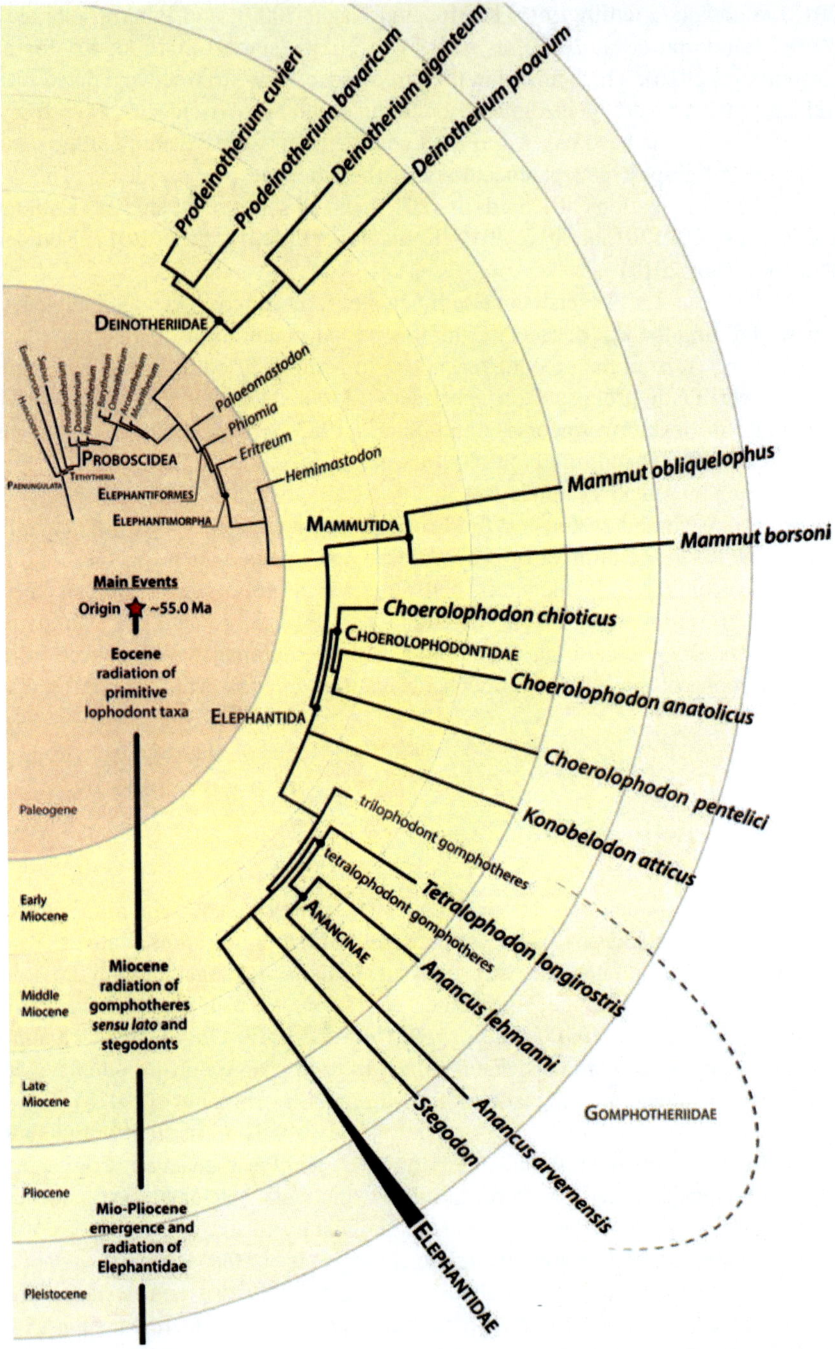

Fig. 1 Simplified cladogram showing the relationships of Paenungulata and Proboscidea based on Gheerbrant and Tassy (2009), Seiffert et al. (2012), and Gheerbrant et al. (2018)

tooth displacement, the reduction in the number of teeth with regard to the complete eutherian dentition, the enlargement and specialization of the cheek teeth, and the increase in the value of the encephalization quotient, constitute evolutionary trends within Proboscidea that evolved progressively since the Eocene (Shoshani and Tassy 1996a; Shoshani 1998, 2002; Gheerbrant and Tassy 2009; Sanders et al. 2010).

In this chapter, we mainly follow the classification proposed by Shoshani and Tassy (2005). However, we include choerolophodonts and amebelodonts in the families Choerolophodontidae and Amebelodontidae, respectively (Gheerbrant and Tassy 2009), and the tetralophodont gomphotheres *Tetralophodon* and *Anancus* of the family Gomphotheriidae in the subfamilies Tetralophodontinae and Anancinae, respectively (Sanders et al. 2010).

Focusing on the proboscidean taxa that are present during the Neogene of Greece, Deinotheriidae are more closely related to Elephantiformes, than to the more basal proboscideans (Fig. 1) based on dental features (e.g., presence of hypocone in P3 and P4 and the higher hypolophid than the protolophid in the lower molars) and postcranial traits (e.g., astragalus with long neck, enlarged lateral calcaneal facet, and reduced fibular facet) (Tassy 1994, 1996a; Gheerbrant et al. 2005). Deinotheres are unique within Proboscidea in lacking upper tusks and possessing strong and downward curved lower tusks (Gheerbrant and Tassy 2009). Apart from Deinotheriidae, all other proboscideans of the Greek fossil record belong to the monophyletic clade Elephantimorpha (Fig. 1). The monophyly of Elephantimorpha (*Eritreum*, *Hemimastodon*, Mammutida, Elephantida) is based among others on the "horizontal tooth displacement", i.e., a pattern of dental eruption, in which the cheek teeth during their use move anteriorly along the jaws and when they are completely worn, they are succeeded by teeth that emerge more distally, in a manner that is reminiscent of a "conveyor belt" (Tassy 1994, 1996a; Shoshani et al. 2006; Sanders 2018; Tassy 2018). Within Elephantimorpha, the family Mammutidae is defined by the zygolophodont molars, with sharper loph(id)s due to anteroposterior compression, and with presence of pretrite and posttrite zygodont crests instead of central conules (Tassy 1996a; Shoshani 1996).

The clade Elephantida includes taxa with bunodont molars, which are transformed into molars with lamellae in the families Stegodontidae and Elephantidae (Gheerbrant and Tassy 2009). The family Choerolophodontidae is distinguished by the fused mesoconelets and central conules and by the mesially pointed chevrons of the intermediate and third molars ("V," chevroning) formed by the more mesial position of the conelets in regard to the main cusps, especially in the distal lophids (Tassy 1996a; Sanders 2003; Konidaris et al. 2016). The straight medial border of the lower tusks and the posttrite conules on the molars are the distinguishing traits of the family Amebelodontidae; their most distinctive feature, the flattened lower tusks, is a plesiomorphic character present in *Phiomia* (Tassy 1996a; Shoshani 1996). The paraphyletic family Gomphotheriidae includes the trilophodont gomphotheres [with three loph(id)s in the intermediate molars] represented in Europe by the genus *Gomphotherium* and the tetralophodont gomphotheres [with four loph(id) s in the intermediate molars], which in Europe include the longirostrine

Tetralophodon (with rounded to pyriform lower tusks) and the brevirostrine *Anancus* (lacking lower tusks) (Tassy 1996a; Shoshani 1996).

4 Distribution

Neogene proboscidean remains are present in almost all the large mammal fossiliferous sites of Greece, from the northernmost regions of the country (e.g., Thermopigi, Dikaia) to the southern-most ones (Crete) and from the western ones (e.g., Milia, western Peloponnese) to the easternmost ones (Samos, Kos, and Rhodes Islands); they are present in both today's continental and insular Greece (Figs. 2 and 3). The earliest occurrence of Neogene proboscideans in the Greek fossil record is documented in the early Miocene of Gavathas (Lesvos Island) with the primitive deinothere *Prodeinotherium*, whereas their last occurrence is recorded in the Early Pleistocene of Vatera (Lesvos Island) with the gomphothere *Anancus* (Fig. 2). The taxonomic diversity reached its peak during the Turolian (late Miocene), when at least four proboscidean species have been recorded in Pikermi and Samos.

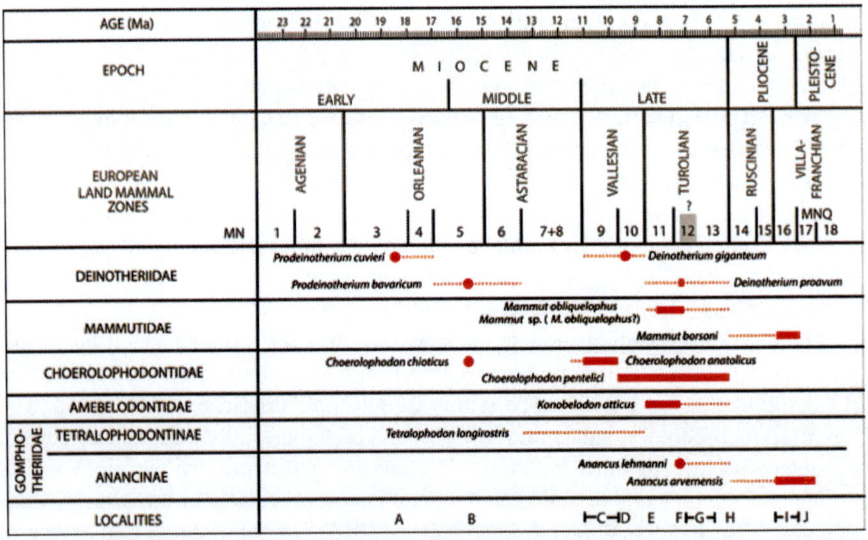

Fig. 2 Biostratigraphic distribution of the Neogene proboscideans in Greece. Solid lines indicate the biostratigraphic range of the species based on their occurrences in Greece; dots indicate single occurrences in Greece; dashed lines indicate the known biostratigraphic range based on data outside Greece; the top of MN12 is controversial and appears in grey. Selected localities with radiometric, biostratigraphical, or magnetostratigraphical data are also shown with letters: **A**, Gavathas; **B**, Thymiana; **C**, Pentalophos-1; **D**, Xirochori-1, Ravin de la Pluie; **E**, Nikiti-2, Ravin des Zouaves-5; **F**, Prochoma-1, Vathylakkos-2, Perivolaki, Pikermi, Pikermi Valley-1, 3, Mytilinii-1A, B, ?Halmyropotamos, ?Kerassia, ?Kryopigi, Chomateri; **G**, Dytiko-2, 3; **H**, Maramena; **I**, Milia, Sesklon, Gephyra; **J**, Vatera. Further details are given in the Appendix

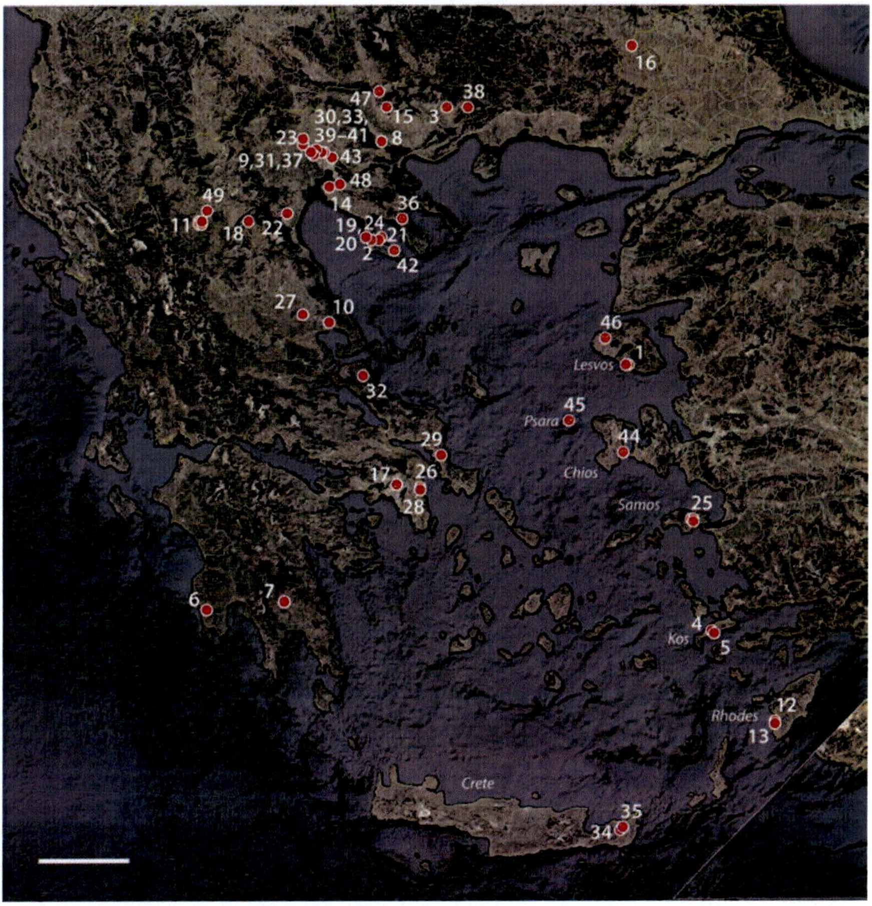

Fig. 3 Map of Greece showing the geographic distribution of the most important localities with Neogene and Early Pleistocene proboscideans. **1**, Vatera localities; **2**, Sani; **3**, Kalliphytos; **4**, Antimachia; **5**, Kardamaena; **6**, Pylos; **7**, Skoura; **8**, Nigrita; **9**, Gephyra-1; **10**, Sesklon; **11**, Milia localities; **12**, Apolakkia I; **13**, Apolakkia II; **14**, Agia Triada; **15**, Maramena; **16**, Dikaia; **17**, Pyrgos Vassilissis; **18**, Servia; **19**, Fourka; **20**, Chelona beach; **21**, sea bed of Kryopigi; **22**, Neokaisareia; **23**, Dytiko localities; **24**, Kryopigi; **25**, Samos localities; **26**, Chomateri; **27**, Perivolaki; **28**, Pikermi localities; **29**, Halmyropotamos; **30**, Prochoma-1; **31**, Vathylakkos localities; **32**, Kerassia localities; **33**, Ravin des Zouaves-5; **34**, Maronia; **35**, Gela; **36**, Nikiti-2; **37**, Ravin X; **38**, Platania; **39**, Ravin de la Pluie; **40**, Xirochori 1; **41**, Ravin des Zouaves 1; **42**, Agia Paraskevi; **43**, Pentalophos 1; **44**, Thymiana; **45**, Psara; **46**, Gavathas; **47**, Thermopigi; **48**, Antonios; **49**, Siatista. See Appendix for more information. Image exported from Google Earth Pro © 2019, map data from US Dept. of State Geographer, SIO, NOAA, U.S. Navy, NGA, GEBCO, image from Landsat/Copernicus. Scale bar equals 80 km, North faces upward

Deinotheres present the widest biostratigraphic range, covering almost the whole Miocene (late MN3 until at least MN12), followed by choerolophodonts, which were present from the middle Miocene until the end of the late Miocene (MN5–13). *Choerolophodon* presents the widest temporal distribution among the genera in Greece, and where present, it is the dominant genus in terms of abundance.

5 Systematic Paleontology

Proboscidea Illiger, 1811
Deinotheriidae Bonaparte, 1845

Comments Deinotheres were specialized browsing proboscideans with low-crowned, tapir-like lophodont and bilophodont cheek teeth—apart from the trilophodont dp4/DP4 and m1/M1—and with a dental formulae 0.0.3/1.0.3 for the deciduous and 0.0.2.3/1.0.2.3 for the permanent teeth (Harris 1973; Sanders et al. 2010) (Figs. 3, 4 and 5). Deciduous dentition is replaced by the permanent one in a vertical manner, and all permanent cheek teeth are in function simultaneously. In this feature, as well as in other cranial, dental, and postcranial traits, deinotheres differ markedly from elephantimorphs. The most distinctive feature of deinotheres is the downcurved mandibular symphysis bearing the lower tusks, the latter emerging almost vertically, whereas they lack upper tusks (Harris 1973). Known from the Late Oligocene of Africa, where they persisted until 1.0 Ma, deinotheres have an evolutionary history of ~25.0 myr (Sanders et al. 2010). In Greece, deinotheres are known from several localities ranging from the early to the late Miocene and belong to the genera *Prodeinotherium* and *Deinotherium*. The original generic name *Deinotherium* Kaup 1829 (consisting of the Greek words δεινός = fearfully great, and θηρίον = wild animal/beast) is valid and not *Dinotherium* used subsequently by several authors (see also comment in Huttunen 2002). Accordingly, Harris (1973) emended the original name *Prodinotherium* Éhik 1930 to *Prodeinotherium*, which is justified and valid.

***Prodeinotherium* Éhik, 1930**

Type Species *Prodinotherium hungaricum* Éhik, 1930.

Other Included Taxa *P. cuvieri* (Kaup, 1832a); *P. hobleyi* (Andrews, 1911); *P. pentapotamiae* (Falconer, 1868); *P. sinense* Qiu et al., 2007.

Distribution Early–early late Miocene of Africa, Europe and Asia.

Comments Based on the conservativeness and the minor evolutionary changes of deinotheres (apart from a gradual increase in size through time) several researchers recognize only *Deinotherium* as the valid European genus (e.g., Gräf 1957; Ginsburg and Chevrier 2001; Pickford and Pourabrishami 2013). Here we follow Harris

(1973), Huttunen (2002), and Aiglstorfer et al. (2014) and regard *Prodeinotherium* as a distinct genus from *Deinotherium*, based on several cranial, dental, and postcranial features. In Europe, *Prodeinotherium* includes the small and primitive deinotheres from the early until the middle Miocene (early Orleanian–early Astaracian, MN3–6). In Greece this genus is recognized with two species, *Prodeinotherium cuvieri* and *Prodeinotherium bavaricum*.

Prodeinotherium cuvieri (Kaup, 1832a)

Nomenclatural and Taxonomical History The species was erected by Kaup (1832a), who distinguished it from *Deinotherium giganteum*, based mainly on the smaller dimensions of the teeth from Chevilly (France) and Eppelsheim Formation (Germany). During the subsequent years and until today, the validity of this species is accepted by several authors (e.g., Ginsburg and Chevrier 2001; Pickford and Pourabrishami 2013), whereas others synonymize it with *Prodeinotherium bavaricum* (e.g., Gräf 1957; Huttunen 2002). The taxonomical status of the early Miocene deinotheres of Europe is not yet clarified and could be briefly summarized into the question whether *P. cuvieri*, *P. hungaricum*, or *P. petenyii* Vörös 1989 are conspecific or not, although the small dental dimensions of all three of them supports the synonymy; in this case, the type species of *Prodeinotherium* would be *P. cuvieri* (Markov 2008a: p. 144, footnote). Regardless of the specific name, dental dimensions and mandibular morphology (and perhaps also cranial and postcranial features) are in favor of a specific distinction from *P. bavaricum* (Markov 2008a).

Type Material MNHN-CHE 13 (lectotype), right and left mandibular fragments (Mayet 1908: pl. 8, figs. 3, 4).

Type Locality Chevilly, France, early Miocene, middle Orleanian, MN4.

Distribution Early Miocene (end of early Orleanian, late MN3) of Greece (Gavathas on Lesvos Island); early Miocene, middle Orleanian (MN4) of Hungary, Germany, France, and Spain.

Remarks This species comprises the smallest, earliest, and most basal deinotheres in Europe, known in Greece from a single occurrence of great importance. In particular, Koufos et al. (2003) studied tooth rows with permanent dentition from the locality Gavathas on Lesvos Island (Fig. 4), well-known for the Petrified Forest of Sigri. The authors noted that the dental morphology and the small dimensions of the Gavathas specimen indicate that it belongs to a primitive form of *P. bavaricum* (following the two European deinothere species concept of *P. bavaricum* and *D. giganteum*, accepted at that time). According to these observations and accepting here the validity of a distinct species from *P. bavaricum*—the valid name for which seems to be *P. cuvieri*—the Gavathas deinothere is provisionally attributed to *P. cuvieri*, pending complete revision of the primitive European deinotheres and new findings. The Gavathas deinothere, dated to as older than 18.4 ± 0.5 Ma (upper part of the early Miocene, late MN3), documents the earliest occurrence of deinotheres in Europe (Koufos et al. 2003). As part of the complex "Proboscidean Datum Event," it marks the penetration of deinotheres into Europe (roughly together with the mammutid *Zygolophodon*, the gomphotheriid *Gomphotherium*, and the amebelodontid *Archaeobelodon*), after the establishment of the so-called *Gomphotherium* landbridge in the middle Burdigalian (~19.0–18.0 Ma, early MN3) (Tassy 1990; Rögl 1999; Antoine et al. 2003; Koufos et al. 2003).

***Prodeinotherium bavaricum* (von Meyer, 1831)**
(=★*Dinotherium bavaricum aegaeum* Paraskevaidis, 1940)

Nomenclatural and Taxonomical History The species was originally coined and described as a member of *Deinotherium* by von Meyer (1831, 1833). Almost a century after, it was transferred to *Prodeinotherium* by Éhik (1930). By being the oldest available name for the small and basal deinotheres, it has been considered by several authors (e.g., Gräf 1957; Bergounioux and Crouzel 1962; Huttunen 2002) as the senior synonym of the species *Prodeinotherium cuvieri*, *P. hungaricum*, and *P. petenyii* (but see above).

Type Material SNSB-BSPG-AS I 220 (lectotype), right p3 (von Meyer 1833: pl. 34, figs. 12–15).

Type Locality Georgensmünd?, Germany, middle Miocene, early Astaracian, MN6.

Distribution Middle Miocene (late Orleanian, MN5) of Greece (Thymiana); middle Miocene (late Orleanian–early Astaracian, MN5–6) of Germany, Austria, France, Hungary, Slovakia, Spain, Bulgaria, Czech Republic.

Remarks This species is well-known with abundant (mainly dental) material from several European localities, including at least two partial skeletons (Huttunen and Göhlich 2002; Huttunen 2004). In Greece its presence is documented with dental material from Thymiana (Paraskevaidis 1940), a vertebrate locality on Chios Island, bio- and magnetostratigraphically dated to ~15.5 Ma (middle Miocene, late Orleanian, MN5; Koufos 2006). Originally attributed to a new subspecies under the original spelling "*Dinotherium bavaricum* var. *Aegäum*," the dimensions and morphology of the teeth indicate that are within the intraspecific variability of *P. bavaricum* without the need for subspecific distinction (Besenecker and Symeonidis 1974). The species is also known from the nearby Psara Island with an isolated M1 (Besenecker and Symeonidis 1974).

***Deinotherium* Kaup, 1829**

Type Species *Deinotherium giganteum* Kaup, 1829.

Other Included Taxa *D. bozasi* Arambourg, 1934; *D. indicum* Falconer, 1845; *D. levius* Jourdan, 1861; *D. proavum* (Eichwald, 1831).

Distribution Middle Miocene–Early Pleistocene of Africa, Europe, and Asia.

Comments The genus *Deinotherium* includes the large-sized deinotheres, and in Europe it is known from the middle until the late Miocene (late Astaracian–Turolian, MN7/8–13).

Deinotherium giganteum **Kaup, 1829**

Nomenclatural and Taxonomical History The species was erected and originally described by Kaup (1829, 1832a). Because it was the oldest available name for the large and derived deinotheres, it has been considered by several authors (e.g., Bergounioux and Crouzel 1962; Huttunen 2002) as the senior synonym of the species *Deinotherium levius* and *D. proavum* (but for *D. levius* see: Gräf 1957; Ginsburg and Chevrier 2001; Böhme et al. 2012; Konidaris and Koufos 2019; and below for *D. proavum*).

Type Material HLMD-Din 466 (holotype), left hemimandible with tusk, m2–m3 and right mandibular fragment with symphysis and tusk fragment (Kaup 1832a: pl. 4, add. pl. 1, fig. 5).

Type Locality Eppelsheim, Germany, late Miocene.

Distribution Late Miocene (late Vallesian, MN 10) of Greece (Ravin de la Pluie); late Miocene, Vallesian (MN9–10) of Germany, Austria, France, Hungary, Spain, Bulgaria, and Turkey.

Remarks This species is well-known with abundant material from several European localities (mainly Germany, Austria, France, and Spain). In Greece it is rare, and known from only few localities and with limited specimens. Konidaris (2013) and Konidaris and Koufos (2013a) described an isolated p4 from the late Vallesian of Ravin de la Pluie in Axios valley, and Tsoukala and Melentis (1994) described an upper cheek tooth row from Agia Paraskevi in Kassandra, Chalkidiki.

Deinotherium proavum **(Eichwald, 1831)**

Nomenclatural and Taxonomical History The species was erected by Eichwald (1831, 1835) based mainly on the large dimensions of some teeth from Rakhny Lesovye (Ukraine). Several years later, Stefanescu (1892) coined another deinothere species with huge dimensions from Găiceana and Mânzaţi (Romania), which he named *D. gigantissimum*. During the subsequent years, although the species name *proavum* was used by several authors, it was mostly neglected in favor of *gigantissimum*, whereas others included these huge-sized deinotheres within *Deinotherium giganteum*. Nowadays, Turolian deinotheres are mostly accepted as a distinct species. Codrea (1994) pointed to the priority of Eichwald's species, which was further supported recently by Pickford and Pourabrishami (2013). Meanwhile, however, Kovachev and Nikolov (2006) erected *D. thraceiensis*, for the Ezerovo (Turolian;

Bulgaria) skeleton, yet its specific distinction is not supported by morphological and metrical data (Markov 2008b).

Type Material left p4 and m1 (holotype) (Eichwald 1835: pl. 60, figs. 1–5).

Type Locality Rakhny Lesovye in Podolia, Ukraine.

Distribution Late Miocene (Turolian) of Greece (Pikermi, Samos Island, Andriano, Halmyropotamos, Perivolaki, Maronia, Gela, Zakros, ?Kerassia); late Miocene (Turolian) of Austria, Spain, Germany, Hungary, Slovakia, Moldova, Ukraine, Romania, North Macedonia, Bulgaria, South Russia, Turkey, Afghanistan, Iran, Iraq (Konidaris et al. 2017).

Remarks During the Turolian, European deinotheres reached huge dimensions and are attributed to *D. proavum*. This species represents the terminal stage of the gradual increase in size characterizing the evolutionary history of deinotheres during the Miocene in western Eurasia. Diagnostic characters include, among others (see Tarabukin 1974; Markov 2008b; Pickford and Pourabrishami 2013; Konidaris and Koufos 2019), the large-sized cheek teeth (juvenile and permanent), the morphology of the dp2, and the strongly developed mandibular angle protruding beyond the ventral border of the horizontal ramus (Figs. 4 and 5). The species is well-known from eastern-southeastern Europe, from where several skeletons are documented (Romania, Moldova, Bulgaria, Russia, Greece). In Greece, *D. proavum* is known with the most abundant material among deinotheres. The most complete material is a partial skeleton from Gela in Crete (Fig. 5; Poulakakis et al. 2005; Iliopoulos et al. 2014). Pikermi, Samos, and Halmyropotamos yielded a very important collection of craniomandibular, dental, and postcranial remains, including a skull from Samos, which is the most complete juvenile deinotheriid skull so far in Eurasia and Africa (Konidaris et al. 2017; Konidaris and Koufos 2019) (Fig. 6).

Elephantimorpha Tassy and Shoshani in Shoshani et al., 1998
Mammutida Tassy and Shoshani in Shoshani et al., 1998
Mammutidae Hay, 1922

Comments The family includes elephantimorphs, whose cheek teeth are characterized by zygolophodonty (presence of yoke-like transverse crests), mesiodistally compressed and sharp transverse ridges, absence of accessory conules, and presence of zygodont crests (Tassy 1996a; Shoshani 1996). The intermediate molars remain trilophodont throughout the evolutionary history of the family (Tobien 1996). In Europe, two genera are present: *Zygolophodon* and *Mammut* (subfamily Mammutinae). The more basal *Zygolophodon* retains a more bunodont character on its cheek teeth, whereas in the more derived *Mammut*, the zygodont character is strongly developed (Tobien 1996). The family originated during the Late Oligocene in Africa, and migrated to Europe in the early Miocene, where it existed until the Pliocene. Mammutids migrated via the Bering Strait to North America during the

Fig. 4 Left tooth row with p3–m3 (cast of GVT-1) of *Prodeinotherium cuvieri* from the early Miocene of Gavathas, Lesvos Island (under permission from N. Zouros, copyright Natural History Museum of the Lesvos Petrified Forest). Scale bar equals 5 cm

Fig. 5 Right mandible with the lower tusk and p3–m3 (NHMC 21.4.2.27) of *Deinotherium proavum* from the late Miocene of Gela, Aghia Photia (under permission from C. Fassoulas, copyright Natural History Museum of Crete). Maximum anteroposterior diameter of the mandible: 1040 mm (Iliopoulos et al. 2014)

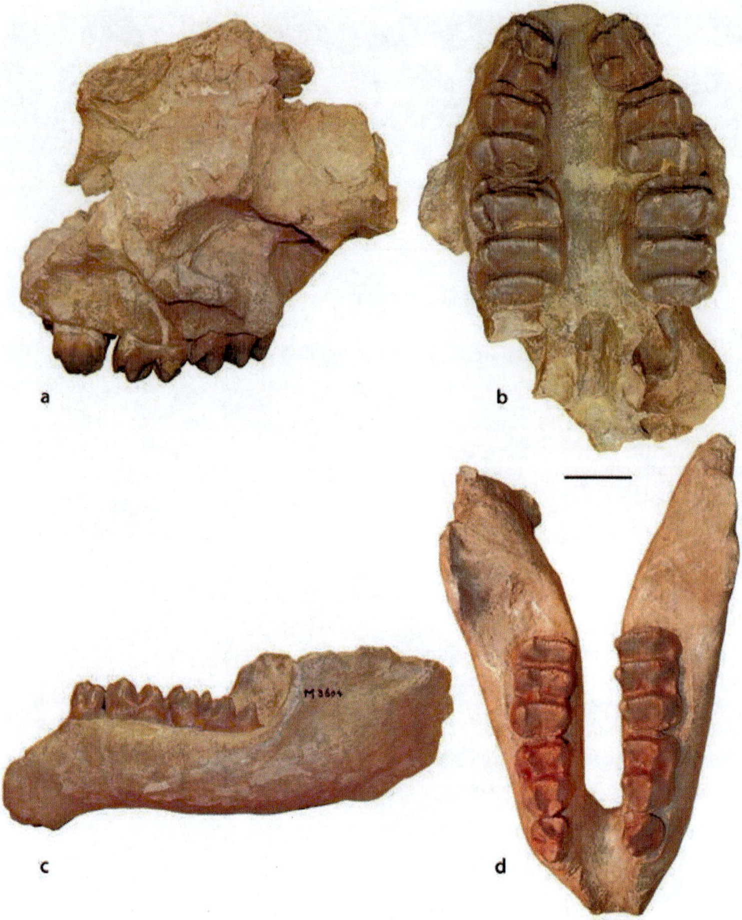

Fig. 6 Partial skull of *Deinotherium proavum* from the late Miocene (Turolian) of Samos (SMF-M 3604; copyright G. Konidaris). (**a–b**), cranium with DP2–DP4 of both sides, in left lateral and ventral view; (**c–d**), mandible with dp2–dp4 of both sides, in left lateral and dorsal view. Scale bar equals 5 cm

middle Miocene and persisted there until the Late Pleistocene (Saunders 1996). In Greece, although some occurrences have been previously attributed to *Zygolophodon*, the only so far representative of Mammutidae is *Mammut*. This genus is present in several localities ranging from the late Miocene to the late Pliocene.

***Mammut* Blumenbach, 1799**

Type Species *Elephas americanus* Kerr, 1792.

Other Included Taxa *M. borsoni* (Hays, 1834); *M. matthewi* (Osborn, 1921); *M. obliquelophus* (Mucha, 1980); *M. pacificus* Dooley et al., 2019.

Distribution Late Miocene–Late Pleistocene of Europe, Asia and North America.

Comments Apart from the well-expressed zygolophodonty on the cheek teeth, the genus *Mammut* is characterized by its shortened mandibular symphysis with relatively small or even vestigial lower tusks and by its straight or upturned upper tusks that lack an enamel band (Fig. 6). In these features, it differs from *Zygolophodon*, which has a longer mandible with well-developed lower tusks, and downwardly curved upper tusks with an enamel band (Tobien 1996). In Europe, the genus is well-known from the Pliocene; however, its first representatives are recorded during the Turolian (late Miocene). It should be noted, however, that the generic name *Mammut* for the Eurasian representatives is in question, pending revision of the whole sample and comparison with the North American specimens (see discussion in Markov 2008b, and von Koenigswald and Göhlich 2019).

***Mammut* sp. [*M. obliquelophus*? (Mucha, 1980)]**

Nomenclatural and Taxonomical History The species was erected by Mucha (1980) for a mammutid mandible from Romanovka (Ukraine). Markov (2008b) validated the species and included into it the skull remains from the Balta Sands in Podolia (Ukraine), which were attributed by Kubiak (1972) to *Mammut praetypicum* as well as material from other late Miocene localities (see discussion in Markov 2008b).

Type Material Paleontological Museum of Odessa State University, Nr. 3347, mandible with m2–m3 (holotype) (Mucha 1980: pl. 1).

Type Locality Romanovka, Ukraine.

Distribution Late Miocene (Turolian) of Greece (Pikermi, Mytilinii-1A-Samos, Ravin des Zouaves-5, Halmyropotamos, Neokaisareia, Palaio Keramidi), and from the late Miocene (Turolian) of Bulgaria, Hungary, Spain, North Macedonia, Moldova, Ukraine, Southern Russia, Spain, and China.

Remarks The late Miocene representative of European (and possibly also Asian) mammutids, *M. obliquelophus*, shows similar cheek tooth morphology as *Mammut borsoni*; however, the length of the symphysis and the size of the lower tusks are distinguishing features between the species: *M. obliquelophus* is characterized by a symphysis that is longer than the tooth row and equipped with well-developed tusks, whereas in *M. borsoni* the symphysis is much reduced and bears small tusks (Markov 2008b). Apart from the Neokaisareia partial skeleton (including an upper tusk) and the Palaio Keramidi molar of this taxon (Konidaris and Tsoukala 2020), the Greek material includes predominantly juvenile specimens (Pikermi, Ravin des Zouaves-5, Mytilinii-1A, Halmyropotamos), but it is very important, including two crania, two maxillae, two mandibles, and isolated teeth (Melentis 1967; Koufos 1980; Tassy 1985; Konidaris and Koufos 2009, 2013a; Konidaris 2013). Their complete study is in progress, and we tentatively attribute here this material to *Mammut* sp. (*M. obliquelophus*?).

Mammut borsoni (Hays, 1834)

Nomenclatural and Taxonomical History Borson (1823) described and figured a tooth from Villanova d'Asti (Italy). Later on, Hays (1834), who had in his possession a cast of this tooth, erected the new species *Mastodon borsoni*. Osborn (1926, 1936) refers to the species as *Zygolophodon borsoni*, a binomen already used by Pohlig (1988). Meanwhile, however, Schlesinger (1922) already included the species within subgenus *Mastodon* (*Mammut*), currently elevated to the genus level.

Type Material MGPT-PU 14896 (holotype), right M3 (Borson 1823: pl. 2).

Type Locality Villanova d'Asti, Piedmont, Italy.

Distribution Late Pliocene (early Villafranchian, MN16) of Greece (Milia, Grevena) and various Pliocene localities (Ruscinian–early Villafranchian, MN14–MN16) of Europe and Asia; its presence during the middle Villafranchian (MNQ 17; Early Pleistocene) of Europe is insecure based on the current evidence.

Remarks The Borson's mastodon is best known from Milia, a locality which has yielded the richest and most important material of this species. The material consists of several partial skeletons and isolated skeletal elements, including two complete pairs of the longest upper tusks ever recorded in the world and the most complete mandible in Europe (Tsoukala 2000; Tsoukala and Mol 2016). The upper tusks are almost straight with a slight upward curvature and torsion, long but slender, they have almost circular cross-section, and they lack enamel (Fig. 7). The brevirostrine mandible bears rudimentary lower tusks of oval section and the molars show well-expressed zygolophodonty (Fig. 7a). *Mammut borsoni* is mentioned also from other localities in Greece, albeit with no precise stratigraphic information (see Appendix).

Elephantida Tassy and Shoshani in Shoshani et al., 1998
Choerolophodontidae Gaziry, 1976

Comments The family Choerolophodontidae includes bunodont trilophodont elephantimorphs, whose most distinctive dental features are the chevroning of the half-loph(id)s (mesially pointing chevrons, "V"), the choerodonty (multiplication of accessory conules), the ptychodonty (corrugated enamel), and the cementodonty (cement cover), with all these traits more-expressed in the later and more derived species. Other choerolophodont features include the upward curvature of the upper tusks that lack enamel, the long mandibular symphysis without tusks, and the strong development of the facial region of the cranium, which is elongated in comparison with the cerebral region. Choerolophodonts originated during the early Miocene, a period when they were distributed in northern/eastern Africa and South Asia.

Fig. 7 *Mammut borsoni* remains from the late Pliocene of Milia. (**a**), Mandible with lower tusks and m2–m3 of both sides (MIL 143), left ulna (MIL 141) and right tibia (MIL 142) from Milia 1; all belong to the same individual; (**b**), partial skeleton from Milia 5; (**c**), close view of the left (MIL 561) and right (MIL 560) upper tusks from Milia 5

Subsequently, during the end of the early Miocene–middle Miocene, they dispersed to Southeastern Europe and China, and possibly to North America. The last appearances of choerolophodonts are traced at the latest Miocene. In Greece, they are known from several localities ranging from the middle to the late Miocene.

★*Choerolophodon* Schlesinger, 1917

Type Species *Mastodon pentelicus* Gaudry and Lartet, 1856.

Other Included Taxa *C. anatolicus* (Ozansoy, 1965); *C. chioticus* Tobien, 1980; *C. corrugatus* (Pilgrim, 1913); *C. guangheensis* Wang and Deng, 2011; *C. kisumuensis* (MacInnes, 1942); *C. ngorora* (Maglio, 1974); *C. palaeindicus* (Lydekker, 1884); *C. zaltaniensis* Gaziry, 1987a.

Distribution Miocene of Europe, Africa, and Asia.

★*Choerolophodon chioticus* Tobien, 1980

Nomenclatural and Taxonomical History A complete skull from the middle Miocene of Thymiana on Chios Island (Greece), originally attributed to *Gomphotherium angustidens* (Tobien 1973), constitutes the holotype of *Choerolophodon chioticus*, erected by Tobien (1980). Pickford (2001), who studied cranial material from Kenya, erected the genus *Afrochoerodon*, in which he included also the Thymiana specimen. According to Shoshani and Tassy (2005) *Afrochoerodon* constitutes a paraphyletic taxon placed within *Choerolophodon*.

Type Material AMPG-937 (holotype), cranium with I2, M1–M2, and erupting M3 of both sides, and associated mandible with m1–m2 and erupting m3 of both sides (Fig. 8).

Type Locality Thymiana, Chios Island, Greece, middle Miocene, late Orleanian, MN5.

Distribution Middle Miocene (late Orleanian, MN5) of Greece (Thymiana), and possibly from the middle Miocene of Turkey and Bulgaria.

Remarks *Choerolophodon chioticus* from Thymiana represents the earliest known choerolophodont in Europe. The skull is characterized, among others, by: moderate elongation, steeply inclined facial region and high cerebral region of the cranium; upper tusks that emerge downwards, then curve upward and at the tip inward; long and downwards deflected mandibular symphysis (Fig. 8); and, weak to moderate expressed choerodonty, ptychodonty, and cementodonty on the molars. The Thymiana cranium is less derived than the late Miocene choerolophodonts and closer to *C. kisumuensis* (Tobien 1980; Tassy et al. 1989; Pickford 2001; Konidaris et al. 2016). The distal fragment of an m3 from Thymiana, attributed by Paraskevaidis (1940: pl. 14, fig. 6) to *Trilophodon (Mastodon) angustidens*, possibly also belongs

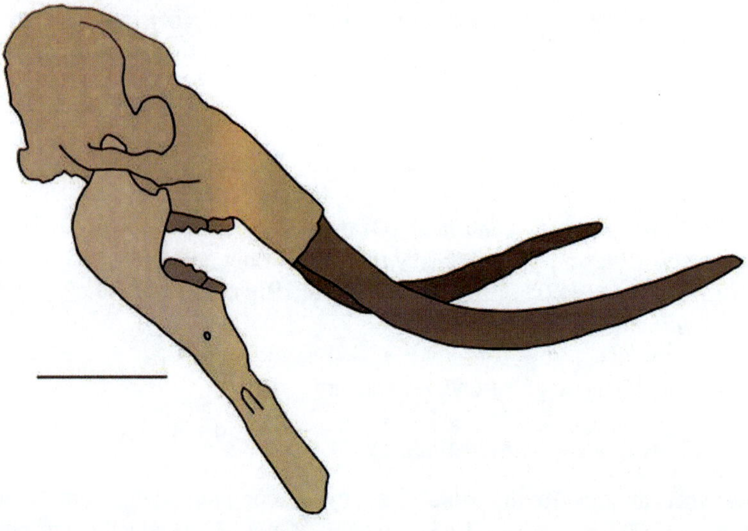

Fig. 8 Cranium with upper tusks, M1–M2 and erupting M3 of both sides, and associated mandible with m1–m2 and erupting m3 of both sides (holotype; AMPG-937) of *Choerolophodon chioticus* from the middle Miocene of Thymiana, Chios Island (copyright G. Konidaris). Scale bar equals 50 cm

to *Choerolophodon*. Study of additional material from Thymiana is in progress. Thymiana is bio- and magnetostratigraphically correlated to the middle Miocene (late Orleanian, MN 5), at ~15.5 Ma (Koufos 2006).

Choerolophodon anatolicus (Ozansoy, 1965)

Nomenclatural and Taxonomical History *Trilophodon* (*Choerolophodon*) *anatolicus* was erected as a new choerolophodont species by Ozansoy (1965) based on material from Yassiören (Middle Sinap, Turkey). However, in the subsequent years, the species was regarded as synonymous with *C. pentelici*, until Sanders (2003), who studied new material from the Sinap localities, re-established its validity. Sanders attributed to this species specimens from several other Turkish localities, including some of the material that had been attributed by Gaziry (1976) to *C. pentelici* and a mandible from Eşme Akçaköy assigned to *C. pentelici lydiensis* by Tassy et al. (1989) (see below). Recently, Konidaris et al. (2016) revised *C. anatolicus* and marked its presence for the first time outside Turkey in the Greek locality Pentalophos-1 of Axios valley.

Type Material MNHN-TRQ-1000 (lectotype), maxilla with right DP2–DP4 and left DP3–DP4.

Type Locality Yassiören, Middle Sinap, Turkey, early Vallesian, MN9.

Distribution Late Miocene (early Vallesian, MN9) of Greece (Pentalophos-1), and the late Miocene (early Vallesian–?late Vallesian, MN9–?10) of Turkey, and possibly of Moldova and Romania.

Remarks *Choerolophodon anatolicus* is diagnosed and differentiated from *C. pentelici*, among other characters, by the moderate retracted perinasal area in the cranium, the ventrally deflected mandibular symphysis in adult individuals, the small size of the deciduous teeth and the weak development of the distal cingulum in the DP3/dp3, which is connected to the second loph(id) (Sanders 2003; Konidaris and Koufos 2013a; Konidaris et al. 2016). The first and so far only record of *C. anatolicus* in Greece is traced in the early Vallesian locality Pentalophos-1 of Axios valley (Konidaris and Koufos 2013a; Konidaris et al. 2016). The material consists of juvenile and adult specimens (Fig. 9), which present more primitive morphology and smaller dimensions than the rest of the choerolophodont material from Axios valley, as well as from Pikermi and Samos, and are similar to those of *C. anatolicus* from Yassiören, Sinap 12 and Eşme Akçaköy (Turkey).

★*Choerolophodon pentelici* (Gaudry and Lartet, 1856)

Nomenclatural and Taxonomical History Gaudry and Lartet (1856) erected the species *Mastodon pentelicus* based on proboscidean remains from the Turolian locality Pikermi. This material was subsequently studied in detail by Gaudry (1862–1867) (specimens of *Konobelodon atticus* were included in this study as

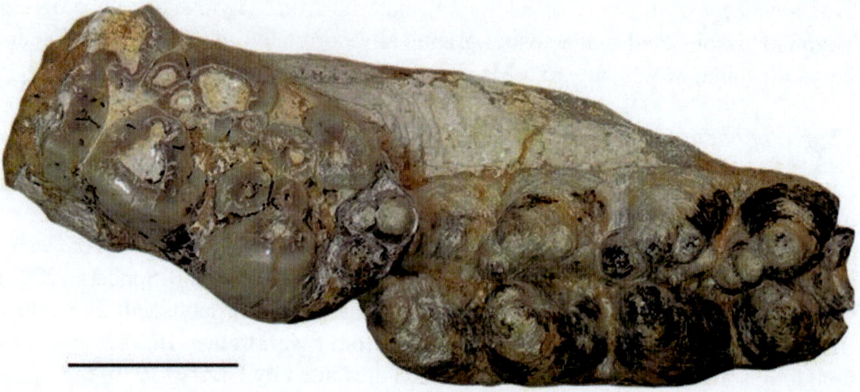

Fig. 9 Right mandibular fragment with m2–m3 (LGPUT-PNT-154) of *Choerolophodon anatolicus* from the late Miocene (Vallesian) of Pentalophos-1, Axios valley (copyright G. Konidaris). Scale bar equals 5 cm

well; see below), who emended also the species name *Mastodon pentelici*, although such a correction was not necessary (Tassy 1985: p. 617). Later on, Schlesinger (1917, 1922) studied choerolophodonts from Samos Island and Maragheh (Iran) and erected the subgenus *Mastodon (Choerolophodon)*, while Arambourg and Piveteau (1929) studied specimens from Axios valley. Since then, all the late Miocene choerolophodont material has been referred to this species without any distinction. Gaziry (1976) was the first, who recognized different evolutionary dental features within the late Miocene choerolophodonts of Turkey. Subsequently, Tassy et al. (1989) erected the subspecies *C. p. lydiensis* for an adult mandible from Eşme Akçaköy, a specimen, which was later transferred by Sanders (2003) to *C. anatolicus* (see above). From the fossiliferous sites of Kemiklitepe, Tassy (1994) identified two different evolutionary morphs, a primitive one in Kemiklitepe-D and a more advanced one in Kemiklitepe A-B. Recently, Konidaris (2013) examined all the choerolophodont material from Greece (old and new collections) and revised specimens from western Eurasia, and recognized the presence of two evolutionary stages within *C. pentelici*. In particular for Greece, "*C. pentelici* primitive" is correlated to the late Vallesian until possibly the early Turolian (MN10–?11; Xirochori-1, Ravin de la Pluie, Ravin des Zouaves-1, and from an unknown locality from the Turolian of Samos Island) and "*C. pentelici* advanced" from the Turolian (MN11–13; Pikermi, Samos Island, Mytilinii-1B-Samos, Andriano-Samos, Nikiti-2, Ravin des Zouaves-5, Prochoma-1, Vathylakkos-2, Ravin X and Dytiko-2, 3 (Konidaris and Koufos 2016, 2019; Konidaris et al. 2016).

Type Material MNHN-PIK-3665 (lectotype), juvenile cranium with right DP2–DP4, left I2 and left DP4, and associated mandible with right dp3–dp4 and left dp4 (Fig. 10a–d).

Type Locality Pikermi, Attica, Greece, late Miocene, middle Turolian, MN12.

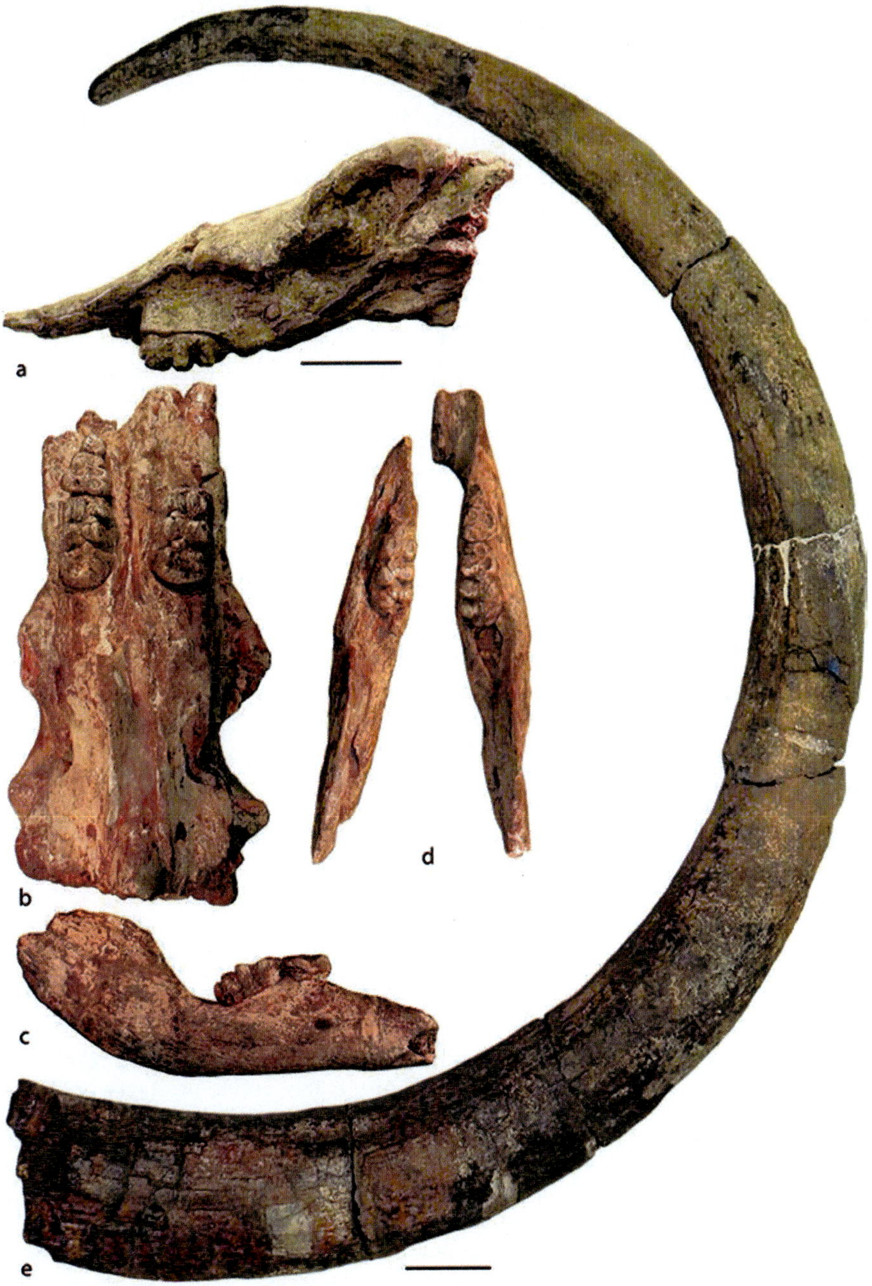

Fig. 10 (**a–d**), Juvenile cranium with right DP2–DP4, left upper tusk and left DP4, and associated mandible with right dp3–dp4 and left dp4 (lectotype; MNHN-PIK-3665) of *Choerolophodon pentelici* from the late Miocene (Turolian) of Pikermi; (**a**), left lateral view of the cranium; (**b**), ventral view of the cranium; (**c**), lateral view of the right hemimandible; (**d**), dorsal view of the mandible (copyright G. Konidaris); (**e**), right upper tusk (LGPUT-NIK-1776) in lateral view of *C. pentelici* from the late Miocene (Turolian) of Nikiti-2 (copyright G. Konidaris). Scale bar equals 10 cm

Distribution Late Miocene (late Vallesian–late Turolian, MN10–13) of Greece (Xirochori-1, Ravin de la Pluie, Ravin des Zouaves-1, Ravin des Zouaves-5, Nikiti-2, Prochoma-1, Vathylakkos-2, Pikermi, Mytilinii-1B, Samos Island-old collections, Andriano, Kerassia, Ravin X, Kryopigi, Dytiko-2, 3, Dikaia, Sani, ?Maramena, ?Pyrgos Vassilissis, ?Servia). Outside Greece it is known from the late Miocene (?late Vallesian–middle Turolian, ?MN10, MN11–12) of Turkey, Bulgaria, North Macedonia, Romania, Moldova, Iran, Iraq, and possibly in Ukraine and Azerbaijan.

Remarks The skull of *C. pentelici* is characterized, among other traits, by an elongated cranium with moderately inclined facial region and low cerebral region, orbits situated at the top of the cranium and far behind the last molar in function, retracted perinasal area and narrow nasal aperture, and mandibular symphysis situated ventrally at the extension of the corpus in both juvenile and adult specimens (Fig. 10a–d). The upper tusks emerge sub-horizontally and outwards and then curve upward (Fig. 10e). The deciduous teeth are large, the distal cingulum of the dp3/DP3 is well developed, and it is separated from the second loph(id) in the more advanced morphs. Choerodonty, ptychodonty, and cementodonty are well-expressed in the cheek teeth. The last appearance of *C. pentelici* is traced in the late Turolian (MN 13) localities of Dytiko in Axios valley (Konidaris et al. 2016). Schmidt-Kittler et al. (1995) note the occurrence of this species in the faunal list of Maramena (Serres basin), dated to the Miocene/Pliocene (MN13/14) boundary. If this occurrence is indeed *C. pentelici*, then the species survived until the uppermost Miocene.

Amebelodontidae Barbour, 1927

Comments The family includes predominantly bunodont trilophodont elephantimorphs (tetralophodonty in all intermediate molars has reached only the genus *Konobelodon*), commonly called "shovel-tuskers" due to the broadening and flattening of their lower tusks. However, the monophyly of the family is based on the straight medial border of the lower tusks and the posttrite conules on the molars (Tassy 1996a; see also Shoshani 1996). Based on the internal structure of the lower tusks, two groups of amebelodontids are distinguished: one with only concentric laminated dentine (*Archaeobelodon*, *Protanancus*, *Amebelodon*, *Serbelodon*, *Aphanobelodon*) and another with presence of tubular dentine or otherwise called dentinal rods (*Platybelodon*, *Torynobelodon*, *Konobelodon*) (Tassy 1986, 1996a; Konidaris et al. 2014; Wang et al. 2017a). The family originated during the early Miocene in Africa, but rapidly dispersed into Europe and Asia, and subsequently into North America; by the beginning of the Pliocene amebelodontids had vanished (Lambert and Shoshani 1998; Sanders et al. 2010). In Greece, amebelodontids are represented so far only by the late Miocene *Konobelodon*.

Konobelodon Lambert, 1990

Type Species *Amebelodon (Konobelodon) britti* Lambert, 1990.

Other Included Species *K. atticus* (Wagner, 1857); *K. cyrenaicus* (Gaziry, 1987b); *K. robustus* Wang et al., 2016.

Distribution Late Miocene of Europe, Africa, Asia, and North America; in Europe is known from the late Vallesian until the late Turolian.

Comments *Konobelodon* was originally erected as a subgenus of *Amebelodon*, which included the shovel-tuskers with tetralophodont first and second molars, and flattened lower tusks bearing internal dentinal rods from the late Miocene of North America (Barbour and Hibbard 1941; Gregory 1945; Lambert 1990) [we note here that the DP4, associated with fully tetralophodont M1 in the palate KUVP-3477 from the Rhino Hill site, Kansas, U.S.A. (Barbour and Hibbard 1941), albeit considerably worn shows an incipient fourth loph (GK observation on photos provided by C. Beard, University of Kansas); Mebrate (1987: p. 232) described the DP4 as trilophodont with a talon, Lambert (1990) mentioned it as trilophodont, while Lambert (pers. comm. 2020 to GK) as quasi-tetralophodont]. Lambert (1990) included also to the same genus/subgenus the Sahabi *Amebelodon cyrenaicus*—described by Gaziry (1987b) and considered a possible junior synonym of *Mastodon grandincisivus* by Tassy (1999) —whose lower tusk shows tubular dentine; trilophodont m1 and tetralophodont m2 were originally attributed to this taxon; however, it is uncertain whether these molars belong to the same taxon as the lower tusk (holotype) or each other (Markov 2008b; Sanders 2008). Konidaris et al. (2014) described juvenile mandibles from Pikermi belonging to a tetralophodont grade species bearing flat lower tusks with tubular dentine, as well as an adult large lower tusk with flattened-pyriform cross-section and internal dentinal rods. The upper and lower deciduous premolars (some of the lower ones associated with flat lower tusks) are both morphologically and metrically similar to those of *Mastodon atticus* (Wagner 1857; Gaudry 1862–1867; usually attributed to *Tetralophodon atticus*), and the adult lower tusk is morphologically and metrically similar to the holotype of *Mastodon grandincisivus* from Maragheh (Iran; Schlesinger 1917). Konidaris et al. (2014) revised the tetralophodont material from Pikermi (see below), noting the similarities with the lower tusks from North America, and proposed the elevation of *Konobelodon* to generic level, the synonymy between *Mastodon atticus* and *Mastodon grandincisivus*, and the inclusion of all the Turolian tetralophodont shovel-tuskers from western Eurasia to *Konobeledon atticus*. New tetralophodont amebelodontid material from China, including juvenile specimens (with striking morphological resemblance to the deciduous premolars from Pikermi, including the holotype of *Mastodon atticus*) and adult mandibles preserving the flattened lower tusks with internal dentinal tubules (Wang et al. 2016), corroborates further the attribution of the Pikermi tetralophodont material to an amebelodontid. Tassy (2016) suggested that the paratype of *K. britti* is similar to the holotype of *Torynobelodon loomisi* Barbour 1929 [*Torynobelodon* is considered a junior synonym of *Platybelodon* in Shoshani and Tassy (1996b: app. A)] and therefore that *Konobelodon* is a junior synonym of *Torynobelodon*; *Konobelodon atticus* should thus be *Torynobelodon atticus*. However, the tusk of *T. loomisi* has different cross-sectional shape from the tusk of *K. britti*, as well as from those of *K. robustus* and *K. atticus*, e.g., from Pikermi, Maragheh and Pestszentlőrinc (Hungary), having wide and straight medial border, roughly the same width in the middle and lateral parts, and with very shallow dorsal and ventral concavities. Moreover, it is unknown whether the tusk of *T. loomisi*, which is so far the single known specimen of this species, belonged to a trilophodont or tetralophodont grade species. Consequently, the current evidence does not support the synonymy between *Konobelodon* and *Torynobelodon*, and they should be considered distinct genera. Finally, the coexistence between a tetralophodont amebelodontid (here *Konobelodon*) and a derived tetralophodont taxon, such as that found in the Turolian of Crevillente 2 [Spain, Mazo and Montoya 2003; possibly related to *Stegotetrabelodon* (Tassy 2016)], or *Stegotetrabelodon* (present in the Turolian of Cessaniti, Italy; Ferretti et al. 2003), both having oval/sub-circular cross-sections with lamellar dentine in their lower tusks, cannot be excluded for Pikermi or other Turolian localities of the wider region, as discussed by Tassy (2016). Nonetheless, such evidence is so far not recorded in the Turolian of the Greco-Iranian-Afghan (*sensu* Bonis et al. 1992; Balkano-Iranian or Sub-Paratethyan) paleobiogeographic province.

★*Konobelodon atticus* (Wagner, 1857)

Nomenclatural and Taxonomical History The type locality of the tetralophodont amebelodontid *K. atticus* is Pikermi, from where the richest material of this species is known. The species has a long and complicated taxonomical history. The presence of tetralophodonts in Pikermi was recognized early by Wagner (1857), who described a juvenile maxilla and erected the species "*Mastodon atticus* G. and L." However, Wagner (1857) attributed *Mastodon atticus* to Gaudry and Lartet (1856), but as noted by Gaudry (1862: p. 142 footnote), the only bunodont proboscidean referred by Gaudry and Lartet (1856: p. 273) was the trilophodont *Mastodon pentelicus* (= *Choerolophodon pentelici*). Few years later, Lartet (1859) transferred the Pikermi tetralophodonts to *Mastodon longirostris* (= *Tetralophodon longirostris*). Gaudry (1862–1867) included all bunodont proboscideans from Pikermi (tri- and tetralophodonts) to *Mastodon pentelici*. Subsequently, Vacek (1877: p. 32, pl. 7) attributed a third molar from Pikermi to *Mastodon atticus*, which was transferred to *Stegotetrabelodon grandincisivus* by Tobien (1978). The tetralophodont material of Wagner (1857) and Gaudry (1862–1867), as well as that described by the latter author as *Mastodon turicensis*, was included in the zygodont taxon *Turicius atticus* by Osborn (1936). More recently, Tassy (1985) referred the tetralophodonts of Pikermi to *Tetralophodon atticus*, although he later questioned this generic attribution (Tassy 2005). Konidaris et al. (2014) studied all the published tetralophodont material from Pikermi, as well as previously unpublished material from the locality originating from the old excavations, and attributed them to the amebelodontid *Konobelodon atticus* (see discussion above).

Type Material SNSB-BSPG-AS II 182 (holotype), left maxillary fragment with DP2–DP3 (Fig. 10a).

Type Locality Pikermi, Attica, Greece, late Miocene, middle Turolian, MN12.

Distribution Late Miocene (Turolian) of Greece (Pikermi, Samos-old collections, Kerassia-4, and possibly from Platania). Outside Greece, it is known from the late Miocene (Turolian, MN11–13) of Turkey, Bulgaria, Moldova, Ukraine, Hungary, North Macedonia, and Iran.

Remarks The species is characterized, among other features, by the tetralophodont intermediate molars, the enlarged third loph in the DP3 with posttrite-pretrite connection in both transverse valleys and well-marked second ento-/ectoflexus (Fig. 11a), the large-sized and dorsoventrally flattened-pyriform adult lower tusks that have thin concentric laminated dentine layer externally and tubular dentine internally (Fig. 11b), and by the long and high mandibular symphysis (Konidaris et al. 2014). Apart from Pikermi, the species is known from Samos, Kerassia-4 and possibly from Platania (Lehmann 1950; Theodorou et al. 2003; Konidaris et al. 2014; Konidaris and Koufos 2019; Konidaris and Tsoukala 2020). It is well distributed from the early until the late Turolian of eastern Europe-western Asia and in Greece, in particular, is present possibly from the Vallesian/Turolian boundary (Platania) until the middle Turolian (Pikermi).

Gomphotheriidae Hay, 1922

Comments This paraphyletic family includes the bunodont trilophodont and tetralophodont elephantimorphs, excluding the choerolophodontids and amebelodontids, and constitutes the most diverse family within Elephantimorpha. By the early Miocene, the primitive *Gomphotherium* "*annectens* group" had already a cosmopolitan distribution in Africa, Europe, and Asia (Tassy 1996b); during the middle Miocene, gomphotheriids entered North America and persisted in South America until the beginning of the Holocene (Lambert and Shoshani 1998; Mothé et al. 2017). In Europe, gomphotheriids are represented by the trilophodont *Gomphotherium* and the tetralophodonts *Tetralophodon* and *Anancus*. In Greece, only tetralophodont gomphotheres are known so far.

Tetralophodontinae van der Maarel, 1932
Tetralophodon Falconer, 1857

Type Species *Tetracaulodon longirostre* Kaup, 1832b.

Other Included Species *T. euryrostris* Wang et al., 2017b; *T. xialongtanensis* (Chow and Chang, 1974).

Distribution Middle–late Miocene of Europe, Asia, and Africa.

Comments *Tetralophodon* includes tetralophodont gomphotheres with long mandibular symphysis bearing pyriform to oval in cross-section lower tusks (in contrast to the brevirostrine and tuskless *Anancus*), consisting of concentric lamellar dentine

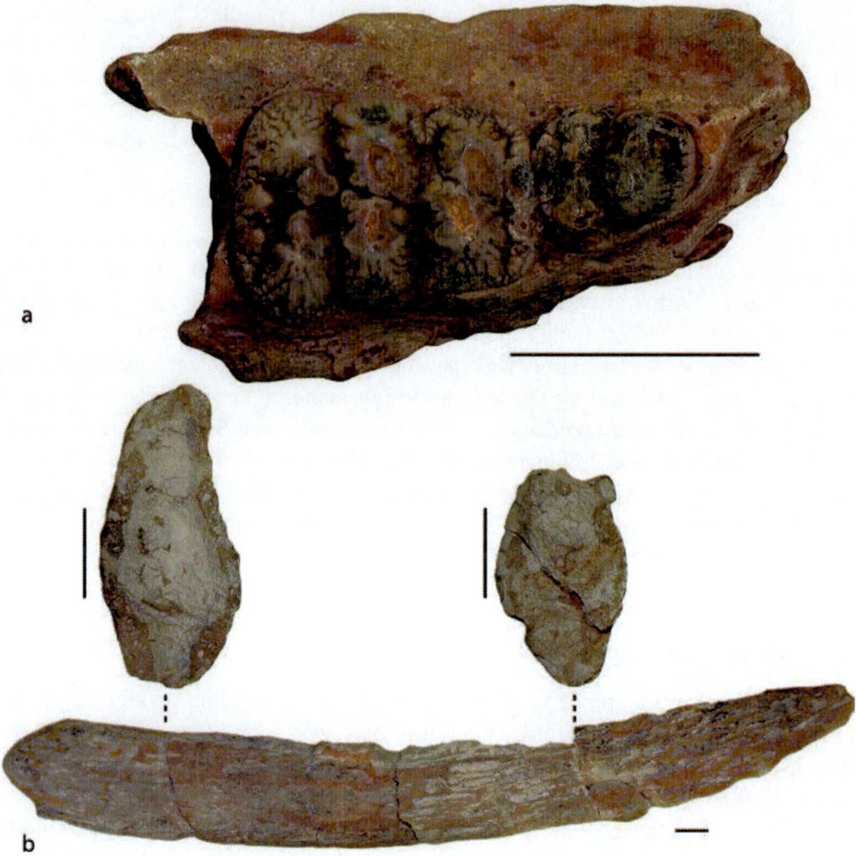

Fig. 11 *Konobelodon atticus* from the late Miocene (Turolian) of Pikermi. (**a**), Left maxilla fragment with DP2–DP3 (holotype; SNSB-BSPG-AS II 182) (copyright G. Konidaris), (**b**), right lower tusk (AMPG-PA1972/61) in ventral view with cross-sections (copyright G. Konidaris). Scale bars equal 5 cm

(no dentinal rods like *Konobelodon*), intermediate molars, and third molars that show trefoil wear patterns (not plate like pattern such as *Stegotetrabelodon*) and rounded upper tusks that lack enamel bands.

Tetralophodon longirostris (Kaup, 1832b)

Nomenclatural and Taxonomical History Tetralophodont proboscidean remains from the Eppelsheim Formation (Dinotheriensande) of the Mainz Basin in Germany were originally allocated to the species *Tetracaulodon longirostre* by Kaup (1832b) (*Tetracaulodon* = *Mammut*), and slightly later they were assigned to *Mastodon longirostris* (Kaup 1835). This species was subsequently included within the subgenus *Mastodon* (*Tetralophodon*) by Falconer (1857).

Type Material HLMD-Din 111 (holotype), mandible fragment with left m2–m3.

Type Locality Eppelsheim, Germany, Miocene.

Distribution ?Late Miocene (?Vallesian) of Greece (Fourka area, seabed of Kryopigi, Chelona beach). Outside Greece, this species is known from the middle–late Miocene (?early Astaracian, late Astaracian–Vallesian, ?MN6, MN7/8–9, ?MN10) of Europe.

Remarks *Tetralophodon longirostris* shows an evident variation in the dental morphology and the shape of the mandible (especially the curvature of the mandibular symphysis), which were attributed to polymorphism (Tassy 1985, 1999). Most of the known specimens originate from the Eppelsheim Formation, which was generally considered to be of Vallesian age. However, recent studies indicate the stratigraphic inhomogeneity due to reworking of the sediments and a chronological range of the fauna from the middle Miocene to the late Miocene (Böhme et al. 2012; Pickford and Pourabrishami 2013). In this aspect, the revision of all known material from Europe and the discovery of new specimens are necessary to clarify the taxonomy of the European *Tetralophodon*, and especially whether *T. curvirostris* Bergounioux and Crouzel 1960 and *T. gigantorostris* (Klähn 1922) are indeed varieties of *T. longirostris*.

In Greece, Lazaridis and Tsoukala (2014b) and Lazaridis (2015) report on the presence of isolated findings of *T. longirostris* from several collection spots (two of them were recovered from the seabed) of provisionally Vallesian age within the Kassandra sand deposits, considering that they originate from the Antonios Formation of Western Chalkidiki and date to the late Miocene. This Formation is biostratigraphically dated from the early/middle Miocene boundary (close to MN4/5) until the late Miocene (pre-middle Turolian but possibly Vallesian) (Syrides 1990; Koufos 2013; Lazaridis et al. 2017).

Anancinae Hay, 1922
***Anancus* Aymard in Dorlhac, 1855**

Type Species *Mastodon arvernensis* Croizet and Jobert, 1828.

Other Included Species *A. capensis* Sanders, 2007; *A kenyensis* (MacInnes, 1942); *A. lehmanni* Gaziry, 1997; *A. osiris* Arambourg, 1945; *A. perimensis* (Falconer and Cautley, 1847); *A. petrocchii* Coppens, 1965; *A. sinensis* (Hopwood, 1935); *A. sivalensis* (Cautley, 1836); *A. ultimus* Sanders, 2011.

Distribution Late Miocene–Early Pleistocene of Europe, Africa, and Asia.

Comments Anancine gomphotheres, represented by the single genus *Anancus*, have tetralophodont (to pentalophodont) intermediate molars, whose main morphological feature is the dislocation of the pretrite and posttrite half-loph(id)s and the resultant alternate arrangement of the successive loph(id)s (anancoidy). In particular, in upper molars, the pretrite half-lophs are mesially offset, whereas in lower molars the pretrite half-lophids are distally dislocated, establishing thus an alternate contact of the successive loph(id)s (Tassy 1986). Other anancine characters include the high and short cranium with domed and elevated vault, the enlarged tympanic bullae, the brevirostrine mandible without tusks, the straight upper tusks (*Anancus* means "without a bend"; although slightly curved upper tusks exist) lacking enamel bands, and the absence of premolars, except in *A. kenyensis* (Tassy 1986; Hautier et al. 2009). The earliest occurrences of *Anancus* are traced during the Turolian (late Miocene). The genus flourished in the Old World during the Pliocene (in Europe commonly co-occurring with the zygolophodont *Mammut borsoni*) and survived until the Early Pleistocene (co-occurring also with *Mammuthus*), being the last gomphothere of the Old World.

***Anancus lehmanni* Gaziry, 1997**

Nomenclatural and Taxonomical History Anancine gomphotheres from the late Miocene of Europe were recognized as *Mastodon* cf. *longirostris* (Schlosser 1907), subsequently to transitional forms between *Mastodon longirostris* and *Mastodon arvernensis* (e.g., Schlesinger 1917; Zapfe 1957) and later on to *Anancus* sp. (Tassy 1986; Markov 2008b). Gaziry (1997) studied the elephantimorphs from Dorn-Dürkheim 1 (Germany) and described four taxa: *Tetralophodon longirostris*, *Anancus arvernensis turoliensis*, *Stegotetrabelodon lehmanni*, and *Stegolophodon caementifer*, the second one a new subspecies and the latter two new species. However, apart from the material referred by Gaziry to "*A. a. turoliensis*," the holotype of "*S. lehmanni*" and other specimens attributed to this species are also included in *Anancus*, as well as specimens of *T. longirostris* (Konidaris and Roussiakis 2019). The species name *lehmanni* has nomenclatural priority over the subspecies name *turoliensis* due to its original higher taxonomic ranking (see also discussion in Markov 2008b). Konidaris and Roussiakis (2019) revised the known anancine specimens from the late Miocene of Europe (see below) and attributed them to *Anancus lehmanni*.

Type Material SMF-DD 3151(holotype), right M3 (Gaziry 1997: pl. 3, fig. 3).

Type Locality Dorn-Dürkheim-1, Germany, Turolian.

Distribution Late Miocene (middle Turolian, MN12) of Greece (Chomateri), and from numerous localities from the late Miocene (middle–late Turolian, MN12–13) of Germany, Austria, Slovakia, Hungary, Spain, Romania, North Macedonia, Bulgaria, Northern Caucasus, and Turkey.

Remarks The late Miocene anancines from Europe are well distinguished from the roughly contemporaneous, but more derived, *A. kenyensis*, and the Plio–Pleistocene *A. arvernensis*, as well as from other anancines, in the morphology of the mandible (longer symphysis, condyle only slightly higher than the coronoid process), the cranium (straight and parallel premaxillary tusk alveoli, more anterior location of the orbit), the upper tusks (downturned, running almost parallel), and the primitive molar features on the cheek teeth (weak anancoidy, simple occlusal morphology, thick and unfolded to coarsely folded enamel on the molars) (Konidaris and Roussiakis 2019). The combination of the above traits is unique among anancines and permits the distinction at the species level, the proper name of which is *Anancus lehmanni* [see discussions in Markov (2008b) and Konidaris and Roussiakis (2019)]. Previously known only from the Pliocene and the Pleistocene of Greece, *Anancus* was recently reported for the first time from the late Miocene of Greece in Chomateri (Fig. 12a), marking its earliest occurrence in the Greek fossil record (Konidaris 2013; Konidaris and Koufos 2013a, b; Konidaris and Roussiakis 2017, 2019). The appearance of *Anancus* in Europe at the second half of the Turolian (~7.2 Ma) coincides with a faunal turnover in both the eastern and western sector of the European Mediterranean region and, in the Southern Balkans in particular, with the decline of the "Pikermian" large mammal fauna (Kostopoulos 2009; Böhme et al. 2017).

Anancus arvernensis (Croizet and Jobert, 1828)

Nomenclatural and Taxonomical History Croizet and Jobert (1828) erected *Mastodon arvernensis* based on juvenile dental specimens from Perrier in Puy-de-Dôme (Auvergne-Rhône-Alpes, France). In 1855 Aymard coined the genus *Anancus* and the species *A. macroplus*, which was slightly later synonymized with *Mastodon*

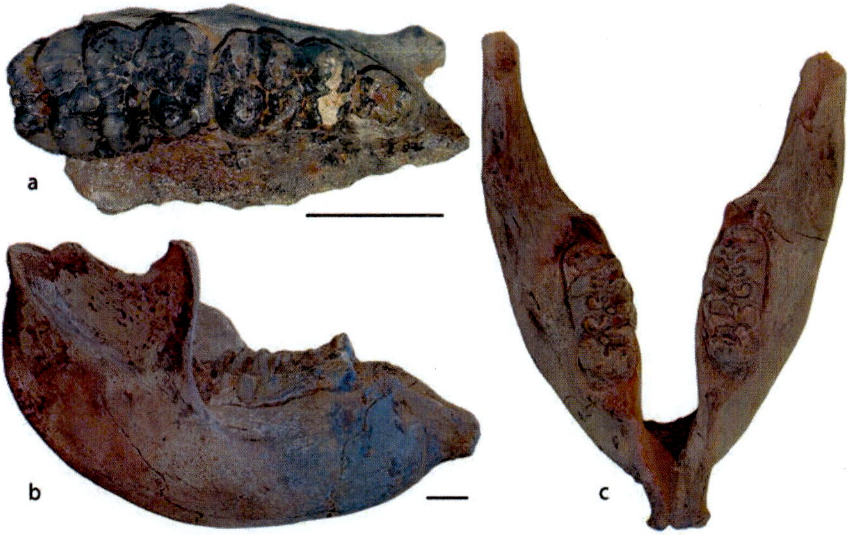

Fig. 12 (**a**), Right maxilla fragment with DP2–DP4 (AMPG-13Π/1972) of *Anancus lehmanni* from the late Miocene (Turolian) of Chomateri (copyright G. Konidaris); (**b–c**), mandible with the m3s (AMPG-1918) of *Anancus arvernensis* from the late Pliocene of Sesklon in b dorsal view and c right lateral view (copyright A. Athanassiou). Scale bars equal 5 cm

arvernensis (Lartet 1859), although the new combination was not utilized. Osborn (1936) recognized the binomen *Anancus arvernensis*. Meanwhile, Schlesinger (1917, 1922) attributed material from Austria and Hungary to *Mastodon* (*Dibunodon*) *arvernense*, a subgenus which is though a junior synonym of *Anancus* (Matsumoto 1927; Gaziry 1976).

Type Material MNHN-A.C. 1830 (lectotype), right maxilla fragment with DP2–DP3 (Croizet and Jobert 1828: pl. 2, fig. 7).

Type Locality Perrier-les-Étouaires, France, late Pliocene, late Ruscinian, MN16.

Distribution Several Pliocene–Early Pleistocene localities of Greece, (e.g., Milia, Gephyra, Sesklon, Vatera, Apolakkia, Nigrita) and in the Pliocene–Early Pleistocene (Ruscinian–middle Villafranchian, MN14–MNQ17) of Europe.

Remarks This species represents the Pliocene–Pleistocene representative of the European anancines and the last gomphothere to have survived in Europe. The richest material originates from Dorkovo (early Pliocene; Bulgaria), but important material is known also from Chilhac (France) and Valdarno (Italy) (Weithofer 1890; Boeuf 1992; Metz-Muller 2000; Rook et al. 2013). The research of Metz-Muller (2000) proved the high intraspecific variability (including the complexity of the cheek teeth and the occasional occurrence of pentalophodont m2) and included the biometrical study of the mandibles resulting in the definition of dental ages; moreover, the study showed that *A. arvernensis* presents a tendency towards shortening of the molars, increase of the hypsodonty, and simplification of the crown.

In Greece, although several *A. arvernensis* specimens are known, most of them are isolated dental findings with uncertain stratigraphic position. However, important and stratified material is known from Sesklon (including a partial cranium, a mandible and an upper tusk, Fig. 12b–c; Symeonidis and Tataris 1983; Athsanassiou 2016, 2018), Nigrita (a partial cranium with the upper tusks and the molars; Athanassiou 2017), Apolakkia (a partial cranium with the upper tusks and the molars; Theodorou et al. 2000), Milia (mandible and teeth; Tsoukala and Mol 2016), Gephyra (mandibles and postcranials; Crégut-Bonnoure and Tsoukala 2017), Kalliphytos (mandible; Athanassiou 2016), and Vatera (teeth; de Vos et al. 2002).

Stegodontidae Osborn, 1918
***Stegodon* Falconer and Cautley, 1847**

Type Species *Mastodon elephantoides* Clift, 1828.

Other Included Species See Saegusa et al. (2005) and Aiba et al. (2010) for a complete list.

Distribution Late Miocene–Late Pleistocene of Asia, Africa, and Europe.

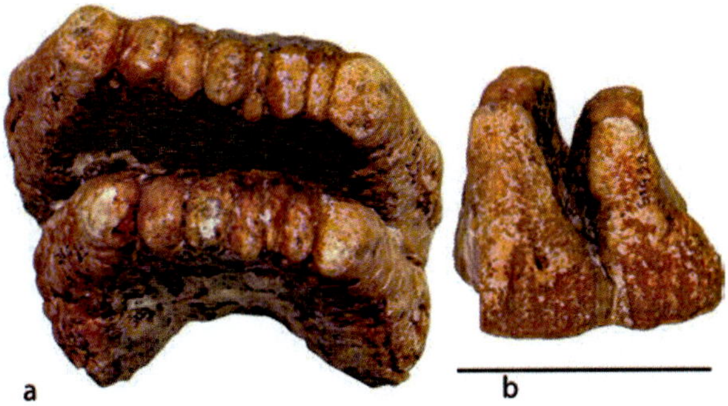

Fig. 13 Right m3 fragment (SIA 22) of *Stegodon* sp. possibly from the wider area of Siatista (Kozani), without information on the exact locality; (**a**), occlusal view; (**b**), lingual view (photo Hans Wildschut, Dick Mol). Scale bar equals 5 cm

Comments The genus *Stegodon* includes the stegodontids, which are characterized and differentiated from the more archaic *Stegolophodon* by their intermediate molars with five or more loph(id)s, third molars with no distinct central conule, absence of lower tusks, and two lophids above the mesial root of the lower third molar (Saegusa et al. 2005).

***Stegodon* sp.**

Remarks A fragment of a third molar of *Stegodon* sp. (Fig. 13) from an unknown locality (and therefore no information about geological age), but possibly from the wider area of Siatista (Kozani), constitutes the first evidence of the presence of stegodontids in Europe, extending their previously known geographical distribution (Mol et al. 2010).

6 Concluding Remarks

Proboscideans are common in the Neogene faunas of Greece with relatively abundant material, and they are present in most of the Neogene fossiliferous sites, both in today's continental and insular Greece. They have been the largest terrestrial mammals in the Neogene faunas of the country, dominating the ecosystems of that time. Recent investigations on the Miocene and Pliocene proboscideans, including the revision and study of old collections, and the study of new specimens from excavations carried out in Greece during the last years, have resulted in the update of their taxonomy and biostratigraphy. Based on our current knowledge, we can reach several conclusions.

During the early Miocene, the only proboscidean that is known in Greece is the deinothere *Prodeinotherium cuvieri*, documented at Gavathas (Lesvos Island), where the find-bearing locality is dated as older than 18.4 Ma (late MN3). The Gavathas deinothere marks the first appearance of deinotheres in Europe and documents their penetration into the continent as part of the complex "Proboscidean

Datum Event," after the establishment of the so-called *Gomphotherium* landbridge in the middle Burdigalian (~19.0–18.0 Ma, early MN3).

In the middle Miocene, the deinothere *Prodeinotherium bavaricum* and the trilophodont choerolophodontid *Choerolophodon chioticus* are recognized in Thymiana (Chios Island), dated to ~15.5 Ma (MN5). *Choerolophodon chioticus* is the most primitive choerolophodontid in Europe and marks their first penetration into the continent, being part of the third European phase of the "Proboscidean Datum Event" (Tassy 1990). This phase is probably part of the first middle Miocene migrational event that took place in the late Orleanian (17.0–15.0 Ma, MN5), involving also the arrival in Europe of the amebelodontid *Protanancus* (Rögl 1999; Koufos et al. 2005; Markov and Vergiev 2010).

During the late Miocene, proboscideans are more diverse and well documented in the Greek fossil record due to a high number of fossiliferous localities; however, *Choerolophodon* is in most of the localities the dominant proboscidean in terms of abundance. In the early Vallesian (MN9), *Choerolophodon anatolicus* is identified in Pentalophos-1 (Axios valley). During the late Vallesian (MN 10), the deinothere *Deinotherium giganteum* and the primitive morph of *Choerolophodon pentelici* co-existed in Ravin de la Pluie (Axios valley). In the early Turolian (MN11) appeared the zygodont *Mammut* sp. (*M. obliquelophus*?), which co-existed with the advanced morph of *C. pentelici* in Ravin des Zouaves-5 (Axios valley). In the middle Turolian *C. pentelici*, the tetralophodont amebelodontid *Konobelodon atticus*, *Mammut* sp. (*M. obliquelophus*?), and the huge-sized deinothere *D. proavum* are recorded, the co-occurrence of which is documented in Pikermi. These four species are recognized also in the Mytilini Formation of Samos Island. During the middle Turolian, but postdating the classical Pikermi, appeared the tetralophodont gomphothere *Anancus lehmanni*, the most primitive anancine of Europe. In the late Turolian (MN13) are traced the last occurrences of *C. pentelici*, present with certainty in the Dytiko localities (Axios valley) and perhaps in Maramena.

During the Pliocene, *Anancus arvernensis* and *Mammut borsoni* are recognized. The fossiliferous localities of this period are few and the lack of stratigraphic information for several findings makes the determination of their first appearances problematic. However, *M. borsoni* survived until the end of the Pliocene (MN16) and *A. arvernensis* until the beginning of the Early Pleistocene (MNQ17). The co-occurrence of these species is documented in Milia (early late Pliocene, MN16a). Finally, the first recorded occurrence of *Stegodon* in Greece represents also the first evidence of the presence of stegodontids in Europe.

Acknowledgments We thank all persons that allowed us access to the several proboscidean collections under their disposal in Greece and the rest of Europe. We are grateful for providing us photos of specimens and permission to use them to N. Zouros, K. Vasileiadou (Natural History Museum of the Lesvos Petrified Forest) and to G. Koufos (University of Thessaloniki) for the Lesvos specimen, A. Athanassiou (Hellenic Ministry of Culture) for the Sesklon one, and C. Fassoulas (Natural History Museum of Crete) and G. Iliopoulos (University of Patras) for the Gela one. We greatly thank G. Markov (National Museum of Natural History, Sofia) and A. Athanassiou for their comments and suggestions that improved the manuscript. GK thanks for providing important data V. Titov (Southern Scientific Centre RAS, Rostov-on-Don) about the holotype of *Mammut obliquelophus*, M. Pavia (University of Torino) about the holotype of *M. borsoni*, and D. Lambert (Holy Innocents' Episcopal School, Atlanta), C. Beard, and M. Sims (University of Kansas) about *Konobelodon britti*. GK thanks G. Koufos and D. Kostopoulos (University of Thessaloniki) for all the scientific support during his research on Neogene proboscideans. GK is supported by the ERC CoG no. 724703 (CROSSROADS).

Appendix

List of Greek localities containing Neogene and Early Pleistocene proboscidean fossils. Type localities are marked with bold. Locality numbers refer to the collection numbers of the PaleoBiology Database (PBDB)

Localities[PBDB No]	Age (MN; GPTS in Ma)	Taxon
Vatera F[183341]	Early Pleistocene (MNQ17)	*Anancus* cf. *arvernensis*[1]
Vatera H[183343]	Early Pleistocene (MNQ17)	*Anancus arvernensis*[1]
Chorigos	Pliocene–Early Pleistocene	*Mammut borsoni*[2]
Neapolis-Grevena basin	Pliocene–Early Pleistocene	*Mammut borsoni*[3]
Sani[202529]	Pliocene–Early Pleistocene	*Anancus arvernensis*[4]
Axios valley ("Dormislou", perhaps close to Gephyra)	Pliocene–Early Pleistocene	*Anancus arvernensis*[5]
Kalliphytos[202532]	Pliocene–Early Pleistocene	*Anancus arvernensis*[6]
Klima	Pliocene–Early Pleistocene	*Anancus arvernensis*[7]
Antimachia[207130]	Pliocene–Early Pleistocene	*Anancus arvernensis*[8]
R. Almyri	Pliocene–Early Pleistocene	*Anancus arvernensis*[9]
Kardamaena[204662]	Pliocene–Early Pleistocene	*Anancus arvernensis*[10]
Pylos	Pliocene–Early Pleistocene	*Anancus arvernensis*[11]
Skoura	Pliocene–Early Pleistocene	*Anancus arvernensis*[12]
Spaides (Eleonas)	Pliocene–Early Pleistocene	*Anancus arvernensis*[13,14]
Nigrita[202533]	Pliocene–Early Pleistocene	*Anancus arvernensis*[15]
Chilia Dendra, Koufalia	Pliocene–Early Pleistocene	*Anancus arvernensis*[b]
Gephyra-1[182685]	Late Pliocene (MN16)	*Anancus arvernensis*[16]

Localities[PBDB No]	Age (MN; GPTS in Ma)	Taxon
Sesklon[34614]	Late Pliocene (MN16)	*Anancus arvernensis*[6,17–18]
Milia-1[187636]	Late Pliocene (MN 16a)	*Mammut borsoni*[19–20]
Milia-2[185862]	Late Pliocene (MN 16a)	*Mammut borsoni*[20], *Anancus arvernensis*[20]
Milia-3[195291]	Late Pliocene (MN 16a)	*Mammut borsoni*[20]
Milia-4[182686]	Late Pliocene (MN 16a)	*Mammut borsoni*[20]
Milia-5[185859]	Late Pliocene (MN 16a)	*Mammut borsoni*[20], *Anancus arvernensis*[20]
Milia-6[195292]	Late Pliocene (MN 16a)	*Mammut borsoni*[20]
Milia-7[195294]	Late Pliocene (MN 16a)	*Mammut borsoni*[20]
Milia-8[195295]	Late Pliocene (MN 16a)	*Mammut borsoni*[20], *Anancus arvernensis*[20]
Milia-9[195296]	Late Pliocene (MN 16a)	*Mammut borsoni*[20]
Milia-10[185860]	Late Pliocene (MN 16a)	*Mammut borsoni*[20]
Milia-11[195299]	Late Pliocene (MN 16a)	*Mammut borsoni*[20]
Milia[202530]	Late Pliocene	*Anancus arvernensis*[4,a]
Apolakkia I[184242]	Pliocene	*Anancus arvernensis*[14]
Apolakkia II[202499]	Pliocene	Proboscidea indet.[14]
Agia Triada[202531]	Pliocene	*Anancus arvernensis*[4]
Angelochori, Angelon beach	Pliocene	*Anancus arvernensis*[c]
Maramena[32189]	Miocene/Pliocene (MN13/14)	*Choerolophodon pentelici*[21]
Dikaia[210641]	Late Miocene	*Choerolophodon pentelici*[22]
Pyrgos Vassilissis[195555]	Late Miocene	*Choerolophodon pentelici*[23]
Sani[202528]	Late Miocene	*Choerolophodon pentelici*[4]
Servia[202257]	Late Miocene	*Choerolophodon pentelici*[24], *Deinotherium* sp.[24]
Fourka[202330]	Late Miocene	*Tetralophodon longirostris*[25]
Chelona beach[202500]	Late Miocene	*Tetralophodon longirostris*[25]
seabed of Kryopigi[202500]	Late Miocene	*Tetralophodon longirostris*[25]
Neokaisareia[206458]	Turolian	*Mammut* sp. (*M. obliquelophus*?)[26]
Palaio Keramidi	Turolian	*Mammut* sp. (*M. obliquelophus*?)[26]
Dytiko 2[32375]	Late Turolian (MN13)	*Choerolophodon pentelici*[27]
Dytiko 3[32376]	Late Turolian (MN13)	*Choerolophodon pentelici*[27]
Kryopigi[157582]	Late Miocene (MN12–13)	*Choerolophodon pentelici*[28]

Localities[PBDB No]	Age (MN; GPTS in Ma)	Taxon
Mytilinii 1A-Samos[202215]	Middle Turolian (MN12; ~7.1)	*Mammut* sp. (*M. obliquelophus*?)[29–30]
Mytilinii 1B-Samos[202216]	Middle Turolian (MN12; ~7.1)	*Choerolophodon pentelici*[27,29,30]
Andriano-Samos	Middle Turolian (MN12; ~7.1)	*Choerolophodon pentelici*[30,32], *Deinotherium proavum*[30,32]
Chomateri[195562]	Middle Turolian (MN12; ~7.16)	*Anancus lehmanni*[33–34]
Perivolaki[194879]	Middle Turolian (MN12; 7.3–7.1)	*Deinotherium proavum*[33,35]
Pikermi[182754]	**Middle Turolian (MN12; ~7.3)**	***Choerolophodon pentelici*[27,30,33,36–38]**, *Deinotherium proavum*[33,36–37,39–44], *Mammut* sp. (*M. obliquelophus*?)[33,36–38], ***Konobelodon atticus*[33,36–38,40,45]**
Pikermi Valley-1[202630]	Middle Turolian (MN12)	*Mammut* sp. (*M. obliquelophus*?)[46]
Pikermi Valley-3[202631]	Middle Turolian (MN12)	*Deinotherium proavum*[44]
Halmyropotamos[202213]	?middle Turolian (?MN12)	*Deinotherium proavum*[44,47], *Mammut* sp. (*M. obliquelophus*?)[33,48]
Prochoma-1[202222]	Middle Turolian (MN12)	*Choerolophodon pentelici*[31,33]
Vathylakkos-2[202703]	Middle Turolian (MN12)	*Choerolophodon pentelici*[31,33]
Kerassia-1	?middle Turolian (?MN 12)	*Choerolophodon*? sp.[49]
Kerassia-4[195435]	?middle Turolian (?MN 12)	*Konobelodon atticus*[45,49]
Kerassia-6[195437]	?middle Turolian (?MN 12)	*Choerolophodon* sp.[49]
Kerassia[195431]	?middle Turolian (?MN 12)	*Deinotherium* sp.[50]
Ravin des Zouaves 5[195489]	Early Turolian (MN11; ~8.2)	*Mammut* sp. (*M. obliquelophus*?)[30,33,51], *Choerolophodon pentelici*[30,33,51]
Maronia, Crete[202722]	Turolian (MN11–13)	*Deinotherium proavum*[52]
Gela, Crete[202723]	Turolian (MN11–13)	*Deinotherium proavum*[53–54]
Zakros, Crete	Turolian (MN11–13)	*Deinotherium proavum*[54–55]
Samos Island, NHMW collection[182751] MGL collection[202120] HGI collection HLMD collection[202724] SMF collection[202725]	Turolian (MN11–13)	*Choerolophodon pentelici*[29–33,56–58], *Konobelodon atticus*[30,32,33,45,58], *Deinotherium proavum*[30,32,33]
Nikiti-2[73869]	Early Turolian (MN11; 8.7–8.2)	*Choerolophodon pentelici*[30,31,33,59]

Localities[PBDB No]	Age (MN; GPTS in Ma)	Taxon
Ravin X[182745]	?early Turolian (?MN11)	*Choerolophodon pentelici*[30–31]
Platania, Drama[182682]	Late Vallesian/early Turolian (MN10/ MN11)	*Konobelodon* cf. *atticus*[26]
Ravin de la Pluie[191070]	Late Vallesian (MN10; ~9.3)	*Deinotherium giganteum*[30,33], *Choerolophodon pentelici*[30–31,33]
Xirochori 1[195490]	Late Vallesian (MN10; ~9.6)	*Choerolophodon pentelici*[30–31,33]
Ravin des Zouaves 1[182746]	Late Vallesian (MN10)	*Choerolophodon pentelici*[30–31,33]
Agia Paraskevi, Kassandra[202726]	Late Miocene	*Deinotherium giganteum*[60]
Pentalophos 1[202119]	Early Vallesian (MN9)	*Choerolophodon anatolicus*[30–31,33]
Thymiana[182752]	**Middle Miocene (MN5; >15.5)**	***Choerolophodon chioticus*[27,61–62]**, *Prodeinotherium bavaricum*[61]
Psara Island	?middle Miocene	*Prodeinotherium bavaricum*[63]
Gavathas, Lesvos Island[195540]	Early Miocene (MN3; >18.4)	*Prodeinotherium cuvieri*[64,a]
Thermopigi[73553]	Turolian (MN11–13)	*Deinotherium* sp.[26], Elephantimorpha indet.[26]
Nikiti-1[202729]	Late Vallesian (MN10)	Proboscidea indet.[59]
Antonios[73861]	Early/middle Miocene (MN4/5)	Proboscidea indet.[65]
Central Macedonia	?	*Deinotherium giganteum*[66]
Vathylakkos area	?	*Mammut borsoni*[11]
possibly from the wider area of Siatista (Kozani)[202732]	?	*Stegodon* sp.[67]

[a]This study

[b]A right hemimandible fragment with the m3 (partially broken), stored at the Ephorate of Palaeoanthropology and Speleology, Office of Northern Greece (Thessaloniki) (pers. comm. Athanassiou 2019)

[C]A small fragment of a molar (LGPUT-AAT 17) and a medial condyle of a right femur (LGPUT-AAT 16) were collected (surface findings/unknown precise stratigraphic position) from an area near the fossil findspot of Angelochori, Angelon beach (Thessaloniki), for which Tsoukala (2018) proposed a correlation to the late Pliocene (MN 16)

[1]de Vos et al. 2002, [2]Dermitzakis et al. 1982, [3]Brunn 1956, [4]Koufos 1977, [5]Sakellariou-Mane 1972, [6]Athanassiou 2016, [7]Steensma 1988, [8]Forsyth Major 1887, [9]Desio 1931, [10]Charrier and Giglio 1969, [11]Mitzopoulos 1967, [12]Georgalas 1941, [13]Gidarakos 1938, [14]Theodorou et al. 2000, [15]Athanassiou 2017, [16]Crégut-Bonnoure and Tsoukala 2017, [17]Symeonidis and Tataris 1983, [18]Athanassiou 2018, [19]Tsoukala 2000, [20]Tsoukala and Mol 2016, [21]Schmidt-Kittler et al. 1995, [22]Tsoukala et al. 2007, [23]Freyberg 1951, [24]Paraskevaidis 1977, [25]Lazaridis and Tsoukala 2014b, [26]Konidaris and Tsoukala 2020, [27]Konidaris et al. 2016, [28]Lazaridis and Tsoukala 2014a, [29]Konidaris and Koufos 2009, [30]Konidaris and Koufos 2013a, [31]Konidaris et al. 2016, [32]Konidaris and Koufos 2019, [33]Konidaris 2013; [34]Konidaris and Roussiakis 2019, [35]Koufos 2006, [36]Gaudry 1862, [37]Gaudry 1867; [38]Tassy 1985, [39]Wagner 1848, [40]Wagner 1857, [41]Weinsheimer 1883, [42]Dietrich 1916, [43]Weithofer 1888, [44]Konidaris et al. 2017, [45]Konidaris et al. 2014, [46]Roussiakis et al. 2014, [47]Melentis 1969, [48]Melentis 1967, [49]Theodorou et al. 2003, [50]van der Made and Moyà-Solà 1989, [51]Koufos 1980, [52]Athanassiou 2004, [53]Poulakakis et al. 2005, [54]Iliopoulos et al. 2014, [55]Fassoulas and Iliopoulos 2011, [56]Schlesinger 1917, [57]Schlesinger 1922, [58]Lehmann 1950, [59]Konidaris and Koufos 2016, [60]Tsoukala and Melentis 1994, [61]Paraskevaidis 1940, [62]Tobien 1980, [63]Besenecker and Symeonidis 1974, [64]Koufos et al. 2003, [65]Koufos and Syrides 1997, [66]Symeonidis 1969, [67]Mol et al. 2010

References

Aiba H, Baba K, Matsukawa M (2010) A new species of *Stegodon* (Mammalia, Proboscidea) from the Kazusa Group (lower Pleistocene), Hachioji City, Tokyo, Japan and its evolutionary morphodynamics. Palaeontology 53:471–490

Aiglstorfer M, Göhlich UB, Böhme M, Gross M (2014) A partial skeleton of *Deinotherium* (Proboscidea, Mammalia) from the late Middle Miocene Gratkorn locality (Austria). Palaeobiodivers Palaeoenviron 94:49–70

Andrews CW (1911) On a new species of *Dinotherium* (*Dinotherium hobleyi*) from British East Africa. Proc Zool Soc London 81:943–945

Antoine PO, Welcomme JL, Marivaux L, Baloch I, Benammi M, Tassy P (2003) First record of Paleogene Elephantoidea (Mammalia, Proboscidea) from the Bugti Hills of Pakistan. J Vertebr Paleontol 23:977–980

Arambourg C (1934) Le *Dinotherium* des gisements de l'Omo. C R Soc Géol Fr 1934:86–87

Arambourg C (1945) *Anancus osiris*, un mastodonte nouveau du Pliocène inférieur d'Egypte. Bull Soc Géol Fr 15:479–495

Arambourg C, Piveteau J (1929) Les Vertébrés du Pontien de Salonique. Ann Paléontol 18:59–138

Athanassiou A (2004) On a *Deinotherium* (Proboscidea) finding in the Neogene of Crete. Carnets Géol 2004(05):1–7

Athanassiou A (2016) Craniomandibular remains of *Anancus arvernensis* (Proboscidea, Mammalia) from Greece: The samples from Kallíphytos (E. Macedonia) and Sésklo (Thessaly). Quat Int 406:25–34

Athanassiou A (2017) A cranial specimen of *Anancus arvernensis* from Nigríta, Northern Greece. 7th International Conference on Mammoths and their Relatives, Taichung

Athanassiou A (2018) A Villafranchian *Hipparion*-bearing mammal fauna from Sésklo (E. Thessaly, Greece): implications for the question of *Hipparion*–*Equus* sympatry in Europe. Quaternary 1:1–24

Athanassiou A (this volume) The fossil record of non-endemic elephants and mammoths (Mammalia: Proboscidea: Elephantidae) in Greece. In: Vlachos E (ed) The fossil vertebrates of Greece vol. 1 – basal vertebrates, amphibians, reptiles, afrotherians, glires, and primates. Springer, Cham

Barbour EH (1927) Preliminary notice of a new proboscidean *Amebelodon fricki*, gen. et sp. nov. Bull Nebr State Museum 13:131–134

Barbour EH (1929) *Torynobelodon loomisi*, gen. et sp. nov. Bull Nebr State Museum 16:147–153

Barbour EH, Hibbard CW (1941) A shovel-tusked mastodon, *Amebelodon fricki*, from Kansas. Bull Univ Nebr State Museum 2:37–46

Bergounioux FM, Crouzel F (1960) *Tetralophodon curvirostris* n. sp. (Mamm., Proboscidea) aus dem Unterpliozän (Pontien) von Esselborn (Rheinhessen). Jahresber Mitt Oberrhein Geol Ver 42:109–121

Bergounioux FM, Crouzel F (1962) Les Déinothéridés d'Europe. Ann Paléontol 48:1–56

Besenecker H, Symeonidis NK (1974) Der erste Säugetierfund aus dem Neogen der griechischen Insel Psara (Ostägäis). Ann Géol Pays Hellén 26:109–117

Blumenbach JF (1799) Handbuch der Naturgeschichte, 6th edn. Dieterich, Göttingen

Boeuf O (1992) *Anancus arvernensis chilhiacensis* nov. subsp. (Proboscidea, Mammalia), un Mastodonte du Plio-Pléistocène de Haute-Loire, France. Geobios 14:179–188

Böhme M, Aiglstorfer M, Uhl D, Kullmer O (2012) The antiquity of the Rhine River: stratigraphic coverage of the *Dinotheriensande* (Eppelsheim Formation) of the Mainz Basin (Germany). PLoS One 7:e36817

Böhme M, Spassov N, Ebner M, Geraads D, Hristova L, Kirscher U, Kötter S, Linnemann U, Prieto J, Roussiakis S, Theodorou G, Uhlig G, Winklhofer M (2017) Messinian age and savannah environment of the possible hominin *Graecopithecus* from Europe. PLoS One 12:e0177347

Bonaparte CL (1845) Catalogo Metodico dei Mammiferi Europei. Luigi di Giacomo Pirola, Milan

Borson E (1823) Note sur des dents du grand mastodonte trouvées en Piémont et sur des machoires et dents fossiles prises dans la mine de houille de Cadibona proche Savone. Mem Reale Acad Sci Torino 27:31–42

Brunn JH (1956) Contribution à l'étude géologique du Pinde septentrional et d'une partie de la Macédoine occidentale. Ann Géol Pays Hellén 7:1–358

Cautley PT (1836) Note on the teeth of the mastodon à dents etroites of the Siwálik Hills. J Asiat Soc Bengal 5:294–296

Charrier G, Giglio A (1969) Primi risultati di una campagna di rilevamento geologico nell'isola di Coo (Sporadi meridionali – Mare Egeo). Boll Assoc Min Subalp 6:482–516

Chow MC, Chang YP (1974) Chinese fossil elephantoids (in Chinese). Science Press, Beijing

Clift W (1828) On the fossil remains of two new species of *Mastodon*, and of other vertebrated animals, found on the left bank of the Irawadi. Trans Geol Soc Lond 2:369–375

Codrea V (1994) A priority issue: *Deinotherium proavum* Eichwald or *Deinotherium gigantissimum* Ştefănescu? In: Nicorici E (ed) The Miocene from the Transylvanian Basin-Romania, Cluj-Napoca, pp 105–110

Cooper LN, Seiffert ER, Clementz M, Madar SI, Bajpai S, Hussain ST, Thewissen JGM (2014) Anthracobunids from the Middle Eocene of India and Pakistan are stem perissodactyls. PLoS One 9:e109232

Coppens Y (1965) Les proboscidiens du Tchad. In: Actes du Ve Congrés Panafricain de Préhistoire et de l'Étude du Quaternaire, Santa Cruz de Tenerife, pp 331–387

Crégut-Bonnoure E, Tsoukala E (2017) The Pliocene Artiodactyla and Proboscidea (Mammalia) from Gephyra (lower Axios valley, Macedonia, Greece). Discovery of a new boselaphine. Quat Int 445:200–214

Croizet JB, Jobert ACG (1828) Recherches sur les ossemens fossiles du département du Puy-de-Dôme. Principaux Libraries, Paris

de Bonis L, Brunet M, Heintz E, Sen S (1992) La province greco-irano-afghane et la réparition des faunes mammaliennes au Miocène supérieur. Paleontol i Evol 24–25:103–112

de Vos J, van der Made J, Athanassiou A, Lyras G, Sondaar PY, Dermitzakis MD (2002) Preliminary note on the Late Pliocene fauna from Vatera (Lesvos, Greece). Ann Géol Pays Hellén 39:37–70

Dermitzakis MD, Symeonidis NK, De Boer LEM, Sondaar PY (1982) The evolution of the elephants (in Greek). Edit Lab Geol Palaeont, Univ Athens

Desio A (1931) Le isole Italiane dell'Egeo. Mem Descrittive Carta Geol Ital 24:1–534

Dietrich WO (1916) Über die Hand und den Fuss von *Dinotherium*. Zeitschr D Geol Ges 68:44–53

Dooley ACJ, Scott E, Green J, Springer KB, Dooley BS, Smith GJ (2019) *Mammut pacificus* sp. nov., a newly recognized species of mastodon from the Pleistocene of western North America. Peer J 7:e6614

Dorlhac MJ (1855) Notice géologique sur le cratère de Coupet et sur son gisement de gemmes et d'ossements fossiles. Ann Soc Agric Sci Arts Commer Puy 19:497–517

Éhik J (1930) *Prodinotherium hungaricum* n. g., sp., with an appendix by Szalay, T.: on the geological occurrence of *Prodinotheirum hungaricum* ÉHIK. Geol Hung, Ser Palaeontol 6:1–24

Eichwald E (1831) Zoologia specialis quam expositis animalibus tum vivis, tum fossilibus potissimum Rossiae in universum, et Poloniae in specie. Josephi Zawadzki, Vilnae

Eichwald E (1835) De pecorum et pachydermorum reliquiis fossilibus in Lithuania, Volhynia et Podolia repertis. Nova Acta Acad Caesareae Leopoldino-Carolinae Ger Nat Curiosorum 17:677–780

Falconer H (1845) Description of some fossil remains of *Dinotherium*, Giraffe, and other Mammalia from the gulf of Cambay, western Coast of India. Q J Geol Soc Lond 1:356–372

Falconer H (1857) On the species of mastodon and elephant occurring in the fossil state in Great Britain. Part I. *Mastodon*. Q J Geol Soc Lond 13:307–360

Falconer H (1868) Description of some fossil remains of *Dinotherium*, Giraffe, and other Mammalia, from Perim Island, Gulf of Cambay, western coast of India. In: Murchison C (ed) Palaeontological memoirs and notes of the late Hugh Falconer, A.M., M.D, London, pp 391–411

Falconer H, Cautley PT (1847) Fauna antiqua sivalensis, being the fossil zoology of the Sewalik Hills, in the North of India. Smith, Elder and Co., London

Fassoulas C, Iliopoulos G (2011) The excavations of *Deinotherium giganteum* from Siteia: life and environment in Crete during the Miocene. In: Proceedings of the 10th international cretological congress, Chania, pp 11–25

Ferretti MP, Rook L, Torre D (2003) *Stegotetrabelodon* (Proboscidea, Elephantidae) from the late Miocene of southern Italy. J Vertebr Paleontol 23:659–666

Forsyth Major CJ (1887) Faune mammalogische delle isole di Cos et di Samos. Atti Soc Tosc Sci Nat 5:272–275

Forsyth Major CJ (1888) Sur un gisement d'ossements fossiles dans l'île de Samos, contemporains de l'âge de Pikermi. C R Hebd Séances Acad Sci 107:1178–1181

Forsyth Major CJ (1894) Le gisement ossifère de Mytilinii et catalogue d'ossements fossiles recueillis à Mitylini, île de Samos, et déposés au Collège Galliard à Lausanne. Georges Bridel and Cie, Lausanne

Freyberg BV (1951) Die Pikermifauna von Tour la Reine. Ann Géol Pays Hellén 3:7–10

Gaudry A (1862) Animaux fossiles et géologie de l'Attique. Savy, Paris

Gaudry A (1867) Animaux fossiles et géologie de l'Attique. Atlas, Savy, Paris

Gaudry A, Lartet E (1856) Résultats des recherches paléontologiques entreprises dans l'Attique sous les auspices de l'Académie. C R Acad Sci Paris 43:271–274

Gaziry AW (1976) Jungtertiäre Mastodonten aus Anatolien (Türkei). Geol Jb 22:3–143

Gaziry AW (1987a) New mammals from the Jabal Zaltan site, Libya. Senckenb Lethaea 68:69–89

Gaziry AW (1987b) Remains of Proboscidea from the early Pliocene of Sahabi, Libya. In: Boaz NT, El-Arnauti A, Gaziry AW, Heinzelin JD, Boaz DD (eds) Neogene paleontology and geology of Sahabi. A. R. Liss, New York, pp 183–203

Gaziry AW (1997) Die Mastodonten (Proboscidea, Mammalia) aus Dorn-Dürkheim 1 (Rheinhessen). Cour Forsch-Inst Senck 197:73–115

Georgalas CG (1941) Über das Vorkommen von *Anancus* (*Mastodon*) *arvernensis* Croiz. und Job. in der Umgebung von Skoura (SÖ von Sparta). Prakt Akad Ath 16:94–100

Gheerbrant E (2009) Paleocene emergence of elephant relatives and the rapid radiation of African ungulates. PNAS 106:10717–10721

Gheerbrant E, Tassy P (2009) L'origine et l'évolution des éléphants. C R Palevol 8:281–294

Gheerbrant E, Sudre J, Cappetta H (1996) A Palaeocene proboscidean from Morocco. Nature 383:68–70

Gheerbrant E, Sudre J, Cappetta H, Iarochéne M, Amaghzaz M, Bouya B (2002) A new large mammal from the Ypresian of Morocco: evidence of surprising diversity of early proboscideans. Acta Palaeontol Pol 47:493–506

Gheerbrant E, Sudre J, Tassy P, Amaghzaz M, Bouya B, Iarochène M (2005) Nouvelles données sur *Phosphatherium escuilliei* (Mammalia, Proboscidea) de l'Éocène inférieur du Maroc, apports à la phylogénie des Proboscidea et des ongulés lophodontes. Geodiversitas 27:239–333

Gheerbrant E, Schmitt A, Kocsis L (2018) Early African fossils elucidate the origin of embrithopod mammals. Curr Biol 28:2167–2173

Gidarakos D (1938) Trouvailles paléontologiques des mammifères en Béotie. Prakt Akad Ath 13:418–422

Ginsburg L, Chevrier F (2001) Les Dinothères du bassin de la Loire et l'évolution du genre *Deinotherium* en France. Symbioses 5:9–24

Gräf IE (1957) Die Prinzipien der Artbestimmung bei *Dinotherium*. Palaeontographica 108:131–185

Gregory JT (1945) An *Amebelodon* jaw from the Texas Panhandle. Univ Texas Publ 4401:477–484

Harris JM (1973) *Prodeinotherium* from Gebel Zelten, Libya. Bull Brit Mus (Nat Hist) (Geol) 23:285–348

Hautier L, Mackaye HT, Lihoreau F, Tassy P, Vignaud P, Brunet M (2009) New material of *Anancus kenyensis* (Proboscidea, Mammalia) from Toros-Menalla (Late Miocene, Chad): contribution to the systematics of African anancines. J Afr Earth Sci 53:171–176

Hay OP (1922) Further observations on some extinct elephants. Proc Biol Soc Wash 35:97–101
Hays I (1834) Descriptions of the specimens of inferior maxillary bones of mastodons in the cabinet of the American Philosophical Society, with remarks on the genus *Tetracaulodon* (Godman), etc. Trans Am Philos Soc 4:317–339
Hopwood AT (1935) Fossil Proboscidea from China. Palæont Sin 9:1–108
Huttunen K (2002) Systematics and taxonomy of the European Deinotheriidae (Proboscidea, Mammalia). Ann Naturhist Mus Wien 103(A):237–250
Huttunen K (2004) On a *Prodeinotherium bavaricum* (Proboscidea, Mammalia) skeleton from Franzensbad, Czech Republic. Ann Nat Mus Wien 105(A):333–361
Huttunen K, Göhlich UB (2002) A partial skeleton of *Prodeinotherium bavaricum* (Proboscidea, Mammalia) from the Middle Miocene of Unterzolling (Upper Freshwater Molasse, Germany). Geobios 35:489–514
Iliopoulos G, Fassoulas C, Tzortzi M (2014) An almost complete skeleton of a large *Deinotherium* (Proboscidea, Mammalia) from the Late Miocene of Aghia Photia, Siteia (Crete Island, Greece). In: Kostopoulos DS, Vlachos E, Tsoukala E (eds) 6th International Conference on Mammoths and their Relatives, Grevena-Siatista. Sci Annals, School Geol, Aristotle Univ Thessaloniki, Greece, sp vol 102: 72–73
Illiger C (1811) Prodromus systematis mammalium et avium additis terminis zoographicis utriusque classis. C. Salfield, Berlin
Jourdan M (1861) Des terrains sidérolitiques. C R Hebd Séances Acad Sci 53:1009–1014
Kaup JJ (1829) *Deinotherium giganteum.* Eine Gattung der Vorwelt aus der Ordnung der Pachydermen. Isis 22:401–404
Kaup JJ (1832a) Description d'ossements fossiles de mammifères inconnus jusqu'à présent, qui se trouvent au Muséum grand-ducal de Darmstadt. J.G. Heyer, Darmstadt
Kaup J (1832b) Ueber zwei Fragmente eines Unterkiefers von *Mastodon angustidens* Cuv., nach welchen diese Art in die Gattung Tetracaulodon Godmann gehört. Isis 25:628–631
Kaup JJ (1835) Description d'ossements fossiles de mammifères inconnus jusqu'à-présent qui se trouvent au Muséum grand-ducal de Darmstadt. J.P. Diehl, Darmstadt
Kerr R (1792) The animal kingdom, or zoological system, of the celebrated Sir Charles Linnaeus, Edinburgh
Klähn H (1922) Die badischen Mastodonten und ihre süddeutschen Verwandten. Borntraeger, Berlin
Konidaris GE (2013) Palaeontological and biostratigraphical study of the Neogene Proboscidea from Greece. Ph.D. thesis, Sci Annals, School Geol, Aristotle Univ Thessaloniki, Greece 153, 1–326
Konidaris GE, Koufos GD (2009) The late Miocene mammal faunas of the Mytilinii Basin, Samos Island, Greece: new collection. 8. Proboscidea. In: Koufos GD, Nagel D (eds) The late Miocene Mammal Faunas of Samos, vol 31. Beit Paläontol Wien, pp 139–155
Konidaris GE, Koufos GD (2013a) Late Miocene Proboscidea (Mammalia) from Macedonia and Samos Island, Greece: preliminary results. Paläont Z 87:121–140
Konidaris GE, Koufos GD (2013b) The late Miocene proboscideans of Greece: taxonomy, biochronology, palaeoecology. 14th Congress of the Regional Committee on Mediterranean Stratigraphy (RCMNS), Istanbul
Konidaris GE, Koufos GD (2016) Proboscidea. In: Koufos GD, Kostopoulos DS (eds) Paleontology of the upper Miocene vertebrate localities of Nikiti (Chalkidiki Peninsula, Macedonia, Greece). Geobios 49:37–44
Konidaris GE, Koufos GD (2019) Late Miocene proboscideans from Samos Island (Greece) revisited: new specimens from old collections. Paläont Z 93:115–134
Konidaris GE, Roussiakis SJ (2017) *Anancus* (Proboscidea, Mammalia) from Chomateri – the first record of the genus in the late Miocene of Greece. 15th Congress of the Regional Committee on Mediterranean Neogene Stratigraphy (RCMNS), Athens
Konidaris GE, Roussiakis SJ (2019) The first record of *Anancus* (Mammalia, Proboscidea) in the late Miocene of Greece and reappraisal of the primitive anancines from Europe. J Vertebr Palaeontol:e1534118

Konidaris GE, Tsoukala E (2020) Proboscideans from the upper Miocene localities of Thermopigi, Neokaisareia and Platania (Northern Greece). Ann Paléontol 106:102380

Konidaris GE, Roussiakis SJ, Theodorou GE, Koufos GD (2014) The Eurasian occurrence of the shovel-tusker *Konobelodon* (Mammalia, Proboscidea) as illuminated by its presence in the late Miocene of Pikermi (Greece). J Vertebr Paleontol 34:1437–1453

Konidaris GE, Koufos GD, Kostopoulos DS, Merceron G (2016) Taxonomy, biostratigraphy and palaeoecology of *Choerolophodon* (Proboscidea, Mammalia) in the Miocene of SE Europe-SW Asia: implications for phylogeny and biogeography. J Syst Palaeontol 14:1–27

Konidaris GE, Roussiakis SJ, Athanassiou A, Theodorou GE (2017) The huge-sized deinothere *Deinotherium proavum* (Proboscidea, Mammalia) from the Late Miocene localities Pikermi and Halmyropotamos (Greece). Quat Int 430:5–21

Kostopoulos DS (2009) The Pikermian Event: temporal and spatial resolution of the Turolian large mammal fauna in SE Europe. Palaeogeogr Palaeoclimatol Palaeoecol 274:82–95

Koufos GD (1977) New finds of mastodonts from Macedonia (Greece). Sci Ann Fac Phys Mathem Univ Thessaloniki 17:97–115

Koufos GD (1980) Palaeontological and stratigraphical study of the continental Neogene deposits of Axios Basin. Ph.D. thesis, Sci Annals, Fac Phys & Mathem, Univ Thessaloniki, Greece 19:1–322

Koufos GD (2006) The Neogene mammal localities of Greece: Faunas, chronology and biostratigraphy. Hellenic J Geosci 41:183–214

Koufos GD (2013) Neogene mammal biostratigraphy and chronology of Greece. In: Wang X, Flynn LJ, Fortelius M (eds) Fossil mammals of Asia. Neogene biostratigraphy and chronology. Columbia University Press, New York, pp 596–621

Koufos GD, Kostopoulos DS (eds) (2016) Palaeontology of the upper Miocene vertebrate localities of Nikiti (Chalkidiki Peninsula, Macedonia, Greece). Geobios 49(1–2):1–154

Koufos GD, Nagel D (eds) (2009) The Late Miocene mammal faunas of Samos, vol 31. Beit Paläontol Wien

Koufos GD, Syrides GE (1997) A new Early/Middle Miocene mammal locality from Macedonia, Greece. C R Acad Sci Paris 325:511–516

Koufos GD, Zouros N, Mourouzidou O (2003) *Prodeinotherium bavaricum* (Proboscidea, Mammalia) from Lesvos island, Greece; the appearance of deinotheres in the eastern Mediterranean. Geobios 36:305–315

Koufos GD, Kostopoulos DS, Vlachou TD (2005) Neogene/Quaternary mammalian migrations in Eastern Mediterranean. Belgian J Zool 135:181–190

Kovachev D, Nikolov I (2006) *Deinotherium thraceiensis* sp. nov. from the Miocene near Ezerovo, Plovdiv District. Geol Balc 35:5–40

Kubiak H (1972) The skull of *Mammut praetypicum* (Proboscidea, Mammalia) from the collection of the Jagiellonian University in Cracow. Poland Acta Zool Cracoviensia 17:305–324

Lambert WD (1990) Rediagnosis of the genus *Amebelodon* (Mammalia, Proboscidea, Gomphotheriidae), with a new subgenus and species, *Amebelodon* (*Konobelodon*) *britti*. J Paleontol 64:1032–1040

Lambert WD, Shoshani J (1998) Proboscidea. In: Janis CM, Scott KM, Jacobs LL (eds) Evolution of tertiary mammals of North America. Cambridge University Press, Cambridge, pp 606–621

Lartet E (1859) Sur la dentition des proboscidiens fossiles (*Dinotherium*, Mastodontes et Éléphants) et sur la distribution géographique et stratigraphique de leurs débris en Europe. Bull Soc Géol Fr 16:469–515

Lazaridis G (2015) Study of the Late Miocene vertebrate locality of Kryopigi and other localities of Kassandra Peninsula, Chalkidiki (Greece). Systematics, taphonomy, paleoecology, biochronology. Ph.D. thesis, Sci Annals, School Geol, Aristotle Univ Thessaloniki, Greece 174, 1–355

Lazaridis G, Tsoukala E (2014a) *Choerolophodon pentelici* (Gaudry and Lartet, 1856) from the Turolian locality of Kryopigi (Kassandra, Chalkidiki, Greece). In: Kostopoulos DS, Vlachos E, Tsoukala E (eds) 6th international conference on mammoths and their relatives, Grevena-Siatista. Sci Annals, School Geol, Aristotle Univ Thessaloniki, Greece, sp vol 102:100

Lazaridis G, Tsoukala E (2014b) *Tetralophodon longirostris* (Kaup, 1832) from Late Miocene of the Kassandra peninsula (Chalkidiki, Greece). In: Kostopoulos DS, Vlachos E, Tsoukala E (eds) 6th international conference on mammoths and their relatives, Grevena-Siatista. Sci Annals, School Geol, Aristotle Univ Thessaloniki, Greece, sp vol 102:101

Lazaridis G, Kostopoulos DS, Lyras G, Roussiakis S (2017) A new Late Miocene ovibovine-like bovid (Bovidae, Mammalia) from the Kassandra Peninsula (Chalkidiki, Northern Greece) and implications to the phylogeography of the group. Paläontol Z 91:427–437

Lehmann U (1950) Über Mastodontenreste in der Bayerischen Staatssammlung in München. Palaeontogr Abt A 99:121–228

Lydekker R (1884) Additional Siwalik Perissodactyla and Proboscidea Memoirs of the Geological Survey of India. Palaeontol Indica 10:1–34

MacInnes DG (1942) Miocene and Post-Miocene Proboscidea from East Africa. Trans Zool Soc Lond 25:33–106

Maglio VJ (1974) A new proboscidean from the late Miocene of Kenya. Palaeontology 17:699–705

Markov GN (2008a) Fossil proboscideans (Mammalia) from the vicinities of Varna: a rare indication of middle Miocene vertebrate fauna in Bulgaria. Hist Nat Bulg 19:137–152

Markov GN (2008b) The Turolian proboscideans (Mammalia) of Europe: preliminary observations. Hist Nat Bulg 19:153–178

Markov GN, Vergiev S (2010) First report of cf. *Protanancus* (Mammalia, Proboscidea, Amebelodontidae) from Europe. Geodiversitas 32:493–500

Matsumoto H (1927) On two new mastodonts and an archetypal stegodont of Japan. Sci Rep Tohoku Imperial Univ 10:1–11

Mayet L (1908) Étude des mammifères miocènes des Sables de l'Orléanais et des Faluns de la Touraine. Ann Univ Lyon 24:1–336

Mazo AV, Montoya P (2003) Proboscidea (Mammalia) from the Upper Miocene of Crevillente (Alicante, Spain). Scr Geol 126:79–109

Mebrate A (1987) The long-jawed gomphotheres. Ph.D. thesis, University of Kansas

Melentis JK (1967) Die Pikermi-Fauna von Halmyropotamos (Euböa, Griechenland), 1. Teil: Odontologie und Kraniologie. Ann Géol Pays Hellén 19:283–411

Melentis JK (1969) Die Pikermi-Fauna von Halmyropotamos (Euböa, Griechenland), 2. Teil: Osteologie. Ann Géol Pays Hellén 28:217–306

Metz-Muller F (2000) La population d'*Anancus arvernensis* (Proboscidea, Mammalia) du Pliocène de Dorkovo (Bulgarie); étude des modalités évolutives d'*Anancus arvernensis* et phylogénie du genre *Anancus*. Ph.D. thesis, Muséum National d'Histoire Naturelle

Mitzopoulos MK (1967) *Zygolophodon borsoni* und *Anancus (Bunolophodon) arvernensis* aus dem Oberpliozän von Griechenland. Ann Géol Pays Hellén 18:436–446

Mol D, Tsoukala E, Batsi A (2010) A *Stegodon* from the area of Siatista (Kozani, Macedonia, Greece), the first evidence of the presence of *Stegodon* in Europe. In: 5th International conference on mammoths and their relatives Le Puy-en-Velay. Quaternaire, sp vol 102:85–87

Mothé D, Avilla LS, Asevedo L, Borges-Silva L, Rosas M, Labarca-Encina R, Souberlich R, Soibelzon E, Roman-Carrion JL, Ríos SD, Rincon AD, de Oliveira GC, Lopes RP (2017) Sixty years after 'The mastodonts of Brazil': The state of the art of South American proboscideans (Proboscidea, Gomphotheriidae). Quat Int 443:52–64

Mucha BB (1980) A new species of yoke-toothed mastodont from the Pliocene of Southwest USSR. In: Negadaev-Nikonov KN (ed) Quaternary and Neogene faunas and floras of Moldavskaya SSR. Shtiintsa, Kishinev, pp 19–26

Osborn HF (1918) A long-jawed mastodon skeleton from South Dakota and phylogeny of the Proboscidea. Bull Geol Soc Am Bull 29:133–137

Osborn HF (1921) First appearance of the true mastodon in America. Am Mus Novit 10:1–6

Osborn HF (1926) Additional new genera and species of the mastodontoid Proboscidea. Am Mus Novit 238:1–16

Osborn HF (1936) Proboscidea. A monograph of the discovery, evolution, migration and extinction of the mastodonts and elephants of the world, Volume I: Moeritherioidea, Deinotherioidea, Mastodontoidea. American Museum Press, New York

Ozansoy F (1965) Étude des gisements continentaux et des mammifères du Cénozoïque de Turquie. Mem Soc Géol Fr 102:1–92

Paraskevaidis I (1940) Eine obermiozäne Fauna von Chios. N Jb Min Geol Paläont 83:363–442

Paraskevaidis I (1977) Säugetierreste aus Griechenland. Proceedings of the 6th Colloquium on the Geology of the Aegean Region 3:1143–1154

Pickford M (2001) Afrochoerodon nov. gen. *kisumuensis* (MacInnes) (Proboscidea, Mammalia) from Cheparawa, Middle Miocene, Kenya. Ann Paléontol 87:99–113

Pickford M, Pourabrishami Z (2013) Deciphering Dinotheriensande deinotheriid diversity. Palaeobiodivers Palaeoenviron 93:121–150

Pilgrim GE (1913) The correlation of the Siwaliks with mammal horizons of Europe. Rec Geol Surv India 43:264–326

Pohlig H (1988) Dentition und Kraniologie des *Elephas antiquus* Falc. mit Beiträgen über *Elephas primigenius* Blum. und *Elephas meridionalis* Nesti. Nova Acta Ksl Leop-Carol Deutsch Akad Naturf 53:1–280

Poulakakis N, Lymberakis P, Fassoulas C (2005) *Deinotherium giganteum* (Proboscidea, Deinotheriidae) from the late Miocene of Crete. J Vertebr Paleontol 25:732–736

Qiu ZX, Wang BJ, Li H, Deng T, Sun Y (2007) First discovery of deinothere in China. Vertebr Palas 45:261–277

Rögl F (1999) Circum-Mediterranean Miocene paleogeography. In: Rössner GE, Heissig K (eds) The Miocene land mammals of Europe. Verlag Dr. Friedrich Pfeil, München, pp 39–48

Rook L, Croitor R, Delfino M, Ferretti MP, Gallai G, Pavia M (2013) The Upper Valdarno Plio-Pleistocene vertebrate record: an historical overview, with notes on palaeobiology and stratigraphic significance of some important taxa. Ital J Geosci 132:104–125

Roussiakis S, Athanassiou A, Michailidis D, Mitsopoulou V, Solomos C, Theodorou G (2014) Remarks on new proboscidean remains from the classical Late Miocene locality of Pikermi and their associated fauna. In: Kostopoulos DS, Vlachos E, Tsoukala E (eds) 6th International Conference on Mammoths and their Relatives. Grevena-Siatista. Sci Annals, School Geol, Aristotle Univ Thessaloniki, Greece, sp vol 102:171–172

Saegusa H, Thasod Y, Ratanasthien B (2005) Notes on Asian stegodontids. Quat Int 126–128:31–48

Sakellariou-Mane H (1972) Presence d'*Anancus arvernensis* dans les couches supérieures du Pliocène de la Valée du Vardar. Folia Biochim Biol Graeca 9:31–36

Sanders WJ (2003) Proboscidea. In: Fortelius M, Kappelman J, Sen S, Bernor RL (eds) Geology and paleontology of the Miocene Sinap formation. Turkey. Columbia University Press, New York, pp 202–219

Sanders WJ (2007) Taxonomic review of fossil Proboscidea (Mammalia) from Langebaanweg, South Africa. Trans Royal Soc S Afr 62:1–16

Sanders WJ (2008) Review of fossil Proboscidea from the Late Miocene-Early Pliocene site of As Sahabi, Libya. In: Boaz NT, El-Arnauti A, Pavlakis P, Salem MJ (eds) Circum-Mediterranean geology and biotic evolution during the Neogene period: the perspective from Libya. University of Garyounis, Benghazi, pp 241–256

Sanders WJ (2011) Proboscidea. In: Harisson T (ed) Paleontology and geology of laetoli: human evolution in context. Volume 2: fossil hominins and the associated fauna, Vertebrate paleobiology and paleoanthropology. Springer, pp 233–262

Sanders WJ (2018) Horizontal tooth displacement and premolar occurrence in elephants and other elephantiform proboscideans. Hist Biol 30:137–156

Sanders WJ, Gheerbrant E, Harris JM, Saegusa H, Delmer C (2010) Proboscidea. In: Werdelin L, Sanders WJ (eds) Cenozoic mammals of Africa. University of California Press, Berkeley, pp 161–251

Saunders J (1996) North American Mammutidae. In: Shoshani J, Tassy P (eds) The proboscidea: evolution and palaeoecology of elephants and their relatives. Oxford University Press, New York, pp 271–279

Schlesinger G (1917) Die Mastodonten des K. K. Naturhistorischen Hofmuseums. Denkschr K K Nat Hofmus 1:1–230

Schlesinger G (1922) Die Mastodonten der Budapester. Sammlungen Geol Hung 2:1–284

Schlosser M (1907) Ueber Säugetiere und Süsswassergastropoden aus Pliocänablagerungen Spaniens und über die natürliche Grenze von Miocän und Pliocän. N Jb Min Geol Paläont 2:1–41

Schmidt-Kittler N, de Bruijn H, Doukas C (1995) 1. General introduction. In: Schmidt-Kittler N (ed) The vertebrate locality of Maramena (Macedonia, Greece) at the Turolian-Ruscinian boundary (Neogene), vol 28. Münchner Geowiss Abh, München, pp 9–18

Seiffert ER, Nasir S, Al-Harthy A, Groenke J, Kraatz B, Stevens N, Al-Sayigh A (2012) Diversity in the later Paleogene proboscidean radiation: a small barytheriid from the Oligocene of Dhofar Governorate, Sultanate of Oman. Naturwiss 99:133–141

Shoshani J (1996) Para- or monophyly of the gomphotheres and their position within Proboscidea. In: Shoshani J, Tassy P (eds) The Proboscidea: evolution and palaeoecology of elephants and their relatives. Oxford University Press, New York, pp 149–177

Shoshani J (1998) Understanding proboscidean evolution: a formidable task. Trends Ecol Evol 13:480–487

Shoshani J (2002) Proboscidea (Elephants). eLS. Wiley

Shoshani J, Tassy P (1996a) Summary, conclusions, and a glimpse into the future. In: Shoshani J, Tassy P (eds) The proboscidea: evolution and palaeoecology of elephants and their relatives. Oxford University Press, New York, pp 335–348

Shoshani J, Tassy P (eds) (1996b) The proboscidea: evolution and palaeoecology of elephants and their relatives. Oxford University Press, New York

Shoshani J, Tassy P (2005) Advances in proboscidean taxonomy and classification, anatomy and physiology, and ecology and behavior. Quat Int 126–128:5–20

Shoshani J, Golenberg EM, Yang H (1998) Elephantidae phylogeny: morphological versus molecular results. Acta Theriologica 5:89–122

Shoshani J, Walter RC, Abraha M, Berhe S, Tassy P, Sanders WJ, Marchant GH, Libsekal Y, Ghirmai T, Zinner D (2006) A proboscidean from the late Oligocene of Eritrea, a "missing link" between early Elephantiformes and Elephantimorpha, and biogeographic implications. Proc Natl Acad Sci 103:17296–17301

Steensma KJ (1988) Plio-/Pleistozäne Großsäugetiere (Mammalia) aus dem Beckem von Kastoria/Grevena, südlich von Neapolis - NW-Griechenland. Ph.D. thesis, Technische Universität Clausthal

Stefanescu G (1892) On the existence of the *Dinotherium* in Roumania. Bull Geol Soc Am 3:81–83

Symeonidis NK, Tataris A (1983) The first results of the geological and palaeontological study of the Sesklo basin and its broader environment (Eastern Thessaly, Greece) (in Greek). Ann Géol Pays Hellén 31:146–190

Syrides GE (1990) Lithostratigraphic, biostratigraphic and palaeogeographic study of the Neogene-Quaternary sedimentary deposits of Chalkidiki peninsula, Macedonia, Greece (in Greek). Ph.D. thesis, Sci Annals, School Geol, Aristotle Univ Thessaloniki, Greece 1:1–243

Tarabukin BA (1974) New data on the systematics, phylogeny and ecology of Suborder Deinotherioidea Osborn (1921). In: Mammals of the late Cenozoic from Southwestern USSR, Kishinev, pp 77–90

Tassy P (1985) La place des mastodontes Miocènes de l'Ancien Monde dans la phylogénie des Proboscidea (Mammalia): Hypothèses et conjectures. Ph.D. thesis, Université Pierre et Marie Curie

Tassy P (1986) Nouveaux Elephantoidea (Mammalia) dans le Miocène du Kenya. CNRS, Paris

Tassy P (1990) The "Proboscidean Datum Event": how many proboscideans and how many events? In: Linsday EH, Mein P, Fahlbusch V (eds) European Neogene mammal chronology. Plenum Press, New York, pp 237–252

Tassy P (1994) Origin and differentiation of the Elephantiformes (Mammalia, Proboscidea). Verh Nat Ver Hamb 34:73–94

Tassy P (1996a) Who is who among the Proboscidea? In: Shoshani J, Tassy P (eds) The Proboscidea: evolution and palaeoecology of elephants and their relatives. Oxford University Press, New York, pp 39–48

Tassy P (1996b) The earliest gomphotheres. In: Shoshani J, Tassy P (eds) The Proboscidea: evolution and palaeoecology of elephants and their relatives. Oxford University Press, New York, pp 89–91

Tassy P (1999) Miocene Elephantids (Mammalia) from the Emirate of Abu Dhabi, United Arab Emirates: palaeobiogeographic implications. In: Whybrow PJ, Hill A (eds) Fossil vertebrates of Arabia. Yale University Press, New Haven/London, pp 209–233

Tassy P (2005) Proboscideans (Mammalia) from the late Miocene of Akkaşdağı, Turkey. In: Sen S (ed) Geology, mammals and environments at Akkaşdağı, late Miocene of Central Anatolia, vol 27. Geodiversitas, Paris, pp 707–714

Tassy P (2016) Proboscidea. In: Sen S (ed) Late Miocene mammal locality of Küçükçekmece, European Turkey, vol 38. Geodiversitas, Paris, pp 261–271

Tassy P (2018) Remarks on the cranium of *Eozygodon morotoensis* (Proboscidea, Mammalia) from the early Miocene of Africa, and the question of the monophyly of Elephantimorpha. Rev Paléobiol 37:593–607

Tassy P, Sen S, Jaeger JJ, Mazin JM, Dalfes N (1989) Une sous-espèce nouvelle de *Choerolophodon pentelici* (Proboscidea, Mammalia) à Esme Akçaköy, Miocène supérieur d'Anatolie occidentale. C R Acad Sci Paris 309:2143–2146

Theodorou G, Spjeldnaes N, Hanken NM, Lauritzen SE, Velitzelos E, Athanassiou A, Roussiakis S (2000) Description and taphonomic investigations of Neogene Proboscidea from Rhodos, Greece. Ann Géol Pays Hellén 38:133–156

Theodorou G, Athanassiou A, Roussiakis S, Iliopoulos G (2003) Preliminary remarks on the Late Miocene herbivores of Kerassiá (Northern Euboea, Greece). In: Reumer JWF, Wessels W (eds) Distribution and migration of Tertiary Mammals in Eurasia. A volume in honour of Hans de Bruijn, vol 10. Deinsea, Rotterdam, pp 519–530

Theodorou GE, Roussiakis SJ, Athanassiou A, Filippidi A (2010) Mammalian remains from a new site near the classical locality of Pikermi (Attica, Greece). In: Christofides G, Kantiranis N, Kostopoulos DS, Chatzipetros AA (eds) 19th CBGA Congress, Thessaloniki. Sci Annals, School Geol, Aristotle Univ Thessaloniki, Greece sp vol 99:109–119

Tobien H (1973) On the evolution of mastodonts (Proboscidea, Mammalia). Part I: The bunodont trilophodont groups. Notizbl hess Landesamt Bodenforsch 101:202–276

Tobien H (1978) On the evolution of mastodonts (Proboscidea, Mammalia). Part 2: The bunodont tetralophodont groups. Geol Jb Hessen 106:159–208

Tobien H (1980) A note on the skull and mandible of a new choerolophodont mastodont (Proboscidea, Mammalia) from the middle Miocene of Chios (Aegean Sea, Greece). In: Jacobs LL (ed) Aspects of vertebrate history: essays in honor of Edwin Harris Colbert. Museum of Northern Arizona Press, Flagstaff, pp 299–307

Tobien H (1996) Evolution of zygodons with emphasis on dentition. In: Shoshani J, Tassy P (eds) The Proboscidea: evolution and palaeoecology of elephants and their relatives. Oxford University Press, New York, pp 76–85

Tsoukala E (2000) Remains of a Pliocene *Mammut borsoni* (Hays, 1834) (Proboscidea, Mammalia), from Milia (Grevena, W. Macedonia, Greece). Ann Paléontol 86:165–191

Tsoukala E (2018) Rhinocerotidae from the Late Miocene and Late Pliocene of Macedonia, Greece. A revision of the Neogene-Quaternary Rhinocerotidae of Greece. Rev Paléobiol 37:609–630

Tsoukala ES, Melentis JK (1994) *Deinotherium giganteum* Kaup (Proboscidea) from Kassandra peninsula (Chalkidiki, Macedonia, Greece). Geobios 27:633–640

Tsoukala E, Mol D (2016) The Proboscidea of the Early Villafranchian site of Milia (Grevena, Macedonia, Greece). Quat Int 406:4–24

Tsoukala E, Vaxevanopoulos M, Lazaridis G (2007) Neogene mammalian remains from Dikaia (Evros, Thrace, Greece). The Archaeological Work in Macedonia and Thrace 2005, 19:1–11

Vacek M (1877) Über österreichische Mastodonten und ihre Beziehungen zu den Mastodon-Arten Europas. Abh Kais-Königlichen geol Reichenstalt 7:1–45

van der Maarel FH (1932) Contributions to the knowledge of the fossil mammalian fauna of Java. Wetenschappelijke Mededelingen Dienst van den Mijnbouw in Nederlandsch-Indië 15:1–208

van der Made J, Moyà-Solà S (1989) European Suinae (Artiodactyla) from the Late Miocene onwards. Boll Soc Paleontol Ital 28:329–339

von Koenigswald W, Göhlich UB (2019) Biogeography of Mammutidae (Proboscidea) or What's your name again? *Mammut*?, 90th Annual Meeting of the Paläontologische Gesellschaft. Staatliche Naturwissenschaftliche Sammlungen Bayerns - Bayerische Staatssammlung für Paläontologie und Geologie, Munich, p 83

von Meyer H (1831) Mittheilung an geheimen Rath von Leonhard. Jb Min Geogn Geol Petrefaktenkunde 2:296–297

von Meyer H (1833) Das *Dinotherium bavaricum*, mit Rückshicht auf die riesenmäßige fossile Thiergattung der Dinotherien überhaupt, und auf die Struktur der Mahlzähne in den Tapiren. Verh Kais Leopoldinisch-Carolinischen Akad Naturforsch 16:487–516

Vörös I (1989) *Prodeinotherium petenyii* sp. n. from the Lower Miocene at Putnok (North Hungary). Fragm Min Palaeont 14:101–110

Wagner A (1848) Urweltliche Säugthier-Ueberreste aus Griechenland. Abh bayerisch Akad Wiss 5:335–378

Wagner A (1857) Neue Beiträge zur Kenntnis der fossilen Säugthier-Ueberreste von Pikermi. Abh bayerisch Akad Wiss 8:109–158

Wang SQ, Deng T (2011) The first *Choerolophodon* (Proboscidea, Gomphotheriidae) skull from China Science. China Earth Sci 54:1326–1337

Wang SQ, Shi QQ, He W, Chen SQ, Yang XW (2016) A new species of the tetralophodont amebelodontine *Konobelodon* Lambert, 1990 (Proboscidea, Mammalia) from the Late Miocene of China. Geodiversitas 38:65–97

Wang SQ, Deng T, Ye J, He W, Chen SQ (2017a) Morphological and ecological diversity of Amebelodontidae (Proboscidea, Mammalia) revealed by a Miocene fossil accumulation of an upper-tuskless proboscidean. J Syst Palaeontol 15:601–615

Wang SQ, Saegusa H, Duangkrayom J, He W, Chen SQ (2017b) A new species of *Tetralophodon* from the Linxia Basin and the biostratigraphic significance of tetralophodont gomphotheres from the Upper Miocene of Northern China. Palaeoworld 26:703–717

Weinsheimer O (1883) Über *Dinotherium giganteum* Kaup. Palaeontol Abh 1:207–281

Weithofer KA (1888) Beiträge zur Kenntnis der Fauna von Pikermi bei Athen. Beitr Paläontol Österr-Ung 6:225–292

Weithofer KA (1890) Die fossilen Proboscidier des Arnothales in Toskana. Beit Palaeontol Oesterr – Ungarns und des Orients 8:107–240

Zapfe H (1957) Ein bedeutender *Mastodon*-Fund aus dem Unterpliozän von Niederösterreich. N Jb Geol Palaeont Abh 104:382–406

The Fossil Record of Continental Elephants and Mammoths (Mammalia: Proboscidea: Elephantidae) in Greece

Athanassios Athanassiou

1 Introduction

The family Elephantidae Gray, 1821, is the evolutionary most derived clade of the order Proboscidea. It comprises very large-sized proboscidean mammals that share in common a suite of morphologic characters, such as the anteroposteriorly shortened skull with pneumatised cranial bones; the long, columnar extremities; and the multiplication of dental cusps, resulting in a lamellar molar structure. Quite like the other derived proboscidean families (see Konidaris and Tsoukala this volume), the Elephantidae also have long tusks, horizontal replacement of the cheek teeth, and a long trunk.

As it is the case with most proboscidean families, the Elephantidae emerged in Africa. This happened in the late Miocene (about 9–7 Ma), during the last major radiation event of the African Proboscidea (Maglio 1973; Todd and Roth 1996; Sanders et al. 2010). Starting in the late Pliocene, members of the family migrated repeatedly out of the continent and dispersed rapidly across Eurasia. During the Early Pleistocene, they colonised North America, but failed to disperse to South America, where the already established—though evolutionary more basal—gomphotheriids continued to thrive. As a consequence of the family's late migration out of Africa, the Elephantidae of Eurasia and America were already highly derived forms within proboscideans. They maintained their nearly ubiquitous presence in the Northern Hemisphere as essential members of the megaherbivore faunas until the latest Pleistocene–early Holocene, when they succumbed widely to the extensive end-Pleistocene megafaunal extinction. At present, the family's geographic

A. Athanassiou (✉)
Ephorate of Palaeoanthropology–Speleology, Hellenic Ministry of Culture and Sports, Athens, Greece

Museum of Paleontology and Geology, National and Kapodistrian University of Athens, Athens, Greece
e-mail: aathanas@geol.uoa.gr

E. Vlachos (ed.), *Fossil Vertebrates of Greece Vol. 1*,
https://doi.org/10.1007/978-3-030-68398-6_13

distribution, taxonomic diversity, and populations are extremely reduced, primarily because of habitat loss due to the expansion of human activities. The present populations are restricted in patchy regions of sub-Saharan Africa and SE Asia and are classified to the African genus *Loxodonta* Cuvier, 1825—with two species, *L. africana* (Blumenbach, 1797) and *L. cyclotis* (Matschie, 1900)—and the Asian *Elephas* Linnaeus, 1758, with one species, *E. maximus* Linnaeus, 1758.

Except for the extant *Elephas*, Eurasia has accommodated three additional elephantid genera during the late Neogene and Quaternary: *Stegotetrabelodon* Petrocchi, 1941; *Mammuthus* Brookes, 1828; and *Palaeoloxodon* Matsumoto, 1924. The last two have an important fossil record in the whole continent, as well as in the territory of Greece in particular. *Stegotetrabelodon* is a less derived taxon, documented in just a few upper Miocene–lower Pliocene localities of Eurasia (Tassy 1999; Ferretti et al. 2017), and it is not considered here, because it is as yet unknown in the fossil record of Greece. Another fossil group not addressed in the present review is the dwarf insular elephant forms, evolved after successful colonisation of remote islands. These are presented by Lyras et al. (volume 2).

Methods Upper and lower molars are abbreviated as M and m respectively; deciduous molars as D/d (e.g., M2, m3, D4, etc.). The measurements and indices were taken and calculated according to Maglio (1973). In particular, the lamellar frequency of molars is calculated as the number of lamellae per 10 cm of molar length, and the hypsodonty index as the percentage of molar height with respect to molar width. All measurements include cement. The length is measured perpendicular to the average direction of the lamellae. The geochronologic and stratigraphic framework follows Pillans and Gibbard (2012). Geochronologic ages are given, where available, in thousand or million years before present, abbreviated as 'ka' or 'Ma', respectively. 'MIS' stands for 'Marine Isotope Stage', 'MNQ' for 'Mammal Neogene and Quaternary Zones' (Guérin 1990), and 'ICZN' for 'International Code of Zoological Nomenclature'. Transliterated local geographic names are typed accented when necessary, in order to help with their correct pronunciation.

2 Phylogenetic Relationships and Taxonomy

Traditionally, the phylogenetic roots of Elephantidae were usually placed among the Stegodontidae Osborn, 1918, close to a form morphologically similar to the genera *Stegodon* or *Stegolophodon* (Osborn 1936; Aguirre 1969). Indeed, both families exhibit similar specialisations, such as a fore-and-aft shearing movement of the jaws during mastication and the dense packing of transverse ridges in molars, and they have been placed together in the past as subfamilies within Elephantidae (e.g., Kalb and Mebrate 1993; Kalb and Froehlich 1995). More recent studies have shown, however, that these characters are rather convergent and do not indicate a phylogenetic relationship. Moreover, the stegodonts were already advanced enough to be considered as ancestors of the elephants (Maglio 1973; Tassy 1996; Todd and Roth

1996). Instead, the ancestry of the Elephantidae (and the Stegodontidae as well) is currently traced to the late Neogene tetralophodont gomphotheres of Africa, i.e., those members of Gomphotheriidae with four lophs/lophids in the intermediate molars (D4–M2 and d4–m2) (see, e.g., Maglio 1973; Tassy 1996). Advanced gomphotheriids already showed trends towards cranial rostrocaudal shortening, that characterises the Elephantidae, while the highly derived elephant molar morphology resulted from the rearrangement of the main and accessory cusps in order to form transverse enamel ridges.

This particular lamellar molar structure, i.e., the merging of the lingual and buccal cusps into a mesiodistally compressed plate (lamella), is the synapomorphy that unites the Elephantidae. Several successive plates (at least six in m3/M3), joined to each other with cement, form the elephantid molar (Kalb and Mebrate 1993; Todd and Roth 1996). With advancing dental wear, this structure forms transverse enamel loops on the occlusal surface. The family is further subdivided in two subfamilies, Stegotetrabelodontinae Aguirre, 1969, and Elephantinae Gray, 1821, which appear to be phylogenetically sister taxa (Kalb and Mebrate 1993; Kalb and Froehlich 1995; Tassy and Debruyne 2001), although a direct ancestor–descendant relationship between them has been also proposed in the past (e.g., Maglio 1973; Coppens et al. 1978). The Stegotetrabelodontinae are monotypic and include only the genus *Stegotetrabelodon* Petrocchi, 1941, which still retains several gomphothere-like characters, such as the long mandibular symphysis bearing a pair of long tusks, the presence of premolars, the presence of a well-developed median sulcus between the lingual and buccal cusps/cuspids (in unworn or little worn molars only), and the occurrence of central conules in the molars (Maglio 1973; Coppens et al. 1978; Sanders et al. 2010). The Elephantinae include all the other genera of the family, namely, *Stegodibelodon* Coppens, 1972; *Primelephas* Maglio, 1970; *Mammuthus* Brookes, 1828; *Elephas* Linnaeus, 1758; *Palaeoloxodon* Matsumoto, 1924; and *Loxodonta* Cuvier, 1825. Collectively, they are characterised by a tuskless mandible with shortened symphysis, true lamellar structure of the molars (totally obliterated median sulcus), and at least seven lamellae in M3/m3 (Kalb and Mebrate 1993; Todd and Roth 1996; Sanders et al. 2010). *Stegodibelodon* and *Primelephas* are the most primitive genera, featuring brachyodont molars with few lamellae. Both are monotypic and differ from *Stegotetrabelodon* in the absence of lower tusks and the more apparent lamellar structure of the molars. Because of its plesiomorphic morphology, *Stegodibelodon* has previously been placed by certain authors in the Stegotetrabelodontinae (for a comprehensive review of the various proposed taxonomic schemes, see Kalb and Mebrate 1993). *Mammuthus*, *Elephas*, *Palaeoloxodon*, and *Loxodonta* constitute the most derived genera of the Elephantidae (note, however, that *Palaeoloxodon* has been often considered until recently as a subgenus of *Elephas*; e.g., Palombo and Ferretti 2005). *Mammuthus*, *Elephas*, and *Loxodonta* emerged in Africa as archaic forms during the late Miocene–early Pliocene and followed similar evolutionary trends across their distinct clades, which mainly involve shortening of the skull and mandible, increasing number of dental lamellae, and increasing hypsodonty (Maglio 1973; Todd and Roth 1996; Sanders et al. 2010). *Palaeoloxodon* is most typically known from the Middle–Late Pleistocene of

Eurasia, but the immediate ancestry of the Eurasian species can be traced to *P. recki* (Dietrich, 1916) from the Plio-Pleistocene of East Africa (Maglio 1973; Beden 1983; Saegusa and Gilbert 2008). *Loxodonta* remained an exclusively African taxon from its first appearance until today. On the contrary, *Mammuthus*, *Elephas*, and *Palaeoloxodon* migrated out of their continent of origin through at least three migration events during the late Pliocene–Early Pleistocene (Maglio 1973).

The phylogenetic relationships among the derived Elephantidae remain as yet not fully resolved, despite the numerous classifications published after Osborn's (1936, 1942) monumental opus (for an overview, see Tassy 1990; Tassy and Shoshani 1996). Morphology-based phylogenies usually place *Elephas* (including *Palaeoloxodon*) close to *Mammuthus* and consider *Loxodonta* as a sister taxon of the former two (Beden 1979; Tassy 1990, 1996; Shoshani 1996; Shoshani et al. 1998, 2007; Shoshani and Tassy 2005; Sanders et al. 2010). However, different phylogenetic trees, also deriving from morphological data, exist as well. Maglio (1973, figs 13, 15) considered that these three genera derived as independent lineages from *Primelephas*. Todd (2010), based on cranial and dental morphology, found that *Loxodonta* and *Elephas* (including *Palaeoloxodon*) derived from *Primelephas*, while the *Mammuthus* lineage was traced back to *Stegotetrabelodon*. Phylogenetic studies involving molecular data still support the traditional *Elephas–Mammuthus* clustering (e.g., Yang et al. 1996; Ozawa et al. 1997). Nevertheless, a *Mammuthus–Loxodonta* clade has been also indicated as most probable as well (Hagelberg et al. 1994; Thomas et al. 2000). Recently, Meyer et al. (2017) and Palkopoulou et al. (2018) clustered *Elephas* with *Mammuthus*, and *Palaeoloxodon* with *Loxodonta*, and noted that *Palaeoloxodon*'s genome has also an *Elephas* and a *Mammuthus* component, illustrating a quite complicated picture of elephantid phylogeny and interrelationships.

3 Historical Overview and Distribution

Fossil remains of elephants were known in Greece since the Antiquity, but they were not recognised as such. Instead, they were thought of as bones of mythical, huge-sized anthropomorphic creatures, such as Giants or Cyclopes. Pausanias, a Greek geographer of the second century AD, mentioned the presence of huge bones in the Megalopolis area (central Peloponnese), which he attributed to fallen Giants, after a fierce battle between them and the Gods. According to a well-known ancient myth, the battle had taken place in this area. The Megalopolis Basin is presently well known as a rich area in fossil mammal sites. Interestingly, the oldest known scientific account to me on fossil elephantids from Greece (Roth 1854) explicitly mentions the presence of numerous elephant remains in this basin, while Kandeloros (1898, p. 8) and Bürchner (1903) recorded information made available to them by local people, according to which huge bones were unearthed in the area in the late 1830s and the 1850s. A few years later, Mitzopoulos et al. (1862) reported very briefly on the excavation of an elephant molar in August 1861 at an undefined findspot near the village Leontári, within the same basin. Later on, the exposure of rich fossil accumulations in the same region due to erosion was the occasion for extensive excavations by Prof. Theodor Skuphos (also transliterated as Skouphos or Skoufos; University of Athens) at two sites, Íssoma and Kalývia Karyón (Bürchner 1903; Skuphos 1905). According to Bürchner (1903), among the finds were numerous elephant fossils, including an allegedly complete skeleton with skull (Fig. 1).

Since the first fossil discoveries in Megalopolis Basin, proboscidean fossils have been recovered in numerous other localities from Greece (see Figs. 2, 3 and Appendix), being one of the richest and widely distributed vertebrate groups in the country. Other early references to elephant fossils were similarly very brief and anecdotal, usually providing neither a description, nor any photograph of the finds, and are not mentioned here (see Appendix for a more complete list). During the 1950s and later, there was a marked increase in the vertebrate palaeontological research within the Greek territory, resulting in new occurrences of both mammoths and elephants. Psarianos and Thenius (1954) were the first authors to describe an elephantid fossil from Greece in detail: they studied a lower third molar referred to *Elephas (Archidiskodon) meridionalis* (= *Mammuthus meridionalis*) from Giáltra, Aedipsós region, N. Euboea, a single find collected from a conglomerate layer within a lignite sequence. Four years later, one of these authors (Psarianos 1958) reported on isolated molars from W. Macedonia (localities Tsotýli and Polýlakkos) and E. Macedonia (Phílippi) and attributed them to species of the *Mammuthus*

Fig. 1 A partial skull with tusk(s) referred to *Palaeoloxodon antiquus*, in right lateral view, during the excavation of Prof. Th. Skuphos (University of Athens) in Megalopolis Basin, 1902 (photograph by Th. Skuphos). This is quite possibly the skull from the area of the village Íssoma, mentioned by Bürchner (1903), which is considered lost

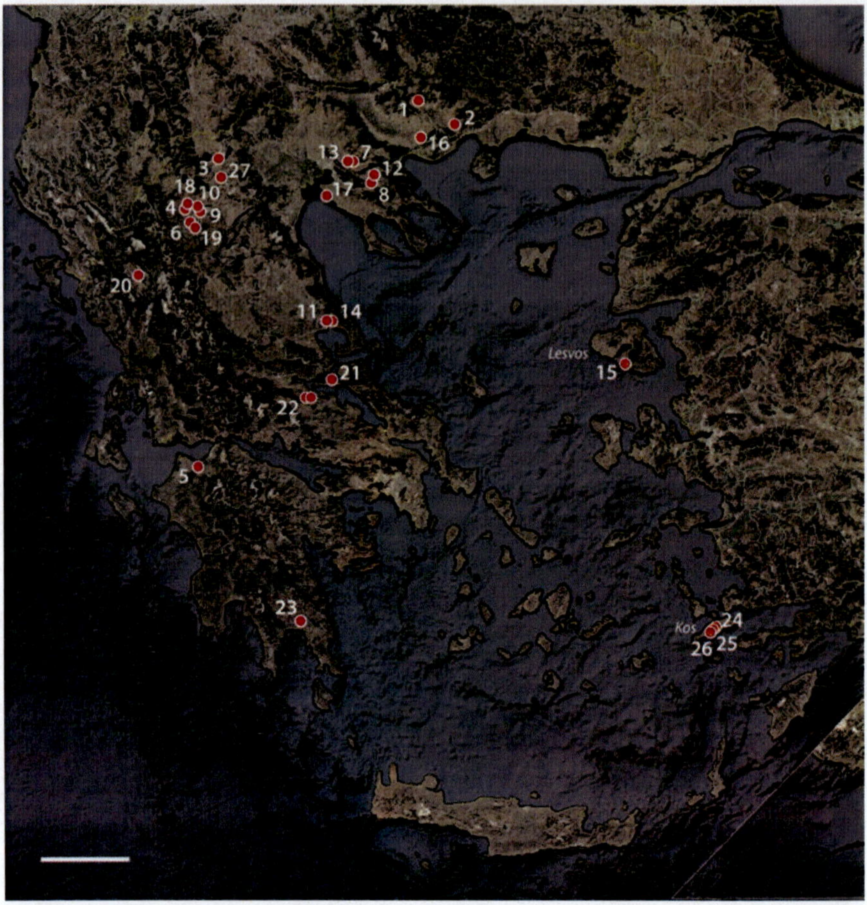

Fig. 2 Map of Greece showing the geographic distribution of localities with continental elephantids referred to the genus *Mammuthus* (including a possible occurrence marked with question mark). Geographically adjacent localities are grouped together under the same numbered bullet: **1**, Angítis; **2**, Phílippi; **3**, Sotíras; **4**, Tsotýli; **5**, Loussiká; **6**, Polýlakkos, Peponiá; **7**, Apollonía; **8**, Kalamotó localities; **9**, Kapetánios; **10**, Libákos; **11**, Sesklo; **12**, Tsiótra Vrýssi; **13**, Gerakaroú; **14**, Halykés(?); **15**, Vaterá localities; **16**, Symbolí; **17**, Epanomí; **18**, Kípi; **19**, Q-Profil; **20**, Ioánnina (exact location unknown); **21**, Giáltra; **22**, Reghínio, Zéli; **23**, Vlachiótis; **24**, Antimáchia; **25**, Kardámaena; **26**, Almyrí; **27**, Kardiá. See Appendix for more information. Image exported from Google Earth Pro © 2019, map data from US Dept. of State Geographer, SIO, NOAA, U.S. Navy, NGA, GEBCO, image from Landsat/Copernicus. Scale bar equals 80 km, North faces upward

lineage (*Archidiskodon meridionalis*, *A.* cf. *meridionalis* and *Mammonteus trogontherii*, respectively). There is no further faunal, geographic, or geologic information about the localities. According to Koulidou (2013), one of the *M. meridionalis* molars actually belongs to *M. rumanus*, due to its very low lamellar frequency. Shortly later, Melentis (1960) published another isolated find, a lower third molar, from an unknown locality in the Ioánnina region, Epirus, which he determined

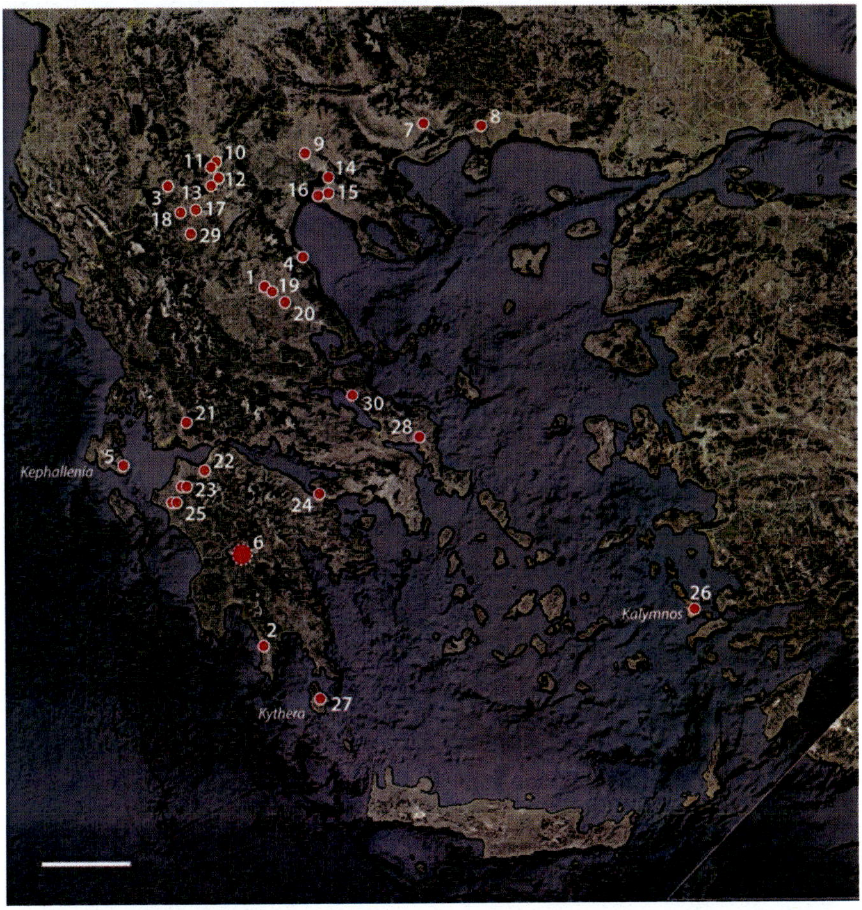

Fig. 3 Map of Greece showing the geographic distribution of localities with continental elephantids referred to the genus *Palaeoloxodon* (including certain insecure occurrences marked with question mark). Geographically adjacent localities are grouped together under the same numbered bullet: **1**, Peniós Valley; **2**, Kalamákia Cave; **3**, Tsákoni(?); **4**, Peniós estuary; **5**, Póros; **6**, various localities in Megalópolis Basin (Kyparíssia, Marathousa, Íssoma, Íssoma KYT, Kalývia Karyón, Léfktro, Leontári); **7**, Symbolí; **8**, Xeriás; **9**, Ravin de l'éléphant(?); **10**, Pétres, Sotíras; **11**, Amýntaio, Philótas; Lágoura(?); **12**, Perdíkkas; **13**, Ptolemaís; **14**, Allatíni(?); **15**, Trílophos(?); **16**, Epanomí(?); **17**, Kalonéri; **18**, Tsotýli; **19**, Terpsithéa; **20**, Néa Léfki, Chálki; **21**, Trichonía; **22**, Pátras; **23**, Nissí, Símiza; **24**, Canal of Corinth; **25**, Roupáki, Ano Olga(?); **26**, Kálymnos; **27**, Kýthera; **28**, Alivéri; **29**, Ambélia (Grevená); **30**, Límni(?). See Appendix for more information. Image exported from Google Earth Pro © 2019, map data from US Dept. of State Geographer, SIO, NOAA, U.S. Navy, NGA, GEBCO, image from Landsat/Copernicus. Scale bar equals 80 km, North faces upward

taxonomically down to the subspecies level: *Elephas (Archidiskodon) meridionalis archaicus* Depéret and Mayet, 1923. The find exhibits indeed all the characters of *M. meridionalis* and can be attributed to this species, though a subspecific classification is not justified according to the current taxonomic practice.

The following year, Melentis initiated a series of publications on the rich fossil material excavated by Skuphos in 1902 in the Megalopolis Basin, of which the elephantid remains constituted a significant component (Melentis 1961, 1963; Fig. 4). Note, however, that the skull mentioned by Bürchner (1903) and photographed by Skuphos (Fig. 1) is not included in the sample studied by Melentis, presumably because meanwhile it had been severely damaged or destroyed. Its whereabouts is currently unknown. Melentis (1961, 1963) described the material in every possible detail and concluded to an extreme taxonomic diversity of elephant species within the uppermost part of the basin's sedimentary sequence (Íssoma section; Melentis 1961, p. 242), recognizing no less than six taxa: *Palaeoloxodon antiquus antiquus*, *P. antiquus italicus*, *P. melitensis*, *Archidiskodon meridionalis meridionalis*, *A. meridionalis cromerensis*, and *Mammonteus primigenius primigenius*. Thus, the sequence was considered by the author as spanning the entire Pleistocene. More recently, it has been documented that the Megalopolis Basin

Fig. 4 Craniodental specimens from the Megalopolis Basin, referred to *Palaeoloxodon antiquus*. Both were published by Melentis (1961), who reported that they come from the locality Íssoma (Skuphos excavations, 1902). (**a**) A maxillary part with both M3s (MEG 1960/70), occlusal view. The rostral end is at the bottom of the figure. (**b**) A right m3 (MEG 1960/77), occlusal view. (**c**) The same specimen in lingual view. Note the characteristic for this species narrow crowns, with plicate enamel that forms median sinuses, and the 'dot-dash-dot' wear pattern in incipiently worn lamellae. Both specimens, as well as their original photographs used here, belong to the collections of the Museum of Paleontology and Geology, University of Athens. Scale bar equals 10 cm

lignite sequence (Marathousa member) was deposited during the Middle Pleistocene (e.g., van Vugt et al. 2000; Tourloukis et al. 2018). The Íssoma section extends (at least mainly) in a detrital sequence that overlies uncomformably the Marathousa member and could be of late Middle–Late Pleistocene age. Consequently, Melentis' elephantid taxa list has to be revised. Indeed, a re-examination of the available material (including unpublished samples stored in the Museum of Paleontology and Geology, University of Athens) shows that the dental specimens referred by Melentis (1961) to *A. meridionalis* are very similar to those he referred to *P. antiquus*, being hypsodont molars with narrow crowns and *P. antiquus*-type wear patterns (see Fig. 4 for a representative sample), as well as tusks that exhibit weak bent and torsion, and well obtuse Schreger angles in their dentine (see Fig. 5 for an example of

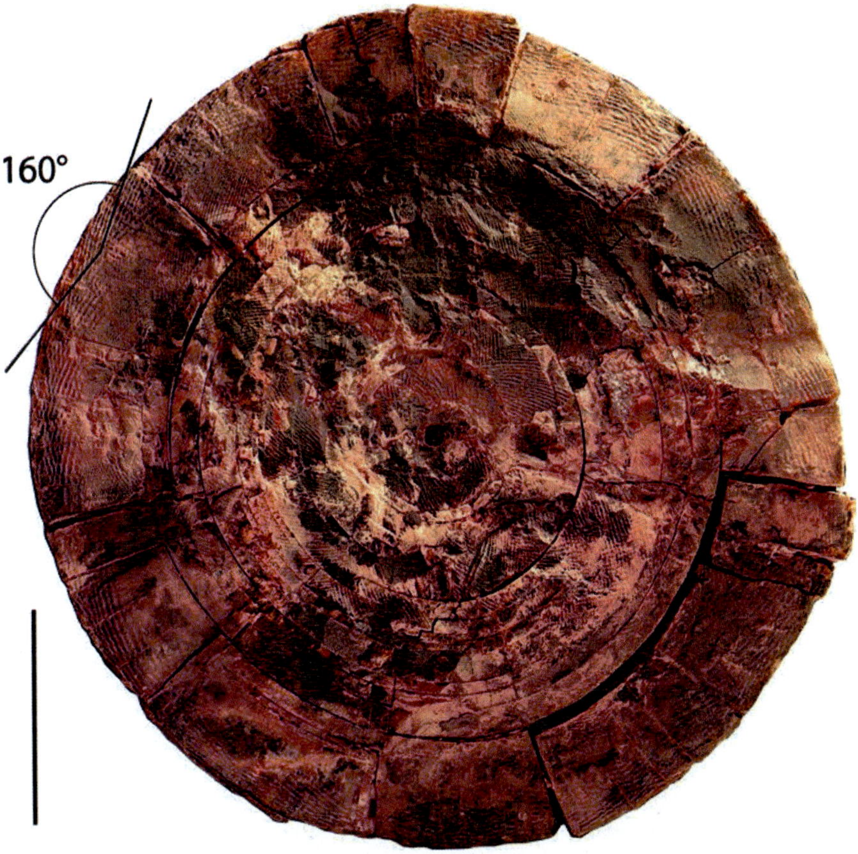

Fig. 5 Natural cross section of a *Palaeoloxodon antiquus* tusk, exhibiting obtuse Schreger angles near its periphery, which distinguish this species from those of the genus *Mammuthus* (see, e.g., Palombo and Villa 2001). This specimen—a 1.8 m-long tusk part—was excavated in September 2011 at Íssoma KYT (Megalopolis Basin, Peloponnese) and is kept in the collections of the Ephorate of Palaeoanthropology–Speleology (Ministry of Culture and Sports, Athens). Graphical scale equals 3 cm

a recently excavated specimen from the Íssoma area). They are morphologically very similar to *P. antiquus*, and they should be assigned to this species. Moreover, the sympatric occurrence of a dwarf species ('*P. melitensis*') among normal-sized elephants is ecologically impossible: dwarf elephant populations occurred only in isolated insular environments, characterised by greatly impoverished endemic faunas. The two dental specimens assigned to this species are better referred to a juvenile individual of *P. antiquus*. These inconsistencies in Melentis' taxonomy were first pointed out by Sondaar and Boekschoten (1967, pp. 562–563), who argued that only *P. antiquus* and *M. primigenius* existed in the Megalopolis Basin. Later research has further shown that the woolly mammoth molar described by Melentis (1961) actually belongs to a sample from the locality Telichka (Kiev, Ukraine), which was acquired by the Natural History Museum of the University of Athens in 1904, and was apparently mixed with the similarly coloured Megalopolis material excavated 2 years before, despite the clear labelling on the Telichka material (Lyras 2007; Iliopoulos et al. 2010). Indeed, recent research at several sites within the Megalopolis Basin never yielded any other elephantid species except for *P. antiquus*, which is currently the only proboscidean known to have existed in the area (Athanassiou 2018; Konidaris et al. 2018). A few proboscidean dental fragments (one of a molar and three of tusks) from undefined sites within the Megalopolis Basin published by Sickenberg (1976) as *Mammuthus (Archidiscodon) meridionalis*, based on the molar fragment morphology, are not in fact inconsistent with an attribution to *P. antiquus*, because of their incomplete preservation. The sample is actually not determinable further than the family level. The same is true for the postcranial finds from the basin referred to various species (Melentis 1963; Sickenberg 1976), since, contrary to what has been stated in older publications (see, e.g., Andrews and Forster Cooper 1928; Melentis 1963), there are no reliable criteria to distinguish between *Palaeoloxodon* and *Mammuthus* postcrania (Lister and Stuart 2010).

Melentis carried out excavations in the Megalopolis Basin as well, yielding most importantly a nice, partially preserved skull, which he assigned to *P. antiquus germanicus* (Stefanescu, 1924) (Melentis 1965). The specimen's morphology (two-bulged parieto-occipital crest, long and narrow hypsodont M3s) is consistent with an attribution to the straight-tusked elephant (Fig. 12a).

During the 1960s, numerous other papers were published describing or reporting on isolated elephantid dental specimens throughout the territory of Greece (Marinos 1964; Mitzopoulos 1964, 1967; Boessneck in Milójčić et al. 1965; Astre 1966; Faugères 1966; Melentis 1966a, b; Schneider 1968; Symeonidis 1970). Most important of these studies is the one dealing with the *Palaeoloxodon* sample found in the Peniós River banks, west of Lárissa (Boessneck in Milójčić et al. 1965), which is associated with Palaeolithic artefacts. The area continues to yield Late Pleistocene fossils when the river water level is low (Athanassiou 2001, 2011; Fig. 6). Three dental specimens from the Peniós River estuary, situated about 50 km NE of Lárissa, are referred by Paraskevaidis (1977) to three distinct species, but they are all referable to *Palaeoloxodon* as well. Unpublished proboscidean samples from the area of Lárissa, which quite probably belong to elephantid species, include a 2.6 m-long and 14 cm in diameter tusk found close to the village Mesorráchi in 1907 (reported

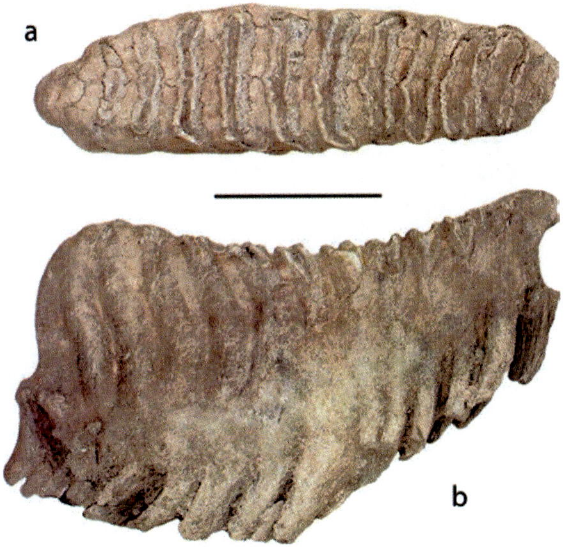

Fig. 6 Lower right third molar, referred to *Palaeoloxodon antiquus*, from Peniós River valley, west of the city of Lárissa (Late Pleistocene). The figured specimen belongs to a sample collected by the author in August 2007, during a dry period that resulted in exposure of large areas of the riverbed, due to very low water level. (**a**) occlusal view, (**b**) buccal view. Graphical scale equals 10 cm

anonymously in the local newspaper 'Mikrá', number 305, June 6, 1907), as well as bone and/or tusk fragments from Megálo Monastíri (findspot 50 in Runnels and van Andel 1993).

The presence of the straight-tusked elephant (*P. antiquus*) is also recorded at Vathýlakkos (Axiós Valley, Central Macedonia) (de Bonis et al. 1973), in an area well known for its late Miocene mammal faunas. The elephant-bearing site—accordingly named 'Ravin de l'éléphant'—has yielded a single elephant find: an unworn m3. Recently, the specimen was described by Koulidou (2013) as *M.* cf. *primigenius*. However, its low number of lamellae for a woolly mammoth (17), as well as its long and narrow shape, makes the original attribution to *P. antiquus* more probable, despite the specimen's high lamellar frequency. Nevertheless, it remains taxonomically enigmatic, requiring certainly a more thorough study.

Another very interesting locality is Pétres (W. Macedonia), where a skull of *Palaeoloxodon antiquus* was found, but never excavated (Velitzelos and Schneider 1973). The authors removed and studied the molars only, while the rest apparently was left in situ and possibly eroded soon thereafter. In the same paper, three other molars from the nearby locality of Sotíras are presented, referred to *M. trogontherii* and *A. meridionalis*, although they exhibit typical *Palaeoloxodon* morphology (see also Tsoukala et al. 2011). Judging from the published material, it seems that both *P. antiquus* and *M. trogontherii* were present at Sotíras (see also Marinos 1964; Tsoukala et al. 2011; Koulidou 2013). New material from this locality was recently published by Tsoukala et al. (2011) (*P. antiquus*) and Koulidou (2013) (*M. trogontherii*).

Poulianos and Poulianos (1980) presented a nearly complete elephant skeleton they excavated in a sand pit near Perdíkkas (Ptolemaís, W. Macedonia) in 1977 (Fig. 7). The authors emphasised the presence of lithic artefacts associated with the

Fig. 7 The almost complete skeleton of *Palaeoloxodon antiquus* from Perdíkkas, W. Macedonia, exhibited in situ (photograph taken on October 20, 1999). The partially preserved skull is to the left. Note the minimal dispersion of the skeletal elements and the anatomical association among most of them. An extraordinary feature of the find is the total lack of the autopodia. This skeleton was excavated in 1977 by A. Poulianos (Poulianos and Poulianos 1980) and described briefly by N. Poulianos (1986)

elephant, and since they referred the find to the species *Archidiskodon meridionalis*, they argued for evidence of hunting in a quite early time period. A few years later, the second author revised the specimen's taxonomic attribution and referred it to *Elephas (Palaeoloxodon)* aff. *antiquus*, though he insisted on an age of about 2.0–2.5 Ma (Poulianos 1986). Nevertheless, the Perdíkkas elephant has clear dental characters of *Palaeoloxodon antiquus* (narrow crowns, plicate enamel, presence of a median sinus in enamel loops), and it should be referred to as such (see also Tsoukala et al. 2011). Its age is Middle–Late Pleistocene. Elephantid fossils continue to emerge in the sand pits of the area, but are usually fragmentary.

In 1981, Koufos described a small fossil fauna of Late Pleistocene age from Angítis river valley, Sérres (E. Macedonia). The fauna is of a cold-steppe character and includes two partially preserved dental specimens referred to *Mammuthus* cf. *primigenius*. Excavations carried out later (1992–1999) by the Ministry of Culture yielded some more mammoth material in a horse-dominated faunal assemblage (Trantalidou 2013). In the same decade of 1980s, two other studies dealt with fossil elephantids. Symeonidis and Theodorou (1986) described a partly preserved upper molar and tusk parts from a locality near Vlachiótis (Laconia, SE Peloponnese) and referred it to *Archidiskodon meridionalis*. A few years later, Steensma (1988) studied the mammal fossils from five sites along the Haliákmon valley, in the region of Neápolis (Kozáni, W. Macedonia), which were collected in the late 1970s during fieldwork by the Technical University of Clausthal (Germany). Elephantidae were identified on dental and osteological specimens in four of the studied sites, Libákos, Kapetánios, Polýlakkos, and Q-Profil, and referred to *Archidiskodon meridionalis* (tentatively in the last two localities). Based on their faunal content, Steensma (1988) places the localities in the Lower Pleistocene (see also Fig. 13). In the same year, Koufos and Pavlides (1988) reported the discovery of a partial elephantine mandible at Sotíras (W. Macedonia), which they preliminarily referred to *Archidiskodon* cf. *meridionalis*. The same specimen was later described by Koulidou (2013) as *M. trogontherii*.

A prominent elephantid find published during the next decade is a partial skeleton from the locality Ambélia, in the town of Grevená (W. Macedonia) (Tsoukala and Lister 1998). The find, its skull and dentition, in particular, show typical characters of *P. antiquus*, such as divergent tusk alveoli, narrow molars with thick, plicate enamel, etc. (Tsoukala and Lister 1998). An interesting point of this study is the dating of the find in a glacial period (MIS 6) of the late Middle Pleistocene, when this species was not present in northern European regions. This indicates that the southern European regions acted as refugia for temperate-climate species, which later recolonised the North during interglacials. Other contemporary studies include Athanassiou (1998), who identified the species *M. meridionalis* at Sésklo (Magnesia, Thessaly) and Elephantidae indet. at the closely situated Halykés, and Tsoukala (1999), who documented the presence of Elephantidae indet. in the cave of Apídima C (Mani, S. Peloponnese). Regarding the sample from Sésklo, it should be noted that a tusk referred by Athanassiou (1998) to *M. meridionalis*, belongs most probably to the gomphothere *Anancus arvernensis* (see Athanassiou 2016). The presence of the southern mammoth at Sésklo is inferred from two carpal bones, which are too large to be assigned to *Anancus*, but it is not well supported on this scarce available material. The presence of *Mammuthus* at Sésklo should be regarded as tentative.

During the 1990s, elephant dental remains were also excavated at Nissí, Símiza and Roupáki areas, all closely situated in the Peniós River basin (Elis region, NW Peloponnese). At Nissí, a tusk excavated by the Museum of Paleontology and Geology, University of Athens, remains unpublished, but its physical characters show that it belongs to *P. antiquus* (see also Agiadi and Theodorou 2005). A very similar tusk was found at Símiza during road construction works (Dermitzakis pers. com.), as yet also unpublished. The same species was documented at Roupáki, based on a tusk, tusk fragments, and two very large M3s, all deriving from an illegal excavation and probably belonging to the same individual (Athanassiou 2000). Proboscidean remains indicated by Dermitzakis and Theodorou on a map (in Dermitzakis et al. 1982, fig. 64) from the nearby locality of Ano Olga may also belong to *P. antiquus*, as does the rest of the proboscidean samples from the region of Elis, but the whereabouts of the corresponding sample is unknown.

In a brief review of the Pliocene–Early Pleistocene Proboscidea of Greece, Athanassiou and Kostopoulos (2001) described elephantid specimens from the Lower Pleistocene localities of Gerakaroú and Apollonía (Central Macedonia), which they tentatively referred to *Mammuthus meridionalis*. New material published quite recently from the latter locality corroborated that insecure attribution

(Konidaris et al. 2020). During the following years scanty elephantid material was published from Vaterá (Lesbos Island) as *M.* cf. *meridionalis* and from Tsákoni (Kastoriá, W. Macedonia) as cf. *Elephas antiquus* (de Vos et al. 2002; Athanassiou 2004, respectively). Other localities, which are rather poor in elephantid material, are Kalamotó with *M. meridionalis* (Tsoukala and Chatzopoulou 2005), Reghínio (Athanassiou 2006a), Kípi (in the area of Neápolis, W. Macedonia; Athanassiou 2006b, fig. 6, 7; Harvati et al. 2008), and Epanomí with *M.* cf. *meridionalis* (Athanassiou and Kostopoulos 2010). An additional locality in the area of Neápolis, Trapezítsa, has yielded numerous tusk fragments exhibiting acute Schreger angles that could be assigned to *Mammuthus*. However, the original position of most identifiable fragments close to the tusk axis makes this feature atypical, and the finds are better referred to Elephantidae indet. Harvati et al. (2008) mentioned the presence of *Elephas antiquus* at Lágoura close to Néo Kostarázi (W. Macedonia), based on a lamella fragment. Although the morphology of the lamella is consistent with *P. antiquus*, an attribution to an early *Mammuthus* such as *M. meridionalis* cannot be excluded, so the find is referred here to Elephantidae indet. as well. The site has also yielded long bones, most notably a distal humerus and a radius.

Similarly taxonomically problematic is a partial left lower molar found in 2015 at Zéli (Central Greece,). The specimen (Fig. 8) preserves just four lamellae that belong quite probably to its distal part. Its maximal width is 91 mm (measured mesially) and its height about 125 mm (measured in the second preserved lamella). A hypsodonty index of 137 is calculated for the preserved part. The mean lamellar frequency is calculated as 5.2 lamellae per 10 cm, ranging between 4.5 and 5.8 from the base to the top of the crown. All lamellae are completely unworn, except for the

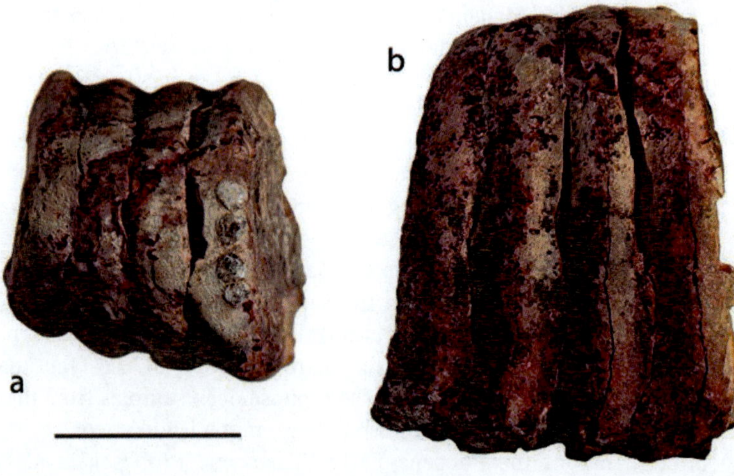

Fig. 8 A partial left lower molar referred to *Mammuthus* sp. from Zéli, Central Greece. (**a**) Occlusal view, (**b**) lingual view (the mesial end is on the right). Graphical scale equals 5 cm. The specimen is kept in the Ephorate of Palaeoanthropology–Speleology (Ministry of Culture and Sports, Athens)

Fig. 9 A part of the *Mammuthus trogontherii* skeleton from the lower Middle Pleistocene of Loussiká, NW Peloponnese (Athanassiou 2012): scapulae and ribs in situ during the excavation carried out by the Ephorate of Palaeoanthropology–Speleology, May 2003. The hammer is used for scale (length: 30 cm). The skeleton is currently curated at the Archaeological Museum of Patras

mesialmost one, which shows signs of incipient wear. More specifically, its occlusal surface has four transversely (linguobuccally) arranged enamel islets of quasi-circular shape. The specimen's metrical characters (proportions and lamellar frequency), as well as the wear pattern observable on the mesialmost lamella, are consistent with an attribution to an early species of *Mammuthus*, like *M. meridionalis*. However, its incomplete preservation does not allow for a taxonomic attribution further than the genus level.

Masseti (2006) figured a partial lower molar of *P. antiquus*, still attached to a mandibular fragment, allegedly coming from Kálymnos Island. The specimen is kept in the Archaeological Museum of Kálymnos, together with some additional elephant and hippopotamus specimens, possibly from the same locality(ies) on the island. Kálymnos was connected to the mainland during most of the Pleistocene, having, as a result, a fossil fauna of continental, non-endemic character.

In the recent years, of particular interest is a partial skeleton of a mammoth, excavated in 2001 and 2003 near Loussiká (NW Peloponnese) by the Ministry of Culture (Fig. 9). The find was referred to the Middle Pleistocene steppe mammoth, *M. trogontherii*, as evidenced from its cranial and dental morphology (Athanassiou 2012), though it had been previously preliminarily assigned to *Elephas antiquus* (Doukas and Athanassiou 2003; Athanassiou 2010). This wrong taxonomic identification has been also cited by Tsoukala et al. (2011). The latter authors, in their review of the occurrences of *P. antiquus* in Greece, described additionally new material of this species from the localities Kalonéri, Xeriás, Terpsithéa, Néa Léfki,

Fig. 10 A left upper third molar from Symbolí (Sérres, E. Macedonia), the holotype of the subspecies *Archidiskodon meridionalis proarchaicus* Melentis, 1966. Occlusal view. The mesial end is at the top. Note the wide crown, with thick lamellae and enamel. The specimen is referred here to *Mammuthus meridionalis*. It belongs to the collections of the Museum of Paleontology and Geology, University of Athens. Original photograph courtesy of the same museum. Scale bar equals 10 cm

and Sotíras (already mentioned above in this section). The Kalonéri (Kozáni, W. Macedonia, located very close to Libákos and Kípi) specimen is a damaged skull, preserving the premaxillaries and both tusks. Xeriás (Kavála, E. Macedonia) yielded only fragmentary material. In Terpsithéa and Néa Léfki, both very close to the city of Lárissa (Thessaly), a fragmentary mandible and tusk parts (respectively) have been recovered from the terrace deposits of the river Peniós. A young-adult mandible from Sotíras with m1s and m2s is an important find, enriching the *P. antiquus* sample that was already known from this locality (Velitzelos and Schneider 1973).

Koulidou (2013) reviewed the museum collections (Geology–Paleontology Museum, Aristotle University of Thessaloniki) of Plio-Pleistocene Proboscidea from Northern Greece and concluded that *P. antiquus* occurs at Tsotýli and Sotíras, *M. meridionalis* at Tsotýli and Phílippi, *M. trogontherii* at Sotíras, Phílippi and possibly (cf.) at Tsotýli, and *M.* cf. *primigenius* at Phílippi and Vathýlakkos (site 'Ravin de l'éléphant', based on a m3 originally referred to *P. antiquus* by de Bonis et al. 1973). She also documented the presence of *M. rumanus* at Tsotýli and possibly (cf.) at Kardiá, for the first time in Greece. The Tsotýli specimen was described soon after in detail by Kostopoulos and Koulidou (2015), who, although recognised its morphologic affinities with dental specimens referred to *M. rumanus*, refrained from assigning it to a species.

Two other important finds published recently are two partial skeletons of *Palaeoloxodon antiquus*, discovered in W. Macedonia and in Peloponnese, respectively. The former was unearthed in 2005 during mining works in Amýntaio Lignite Mine and was described in detail by Kevrekidis and Mol (2016). The latter was

Fig. 11 The Marathousa 1 skeleton of a male *Palaeoloxodon antiquus* (Megalopolis Basin, central Peloponnese; view to the East) during an early excavation stage (November 3, 2014), when several skeletal elements were still buried. The skeleton was found scattered mainly in a ENE–WSW direction, but generally well preserved. The plaster jacket at the eastern end of the excavation covers the skull (figured after its preparation in Fig. 12b). The find was published preliminarily by Konidaris et al. (2018)

located in 2013 during archaeological prospecting in Marathoúsa Lignite Mine, Megalopolis Basin, and was excavated during the years 2013–2019 (Konidaris et al. 2018; Panagopoulou et al. 2018; Figs. 11 and 12b). A significant feature of the Marathousa skeleton is that it is associated with lithic artefacts and bears cut marks, documenting butchering activities by humans on the elephant corpse (Konidaris et al. 2018; Panagopoulou et al. 2018). At about the same time, the Megalopolis

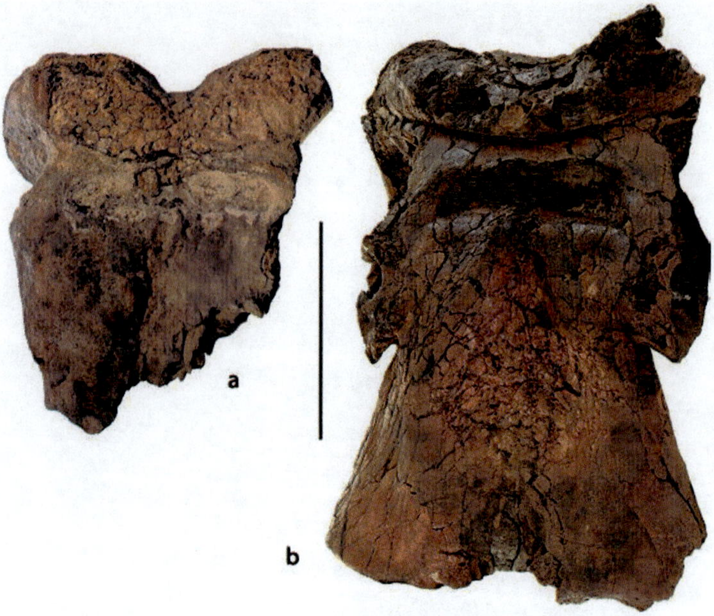

Fig. 12 Cranial specimens of *Palaeoloxodon* from Greek localities. (**a**) Partial skull from Léfktro, Megalopolis Basin, in frontal view, published by Melentis (1965) (Museum of Paleontology and Geology, National and Kapodistrian University of Athens), (**b**) the complete skull of the Marathousa 1 skeleton, Megalopolis Basin, in frontal view (Ephorate of Palaeoanthropology–Speleology, currently under study; see also Konidaris et al. 2018). Scale bar equals 50 cm

Basin yielded additional fossil samples: in September 2011 a tusk and a scapula, possibly associated with a lithic artefact, were found during the construction of an electricity distribution centre, not far from the village Íssoma (locality 'Íssoma KYT'). A few days later, two more tusk parts were found in the same locality, 200 m to the North of the previous finds, but both were reduced to small fragments, presumably because of long aerial exposure before burial. Íssoma KYT yielded some more tusk parts in October of the same year. The following year, 2012, a team of the University of Athens excavated a ravine at Íssoma (which is allegedly one of the spots exploited by Skuphos during his 1902 expedition) and unearthed a few mammal fossils, most notably a complete elephant tusk (Theodorou 2014). All these recent finds in the broader area of Íssoma (localities 'Íssoma' and 'Íssoma KYT') are referred to *P. antiquus*, based on the tusk morphology (weak curve and torsion) and the obtuse Schreger angles observed in the fragment cross sections.

A recent (2015) discovery of an elephant maxilla on the island of Kephallenía led Theodorou et al. (2018) to erect a new species, *Elephas (Palaeoloxodon) cephallonicus*, accepting that the find belonged to an endemic, incipiently dwarfing population. Following Athanassiou et al. (2019), *P. cephalonicus* is considered here as a junior synonym of *P. antiquus*, because it does not differ diagnostically from the mainland *Palaeoloxodon* of Greece (see Sect. Invalid Taxa).

4 Systematic Paleontology

This section considers only non-endemic, normal-sized elephants, excluding the several species of insular dwarfs, which have been described from the islands of Greece and presented separately (Lyras et al. volume 2).

Valid Taxa

Elephantidae Gray, 1821

Type Genus *Elephas* Linnaeus, 1758.

***Mammuthus* Brookes, 1828**

Type Species *Elephas primigenius* Blumenbach, 1799.

Included Taxa The genus is known with four continental species in Greece, *M.* ex gr. *rumanus*, *M. meridionalis*, *M. trogontherii*, and *M.* cf. *primigenius*.

Remarks *Mammuthus* is an elephantid genus that reached Eurasia during the late Pliocene (at about 3.5 Ma, or somewhat later), after having evolved in Africa. The oldest representatives of the genus in Eurasia are, though, quite primitive, featuring brachyodont or barely hypsodont molars that consist of just a few lamellae. Yet, during the Pleistocene the mammoth lineage underwent impressive evolutionary modifications in response to the recession of the forests and the expansion of open environments. These included primarily a great increase in hypsodonty, multiplication of the molar lamellae, and increase of their frequency, thinning of the enamel, and rostrocaudal shortening and dorsoventral deepening of the skull and the mandible. These morphological changes had been traditionally considered as a prime example of gradual, orthogenetic evolution, a conclusion biased by the fact that practically only the European fossil samples were well known. In a Eurasian-wide perspective, Lister et al. (2005) illustrated a quite different evolutionary setting, according to which long periods of stasis, corresponding to the biochronological ranges of species, were interrupted by introductions of new morphotypes, apparently originating in NE Asia and spreading to the west possibly by means of genetic input to geographically adjacent populations.

Mammuthus exhibit a unique suite of morphological characters that include a single-domed skull, subparallel tusk alveoli, curved and twisted tusks, and broad molars with usually weakly plicate enamel. Early and Middle Pleistocene species evolved very large body sizes, but a considerable gradual body size reduction is observed from the late Middle Pleistocene onwards. The Eurasian species currently classified in this genus are the following (from the oldest to most recent), *M. rumanus*, *M. meridionalis*, *M. trogontherii*, and *M. primigenius*, the first being by far the least well known and taxonomically documented, due to its scarcity. All were originally placed in *Elephas*, considered in a much wider sense, usually equivalent to the current subfamily Elephantinae. Other taxonomic names have been applied to members of this clade as well, usually in order to split it in taxa of genus or subgenus rank, such as *Archidiskodon* Pohlig, 1888, *Parelephas* Osborn, 1924, and *Mammonteus* Osborn, 1924. Maglio (1973) showed the phylogenetic continuity of the relevant species and adopted their classification under a single genus, *Mammuthus*. The same opinion had been expressed before in Simpson's classification of mammals (Simpson 1945), while Osborn (1942), though splitting the lineage into three genera, recognised their close relationship and placed them together in a single subfamily, Mammontinae Osborn, 1921. The classification of these species under the genus *Mammuthus* is followed today by the majority of authors, except for most Eastern European ones, who generally prefer to classify the late Pliocene–Early Pleistocene mammoth species to *Archidiskodon* (with *Elephas meridionalis* as the type species).

The species of the mammoth lineage shared the continent with other proboscideans, such as the mammutid *Mammut borsoni* during the late Pliocene, the gomphotheriid *Anancus arvernensis* during the late Pliocene and the Early Pleistocene, and the elephantid *Palaeoloxodon antiquus* during the Middle and the Late Pleistocene.

***Mammuthus* ex gr. *rumanus* (Stefanescu, 1924)**

Nomenclatural and Taxonomical History The species was erected by its author, without any description, as a primitive 'mutation' of *Elephas antiquus* (i.e., a subspecies in the current taxonomic practice). Stefanescu (1924) referred to an older publication (Athanasiu 1915) for a description of the type specimen (a partial m3 from Tulucești, Romania), where it was assigned to *Elephas* cf. *meridionalis*. More recently, Radulescu and Samson (1995) described a molar from Cernătești, Romania, as *M. rumanus*. The late Pliocene and earliest Pleistocene elephantid material from Europe consists of scarce dental samples, which, due to their plesiomorphic characters, are difficult to assign to a genus. They were habitually referred to the Asian species *Elephas planifrons* Falconer and Cautley, 1846 (e.g., Depéret and Mayet 1923; Osborn 1942, p. 969; Maglio 1973), to *Archidiskodon gromovi* Garutt and Alexeeva, 1964 (e.g., Azzaroli 1977), or considered primitive forms of *M. meridionalis*. In addition, Maschenko (2010) coined *Archidiskodon garutti* for a small Early Pleistocene dental sample from the Stavropol region, Russia (its provenance locality is disputed), which, however, except for its more pronounced brachyodonty, does not seem to differentiate significantly from *M. rumanus* on the basis of the author's metrical and morphological data (see also Baygusheva et al. 2011).

Type Material Stefanescu (1924) apparently considered as the holotype a partially preserved lower third molar (m3) figured by Athanasiu (1915, pl. XVII, fig. 4) and referred to as *Elephas* cf. *meridionalis* by the same author. Fairly recently, Lister and van Essen (2003) proposed as a neotype the complete upper third molar from

Cernăteşti (Romania), because of the false assumption that the holotype had been lost (Lister et al. 2005).

Type Locality Upper Pliocene of Tuluceşti (Galaţi region, Romania), biochronologically dated to the zone MN16a. Regional stratigraphic correlations and palaeomagnetic data place this locality to the middle Gauss (Kaena event, about 3.0–3.1 Ma) (Radulescu et al. 2003).

Distribution Primitive elephantine dental specimens attributable or similar to *M. rumanus* have been described (apart from the two Romanian localities mentioned above) from Bossilkovtsi (Bulgaria; Markov and Spassov 2003), the Stavropol area (SW Russia, Maschenko 2010), Kale Tepe-3 (Turkey; Albayrak 2017; though morphologically closer to *Elephas*, according to the author), Bethlehem and Erq el Ahmar (Israel; Markov 2012; Rabinovich and Lister 2017; Rabinovich et al. 2019), and two localities in Shanxi (NE China; Wei et al. 2006). Lister and van Essen (2003) also referred to this species the samples from Montopoli and Laiatico (Italy) and Red Crag (England), but this opinion was moderated as provisional shortly later (Lister et al. 2005). Obada (2010) added another Romanian locality to the species' distribution, Orodelu (Dolj, S. Romania), but considered the Cernăteşti specimen as a new distinct species (*Archidiskodon stefanescui*). An m3 from Podari (Dolj, S. Romania), described by the same author as *Loxodonta* sp., may also be attributable to *M. rumanus*, on the basis of the morphologic and metrical characters presented by Obada (2010).

In Greece, mammoth finds with archaic dental characters are known from Tsotýli (two specimens) and Kardiá (one specimen) in W. Macedonia (Koulidou 2013). Their stratigraphic origin is unknown, because they derive from old incidental collections, not systematic excavations. Koulidou (2013) referred them to *M. rumanus*. One of the Tsotýli specimens had been already published by Psarianos (1958) as *Archidiskodon* cf. *meridionalis*. The second specimen from the same locality is the most important of them and has been re-described in detail by Kostopoulos and Koulidou (2015). It is a part of a maxilla, bearing two rather brachyodont molars (M2 and M3) that consist of very few lamellae. Kostopoulos and Koulidou (2015) do not assign it to a species, due to our incomplete knowledge of the early mammoth morphology and variation. Here, however, it is included provisionally to *M.* ex gr. *rumanus*, together with the other two primitive elephantid finds from Greece, for the sake of simplicity, until more material becomes available.

Taxonomic Remarks The species *M. rumanus* was reviewed and described in detail by Lister and van Essen (2003), who essentially resurrected it internationally after many years of local use (e.g., Radulescu and Samson 1995) and expanded its distribution out of Eastern Europe. The authors pointed out that this species is distinctly different from *M. meridionalis*, in contrast with other names proposed for primitive mammoths, such as *M. gromovi*, whose type material is similar to the typical *M. meridionalis* (at least dentally), thus not separable from it at the species level.

Moreover, in African localities there is evidence for populations very similar to *M. rumanus*, which may indicate that this species had a transcontinental distribution, and was involved in the mammoths' migration out of Africa (Markov 2012). Certain authors, however, prefer to classify *rumanus* as a basal subspecies of *'Archidiskodon' meridionalis*, immediately preceding *gromovi* in the biostratigraphic subspecies succession of this species (e.g., Vislobokova 2005; Baygusheva and Titov 2012). The name is used here as *M.* ex gr. *rumanus* to include the primitive elephantid finds from Greece and similar samples from the wider European region, pending the availability of more material and a systematic revision of this group. Currently, the scarcity of well-dated dental finds and the lack of cranial material make even a generic attribution insecure in most cases, while the true metrical and morphological variation within the species is incompletely known. *Mammuthus rumanus* is characterised dentally by a low number of lamellae (8–10 in M3/m3) and low lamellar frequency (3–5 lamellae per 10 cm of molar length), while their hypsodonty index is comparable to the minimum values calculated for *M. meridionalis* (Lister and van Essen 2003; Lister et al. 2005). Occlusally, the enamel loops often form prominent, distally directed median folds, a feature homologous to the gomphothere central conules, which is reduced in later mammoth species.

***Mammuthus meridionalis* (Nesti, 1825)**

Nomenclatural and Taxonomical History This species was defined by Nesti (1825) on an Early Pleistocene elephantine sample collected in the region of Upper Valdarno, N. Italy. Part of this material had been previously referred to the Asian elephant, *Elephas maximus*, and to the woolly mammoth, *M. primigenius* (see a historical account in Palombo and Ferretti 2005). Because of the species' long biostratigraphic range, numerous subspecies (usually chrono-subspecies) have been named to account for different evolutionary levels. Common names found in the literature are *M. m. gromovi* (Garutt and Alexeeva, 1964), *M. m. meridionalis*, *M. m. vestinus* (Azzaroli, 1972), *M. m. tamanensis* (Dubrovo, 1963), and *M. m. depereti* Coppens and Beden, 1982 (roughly from the stratigraphically oldest to more recent ones). Subspecies names that have been used to identify Greek fossils include *M. m. meridionalis*, *M. m. cromerensis* Depéret and Mayet, 1923, *M. m. archaicus* Depéret and Mayet, 1923, *M. m. proarchaicus* Melentis, 1966, and *M. m. vestinus* (Azzaroli, 1972) (Melentis 1960, 1961, 1966a; Mitzopoulos 1967; Konidaris et al. 2020).

Type Material No holotype was designated by Nesti (1825), who, however, figured two skulls and several limb bones in his study. One of these skulls (IGF-1054; Palaeontological Museum of Florence, Italy) was later designated as the lectotype by Depéret and Mayet (1923). The type locality has yielded until today a rich hypodigm that includes skeletons, several skulls and mandibles, and hundreds of molars and postcranial elements (Palombo and Ferretti 2005).

Type Locality Lower Pleistocene fluvio-lacustrine deposits of the Montevarchi group, Upper Valdarno, Italy. This sequence includes the Matassino (older) and the Tasso (newer) faunas, and it is dated to the time interval 2.0–1.77 Ma (Palombo and Ferretti 2005).

Distribution The species is quite common, known from hundreds of localities widespread across Eurasia, except for its northernmost regions. It is also tentatively reported from Northern Africa (Maglio 1973; Sanders et al. 2010). Its presence in North America, assumed in the past by some authors (e.g., Osborn 1942; Maglio 1973), is not supported by recent research (Lister and Sher 2015). Geochronologically, it appeared at about the beginning of the Pleistocene (2.6 Ma) and persisted till the early part of the Middle Pleistocene (possibly as late as 0.6 Ma) (Lister et al. 2005). In Greece, the species occurs usually as single dental finds currently in more than 20 localities (see Appendix) situated throughout the country. Most of them are not datable further than a general placing in the Lower Pleistocene.

Taxonomic Remarks This is the oldest well-known species of the genus. It is of large body size, with moderately rostro-caudally shortened skull and mandible. The skull has a concave forehead in lateral view and forms dorsally a single dome due to the dorsal expansion of the parietal and occipital bones. With regard to more advanced species of the genus, the mandibular symphysis is long. The tusks are robust, curved, and twisted, and the molars are wide, weakly to moderately hypsodont (hypsodonty index: 99–146 in M3s, 110–143 in m3s; Lister and Stuart 2010), with thick lamellae and moderately thick enamel. The third molars (M3/m3) consist of 12–14 lamellae (in rare cases 11 or 15; Lister et al. 2005).

The species exhibited a gradual evolution during its long history mainly towards a higher cranial dome, higher lamellae number, increased hypsodonty, and thinner enamel. This guided certain authors to assign the known samples to distinct evolutionary stages. Depéret and Mayet (1923) distinguished four 'mutations' (archaic, typical, Saint-Prestien, and *cromerensis*), the last of which is presently referable to *M. trogontherii*, at least partly. Maglio (1973) designated three evolutionary stages, 'Laiatico', 'Montevarchi', and 'Bacton' (from less to more derived), the second of which is the one observed in the material of the type locality. The Laiatico stage samples have been later related to *'Archidiskodon' gromovi* (see Azzaroli 1977) and to *M. rumanus* (see Lister and van Essen 2003; Lister et al. 2005). However, the scarce available material of these archaic mammoth forms is still inadequate for an unambiguous species-level taxonomy (Palombo and Ferretti 2005). The Bacton stage includes the most advanced samples of the species that reached the highest lamellae number and hypsodonty and corresponds roughly to the local subspecies *M. m. vestinus*, *M. m. depereti*, and *M. m. tamanensis* mentioned above.

The Greek fossil record of *M. meridionalis* remains still rather poor. Despite this fact, certain isolated dental specimens were determined taxonomically down to the subspecies level, without considering the wide morphological variability of mammoth molar characters, which are also wear dependent (see, e.g., Lister and Sher 2015). This taxonomic practice, followed during the 1960s, is certainly not appropriate, and the subspecific part of the name should be dropped. The molars studied by Melentis (1960, 1966a) form an example (Fig. 10; see also above). Moreover, in some cases, the samples are referable to *Palaeoloxodon antiquus* (see below in this section), not to a mammoth. The case of the Megalopolis sample (Melentis 1961) has been already mentioned above. Another example are the two molars from Néa Kómi or Perdíkkas (exact locality data are not given), which were referred to *Archidiskodon meridionalis archaicus* by Mitzopoulos (1964, 1967): the high and narrow crowns, as well as the occlusal wear pattern point to the straight-tusked elephant, which is frequently found in the area, and not to an archaic mammoth. The same is quite probably true for a molar from Chálki referred by Schneider (1968) to *Elephas (Archidiskodon) meridionalis* cf. *cromerensis* (see also Athanassiou 2002). A taxonomic determination at the subspecies level has been published again very recently (*M. m. vestinus* from Apollonía; Konidaris et al. 2020), but this time it was employed to classify a richer sample, which is clearly evolutionary more advanced than the typical samples of the species.

***Mammuthus trogontherii* (Pohlig, 1885)**

Nomenclatural and Taxonomical History The recognition of this taxon was the result of an extensive study on the Pleistocene elephants of Europe elaborated by Pohlig (1888, 1891). However, a short description of the species as an intermediate form between *M. meridionalis* and *M. primigenius*, accompanied by a statement of the name *Elephas trogontherii*, is to be found in a letter of the author referring to this study, which was published in 1885. The use of this name is widespread since then. Maglio (1973) preferred to use an older name, *M. armeniacus* (Falconer, 1857), which he considered as a senior synonym of *M. trogontherii*. This opinion was not followed by most later authors, however, because *M. armeniacus* is insufficiently documented on the basis of poor material from a difficult to relocate and undated type locality (Adam 1988; Lister 1996). The use of the name *M. trogontherii* is currently ubiquitous. Other junior synonyms of *M. trogontherii* include *Elephas intermedius* Jourdan, 1891; *Elephas nestii* Pohlig, 1891; and *Elephas wuesti* Pavlow, 1910.

Type Material No type material was designated by the species author, so that the total molar sample from Süßenborn (Thuringia, Germany), described in detail by Pohlig (1888), can be considered as syntypes. Later, Osborn (1942) chose an upper and a lower third molar out of the type series as a lectotype, despite the fact that the two specimens, though belonging to the same fossil collection, do not necessarily belong to the same individual. To correct for this error, Kahlke (1990) fixed one of Osborn's specimens (the upper third molar) as the lectotype, based on the same choice made by Dubrovo in a series of papers (e.g., Dubrovo 1965), who, however, repeatedly referred to the molar as a lower third one. The type locality has yielded more than a thousand molars of this species (Guenther 1969), which constitute a very good statistical sample to characterise the species dentally.

Type Locality Fluvial sediments of Süßenborn, upper terrace system of the river Ilm (Thuringia, central Germany), dated to early Middle Pleistocene (about 600 ka) (Lister and Sher 2001).

Distribution This species is common, known from many localities across Eurasia. In Europe it is first recorded at about 1.0 Ma, initially coexisting in certain parts of the continent with *M. meridionalis* (see Lister et al. 2005) and becomes a common faunal element after about 0.8 Ma. However, fossils with *M. trogontherii* morphology are already known from much older, Lower Pleistocene deposits in N. China, dated at 1.66 Ma (Wei and Lister 2005), and in E. Siberia at about 1.2 Ma (Lister and Sher 2001), suggesting that E. Asia was the centre of evolution and dispersion of this species. *M. trogontherii* persisted in Europe until about 200 ka, but, again, Chinese finds postdate the species' last occurrence in Europe, surviving until the latest Pleistocene (Wei et al. 2010).

Taxonomic Remarks Morphologically intermediate between *M. meridionalis* and *M. primigenius*, *M. trogontherii* features, with respect to older mammoth forms, a rostro-caudally shorter and higher skull, a deeper mandible with shorter symphysis, and more advanced molars, with more lamellae, thinner enamel and distinctly increased hypsodonty. Some early finds from Europe exhibit a less derived, mosaic dental morphology, characterised by advanced hypsodonty, but rather low lamellar count (15–17), or vice versa, high lamellar count, but moderate hypsodonty (Lister et al. 2005; Lister and Stuart 2010). Typical *M. trogontherii* have a lamellar number of 17–22 and a hypsodonty index of 153–206 in M3 and 141–177 in m3 (Lister and Stuart 2010). *M. trogontherii* has been credited the title of the largest-bodied elephantid (possibly also proboscidean), with exceptionally large males estimated to weigh more than 20,000 kg (Christiansen 2004). However, the species underwent a progressive body size reduction during the Middle Pleistocene, with later and smaller forms often referred to the subspecies *M. trogontherii chosaricus* Dubrovo, 1966 (in some cases elevated to the species level), approaching the morphology of *M. primigenius*. Due to the complexity of the *M. trogontherii–M. primigenius* transition, this taxon (in subspecies or species rank) is rarely used in systematics anymore (see Lister 1996).

The species' fossil record in Greece is limited (see Appendix), as it is not so common in southern latitudes. Psarianos (1958) and Marinos (1964) mention its presence in Macedonia (localities Sotíras, W. Macedonia, and Phílippi, E. Macedonia) on the basis of three isolated molars. Later, Velitzelos and Schneider (1973) described two additional molars from Sotíras as *M. trogontherii*, but one of them (in their figs. 6 and 7) is better attributed to *P. antiquus*, which is also present in the site. More recently, Koulidou (2013), in a comprehensive study of the available proboscidean material from Northern Greece, described dental specimens referable to this species from the localities Tsotýli, Sotíras, and Phílippi.

An exceptional find, compared to the generally poor samples of the species throughout the country, is a partial skeleton excavated in 2001 and 2003 at Loussiká, NW Peloponnese (Athanassiou 2012; Fig. 9). The recovered anatomical parts include the skull and mandible, part of the axial skeleton, and many limb bones that document a large 45-year-old male individual, estimated to have stood about 3.5–3.8 m high and have weighed 8 tons (Athanassiou 2012; Larramendi 2016). The Loussiká mammoth is also important in a European perspective, due to the rarity of skeletons of this species, and its location, which is comprised among the southernmost recorded.

***Mammuthus* cf. *primigenius* (Blumenbach, 1799)**

Nomenclatural and Taxonomical History The woolly mammoth was one of the first fossil proboscidean species to be recorded in the scientific literature. Thanks to its abundant remnants discovered throughout the Northern Hemisphere, which include not only fossil skeletal elements but also well-preserved carcasses with soft tissues often found in the permafrost of the northernmost territories, *M. primigenius* is likewise one of the best-known extinct animals. Quite like most derived Proboscidea, this species was initially classified in the genus *Elephas*. Osborn (1942) used the name *Mammonteus* Osborn, 1924, but Maglio (1973) restored *Mammuthus* as the valid genus name according to the rules of zoological nomenclature. Many woolly mammoth fossils discovered during the nineteenth and early twentieth centuries were given a plethora of different names of species or subspecies rank, most of which were not accompanied by a diagnosis or even a description and were soon forgotten (Garutt 1964; see also the synonymy list in Maglio 1973, p. 60).

Type Material The species' original publication did not include a description, nor a designation of type material. Osborn (1942, p. 1122) selected as lectotypes two molars from Blumenbach's personal collection, deriving from Germany and Siberia, respectively. Gromova argued in 1965 (cited in Garutt et al. 1990) that only one of these specimens, the molar from Siberia, should be designated as the lectotype of the species. Maglio (1973, p. 60) shared the same opinion. However, the Blumenbach's collection seems to have been lost or destroyed during World War II. To address these problems, Garutt et al. (1990) proposed to designate an adult nearly complete male skeleton discovered in 1948 on the Taimyr Peninsula, N. Siberia, as the neotype of the species.

Type Locality Uppermost Pleistocene of Taimyr Peninsula, N. Siberia (place of origin of the species' neotype; Garutt et al. 1990).

Distribution As in the case of its immediate ancestor, *M. primigenius* was first emerged in the E. Asia and subsequently dispersed west to the rest of the Eurasia, and east to N. America through Beringia. In Europe, it is first recorded at about 200 ka (Lister et al. 2005) and became widespread, particularly in periods of cold climate. By the end of the Pleistocene the species populations started to recede to the North, probably because of the climate warming and expansion of forests, combined with hunting pressure by humans. The last *M. primigenius* finds from the European North have been ^{14}C dated to the Younger Dryas / Holocene boundary, about 11.4–11.8 ka (Ukkonen et al. 2011). The species' palaeogeographic range continued to shrink rapidly, until its ultimate extinction at about 3.7 ka on Wrangel Island, NE Siberia (Stuart et al. 2002).

Since this is a cold-adapted species, it is very rare in Greece, limited geographically to the country's northern part, and quite possibly also geochronologically to periods of cold climate. It has been tentatively identified (as *M.* cf. *primigenius*) on dental and osteological material in Angítis (E. Macedonia), dated to the Late Pleistocene (34–28 ka) (Koufos 1981; Trantalidou 2013) and Phílippi (E. Macedonia) on a molar fragment (Koulidou 2013). A third potential locality, according to Koulidou (2013), is Ravin de l'éléphant (Vathýlakkos, Central Macedonia), based on a molar with enigmatic morphology originally assigned to *P. antiquus* (de Bonis et al. 1973; see also below).

Taxonomic Remarks *Mammuthus primigenius* represents the last stage in mammoth evolution in Eurasia, a species adapted to life in cold and dry steppe–tundra habitats. Compared to its ancestral species *M. trogontherii*, it is characterised by a reduction in body size and the evolution of more advanced molars, with more lamellae and thinner enamel. Typical *M. primigenius* have third molars with 23–28 lamellae, though specimens with fewer lamellae (as few as twenty) have been recorded in Europe (Lister and Sher 2001; Lister and Stuart 2010). The hypsodonty index does not seem to have changed, with the range of 160–209 calculated for the rich Upper Pleistocene locality Předmostí (Czechia) (Lister and Sher 2001; Lister and Stuart 2010).

Several fossils from Greece have been referred to *M. primigenius*, but their identity is often doubtful. The dental specimen allegedly from Megalopolis Basin (Melentis 1961) clearly belongs to a woolly mammoth, but comes in fact from a Ukrainian locality (see Sect. 3). A molar from the Peniós estuary (Thessaly) (Paraskevaidis 1977, pl. 3, fig. 3) seems to have thick lamellae and enamel (the author did not provide a scale bar or measurements, while the whereabouts of the specimen is unknown). It is rather attributable to a less advanced mammoth species, or more probably to *P. antiquus*, which is common at several sites along the Peniós valley. Angítis (E. Macedonia) is the only site where the species seems to have been more frequent, though more diagnostic material is needed for a secure species-level determination (Koufos 1981; Trantalidou 2013). Recently, Koulidou (2013) referred to the woolly mammoth a third molar from Ravin de l'éléphant (Vathýlakkos, Central Macedonia), reported as *P. antiquus italicus* by de Bonis et al. (1973). The specimen consists of only 17 lamellae according to Koulidou (2013), a feature which differentiates it strongly from the *M. primigenius* hypodigm. Its completely unworn occlusal surface hampers its taxonomic identification. Its long and narrow crown may indicate an affinity to *P. antiquus* (in accordance to the original assignment by de Bonis et al. 1973), but the lamellar frequency is quite high (about 9 based on Koulidou's figures), consistent with the morphology of an advanced mammoth. The Vathýlakkos find is referred here to cf. *P. antiquus*, until more data become available. A partial molar from Phílippi referred by the same author (Koulidou 2013) to *M.* cf. *primigenius* is rather too incompletely preserved to be identified at the species level, though, also in this case, it is characterised by very high lamellar frequency, as in *M. primigenius*.

***Palaeoloxodon* Matsumoto, 1924**

Type Species *Elephas namadicus naumanni* Makiyama, 1924 (raised to the species level since the early 1970s).

Included Taxa The genus is known with a single species in continental Greece, *P. antiquus*.

Remarks Although *Palaeoloxodon* is a genus mainly occurring and studied in Europe and E. Asia, it is of African origin, just as the other elephantine genera (Osborn 1942; Aguirre 1969; Maglio 1973; Saegusa and Gilbert 2008). It is known to have existed in E. Africa at least since the early Pliocene with the species *P. ekorensis* (Maglio, 1970) and *P. recki* (Dietrich, 1916), which already exhibited the apomorphic characters of the genus, though in more ancestral states (Maglio 1973). *Palaeoloxodon* probably migrated to Eurasia during the terminal Early Pleistocene, at about 1.0 Ma (Saegusa and Gilbert 2008) and enjoyed a wide dispersal in the continent in habitats of temperate climate. *Palaeoloxodon* never migrated to America, quite probably because of the cold and arid steppe environment of NE Asia and Beringia, which was hostile to the populations of this genus. In Eurasia it became extinct during the last glacial period of the Late Pleistocene (Stuart 2005; Saegusa and Gilbert 2008).

Palaeoloxodon are readily distinguished from the mammoths on cranial characters, while there are also distinctive dental features. The skull is very large, high and rostro-caudally short, with a double-domed vertex dorsally and concave frontoparietal surface. The tusk alveoli are divergent (at least in advanced species) not closely positioned like *Mammuthus*. The tusks are weakly curved and twisted, in contrast to the spirally twisted and more curved tusks of *Mammuthus*. The molars exhibit low hypsodonty and low lamellae number in early forms, but have more lamellae and become very hypsodont in Pleistocene species. However, *Palaeoloxodon* never reached the level of lamellar multiplication observed in advanced mammoths. Moreover, *Palaeoloxodon* is characterised by the presence of a 'loxodont sinus', a thickening of the lamellae in their central area, observed as a large median fold on the occlusal surface, a persisting plesiomorphic feature not retained in advanced *Mammuthus*.

***Palaeoloxodon antiquus* (Falconer and Cautley, 1847)**

Nomenclatural and Taxonomical History This very common European elephant species was initially published as *Elephas antiquus* Falconer and Cautley, 1847, together with a very similar Asian species, *E. namadicus* Falconer and Cautley, 1847. Matsumoto (1924) introduced a new name for the Japanese samples, *Palaeoloxodon*, and placed it in subgeneric rank under the genus *Loxodonta*. Later in the same year, Osborn (1924) coined the genus names *Sivalikia* for the Asian samples and *Pilgrimia* for the Mediterranean dwarf forms and the African species

E. recki. The same author soon created a new genus for the European species, *Hesperoloxodon* Osborn, 1931, based on a skull from Italy. Finally, Osborn (1942, p. 1179) accepted the synonymy of *Sivalikia* and *Pilgrimia* with *Palaeoloxodon*, but retained *Hesperoloxodon* for the European non-insular samples. Aguirre (1969) considerably reduced Osborn's taxonomic splitting, using *Palaeoloxodon* as a genus name for all African and Eurasian species of the relevant morphology (*recki*, *namadicus*, *antiquus*, *falconeri*, etc.). Maglio (1973) synonymised *Palaeoloxodon* with *Elephas*, and *antiquus* with *namadicus*, basing his arguments on cranial morphology and accepting a certain degree of intraspecific variation. However, these suggestions were generally not followed by later authors, who continued to use both *Palaeoloxodon* (in genus or subgenus rank under *Elephas*) and *antiquus*, pending a large-scale revision of the *namadicus–antiquus* group. In the recent literature, *Palaeoloxodon* is consistently employed in genus rank (following Inuzuka and Takahashi 2004, Shoshani et al. 2007 and Todd 2010), and *antiquus* continues to be used exclusively for the European and Middle East populations.

Type Material The species is defined on a partially preserved mandibular part bearing the m2, which was designated as the lectotype by Osborn (1942, p. 1218).

Type Locality The type locality is unknown, but Palombo and Ferretti (2005) cited a personal communication with P. Davies and A. Lister, according to which the lectotype probably comes from the upper Middle Pleistocene deposits at Grays, a site in the terrace system of the river Thames (England).

Distribution *Palaeoloxodon antiquus* is by far the most common elephantine species in Southern Europe, expanding its range to the North during periods of warmer climate. The species occurs in hundreds of localities throughout Europe, particularly in Italy, Germany, and France (see, e.g., Palombo et al. 2010, fig. 6). In Greece, it is quite widespread geographically, occurring in numerous localities, from Western and Central Macedonia in the North to Southern Peloponnese and Kýthera Island in the South and from Kephallenía Island in the West to Kálymnos Island in the East. Yet unpublished material includes tusks from Alivéri (Euboea) and Nissí (NW Peloponnese) (Museum of Paleontology and Geology, National and Kapodistrian University of Athens). The former was believed to belong to *M. meridionalis* (see Doukas and Athanassiou 2003), but has clear *P. antiquus* morphology. A left m3 from Philótas (NW Macedonia), curated in the Petralona Museum (Chalidikí) may also belong to this species.

Palaeoloxodon antiquus first appeared in Europe just before the beginning of the Middle Pleistocene, at about 0.9–0.8 Ma and became extinct in Northern Europe at the end of the Last Interglacial (Eemian), as its range contracted to the South following the retreat of the temperate vegetation to the same direction (Stuart 2005)—note, nevertheless, that *P. antiquus* molars from the Netherlands have been dated to as late as 32.5 ka (Bosscha Erdbrink et al. 2001; Mol et al. 2007). Southern Europe acted as a refugium for temperate-adapted floras and faunas during the glacial periods of the Pleistocene, and this was also the case during the Last Glacial, allowing for a later extinction date in this region. On the Italian Peninsula, this species continued to constitute a common element of the Late Pleistocene faunas and became extinct sometime during the MIS 3 (about 60–27 ka) (Palombo and Ferretti 2005). On the Iberian Peninsula, *P. antiquus* is reported to have persisted until about 33 ka (Sousa and Figueiredo 2001), while a more recent date of 23.5 ka appears to be controversial (Stuart 2005).

The Greek fossil record of *P. antiquus* is in most cases not dated using physical methods (either directly on the elephant fossils, or on associated samples). Only three samples have been dated so far. The geochronologically oldest dated find is the skeleton from Marathousa 1 (Megalopolis Basin; Fig. 11), which yielded an age estimation of 0.50–0.42 Ma (MIS 12), resulting from combining palaeomagnetic, radiometric, and optical methods (Blackwell et al. 2018; Jacobs et al. 2018; Tourloukis et al. 2018). A more recent elephant skeleton from Ambélia, Grevená, was dated to MIS 6 (165–170 ± 25 ka) using the Electron Spin Resonance method (Bassiakos in Tsoukala and Lister 1998). Last, a maxilla from Póros, Kephallenía Island, was dated by Theodorou et al. (2018) at 104.2 ± 18.5 ka, using Infrared Optically Stimulated Luminescence Dating of feldspar. An additional locality is dated indirectly. This is Peniós Valley, located west of the city of Lárissa (Thessaly), where the fossils are found in the river's lower terrace. The age of these deposits has been estimated to 45–30 ka, based on ^{14}C dating of freshwater molluscs and U/Th dating of pedogenic carbonates from several sites in the river banks (Demitrack 1986), while Runnels and van Andel (1993) extended this range back to 60–30 ka, based on archaeological correlations. This is the most recent date obtained for *P. antiquus* samples in Greece (Athanassiou 2011).

Taxonomic Remarks *Palaeoloxodon antiquus* is a very large-bodied elephant with largest male individuals estimated to have stood 4 m high at the shoulders (Larramendi et al. 2017). It is characterised by a massive rostrocaudally compressed skull, with a two-bulged vertex. Dorsally the skull often forms a strong parieto-occipital crest that folds rostrally over the frontal surface and overhangs the external nares. Another prominent feature of this species is the long and fan-shaped premaxillaries, which widen distally in accord to the strongly divergent tusks (see, e.g., Fig. 12b). The mandible is deep, to accommodate the hypsodont molars, and has a short symphysis. The tusks are weakly curved and exhibit only slight torsion. Their dentine structure is manifested in cross sections as curved Schreger lines that form obtuse angles near the tusk periphery (Palombo and Villa 2001; see also Fig. 5). The molars are narrow and hypsodont, with coarsely folded, rather thick enamel that forms a 'loxodont sinus', i.e., a central thickening of the lamellae expressed on the occlusal surface as a median widening of the enamel loops. Incipiently worn lamellae usually present a 'dot-dash-dot' wear pattern, i.e., a central wide elliptical enamel loop, and two smaller of more circular shape on the lingual and buccal side. The third molars (M3/m3) normally have 12–18 plates (Maglio 1973) but may reach 20 in exceptional cases (Palombo and Ferretti 2005). The lamellar frequency ranges between 4.4 and 7.7 lamellae per 10 cm of molar length (Maglio 1973), usually about 5–6.

The question on the taxonomic relationship between the European *P. antiquus* and the Asian *P. namadicus* remains still open, because of the need for a wide-scale revision of the available cranial material. Maglio (1973) considered the two species as synonyms, with *P. namadicus* having priority, based on the absence of dental differences and the considerable variability observed in the Indian cranial samples, which incorporate the morphologies found in Europe. Indeed, the characteristic development of a parieto-occipital crest observed in Indian samples (the so-called '*namadicus* morph' of Saegusa and Gilbert 2008) is also found in Italy (e.g., in Pian dell'Olmo, La Polledrara di Cecanibbio, and Pignataro Interamna; Maccagno 1962; Palombo and Ferretti 2005; Anzidei et al. 2012). Skulls with a weak parieto-occipital crest (the so-called 'Stuttgart morph' of Saegusa and Gilbert 2008) are more common in Germany (e.g., in Steinheim an der Murr and Neumark-Nord; Osborn 1942, p. 1254; Palombo et al. 2010), but this morphology is also present in India (Saegusa and Gilbert 2008). Saegusa and Gilbert (2008) considered three alternative hypotheses to explain the occurrence of both these morphologies: (a) they reflect pronounced intraspecific variation; (b) they represent two distinct taxa, of which the more advanced (*namadicus* morph) replaced the ancestral one; and (c) they represent two distinct taxa occurring contemporaneously. The authors considered option (a) as the least plausible and discussed the other two. In the case of option (b), they place the taxonomic replacement sometime during, or slightly before, MIS 9.

Concerning the Greek samples, the morphology of the parietofrontal region is observable in two skulls, one from Léfktro (Melentis 1965) and one from Marathousa 1 (Konidaris et al. 2018). Both bear prominent parieto-occipital crests (Fig. 12) that bend rostrally over the nasal opening. The skull from Léfktro (Fig. 12a) seems to have had a somewhat stronger crest that bends down closer to the nares (*contra* Melentis 1965, who stated that there is no such crest). A more accurate assessment is not possible, however, because the region is partially eroded before burial and partially damaged, possibly during excavation or preparation. Nevertheless, none of the Megalopolis Basin skulls exhibits the extreme infolding of the crest observed in the specimens from Godávari, India (Osborn 1942, p. 1250) and La Polledrara di Cecanibbio, Italy (Palombo and Ferretti 2005; Anzidei et al. 2012; but see also Larramendi et al. 2020 for a re-evaluation of the condition seen in this skull). Léfktro and Marathousa 1 are geographically adjacent, the former being located about 8 km SSE of the latter. Their exact relative stratigraphic position is unknown, but both derive from organic-rich lacustrine sediments of the Marathousa Member of the Choremi Formation within the Megalopolis Basin (Vinken 1965; Tourloukis et al. 2018) and are expected to have a similar geological age. The development of the parieto-occipital crest in other skulls from Greek localities, like those from Grevená (Tsoukala and Lister 1998) and Amýntaio (Kevrekidis and Mol 2016) is unknown because the relevant cranial part is not preserved. The Perdíkkas elephant skull apparently preserves this region, but it is not observable, since the skull remains in situ lying on its frontal plane (Fig. 7).

The Greek fossil record sheds some light to the problem of the relationship between the '*namadicus*' and the 'Stuttgart' morphs. The early presence of '*namadicus* morph' skulls in Greece (Marathousa 1 is dated to MIS 12; Blackwell et al. 2018; Jacobs et al. 2018; Tourloukis et al. 2018), with respect to the more recent dating of 'Stuttgart morph' specimens from Germany (MIS 11 or 9 in Steinheim, van Asperen 2013; MIS 7 in Neumark-Nord, Palombo et al. 2010), is not consistent with the hypothesis (b) of Saegusa and Gilbert (2008), according to which the '*namadicus* morph' succeeded the 'Stuttgart morph' during MIS 9. Thus, the two morphologies are not evolutionarily related and must be referred either to different conspecific populations or to different species [hypotheses (a) and (c), respectively, in Saegusa and Gilbert 2008]. According to an additional option, favoured here, the two morphs and their intermediate stages might be an effect of intra-population variation, occurring mainly as a result of sexual dimorphism and ontogeny. If the parieto-occipital crest primarily functioned as a muscle insertion area, then its development might reflect the relative size of the tusks, which is expected to vary significantly among male, female, young, and old individuals. Quite recently, Larramendi et al. (2020) revisited the issue, examining the development of the parieto-occipital crest in cranial specimens from Africa, Asia, and Europe, and concluded that, indeed, the European samples cannot be divided into two species based on this feature. Evaluating a wider range of cranial and postcranial characters, they retained, though, the specific distinction between the European and the S. Asian straight-tusked elephants (*P. antiquus* and *P. namadicus*, respectively).

Unavailable Taxa

Elephas gortynius Skuphos, *Elephas antiquus* var. *gortyniensis* Depéret, 1913, and *Elephas minor* Skuphos

These three species-group names appear in the literature, being attributed directly or indirectly to Th. Skuphos, and referring to the elephant fossils he excavated in 1902 in the Megalopolis Basin (see Sect. 3). In particular, Bürchner (1903) and Georgalas (1929) mentioned *E. gortynius* as a species-rank name coined and used in lectures by Skuphos. According to these authors, the name *gortynius* referred to an allegedly complete skeleton from the area of Íssoma, which is not available presently, at least as a whole. The find was not described, except for a vague mentioning of its large size. Georgalas (1929) added *E. minor* as well, but in its Greek version, not latinised. According to him, the name *minor* was based on a very small molar (not described), which apparently belongs to a juvenile individual. Both names are considered herein as *nomina nuda*. This is because, to my knowledge, none of them has been ever published by Skuphos in any form of scientific or popular publication (that is why their authorships lack a publication year), and consequently they do not fulfil the criteria of availability (Art. 11 of the ICZN). Moreover, although they were treated as valid names by the two authors mentioned above, the corresponding samples remained undefined and undescribed by them. Thus, their publications do not meet the criteria of Art. 12 of the ICZN, and the names can neither be attributed to Bürchner (1903) nor to Georgalas (1929), who used them.

Depéret (1913), when referring to the material excavated by Skuphos, also uses the first name (as a 'variation' of *E. antiquus*, that is of subspecies rank), but with the different spelling '*gortyniensis*', followed by the designation 'n. var.'. Supposedly, Depéret was aware of the unpublished status of this name and published it, but also in this case without any description of any kind. Since *E. antiquus gortyniensis* lacks a description, it is also a *nomen nudum*, according to the Art. 12 of the ICZN.

Invalid Taxa

In two cases, authors erected new species-group names for elephantid samples from Greece. Both names conform to the requirements of the ICZN, thus they are available, but are considered herein as junior synonyms of pre-existing names.

***Archidiskodon meridionalis proarchaicus* Melentis, 1966**
(= *Mammuthus meridionalis*)

Type Material AMPG 1964/449, left upper third molar (M3); Museum of Paleontology and Geology, University of Athens. Described by Melentis (1966a).

Type Locality Symbolí, (Sérres, E. Macedonia); Early Pleistocene.

Remarks This subspecies was defined on a single molar, described as presenting archaic morphological characters, which allegedly distinguish it as a basal subspecies of *M. meridionalis*. Indeed, the specimen is characterised by low lamellar frequency and thick enamel. However, given the considerable individual and wear-dependent morphological variability of the elephantid molars, a single specimen cannot be considered as a representative of a taxonomically distinct population. Moreover, despite the presence of characters that can be considered ancestral, others are derived (e.g., the molar's height). In general, the molar AMPG 1964/449 is morphologically and metrically comparable to published samples of *M. meridionalis* (e.g., Maglio 1973; Lister and Sher 2015) and does not present any special features to justify a taxonomic separation. As such, *A. meridionalis proarchaicus* is considered a junior synonym of *Mammuthus meridionalis*. To my knowledge, this species-group name has never been used again for other samples.

***Elephas (Palaeoloxodon) cephallonicus* Theodorou et al., 2018**
(= *Palaeoloxodon antiquus*)

Type Material AMPG 900, fragment of maxilla with molars on both sides; Museum of Paleontology and Geology, University of Athens. Described by Theodorou et al. (2018).

Type Locality North coast of Póros, SE Kephallenía Island; early Late Pleistocene (104.2 ± 18.5 ka) (Theodorou et al. 2018).

Remarks This species was erected as an endemic, incipiently dwarfing *Palaeoloxodon* elephant taxon. The single specimen studied by the authors is dimensionally smaller than some continental samples of *P. antiquus*, but still remains large sized, having molar widths very close to the mean values of *P. antiquus* (according to Maglio 1973). As shown by Athanassiou et al. (2019), the Kephallenía specimen does not differentiate metrically from the mainland *Palaeoloxodon* of Greece, being larger than certain specimens from the Megalopolis Basin. Thus, AMPG 900 is not sufficient for the documentation of a new endemic species, which would require a better statistical sample to show the alleged size reduction trend. The palaeogeography of the area is also not favourable for the isolation of a *Palaeoloxodon* population on the island. During the early Late Pleistocene, when this individual was living, the sea strait separating Kephallenía from the mainland would be less than 8 km wide, a small distance considering the swimming capabilities of the elephants. For these reasons, *P. cephalonicus* is considered herein as a junior synonym of *P. antiquus*.

5 Conclusions

The elephantid fossil record of Greece is known from dental and osteological samples deriving from about 90 sites throughout the country's territory (see Appendix) and spanning geochronologically from the latest Pliocene to the Late Pleistocene (Fig. 13). The best-represented species is *Palaeoloxodon antiquus*, well known from skeletons, cranial, and postcranial material. The genus *Mammuthus* occurs less frequently, but is diverse taxonomically, represented by four species: *M.* ex gr. *rumanus*, *M. meridionalis*, *M. trogontherii*, and *M.* cf. *primigenius*. The relative abundance of *Palaeoloxodon* over *Mammuthus* during the Middle and Late Pleistocene is in accordance with the milder climatic conditions of S. Europe, which favoured the temperate species over the cold-adapted ones. Ongoing research in elephantid-bearing sites, particularly those of the Megalopolis Basin, is expected to yield additional data on the morphology and palaeoecology of the family in the SE extremities of Europe.

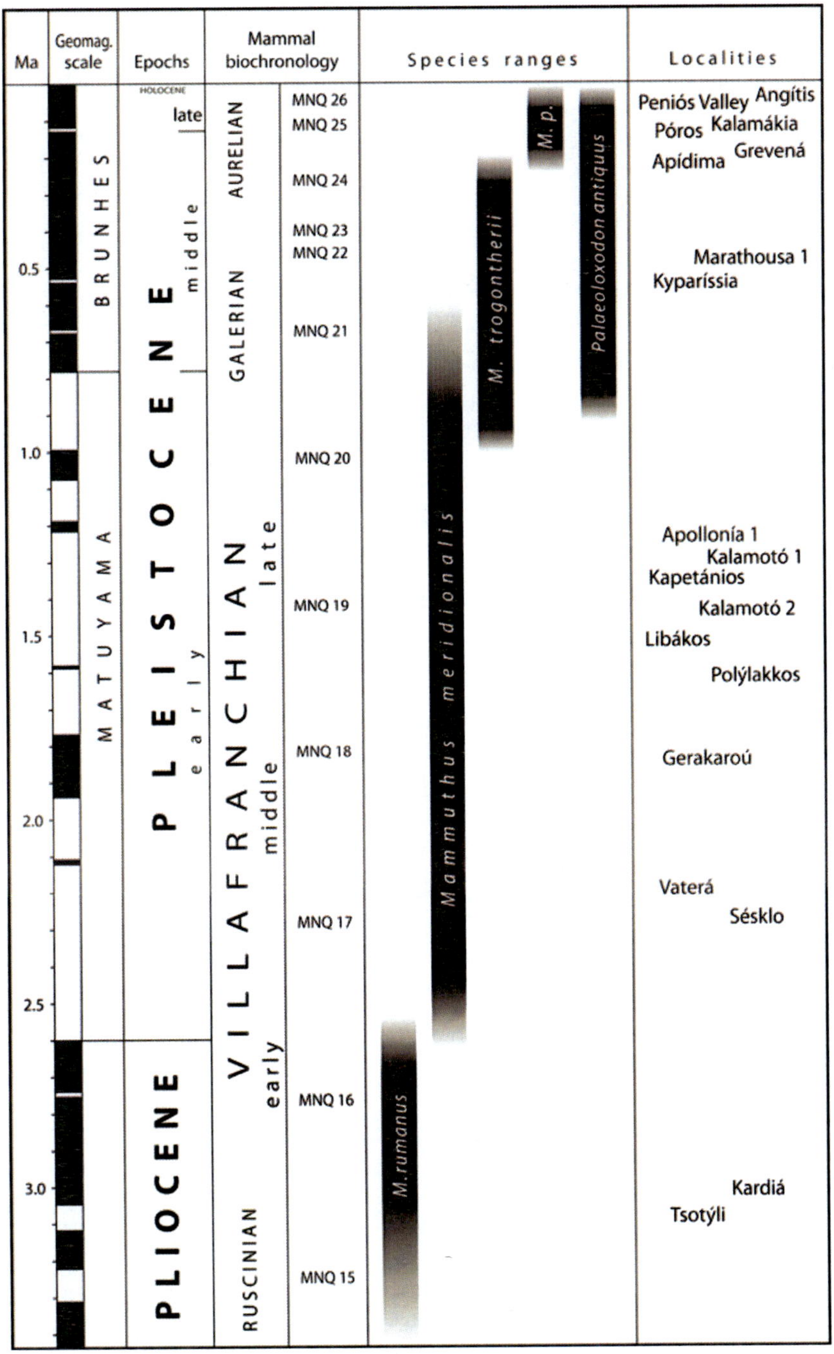

Fig. 13 Geochronologic chart showing the distributions of the elephantid species that occur in Greece and the stratigraphic positions of well-dated *Mammuthus*- and *Palaeoloxodon*-bearing Greek localities. Geochronologic and magnetostratigraphic subdivisions according to Pillans and Gibbard (2012). Mammal biozonation (MNQ zones) according to Guérin (1990). The biochronologic ranges of the *Mammuthus* species follow the concept introduced by Lister et al. (2005). '*M. p.*' stands for *Mammuthus primigenius*

Acknowledgments S. Roussiakis and G. Lyras (National and Kapodistrian University of Athens), G. Konidaris (Eberhard Karls University of Tübingen), and E. Tsoukala (Aristotle University of Thessaloniki) are thanked for providing access to fossil samples and for useful discussions on species distributions. L. Bellucci helped with hard-to-find Italian literature. G. Konidaris and R. Rabinovich reviewed the original manuscript, making significant comments and remarks, which essentially improved the content of this chapter.

Appendix

List of Greek localities with known occurrences of fossil elephants and mammoths (including some highly tentative ones), arranged in approximate geochronological order. The cited ages are in most cases broadly estimated, based on associated faunas (if any) and the current biostratigraphic framework. The taxonomic identification follows the present revision. Locality numbers refer to the collection numbers of the PaleoBiology Database (PBDB)

Localities[PBDB No]	Age (biozone or chronometric dating)	Taxon
Angítis[204391]	Late Pleistocene (34–28 ka)	*Mammuthus* cf. *primigenius*[1]
[a]Peniós Valley[204397]	Late Pleistocene (45–30 ka)	*Palaeoloxodon antiquus*[2]
Kalamákia Cave[184245]	Late Pleistocene (100–40 ka)	*Palaeoloxodon antiquus*[3]
Phílippi[204747]	Late Pleistocene	*Mammuthus* cf. *primigenius*[4]
Tsákoni[204402]	Late Pleistocene	cf. *Palaeoloxodon antiquus*[5]
Peniós estuary	Late Pleistocene	*Palaeoloxodon antiquus*[6]
Póros (Kephallenía)[204403]	early Late Pleistocene (104 ka)	*Palaeoloxodon antiquus*[7]
[b]Íssoma[204546]	late Middle (–Late?) Pleistocene	*Palaeoloxodon antiquus*[8, *]
Íssoma KYT	late Middle (–Late?) Pleistocene	*Palaeoloxodon antiquus*[*]
Symbolí[204807]	Middle–Late Pleistocene	*Palaeoloxodon antiquus*[9]
Xeriás[182679]	Middle–Late Pleistocene	*Palaeoloxodon antiquus*[10]
Ravin de l'éléphant[204549]	Middle–Late Pleistocene	cf. *Palaeoloxodon antiquus*[4,11,*]
Philótas	Middle–Late Pleistocene	*Palaeoloxodon antiquus*[*]
Pétres	Middle–Late Pleistocene	*Palaeoloxodon antiquus*[10,12]
Amýntaio[204550]	Middle–Late Pleistocene	*Palaeoloxodon antiquus*[13]
Sotíras[182678]	Middle–Late Pleistocene	*Palaeoloxodon antiquus*[10]
Néa Kómi[204754]	Middle–Late Pleistocene	*Palaeoloxodon antiquus*[14,*]
Perdíkkas[204753]	Middle–Late Pleistocene	*Palaeoloxodon antiquus*[10,15]
Pentávryssos	Middle–Late Pleistocene	Elephantidae indet.[16, *]
Ptolemaís[204552]	Middle–Late Pleistocene	*Palaeoloxodon antiquus*[17]
Allatíni	Middle–Late Pleistocene	?*Palaeoloxodon* sp.[18]
Trílophos[204743]	Middle–Late Pleistocene	?*Palaeoloxodon* sp.[18]
Epanomí[204744]	Middle–Late Pleistocene	?*Palaeoloxodon* sp.[18]

Localities[PBDB No]	Age (biozone or chronometric dating)	Taxon
Petrálona Cave[183123]	Middle–Late Pleistocene	Elephantidae indet.[*]
Kalonéri[182677]	Middle–Late Pleistocene	*Palaeoloxodon antiquus*[10]
[c]Tsotýli	Middle–Late Pleistocene	*Palaeoloxodon antiquus*[19]
Siátista	Middle–Late Pleistocene	*Palaeoloxodon antiquus*[20]
Terpsithéa[182680]	Middle–Late Pleistocene	*Palaeoloxodon antiquus*[10]
Mesorráchi	Middle–Late Pleistocene	Elephantidae indet.[21]
Chálki	Middle–Late Pleistocene	*Palaeoloxodon antiquus*[22]
Néa Léfki[182681]	Middle–Late Pleistocene	*Palaeoloxodon antiquus*[10]
Límni (Euboea)	Middle–Late Pleistocene	*Palaeoloxodon antiquus*[20]
Trichonía	Middle–Late Pleistocene	*Palaeoloxodon antiquus*[20]
Patras[204554]	Middle–Late Pleistocene	*Palaeoloxodon antiquus*[23]
Nissí[204555]	Middle–Late Pleistocene	*Palaeoloxodon antiquus*[24,*]
Canal of Corinth[204553]	Middle–Late Pleistocene	*Palaeoloxodon antiquus*[23]
Roupáki[204556]	Middle–Late Pleistocene	*Palaeoloxodon antiquus*[25]
Símiza	Middle–Late Pleistocene	*Palaeoloxodon antiquus*[*]
Kálymnos[34770]	Middle–Late Pleistocene	*Palaeoloxodon antiquus*[26]
Apídima Cave C[32082]	Middle–Late Pleistocene	Elephantidae indet.[27]
Cape Cheládi, Kýthera[204298]	Middle–Late Pleistocene	*Palaeoloxodon antiquus*[28]
Alivéri	Middle–Late Pleistocene	*Palaeoloxodon antiquus*[*]
Ambélia (Grevená)[204557]	late Middle Pleistocene	*Palaeoloxodon antiquus*[29]
Phílippi[204749]	Middle Pleistocene	*Mammuthus trogontherii*[4]
Sotíras[182678]	Middle Pleistocene	*Mammuthus trogontherii*[4,18]
Lágoura (Néo Kostarázi)[204558]	Middle Pleistocene	Elephantidae indet.[30]
Tsotýli[204751]	Middle Pleistocene	*Mammuthus* cf. *trogontherii*[4]
Megálo Monastíri[204560]	Middle Pleistocene	Elephantidae indet.[31]
Megalópolis Basin (undefined sites)[182721]	Middle Pleistocene	Elephantidae indet.[32]
Marathoúsa 1[187637]	Middle Pleistocene (0.48–0.42 Ma)	*Palaeoloxodon antiquus*[33]
Kyparíssia 1[194472]	Middle Pleistocene	*Palaeoloxodon antiquus*[34]
Kyparíssia 3[194474]	Middle Pleistocene	*Palaeoloxodon antiquus*[34]
Kyparíssia 4[194475]	Middle Pleistocene	*Palaeoloxodon antiquus*[34]
[d]Kalývia Karyón	Middle Pleistocene	*Palaeoloxodon antiquus*[35]
[e]Léfktro	Middle Pleistocene	*Palaeoloxodon antiquus*[36]
Leontári	Middle Pleistocene?	Elephantidae indet.[37]
Loussiká[204561]	early Middle Pleistocene	*Mammuthus trogontherii*[38]
Polýlakkos	early Middle Pleistocene?	*Mammuthus trogontherii*[18]
Apollonía[34784]	latest Early Pleistocene (MNQ19)	*Mammuthus meridionalis*[39]
Kalamotó-1[200083]	latest Early Pleistocene (MNQ19)	*Mammuthus meridionalis*[40]
Kalamotó-2[200084]	latest Early Pleistocene (MNQ19)	*Mammuthus meridionalis*[40]

Localities[PBDB No]	Age (biozone or chronometric dating)	Taxon
Platanochóri-1[204658]	latest Early Pleistocene (MNQ19)	Elephantidae indet.[41]
Kapetánios[34781]	Early Pleistocene (MNQ19)	*Mammuthus meridionalis*[42]
Libákos[34764]	Early Pleistocene (MNQ19)	*Mammuthus meridionalis*[42]
Halykés[34782]	Early Pleistocene (MNQ19)	Elephantidae indet.[43]
Tsiótra Vrýssi	Early Pleistocene (MNQ18–19)	*Mammuthus meridionalis*[44]
Gerakaroú[34617]	Early Pleistocene (MNQ18)	*Mammuthus meridionalis*[45]
Polýlakkos[34763]	Early Pleistocene (MNQ18?)	*Mammuthus* cf. *meridionalis*[43]
Sésklo[34614]	Early Pleistocene (MNQ17)	*Mammuthus meridionalis*[44]
Vaterá-DS[183344]	Early Pleistocene (MNQ17)	*Mammuthus* cf. *meridionalis*[46]
Vaterá-U[183346]	Early Pleistocene (MNQ17)	*Mammuthus* cf. *meridionalis*[46]
Symbolí[204808]	Early Pleistocene	*Mammuthus meridionalis*[9,*]
Phílippi[204750]	Early Pleistocene	*Mammuthus meridionalis*[4]
Epanomí[204660]	Early Pleistocene	*Mammuthus* cf. *meridionalis*[47]
[f]Kípi[204933]	Early Pleistocene	*Mammuthus* cf. *meridionalis*[48]
Peponiá	Early Pleistocene	*Mammuthus meridionalis*[49]
Tsotýli	Early Pleistocene	*Mammuthus meridionalis*[18]
Trapezítsa	Early Pleistocene	Elephantidae indet.[*]
Q-Profil[34812]	Early Pleistocene	*Mammuthus* cf. *meridionalis*[43]
Ioánnina[204740]	Early Pleistocene	*Mammuthus meridionalis*[50]
Giáltra[204804]	Early Pleistocene	*Mammuthus meridionalis*[51]
Reghínio[204661]	Early Pleistocene	*Mammuthus* cf. *meridionalis*[52]
Zéli	Early Pleistocene?	*Mammuthus* sp.[*]
Lefkóchoma[204741]	Early Pleistocene	Elephantidae indet.[53]
Vlachiótis[204742]	Early Pleistocene	*Mammuthus meridionalis*[54]
Antimáchia[207130]	Early Pleistocene	*Mammuthus meridionalis*[55]
Kardámaena[204662]	Early Pleistocene	*Mammuthus meridionalis*[56]
Almyrí	Early Pleistocene	*Mammuthus meridionalis*[56]
Límni (Vromonéra)[204934]	Pleistocene	Elephantidae indet.[57]
Tsotýli[204400]	late Pliocene	*Mammuthus* ex gr. *rumanus*[4,58,*]
Kardiá[204752]	late Pliocene	*Mammuthus* ex gr. *rumanus*[4,*]

GPTS Geomagnetic Polarity Time Scale, *ELMA* European Land Mammal Age, *MNQ* Mammal Neogene–Quaternary Zone

[a]Also mentioned as 'Lárissa' in old publications (e.g., Georgalas 1929; Paraskevaidis 1956)

[b]Also appears in the literature under the names 'Hágios Ioánnis tis Vathiás Choúnis' (e.g., Bürchner 1903; Melentis 1961) and 'Grána tou Skoufou' (Theodorou 2014)

[c]According to the sketchy map provided by Melentis (1966b), this locality is rather in the area of Kalonéri or Polýlakkos, than Tsotýli
[d]Also appears in the literature as 'Musaklá' (Bürchner 1903; Melentis 1961)
[e]Reported as 'Biláli' in the original publication
[f]This locality name is recorded here for the first time, as in the original publication it was referred to vaguely as 'Neápolis area'
[*]This study
[1]Koufos (1981), [2]Boessneck in Milójčić et al. (1965), Athanassiou (2001, 2011), [3]Harvati et al. (2013), [4]Koulidou (2013), [5]Athanassiou (2004), [6]Paraskevaidis (1977), [7]Theodorou et al. (2018), Athanassiou et al. (2019), [8]Melentis (1961, 1963), Theodorou (2014), [9]Melentis (1966a), [10]Tsoukala et al. (2011), [11]de Bonis et al. (1973), [12]Velitzelos and Schneider (1973), [13]Kevrekidis and Mol (2016), [14]Mitzopoulos (1967), [15]Poulianos (1986), [16]Stratigopoulos (2008), [17]Astre (1966), [18]Marinos (1964), [19]Melentis (1966b), [20]Georgalas (1929), [21]reported anonymously in the local newspaper 'Mikrá', number 305, June 6, (1907), [22]Schneider (1968), [23]Depéret (1913), [24]Agiadi and Theodorou (2005), [25]Athanassiou (2000), [26]Masseti (2006), [27]Tsoukala (1999), [28]Kuss (1967), Athanassiou et al. (2019), [29]Tsoukala and Lister (1998), [30]Harvati et al. (2008), [31]Runnels and van Andel (1993), [32]Sickenberg (1976), [33]Konidaris et al. (2018), [34]Athanassiou (2018), [35]Melentis (1961, 1963), [36]Melentis (1965), [37]Mitzopoulos et al. (1862), [38]Athanassiou (2012), [39]Konidaris et al. (2020), [40]Tsoukala and Chatzopoulou (2005), [41]Konidaris et al. (2015), [42]Steensma (1988), [43]Athanassiou (1998), [44]Konidaris et al. (2016), [45]Athanassiou and Kostopoulos (2001), [46]de Vos et al. (2002), [47]Athanassiou and Kostopoulos (2010), [48]Athanassiou (2006b), [49]Hilber (1894), [50]Melentis (1960), [51]Psarianos and Thenius (1954), [52]Athanassiou (2006a), [53]Symeonidis (1970), [54]Symeonidis and Theodorou (1986), [55]Forsyth Major (1887), Airaghi (1928), Desio (1931), [56]Desio (1931), Charrier and Giglio (1969), [57]Paraskevaidis (1956), [58]Kostopoulos and Koulidou (2015)

References

Adam KD (1988) Über pleistozäne Elefanten-Funde im Umland von Erzurum in Ostanatolien. Ein Beitrag zur Namengebung von *Elephas armeniacus* und *Elephas trogontherii*. Stuttgarter Beitr Naturkd 146:1–89

Agiadi K, Theodorou G (2005) Tusk paleohistology as a tool in the discrimination of fossil tusks from Greece. In: Alcover JA, Bover P (eds) Proceedings of the International Symposium "Insular Vertebrate Evolution: the Palaeontological Approach", Palma de Mallorca, 2003. Monografies de la Societat d'Història Natural de les Balears, pp 1–8

Aguirre E (1969) Evolutionary history of the elephant. Science 164:1366–1376

Airaghi C (1928) Mammiferi pliocenici dell'isola di Coo (Dodecaneso). Atti Soc It Sci Nat Mus Civ St Nat Milano 67:125–135

Albayrak E (2017) A new primitive elephantid from Turkey. Quat Int 445:43–49

Andrews CW, Forster Cooper C (1928) On a specimen of *Elephas antiquus* from Upnor. British Museum (Natural History), London

Anzidei AP, Bulgarelli GM, Catalano P, Cerilli E, Gallotti R, Lemorini C, Milli S, Palombo MR, Pantano W, Santucci E (2012) Ongoing research at the late Middle Pleistocene site of La Polledrara di Cecanibbio (central Italy), with emphasis on human–elephant relationships. Quat Int 255:171–187

Astre G (1966) *Elephas antiquus* du bassin de Kozani–Ptolémais en Macédoine. Bull Soc Hist Nat Toulouse 102:292–294

Athanasiu S (1915) Resturile de mamifere pliocene superioare dela Tuluceşti în Districtul Covurlui. An Inst Geol Rom 6:408–415

Athanassiou A (1998) [Contribution to the study of the fossil mammals of Thessaly]. PhD thesis (1996), National and Kapodistrian University of Athens. Gaia 5:1–393. (in Greek)

Athanassiou A (2000) Presence of fossil elephants in the area of Peniós valley (NW Peloponnesus, Greece). Ann Géol Pays Hellén 38:63–76

Athanassiou A (2001) New findings of fossil large mammal remains in the Peniós Valley (area of Larissa, Thessaly, Greece). Bull Geol Soc Greece 34:533–539

Athanassiou A (2002) Neogene and quaternary faunas of Thessaly. Ann Géol Pays Hellén 39:279–293

Athanassiou A (2004) A new elephant site in the Haliákmon River valley (W. Macedonia, Greece). In: Proceedings of the 5th international symposium on Eastern Mediterranean Geology, Thessaloniki, vol 2004, pp 299–302

Athanassiou A (2006a) Reghínio, a new mammal locality from the Plio-Pleistocene of Central Greece. N Jb Geol Pal Mh 2006:116–128

Athanassiou A (2006b) Biostratigraphical implications of the fossil tusks' structure. Bull Geol Soc Greece 39:9–16. (in Greek)

Athanassiou A (2010) An elephant (*Elephas antiquus* Falconer & Cautley, 1847) skeleton from Lousiká (NW Peloponnese, Greece). Quaternaire hors-série 3:74

Athanassiou A (2011) The Late Pleistocene fauna of Peneiós Valley (Lárissa, Thessaly, Greece): new collected material. In: van der Geer A, Athanassiou A (eds) 9th European Association of Vertebrate Palaeontologists meeting, Heraklion, 2010. Abstracts, p 14

Athanassiou A (2012) A skeleton of *Mammuthus trogontherii* (Proboscidea, Elephantidae) from NW Peloponnese, Greece. Quat Int 255:9–28

Athanassiou A (2016) Craniomandibular remains of *Anancus arvernensis* (Proboscidea, Mammalia) from Greece: The samples from Kallíphytos (E. Macedonia) and Sésklo (Thessaly). Quat Int 406:25–34

Athanassiou A (2018) Pleistocene vertebrates from the Kyparíssia lignite mine, Megalopolis Basin, S. Greece: Rodentia, Carnivora, Proboscidea, Perissodactyla, Ruminantia. Quat Int 497:198–221

Athanassiou A, Kostopoulos DS (2001) Proboscidea of the Greek Pliocene-Early Pleistocene faunas; biostratigraphic and palaeoecological implications. In: Cavarretta G, Gioia P, Mussi M, Palombo MR (eds) Proceedings of the 1st international congress "The World of Elephants", Rome, vol 2001, pp 85–90

Athanassiou A, Kostopoulos DS (2010) On mammal remains in the area of Epanomi (Macedonia, Greece). In: Christofides G, Kantiranis N, Kostopoulos DS, Chatzipetros AA (eds) Proceedings of the XIX Congress of the Carpathian-Balkan Geological Association, Thessaloniki, 2010. Sci Ann School Geol, Fac Sci, Aristotle Univ Thessaloniki, pp 63–68

Athanassiou A, van der Geer AAE, Lyras G (2019) Pleistocene insular Proboscidea of the Eastern Mediterranean: a review and update. Quat Sci Rev 218:306–321

Azzaroli A (1977) Evolutionary patterns of Villafranchian elephants in Central Italy. Atti Accad Naz Lincei, Mem 14:149–168

Baygusheva V, Titov V (2012) The evolution of Eastern European meridionaloid elephants' dental characteristics. Quat Int 255:206–216

Baygusheva VS, Vangengeim EA, Tesakov AS, Titov VV, Shvyreva AK (2011) Comments on the article by E.N. Mashchenko "Evolution of the elephants of the mammoth lineage in the northern Caucasus (Russia) in the Pliocene and Pleistocene". Proc Zool Inst RAS 315:483–493

Beden M (1979) Les éléphants (*Loxodonta* et *Elephas*) d'Afrique orientale: systématique, phylogénie, intéret biochronologique. Thèse, Université de Poitiers

Beden M (1983) Family Elephantidae. In: Harris JM (ed) Koobi Fora Research Project, vol. 2: The fossil ungulates. Proboscidea, Perissodactyla, and Suidae, Oxford, pp 40–129

Blackwell BAB, Sakhrani N, Singh IK, Gopalkrishna KK, Tourloukis V, Panagopoulou E, Karkanas P, Blickstein JIB, Skinner AR, Florentin JA, Harvati K (2018) ESR dating ungulate teeth and molluscs from the Paleolithic site Marathousa 1, Megalopolis Basin, Greece. Quaternary 1:22. (1–25)

Bonis L de, Bouvrain G, Keraudren B, Melentis J (1973) Premiers résultats des fouilles récentes en Grèce septentrionale (Macédoine). C R Hebd Séances Acad Sci 277:1431–1434

Bosscha Erdbrink DP, Brewer JG, Mol D (2001) Some remarkable Weichselian elephant remains. Deinsea 8:21–26
Bürchner L (1903) Wichtige Funde fossiler Knochen in Arkadien. Ber Naturwiss Ver Regensburg 9:119–123
Charrier G, Giglio A (1969) Primi risultati di una campagna di rilevamento geologico nell'isola di Coo (Sporadi meridionali – Mare Egeo). Boll Assoc Min Subalp 6:482–516
Christiansen P (2004) Body size in proboscideans, with notes on elephant metabolism. Zool J Linn Soc 140:523–549
Coppens Y, Maglio VJ, Madden CT, Beden M (1978) Proboscidea. In: Maglio VJ, Cooke HBS (eds) Evolution of African mammals. Harvard University Press, Cambridge, pp 336–367
de Vos J, van der Made J, Athanassiou A, Lyras G, Sondaar PY, Dermitzakis MD (2002) Preliminary note on the Late Pliocene fauna from Vatera (Lesvos, Greece). Ann Géol Pays Hellén 39:37–70
Demitrack A (1986) The Late Quaternary geologic history of the Larissa plain, Thessaly, Greece: Tectonic, climatic, and human impact on the landscape. PhD thesis, Stanford University.
Depéret C (1913) Observations sur l'histoire géologique pliocène et quaternaire du golfe et de l'isthme de Corinthe. C R Hebd Séances Acad Sci 156:427–431, 659–663, 1048–1052
Depéret C, Mayet L (1923) Monographie des éléphants pliocènes de l'Europe et de l'Afrique du Nord. Ann Univ Lyon 42:89–224
Dermitzakis MD, Symeonidis NK, de Boer LEM, Sondaar PY (1982) [The evolution of the elephants]. Editions of the Laboratory of Geology and Palaeontology, University of Athens, Athens. (in Greek)
Desio A (1931) Le isole Italiane dell'Egeo. Mem Descr Carta Geol Italia 24:1–534
Doukas CS, Athanassiou A (2003) Review of the Pliocene and Pleistocene Proboscidea (Mammalia) from Greece. Deinsea 9:97–110
Dubrovo IA (1965) Systematic position of *Elephas wüsti*. Int Geol Rev 7:1110–1115
Faugères L (1966) Découverte d'une molaire d'*Elephas antiquus* dans le bassin de Kozani–Ptolemaïs (Macédoine occidentale). C R Som Séanc Soc Géol Fr 1966:184–185
Ferretti MP, Rook L, Carone G, Marra AC (2017) New findings of *Stegotetrabelodon syrticus* from the Late Miocene of Cessaniti, southern Italy. Bol Soc Pal It 56:89–92
Forsyth Major CJ (1887) Faune mammalogiche delle isole di Kos e di Samos. Atti Soc Tosc Sc Nat, Proc Verb 5:272–275
Garutt WE (1964) Das Mammut *Mammuthus primigenius* (Blumenbach). Ziemsen, Wittenberg Lutherstadt
Garutt WE, Gentry A, Lister AM (1990) *Mammuthus* Brookes, 1828 (Mammalia, Proboscidea): proposed conservation, and *Elephas primigenius* Blumenbach, 1799 (currently *Mammuthus primigenius*): proposed designation as the type species of *Mammuthus*, and designation of a neotype. Bull Zool Nomencl 47:38–44
Georgalas GC (1929) Elephas (palaeozoology). Megáli Hellinikí Egkyklopaédia 9:945–946. (in Greek)
Guenther EW (1969) Die Elefantenmolaren aus den Kiesen von Süßenborn bei Weimar. Paläont Abh 3:711–734
Guérin C (1990) Biozones or mammal units? Methods and limits in biochronology. In: Lindsay EH, Fahlbusch V, Mein P (eds) European Neogene mammal chronology, NATO ASI Series (A: Life Sciences). Plenum, New York, pp 119–130
Hagelberg E, Thomas MG, Cook CE, Sher AV, Baryshnikov GF, Lister AM (1994) DNA from ancient mammoth bones. Nature 370:333–334
Harvati K, Darlas A, Bailey SE, Rein TR, El Zaatari S, Fiorenza L, Kullmer O, Psathi E (2013) New Neanderthal remains from Mani peninsula, Southern Greece: The Kalamakia Middle Paleolithic cave site. J Hum Evol 64:486–499
Harvati K, Panagopoulou E, Karkanas P, Athanassiou A, Frost SR (2008) Preliminary results of the Aliakmon Paleolithic/Paleoanthropological Survey, Greece, 2004–2005. In: Darlas A, Mihailović D (eds The Palaeolithic of the Balkans. BAR International Series, vol 1819. Archaeopress, Oxford, pp 15–20
Hilber V (1894) Geologische Reise in Nordgriechenland und Makedonien 1894. Sitzungsber Kaiserl Akad Wiss – Math-Naturwiss Klasse 103:616–623

Iliopoulos G, Lyras G, Roussiakis S, Athanassiou A (2010) Woolly rhinos and woolly mammoths in Southern Greece? Using REE elements to trace the provenance of old museum collections. In: 8th Annual Meeting of the European Association of Vertebrate Palaeontologists, Aix-en-Provence, 2010. Abstracts, p 46

Inuzuka N, Takahashi K (2004) Discrimination between the genera *Palaeoloxodon* and *Elephas* and the independent taxonomical position of *Palaeoloxodon* (Mammalia: Proboscidea). Zona Arqueol 4:428–463

Jacobs Z, Li B, Karkanas P, Tourloukis V, Thompson N, Panagopoulou E, Harvati K (2018) Optical dating of K-feldspar grains from Middle Pleistocene lacustrine sediment at Marathousa 1 (Greece). Quat Int 497:170–177

Kahlke RD (1990) Zur Festlegung des Lectotypus von *Mammuthus trogontherii* (Pohlig, 1885) (Mammalia, Proboscidea). Quartärpaläont 8:119–124

Kalb JE, Froehlich DJ (1995) Interrelationships of Late Neogene Elephantoids: new evidence from the Middle Awash Valley, Afar, Ethiopia. Geobios 28:727–736

Kalb JE, Mebrate A (1993) Fossil elephantoids from the hominid-bearing Awash Group, Middle Awash Valley, Afar Depression, Ethiopia. Trans Am Phil Soc 83:1–114

Kandeloros TC (1898) [Gortynía. History from the ancient times until our days]. Patras (in Greek)

Kevrekidis C, Mol D (2016) A new partial skeleton of *Elephas (Palaeoloxodon) antiquus* Falconer and Cautley, 1847 (Proboscidea, Elephantidae) from Amyntaio, Macedonia, Greece. Quat Int 406:35–56

Konidaris GE, Tsoukala E (this volume) The fossil record of Neogene proboscideans in Greece. In: Vlachos E (ed) , The Fossil Vertebrates of Greece Vol. 1 – Basal vertebrates, basal tetrapods, afrotherians, glires, and primates. Springer, Cham

Konidaris GE, Tourloukis V, Kostopoulos DS, Thompson N, Giusti D, Michailidis D, Koufos GD, Harvati K (2015) Two new vertebrate localities from the Early Pleistocene of Mygdonia Basin (Macedonia, Greece): Preliminary results. C R Palevol 14:353–362

Konidaris GE, Kostopoulos DS, Koufos GD, Tourloukis V, Harvati K (2016) Tsiotra Vryssi: a new vertebrate locality from the Early Pleistocene of Mygdonia Basin (Macedonia, Greece). In: XIV annual meeting of the European Association of Vertebrate Palaeontologists, Haarlem 2016, Programme and Abstract Book, p 38

Konidaris GE, Athanassiou A, Tourloukis V, Thompson N, Giusti D, Panagopoulou E, Harvati K (2018) The skeleton of a straight-tusked elephant (*Palaeoloxodon antiquus*) and other large mammals from the Middle Pleistocene butchering locality Marathousa 1 (Megalopolis Basin, Greece): preliminary results. Quat Int 497:65–84

Konidaris GE, Kostopoulos DS, Koufos GD (2020) *Mammuthus meridionalis* (Nesti, 1825) from Apollonia-1 (Mygdonia Basin, Northern Greece) and its importance within the Early Pleistocene mammoth evolution in Europe. Geodiversitas 42:69–91

Kostopoulos DS, Koulidou I (2015) An early mammoth maxilla from north-western Greece. Quat Int 379:155–163

Koufos GD (1981) A new Late Pleistocene (Würmian) mammal locality from the basin of Drama (Northern Greece). Sci Annals, Fac Phys & Mathem, Univ Thessaloniki 21:129–148

Koufos GD, Pavlides SB (1988) Correlation between the continental deposits of the lower Axios valley and Ptolemais basin. Bull Geol Soc Greece 20:9–19

Koulidou IK (2013) [Study of Plio-Pleistocene Proboscidea from Northern Greece]. Master thesis, Aristotle University of Thessaloniki (in Greek)

Kuss SE (1967) Pleistozäne Säugetierfunde auf den ostmediterranen Inseln Kythera und Karpathos. Ber Naturf Ges Freiburg i Br 57:207–216

Larramendi A (2016) Shoulder height, body mass and shape of proboscideans. Acta Palaeont Pol 61:537–574

Larramendi A, Palombo MR, Marano F (2017) Reconstructing the life appearance of a Pleistocene giant: size, shape, sexual dimorphism and ontogeny of *Palaeoloxodon antiquus* (Proboscidea: Elephantidae) from Neumark-Nord 1 (Germany). Bol Soc Pal It 56:299–317

Larramendi A, Zhang H, Palombo MR, Ferretti MP (2020) The evolution of *Palaeoloxodon* skull structure: Disentangling phylogenetic, sexually dimorphic, ontogenetic, and allometric morphological signals. Quat Sci Rev 229:106090

Lister AM (1996) Evolution and taxonomy of Eurasian mammoths. In: Shoshani J, Tassy P (eds) The Proboscidea: evolution and paleoecology of elephants and their relatives. Oxford University Press, New York, pp 203–213
Lister AM, Sher AV (2001) The origin and evolution of the woolly mammoth. Science 294:1094–1097
Lister AM, Sher AV (2015) Evolution and dispersal of mammoths across the Northern Hemisphere. Science 350:805–809
Lister AM, Stuart AJ (2010) The West Runton mammoth (*Mammuthus trogontherii*) and its evolutionary significance. Quat Int 228:180–209
Lister AM, van Essen H (2003) *Mammuthus rumanus* (Ştefănescu), the earliest mammoth in Europe. In: Petculescu A, Stiucă E (eds) Advances in Vertebrate Paleontology: Hen to Panta. Romanian Academy "Emil Racoviţă" Institute of Speleology, Bucharest, pp 47–52
Lister AM, Sher AV, van Essen H, Wei G (2005) The pattern and process of mammoth evolution in Eurasia. Quat Int 126–128:49–64
Lyras G (2007) Museum labels and fieldnotes. Cranium 24:45–47
Lyras GA, Athanassiou A, van der Geer AAE (volume 2) The fossil record of insular endemic mammals from Greece. In: Vlachos E (ed) The Fossil Vertebrates of Greece Vol. 2 – Laurasiatherians, artiodactyles, perissodactyles, carnivorans, and island endemics. Springer, Cham
Maccagno AM (1962) Gli elefanti fossili di Riano (Roma). Geol Rom 1:33–131
Maglio VJ (1973) Origin and evolution of the Elephantidae. Trans Am Philos Soc 63:1–149
Marinos G (1964) Beiträge zur Kenntnis der Verbreitung des Pleistozäns in Mazedonien. Sci Annals, Fac Phys & Mathem, Univ Thessaloniki 9:95–111. (in Greek)
Markov GN (2012) *Mammuthus rumanus*, early mammoths, and migration out of Africa: some interrelated problems. Quat Int 276–277:23–26
Markov GN, Spassov N (2003) Primitive mammoths from Northeast Bulgaria in the context of the earliest mammoth migrations in Europe. In: Petculescu A, Stiucă E (eds) Advances in Vertebrate Paleontology: Hen to Panta. Romanian Academy "Emil Racoviţă" Institute of Speleology, Bucharest, pp 53–58
Maschenko EN (2010) Evolution of the elephants of the mammoth lineage in the Northern Caucasus (Russia) in the Pliocene and Pleistocene. Proc Zool Inst RAS 314:197–210
Masseti M (2006) Deer and dwarf elephants of the Dodecanese. The Holocene redefinition of the faunal ecosystems of Rhodes and other nearby islands. In: Karageorghis V, Giannikouri A (eds) Conservation and presentation of the cultural and natural heritage of the large islands of the Mediterranean. Proceedings of the International Symposium, Rhodes 2005. Ministry of Culture, Archaeological Institute of Aegean Studies. Foundation Anastasios G. Leventis, Athens, pp 113–122
Matsumoto H (1924) Preliminary note on fossil elephants in Japan. J Geol Soc Tokyo 31:255–272. (in Japanese)
Melentis JK (1960) Ein Beitrag zur Kenntnis der Verbreitung von *Elephas (Archidiskodon) meridionalis archaicus* Depéret und Mayet, 1923. Ann Géol Pays Hellén 11:266–284
Melentis JK (1961) Die Dentition der pleistozänen Proboscidier des Beckens von Megalopolis im Peloponnes (Griechenland). Ann Géol Pays Hellén 12:153–262
Melentis JK (1963) Die Osteologie der pleistozänen Proboscidier des Beckens von Megalopolis im Peloponnes (Griechenland). Ann Géol Pays Hellén 14:1–107
Melentis JK (1965) Erster Fund von *Palaeoloxodon antiquus germanicus* in den jungpleistozänen Ablagerungen des Beckens von Megalopolis (Peloponnes). Prakt Akad Ath 40:197–207. (in Greek)
Melentis JK (1966a) *Archidiskodon meridionalis proarchaicus* n. ssp., die geologisch ältesten Elefantenreste aus Griechenland. Ann Géol Pays Hellén 17:211–220
Melentis JK (1966b) Die pleistozäne Säugetierfauna des Beckens von Haliakmon (Griechenland). Ann Géol Pays Hellén 17:247–265
Meyer M, Palkopoulou E, Baleka S, Stiller M, Penkman KEH, Alt KW, Ishida Y, Mania D, Mallick S, Meijer T, Meller H, Nagel S, Nickel B, Ostritz S, Rohland N, Schauer K, Schüler T, Roca AL, Reich D, Shapiro B, Hofreiter M (2017) Palaeogenomes of Eurasian straight-tusked elephants challenge the current view of elephant evolution. eLife 6:e25413

Milójčić V, Boessneck J, Jung D, Schneider H (1965) Paläolithikum um Larissa in Thessalien. Beiträge zur Ur- und Frühgeschichtlichen Archäologie des Mittelmeerkulturräumes, vol 1. Rudolf Habelt, Bonn

Mitzopoulos MK (1964) Über das Vorkommen von *Archidiskodon meridionalis archaicus* im Becken von Ptolemais (Griechisch-Mazedonien). Prakt Akad Ath 39:381–388. (in Greek)

Mitzopoulos MK (1967) Über das Vorkommen von *Archidiskodon meridionalis archaicus* im Becken von Ptolemais (Griechisch-Mazedonien). Ann Géol Pays Hellén 18:463–470

Mitzopoulos H, Orphanides TG, von Heldreich T (1862) [General report on the accessions in the collections of the Physiographic Museum in Athens during the academic year 1860–61]. Athens (in Greek)

Mol D, de Vos J, van der Plicht J (2007) The presence and extinction of *Elephas antiquus* Falconer and Cautley, 1847, in Europe. Quat Int 169–170:149–153

Nesti F (1825) Lettere sopra alcune ossa fossili del Valdarno non per anco descritte sulla nuova specie di elefante fossile del Valdarno all'illustrissimo Sig. Dott . Prof. Ottaviano Targioni Tozzetti. Nistri, Pisa

Obada T (2010) The remarks on the systematic atrribution of the most ancient Elephantinae Gray, 1821 (Mammalia, Proboscidea) of Europe. In: Lazarev P, Boeskorov G, Maschenko E (eds) Proceedings of the IV Internatioinal Mammoth Conference, Yakutsk, 2007. Federal State Scientific Institution "Institute of Applied Ecology of the North", pp 80–106

Osborn HF (1924) *Parelephas* in relation to phyla and genera of the family Elephantidae. Am Mus Nov 152:1–7

Osborn HF (1936) Proboscidea – a monograph of the discovery, evolution, migration and extinction of the mastodonts and elephants of the World, Volume I: Moeritherioidea, Deinotherioidea, Mastodontoidea. American Museum Press, New York

Osborn HF (1942) Proboscidea – a monograph of the discovery, evolution, migration and extinction of the mastodonts and elephants of the World, , Volume II: Stegodontoidea and Elephantoidea. American Museum Press, New York

Ozawa T, Hayashi S, Mikhelson VM (1997) Phylogenetic position of mammoth and Steller's sea cow within Tethytheria demonstrated by mitochondrial DNA sequences. J Mol Evol 44:406–413

Palkopoulou E, Lipson M, Mallic S, Nielsen S, Rohland N, Baleka S, Karpinski E, Ivancevic AM, To T-H, Kortschak RD, Raison JM, Qu Z, Chin T-J, Alt KW, Claesson S, Dalén L, MacPhee RDE, Meller H, Roca AL, Ryder OA, Heiman D, Young S, Breen M, Williams C, Aken BL, Ruffier M, Karlsson E, Johnson J, Di Palma F, Alfoldi J, Adelson DL, Mailund T, Munch K, Lindblad-Toh K, Hofreiter M, Poinar H, Reich D (2018) A comprehensive genomic history of extinct and living elephants. Proc Natl Acad Sci USA 115:E2566–E2574

Palombo MR, Ferretti MP (2005) Elephant fossil record from Italy: knowledge, problems, and perspectives. Quat Int 126–128:107–136

Palombo MR, Villa P (2001) Schreger lines as support in the Elephantinae identification. In: Cavarretta G, Gioia P, Mussi M, Palombo MR (eds) Proceedings of the 1st International Congress "The World of Elephants", Rome, 2001, pp 656–660

Palombo MR, Albayrak E, Marano F (2010) The straight-tusked elephants from Neumark-Nord. A glance into a lost world. In: Meller H (ed) Elefantenreich – Eine Fossilwelt in Europa. Landesamt für Denkmalpflege und Archäologie Sachsen-Anhalt, Landesmuseum für Vorgeschichte, Halle, pp 218–247

Panagopoulou E, Tourloukis V, Thompson N, Konidaris G, Athanassiou A, Giusti D, Tsartsidou G, Karkanas P, Harvati K (2018) The Lower Palaeolithic site of Marathousa 1, Megalopolis, Greece: overview of the evidence. Quat Int 497:33–46

Paraskevaidis I (1956) Observations sur le Quaternaire de la Grèce. Actes de IV Congrès International du Quaternaire, Rome–Pise 1953:167–178

Paraskevaidis E (1977) Säugetierreste aus Griechenland. In: VIth Colloquium on the Geology of the Aegean Region, Athens, 1977. Proceedings, pp 1143–1154

Pillans B, Gibbard P (2012) The quaternary period. In: Gradstein FM, Ogg JG, Schmitz MD, Ogg GM (eds) The Geologic Time Scale 2012. Elsevier, Oxford, pp 979–1010

Pohlig H (1888) Dentition und Kranologie des *Elephas antiquus* Falc., mit Beiträgen über *Elephas primigenius* Blum. und *Elephas meridionalis* Nesti. Nova Acta Ksl Leop-Carol Deutsch Akad Naturf 53:1–280

Pohlig H (1891) Dentition und Kranologie des *Elephas antiquus* Falc., mit Beiträgen über *Elephas primigenius* Blum. und *Elephas meridionalis* Nesti. Zweiter Abschnitt. Nova Acta Ksl Leop-Carol Deutsch Akad Naturf 57:267–466

Poulianos NA (1986) Osteological data of the Late-Pliocene elephant of Perdikkas. Anthropos 11:49–80

Poulianos AN, Poulianos NA (1980) Pliocene elephant hunters in Greece. Anthropos 7:108–121. (in Greek)

Psarianos P (1958) Neue Proboscidierreste aus dem Pleistozän von Mazedonien (Griechenland). Ann Géol Pays Hellén 9:221–226

Psarianos P, Thenius E (1954) Über *Elephas (Archidiskodon) meridionalis* (Elephant., Mammal.) von Euboea (Griechenland). Prakt Akad Ath 28:413–424

Rabinovich R, Lister AM (2017) The earliest elephants out of Africa: taxonomy and taphonomy of proboscidean remains from Bethlehem. Quat Int 445:23–42

Rabinovich R, Herzlinger G, Calvo R, Rivals F, Mischke S, Beiner G (2019) Erq el Ahmar Elephant Site – A mammoth skeleton at a rare and controversial Plio-Pleistocene site along the mammal migration route out of Africa. Quat Sci Rev 221:105885

Radulescu C, Samson P (1995) Review of the Villafranchian *s.s.* faunas of Romania. Il Quaternario 8:377–381

Radulescu C, Samson PM, Petculescu A, Stiucă E (2003) Pliocene large mammals of Romania. Coloq Paleont vol. ext. 1:549–558

Roth J (1854) Über seine Reise nach Griechenland und Syrien. Gelehrte Anz, Bull Königl Bayer Akad Wiss 7:234–240

Runnels C, van Andel TH (1993) The Lower and Middle Paleolithic of Thessaly, Greece. J Field Arch 20:299–317

Saegusa H, Gilbert WH (2008) Elephantidae. In: Gilbert WH, Asfaw B (eds) *Homo erectus*: Pleistocene evidence from the Middle Awash. Ethiopia. The Middle Awash Series. University of California, Berkeley, pp 193–226

Sanders WJ, Gheerbrant E, Harris JM, Saegusa H, Delmer C (2010) Proboscidea. In: Werdelin L, Sanders WJ (eds) Cenozoic mammals of Africa. University of California Press, Berkerley, pp 161–251

Schneider HE (1968) Zur quartärgeologischen Entwicklungsgeschichte Thessaliens (Griechenland). Beiträge zur Ur- und Frühgeschichtlichen Archäologie des Mittelmeerkulturräumes, vol 6. Rudolf Habelt, Bonn

Shoshani J (1996) Para- or monophyly of the gomphotheres and their position within Proboscidea. In: Shoshani J, Tassy P (eds) The Proboscidea: evolution and paleoecology of elephants and their relatives. Oxford University Press, New York, pp 149–177

Shoshani J, Tassy P (2005) Advances in proboscidean taxonomy & classification, anatomy & physiology, and ecology & behavior. Quat Int 126–128:5–20

Shoshani J, Golenberg EM, Yang H (1998) Elephantidae phylogeny: morphological versus molecular results. Acta Theriol suppl. 5:89–122

Shoshani J, Ferretti MP, Lister AM, Agenbroad LD, Saegusa H, Mol D, Takahashi K (2007) Relationships within the Elephantinae using hyoid characters. Quat Int 169–170:174–185

Sickenberg O (1976) Eine Säugetierfauna des tieferen Bihariums aus dem Becken von Megalopolis (Peloponnes, Griechenland). Ann Géol Pays Hellén 27:25–73

Simpson GG (1945) The principles of classification and a classification of mammals. Bull Am Mus Nat Hist 85:i–xvi, 1–350

Skuphos TG (1905) Über die palæontologischen Ausgrabungen in Griechenland in Beziehung auf das Vorhandensein des Menschen. Comptes Rendus du Congrès International d'Archéologie, Athènes 1905:231–236

Sondaar PY, Boekschoten GJ (1967) Quaternary Mammals in the South Aegean island arc, with notes on other fossil Mammals from the coastal regions of the Mediterranean. Proc Konin Nederl Akad Wet 70:556–576

Sousa MF, Figueiredo SM (2001) The Pleistocene elephants of Portugal. In: Cavarretta G, Gioia P, Mussi M, Palombo MR (eds) Proceedings of the 1st International Congress "The World of Elephants", Rome, 2001, pp 611–616

Steensma KJ (1988) Plio-/Pleistozäne Großsäugetiere (Mammalia) aus dem Becken von Kastoria / Grevena, südlich von Neapolis — NW Griechenland. Inaugural-Dissertation, Technische Universität Clausthal

Stefanescu S (1924) Sur la présence de l'*Elephas planifrons* et de trois mutations de l'*Elephas antiquus* dans les couches géologiques de Roumanie. C R Hebd Séances Acad Sci 179:1418–1419

Stratigopoulos SA (2008) Mammoth remains from the Pleistocene of Ptolemais Basin in W. Macedonia. In: Abstracts of the 33rd International Geological Congress, Oslo, August 2008

Stuart AJ (2005) The extinction of woolly mammoth (*Mammuthus primigenius*) and straight-tusked elephant (*Palaeoloxodon antiquus*) in Europe. Quat Int 126–128:171–177

Stuart AJ, Sulerzhitsky LD, Orlova LA, Kuzmin YV, Lister AM (2002) The latest woolly mammoths (*Mammuthus primigenius* Blumenbach) in Europe and Asia: a review of the current evidence. Quat Sci Rev 21:1559–1569

Symeonidis NK (1970) Stratigraphisch-paläontologische Untersuchungen über die Neogenablagerungen von Lakonien, insbesondere der Gebiete von Neapolis (Eparchia Epidaurus Limira) und gegenüber liegenden Insel Elafonisos. Ann Géol Pays Hellén 21:531–553. (in Greek)

Symeonidis NK, Theodorou GE (1986) On a new occurrence of Lower Pleistocene Proboscidea in South Peloponnese. Ann Géol Pays Hellén 33:251–261. (in Greek)

Tassy P (1990) Phylogénie et classification des Proboscidea (Mammalia): historique et actualité. Ann Paléont 76:159–224

Tassy P (1996) Who is who among the Proboscidea? In: Shoshani J, Tassy P (eds) The Proboscidea: evolution and paleoecology of elephants and their relatives. Oxford University Press, New York, pp 39–48

Tassy P (1999) Miocene elephantids (Mammalia) from the Emirate of Abu Dhabi, United Arab Emirates: palaeobiogeographic implications. In: Whybrow PJ, Hill A (eds) Fossil vertebrates of Arabia. Yale University Press, New Haven, pp 209–233

Tassy P, Debruyne R (2001) The timing of early Elephantinae differentiation: the palaeontological record, with a short comment on molecular data. In: Cavarretta G, Gioia P, Mussi M, Palombo MR (eds) Proceedings of the 1st International Congress "The World of Elephants", Rome, 2001, pp 685–687

Tassy P, Shoshani J (1996) Historical overview of classification and phylogeny of the Proboscidea. In: Shoshani J, Tassy P (eds) The Proboscidea: evolution and paleoecology of elephants and their relatives. Oxford University Press, New York, pp 3–8

Theodorou G (2014) Megalopolis – 112 years after the first excavation by National and Kapodistrian University of Athens (NKUA) – and the post lignite era. In: Kostopoulos DS, Vlachos E, Tsoukala E (eds) Abstract Book of the VIth International Conference on Mammoths and Their Relatives, Grevena–Siatista 2014. Sci Ann, School Geol, Aristotle Univ Thessaloniki, pp 195–196

Theodorou G, Bassiakos Y, Tsakalos E, Yiannouli E, Maniatis P (2018) The use of CT scans and 3D modeling as a powerful tool to assist fossil vertebrate taxonomy. In: Ioannides M, Fink E, Brumana R, Patias P, Doulamis A, Martins J, Wallace M (eds) Digital Heritage. Progress in Cultural Heritage: Documentation, Preservation, and Protection. 7th International Conference, EuroMed 2018, Nicosia, Cyprus, October 29 – November 3, 2018, Proceedings, Part I. Lecture Notes in Computer Science vol 11196. Springer, Cham, pp 79–89

Thomas MG, Hagelberg E, Jones HB, Yang Z, Lister AM (2000) Molecular and morphological evidence on the phylogeny of the Elephantidae. Proc R Soc London 267:2493–2500

Todd NE (2010) New phylogenetic analysis of the family Elephantidae based on cranial-dental morphology. Anat Rec 293:74–90

Todd NE, Roth VL (1996) Origin and radiation of the Elephantidae. In: Shoshani J, Tassy P (eds) The Proboscidea: evolution and paleoecology of elephants and their relatives. Oxford University Press, New York, pp 193–202

Tourloukis V, Muttoni G, Karkanas P, Monesi E, Scardia G, Panagopoulou E, Harvati K (2018) Magnetostratigraphic and chronostratigraphic constraints on the Marathousa 1 Lower Palaeolithic site and the Middle Pleistocene deposits of the Megalopolis basin, Greece. Quat Int 497:154–169

Trantalidou K (2013) On the banks of the river Angitis: environmental data for the basin of Drama in Prehistoric times. In: Proceedings of the 5th Scientific Meeting, Drama, 2006, pp 41–156 (in Greek)

Tsoukala E (1999) Quaternary large mammals from the Apidima Caves (Lakonia, S. Peloponnese, Greece). Beitr Paläontol 24:207–229

Tsoukala E, Chatzopoulou K (2005) A new Early Pleistocene (latest Villafranchian) site with mammals in Kalamotó (Mygdonia Basin, Macedonia, Greece) – preliminary report. Mitt Komm Quartärforsch Österr Akad Wiss 14:213–233

Tsoukala E, Lister A (1998) Remains of straight-tusked elephant, *Elephas (Palaeoloxodon) antiquus* Falc. & Caut., 1847 ESR-dated to oxygen isotope Stage 6 from Grevena (W. Macedonia, Greece). Bol Soc Pal It 37:117–139

Tsoukala E, Mol D, Pappa S, Vlachos E, van Logchem W, Vaxevanopoulos M, Reumer J (2011) *Elephas antiquus* in Greece: New finds and a reappraisal of older material (Mammalia, Proboscidea, Elephantidae). Quat Int 245:339–349

Ukkonen P, Aaris-Sørensen K, Arppe L, Clark PU, Daugnora L, Lister AM, Lõugas L, Seppä H, Sommer RS, Stuart AJ, Wojtal P, Zupiņš I (2011) Woolly mammoth (*Mammuthus primigenius* Blum.) and its environment in northern Europe during the last glaciation. Quat Sci Rev 30:693–712

van Asperen EN (2013) Position of the Steinheim interglacial sequence within the marine oxygen isotope record based on mammal biostratigraphy. Quat Int 292:33–42

van Vugt N, de Bruijn H, van Kolfschoten T, Langereis CG (2000) Magneto- and cyclostratigraphy and mammal-fauna's of the Pleistocene lacustrine Megalopolis Basin, Peloponnesos, Greece. Geol Ultraject 189:69–92

Velitzelos E, Schneider HE (1973) Beiträge zur Geologie West-Mazedoniens. 1. Elephantiden-Reste aus dem Pleistozän der Provinz Florina. Ann Mus Goulandris 1:251–256

Vinken R (1965) Stratigraphie und Tektonik des Beckens von Megalopolis (Peloponnes, Griechenland). Geol Jb 83:97–148

Vislobokova I (2005) On Pliocene faunas with Proboscideans in the territory of the former Soviet Union. Quat Int 126–128:93–105

Wei GB, Lister AM (2005) Significance of the dating of the Majuangou site for understanding Eurasian mammoth evolution. Vert Palas 43:243–244

Wei GB, Taruno H, Kawamura Y, Jin CZ (2006) Pliocene and Early Pleistocene primitive mammoths of Northern China: Their revised taxonomy, biostratigraphy and evolution. J Geosci, Osaka City Univ 49:59–101

Wei G, Hu S, Yu K, Hou Y, Li X, Jin C, Wang Y, Zhao J, Wang W (2010) New materials of the steppe mammoth, *Mammuthus trogontherii*, with discussion on the origin and evolutionary patterns of mammoths. Sci China Earth Sci 53:956–963

Yang H, Golenberg EM, Shoshani J (1996) Phylogenetic resolution within the Elephantidae using fossil DNA sequence from the American mastodon (*Mammut americanum*) as an outgroup. Proc Natl Acad Sci USA 93:1190–1194

The Fossil Record of Sea Cows (Mammalia: Sirenia) in Greece

George Iliopoulos, Kaliana Svana, and Socrates Roussiakis

1 Introduction

Sirenians, commonly known as sea cows, are a group of aquatic mammals belonging to the order Sirenia. Their name derives from the sirens, the mermaids of the Greek mythology. Modern sea cows are large and robust mammals that live in shallow, protected, tropical, and subtropical marine, riverine, and estuarine environments, and feed preferentially on aquatic angiosperm grasses belonging mainly to the families Hydrocharitaceae and Potamogetonaceae (Berta et al. 2015; Domning 2001). Extant Sirenia are affiliated to two families, Trichechidae and Dugongidae. The three living trichechid species are the West Indian manatee (*Trichechus manatus*), the West African manatee (*Trichechus senegalensis*), and the Amazon manatee (*Trichechus inunguis*). Dugongidae are represented by the living dugong, *Dugong dugon*, and the extinct Steller's sea cow, *Hydrodamalis gigas*, which became extinct in 1768 (Forsten and Youngman 1982). They have a fuselage-shaped body with characteristic downturned snouts, their forelimbs have evolved into short and rounded flippers, their hindlimbs have been lost and only a vestigial pelvic bone has been maintained (in some species it is indicative of sexual dimorphism), and they possess a horizontal tail fluke (Berta et al. 2015).

G. Iliopoulos (✉)
Department of Geology, University of Patras, Rio Patras, Greece
e-mail: iliopoulosg@upatras.gr

K. Svana
Limassol, Cyprus

S. Roussiakis
Faculty of Geology and Geoenvironment, National & Kapodistrian University of Athens, Athens, Greece
e-mail: srousiak@geol.uoa.gr

E. Vlachos (ed.), *Fossil Vertebrates of Greece Vol. 1*,
https://doi.org/10.1007/978-3-030-68398-6_14

Another typical feature of Sirenia is the pachyostosis and osteosclerosis that their bones present (Domning et de Buffrénil 1991). Pachyostosis is the observed increased volume of the bones and osteosclerosis refers to their high density due to the absence of cancellous bone tissue cavities. Sirenian ribs are particularly pachyo-steosclerotic, whereas their lumbar vertebrae possess elongated transverse processes to support the body. Pachyostosis and osteosclerosis increase the specific weight of the bones forming a kind of ballast that helps the animals to weigh down their body weight and dive and in addition maintain neutral or negative buoyancy overcoming the buoyant gases that are produced during plant digestion, keeping their bodies efficiently stable above the sea bottom to feed with aquatic plants (Uhen 2007). Moreover, anatomical changes such as the downturned jaws evince their adjustment to bottom feeding on the rhizomes or leaves of aquatic plants and mainly seagrasses (Domning 1977, 2001; Uhen 2007).

The first sirenians, *Prorastomus* and *Pezosiren*, appeared in the fossil record during the early to middle Eocene (Berta et al. 2015) possessing anatomical features that already indicated aquatic adaptations. Sirenians originated and evolved around the coastlines of the former Tethys Ocean, thus their fossil remains are found in riverine, estuarine, and shallow marine deposits around a zone from southeast Asia to northern Africa, central and southern Europe, the Caribbean, south North America, and north South America.

In Greece, until now, Sirenian fossils have been reported only from upper Miocene and mainly Tortonian deposits from Crete Island (Svana et al. 2010). Hence, 18 localities with Sirenian remains have been reported from Crete Island (see below). Most of the findings consist of scattered and fragmented skeletal elements, mostly the typical pachyosteosclerotic ribs found in shallow shelf deposits. Nevertheless, in a small number of localities more complete findings have been recovered, with articulated postcranial elements (Stylos and Trypitos), scattered postcranial elements (Panassos 1, Panassos 2, and Ampelouzos), and cranial elements (Tympana, Kefala, and Panassos 1).

2 Historical Overview

The first report of sirenian findings from Greece came from Symeonidis and Schultz (1973) reporting the presence of ribs from shallow shelf deposits at Kotsiana, western Crete. The next report was published more than 25 years later by Markopoulou-Diakantoni (2001) mentioning sirenian findings from the locality Machairidi at the village Gdochia, close to Ierapetra. In the next decade, the number of localities with Sirenian findings on the island of Crete increased drastically mainly by the systematic work done by the paleontology laboratory of the Natural History Museum of Crete. In 2004, Markopoulou-Diakantoni and Logos mentioned the discovery of cranial material from two localities, Tympana and Kefala, near Sitia, as well as the existence of one more locality, Mouchli Lakkos from Chania. The same year, Kröger (2004) reported in his PhD thesis the presence of sirenian ribs in conglomerate deposits south of Vasiliki in the Messara basin. In 2006, Reuter provided data for the presence of sirenian remains from two more localities in the Messara basin, Apomarma and Vasiliki.

At the same time, Kaliana Svana for her MSc thesis started a systematic work on sirenian remains from different localities across Crete that were provided from systematic fieldwork and excavations of the paleontology laboratory of the Natural History Museum of Crete (Svana 2007). During this study, the number of sirenian localities was doubled, providing articulated (Stylos and Trypitos) and associated (Panassos 1 and 2 and Ampelouzos) skeletal elements (Svana 2007; Svana et al. 2010). The reported localities expand all over Crete with the localities of Stylos, Afrata, and seven additional sites at the locality of Kotsiana in the area of Chania; the localities of Panassos 1, Panassos 2, Ampelouzos, and Kefala Gergeri in the Messara basin; and finally the locality of Trypitos near Sitia (Svana et al. 2010). One more locality, Filippoi, is mentioned in this work again from the Messara basin, which came through personal communication with Efterpi Koskeridou (Svana et al. 2010). Herein, two more localities that have yielded sirenian findings are reported for the first time: Arkalies, close to Leukochori at the Messara basin, and Gavdos Island. A few months ago, one of the authors (SR) discovered in the old collections of the Museum of Paleontology and Geology of the National and Kapodistrian University of Athens a few sirenian remains (rib fragments and a tusk) that had been collected from the locality of Skineas at Paliki peninsula on Cephalonia Island some decades ago by old staff members of the Geology Department; this is also reported for the first time herein.

3 Phylogenetic Relationships

Based on anatomical and molecular data, Sirenia have been placed together with Proboscidea and Desmostylia, forming the monophyletic group Tethytheria (McKenna 1975; Berta et al. 2015; Uhen 2007). They constitute a monophyletic group that appeared in the fossil record for the first time during the early Eocene, and this monophyly is based on the following synapomorphies (Domning 1994; Vélez-Juarbe and Domning 2014): enlarged and retracted nasal bones, premaxillar and frontal bones in contact, absence of sagittal crest, laterally compressed symphysis of premaxillae that forms an elongated rostrum, enlarged mastoid processes visible through the occipital fenestrae, five premolars (the front ones are lost in later forms), enlarged tear-shaped ectotympanic bone, and bones that present pachyostosis and osteosclerosis.

The order Sirenia reached its peak diversity during the Miocene (Domning et al. 2010). Four families are currently recognized, namely, Prorastomidae,

Protosirenidae, Dugongidae, and Trichechidae. The family Dugongidae comprises the most successful and diverse sirenian group and is separated into three subfamilies: Halitheriinae (Eocene to Pliocene), Hydrodamalinae (Miocene to eighteenth century), and Dugonginae (late Oligocene–today). The most common representative of the family during the Neogene was the genus *Metaxytherium*. *Metaxytherium* consists of two stocks, an American and a European one. The European species of *Metaxytherium* consists a monophyletic group which includes *Metaxytherium krahuletzi* (early Miocene–Burdigalian), *M. medium* (Serravallian–Tortonian), *Metaxytherium serresii* (late Tortonian–early Pliocene), and *Metaxytherium subapenninum* (Pliocene). These four *Metaxytherium* species are considered members of the same lineage, being different chronospecies, while *M. serresii*, a small-sized *Metaxytherium*, and *M. subapenninum* are characterized as Mediterranean endemics (Bianucci et al. 2008; Sorbi et al. 2012). In Europe during the lower Miocene, the dugongine species *Rytiodus capgrandi* and *Rytiodus heali* were sympatric with *Metaxytherium* (Domning and Sorbi 2011). During the middle and late Miocene, the species *Metaxytherium crataegense*, *Metaxytherium floridanum*, and *Metaxytherium arctodites* were found in America.

Concerning the sirenian remains that have been found in Greek localities, most of the findings, and basically those that could be determined, have been attributed to the genus *Metaxytherium*. More specifically, where cranial elements or more complete articulated or associated skeletal parts have been found it has been possible to attribute the material to *Metaxytherium medium* (Markopoulou-Diakantoni and Logos 2004; Svana 2007; Svana et al. 2010).

4 Distribution

As mentioned above, until now fossil sirenian remains in Greece had been reported only from Crete Island. As seen in Fig. 1 and the information listed in the Appendix, the localities where sirenian remains have been found in Crete extend all over the island from western to eastern Crete, with the majority of the localities found in the southern part of central Crete and more specifically in the Messara plain, in the north part of western Crete, and in another small cluster in the north part of eastern Crete. A new locality with sirenian remains from Gavdos Island is mentioned for the first time herein. These localities indicate that during the Tortonian, shallow shelf marine environments rich in marine angiosperms prevailed at these parts of Crete. However, another new locality on the Island of Cephalonia reported herein also for the first time extends the distribution of the group further north. The actual locality is situated close to the village Skineas at Paliki peninsula comprising the first sirenian locality in Greece outside Crete. The age of the sirenian bearing deposits from Skineas is late Tortonian (Symeonidis and Schultz 1970; Georgiadou-Dikeoulia 1967).

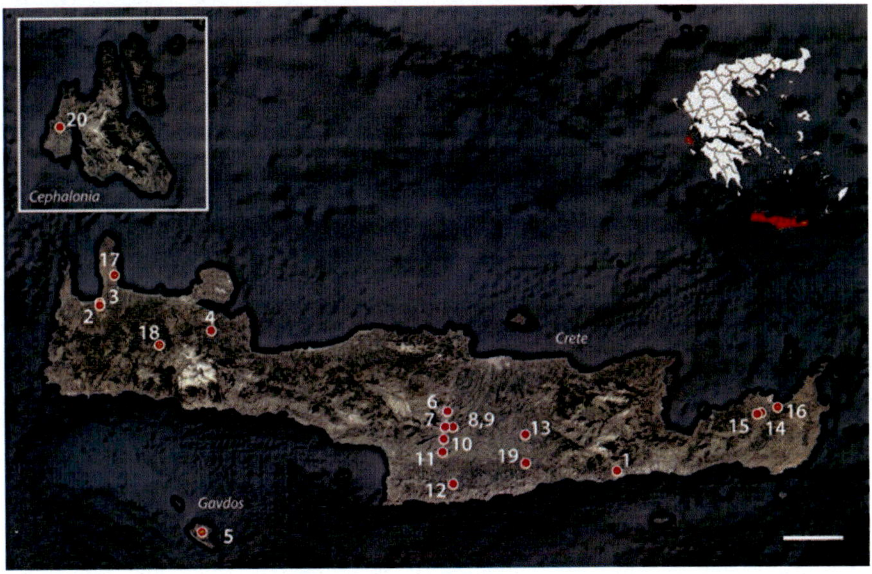

Fig. 1 Map of Crete, Gavdos, and Cephalonia Islands, showing the Miocene occurrences of fossil sea cows (Sirenia). **1**, Machairidi; **2**, Kotsiana; **3**, Kotsiana 7 localities; **4**, Stylos; **5**, Gavdos Island; **6**, Psalida; **7**, Kefala Gergeri; **8**, Panassos; **9**, Panassos 2; **10**, Apomarma; **11**, Voulismata; **12**, Vasiliki 1 and 2 localities; **13**, Arkalies; **14**, Tympana; **15**, Kefala; **16**, Tripitos; **17**, Afrata; **18**, Mouchli Lakkos; **19**, Filippoi; **20**, Skineas (Kefalonia). See Appendix for more information. Image exported from Google Earth Pro © 2019. Map data from SIO, NOAA, US Navy, NGA, GEBCO, image from Landsat/Copernicus. Additional map data for Cephalonia LDEO-Columbia, NSF. Scale bar equals 20 km, north faces upward

5 Systematic Paleontology

Sirenia Illiger, 1811
Dugongidae Gray, 1821

Type Genus *Dugong* Lacépède, 1799.

Halitheriinae (Carus, 1868)
***Metaxytherium* de Christol, 1840**

Type Species *Hippopotamus medius* (Desmarest, 1822).

Included Taxa *Metaxytherium arctodites* Aranda-Manteca et al., 1994; *Metaxytherium crataegense* (Simpson, 1932); *Metaxytherium floridanum* Hay, 1922; *Metaxytherium krahuletzi* Deperet, 1895; *Metaxytherium medium* (Desmarest, 1822); *Metaxytherium serresii* (Gervais, 1847); Deperet 1895; *Metaxytherium subapenninum* (Bruno, 1839); *Metaxytherium albifontanum* Velez-Juarbe and Domning, 2014.

Distribution Late Oligocene of southeastern United States; early–late Miocene of North and South America, Caribbean, Portugal, Spain, France, Italy, Malta, Switzerland, the Netherlands, Germany, Austria, Hungary, Slovakia, Croatia, Greece, Libya, Tunisia, and Turkey; Pliocene of Morocco, Spain, and Italy (Sorbi 2007; Carone and Domning 2007; Domning et al. 2010; Sorbi et al. 2012; Bianucci et al. 2003; Velez-Juarbe and Domning 2014; Landini et al. 2005; Šuklje 1938; Tekkaya 1973).

Comments In most cases, the Greek findings consist mainly of scattered ribs and vertebrae that do not allow identification beyond the generic level (Fig. 2).

***Metaxytherium medium* (Desmarest, 1822)**

Nomenclatural and Taxonomical History *Hippopotamus medius* in Desmarest 1822 (new species); *Metaxytherium cuvieri* in de Christol 1840 (new combination); *Metaxytherium medium* in Hooijer 1952 (new combination and mandatory change of the specific epithet).

Type Material MNHN Fs 2706, partial left mandible with M2–3 and roots of the M1, Muséum National d' Histoire Naturelle (p. 388 in Desmarest 1822).

Type Locality Saint-Michel en Chaisine, Maine et Loire, France.

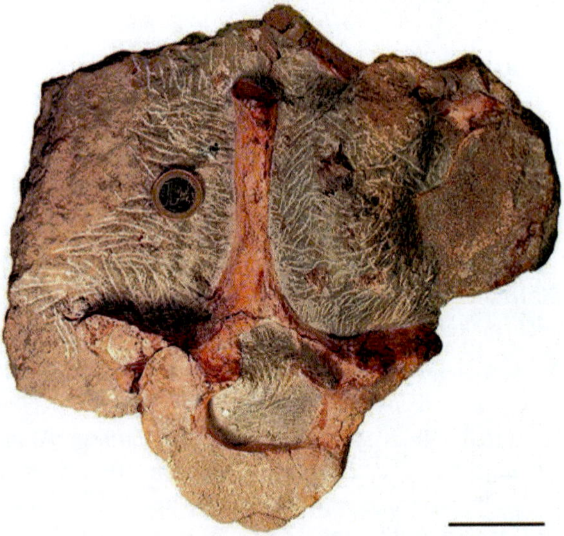

Fig. 2 Two *Metaxytherium* thoracic vertebrae from the locality of Apomarma, Messara, Heraklion (courtesy Markus Reuter). The diameter of the 1 euro coin equals 2.3 cm

Distribution Besides its type locality, this species is known from the middle to late Miocene (Serravalian–Tortonian) of Portugal, Spain, France, Italy, the Netherlands, Austria, Hungary, and Slovakia. In Greece, it is known from the late Miocene (Tortonian) of Crete Island (southern Greece).

Taxonomic Remarks *Metaxytherium medium* is a halitheriine species of *Metaxytherium* characterized by the loss of permanent premolars 1–4, the reduced supraorbital process of the frontal bone, the separated in the midline nasal bones, the wider ventrally supraoccipital bone, the ventrally strong concave horizontal mandible ramus, and the exoccipitals that do not meet usually along the midline suture above the foramen magnum. *Metaxytherium medium* differs from *M. krahuletzi* by the fact that its jugal ventral extremity is found ventrally to the orbit; compared to *M. serresii* it has a deeper supracondylar fossa and a smaller tusk alveolus, and compared to *M. subapenninum*, it differs in the supracondylar fossa, which is even shallower or lost, and in the even longer tusk alveolus. Its tusk size is reduced compared to *M. serresii* and *M. subapenninum*.

Comments on the Greek Material Markopoulou-Diakantoni and Logos (2004) identified two partial skulls from the localities of Kefala (Fig. 3a) and Timpana (Fig. 3b) as *Metaxytherium cuvieri*, which is currently considered a junior synonym of *M. medium*. The anterior and posterior parts of the skulls are broken, and despite the fact that maxillae are generally preserved, the teeth in both samples are broken. The specimen from Kefala is better preserved but not well prepared, where the three molars and the fifth deciduous premolar in both toothrows are found in place, but they were broken prior to the extraction of the skull from the sediment. Cranial material consisting of one well-preserved left tusk (Fig. 3c) has been found in the locality of Panassos. The total length of the tooth is 114.5 mm and the length of the crown is 23.7 mm. The enamel of the tooth presents typical parallel and perpendicular striations from the base of the crown to the apex. The short crown is a typical feature of *M. medium* with respect to the longer tusks of *M. serresii* and *M. subapenninum* (Bianucci et al. 2008). In most localities, the findings consist mainly of scattered pachyosteosclerotic ribs. Nevertheless, in the localities Stylos and Trypitos, articulated postcranial elements (Fig. 4) have been found, with the Stylos specimen being more complete, preserving the axial skeleton articulated with the thoracic and lumbar vertebrae and most of the ribs in place—according to information by the local people of Stylos, the skull was also in place but removed shortly after its discovery. Also, the localities Panassos 1, Panassos 2, and Ampelouzos have provided scattered postcranial elements that belong to the same individual consisting mainly of several vertebrae and ribs. The ribs present the typical pachyostosis and osteosclerosis of the sirenian ribs, while the vertebrae are also characteristic and particularly the lumbar ones with fairly elongated and thick lateral apophyses (Fig. 5).

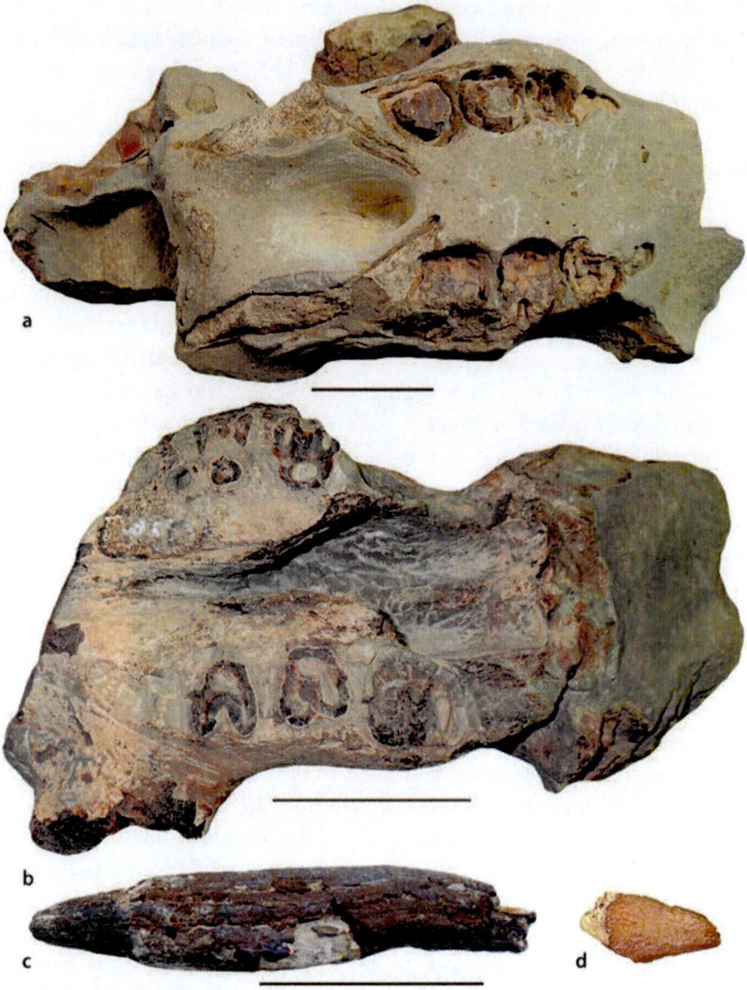

Fig. 3 Cranial remains of the sirenian *Metaxytherium medium* from the Tortonian of Crete Island and Cephalonia Island, Greece. (**a**) The partial skull from Kefala (ventral view), (**b**) the partial skull from Tympana (ventral view), (**c**) the left tusk from Panassos (lateral view), and (**d**) the crown of the tusk from Skineas, Cephalonia Island (lateral view). Scale bars equal 5 cm

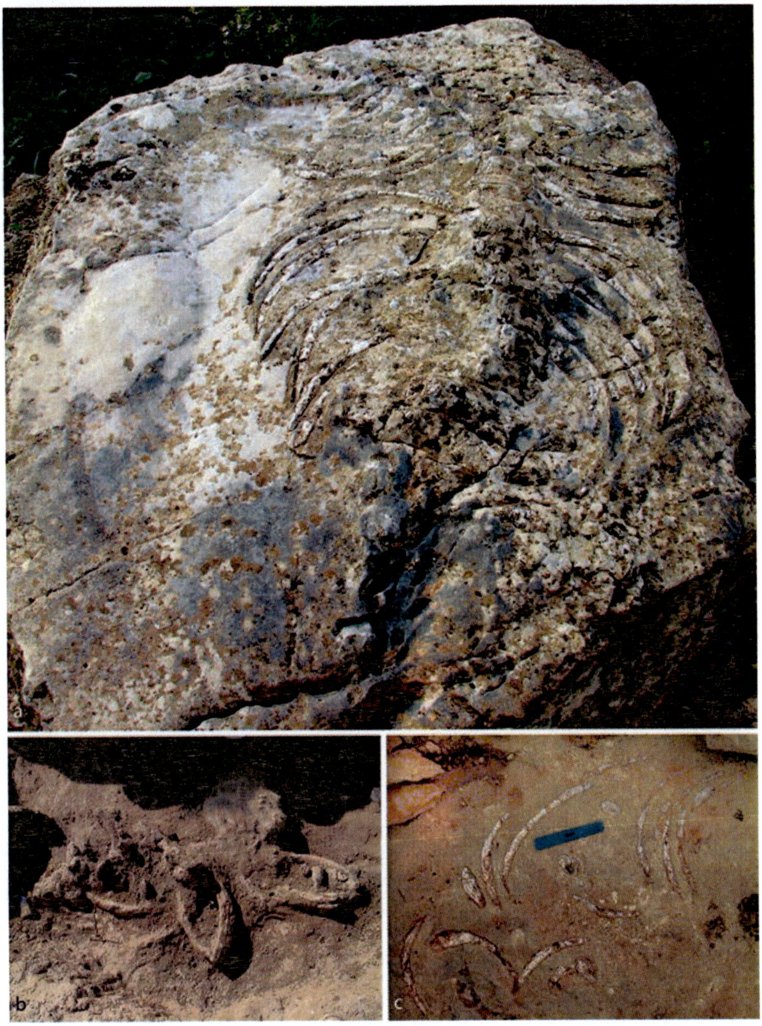

Fig. 4 Postcranial remains of the sirenian *Metaxytherium medium* from the Tortonian of Crete Island. (**a**) The articulated postcranial skeleton from Stylos with vertebrae and ribs in place. (**b**) The partly articulated postcranial skeleton from Trypitos, Sitia, with articulated thoracic vertebrae and ribs. (**c**) Scattered ribs (most common findings) from Kotsiana

Fig. 5 Lumbar vertebra of the sirenian *Metaxytherium medium* from the Tortonian locality of Panassos. Scale bar equals 5 cm

Acknowledgments We would like to thank Charalampos Fassoulas, curator of the Natural History Museum of Crete for providing to us the access to the paleontological collections of the museum to study the Sirenian material. We would also like to thank Apostolos Alexopoulos, Euterpi Koskeridou, and Markus Reuter for providing us information about the fossiliferous Sirenian sites of Gavdos, Filippoi, Psalida, and Apomarma. We deeply thank M. Uhen and A. Collareta for their comments that helped improve this manuscript.

Appendix

List of the Greek localities containing sea cow fossils. Locality numbers refer to the collection numbers of the PaleoBiology Database (PBDB)

Localities[PBDB No]	Age	Taxon
Machairidi Gdochia, Lassithi[204401]	Late Tortonian, Miocene	*Metaxytherium* sp.[1]
Kotsiana, Chania[110125]	Tortonian, Miocene	*Metaxytherium* sp.[2,3]
Kotsiana 7 further sites, Chania[204403]	Tortonian, Miocene	*Metaxytherium* sp.[3]
Stylos, Chania[110123]	Tortonian, Miocene	*Metaxytherium medium*[3]
Gavdos, Chania[**]	Tortonian, Miocene	*Metaxytherium medium*[*]
Psalida, Heraklion[110130]	Tortonian, Miocene	*Metaxytherium* sp.[4]
Kefala Gergeri, Heraklion[110127]	Tortonian, Miocene	*Metaxytherium* sp.[3]
Panassos, Heraklion[106474]	Tortonian, Miocene	*Metaxytherium medium*[3]
Panassos 2, Heraklion[106474]	Tortonian, Miocene	*Metaxytherium medium*[3]
Apomarma, Heraklion[110131]	Tortonian, Miocene	*Metaxytherium* sp.[4]
Voulismata Ampelouzos, Heraklion[110126]	Tortonian, Miocene	*Metaxytherium* sp.[3]
Vasiliki 1, Heraklion[110132]	Tortonian, Miocene	*Metaxytherium* sp.[4]
Vasiliki 2, Heraklion[110132]	Tortonian, Miocene	*Metaxytherium* sp.[5]
Arkalies Leukochori, Heraklion	Tortonian, Miocene	*Metaxytherium* sp.[*]
Timpana, Lassithi[110174]	Tortonian, Miocene	*Metaxytherium medium*[6]
Kefala, Lassithi[110175]	Tortonian, Miocene	*Metaxytherium medium*[6]

Localities[PBDB No]	Age	Taxon
Tripitos, Lassithi[110128]	Tortonian, Miocene	*Metaxytherium medium*[3]
Afrata, Chania[110124]	Miocene	*Metaxytherium* sp.[3]
Mouchli Lakkos, Chania[204042]	Miocene	*Metaxytherium* sp.[6]
Filippoi, Heraklion	Miocene	*Metaxytherium* sp.[***]
Skineas, Cephalonia	Late Tortonian, Miocene	*Metaxytherium* cf. *medium*[*]

[*]New locality reported herein
[**]A. Alexopoulos, personal communication
[***]E. Koskeridou, personal communication
[1]Markopoulou-Diakantoni (2001), [2]Symeonidis and Schultz (1973), [3]Svana et al. (2010), [4]Reuter (2006), [5]Kröger (2004), [6]Markopoulou-Diakantoni and Logos (2004)

References

Aranda-Manteca FJ, Domning DP, Barnes LG (1994) A new middle Miocene sirenian of the genus *Metaxytherium* from Baja California and California: relationships and palaeobiogeographic implications. Proc San Diego Soc Nat Hist 29:191–204

Berta A, Sumich JL, Kovacs KM (2015) Marine mammals: evolutionary biology, 3rd edn. Academic Press, Amsterdam

Bianucci G, Landini W, Varola A (2003) New records of *Metaxytherium medium* (Mammalia: Sirenia) from the late Miocene of Cisterna quarry (Apulia, southern Italy). Boll Soc Paleontol Ital 42(1–2):59–63

Bianucci G, Carone G, Domning DP, Landini W, Rook L, Sorbi S (2008) Peri-Messinian dwarfing in Mediterranean *Metaxytherium* (Mammalia, Sirenia): evidence of habitat degradation related to the Messinian salinity crisis. In: El-Arnauti A, Salem M, Pavlakis P, Boaz N (eds) Garyounis scientific bulletin, Special issue 5, pp 145–157

Bruno GD (1839) Illustrazione di un nuovo cetaceo fossile. Memorie dell' Accademia di Scienze Torino 2:143–160

Carus JV (1868) Handbuch der Zoologie. 1ster Band, Wirbelthiere, Mollusken und Molluscoiden. Wilhelm Engelmann, Leipzig, pp 894

Carone G, Domning DP (2007) *Metaxytherium serresii* (Mammalia: Sirenia): new pre-Pliocene record, and implications for Mediterranean paleoecology before and after the Messinian salinity crisis. Boll Soc Paleontol Ital 46(1):55–92

Depéret C (1895) Uber die Fauna von miocänen Wirbelthieren aus der ersten Mediterranstufe von Eggenburg. Sitzungsber Akad Wiss Wien, math naturwissen Cl 104:395–416

Desmarest AG (1822) Mammalogie ou description des espèces de mammifères. Seconde partie, contenant les ordres des rongeurs, des édentés, des pachydermes, des ruminants et des cétacés. Chez Mme. Veuve Agasse, Imprimeur-Libraire, Paris, pp 277–556

de Christol J (1840) Recherches sur divers ossements fossiles attribués par Cuvier à deux phoques, au lamantin, et à deux espèces d'hippopotames, et rapportés au Metaxytherium, nouveau genre de cétacé de la famille des dugongs. L 'Institut 8:322–323

Domning DP (1977) Observations on the myology of *Dugong dugon* (Muller). Smithsonian Contrib Zool 226:1–57

Domning DP (1994) A phylogenetic analysis of the Sirenia. Proc San Diego Soc Nat Hist 29:177–189

Domning DP (2001) Sirenians, seagrasses, and Cenozoic ecological change in the Caribbean. Palaeogeogr Palaeoclimatol Palaeoecol 166(1–2):27–50

Domning DP, de Buffrénil V (1991) Hydrostasis in the Sirenia: quantitative data and functional interpretations. Mar Mamm Sci 7:331–368

Domning DP, Sorbi S (2011) *Rytiodus heali*, sp. nov., a new Sirenian (Mammalia, Dugonginae) from the Miocene of Libya. J Vertebr Paleontol 31(6):1338–1355

Domning DP, Zalmut IS, Gingerich PD (2010) Sirenia. In: Werdelin L, Sanders WJ (eds) Cenozoic mammals of Africa. University California Press, Berkeley, pp 147–160

Forsten A, Youngman PM (1982) *Hydrodamalis gigas*. Mamm Species 165:1–3

Georgiadou-Dikeoulia E (1967) Beiträge zur Kenntnis des Neogen auf Kephallinia. Ann Géol Pays Hellén 18:43–106. (in Greek, German summary)

Gervais FLP (1847) Observations sur les mammifères du midi de la France. Deuxième partie. Annales des Sciences Naturelles (Zoologie) 8:203–224

Gray JE (1821) On the natural arrangement of vertebrose animals. Lon-don Medical Repository 15:296–310

Hay OP (1922) Description of a new fossil sea cow from Florida, *Metaxytherium floridanum*. Proceedings of the U.S. National Museum 61:1–4

Hooijer DA (1952) Fact and fiction in hippopotamology (sampling the history of scientific error). Osiris 10:109–116

Illiger JKW (1811) Prodomus Systematis Mammalium et Avium: Additis Terminis Zoographicis Utriusque Classis, Eorumque Versione Germanica. Sumptibus C. Salfeld, Berlin, pp 302

Kröger KF (2004) Sedimentary environments and climate change: a case study (late Miocene, central Crete). PhD thesis, Johannes Gutenberg-Universitat, Mainz

Lacepède BGE (1799) Tableau des divisions, sous-divisions, ordres et genres des mammifères 1–18

Landini W, Bianucci G, Bisconti M, Carnevale G, Sorbini C, Varola A (2005) Il Miocene: I vertebrati marini. In: Bonfiglio L (ed) Paleontologia dei Vertebrati in Italia. Evoluzione biologica, significato ambientale e paleogeografia, Mem Mus Civico Storia Nat, sezione Sci della Terra 6, pp 145–154

Markopoulou-Diakantoni A (2001) Paleoecological observations on the coral reefs of the upper Tortonian between Viannos and Ierapetra provinces (SE Crete). Bull Geol Soc Greece 34(2):495–502

Markopoulou-Diakantoni A, Logos E (2004) The occurrence of the *Metaxytherium cuvieri* Christol in the late Miocene sediments of Sitia, Crete. Bull Geol Soc Greece 36:764–771

McKenna MC (1975) Toward a phylogenetic classification of the mammalia. In: Luckett WP, Szalay FS (eds) Phylogeny of the primates. Plenum Press, New York, pp 21–45

Reuter M, Brachert TC, Kröger KF (2006) Shallow-marine carbonates of the tropical–temperate transition zone: effects of hinterland climate and basin physiography (late Miocene, Crete, Greece). In: Pedley HM, Carannante G (eds) Cool-water carbonates: depositional systems and palaeoenvironmental controls, Geological Society special publications 255. Geological Society, London, pp 157–178

Simpson GG (1932) Fossil Sirenia of Florida and the evolution of the Sirenia. Bull Am Mus Nat Hist 59:419–503

Sorbi S (2007) The Neogene Euro-North African Sirenia: phylogenetic and systematic study, palaeoecological and palaeobiogeographic considerations. 363 pp. Unpublished thesis. Dipartimento di Scienze della Terra, Università di Pisa, Italy

Sorbi S, Domning DP, Vaiani SC, Bianucci G (2012) *Metaxytherium subapenninum* (Bruno, 1839) (Mammalia, Dugongidae), the latest sirenian of the Mediterranean Basin. J Vertebr Paleontol 32:686–707

Šuklje F (1938) Mediteranska sirena iz Vrapča kod Zagreba i Otruševca kod Samobora. [Mediterranean sirenians from Vrapce near Zagreb and Otrusevac near Samobor]. Glasnik Hrvatskoga Prirodoslovnoga Društva Zagreb 49–50:87–93

Svana K (2007) The late Miocene Sirenians of Crete. MSc thesis, University of Crete, Heraklion, Greece

Svana K, Iliopoulos G, Fassoulas C (2010) New Sirenian findings from Crete Island. Bull Geol Soc Greece 43(2):746–753

Symeonidis NK, Schultz O (1970) Eine miozäne Selachierfauna der Halbinsel Paliki (Kephallinia, Griechenland). Ann Géol Pays Hellén 21:153–162

Symeonidis NK, Schultz O (1973) Bemerkungen zur neogenen Fischfauna Kretas is und Beschrebung zweier Fundstellen mit miozänen Mollusken, Echiniiden etc., Insel Kreta, Griechenland. Ann Naturhist Mus Wien 77:141–147

Tekkay I (1973) Preliminary note on the vertebrate fauna of Bayraktepe, southeast of Canakkale (Dardanelles). Maden Tetkik Arama Enst. (Mineral Research & Exploration Institute of Turkey), Bulletin, Foreign Ed. 81: 118–122

Uhen MD (2007) Evolution of marine mammals: Back to the sea after 300 million years. Anat Rec 290:514–522

Vélez-Juarbe J, Domning DP (2014) Fossil Sirenia of the West Atlantic and Caribbean region. Ix. *Metaxytherium albifontanum*, sp. nov. J Vertebr Paleontol 34(2):444–464

The Fossil Record of Rodents (Mammalia: Rodentia) in Greece

Katerina Vasileiadou and Ioanna Sylvestrou

1 Introduction

Rodents are today one of the most successful and diverse mammalian orders, since they include almost half of the modern placental species (2277 species), occupying almost all habitats worldwide except for Antarctica (Wilson and Reeder 2005; Macdonald 2009). The order is characterized by the presence of one pair of large, sharp and ever-growing incisors on each of the upper and lower jaw, used for gnawing, followed by a diastema, a reduced number of premolars (if any) and usually three molars. Together with the order Lagomorpha, they form the clade Glires, based on various morphological similarities, the main of which is the presence of large ever-growing incisors on both jaws with enamel only on their buccal side (Meng and Wyss 2005). Rodents are separated from lagomorphs by several cranial and postcranial features, the most essential being the presence in the latter of a second pair of incisors.

The sampling of rodent fossils differs greatly from that of large animals, as in their vast majority the fossils are isolated skeletal and dental elements inside lacustrine, karstic, fluvial deposits that cannot be distinguished by the naked eye. The elaborate procedure includes sampling of large quantities of sediments, screen-washing and sieving followed by picking under a stereoscopic microscope. The isolated bones are generally not diagnostic, in contrast with the dental elements, which, due to the specialization of each rodent species in its food, are highly diagnostic.

K. Vasileiadou (Deceased)
Natural History Museum of the Lesvos Petrified Forest, Mytilene, Lesvos, Greece

I. Sylvestrou (✉)
School of Geology, Aristotle University of Thessaloniki, Thessaloniki, Greece
e-mail: sylvest@geo.auth.gr

E. Vlachos (ed.), *Fossil Vertebrates of Greece Vol. 1*,
https://doi.org/10.1007/978-3-030-68398-6_15

The fairly rapid evolution of many rodent groups, mainly based on the short lifespan of the individuals, combined with the strong habitat and dietary preference (observed in extant rodent species and presumed for their extinct relatives), leads to the use of fossil rodents for biostratigraphy and also palaeoenvironmental and palaeoclimatic reconstructions. The high-quality taxonomic framework is thus needed and can be used as a valuable tool in many aspects of the geological and biological research.

The Greek rodent fossil record, which starts from the late Oligocene, is reasonably complete, mainly during the late Miocene, due to the large number of localities. However, there are only a very few areas where the temporal sequence of rodent faunas can be followed (e.g., the Ptolemais Basin in North Greece); it is thus difficult to reconstruct local rodent evolution and presume migration events.

The sequence of rodent families in the text follows McKenna and Bell (1997) though subfamilies and genera are roughly in chronological order. Fossil Greek occurrences of taxa surviving today are not treated separately and take part of the more general genus remarks. Taxa originally described from Greece are marked with a star.

2 Historical Overview

The first Greek rodent fossil remains were reported in the nineteenth century, by Wagner (1848, 1857), Roth and Wagner (1854), Gaudry (1862–1867) and Gaudry and Lartet (1856), from the Upper Miocene locality Pikermi (Attica); all above authors referred to the presence of *Hystrix primigenia* remains (some under other species names; see paragraph 5.8). Later, in the first quarter of the twentieth century, late Miocene murids from the island of Samos were added to the Greek rodent fossil record by Schaub (1926). It was only in the 1960s that Greek researchers started reporting the presence of rodent fossils in various localities (Melentis 1966, 1967; Melentis and Schneider 1966).

The studies significantly increased in the 1970s and, mostly, in the 1980s, mainly due to the Dutch–Greek collaboration among the Faculty of Earth Sciences of the University of Utrecht, the Geology Department of the University of Athens and the Greek Institute of Geology and Mineral Exploration. This collaboration led to the description of rich Neogene rodent assemblages including several new rodent species from Maritsa 1 (Rhodes Island), Aliveri (Evia Island) and Attica (localities Biodrak, Pikermi, Tourkobounia) (e.g., de Bruijn et al. 1970, 1980, 1987; de Bruijn and van der Meulen 1975; de Bruijn 1976; van der Meulen and de Bruijn 1982; Klein Hofmeijer and de Bruijn 1985, 1988; Alvarez-Sierra et al. 1987). In the following years, the intense sampling at various Neogene localities (of various ages) in the Ptolemais and Strimon Basins (Macedonia, North Greece) by large teams from the Universities of Utrecht and Athens (and Mainz at Strimon Basin) resulted in publications of huge importance for the knowledge of the Neogene rodent faunas and the understanding of their evolution in south-east Europe (e.g., van de Weerd 1979; de Bruijn and van der Meulen 1979; Bouwens and de Bruijn 1986; de Bruijn 1989, 1995; Theocharopoulos 1991; Daxner-Höck 1992, 1995; Storch and Dahlmann 1995; Hordijk and de Bruijn 2009).

Since the 1990s and until today, studies carried out in the Greek Geology Departments of the Universities of Athens, Thessaloniki and Patras have led to the discovery of more Neogene and also Quaternary rodent remains (e.g., Koliadimou and Koufos 1991; Syrides and Koliadimou 1994; Koliadimou 1996; Athanassiou 1998; Doukas and Theocharopoulos 1999; Theocharopoulos 2000; Koufos et al. 2001; Sylvestrou 2002; Vasileiadou et al. 2003, 2012; Tsoukala and Chatzopoulou 2005; Tsoukala et al. 2006; Sylvestrou and Kostopoulos 2007; Vasileiadou and Sylvestrou 2009; Chatzopoulou 2014; Skandalos 2017; Konidaris et al. 2018; Doukas et al. 2018; Kokotini et al. 2019).

The study of the rich Greek Neogene fossil rodent material over the past almost two centuries has led to the creation of one sciurid *Aliveria* and seven murid new genera, *Heramys*, *Byzantinia*, *Karydomys*, *Hypsocricetus*, *Parapodemus*, *Hansdebruijnia* and *Kalymnomys*, as well as 53 new species, four glirids, seven sciurids, 41 murids and one hystricid, and one new subspecies, the murid *Rhagapodemus frequens athensis*.

3 Phylogenetic Relationships

The monophyly, divergence dates and the phylogenetic relationships of rodents with other orders, as well as relationships among the order's clades, have been matters of long-lasting debates (see Meng and Wyss 2005 for a brief summary). In general, morphological data mostly support the monophyly of the order, but molecular data indicate both monophyly and paraphyly; the monophyly of the rodentia is, however, in general more commonly accepted (see Blanga-Kanfi et al. 2009; Ryu et al. 2013 and references therein). The oldest rodent fossils seem to come from the late Paleocene (57 Ma) of N America (Meng and Wyss 2005). However, several studies based on molecular data place the divergence date for rodents much earlier, e.g., 109–112 Ma or at around 75 Ma (Adkins et al. 2001; Meng and Wyss 2005 and references therein).

Traditionally, rodents were divided into three suborders, Sciuromorpha, Myomorpha and Hystricomorpha, based on the position of the masticatory muscles (masseters); however more recent works divide rodents into two suborders, Sciurognathi and Hystricognathi, based on the position of the incisors and the angle of the jaw (Blanga-Kanfi et al. 2009; Macdonald 2009). The monophyly of Hystricognathi is well established, but that of Sciurognathi as well as the phylogenetic relationships between the two suborders have been debated by many taxonomists and molecular biologists (Blanga-Kanfi et al. 2009). The most recent works on rodent classification, based on both morphological and molecular data, accept the division of rodents in five suborders, Sciuromorpha, Castorimorpha, Myomorpha, Anomaluromorpha and Hystricomorpha; reports based on mitochondrial and/or nuclear genes support the monophyly of these suborders, but phylogenetic relationships among them are still unclear (Ryu et al. 2013 and references therein).

4 Distribution

Rodent fossil remains have been recovered from 108 Upper Oligocene to Upper Pleistocene localities in Greece (and 1 sub-recent locality): 39 are correlated with the Pleistocene, 22 with the Pliocene, four with the Miocene/Pliocene boundary, 42 with the Miocene and one with the Upper Oligocene. 53 of the localities are located in North Greece, 12 in Attica, six in Central Greece, 10 in Peloponnese and 28 in insular Greece (Kos, Naxos, Chios, Kalymnos, Evia, Rhodes, Karpathos, Samos, Crete and Lesvos). The fact that their vast majority can be found in North Greece clearly has to do with the geographical and temporal distribution of the fossiliferous sediments in Greece (Fig. 1 and Appendix).

The only Upper Oligocene locality in Greece is Kyprinos, in Thrace (N Greece), which has yielded five rodent species (Doukas and Theocharopoulos 1999). Aliveri (Evia Island) and Karydia (Thrace, N Greece) are the most diverse Lower Miocene (MN4) localities, with 21 (de Bruijn et al. 1980; van der Meulen and de Bruijn 1982; Klein Hofmeijer and de Bruijn 1985; Ziegler and Fahlbusch 1986; de Bruijn et al. 1987; Alvarez-Sierra et al. 1987; Klein Hofmeijer and de Bruijn 1988; Ünay 1994; Theocharopoulos 2000) and 24 rodent species (Wessels 2020 personal communication), respectively, whereas Thymiana A in Chios Island is the most diverse middle Miocene (MN5) locality with 10 rodent species (de Bonis et al. 1997; López-Antoñanzas et al. 2005). The best known Upper Miocene (MN13) localities are Maritsa 1 (Rhodes Island) with 15 (de Bruijn et al. 1970; de Bruijn and van der Meulen 1975; Şen 1977; Şen et al. 1989) and Lava 2 (N Greece) with 10 rodent species (de Bruijn et al. 1999), whereas Maramena (N Greece) is the most diverse locality correlated with the Turolian/Ruscinian (MN13/14) boundary with 20 rodent species (Schmidt-Kittler et al. 1995). The most diverse Lower Pleistocene locality, Rema Voulgarakis (MNQ20; N Greece), has revealed 12 rodent species (Koliadimou 1996), and the most diverse Middle Pleistocene one, Latomi 1 (Chios Island), provided 18 species (Storch 1975). The best known Upper Pleistocene localities are Loutra Almopias Cave LAC and LAC Ia, with 17 and 15 species, respectively (Chatzopoulou 2014), as well as that of Arnissa with 16 species (Mayhew 1978), all three of them in North Greece.

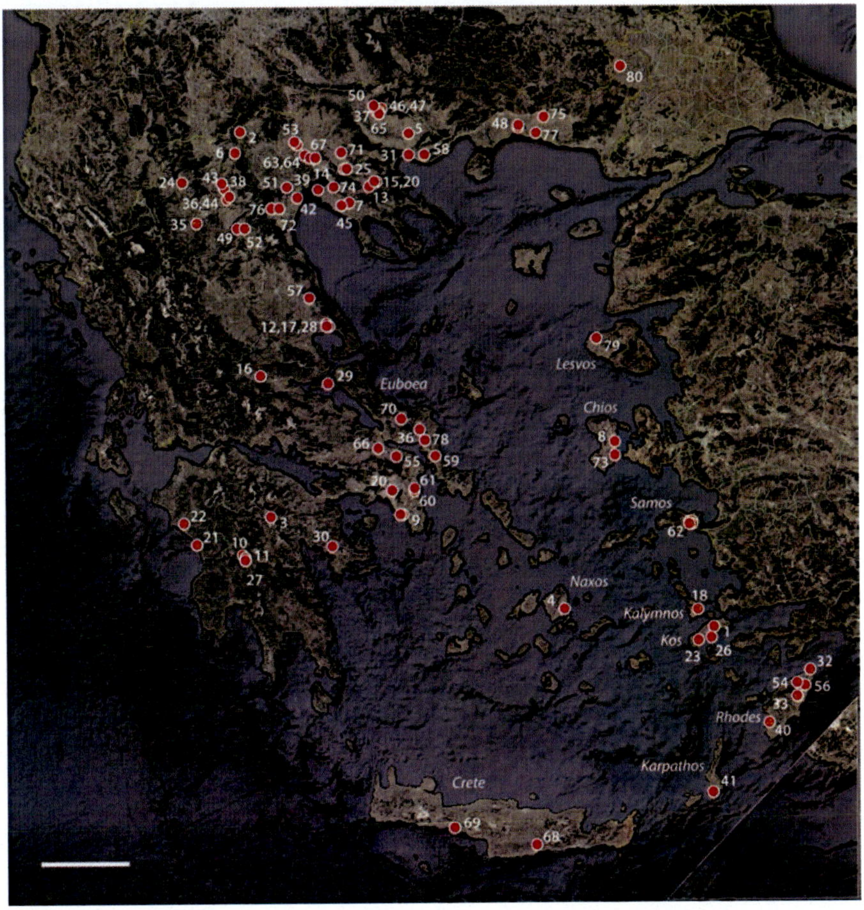

Fig. 1 Map of Greece showing the geographic distribution of the most important localities with rodent fossils. **1**, Pili; **2**, Loutra Almopias Cave; **3**, Kitseli Pothole; **4**, Trypiti; **5**, Dimitra 2; **6**, Arnissa; **7**, Petralona Cave; **8**, Latomi 1; **9**, Varkiza localities; **10**, Marathousa localities; **11**, Choremi localities; **12**, Volos; **13**, Kalamoto-1; 14**,** Apollonia-1; **15**, Ravin Voulgarakis; **16**, Zeli localities; **17**, Alikes; **18**, Kalymnos-Xi; **19**, Marathoussa; **20**, Tourkobounia localities; **21**, Kaiafas; **22**, Pyrgos; **23**, Lagkada; **24**, Kastoria 1 and 2; **25**, Gerakarou 1; **26**, Kardamena; **27**, Thoknia; **28**, Sesklo; **29**, Yialtra; **30**, Karnezeika; **31**, Rema Aslan; **32**, Damatria; **33**, Archipolis; **34**, Milia-5; **35**, Limni localities; **36**, Notio localities; **37**, Spilia localities; **38**, Komanos localities; **39**, Megalo Emvolon; **40**, Apolakkia; **41**, Kato Afiartis; **42**, Makrygialos; **43**, Vorio, Ptolemais, and Kardia localities; **44**, Tomea Eksi localities; **45**, Nea Silata; **46**, Maramena; **47**, Ano Metochi localities; **48**, Kessani 1 and 2; **49**, Prosilion–Mercurion; **50**, Monasteri; **51**, Paliambela localities; **52**, Lava 2; **53**, Dytiko localities; **54**, Maritsa 1; **55**, Chalkoutsi; **56**, Kalithies; **57**, Alifakas; **58**, Rema Marmara; **59**, Halmyropotamos; **60**, Pikermi; **61**, Chomateri; **62**, various localities in Mytilinii basin, Samos Island; **63**, Vathylakkos-3; **64**, Ravin des Zouaves 5; **65**, Lefkon; **66**, Biodrak; **67**, Ravin de la Pluie; **68**, Kastellios localities; **69**, Plakias; **70**, Katheni; **71**, Chryssavgi 1; **72**, Rema Patsiaris; **73**, Thymiana localities; **74**, Antonios; **75**, Komotini; **76**, Moschopotamos; **77**, Karydia; **78**, Aliveri; **79**, Lapsarna; **80**, Kyrpinos. See Appendix for more information. (Image exported from Google Earth Pro © 2019, map data from US Dept. of State Geographer, SIO, NOAA, U.S. Navy, NGA, GEBCO, image from Landsat/Copernicus. Scale bar equals 80 km, North faces upwards)

5 Systematic Paleontology

Rodentia Bowdich, 1821

Family Sciuridae Fischer de Waldheim, 1817

Type Genus *Sciurus* Linnaeus, 1758.

Remarks Remains of Sciuridae indet. have been listed in the faunal lists for Komotini (de Bruijn and van der Meulen 1979), Thymiana C (de Bonis et al. 1997) and Tourkobounia 3 and 5 (Reumer and Doukas 1985), but without any details.

Subfamily Sciurinae Hemprich, 1820

Tribe Pteromyini Brandt, 1855

Type Genus *Pteromys* Cuvier, 1800.

Remarks The flying squirrels used to be grouped in the Sciuridae subfamily Petauristinae Simpson, 1945, but are now classified in the tribe Pteromyini within the subfamily Sciurinae (Macdonald, 2009). Daams and van de Weerd (1980) reported the presence in the early Ruscinian fauna from Agios Ioannis (Karpathos Island) of one fragmented left D4 with a large metaconule and a small metacone, which they assigned to Petauristinae gen. et sp. indet. (now Pteromyini indet.).

***Aliveria* de Bruijn, van der Meulen and Katsikatsos, 1980**

Type Species *A. brinkerinki* de Bruijn, van der Meulen and Katsikatsos, 1980.

Other Taxa Included *A. luteyni* de Bruijn, van der Meulen and Katsikatsos, 1980.

Distribution Early Miocene of Greece, Serbia and Turkey.

★*Aliveria brinkerinki* de Bruijn, van der Meulen and Katsikatsos, 1980

Type Material Right m1, no. 171(holotype) (de Bruijn et al. 1980: Pl. 1.8).

Type Locality Aliveri, Greece, Lower Miocene [MN4].

Distribution Early Miocene of Greece and Serbia (Sibnica 4).

Occurrences in Greece Aliveri and Moschopotamos.

Remarks De Bruijn et al. (1980) created the species based on material from Aliveri (Fig. 2a–f) consisting of two d4, nine p4, five m1, seven m2, four m3, five D4, two P4, 11 M1/2 and two M3 (including the holotype). The diagnosis mentions that this is a medium-sized flying squirrel with estimated p4–m3 length of 9.4 mm, characterized by the presence of small protoconule and metaconule and distinct mesostyle on the M1–2, a large round anteroconid connected with the protoconid and the metaconid and small mesoconid and mesostylid on the m1–2, whereas the unworn lower molars show a peculiar pit labially to the entoconid (de Bruijn et al. 1980). Benda and Steffens (1981) reported the presence in the locality Moschopotamos of one M3 of *A.* cf. *brinkerinki*, which is a little larger than the M3 from Aliveri.

★*Aliveria luteyni* de Bruijn, van der Meulen and Katsikatsos, 1980

Type Material Left m1, no. 253 (holotype) (de Bruijn et al. 1980: Pl. 2.7).

Type Locality Aliveri, Greece, Lower Miocene [MN4].

Distribution Early Miocene of Greece and Turkey, plus a doubtful presence in Serbia (*A.* aff. *luteyni* in Sibnica 1, 2, 4).

Occurrences in Greece Aliveri and Karydia.

Remarks This species is smaller than *A. brinkerinki*, with estimated p4–m3 length of 7.6 mm (de Bruijn et al. 1980). According to the diagnosis, *A. luteyni* is characterized by a small or absent protoconule, an elongate or double metaconule and a small but distinct mesostyle, often connected to the paracone on the M1–2, and an anteroconid incorporated within the anterolophid, a large mesoconid and a small or absent mesostylid on the m1–2 (de Bruijn et al. 1980). The type locality has yielded one d4, six p4, six m1, five m2, four m3, one D4, three P4, 16 M1/2 and five M3 (including the holotype; (Fig. 2g–l). The species has also been found in Karydia, as reported in the faunal list for the locality by Theocharopoulos (2000) and verified in the most recent review of the material (Wessels 2020 personal comminication), but there is no additional information published.

***Pliopetaurista* Kretzoi, 1962**

Type Species *Sciuropterus pliocaenicus* Depéret, 1890.

Other Taxa Included *P. dehneli* Sulimski, 1964; *P. bressana* Mein, 1970; *P. meini* Black and Kowalski, 1974; *P. rugosa* Qiu, 1990; *P. raui* Dahlmann, 2001; *P. kollmanni* Daxner-Höck, 2004; *P. speciosa* Qiu and Ni, 2006.

Distribution Late Miocene–Early Pleistocene of Europe and Asia.

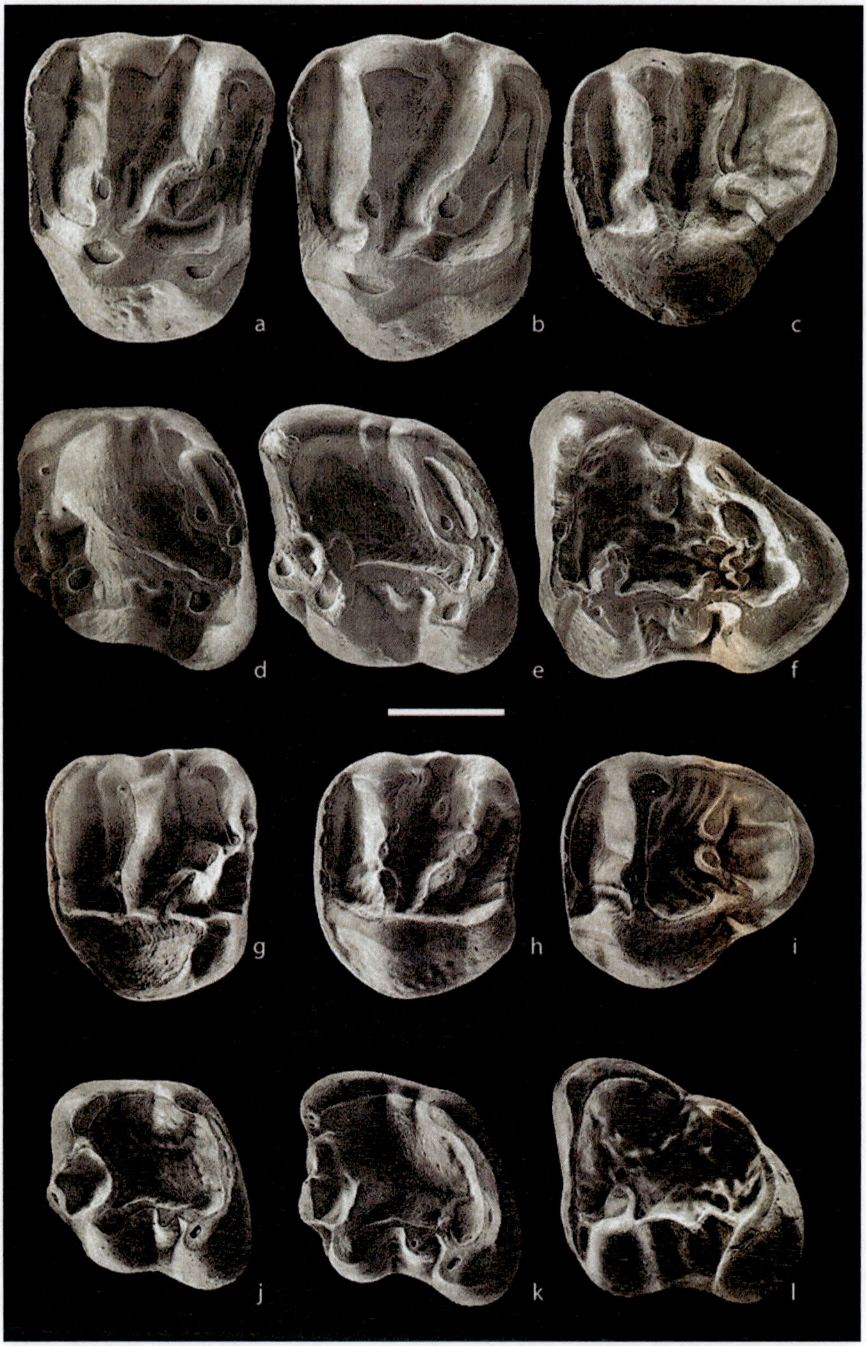

Fig. 2 *Aliveria brinkerinki* from Aliveri (type locality). **a** M1, **b** M1/2, **c** M3, **d** m1 (holotype), **e** m2, **f** m3. *Aliveria luteyni* from Aliveri (type locality). **g** M1/2, **h** M1/2, **i** M3, **j** m1 (holotype), **k** m2, **l** m3. Images courtesy of H. de Bruijn. All specimens figured as left; d and g are reversed. Scale bar equals 1 mm

Remarks *Pliopetaurista* sp. is reported in the faunal list from Lava 2 by de Bruijn et al. (1999), without description or further information on the material, except that the teeth are larger than those of *P. dehneli*.

Pliopetaurista bressana **Mein, 1970**

Type Material Right M1/2, F.S.L. 65438 (holotype) (Mein 1970: Fig. 49).

Type Locality Soblay, France, Upper Miocene [MN10].

Distribution Late Miocene of Europe and Turkey.

Occurrences in Greece Lefkon.

Remarks This species is diagnosed by the absence of a mesostylar crest from the upper molars, the presence of an anteroconid on the p4 and of an anterosinusid on the lower molars and the not very long m3 (Mein 1970). Mein (1970) mentioned in his diagnosis that this is the oldest and smallest *Pliopetaurista* species; however, after the publication of this species, even smaller and older species have been named (i.e., *P. meini*, *P. raui* and *P. kollmanni*). The species is present in the faunal list of the latest Vallesian fauna of Lefkon (de Bruijn 1989), but without further information.

Pliopetaurista dehneli **(Sulimski, 1964)**

Nomenclatural and Taxonomical History *Pliosciuropterus dehneli* in Sulimski 1964 (new species); *Pliopetaurista dehneli* in Mein 1970 (new combination). *P. schaubi* (Sulimski, 1964) is considered a junior synonym by de Bruijn (1995).

Type Material Fragment of a right upper jaw with P4–M1 and the alveoli of P3 and M2, No. VM/I–1 (holotype), Institute of Palaeozoology of the Polish Academy of Sciences in Warsaw (Sulimski 1964: Pl. IV.8).

Type Locality Węże 2, Poland, Pliocene [MN15–16].

Distribution Late Miocene–Pliocene of Europe and Turkey.

Occurrences in Greece Kessani-2, Kessani-1, Ano Metochi 2, Maramena, Nea Silata, Komanos 1 high A and B, and Vorio 3 and 3A.

Remarks According to the diagnosis by Sulimski (1964), the characteristics of the species are the metastylid of the p4–m2 is small and connected with the metaconid, the central field of the p4–m2 crown bears weak proto- and hypolophid, the p4 has three roots, the talonid of the m3 is tapering posteriorly, the mental foramen is large and situated just in front of the anterior root of the p4, the lower masseteric crest is high and sharp, the P4–M2 show open posterior re-entrant folds (the post-metacone flexus, the posteromedial diagonal flexus and the postero-lingual diagonal flexus) and the M3 lacks a metaloph. The species of *Pliopetaurista* differ also in the size of their dental elements; the size of the teeth of *P. dehneli* is intermediate between that of *P. bressana* and *P. pliocaenica* (largest species reported until now).

P. dehneli has been found in many Greek localities. Maramena has yielded the richest material, consisting of 10 d4, 12 p4, 32 m1/2, 15 m3, 14 D4, 17 P4, 37 M1/2 and 15 M3 (de Bruijn 1995). Nea Silata has only yielded one fragmented M1/2, but its morphology and size agree well with this species (Vasileiadou et al. 2003). Material from Kessani includes one d4, two m1/2, one m3, one P4, one M1/2 and one M3 (Vasileiadou et al. 2012). The species is also present in Komanos 1 high A and B with one fragmented m3 and in Vorio 3 and 3A with one fragmented M1/2 and one m3 (Hordijk and de Bruijn 2009). Finally, the species is in the faunal list for Ano Metochi 2 (de Bruijn 1989), but with no information on the material.

***Miopetaurista* Kretzoi, 1962**

Type Species *Sciurus gibberosus* Hofmann, 1893.

Other Taxa Included *M. lappi* (Mein, 1958); *M. gaillardi* (Mein, 1970); *M. neogrivensis* (Mein, 1970); *M. crusafonti* (Mein, 1970); *M. thaleri* (Mein, 1970); *M. diescalidus* Daams, 1977; *M. dehmi* de Bruijn, van der Meulen and Katsikatsos, 1980; *M. asiatica* Qiu, 2002.

Distribution Early Miocene–Pliocene of Europe and Asia.

Miopetaurista dehmi **de Bruijn, van der Meulen and Katsikatsos, 1980**

Type Material Left m1, no. 10997 (1937 II) (holotype) (Dehm 1950: Fig. 36).

Type Locality Wintershof West, Germany, Lower Miocene [MN3].

Distribution Early–middle Miocene of Europe.

Occurrences in Greece Aliveri and Karydia.

Remarks This is a small *Miopetaurista* species, of which the p4–m3 length is 10.4 mm (de Bruijn et al. 1980). The diagnostic features are the following: the P4 is slightly longer but narrower than the M1–2, the parastyle of the P4 is unicuspulate, the mesostyle is absent from the P4 and M3 and forms a tiny ridge on the M1–2, a small extra loph is present behind the protoloph on all upper cheek teeth and the mesostylid is well surrounded in all lower cheek teeth (de Bruijn et al. 1980). The dental elements of this species are almost identical with those of *M. lappi*, but they are much smaller (de Bruijn et al. 1980). *M. dehmi* has been found in Aliveri, represented only by two p4 (de Bruijn et al. 1980). It also appears in the most recent faunal list for Karydia (Wessels 2020 personal communication).

Miopetaurista thaleri (Mein, 1970)

Nomenclatural and Taxonomical History *Cryptopterus thaleri* in Mein 1970 (new species); *Miopetaurista thaleri* in Daxner-Höck and Mein 1975 (new combination). *Cryptopterus tobieni* Mein, 1970 is considered a junior synonym by de Bruijn (1995).

Type Material Partial left maxilla with P4–M3, M.n.P. 1922–1 (holotype) (Mein 1970: Fig. 35).

Type Locality Marnes de Celleneuve, France, Lower Pliocene [MN14].

Occurrences in Greece Maramena.

Distribution Late Miocene–early Pliocene of France, Poland, Slovakia and Greece.

Remarks The species has very large dental elements, and their diagnostic features are the following: the abrupt decrease in height of the endoloph in contact with the posteroloph, the simple parastyle on the P4, the gradually weaker mesostyle from the P4 to the M3, the presence of a lophule on the anterior face of the protoloph and the relatively high average portion of the posteroloph (Mein 1970). The Greek locality Maramena has revealed a poor collection of one p4, one m1, one m2, one m3 and two M1/2 assigned to this species; the two upper molars seem to belong to the same individual and the lower teeth to a second one (de Bruijn 1995).

Blackia Mein, 1970

Type Species *Blackia miocaenica* Mein, 1970.

Other Taxa Included [according to Marković et al. (2016)] *B. polonica* Black and Kowalski, 1974; *B. ulmensis* Werner, 1994.

Distribution Late Oligocene–late Pliocene of Europe and Asia.

Remarks de Bruijn et al. (2012) reported the presence of a single m2 in Plakias, which they identified as cf. *Blackia* sp. – the specimen had been previously described as *Forsythia*? sp. by de Bruijn and Meulenkamp (1972).

Blackia miocaenica Mein, 1970

Taxonomical History *B. wölfersheimensis* Mein, 1970 and *B. parvula* Baudelot, 1972 are junior synonyms, according to Dahlmann (2001) and Engesser (1972), respectively.

Type Material Left P4, F.S.L. 65463 (holotype) (Mein 1970: Fig. 72).

Type Locality La Grive L7, France, middle Miocene [MN7/8].

Distribution Early Miocene (MN4)–Pliocene (MN15) of Europe.

Occurrences in Greece Aliveri, Karydia and Lava 2.

Remarks The diagnosis of the species by Mein (1970) mentions (translated from French): "*The triangular P4 has a narrow internal border, convergent lophs, a strong parastyle and a very prominent protocone; the upper molars show a metacone as high as the paracone; the internal wall decreases strongly at the front and rear of the protocone; the lower molars have an antero–external cingulum and an anterosinusid*". The general morphology of the dental elements remained virtually unchanged during the long existence of this species, and apparently this leads to its synonymies with other *Blackia* species. De Bruijn et al. (1980) assigned to this species a collection of dental elements from Aliveri: five d4, one p4, five m1, four m2, two D4, one P4, seven M1/2 and two M3. Theocharopoulos (2000) listed the species in the faunal list for Karydia and after a review of the material, Wessels (2020 personal communication) verified its presence, but no details on the material have been published yet. De Bruijn et al. (1999) included *B. miocaenica* (as *B. wölfersheimensis*) in the faunal list for the locality Lava 2, without description and details on the material.

***Hylopetes* Thomas, 1908**
(= *Pliopetes* Kretzoi, 1959, according to Bouwens and de Bruijn 1986)

Type Species *Sciuropterus everetti* Thomas, 1908.

Other Taxa Included The genus *Hylopetes* currently includes 10 extant species of flying squirrels, all restricted in Asia (Macdonald 2009). The extinct species of *Hylopetes* include (according to Bosma et al. 2013) *H. hungaricus* (Kretzoi, 1959); *H. macedoniensis* Bouwens and de Bruijn, 1986; *H. auctor* Qiu, 1991; *H. höckarum* de Bruijn, 1998; *H. debruijni* Mein and Ginsburg, 2002; *H. magistri* Reumer and van den Hoek Ostende, 2003.

Distribution Early Miocene–present of Europe and Asia.

Remarks *Hylopetes* is usually poorly represented in fossil localities, a fact that hampers specific identification. Van de Weerd (1979) identified one P4 and one M1 from the early Ruscinian fauna of Ptolemais 1 as *Hylopetes* sp. Furthermore, de Bruijn (1989) mentions the presence of *Hylopetes* sp. in the Turolian/Ruscinian fauna from Monasteri, without giving information on the material. Finally, de Bruijn et al. (2012) described one D4 and one M3 from the early Vallesian fauna of Plakias as cf. *Hylopetes* sp.

Hylopetes hungaricus (Kretzoi, 1959)

Nomenclatural and Taxonomical history *Pliopetes hungaricus* in Kretzoi 1959 (new species); *Hylopetes hungaricus* in Bouwens and de Bruijn 1986 (new combination).

Type Material Left m2, Csarnóta 2/13(holotype) (Kretzoi 1962: Pl. 3.1, designated by van de Weerd 1979:132).

Type Locality Csarnóta 2, Hungary, Lower Pliocene [Ruscinian (MN15)].

Distribution Late Miocene–Pliocene of Europe and Turkey.

Occurrences in Greece Lava 2.

Remarks Kretzoi (1959) created a new genus (*Pliopetes*) with *P. hungaricus* as its type species based on about 30 molars, but without giving detailed descriptions, without designating a holotype or figuring any elements; he mentioned that the molars seem to be identical at first sight to the extant species and the lower molars have a rhombic shape, two roots and wrinkled enamel reminiscent of *Hylopetes*. Later, he figured a series of dental elements from the type locality of the species (Kretzoi 1962:Pl. 3.1–3), but van de Weerd (1979:132) and later Bouwens and de Bruijn (1986:118) verified that the figured M1 (Kretzoi 1962:Pl. 3.3) belongs to *Blackia* and the figured m2 and m3 (Kretzoi 1962: Pl. 3.1–2) belong to a species very similar to extant *Hylopetes*. Van de Weerd (1979:132) mentioned that "*there seems to be little reason to maintain the generic name* Pliopetes … *as the difference between* Hylopetes *and* Pliopetes *is slight*". Bouwens and de Bruijn (1986) synonymized the two genera and included the type species of *Pliopetes*, *P. hungaricus*, to *Hylopetes*. De Bruijn et al. (1999) included *H.* cf. *hungaricus* in the faunal list for Lava 2. Later, Bosma et al. (2013) mentioned that the Lava 2 material constitutes of only one "*little worn right P4 (length 1.42, width 1.47)*".

★*Hylopetes macedoniensis* Bouwens and de Bruijn, 1986

Type Material Right M1–2, no 1651 (holotype) (Bouwens and de Bruijn 1986: Pl. 3.2).

Type Locality Maramena, Greece, Miocene/Pliocene [Turolian/Ruscinian boundary (MN13/14)].

Distribution Late Miocene/Pliocene of Greece and Turkey, possible presence in Moldova (*H.* cf. *macedoniensis*, Nicoara 2011).

Occurrences in Greece Maramena.

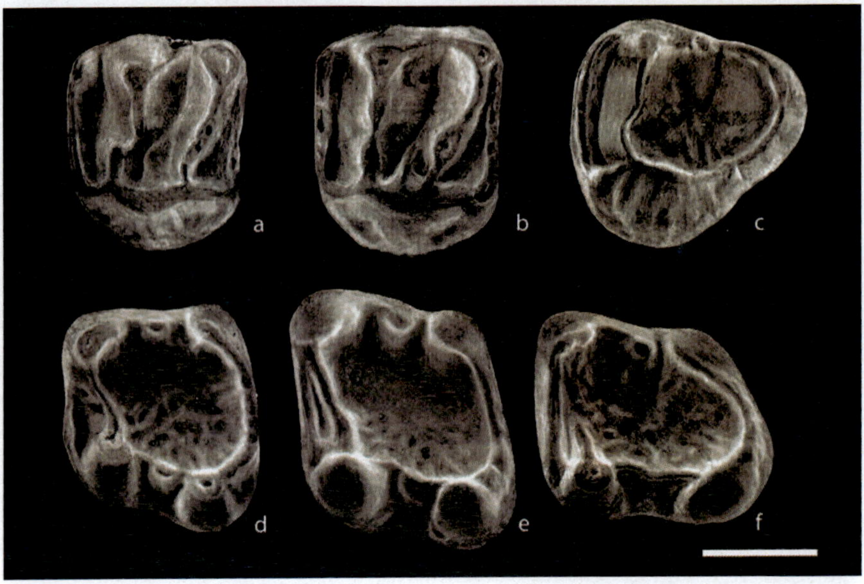

Fig. 3 *Hylopetes macedoniensis* from Maramena (type locality). **a** M1/2, **b** M1/2, **c** M3, **d** ml, **e** m2, **f** m3. Images courtesy of H. de Bruijn. All specimens figured as left; b, c, f are reversed. Scale bar equals 1 mm

Remarks This is a small *Hylopetes* species, the diagnosis of which includes the following features: the enamel of unworn cheek teeth is not much sculptured, the four-rooted lower molars bear a deep notch between the entoconid and the mesostylid, as well as a low and narrow anterolophid is not connected to the protoconid and not extended to the antero-labial cingulum (Bouwens and de Bruijn 1986). After the establishment of the species, more material was found in the type locality Maramena (Fig. 3), which now includes 11 d4, eight p4, 33 m1–2, 10 m3, 11 D4, five P4, 30 M1–2 and 10 M3, making this the richest fossil *Hylopetes* collection up to now (de Bruijn 1995). The species dental elements are very similar metrically and morphologically to those of the extant *H. sagitta*, apart from minor details (Bouwens and de Bruijn 1986).

Tribe Sciurini Fischer de Waldheim, 1817

Type Genus *Sciurus* Linnaeus, 1758.

Palaeosciurus Pomel, 1853

Type Species *Sciurus* (*Palaeosciurus*) *feignouxi* Pomel, 1853.

Other Taxa Included [according to Bosma et al. (2019)] *P. fissurae* (Dehm, 1950); *P. goti* Vianey-Liaud, 1974; *P. sutteri* Ziegler and Fahlbusch, 1986; *P. ultimus* Mein and Ginsburg, 2002 [Fejfar and Kaiser (2005) mentioned the presence of

Palaeosciurus nov. sp. in the Oligocene Doupov Mountains, in northwest Bohemia, but have not yet officially named the species or given any details on the material].

Distribution Early Oligocene–middle Miocene of Europe and Asia.

***Palaeosciurus fissurae* (Dehm, 1950)**

Nomenclatural and Taxonomical History *Sciurus fissurae* in Dehm 1950 (new species); *Palaeosciurus fissurae* in Aguilar 1974 (new combination).

Type Material Right mandible with full dentition and part of the vertical ramus, Sammlung München 1937 II 10154 (holotype) (Dehm 1950: Fig. 13a–c).

Type Locality Wintershof–West, Germany, Lower Miocene [MN3].

Distribution Early–middle Miocene of Europe and Turkey.

Occurrences in Greece Aliveri and Karydia.

Remarks According to Dehm's (1950) diagnosis, the species is of similar size with the extant *Sciurus vulgaris*, but with a lower mandible, weaker masseteric ridge, with no parastyle (Vorderlobus) on the P4 and with longitudinally striped Schmelzband in the lower incisors. One p4, one m1, one m2, one m3 and one M1/2 from Aliveri have been described as *P.* aff. *fissurae* by de Bruijn et al. (1980), who noticed that the Aliveri material "*seem to represent a new species of* Palaeosciurus", as the dental elements are more robust than those from the type locality of the species, the mesoconids of the lower molars are larger, the postero-internal outline of the m3 is more oblique and the enamel of the central basins of unworn lower molars is more rugose (de Bruijn et al. 1980:259). Ziegler and Fahlbusch (1986) included the Aliveri material in their new species *P. sutteri*. However, van den Hoek Ostende et al. (2015), in their review on the Aliveri fauna, listed *P.* aff. *fissurae*. *P.* aff. *fissurae* also appears in the faunal list for Karydia (Theocharopoulos 2000); there is no additional information on the material, but as the Karydia material has been recently revisited, it seems that the material indeed belongs to *P. fissurae* (Wessels 2020, personal communication).

***Palaeosciurus sutteri* Ziegler and Fahlbusch, 1986**

Type Material Left m1, Inv.–Nr.1959 XXVII 225 (holotype), Bavarian State Collection for Paleontology and Historical Geology (Ziegler and Fahlbusch 1986: Tab. 3.16).

Type Locality Forsthart bei Vilshofen, Germany, Lower Miocene [MN4].

Distribution Early–middle Miocene of Europe.

Occurrences in Greece Antonios.

Remarks This is a sciurid with medium-sized dental elements; the entoconid and mesostylid are clearly separated on the m1/2; the metalophid of the lower molars usually forms a lingual spur on the protoconid; the P4–M2 show a distinct, isolated mesostyle and a well-developed metaloph; the proto- and metaloph on the upper molars are connected; the protoconule appears often and the metaconule is mostly present (Ziegler and Fahlbusch 1986). According to Ziegler and Fahlbusch (1986:24), the evolution from *P. fissurae* to *P. sutteri* is evident through the stressing of the entoconid and the mesostylid, the development of a conical mesostyle on the P4, M1 and M2 and the constriction of the protoloph and the metaloph. A small collection of *P. sutteri* dental elements have been found in Antonios, one m2, two P4, one M1/2 and one M3; their morphology is very close to that of the material from Forsthart, but they are larger, within the range of *P. sutteri* from Oberdorf 3 and 4 in Austria described by de Bruijn (1998) (Vasileiadou and Koufos 2005).

***Sciurus* Linnaeus, 1758**

Type Species *Sciurus vulgaris* Linnaeus, 1758.

Other Taxa Included There are at least 27 extant species in Eurasia and America (Macdonald 2009). Of these, *S. anomalus* Güldenstädt, 1785 and *S. vulgaris* have also been found in fossil assemblages. The extinct species of the Old World are *S. whitei* Hinton, 1914; *S. warthae* Sulimski, 1964; *S. maltei* Dahlmann, 2001; and *S. lii* Qiu and Yan, 2005.

Distribution Early–middle Miocene of China to present in Eurasia and America.

Remarks *Sciurus* sp. has been reported from the Lower Pleistocene localities Kalymnos-Xi (one M3; Kuss and Storch 1978) and Kaiafas (van der Meulen and van Kolfschoten 1986). One m1 of the extant species *S. anomalus* has been found in the late Ruscinian fauna of Notio 1, showing great similarity with *S. anomalus* material from Upper Pleistocene cave deposits in Israel (Hordijk and de Bruijn 2006). The Villanyian fauna of Tourkobounia 1 has yielded one d4 and one M3 identified as *S.* cf. *anomalus* by de Bruijn and van der Meulen (1975). Although the material from Tourkobounia 2 is relatively rich (one d4, one D4, four P4, six M1/2 and two M3), the absence of lower molars and the differences between the P4 from Tourkobounia 2 and that of *S. anomalus* from Kebara Cave led to the assignment of the material to *S.* cf. *anomalus* (van der Meulen and Doukas 2001). Van der Meulen and Doukas (2001) mentioned that additional *S.* cf. *anomalus* material was collected from Tourkobounia 1 after the publication of de Bruijn and van der Meulen (1975), including also one M1/2 and one M3 morphologically similar to the Tourkobounia 2 elements, but smaller. One dental element of *S.* cf. *vulgaris* is listed for the Middle Pleistocene locality Choremi 3 without further information (van Vugt et al. 2000).

Subfamily Xerinae Osborn, 1910

Tribe Xerini Osborn, 1910

Type Genus *Xerus* Hemprich and Ehrenberg, 1833.

***Atlantoxerus* Forsyth Major, 1893**

Type Species *Sciurus getulus* Linnaeus, 1758 (only extant species of genus).

Other Taxa Included *A. tadlae* (Lavocat, 1961); *A. blacki* (de Bruijn, 1967); *A. adroveri* (de Bruijn and Mein, 1968); *A. rhodius* de Bruijn, Dawson and Mein, 1970; *A. huvelini* Jaeger, 1977; *A. giganteus* Wu, 1988; *A. junggarensis* Wu, 1988; *A. margaritae* Adrover et al., 1993; *A. cuencae* Aguilar et al., 1995; *A. orientalis* Qiu, 1996; *A. martini* Aguilar, 2002; *A. humboldti* Kordikova et al., 2004; *A. xiyuensis* Wei, 2010.

Distribution Early Miocene–Pleistocene of N Africa and Eurasia. The single extant species, *A. getulus*, lives in the Atlas Mountains (N Africa).

★*Atlantoxerus rhodius* de Bruijn, Dawson and Mein, 1970

Type Material Left m1/2, no. 1183 (de Bruijn et al. 1970: Pl. 7.11).

Type Locality Maritsa 1, Rhodes Island, Greece, Upper Miocene [Turolian (MN13)].

Distribution Apart from the type locality, it is known from Morocco (Mahboubi 2014). *A.* cf. *rhodius* is mentioned from Villaba–Alta in Spain (Adrover et al. 1976), Brisighella in Italy (de Giuli 1989) and Algeria (Coiffait et al. 1985).

Occurrences in Greece Maritsa 1.

Remarks This is an average-sized *Atlantoxerus* species, diagnosed through the absence of a protoconule from the upper molars, the presence of a small mesostyle on all M1–2 and some P4 and a very small metaconule on the M3 (de Bruijn et al. 1970). Additional morphological features include a metalophid that is absent from the p4, complete on the m1–2 and interrupted on the m3, an absent or indistinct entolophid on the p4, weak anteroconids, a small metacone on the M3 and a sharp narrow entolophid on the m1–3 (de Bruijn et al. 1970). The material from the type locality Maritsa 1 (Fig. 4) includes one partial mandible with m1–m2, three d4, four p4, seven m1/2, seven m3, two partial maxillae with M1–M3, five D4, eight P4, 10 M1/2 and three M3.

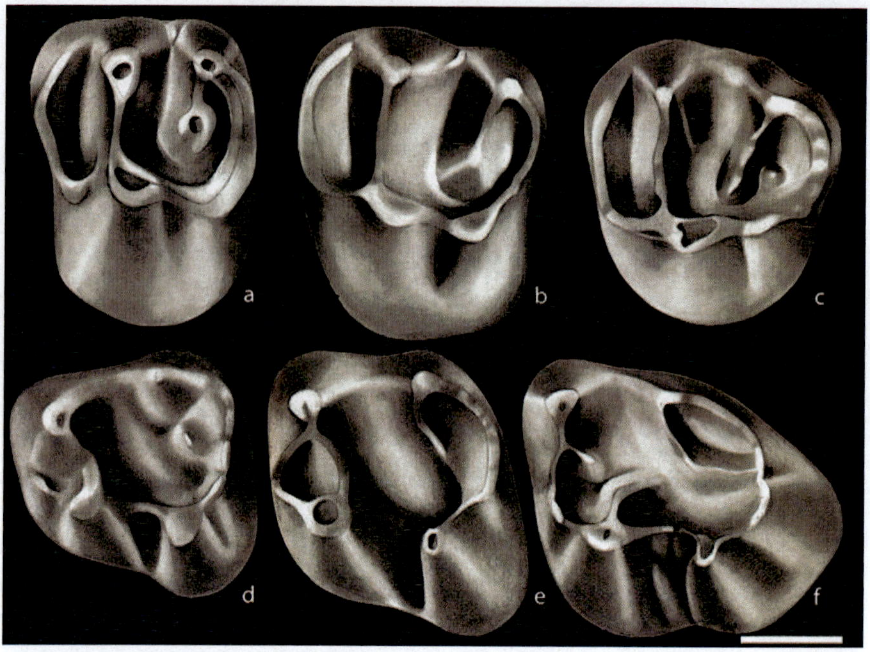

Fig. 4 *Atlantoxerus rhodius* from Maritsa 1 (type locality). **a** P4, **b** M1/2, **c** M3, **d** p4, **e** m1/2 (holotype), **f** m3. Images courtesy of H. de Bruijn. All specimens figured as left; b is reversed. Scale bar equals 1 mm

Tribe Marmotini Pocock, 1923

Type Genus *Marmota* Blumenbach, 1779.

Tamias Illiger, 1811

Type Species *Sciurus striatus* Linnaeus, 1758 (extant).

Other Taxa Included [Only extinct species of the Old World, including the subgenus *Eutamias* Trouessart, 1880, according to Bosma et al. (2019)] *T.* (*E.*) *orlovi* Sulimski, 1964; *T.* (*E.*) *urialis* Munthe, 1980; *T. eviensis* de Bruijn, van der Meulen and Katsikatsos, 1980; *T.* (*E.*) *sihongensis* Qiu and Liu, 1986; *T.* (*E.*) *ertemtensis* Qiu, 1991; *T. atsali* de Bruijn, 1995; *T. allobrogensis* Mein and Ginsburg, 2002; *T.* (*E.*) *lishanensis* Qiu, Zheng and Zhang, 2008; *T. anatoliensis* Bosma, de Bruijn and Wessels, 2013.

Distribution Late Oligocene–present of N America and Eurasia.

Remarks The rare presence of *Tamias* sp. in the localities Komanos 1 low A and B is mentioned by Hordijk and de Bruijn (2009). *Tamias* sp. is also mentioned in the faunal list for Komotini (de Bruijn and van der Meulen 1979).

★*Tamias eviensis* de Bruijn, van der Meulen and Katsikatsos, 1980

Type Material Left M1/2 (holotype) (de Bruijn et al. 1980: Pl. 4.9).

Type Locality Aliveri, Greece, Lower Miocene [MN4].

Distribution Early Miocene of Greece and possibly early–late Miocene of Turkey [*T.* cf. *eviensis* in the Lower Miocene locality Yapıntı (Bosma et al. 2019) and the Upper Miocene localities Karaözü and Hayranlı 1 (Bosma et al. 2013)].

Occurrences in Greece Aliveri.

Remarks The species is based on a small collection of two p4, two m1, three m2, three m3, one P4, nine M1/2 and five M3 from Aliveri (Fig. 5), which have size and general proportions similar with those of the extant *T. sibericus* and the Pliocene species *T. orlovi* (de Bruijn et al. 1980). Its diagnostic features include the absence of a parastyle from the P4, the rather well-developed hypocone on the M1–2, the presence of a small protoconule on the M1–3 and the well-developed anteroconid on the m1–3. The presence of *Tamias* in Aliveri seems to be the earliest occurrence of the genus in the Old World and almost coincides with the earliest occurrence of the genus in N America (de Bruijn et al. 1980).

★*Tamias atsali* de Bruijn, 1995

Type Material Right M1/2, no. 5014 (holotype) (de Bruijn 1995: Pl. 3.3).

Type Locality Maramena, Greece, Miocene/Pliocene [Turolian/Ruscinian boundary (MN13/14)].

Distribution Late Miocene/early Pliocene of Greece and Turkey [in Kangal 1 and *T.* aff. *atsali* in Süleymanlı 2 (Bosma et al. 2013)].

Occurrences in Greece Ano Metochi 3 and Maramena.

Remarks The material from the type locality Maramena (Fig. 5) consists of two d4, four p4, 28 m1/2, 19 m3, two D4, five P4, 13 M1/2 and 10 M3, whereas that from Ano Metochi 3 of one d4, one p4, one m1/2, one D4 and one partial maxilla with the P4–M3 and the alveolus of the P3 (de Bruijn 1995). The species is diagnosed through the average size of its dental elements, the presence of a P3, the large metaconule on the M1–2, the constricted or incomplete metaloph situated between the metaconule and the protocone, the absence of a protoconule from the M1–2, the small parastyle on the P4, rounded occlusal surface of the M3, the usual presence of a notch between the protocone and the posteroloph on the M3 and the absence of mesoconid and mesostylid from the m1–3 (de Bruijn 1995).

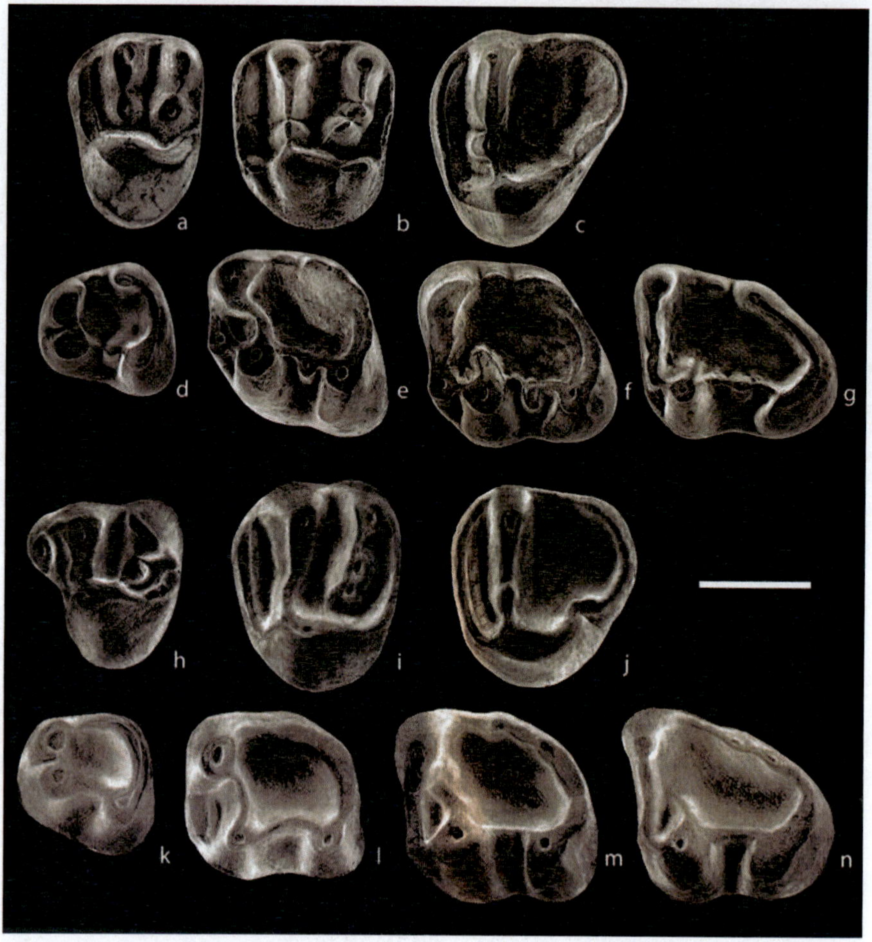

Fig. 5 *Tamias eviensis* from Aliveri (type locality). **a** P4, **b** M1/2 holotype, **c** M3, **d** p4, **e** m1, **f** m2, **g** m3. *Tamias atsali* from Maramena (type locality). **h** P4, **i** M1/2 holotype, **j** M3, **k** p4, **l** ml, **m** m2, **n** m3. Images courtesy of H. de Bruijn. All specimens figured as left; i, k, l are reversed. Scale bar equals 1 mm

Spermophilinus de Bruijn and Mein, 1968

Type Species *Sciurus bredai* von Meyer, 1848.

Other Taxa Included *S. turolensis* de Bruijn and Mein, 1968; *S. giganteus* de Bruijn, Dawson and Mein, 1970; *S. besana* Cuenca Bescós, 1988.

Distribution Early Miocene–early Pliocene of Eurasia.

Remarks The distinction of the *Spermophilinus* species is mainly based on the size of the dental elements, as their morphology is quite similar (de Bruijn 1995).

One P4 and one M3 found in the locality Ravin de la Plui have been described by Aguilar et al. (1995) as *Spermophilinus* sp.; the teeth are clearly larger than those of *S. turolensis* and *S. besana*; they are closer to *S. giganteus*, with which, however, there are morphological differences. Aguilar et al. (1995) believed the material belongs to a species that could be ancestral to *S. giganteus* from Maritsa 1; thus two lineages of *Spermophilinus* could be present in the area of Greece during the late Miocene. *Spermophilinus* sp. is also in the faunal list of Lava 2, with the only information available that the teeth are larger than those of *S. turolensis* (de Bruijn et al. 1999). De Bruijn (1995) suggested that there is a gradual increase in tooth size through time for the *Spermophilinus* species. However, the current data show that the *Spermophilinus* evolutionary history might be more complicated than this. According to Bosma et al. (2019), it is probable that the MN10–MN12 *S.* (cf.) *bredai* from Greece and Turkey is a continuation of the middle Miocene *S. besana* lineage, and the MN9–MN13/14 *S. turolensis* from Greece and Turkey has its roots in the middle Miocene *S. bredai*, whereas *S. giganteus* may have originated from *S. turolensis*. This hypothesis does not, however, fit with the presence of a large *Spermophilinus* in the MN10 Ravin de la Plui and the MN13 Lava 2 faunas.

***Spermophilinus besana* Cuenca Bescós, 1988**

Nomenclatural History The species was named *Spermophilinus besana* by Cuenca Bescós (1988), but many authors have been referring to it as *S. besanus*, marking that since the name *Spermophilinus* is of masculine gender, the species name should be changed from feminine to masculine as well (e.g., de Bruijn 1998, Vasileiadou and Koufos 2005). However, Ruiz-Sánchez et al. (2013) clarify that "besana" is used as a noun in apposition and does not need to agree in gender.

Type Material Left m2, RGM 300 245 (holotype) (Cuenca Bescós 1988: Pl. 1.F).

Type Locality Vargas 1A, Zaragoza, Spain, Lower Miocene [MN4].

Distribution Early–middle Miocene [MN4–5] of Europe and Turkey.

Occurrences in Greece Antonios.

Remarks It is the smallest and oldest *Spermophilinus* species, diagnosed through its rectangular upper molars bearing an anteroloph isolated from the protocone by a shallow furrow and a narrow metaloph with no metaconule, its also rectangular lower dental elements with a rounded postero-lingual angle, with an anterolophid separated from the protoconid by a deep groove, with no evident entoconid but incorporated within the posterolophid, with a residual metalophid and a well-developed sinusid (Cuenca Bescós 1988). The locality Antonios has yielded one m1/2, one m3 and five M1/2 assigned to *S. besana* (erroneously spelled *S. besanus*), which are slightly larger than the dental elements from the type locality but morphologically identical to them (Vasileiadou and Koufos 2005).

***Spermophilinus bredai* (von Meyer, 1848)**

Nomenclatural and Taxonomical History *Sciurus bredai* in von Meyer 1848 (new species); *Spermophilinus bredai* in de Bruijn and Mein 1968 (new combination). Considered by Schlosser (1890) as the senior synonym of *Sciurus spermophilinus* Depéret 1892 and *Sciurus gervaisianus* Lartet 1851.

Type Material von Meyer did not designate a holotype, but the only material coming from the type locality is a crashed skeleton consisting of one mandible with p4–m2, one scapula, one humerus and the hind limbs, to which Schlosser (1884) and later Depéret (1892) refer to as the holotype. The specimen, numbered 42,826, was purchased in 1871 from the Museum of Haarlem by the Natural History Museum of London (Forsyth Major 1899: Pl. XVII).

Type Locality Öhningen, Germany, middle Miocene [MN6 / MN7+8].

Distribution Middle–late Miocene of Eurasia.

Occurrences in Greece Kastellios 1, Samos S3, Mytilinii 1A and B.

Remarks The only comment in von Meyer (1848:472) referring to his new species from Öhningen was that he deposited its remains in the collection of Prof. van Breda (Haarlem Museum, the Netherlands). Later, Schlosser found rough sketches of the specimens deposited in Munich by von Meyer, on which he measured the specimens (Schlosser 1884:88) and he mentioned that the limbs are remarkably short in relation to the length of the tooth row. The holotype was recovered by Forsyth Major in the Natural History Museum of London; he measured the specimens and stated that the lower incisor has a "*longitudinal striation of the enamel*", whereas the cheek teeth present are very worn and he did not describe them (Forsyth Major 1899:365). As the species *Sciurus spermophilinus* Depéret, 1892 from La Grive (France, MN7–8) and *Sciurus gervaisianus* Lartet, 1851 from Sansan (France, MN6) were synonymized with *S. bredai* by Schlosser (1890), most of our knowledge on the species comes from these localities (see Forsyth Major 1899 and Black 1966 for details).

One D4 has been found in the locality Kastellios 1, with short protoloph with no protoconule, metacone connected to the protocone showing a constriction just labially to the connection, well-defined metaconule and a small mesostyle near the metacone; after comparisons with numerous samples of the species from La Grive, the tooth was assigned to *S. bredai* (de Bruijn et al. 1971:11). A few squirrel specimens have also been found in Samos Island: one right M1/2 in the locality S3 (Black et al. 1980) and one m2 and one M1/2 in the localities Mytilinii 1A and B, respectively (Vasileiadou and Sylvestrou 2009); they have all been assigned to *S.* cf. *bredai*, as their size and morphology are close to this species but the specimens too few to allow certain identification.

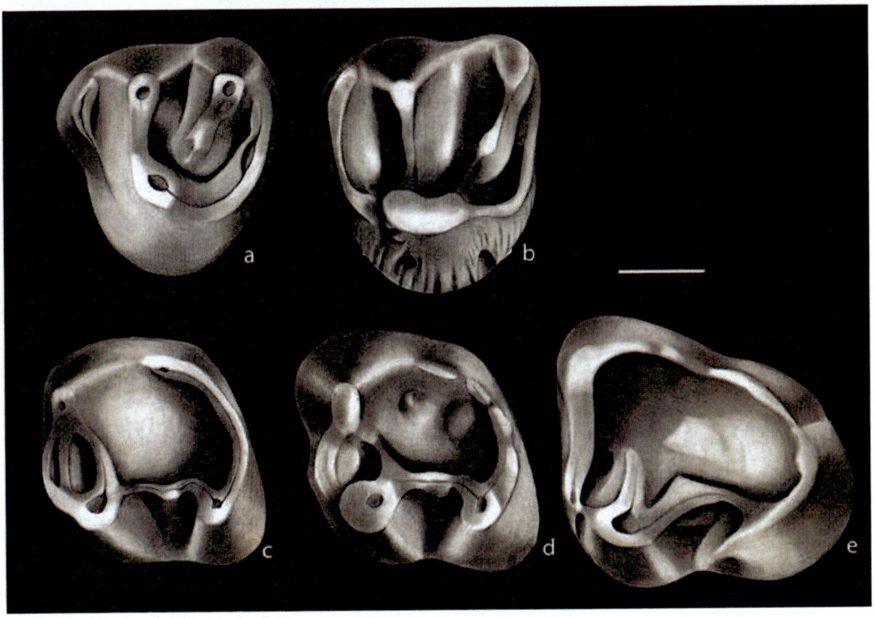

Fig. 6 *Spermophilinus giganteus* from Maritsa 1 (type locality). **a** P4, **b** M1/2, **c** m1/2 (holotype), **d** m1/2, **e** m3. Images courtesy of H. de Bruijn. All specimens figured as left; b, c, d are reversed. Scale bar equals 1 mm

★*Spermophilinus giganteus* de Bruijn, Dawson and Mein, 1970

Type Material Right m1/2, no. 1197 (holotype) (de Bruijn et al. 1970: Pl. 9.2).

Type Locality Maritsa 1, Rhodes Island, Greece, Upper Miocene [Turolian (MN13)].

Distribution The species is only known from the type locality.

Remarks This is the largest *Spermophilinus* species. The diagnosis given by de Bruijn et al. (1970) includes unusually weak for the genus mesoconids on the lower molars, triangular P4, a well-developed metaconule on the M1/2 metalophs and an enlarged m3. The material from Maritsa 1 (Fig. 6) includes one d4, one p4, two m1/2, one m3, one P4 and five M1/2.

Spermophilinus turolensis de Bruijn and Mein, 1968

Type Material Right m1/2, LM 92 (holotype) (de Bruijn and Mein 1968: Pl. 2.4).

Type Locality Los Mansuetos, Spain, Upper Miocene [MN12].

Distribution Late Miocene (late Vallesian–Turolian) of Eurasia.

Occurrences in Greece Maramena and Nea Silata.

Remarks The species, based on two d4, one m1/2, two m3, three D4, three P4, three M1/2 and two M3 from Los Mansuetos, is diagnosed through the rugose talonid basin in unworn specimens, the rounded postero-lingual corner and the absence of a well-developed entoconid on the m1–2, the absence of anteroconid and mesostylid, the strong mesoconid on the m1–3, the much wider than long m1–2, the converging towards the protocone protoloph and metaloph, the constriction of the metaloph near the protocone, the incorporated in the lingual border hypocone and the presence of a mesostyle on the M1–2 but absence on the M3 (de Bruijn and Mein 1968). A rather large collection of *S. turolensis* dental elements was found in Maramena, consisting of six p4, four m1/2, seven m3, two D4, 12 M1/2 and three M3 (de Bruijn 1995). One m3, one D4 and one P4 of *Spermophilinus* from Nea Silata were described by Vasileiadou (2001) in her unpublished MSc thesis; the elements have the size and morphology of *S. turolensis*, with the only difference being the strong mesostyle on the two upper elements, which led Vasileiadou et al. (2003) to the assignment of the material to *S.* cf. *turolensis*.

***Spermophilus* Cuvier, 1825**

Type Species *Mus citellus* Linnaeus, 1766.

Other Taxa Included The genus contains 12 extant species (Sinitsa and Pogodina 2019), of which *S. citellus* (Linnaeus, 1766); *S. suslicus* (Güldenstädt, 1770), *S. major* (Pallas, 1779) and *S. xanthoprymnus* (Bennett, 1835) have also been found in Late Pleistocene faunas. The extinct species include *S. superciliosus* (Kaup, 1839); *S. citelloides* Kormos, 1916; *S. primigenius* (Kormos, 1934); *S. nogaici* (Topachevski, 1957); *S. severskensis* Gromov, 1958; *S. polonicus* (Gromov, 1965); *S. dietrichi* Kretzoi, 1965; *S. taurensis* Gündüz et al., 2007; and *S. praecox* Sinitsa and Pogodina, 2019.

Distribution Latest Pliocene–present of Eurasia, from Austria to China.

Remarks The genus name *Citellus* (Oken 1816) has been used for a long time, until the International Commission on Zoological Nomenclature adopted the opinion that *Spermophilus* (Cuvier 1825) was the correct name for the genus (1956; Opinion 417; see Ramos-Lara et al. 2014 for details). Following this, the upper molar from Petralona assigned to *Citellus* sp. by Sickenberg (1971) is here mentioned as *Spermophilus* sp. One m3, one P4 and two M1/2 from the Biharian fauna of Marathoussa (N Greece) have been assigned to *Spermophilus* sp. (Koufos et al. 2001). *S.* cf. *xanthoprymnus* have been described from the fissure filling Latomi 1 in Chios Island (Storch 1975). Fossil assemblages of the extant European ground squirrel, *S. citellus*, have been described from the Late Pleistocene localities Arnissa (Mayhew 1978) and Loutra Almopias Cave LAC and LAC Ia (Chatzopoulou 2014).

***Spermophilus nogaici* (Topachevski, 1957)**

Nomenclatural and Taxonomical History *Citellus* (*Urocitellus*) *nogaici* in Topachevski 1957 (new species); *Spermophilus nogaici* in accordance with the Opinion 417 of the International Commission on Zoological Nomenclature (1956).

Type Material Lectotype: right maxilla with P4, NMNHU-P 27–190 (formerly NMNHU-P 27–189) (Topachevski 1957: Fig. 3).

Type Locality Nogajsk, Ukraine, Lower Pleistocene [middle Biharian].

Distribution Late Pliocene–Middle Pleistocene of E Europe.

Occurrences in Greece Ravin Voulgarakis.

Remarks A rich material consisting of one d4, five p4, 10 m1/2, two m3, three P3, six P4, six M1/2 and nine M3 from the Biharian fauna of Ravin Voulgarakis has been attributed to *S. nogaici* by Koliadimou (1996). The material is characterized by a well-developed P3 with a low anterior cingulum, a triangular P4, a weak posterior cingulum, weak mesostyle, an Λ-shaped continuous ridge connecting the paracone, protocone, metaconule and metacone, a strong protocone and the absence of a parastyle on the M1–2, the absence of a metacone and the presence of a weak metaconule on the M3, the closed central basin and the presence of a divided posterior root on the p4, the presence of a mesostylid on the m1–2 and the stronger metalophulid on the m2 (Koliadimou 1996).

***Spermophilus citelloides* Kormos, 1916**

Type Material See Kormos 1916.

Type Locality Felsnische Pilisszántó, Hungary, Upper Pleistocene [late Würm].

Distribution Late Pleistocene and Holocene of C and E Europe.

Occurrences in Greece Dimitra 2.

Remarks Eight complete teeth and many tooth fragments of *S. citelloides* have been found in the Upper Pleistocene locality Dimitra 2 (Strimon Basin); the teeth are small and hypsodont, the upper molars are triangular, the lower ones rhomboid; there are well-developed intermediate tubercles on the crests, the mesostyle is absent from the upper molars and the ectostylid is absent from the lower molars; there are two large roots on the p4, the posterior of which extends along the border of the tooth (Syrides and Koliadimou 1994).

Family Gliridae Muirhead, 1819

Type Genus *Glis* Brisson, 1762.

Remarks One small worn tooth with no roots preserved at the locality Plakias has been identified as a p4 or d4 of Gliridae indet. by de Bruijn et al. (2012:69). Gliridae indet. have also been reported in the faunal lists of Tourkobounia 4 and 5 (Reumer and Doukas 1985) and of Karnezeika (Kokotini et al. 2019).

Subfamily Bransatoglirinae Daams and de Bruijn, 1995

Type Genus *Bransatoglis* Hugueney, 1967.

***Bransatoglis* Hugueney, 1967**
(= *Paraglis* Baudelot, 1970, according to Daams 1976; = *Oligodyromys* Bahlo, 1975, according to Bosma and de Bruijn 1982)

Type Species *Bransatoglis concavidens* Hugueney, 1967.

Other Taxa Included (according to de Bruijn et al. 2013a) *B. spectabilis* (Dehm, 1950); *B. fugax* (Hugueney, 1967); *B. astaracensis* (Baudelot, 1970); *B. planus* (Bahlo, 1975); *B. cadeoti* Bulot, 1978; *B. infralactorensis* (Baudelot and Collier, 1982); *B. bahloi* Bosma and de Bruijn, 1982; *B. mayri* Rabeder, 1984; *B. sjeni* Ünay-Bayraktar, 1989; *B. moyai* (Hugueney and Adrover, 1990); *B. complicatus* Ünay, 1994; *B. ingens* (Werner, 1994); *B. parvus* Freudenthal, 1996; *B. adroveri* Hugueney, 1997; *B. attenuatus* Peláez-Campomanes, 2000; *B. bosniensis* de Bruijn et al., 2013a.

Distribution Late Eocene (MP17)–middle Miocene (MN6) of Europe and Turkey.

***Bransatoglis sjeni* Ünay–Bayraktar, 1989**

Type Material Right M2 (holotype) (Ünay-Bayraktar 1989: Pl. 9.8).

Type Locality Kocayarma, Turkey, middle Oligocene.

Distribution Oligocene of Greece and Turkey.

Occurrences in Greece Kyprinos.

Remarks According to the diagnosis of this medium-sized species, the occlusal surface of the cheek teeth is strongly concave, and the cross section of the valleys is U-shaped, whereas that of the ridges V-shaped, the P4 bears a short transverse protoloph and usually two centrolophs, the M1–2 show an interrupted endoloph and the

M3 shows a complex pattern of crests within the trigon (Ünay-Bayraktar 1989). One M1/2 from the late Oligocene of Kyprinos has been assigned to this species, based on its similarity in morphology and size with *B. sjeni* from the type locality: it shows a strongly concave occlusal surface, eight ridges and isolated anteroloph and posteroloph connected to the protocone (Doukas and Theocharopoulos 1999).

***Bransatoglis complicatus* Ünay, 1994**

Type Material Left M1(holotype) (Ünay 1994: Pl. 1.1).

Type Locality Harami 1, Turkey, Lower Miocene [MN2].

Distribution Early Miocene of Anatolia and Greece.

Occurrences in Greece Aliveri.

Remarks This species has medium-sized cheek teeth, with strongly concave occlusal surfaces and complex occlusal pattern, showing strong extra ridges outside the trigon; the P4 bears two centrolophs, and the M1–2 has a complete endoloph connected with the long anterior centroloph (Ünay 1994). Material from Aliveri, two M1, one M2 and one M3, was originally assigned to *B.* cf. *fugax* by van der Meulen and de Bruijn (1982). The size of the M1–2 overlaps with that of the smallest *B. fugax* from the type locality Coderet and other French localities, but the ridges on these molars seem stronger, the valleys less wide and the paracone and metacone less distinguished, whereas the M3 is longer and less triangular than the M3 of *B. fugax* from the type locality (van der Meulen and de Bruijn 1982). The Aliveri molars were transferred to *B. complicatus* by Ünay (1994), due to the continuous endoloph, its connection to the long anterior centroloph and the connections between the protoloph and the metaloph at the endoloph on the M1–2, as well as the complex ridge pattern and the strength of the extra ridges.

Subfamily Dryomyinae de Bruijn, 1967

Type Genus *Dryomys* Pallas, 1778.

***Glirulus* Thomas, 1906**
(= *Amphidyromys* Heller, 1936, according to Mayr 1979)

Type Species Extant Japanese dormouse *Myoxus japonicus* Schinz, 1845 (= *Glirulus elegans* Temminck, 1845, *Glirulus lasiotis* Thomas, 1880).

Other Fossil Taxa Included [including subgenus *Paraglirulus* Engesser, 1972, which Marković et al. (2016) consider synonymous to *Glirulus*] *G. pusillus* (Heller, 1936); *G. gemmula* Kretzoi, 1962; *G. lissiensis* (Hugueney and Mein, 1965); *G.* (*P.*) *werenfelsi* (Engesser, 1972); *G. diremptus* (Mayr, 1979); *G. conjunctus* (Mayr,

1979); *G.* (*P.*) *agelakisi* van der Meulen and de Bruijn, 1982; *G. minor* Wu, 1993; *G. ekremi* Ünay, 1994; *G. daamsi* de Bruijn et al., 2003a; *G.* (*P.*) *schultzi* Daxner-Höck and Höck, 2009.

Distribution Miocene–Pliocene of Europe and Turkey, present of Japan.

***Glirulus* (*Glirulus*) *diremptus* (Mayr, 1979)**

Nomenclatural and Taxonomical History *Paraglirulus diremptus* in Mayr 1979 (new species); *Glirulus* (*Glirulus*) *diremptus* in van der Meulen and de Bruijn 1982 (new combination).

Type Material Right M1 (holotype), Erkertshofen 1962 XIX 3331 (Mayr 1979: Pl. VIII.6).

Type Locality Erkertshofen, South Germany, Lower Miocene [MN4].

Distribution Early Miocene of Europe.

Occurrences in Greece Lapsarna, Aliveri and Karydia.

Remarks According to Mayr's (1979) diagnosis, the P4 bears an anterior centroloph and two extra ridges outside the trigon, the M1–2 always has three additional ridges, two of them outside the trigon and the third situated between the protoloph and the anterior centroloph; the endoloph is continuous and the main ridges tend to connect on the labial edge; the posterior centroloph is shorter than the anterior one, and both are separated from the endoloph; the p4–m2 bear five main ridges and four additional ones; the transverse valley between the centrolophid and the mesolophid is present but narrow; the m1–2 are two-rooted.

A rich collection of *G. diremptus* dental elements has been found in Aliveri: 11 p4, 14 m1, 10 m2, seven m3, one P4, nine M1, 16 M2 and eight M3; the Aliveri material has very few differences from that from the type locality: the Aliveri M3 and m1 are longer, the M1–2 and m2 narrower, an extra ridge in the posterior valley is always present on the M1/2, and there are more extra ridges in the Aliveri lower molars (van der Meulen and de Bruijn 1982:494). *G. diremptus* is also mentioned in the Karydia faunal list by Theocharopoulos (2000) without further information. A small collection from the Lower Miocene locality Lapsarna, one m2, one m3, one P4 and one M3, has been assigned to *G.* cf. *diremptus* by Vasileiadou and Zouros (2012). The Lapsarna teeth are very similar to the ones from Aliveri, the only difference being on the M3–m3 from Lapsarna, where the extra ridges are slightly shorter than on the Aliveri molars, whereas the Lapsarna P4 is smaller and the M3 larger than the Aliveri teeth (Vasileiadou and Zouros 2012:259). Additional material from Lapsarna, consisting of one p4, one m1, four m2, one m3, one M2, one M1/2 and one M3, has confirmed the allocation to *G. diremptus* (Vasileiadou et al. 2017).

★*Glirulus (Paraglirulus) agelakisi* van der Meulen and de Bruijn, 1982

Type Material Left M2, Al 1875 (holotype) (van der Meulen and de Bruijn 1982: Pl. 3.2).

Type Locality Aliveri, Greece, Lower Miocene [MN4].

Distribution Early Miocene of Greece.

Occurrences in Greece Aliveri and Karydia.

Remarks The material from the type locality Aliveri (Fig. 7) consists of two d4, three p4, eight m1, 12 m2, six m3, three P4, 11 M1, eight M2 and one M3. The species is diagnosed based on the common lack of connection between the anterior centroloph and the endoloph in the upper molars (van der Meulen and de Bruijn 1982:495), whereas in *G.* (*P.*) *werenfelsi* there is always a connection between them. Another difference is that in the *G.* (*P.*) *agelakisi* upper molars the ridges are less parallel to each other than in *G.* (*P.*) *werenfelsi* (van der Meulen and de Bruijn 1982:499). *Glirulus* (*P.*) *agelakisi* is also mentioned in the faunal list of Karydia without further information on the material (Theocharopoulos 2000). This species is considered to be the ancestor of *G.* (*P.*) *werenfelsi* by van der Meulen and de Bruijn (1982:499).

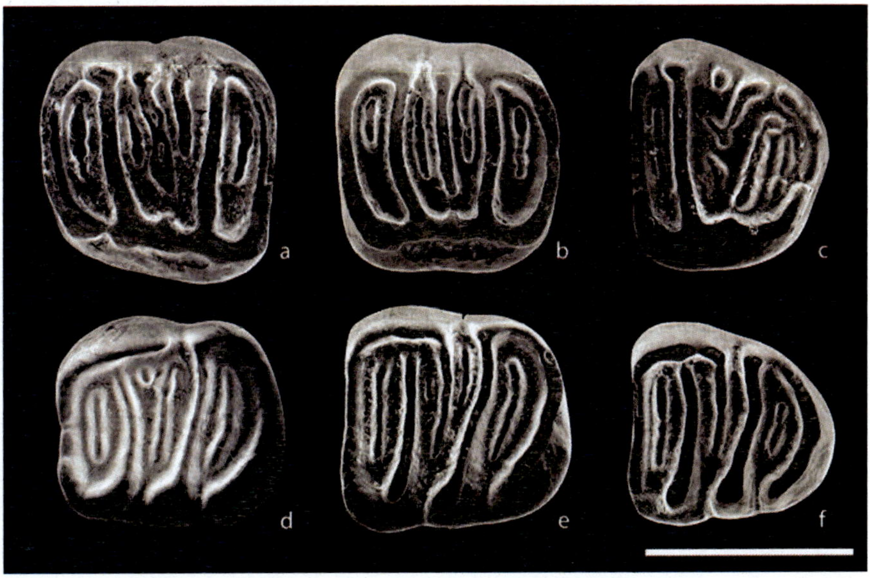

Fig. 7 *Glirulus* (*Paraglirulus*) *agelakisi* from Aliveri (type locality). **a** M1, **b** M2 (holotype), **c** M3, **d** m1, **e** m2, **f** m3. Images courtesy of H. de Bruijn. All specimens figured as left; c, d, f are reversed. Scale bar equals 1 mm

***Glirulus* (*Paraglirulus*) *werenfelsi* Engesser, 1972**

Type Material Left M1, Al. 274 (holotype) (Engesser 1972: Abb. 75.1).

Type Locality Anwil, Switzerland, middle Miocene [MN7/8].

Distribution Early–late Miocene of Europe.

Occurrences in Greece Plakias.

Remarks According to Engesser's (1972) diagnosis, this is a middle-sized glirid, with increasingly concave occlusal surface from the m3 to the m1; two-rooted lower molars, with strong main ridges and an average of four secondary ridges, with a thick ectolophid but no endolophid, with a centrolophid never connected to the mesolophid; upper teeth with an almost square outline, continuous endoloph, main ridges higher than the – usually four – secondary ones, the anterior centroloph always connected with the endoloph but ends freely on the labial side, the two anterior main ridges and the two posterior ones almost always connected. One m3 from an old collection from Plakias had been identified as *Glirudinus* sp. (de Bruijn and Meulenkamp 1972), but a subsequent collection contains two more teeth, one m2 and one M2, which lead to the assignment of the Plakias three glirid teeth to *G.* (*P.*) *werenfelsi*, even though the molars are somewhat larger than those of other European representatives of the species (de Bruijn et al. 2012).

***Microdyromys* de Bruijn, 1966**
[= *Myolidus* Álvarez-Sierra in Álvarez-Sierra and García-Moreno, 1986, according to Daams (1999)]

Type Species *Microdyromys koenigswaldi* de Bruijn, 1966.

Included Taxa *M. praemurinus* (Freudenberg, 1941); *M. complicatus* de Bruijn, 1966; *M. monspeliensis* Aguilar, 1977; *M. legidensis* Daams, 1981; *M. orientalis* Wu, 1986; *M. sinuosus* (Álvarez-Sierra, 1986); *M. misonnei* (Vianey-Liaud, 1994); *M. hidebrandti* Werner, 1994; *M. wuae* Qiu, 1996; *M. heissigi* (Uhlig, 2001); *M. puntarronensis* Freudenthal and Martín-Suárez, 2007; *M. remmerti* García-Paredes, Peláez-Campomanes and Álvarez-Sierra, 2010.

Distribution Late Eocene–Miocene of Europe and Turkey.

Remarks According to de Bruijn's (1966) diagnosis, *Microdyromys* is a small glirid with concave occlusal surface of the cheek teeth, ornamented lingual side of upper cheek teeth, four main ridges connected via an endoloph on the upper molars, anterior centroloph longer than the posterior one on the M1–2, P4 and M3 relatively large, additional ridges lower than the main ones. Van der Meulen and de Bruijn (1982) assigned one p4, one m1 and one m2 from Aliveri to *Microdyromys* sp.: the p4 has five ridges and no centrolophid, whereas the two molars have seven ridges each; metrically and morphologically, the teeth are comparable to *M. legidensis* and *M. monspeliensis* (van der Meulen and de Bruijn 1982:486). De Bonis et al. (1997) have reported the presence of *Microdyromys* sp. in the middle Miocene locality Thymiana A (Chios Island), with no details. Vasileiadou and Koufos (2005) described one m2 from the locality Antonios as cf. *Microdyromys* sp.; even though the size and morphology remind those of *Microdyromys*, the material is extremely poor to allow for a definite identification.

***Eliomys* Wagner, 1840**

Type Species *Eliomys melanurus* Wagner, 1840 (extant).

Other Taxa Included Extant: *E. quercinus* (Linnaeus, 1766); *E. munbyanus* (Pomel, 1856). Extinct: *E. intermedius* Friant, 1953; *E. truci* Mein and Michaux, 1970; *E. reductus* Mayr, 1979; *E. assimilis* Mayr, 1979; *E. darelbeidae* Geraads, 1994; *E. briellensis* Reumer, 2001; *E. lafarguei* Aguilar, Michaux and Lazzari, 2007; *E. yevesi* Masino, Garcia-Alix, Ruiz-Sánchez and Montoya, 2015.

Distribution Middle Miocene–present, in Europe, Turkey, Israel, Egypt and Morocco.

Remarks A fossil assemblage of the extant *E. quercinus* has been described from the Biharian fauna of Tourkobounia 2, consisting of three m1, three m2, three m3, four P4 and two M1/2 (van der Meulen and Doukas 2001). The P4 has four ridges and two fused roots and shows the typical *Eliomys* outline, straight anteriorly and convex posteriorly. The M1/2 has three roots and, apart from the four main ridges, a rather long anterior centroloph and a smaller posterior one. The m1–2 have five ridges, whereas the centrolophid is shorter on the m2 than on the m1 (van der Meulen and Doukas 2001:261). *Eliomys* sp. has been mentioned from Limni 6 (van den Meulen and van Kolfschoten 1986), with no details on the available material.

***Eliomys intermedius* Friant, 1953**

Type Material Unnumbered M1 (Friant 1953: Fig. 2).

Type Locality Sète, France, Lower Pleistocene.

Distribution Late Miocene (Turolian)–Early Pleistocene of Spain, France, W Asia and Greece.

Occurrences in Greece Maritsa 1.

Remarks The holotype from Sète is a small M1 with three roots (one lingual and two labial), four main transverse ridges and a concave occlusal surface with two sharp cusps on the labial side (Friant 1953). Several dental elements from Maritsa 1 (two p4, two m1, two m2, two m3, two partial P4, two M1/2, and four M3) have been assigned to *E.* aff *intermedius* by de Bruijn et al. (1970), after comparison with Turolian *Eliomys* from Spain and *E. intermedius* from the French sites Sète and Balaruc II. The Maritsa 1 teeth are a little smaller than the Sète ones and much smaller than the Balaruc II ones; they also differ from the latter teeth due to the longer extra ridge on the m2 and the presence of an extra ridge on the M1–2; the Turolian species is very different from the Maritsa 1 one (de Bruijn et al. 1970).

***Dryomys* Thomas, 1906**
(= *Afrodryomys* Jaeger, 1975, according to Daams and de Bruijn 1995)

Type Species *Mus nitedula* Pallas, 1778 (extant).

Other Taxa Included Extant: *D. laniger* Felten and Storch, 1968; *D. niethammeri* Holden, 1996. Extinct: *D. ambiguous* Lavocat, 1961; *D. chaabi* Jaeger, 1977; *D. sichuanensis* Wang, 1985; *D. tosyaensis* Ünay and de Bruijn, 1998; *D. apulus* Freudenthal and Martín-Suárez, 2006.

Distribution Middle Miocene of Africa, late Miocene of Germany and Italy, Pliocene–present of C Europe, Balkans, Turkey, Middle East, Russia, Pakistan, China.

Remarks *Dryomys* sp. is mentioned in the faunal list of Limni 6 (van der Meulen and van Kolfschoten 1986) with no information on the material. De Bruijn and van der Meulen (1975:336) assigned one m1 and one M3 from Tourkobounia 1 to Gliridae gen. and sp. indet.; Daams (1981:64) mentioned that these molars resemble both *Microdyromys* and *Dryomys*, but he assigned them to the latter genus, only because they are relatively large (Daams 1981:65). Fossil remains of the extant *D. nitedula* have been described from the Middle Pleistocene locality Latomi 1; the material, consisting of two m1, two m2, one m3 and one M3 (with a short anterior and a long posterior centroloph), falls within the variation range of the recent western Anatolian and Balkan species, in both size and morphology (Storch 1975:171). *D. nitedula* has also been found in Arnissa, with two m1, four m1/2, one M1 and two M2; the M1–2 have a short posterior and a long anterior centroloph, with one extra ridge between the two on the M2, whereas on the one m1 the centrolophid is isolated and on the other connected to the metalophid (Mayhew 1978:305). Finally, the species has also been described from the Late Pleistocene faunas Loutra Almopias Cave LAC and LAC Ia (Chatzopoulou 2014) and the sub-recent cave deposits of Pili (Daams 1981).

Subfamily Glirinae Thomas, 1897

Type Genus *Glis* Brisson, 1762.

***Glirudinus* de Bruijn, 1966**

Type Species *Glirulus gracilis* Dehm, 1950.

Other Taxa Included *G. glirulus* (Dehm, 1935); *G. modestus* (Dehm, 1950); *G. bouzigensis* (Thaler, 1966); *G. undosus* Mayr, 1979; *G. euryodon* van der Meulen and de Bruijn, 1982; *G. minutus* Wu, 1993; *G. engesseri* Ünay, 1994; *G. haramiensis* Ünay, 1994; *G. eggingensis* Werner, 1994; *G. antiquus* Vianey-Liaud, 2003; *G. intermedius* Aguilar and Lazzari, 2006; *G. magnus* Aguilar and Lazzari, 2006; *G. matusi* Joniak et al., 2017.

Distribution Late Oligocene–middle Miocene of C and SW Europe and Anatolia.

Remarks Marković et al. (2016) reported that "*the cheek teeth of* G. minutus, G. gracilis, G. modestus, G. bouzigensis *and* G. undosus *show so much variation in size and morphology that samples overlap, which makes it difficult, if not impossible, to identify these species with certainty*". However, they retain the individual species, and we here follow their opinion.

★*Glirudinus euryodon* van der Meulen and de Bruijn, 1982

Type Material Left M2, AL 1858 (holotype) (van der Meulen and de Bruijn 1982: Pl. 2.2)

Type Locality Aliveri, Greece, Lower Miocene [MN4].

Distribution Early Miocene of Greece (Aliveri). *G.* cf. *euryodon* in the Turkish Lower Miocene localities Semsettin (Şen et al. 1998) and Kaplangı (Dangremond 2012) and *G.* aff. *euryodon* in the Lower Miocene Serbian locality Sibnica (Marković et al. 2016).

Occurrences in Greece Aliveri.

Remarks The species was created based on an assemblage from Aliveri (Fig. 8) consisting of one d4, two p4, six m1, five m2, two D4, one P4, nine M1, seven M2 and three M3 (including the holotype). The diagnosis indicates a medium-sized *Glirudinus* with wide molars having seven ridges, complete endoloph on the upper molars and two-rooted M2 (fusion of the two labial roots). Further features include the connection of the anteroloph to the endoloph on the M1. On the upper molars there is a pronounced lingual projection and two loops at the labial side on which all ridges are connected except for the extra ridge between the anterior centroloph and the protoloph, whereas on the lower molars the labial ends of the main ridges end free (van der Meulen and de Bruijn 1982).

Glirudinus gracilis (Dehm, 1950)

Nomenclatural and Taxonomical History *Glirulus gracilis* in Dehm 1950 (new species); *Glirudinus gracilis* in de Bruijn 1966 (new combination).

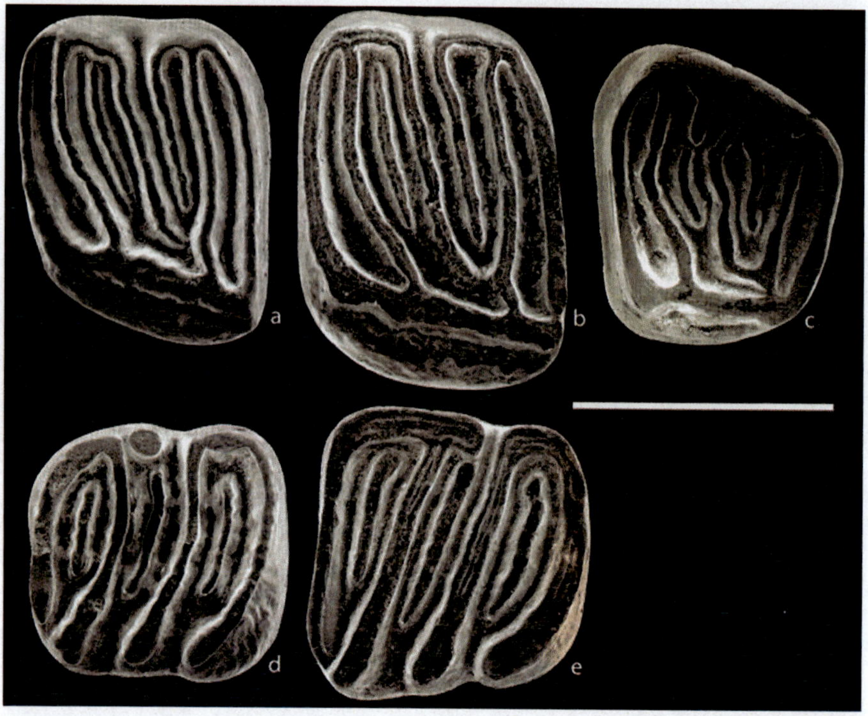

Fig. 8 *Glirudinus euryodon* from Aliveri (type locality). **a** M1, **b** M2 (holotype), **c** M3, **d** m1, **e** m2. Images courtesy of H. de Bruijn. All specimens figured as left; a, c–e are reversed. Scale bar equals 1 mm

Type Material Right lower mandible with incisor and p4–m3, No. 10794 (holotype), Collection München 1937 (Dehm 1950: Abb. 90a–c).

Type Locality Wintershof–West, Bavaria, Germany, Lower Miocene [MN3].

Distribution Early–middle Miocene of Europe.

Occurrences in Greece Karydia.

Remarks Dehm (1950), in his original diagnosis, mentioned that this species is similar to *Glirulus elegans* but smaller, with a molariform p4 bearing seven ridges and m1–3 with numerous ridges and the anterior valley (between anterolophid and metalophid) with three anterior extra ridges. De Bruijn (1966) transferred the species in his genus *Glirudinus*, which differs from *Glirulus* in the absence from the upper cheek teeth of an endoloph and decoration on the lingual side, as well as the ±45° angle formed by the ridges with the longitudinal axis of the teeth. *G. gracilis* is mentioned in the faunal list of the Lower Miocene locality Karydia without additional information (Theocharopoulos 2000).

***Glis* Brisson, 1762**

Type Species *Sciurus glis* Linnaeus, 1766 (only extant species in genus).

Other Taxa Included (extinct) *G. sussenbornensis* Soergel, 1919; *G. sackdillingensis* (Heller, 1930); *G. minor* Kowalski, 1956; *G. major* de Bruijn and Rümke, 1974; *G. truyolsi* Daams, 1976; *G. apertus* Mayr, 1979; *G. vallesiensis* Agustí, 1981; *G. galitopouli* van der Meulen and de Bruijn, 1982; *G. guerbuezi* Ünay, 1990; *G. transversus* Ünay, 1994; *G. mihevci* Aguilar and Michaux, 2011; *G. perkoi* Aguilar and Michaux, 2011.

Distribution Late Oligocene (MP 28)–present of Europe with a gap in MN5–8.

Remarks *Glis glis* fossil material is present in the Upper Pleistocene localities Loutra Almopias Cave LAC and LAC Ia, consisting in total of one d4, three p4, 10 m1, five m2, six m3, two P4, six M1, three M2, three M3, two M1/2 fragments and one partial maxilla with M2–3 (Chatzopoulou 2014). The large size of the teeth, the slightly elevated tooth edges, the square M1–2 with a centroloph, the presence of both anterior and posterior centroloph on the M3 and the strong centrolophid on the lower molars led Chatzopoulou (2014:236) to that identification. Two molars of a large-sized dormouse from Ravin Voulgarakis have been assigned to *Glis* aff. *glis*; the M2 is morphologically similar to the extant species, but the M3 is less complex than that of *G. glis*, which show an anterior and a posterior extra ridge, as well as an anterior centroloph, ridges absent from the Ravin Voulgarakis M3 (Koliadimou 1996).

Glis n. sp. has been reported in the faunal list of the locality Kaiafas by van der Meulen and van Kolfschoten (1986). Van der Meulen and de Bruijn (1982:503) had already mentioned that some of the molars from fissure deposits near Kaiafas are almost as flat as in recent *Glis*, whereas other molars are as concave as in *Glis galitopouli*, and that there are extra ridges on the Kaiafas molars apart from the seven normally present in *Glis*. However, no new species has been erected up to now for the Kaiafas material. A poor glirid assemblage from the Biharian locality Tourkobounia 2, consisting of one p4, one m1, one m2 and two P4, with dimensions intermediate between *G. sackdillingensis* and *G. glis*, has been assigned to *Glis* sp. (van den Meulen and Doukas 2001). On one P4 there are six ridges and on the other one seven; the p4 is two-rooted and shows five ridges; on the m1 there are seven ridges plus an extra small one between the mesolophid and the centrolophid; the m2 has seven ridges (van den Meulen and Doukas 2001:262).

★*Glis galitopouli* van der Meulen and de Bruijn, 1982

Type Material Left M2, AL 1596 (holotype) (van der Meulen and de Bruijn 1982: Pl. 1.5).

Type Locality Aliveri, Greece, Lower Miocene [MN4].

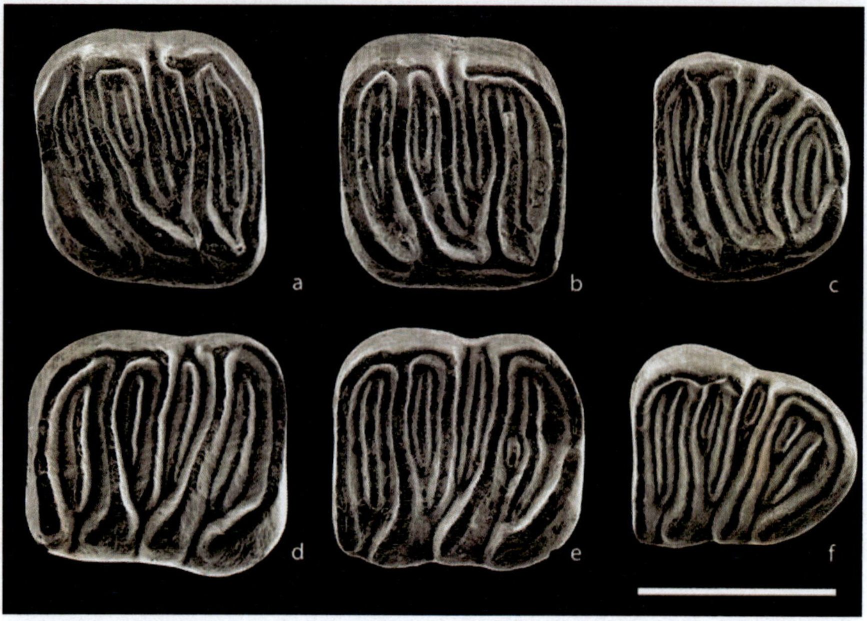

Fig. 9 *Glis galitopouli* from Aliveri (type locality). **a** M1, **b** M2 (holotype), **c** M3, **d** m1, **e** m2, **f** m3. Images courtesy of H. de Bruijn. All specimens figured as left; c is reversed. Scale bar equals 1 mm

Distribution Early Miocene of Greece, Switzerland and Turkey (Keseköy, Sabuncubeli).

Occurrences in Greece Aliveri and Karydia.

Remarks This small *Glis* species has been created based on an assemblage from Aliveri (Fig. 9) consisting of one d4, four p4, 10 m1, 12 m2, 22 m1/2, 10 m3, one D4, five P4, seven M1, 13 M2, 20 M1/2 and 10 M3. According to the species diagnosis, the protoloph, metaloph and posteroloph are usually connected lingually on the upper molars, both upper and lower molars have relatively high crowns, concave occlusal surface and bear nine ridges (van der Meulen and de Bruijn 1982:500). *G. galitopouli* is also mentioned in the faunal list of the locality Karydia (Theocharopoulos 2000), without details on the material.

Glis minor **Kowalski, 1956**

Nomenclatural and Taxonomical History *Glis sackdillingensis minor* in Kowalski 1956 (new subspecies); *Glis minor* in Kretzoi 1959 (elevated to species level).

Type Material Partial mandible with p4 and m1 (holotype) (Kowalski 1956: Pl. IV.8, Fig. 2.f).

Type Locality Podlesice, Poland, Pliocene [MN14].

Distribution Late Miocene–Early Pleistocene of C and E Europe.

Occurrences in Greece Komanos 1 low A and B, Notio 1, Tourkobounia 1.

Remarks Kowalski (1956) did not give a diagnosis for his new species, but he described the type mandible and one isolated M1 from the type locality, possibly of the same individual. All elements are characterized by their small dimensions; the M1 is three-rooted, with the four main ridges slightly oblique and bent backwards on the lingual edge; the p4 bears only the main ridges, which are strongly curved backwards in the middle of the tooth; the m1 shows four main ridges, of which the posterolophid is the weakest (Kowalski 1956:385–386).

G. minor is present in the localities Komanos 1 low A and B and Notio 1, with one m2 and one m1, respectively (Hordijk and de Bruijn 2009). The m1 from Notio 1 has one extra ridge in the anterior and one extra ridge connected to the entoconid in the posterior valley, both of which are as long as the main ridges. The centrolophid is connected to the entoconid and the mesoconid, but not to the metaconid. The m2 from Komanos is similar to the Notio m1, but the centrolophid is short and does not reach the labial and lingual molar edges and the anterior extra ridge is thin (Hordijk and de Bruijn 2009:66).

De Bruijn and van der Meulen (1975) assigned an assemblage from Tourkobounia 1 (one d4, one m1, three m2, two m3, one P4, nine M1–2) to *G. sackdillingensis*. The d4 is flat, with incomplete meso- and metalophid and a low extra ridge between the meso- and posterolophid; the lower molars show the four main ridges, a long centrolophid, one long extra ridge in the anterior and one more in the posterior valley; on the P4 the protoloph and metaloph are separated lingually, a long posterior extra ridge and a short anterior one are present between the protoloph and the metaloph, whereas there is an extra ridge between the protoloph and the anteroloph and one more between the metaloph and the posteroloph; the upper molars show the four main ridges, a long anterior centroloph, one extra ridge between the anteroloph and the protoloph and one more between the metaloph and the posteroloph (de Bruijn and van der Meulen 1975:334). The authors argued that *Glis minor* is a junior synonym of *G. sackdillingensis*. However, the synonymy between the species was never followed by subsequent authors, and Daams and de Bruijn (1995), in their paper on the classification of glirids based on the dental morphology, retain both species as valid. As the size of the dental elements from Tourkobounia 1 falls within the range of *G. minor* (e.g., size of molars from Komanos 1 low and Notio 1 in Hordijk and de Bruijn 2009:66, and from type locality of species Podlesice in de Bruijn and van der Meulen 1975:335) and is much smaller than that of *G. sackdillingensis* (e.g., size of dental elements from type locality of species Sackdillinger Höhle in de Bruijn and van der Meulen 1975:335), the Tourkobounia 1 material should be assigned to *G. minor*, since the morphological features show similarities with both species.

G. cf. *minor* is reported from Lava 2 without further information on the material (de Bruijn et al. 1999) and also from Maramena, with only one M2, which is morphologically similar to *G. minor* (slightly concave, three-rooted, with four long ridges reaching the labial and lingual borders and three short ones that reach only the labial border) but slightly larger than the material from the type locality and from Węże (Daxner-Höck 1995:109).

Glis sackdillingensis (Heller, 1930)

Nomenclatural and Taxonomical History *Myoxus glis sackdillingensis* in Heller 1930 (new subspecies); *Glis sackdillingensis* in Heller 1933 (new combination). *Glis antiques* Kormos 1930 is its junior synonym according to de Bruijn and van der Meulen (1975:335).

Type Material Mandibles and limb bones (Heller 1930: Plate XVI. 5a–b, 6).

Type Locality Sackdillinger Höhle, Germany, Pleistocene.

Distribution Pliocene–Pleistocene of Germany, Hungary and Greece.

Occurrences in Greece Kaiafas.

Remarks The species was based only on lower jaws and a few limb bones from the Early Pleistocene Sackdillinger Höhle, and its only distinctive character is the size of the elements, which are smaller than those of the extant *G. glis* (Kowalski 1963). Van den Meulen and van Kolfschoten (1986) included *G. sackdillingensis* in the faunal list of the Biharian locality Kaiafas, but without any additional information.

Muscardinus Kaup, 1829
(= _Eomuscardinus_ Hartenberger, 1966 = _Pentaglis_ Kretzoi, 1943, according to Daams and de Bruijn 1995)

Type Species *Mus avellanarius* Linnaeus, 1758 (only extant species of genus).

Other Taxa Included (extinct) *M. sansaniensis* (Lartet, 1851); *M. dacicus* Kormos, 1930; *M. pliocaenicus* Kowalski, 1963; *M. davidi* Hugueney and Mein, 1965; *M. vireti* Hugueney and Mein, 1965; *M. hispanicus* de Bruijn, 1966; *M. thaleri* de Bruijn, 1966; *M. vallesiensis* Hartenberger, 1966; *M. heintzi* Aguilar, 1981; *M. cyclopeus* Agustí, Moyà Solà and Pons-Moyà, 1982; *M. helleri* Fejfar and Storch, 1990; *M. topachevskii* Nesin and Kowalski, 1997; *M. meridionalis* García-Alix, Minwer-Barakat, Martín-Suárez and Freudenthal, 2008.

Distribution Middle Miocene–present of Europe and Turkey.

Remarks One p4 from Pikermi (Chomateri) has been assigned to *Muscardinus* sp.; it resembles the p4 of *M. hispanicus*, showing four complete transverse ridges, but is much larger than that (de Bruijn 1976). One P4 and one M1 from Agios Ioannis (Karpathos Island) have been identified as *Muscardinus* sp.: the M1 has four roots and six ridges, like that of *M. vireti*, and the dental pattern of the P4 agrees with that of *M. pliocaenicus* and the extant *M. avellanarius*; both teeth have the same size with the teeth of the extant *M. avellanarius* (Daams and van de Weerd 1980:328). Fossil assemblages of the only extant species *M. avellanarius* have been described from the Upper Pleistocene localities Loutra Almopias Cave LAC (one m3, one M1, one M2) and LAC la (one M1, two M2); the dental elements are similar to those of *M. avellanarius* from the Upper Pleistocene locality Nixloch (Austria) (Chatzopoulou 2014:242).

***Muscardinus hispanicus* de Bruijn, 1966**

Nomenclatural and Taxonomical History *Muscardinus pliocaenicus hispanicus* in de Bruijn 1966 (new subspecies); *Muscardinus hispanicus* in van de Weerd 1976 (elevated to species level). *Muscardinus crusafonti* Hartenberger, 1966 is considered its junior synonym by Daams (1985:12).

Type Material Left m1, no. PEC2008 (holotype) (de Bruijn 1966: Pl. 1.10).

Type Locality Pedregueras II C, Spain, Upper Miocene [Vallesian].

Distribution Late Miocene of Europe.

Occurrences in Greece Kastellios 1.

Remarks According to the diagnosis by de Bruijn (1966:368), the p4 is small and shows four ridges, the m1–2 have six complete ridges, the m1 is two-rooted, the m2–3 are three-rooted, the P4 bears only three ridges and one root and the upper molars are all three-rooted. One m2 from Kastellios 1, with flat occlusal surface and six ridges, was originally identified as *M.* cf. *crusafonti* by de Bruijn and Zachariasse (1979:222), but is here listed as *M.* cf. *hispanicus* due to their synonymy.

***Muscardinus vireti* Hugueney and Mein, 1965**

Type Material Left M1, No. 65228 (holotype) (Hugueney and Mein 1965: Plate 3.64).

Type Locality Lissieu, France, Upper Miocene [MN13].

Distribution Late Miocene of Spain, France, Italy and Greece.

Occurrences in Greece Lava 2.

Remarks The diagnosis by Hugueney and Mein (1965) mentions that this is a small species with relatively wide teeth, the upper molars are rectangular with a number of ridges, the M1 is slightly longer than the M2 and the lower molars are quite similar to those of Quaternary species. *M.* cf. *virety* is mentioned in the faunal list of Lava 2 by de Bruijn et al. (1999), but without details on the material.

***Muscardinus pliocaenicus* Kowalski, 1963**

Type Material A damaged mandible with m1–m2, the alveoli of p4 and m3 and broken incisor, no. MF (636) 63 (holotype), Institute of Systematics and Evolution of Animals of the Polish Academy of Sciences, Krakow (Kowalski 1963: Fig. 4.A–B).

Type Locality Węże, Poland, Upper Pliocene [MN15].

Distribution Late Miocene–Pleistocene of Europe.

Occurrences in Greece Maramena.

Remarks Kowalski (1963) did not give a diagnosis for the species, but he described the holotype in detail: the m1 and m2 show six ridges, all parallel on the m1 and slightly curved anteriorly on the m2; based on its alveoli, the m3 has either three roots with the posterior one bifurcated at its end or four roots; the mandible structure resembles that of the extant *M. avellanarius* but it is finer; the teeth are smaller than those of *M. avellanarius* and have the same width along their whole length, in contrast with *M. avellanarius* (Kowalski 1963:538–539). Further, the material from the type locality is characterized by very short p4 with two ridges forming a horseshoe and an indication of a third one on the anterior margin, three roots on the m1 and four on the m2, m3 with six ridges, M1 with five widely spaced ridges, M2 with four roots, six enamel ridges and a longitudinal ridge on the lingual margin (Kowalski 1963:539). Three m1 and one m2 from Maramena have been assigned to *M. pliocaenicus*; the molars have flat occlusal surface and six transverse ridges running along the whole width; the two middle ridges end free on both labial and lingual borders; the m1 has three and the m2 four roots (Daxner-Höck 1995:109).

***Seorsumuscardinus* de Bruijn, 1998**
(= *Heissigia* Prieto and Böhme, 2007, according to Prieto 2009)

Type Species *Seorsumuscardinus alpinus* de Bruijn, 1998.

Distribution Early Miocene of Switzerland, Austria, Germany, Serbia and Greece.

***Seorsumuscardinus alpinus* de Bruijn, 1998**

Taxonomical History *Heissigia bolligeri* Prieto and Böhme, 2007 is considered as its junior synonym, according to Marković et al. (2016).

Type Material Left M1, NHMW 1997z0201/0002/1(holotype) (de Bruijn 1998: Pl. 5.8).

Type Locality Oberdorf 4, Austria, Lower Miocene [MN4].

Distribution Early Miocene of Switzerland, Austria, Germany, Serbia and Greece.

Occurrences in Greece Karydia 2.

Remarks According to the diagnosis of the species, the P4 has four ridges, the M1 five, the M2 six and the M3 seven. Additional characteristics are P4 with two roots, oval shape and four sub-parallel ridges, M1–2 with three centrally placed roots, p4 with one root and lower molars with two or three roots (de Bruijn 1998:112). *S. alpinus* appears in the faunal list of Karydia 2 (de Bruijn 1998; Theocharopoulos 2000; Doukas 2003), but there is no additional information on the material.

Subfamily Myomiminae Daams, 1981

Type Genus *Myomimus* Ognev, 1924.

Miodyromys Kretzoi, 1943
(= *Ramys* García-Moreno and López Martínez, 1986, according to Marković et al. 2016)

Type Species *Dryomys hamadryas* Forsyth Major, 1899.

Other Taxa Included *M. prosper* (Thaler, 1966); *M. multicrestatus* (de Bruijn, 1966); *M. aegercii* (Baudelot, 1972); *M. biradiculus* Mayr, 1979; *M. vagus* Mayr, 1979; *M. hugueneyae* Agustí and Arbiol, 1989; *M. praecox* Wu, 1993; *M. grycivensis* Nesin and Kowalski, 1997; *M. asiamediae* Maridet et al., 2011.

Distribution Miocene of Europe, Turkey, Kazakhstan and China.

Miodyromys praecox Wu, 1993

Type Material Left mandible with p4–m3, BSP 1980 XXXII 1974 (holotype) (Wu 1993: Abb. 13).

Type Locality Stubersheim 3, Germany, Lower Miocene [MN3].

Distribution Early Miocene of Germany and maybe Greece.

Occurrences in Greece Karydia.

Remarks According to Wu's (1993) diagnosis, the species has two-rooted lower molars and complex morphology, very often with an extra ridge between the centro- and mesolophid of the m1–m2. *M.* cf. *praecox* is in the faunal list for Karydia (Theocharopoulos 2000), but with no additional information.

***Miodyromys multicrestatus* (de Bruijn, 1966)**

Nomenclatural and Taxonomical History *Peridyromys multicrestatus* in de Bruijn 1966 (new species); *Myomimus multicrestatus* in de Bruijn et al. 1970 (new combination); *Vasseuromys multicrestatus* in Mein 1983 (new combination); *Ramys multicrestatus* in García–Moreno and López Martínez 1986 (new combination); *Miodyromys multicrestatus* in Marković et al. 2016 (new combination).

Type Material Left M1/2, Nr. PEA 232 (holotype) (de Bruijn 1966: Pl. II, fig. 22).

Type Locality Pedregueras IIA, Spain, Upper Miocene [MN9].

Distribution Middle–late Miocene of Spain, France, Moldova, Serbia and Greece.

Occurrences in Greece Biodrak.

Remarks According to de Bruijn's (1966) diagnosis, the centrolophid becomes gradually shorter from the m1 to the m3, the m1 has more extra ridges than the m2–3 and all the lower molars have three roots and at least two extra ridges. *M. multicrestatus* has been described from the Vallesian locality Biodrak (as *Myomimus multicrestatus*), represented by two d4, two p4, three m1, one m2, one P4 and five M1/2; the dental elements are very similar to the ones from the species' type locality, one small difference being the higher number of lower molars with an anterior extra ridge in the Biodrak assemblage, but this could also be a misapprehension due to the small number of specimens (de Bruijn 1976).

***Myomimus* Ognev, 1924**
(= *Peridyromys* Stehlin and Schaub, 1951, according to de Bruijn et al. 1970 and Marković et al. 2016)

Type Species *Myomimus personatus* Ognev, 1924 (extant).

Included Taxa Extant: *M. roachi* (Bate, 1937); *M. setzeri* Rossolimo, 1976. Extinct: *M. murinus* (Pomel, 1853); *M. dehmi* (de Bruijn, 1966); *M. judaicus* Tchernov, 1968; *M. maritsensis* de Bruijn, Dawson and Mein, 1970; *M. qafzensis* (Haas, 1973); *M. aquatilis* (de Bruijn and Moltzer, 1974); *M. jaegeri* (Aguilar, 1974); *M. sumbalenwalicus* Munthe, 1980; *M. sinensis* Wu, 1985; *M. columbarii* (Daams, 1989); *M. turbatus* (Daams and de Visser, in Álvarez-Sierra et al., 1990); *M. lavocati* (Peláez-Campomanes and Daams, 2002).

Distribution Late Oligocene–present of Europe and Asia.

Remarks The extant *Myomimus roachi* has been reported from several Greek Plio-Pleistocene localities. Van der Meulen and van Kolfschoten (1986) mentioned its presence at the localities Rema Aslan, Kastoria 2 and Lagkada, without giving information on the assemblages. A rich collection from the Biharian locality Kalymnos-Xi, consisting of two d4, four p4, 10 m1, eight m2, two m3, one D4, four P4, 20 M1/2 and eight M3, has been attributed to *M. roachi* by Kuss and Storch (1978). The molars are very simple; only five of 20 M1/2 and three of eight M3 have short-to-medium-length centroloph; two out of eight m1 show very short and all the m2 a short-to-medium-length centrolophid; only one m2 has a posterior extra ridge (Kuss and Storch 1978:212). The species is also present in Latomi 1, with one d4, 23 m1, six m2, four m3, one D4, four P4, 11 M1, 10 M2 and three M3 (Storch 1975:168) and in the sub-recent cave deposits of Pili, with one m1, two m2, one m3, five M1 and two M2 (Daams 1981). Van der Meulen and van Kolfschoten (1986) included *M.* cf. *roachi* in the faunal list for the locality Limni 6; de Bruijn and van der Meulen (1979) had previously reported *M. roachi* in their faunal list for Limni 6; since details on the assemblage have not been published, the identification is unclear, but we here list the most recent identification. Finally, a small collection of *M. roachi* molars (one m1, two m3, two M1, two M2) is available from the locality Tourkobounia 2: the upper molars lack a complete endoloph; the M1 has three roots, one specimen shows only the four main ridges and the second one shows an additional short anterior centroloph; one M2 has three roots and the other one four, one has a short anterior centroloph additional to the four main ones and the other one has two centrolophs; the m2 has three roots and bears a weak centrolophid reaching the middle of its width; all m3 show a short centrolophid and a long mesolophid reaching the labial crown edge (van der Meulen and Doukas 2001:258).

Fossil assemblages of the extant masked mouse-tailed dormouse *Myomimus personatus* are known from the Pleistocene locality Varkiza 2 (one m1 and one m2) described by van de Weerd (1973) and the Middle Pleistocene Petralona cave (one m3 and one M2) described by Sickenberg (1971).

The oldest *Myomimus* sp. assemblage in Greece has been found in the middle Miocene locality Thymiana A, reported as *Peridyromys* sp. by de Bonis et al. (1997:268), without information on the material. One small m2 with the four main ridges and two secondary ones from the middle Miocene locality Chryssavgi 1 has been assigned to *Myomimus* sp. by Koliadimou (1996). One m2, one D4, one P4 and two M3 from the Biharian locality Ravin Voulgarakis have been attributed to *Myomimus* sp.: the dental elements are small, with a concave occlusal surface and a simple dental pattern, and bear only the four main ridges (D4) or those plus a small one (m2, P4 and M3) (Koliadimou 1996). *Myomimus* sp. is also mentioned by van der Meulen and van Kolfschoten (1986) in the faunal lists of the localities Damatria and Kardamena.

★*Myomimus maritsensis* de Bruijn, Dawson and Mein, 1970

Type Material Right M1–2, no. 1014 (holotype) (de Bruijn et al. 1970: Pl. 7.2).

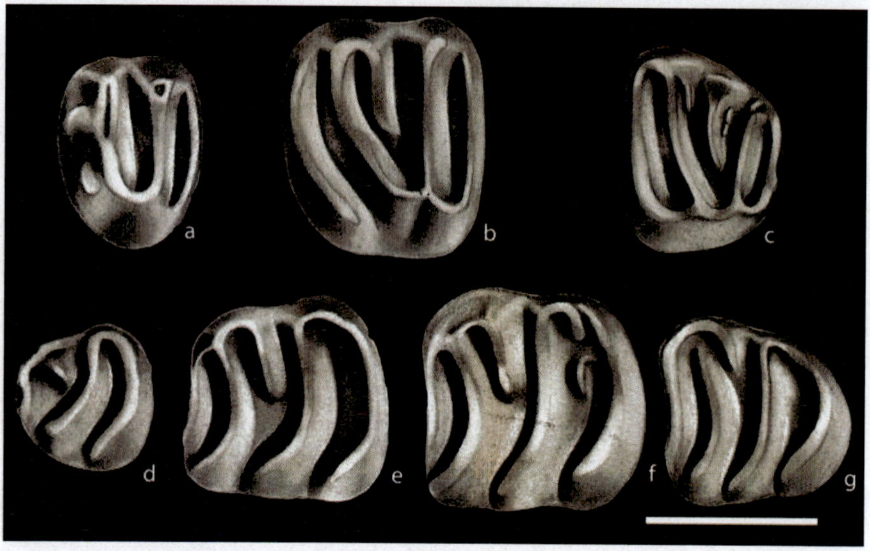

Fig. 10 *Myomimus maritsensis* from Maritsa 1 (type locality). **a** P4, **b** M1/2 (holotype), **c** M3, **d** p4, **e** m1, **f** m2, **g** m3. Images courtesy of H. de Bruijn. All specimens are reversed and figured as left. Scale bar equals 1 mm

Type Locality Maritsa 1, Rhodes Island, Greece, Upper Miocene [Turolian (MN13)].

Distribution Late Miocene–Pliocene of Greece, Ukraine and Turkey.

Occurrences in Greece Maritsa 1, Monasteri, Kessani, Ano Metochi 2–3, Maramena, Nea Silata, Spilia 3–4 and Notio 1.

Remarks The species is represented in its type locality (Fig. 10) by two d4, 10 p4, 29 m1, 30 m2, 26 m3, four D4, 10 P4, 30 M1/2 and 17 M3 (including the holotype) and was erected based on the similarity of its dentition with that of the extant *M. personatus* and its relatively large P4. The P4 has the four main ridges, although the anteroloph looks more like a parastyle, plus incipient centroloph(s) in three out of 10 specimens; the M1–2 always show an anterior centroloph and a posterior one is present in 16 of the 30 specimens; the M3 shows one or two centrolophs; the m1 and m2 bear the four main ridges (antero-, meta-, centro-, posterolophid) plus an extra ridge between the ento- and the posterolophid in 18/29 m1 and 19/30 m2 (de Bruijn et al. 1970:565–566).

M. maritsensis has been described from many Greek localities. The species is in the faunal lists for Ano Metochi 2–3 (*M.* cf. *maritsensis* in de Bruijn and van der Meulen 1979) and Spilia 3–4 (van der Meulen and van Kolfschoten 1986), but there is no additional information on the material from the two localities. Maramena has yielded the richest assemblage, consisting of 2 incomplete dentitions and 247 teeth in total, whereas the assemblage from Monasteri includes 2 incomplete dentitions and 57 teeth (Daxner-Höck 1995:106). In the fauna from Nea Silata there are 30 dental elements that morphologically and metrically resemble the Maritsa 1 material (Vasileiadou et al. 2003). *M. maritsensis* has also been yielded in Kessani, with one m1/2 and two M1/2 (Vasileiadou et al. 2012). *M.* aff. *maritsensis* has been described from Notio 1, represented by one m1, one m2, one m3, one P4, two M1/2 and two M3; the dental elements are very similar to those of *M. maritsensis* but are significantly larger than any teeth ever assigned to this species (Hordijk and de Bruijn 2009:66).

***Myomimus dehmi* (de Bruijn, 1966)**

Nomenclatural and Taxonomical History *Peridyromys dehmi* in de Bruijn 1966 (new species); *Myomimus dehmi* in de Bruijn 1976 (new combination).

Type Material Left m1, no. PEC 2172 (holotype) (de Bruijn 1966: Pl. II.5).

Type Locality Pedregueras IIC, Spain, Upper Miocene.

Distribution Late Miocene of Europe and Turkey.

Occurrences in Greece Katheni, Lefkon and Chomateri.

Remarks According to de Bruijn's (1966) diagnosis, the upper cheek teeth are identical to those of *Peridyromys murinus* (now *Myomimus murinus*), the centrolophid gets successively shorter from the m1 to the m3, the lower molars are three-rooted with an extra ridge in the posterior valley, whereas the upper molars have only the four main ridges and two centrolophs of various length. One m1 and four M1/2 from Lefkon had been previously assigned to *Myomimus maritsensis*, *Vasseuromys pannonicus* and *Eliomys* cf. *truci* by de Bruijn (1989:text–fig. 3), due to morphological and metrical variety, but Daxner-Höck (1995:104) included them all to *Myomimus dehmi*. The species is also reported in the faunal list of Katheni, with no information on the material (de Bruijn and van der Meulen 1979). One p4, five m1, four m2, one m3, one P4 and six M1/2 from Pikermi (Chomateri) have been assigned to *Myomimus* cf. *dehmi* by de Bruijn (1976), as they are very similar to the material from the type locality (three-rooted lower molars with a posterior extra ridge, upper molars with only the four main ridges), but larger.

***Dryomimus* Kretzoi, 1959**

Type Species *Dryomimus eliomyoides* Kretzoi, 1959.

Distribution Pliocene–Pleistocene of Europe and Turkey.

***Dryomimus eliomyoides* Kretzoi, 1959**

Type Material Kretzoi never designated or figured a holotype for the species.

Type Locality Csarnota-4, Hungary, Pliocene [MN15].

Distribution Pliocene–Pleistocene of Europe (Romania, Hungary, Moldova, Bulgaria, Greece) and Turkey.

Occurrences in Greece Kastoria 1 and Tourkobounia 1.

Remarks There is no official diagnosis neither for the genus nor for the species given by Kretzoi; however, he mentioned some important features of his new taxon: P4 rather unreduced, upper molars with *Miodyromys*-like lingually separated protoloph, lower molars with dental features between *Dryomys* and *Myomimus* (Kretzoi 1959:240). De Bruijn and van der Meulen (1975:333) gave some characters for *Dryomimus*: medium-sized glirid with three-rooted lower molars, M1–2 with a posterior centroloph longer than the anterior one, the majority of the M1–2 belonging to Daams' (1974) morphotype D, the majority of the M3 to morphotype Y, the majority of the m1–2 belonging to morphotype M and the majority of the m3 to morphotype R.

D. aff. *eliomyoides* has been described by de Bruijn and van der Meulen (1975) from the locality Tourkobounia 1, represented by four d4, seven p4, 20 m1, 12 m2, 12 m3, four D4, six P4, 24 M1/2 and nine M3. After comparison with a small sample from the type locality of the species, the Csarnota dental elements were found to possess extra ridges more often than the Tourkobounia ones, hence the assignment to *D.* aff. *eliomyoides* (de Bruijn and van der Meulen 1975:334). *D.* cf. *eliomyoides* is mentioned in the faunal list for the locality Kastoria 1 by van der Meulen and van Kolfschoten (1986), with no information on the material.

Family Castoridae Gray, 1821

Type Genus *Castor* Linnaeus, 1758.

Remarks Remains of Castoridae indet. have been reported from the Upper Miocene localities Lefkon and Maramena (de Bruijn 1989 and Schmidt-Kittler et al. 1995, respectively) and from the Pliocene localities Komanos 1 high A and B, Vorio 3 and 3A and Notio 1 (Hordijk and de Bruijn 2009).

Chalicomys Kaup, 1832

Type Species *Chalicomys jaegeri* Kaup, 1832.

Other Taxa Included (according to Casanovas-Vilar and Alba 2011): *C. catalaunicus* (Baraller, 1838); *C. subpyrenaicus* (Lartet, 1851); *C. adroveri* (Aldana Carrasco, 1992); *C. batalleri* Casanovas-Vilar et al., 2008.

Distribution Middle–late Miocene (Astaracian–Turolian) of Europe and Turkey.

Chalicomys jaegeri Kaup, 1832

Taxonomical History Considered the senior synonym of *Palaeomys castoroides* Kaup, 1832, according to Hugueney (1999); of *Chelodus typus* Kaup, 1832, according to Casanovas-Vilar et al. (2008); of *Castor vidali* Crusafont et al., 1948, according to van de Weerd (1976); and of *Palaeomys plassi* (Storch in Franzen and Storch, 1975), according to Casanovas-Vilar and Alba (2011).

Type Material The material from the type locality includes one partial mandible with all cheek teeth, one partial maxilla with the M1–2 and eight isolated cheek teeth (Kaup 1832); no holotype has been designated.

Type Locality Eppelsheim, Germany, Upper Miocene [Vallesian (MN9)].

Distribution Late Miocene (Vallesian–Turolian) of Europe and Turkey.

Occurrences in Greece Serrae.

Remarks The morphological features of the species are given by Hugueney (1999): the dental elements are as large as those of *Castor* and even larger; their height is more than twice their length (high degree of hypsodonty); the upper teeth show four labial ridges, the lower ones show normally four lingual ridges but pro-parafossetids are often present in early wear stages; the hypostria/-striid reach the crown basis; the para-, meso- and metastria/-striid are long but stop far from the crown basis; the mesostria/-striid are longer than the others, and in the P4 and p4 the parastria and parastriid are longer than the metastria and metastriid; the P4 is the longest cheek tooth and the M3 is not longer than the M1–2; the incisor shows a sub-triangular cross section and thin enamel; the synclines are completely filled with cement even in early wear stages; the parastriid of the p4 is long and persists even after the closure of the mesoflexid of the m1–2. One fragmented left maxilla with the P4–M3 found inside a lignite layer in the area north of the city of Serrae (N Greece) has been assigned to *C. jaegeri* by Melentis (1966).

Castor Linnaeus, 1758

Type Species *Castor fiber* Linnaeus, 1758 (extant).

Other Taxa Included (extant) *C. canadensis* Kuhl, 1820; (extinct species of Europe) *C. neglectus* Schlosser, 1902. [*C. praefiber* Depéret, 1897 and *C. plicidens* Major, 1874, both described as extinct species of Europe, are considered to be junior synonyms of *C. fiber* by van de Weerd (1979)].

Distribution Oligocene–present in Asia and N America, late Miocene (Vallesian)–present in Europe; the extant Eurasian beaver *C. fiber* is also found in Pliocene and Pleistocene localities.

Remarks The feature mostly distinguishing *C. fiber* isolated cheek teeth from those of other fossil castorids is the absence of roots and the presence of annual layers of cement below the crown in teeth of old individuals (van de Weerd 1979). Several isolated dental elements, partial jaws and one partly damaged humerus of a young individual from the localities Kardia, Ptolemais 1 and Ptolemais 3 have been described as *C. fiber* by van de Weerd (1979). Furthermore, dental elements of the species from the locality Apolakkia–2 (Rhodes Island) have been described by van de Weerd et al. (1982). The species has also been reported from the localities Choremi 2 (one specimen), Choremi 3 (12 specimens) and Choremi 4 (one specimen); the teeth are smaller than those of the extant species, but similar in size with Middle Pleistocene material from Germany (van Vugt et al. 2000:74). Dental and postcranial elements of *C. fiber* have also been found in the locality Marathousa 1 (Peloponnese); the straight anterior margin, smooth enamel and triangular cross section on the lower incisors, the long and deep striiae/striids on the high-crowned cheek teeth and the wide caput humeri attached to the tuberculum majus characterize the material (Konidaris et al. 2018). The presence of *Castor* sp. has been reported from Sesklon by Symeonidis (1992); however, since the only available material is one lower incisor (no. 75S) and the faunal list as well as the sedimentological data for the locality do not advocate for the presence of a lake, on the contrary show "open and rather dry" environment (Athanassiou 2018), this report is here regarded as inaccurate (see below).

Family Dipodidae Fischer de Waldheim, 1817

Type Genus *Dipus* Zimmermann, 1780.

Subfamily Sicistinae Allen, 1901

Type Genus *Sicista* Gray, 1827.

Sicista Gray, 1827

Type Species *Mus subtilis* Pallas, 1773 (extant).

Other Taxa Included Extant: *S. betulina* Pallas, 1779; *S. concolor* (Büchner,1892); *S. tianshanica* (Salensky, 1903); *S. caudata* Thomas, 1907; *S. napaea* Hollister, 1912; *S. caucasica* Vinogradov, 1925; *S. strandi* (Formozov, 1931); *S. severtzovi* Ognev, 1935; *S. pseudonapaea* Strautman, 1949; *S. kluchorica* Sokolov et al., 1980;

S. kazbegica Sokolov et al., 1986; *S. armenica* Sokolov and Baskevich, 1988. Extinct: *S. praeloriger* Kormos, 1930; *S. bagajevi* Savinov, 1970; *S. pliocaenica* Erbajeva, 1976; *S. wangi* Qiu and Storch, 2000; *S. primus* Kimura, 2011.

Distribution Early Miocene–present in Eurasia and Africa. The extant *S. betulina* and *S. subtilis* are also found in Pleistocene European localities.

Remarks The extant southern birch mouse *S. subtilis* has been located in various Greek Pleistocene localities, with only a few dental elements in each. It is mainly recognized among other *Sicista* species through its brachyodont molars with simple morphology. Koliadimou (1996) described one M1, two M2 and two M3 of *S. subtilis* from the Biharian fauna of Ravin Voulgarakis. One m1 has been found in the Biharian fauna of Marathoussa (N Greece) and has been assigned to *S.* cf. *subtilis* due to the similarity with the species but scarcity of the material (Koufos et al. 2001). One m1, one m2, two m3 and one M2 have been found in the Middle Pleistocene locality Latomi 1; their size and simple morphology are identical to those of the recent species (Storch 1975). The Upper Pleistocene locality Arnissa has yielded one m1, two m2, one M1 and two M2 of *S. subtilis* (Mayhew 1978). Finally, Chatzopoulou (2014) has described one fragmented m1 and one m2 of *S. subtilis* from the Upper Pleistocene locality Loutra Almopias Cave LAC.

Subfamily Allactaginae Vinogradov, 1925

Type Genus *Allactaga* Cuvier, 1836.

Allactaga Cuvier, 1836

Type Species *Dipus major* Kerr, 1792 (extant).

Other Taxa Included Extant: *A. sibirica* (Forster, 1778); *A. tetradactyla* Lichtenstein, 1823; *A. elater* (Lichtenstein, 1828); *A. euphratica* Thomas, 1881; *A. hotsoni* Thomas, 1920; *A. bullata* G. M. Allen, 1925; *A. severtzovi* Vinogradov, 1925; *A. vinogradovi* Argyropulo, 1941; *A. balikunica* Hsia and Fang, 1964; *A. firouzi* Womochel, 1978. Extinct: *A. anderssoni* (Young, 1927); *A. suni* (Teilhard and Young, 1931); *A. varians* (Savinov, 1970); *A. minor* (Zheng, 1982); *A. irgizensis* Zazhigin and Lopatin, 2000; *A. fru* Nesin and Kovalchuk, 2017.

Distribution Late Miocene–present of Eurasia and N Africa. Many extant species can be found in Pleistocene localities.

Remarks The only fossils in Greece that can be attributed to *Allactaga* with certainty have been described from the Middle Pleistocene locality Latomi 1 (two m1, four m2, one P4, five M1, three M2 and two unidentifiable molar fragments; Storch 1975). The size of the teeth falls within the ranges of both *A. euphratica* and *A. ela-*

ter, and their morphology is identical to that of both extant species. Storch (1975) decided to preliminarily assign the Latomi 1 material to *A.* cf. *euphratica* based only on the recent geographical distribution of the two species, since currently *A. euphratica* is found closer to the Aegean area than *A. elater*. One broken molar from the Middle Pleistocene locality Marathousa 1 (Peloponnese) has been assigned to ?*Allactaga* indet., but the state of the molar does not permit a certain identification (Doukas et al. 2018).

Family Muridae Gray, 1821

Note The division of Muridae in subfamilies suggested by de Bruijn (2010) has been used in the present work.

Type Genus *Mus* Linnaeus, 1758.

Subfamily Paracricetodontinae Mein and Freudenthal, 1971a

Type Genus *Paracricetodon* Schaub, 1925.

Paracricetodon Schaub, 1925

Type Species *Cricetodon spectabilis* Schlosser, 1884.

Other Taxa Included *P. cadurcensis* (Schlosser, 1884); *P. confluens* Schaub, 1925; *P. dehmi* Hrubesch, 1957; *P. walgeri* Bahlo, 1975; *P. kavakderensis* Ünay-Bayraktar, 1989; *P. kodjayarmensis* Ünay-Bayraktar, 1989 (according to van de Weerd et al. 2018 *P. kodjayarmensis* considered to be a junior synonym of *P. dehmi*); *P. wentgesi* de Bruijn et al., 2003b (according to de Bruijn et al. 2003b, *P. confluens* seems to be a junior synonym of *P. spectabilis*); *P. stojanovici* van de Weerd et al., 2018; *P. gracilis* van de Weerd et al., 2018.

Distribution Oligocene of Europe and Turkey.

Paracricetodon kavakderensis Ünay-Bayraktar, 1989

Nomenclatural and Taxonomical History Considered as a junior synonym of *Paracricetodon dehmi* Hrubesch, 1957, according to van de Weerd et al. (2018).

Type Material Left Ml (holotype), Mineral Research and Exploration General Directorate (M.T.A.), Ankara, Turkey (Ünay-Bayraktar 1989: Pl. I.4).

Type Locality Kavakdere, Turkey, Lower Oligocene [~MP24–25].

Distribution Oligocene of Turkish Thrace and Greece.

Occurrences in Greece Kyprinos.

Remarks According to Ünay-Bayraktar's (1989) diagnosis, this is a rather small *Paracricetodon* species, with well-developed anterior arms of the protocone and hypocone on the M1–2, a weak hypocone and a shallow sinus on the M3 and a strong posterior arm of the protoconid of the m3. Doukas and Theocharopoulos (1999) assigned three m1, four m2, five m3, five M1, 10 M2 and two M3 from the Upper Oligocene locality Kyprinos to *P.* aff. *kavakderensis*. The features of the Kyprinos species that differentiate it from the Turkish species are the much larger M3, the presence of the posterior arm of the protocone on the M1–2 and the absence of the posterior arm of the hypoconid from all five m3 (Doukas and Theocharopoulos 1999). The authors attributed these differences to the younger age of the Kyprinos species in comparison to that of the Kavakderen one and suggested that the Kyprinos species is taxonomically between *P. kavakderensis* and *P. kodjayarmensis* but closer to the former species. Van de Weerd et al. (2018) demonstrated great size overlapping between *P. kavakderensis* and *P. dehmi* from Bernloch, Germany, and in spite of the minor differences observed by Ünay-Bayraktar (1989), they consider them as possible synonyms.

Subfamily Pseudocricetodontinae Engesser, 1987

Type Genus *Pseudocricetodon* Thaler, 1969.

Pseudocricetodon Thaler, 1969

Type Species *Pseudocricetodon montalbanensis* Thaler, 1969.

Other Taxa Included *P. thaleri* (Hugueney, 1969); *P. philippi* Hugueney, 1971; *P. moguntiacus* (Bahlo, 1975); *P. simplex* Freudenthal, Hugueney and Moissenet, 1994; *P. adroveri* Freudenthal, Hugueney and Moissenet, 1994; *P. nawabi* Marivaux et al., 1999. De Bruijn et al. (2003b) consider *Allocricetodon* Freudenthal, 1994 as a junior synonym of *Pseudocricetodon*, and thus he included the species *incertus* Schlosser, 1884, *landroveri* Daams et al., 1989 and *cornelii* Freudenthal, 1994 to *Pseudocricetodon*, but other authors (e.g., Freudenthal et al. 2010) prefer maintaining *Allocricetodon* as a separate genus.

Distribution Oligocene–early Miocene of Europe and Asia.

Pseudocricetodon moguntiacus orientalis Ünay-Bayraktar, 1989

Type Material Left ml (holotype), Mineral Research and Exploration General Directorate (M.T.A.), Ankara, Turkey (Ünay-Bayraktar 1989: Pl. VI.17).

Type Locality Kocayarma, Turkey, Upper Oligocene [~MP25–MP26].

Distribution Late Oligocene of Turkey and Greece.

Occurrences in Greece Kyprinos.

Remarks The subspecies differs from *P. moguntiacus moguntiacus* through the absence of ridges connecting the mesoloph and metaloph from the M1–2 (Ünay-Bayraktar 1989). Freudenthal et al. (1994) consider *orientalis* Ünay-Bayraktar 1989 to be a species of *Lignitella* Ünay-Bayraktar 1989, which was initially described as a subgenus of *Pseudocricetodon* by Ünay-Bayraktar (1989), but was considered to be a separate genus by Freudenthal et al. (1994). Here, however, we follow the initial taxonomy by Ünay-Bayraktar (1989). The very small late Oligocene Kyprinos assemblage, consisting of only two M1 and one M2 described by Doukas and Theocharopoulos (1999), is rather similar to *P. moguntiacus orientalis* from the type locality Kocayarma (Turkey), but show some differences, hence the assignment to *P. moguntiacus* cf. *orientalis*. The Kyprinos M1 shows a faint crest from the protoloph that ends against a small cuspule in the middle of the anterosinus. Furthermore, on both Kyprinos M1 the anterior arm of the protocone ends against the anterocone, whereas this feature is observed on only 22% of the type locality M1. This difference, according to Doukas and Theocharopoulos (1999), might imply that the Kyprinos species is slightly older than the Kocayarma one. One additional difference is that the Kyprinos M1 bears a transverse protoloph that ends against the posterior arm of the protocone, whereas in the Kocayarma M1 the protoloph usually connects to the entoloph (in 86% of the specimens, according to Ünay-Bayraktar 1989).

Kerosinia Ünay-Bayraktar, 1989

Type Species *Kerosinia variabilis* Ünay-Bayraktar, 1989.

Distribution Late Oligocene of Turkey and Greece.

Kerosinia variabilis Ünay-Bayraktar, 1989

Type Material Left Ml (holotype), Mineral Research and Exploration General Directorate (M.T.A.), Ankara, Turkey (Ünay-Bayraktar 1989: Pl. V.1).

Type Locality Kocayarma, Turkey, Upper Oligocene [~MP25–MP26].

Distribution Late Oligocene of Turkey and possibly Greece.

Occurrences in Greece Kyprinos.

Remarks According to the diagnosis of the species, the cheek teeth are low-crowned, the wear facets of the main cusps are leaning to the central basins, the ectolophids of the lower molars are usually complete, the posterior arm of the protoconid is usually present on the m2, the anterolophid shows a well-developed lingual branch on the m2–3, the hypolophulid is often incomplete on the m3, the anterior arm of the M1 protocone is usually well-developed and ends free, the protoloph on the M3 is complete and there is often a strong lingual cingulum on the M3 (Ünay-Bayraktar 1989:49).

Doukas and Theocharopoulos (1999) assigned six m1, 10 m2, six m3, six M1, seven M2 and seven M3 from the Upper Oligocene locality Kyprinos to *Kerosinia* aff. *variabilis*. In all Kyprinos M1 the anterior arm of the protocone forms the lingual part of the protoloph, whereas this is the case only in some M1 from the type locality. The Kyprinos dental elements are reported by Doukas and Theocharopoulos (1999) to be smaller than the Kocayarma ones. However, the assemblage from the type locality is much larger than the Kyprinos one (about 613 elements vs. 42 ones), and the size variability in Kocayarma is so great that such a statement about the size difference between the molars from the two localities is improper; in most cases, the Kyprinos measurements fall within the size variation of the Kocayarma ones.

Subfamily Eumyarioninae Ünay-Bayraktar, 1989

Type Genus *Eumyarion* Thaler, 1966.

***Mirrabella* de Bruijn et al., 2007**

Type Species *Mirabella tuberosa* de Bruijn et al., 1987.

Other Taxa Included *M. anatolica* (de Bruijn and Saraç, 1992); *M. crenulata* (de Bruijn and Saraç, 1992); *M. hansoulii* (Doukas and Theocharopoulos, 1999).

Distribution Oligocene–middle Miocene of Turkey and Greece.

Remarks The name *Mirabella* de Bruijn et al. 1987 is preoccupied for a homopteran genus; thus de Bruijn et al. (2007) changed it to *Mirrabella*.

★*Mirrabella hansoulii* (Doukas and Theocharopoulos, 1999)

Type Material Left M1 and M2, AMPG KYP–191 and 192 (holotype) (Doukas and Theocharopoulos 1999: Pl. 2.1–2).

Type Locality Kyprinos, Greece, Upper Oligocene.

Distribution The species has been found only in the type locality.

Occurrences in Greece Kyprinos (Fig. 11a, b).

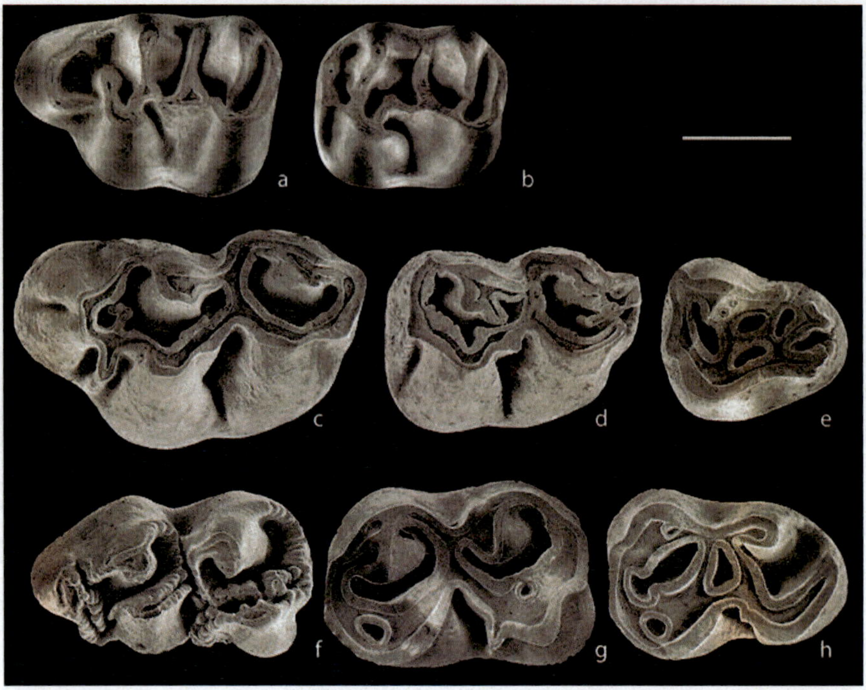

Fig. 11 *Mirrabella hansoulii* from Kyprinos (type locality). **a** M1 (holotype), **b** M2. Images courtesy of C. Doukas. *Mirrabella tuberosa* from Aliveri (type locality). **c** M1, **d** M2, **e** M3, **f** m1 (holotype), **g** m2, **h** m3. Images courtesy of H. de Bruijn. All specimens figured as left; a, c–h are reversed. Scale bar equals 1 mm

Remarks The species has low-crowned and small-sized teeth; in the M1 and M2 the anterocone is not divided, the metalophule is well developed, the posterior paracone-spur is weak and the sinus is wide, whereas there is an anterior arm of the protocone in the M2 (Doukas and Theocharopoulos 1999). The material from the type locality includes the two holotype molars and one additional M2. The size and morphology of the molars show that this species is more primitive than all the other *Mirrabella* species described up to now and more derived than *Paracricetodon* (Doukas and Theocharopoulos 1999).

★*Mirrabella tuberosa* (de Bruijn et al., 1987)

Type Material Right m1, no. 493 (holotype), Institute of Earth Sciences, Utrecht University, The Netherlands (de Bruijn et al. 1987: Pl. 2.5, 5a, 5b).

Type Locality Aliveri, Greece, Lower Miocene [MN4].

Distribution Early Miocene of Greece and Turkey (*M.* cf. *tuberosa* also in German and Swiss MN4 localities).

Occurrences in Greece Aliveri and Karydia.

Remarks The diagnostic characteristics, given by de Bruijn et al. (1987), are the following: the dentition is semi-hypsodont; the ridges on unworn teeth are not much lower than the main cusps; the strong ectomesolophid of the m1–2 is developed as a spur of the hypoconid; the lingual and labial spurs of the anterolophule of the M1 are well developed; there are many tubercles on the edges of the ridges on unworn teeth. Apart from the holotype, the type locality Aliveri has yielded eight m3, eight m2, nine m1, five M1, 10 M2 and eight M3 (Fig. 11c–h). The authors mentioned that the dental patterns from the first to the last molars of both the upper and the lower jaw "*deviate progressively less from the normal cricetid pattern*" (de Bruijn et al. 1987:127). The species is also present in the faunal list of Karydia (Doukas 2003; Wessels 2020 personal comminication), with no details on the material.

Eumyarion Thaler, 1966

Type Species *Cricetodon medius* Lartet, 1851 [= *E. helveticus* (Schaub, 1925)].

Other Taxa Included (according to de Bruijn 2009) *E. latior* (Schaub and Zapfe, 1953); *E. weinfurteri* (Schaub and Zapfe, 1953); *E. bifidus* (Fahlbusch, 1964); *E. leemanni* (Hartenberger, 1965); *E. carbonicus* de Bruijn and Saraç, 1991; *E. montanus* de Bruijn and Saraç, 1991; *E. microps* de Bruijn and Saraç, 1991; *E. intercentralis* de Bruijn and Saraç, 1991; *E. orhani* de Bruijn et al., 2006; *E. lukasi* Joniak et al., 2017.

Distribution Late Oligocene–Miocene of Europe and Anatolia.

Eumyarion latior (Schaub and Zapfe, 1953)

Nomenclatural and Taxonomical History *Cricetodon helveticus latior* in Schaub and Zapfe 1953 (new subspecies); *Cotimus latior* in Cicha, Fahlbusch and Fejfar 1972 (new combination); *Eumyarion latior* in Engesser 1972 (new combination).

Type Material Right mandible with M1–M2 (holotype) (Schaub and Zapfe 1953: Pl. II.4).

Type Locality Neudorf an der March (Spalte 1; also known as Devínská Nová Ves), Slovakia, middle Miocene [MN6].

Distribution Early–middle Miocene of Europe.

Occurrences in Greece Karydia.

Remarks *E. latior* was initially described as a subspecies of *Cricetodon helveticus* based on only two upper molars (the type material). Schaub and Zapfe (1953), in their diagnosis, stated (free translation from German) upper molars with transversal anteriorly leaning cusps and anteriorly leaning sinus; the M1 is significantly wider than in *C. helveticus*. Doukas (2003) gave a faunal list for the Lower Miocene locality Karydia, mentioning the presence of *E.* aff. *latior* (Doukas 2003:table 2), without further information. Later, in her master thesis, Duncan (2012) described the Karydia material (six m1, eight m2, eight m3, nine M1, eight M2, eight M3) and confirmed the assignment to *E.* aff. *latior*. The material shows great morphological variation; its morphological and metrical characters show close similarities with *E. latior*, but there are also important differences from this species, i.e., the long mesoloph with spurs connecting it with the paracone and metacone on the upper molars, the relatively long anterocone in the M1 resulting in a high L/W ratio and the high L/W ratio in the M2 (Duncan 2012).

***Eumyarion weinfurteri* (Schaub and Zapfe, 1953)**

Nomenclatural and Taxonomical History *Cricetodon helveticus weinfurteri* in Schaub and Zapfe 1953 (new subspecies); *Cotimus weinfurteri* in Cicha et al. 1972 (new combination); *Eumyarion weinfurteri* in Engesser 1972 (new combination).

Type Material Right M1(holotype) (Schaub and Zapfe 1953: Pl. II.5).

Type Locality Neudorf an der March (Spalte 1; also known as Devínská Nová Ves), Slovakia, middle Miocene [MN6].

Distribution Early–middle Miocene of Europe.

Occurrences in Greece Aliveri.

Remarks Initially described as a subspecies of *Cricetodon helveticus*, known from its type locality only through two molars, *E. weinfurteri* is very close to *C. helveticus*; there is similarity in features of the tooth crown, such as the transversal position of the cusps and the forward-facing sinus; the overall outline of the M1 is somewhat more compact than in *C. helveticus*, due to the less pronounced constriction behind the anterocone; it is smaller than *C. helveticus*, corresponding in size to *C. brevis*, but morphologically distinct (Schaub and Zapfe 1953; free translation from German). After the initial descriptions of both *E. latior* and *E. weinfurteri* were published, larger assemblages than those from their type locality were studied, showing a very large intraspecific variation in size and morphology for the two species (e.g., Engesser 1972). This fact has led de Bruijn and Saraç (1991) to synonymize *E. weinfurteri* with *E. latior*. Later, Schötz (1993) suggested, and also de Bruijn (2009) agreed, "*... to tentatively maintain the species name* weinfurteri *for associations of small* Eumyarion *teeth...until the homogeneity of the collection from Neudorf is demonstrated*". Following this concept, and due to the limited number of molars from the Lower Miocene locality Lapsarna (two m1, one m2, two m3, one M1, three M2, one M3), Vasileiadou and Zouros (2012) attributed their *Eumyarion* to *E.* aff. *latior/weinfurteri*. The material from the Lower Miocene locality Aliveri (nine m1, seven m2, 10 m3, nine M1, eight M2, 12 M3) initially attributed to *E. weinfurteri* by Klein Hofmeijer and de Bruijn (1988), was later transferred to *E. latior* by de Bruijn and Saraç (1991); in the latest faunal list for the locality is mentioned as *E. weinfurteri* again (van den Hoek Ostende et al. 2015). Since the material is scanty and in view of the possible synonymy of the two species, it is better to refer to the material as *E. latior/weinfurteri*.

Eumyarion leemanni **(Hartenberger, 1965)**

Nomenclatural and Taxonomical History *Cotimus leemanni* in Hartenberger 1965 (new species); *Eumyarion leemanni* in Engesser 1972 (new combination).

Type Material Left M1, CL 3683 (holotype), Museum of Sabadell (Hartenberger 1965: Pl. X.f and Text–fig. 1).

Type Locality Can Llobateres, Spain, Miocene [Vallesian (MN9)].

Distribution Late Miocene of Europe.

Occurrences in Greece Plakias.

Remarks *E. leemanni* is clearly distinguished from the other *Cotimus* species due to its higher degree of hypsodonty (Hartenberger 1965). Since Engesser (1979) proved that the N American genus *Cotimus* Black, 1961 is synonymous to *Leidymys* Wood, 1936 and the European species do not belong to this genus, the genus *Eumyarion* Thaler, 1966 has been used to include the European species. A small assemblage from the Upper Miocene locality Plakias, consisting of one m1 and one M3, both partial, was attributed to *Cotimus* sp. by de Bruijn and Meulenkamp (1972), who mentioned that the M3 possibly resembles *C. leemanni* best. The material was enriched through further sampling and one m3, one M1 and one M2 were added, which were identified as *E. leemanni* by de Bruijn et al. (2012).

Subfamily Anomalomyinae Schaub, 1925

Type Genus *Anomalomys* Gaillard, 1900.

Anomalomys **Gaillard, 1900**

Type Species *Anomalomys gaudryi* Gaillard, 1900.

Other Taxa Included *A. gaillardi* Viret and Schaub, 1946; *A. minor* Fejfar, 1972; *A. aliveriensis* Klein Hofmeijer and de Bruijn, 1985; *A. rudabanyensis* Kordos, 1989; *A. minutus* Bolliger, 1992.

Distribution Miocene of Europe and Turkey.

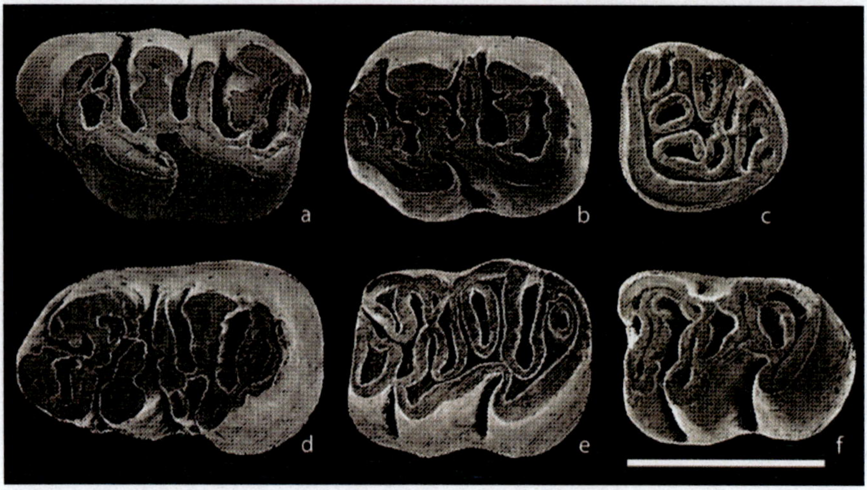

Fig. 12 *Anomalomys aliveriensis* from Aliveri (type locality). **a** M1, **b** M2, **c** M3, **d** m1 (holotype), **e** m2, **f** m3. Images courtesy of H. de Bruijn. All specimens are left. Scale bar equals 1 mm

★*Anomalomys aliveriensis* Klein Hofmeijer and de Bruijn, 1985

Type Material Left m1, no. 1194 (holotype), Institute of Earth Sciences, Utrecht University, The Netherlands (Klein Hofmeijer and de Bruijn 1985: Pl. III.10–10a).

Type Locality Aliveri, Greece, Lower Miocene [MN4].

Distribution Early Miocene of Greece and Turkey.

Occurrences in Greece Aliveri (Fig. 12).

Remarks According to the diagnosis given by Klein Hofmeijer and de Bruijn (1985), the species has a relatively narrow M1 and a long m3, its M1 and M2 bear well-developed and transverse proto- and metaloph, the posterior arms of the proto- and hypoconid are strong and its m1–m2 have always long mesolophid. The species from Aliveri was initially mentioned as *A. minor* by de Bruijn et al. (1980) and Benda and de Bruijn (1982), but it differs from the latter species through its narrower M1, longer m3 and more complex molar pattern (Klein Hofmeijer and de Bruijn 1985). The material from the type locality includes 37 m1, 33 m2, 22 m3, 36 M1, 34 M2 and six M3 (including the holotype; Klein Hofmeijer and de Bruijn 1985). It is the oldest *Anomalomys* species known to date and has the most complex dental morphology (Klein Hofmeijer and de Bruijn 1985).

Anomalomys minor **Fejfar, 1972**

Type Material Left M1 (holotype), Central Geological Institute in Prague, SUUG OF 73303 (Fejfar 1972: Abb. 2.1, 9).

Type Locality Franzensbad, Czech Republic, lower/middle Miocene [MN4/5].

Distribution Early–middle Miocene of Europe and Turkey.

Occurrences in Greece Karydia.

Remarks Fejfar (1972) did not give a formal diagnosis for his species, but he gave detailed descriptions of the molars from the type locality and also of material assigned to this species from Langenmoosen. Based on the most recent study of the fossils from Karydia, the material previously assigned to *A. aliveriensis* by Theocharopoulos (2000) is now assigned to *A. minor* (Wessels 2020 personal communication), but no information on the material has been published yet.

Prospalax **Méhely, 1908**

Type Species *Spalax priscus* Nehring, 1897.

Other Taxa Included (according to Hordijk and de Bruijn 2009) *P. petteri* Bachmayer and Wilson, 1970.

Distribution Late Miocene–Pliocene of Europe.

Prospalax petteri **Bachmayer and Wilson, 1970**

Type Material Partial right lower jaw with m1–m3, no. 1970/1394 (holotype), Museum of Natural History, Vienna (Bachmayer and Wilson 1970: fig. 68).

Type Locality Kohfidisch, Austria, Upper Miocene [MN10].

Distribution Late Miocene of Europe.

Occurrences in Greece Lava 2.

Remarks The diagnosis, given by Bachmayer and Wilson (1970), mentions that this species has cheek teeth with size similar to that of the *P. priscus*' cheek teeth, but the lower jaw is less robust than that of the latter species. Furthermore, its jaws are larger and more specialized than in *Anomalomys gaudryi*. Its m1 is more elongate than in *Pliospalax*, and the degree of development of a sigmoid pattern in the cheek teeth is intermediate between that of *Anomalomys gaillardi* and of *P. priscus*. De Bruijn et al. (1999) mentioned the presence of *P.* aff. *petteri* in the faunal list of the Upper Miocene locality Lava 2 (de Bruijn et al. 1999:Table 10, p. 110), without any further details.

***Prospalax priscus* (Nehring, 1897)**

Nomenclatural and Taxonomical History *Spalax priscus* in Nehring 1897 (new species); *Prospalax priscus* in Méhely 1908 (new combination).

Type Material Fragmented right mandible with well-preserved molars and a large part of the incisor (holotype) (Nehring 1897: Abb. 4.3).

Type Locality "Nagyharsányer Berg in the area of Villány", S Hungary, Pliocene [MN16].

Distribution Pliocene of Europe.

Occurrences in Greece Notio 1.

Remarks Nehring (1897) mentioned that the three molars are very similar to each other, with one lingual and one buccal re-entrant folds and thin enamel layer. The total length at the alveoli is 7.2 mm, whereas the width is 1.8–2 mm. The lower incisor (width: 1.9 mm.) does not show any wrinkling of the enamel but bears three longitudinal ridges running along the anterior face of the enamel. The material available to Nehring for the description of the new species was only one mandible, whereas Méhely acquired a partial left mandible of the same species from Beremend (west of Nagyharsányer Berg) with preserved muscular process and complete incisor but lacking the m2 and m3; based on the two mandibles, Méhely described the new genus *Prospalax* Méhely 1908 as the direct ancestor of all modern *Spalax* species. Three molars (two m2 and one M1) from the Pliocene locality Notio 1 have been attributed to this species by Hordijk and de Bruijn (2009), based on their similarity with *P. priscus* molars from Csarnóta 2 and 4 available to the authors for comparisons. An unusual characteristic of the species, noticed for the first time by these authors and perceived as a derived feature, is that the enamel on the anterior sides of the M1–M2 and the posterior sides of the m1–m2 is much thicker than that on the opposite sides (Hordijk and de Bruijn 2009:63).

Subfamily Spalacinae Gray, 1821

Type Genus *Spalax* Guldenstaedt, 1770.

***Heramys* Klein Hofmeijer and de Bruijn, 1985**

Type Species *Heramys eviensis* Klein Hofmeijer and de Bruijn, 1985.

Other Taxa Included *H. anatolicus* Sarıca and Şen, 2003.

Distribution Early–middle Miocene of Greece and late Miocene of Turkey.

Remarks This genus was created by Klein Hofmeijer and de Bruijn (1985), based on material from Aliveri, to include species that bear relatively longer Ml and M2 than the *Pliospalax* species do and shorter M3 without lingual sinus, as well as long metalophules/-ids in the Ml–2 and m1–2 and an antero-labial sinus in the m1, features that differentiate it from *Spalax*. The presence of *Heramys* sp. in the middle Miocene locality Thymiana A (Chios Island) has been mentioned by de Bonis et al. (1997), but the material has not been published yet.

★*Heramys eviensis* Klein Hofmeijer and de Bruijn, 1985

Type Material Left m1, no. 1289, Institute of Earth Sciences, Utrecht University, The Netherlands (Klein Hofmeijer and de Bruijn 1985: Pl. I.4, 4 a–b).

Type Locality Aliveri, Greece, Lower Miocene [MN4].

Distribution Until now the species has been found only in its type locality.

Remarks Klein Hofmeijer and de Bruijn (1985) gave the diagnosis of the species (same as that of the genus): "*spalacid with semi–hypsodont cheek teeth; Ml and M2 longer than wide; M3 small and without lingual sinus; metaconid of the ml large; lower incisor with two narrow ribs*". The assemblage from the type locality Aliveri (Fig. 13) is rather small (seven m1, seven m2, seven m3, five M1, three M2, two M3), but large enough to confirm its allocation to a new genus and species.

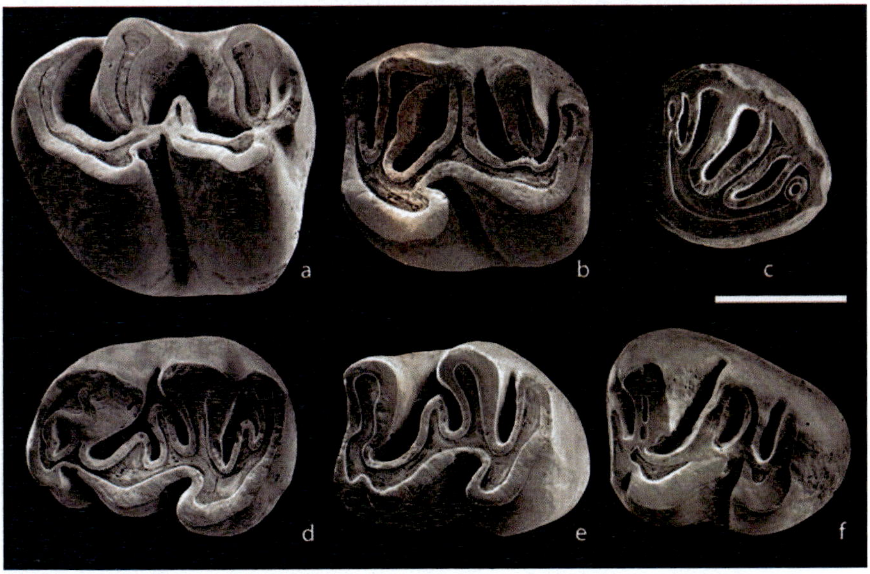

Fig. 13 *Heramys eviensis* from Aliveri (type locality). **a** M1, **b** M2, **c** M3, **d** m1 (holotype), **e** m2, **f** m3. Images courtesy of H. de Bruijn. All specimens are left. Scale bar equals 1 mm

***Debruijnia* Ünay, 1996**

Type Species *Debruijnia arpati* Ünay, 1996.

Other Taxa Included (according to de Bruijn 2016) *Debruijnia* sp. (in Ünay and Göktas 1999); *D. kostakii* de Bruijn 2016.

Distribution Early Miocene of Greece and Turkey.

★*Debruijnia kostakii* de Bruijn, 2016

Type Material Right m1, KR 2, nr. 1014 (holotype), Institute of Earth Sciences, Utrecht University, The Netherlands (de Bruijn 2016: Fig. 3, 1a & 1b).

Type Locality Karydia 2, Greece, Lower Miocene [MN4].

Distribution Until now the species has been found only in its type locality.

Remarks According to de Bruijn (2016), this species has rather large molars. Its M1 is characterized by an anteroloph separated from the protocone through a shallow notch, by an either complete or interrupted anterior arm of the protocone and by a sometimes transverse, or otherwise connected to the posteroloph, metalophule. Its five-lophed M3 bears a sinus that remains open lingually until an advanced stage of wear. Its m1 is long, bears a metaconid that does not fuse with the anteroconid until an advanced stage of wear and when unworn shows a long free-ending posterior arm of the hypoconid. The protosinusid is progressively less developed from the m1 to the m3. The material from the type locality Karydia 2 (Fig. 14) consists of four m1, two m2, three m3, four M1, five M2 and two M3, whereas there is also one M1 found in the synchronous layer Karydia 3 (de Bruijn 2016). The material from Karydia 2 had been mentioned as *Debruijnia* n.sp. (large) in the faunal list for the locality given by Theocharopoulos (2000).

***Pliospalax* Kormos, 1932**

Type Species *Prospalax macoveii* Simionescu, 1930.

Other Taxa Included (including *Sinapospalax* Sarıca and Şen, 2003, according to de Bruijn et al. 2015) *P. compositodontus* Topachevski, 1969; *P. sotirisi* (de Bruijn et al., 1970); *P. tourkobouniensis* de Bruijn and van der Meulen, 1975; *P. primitivus* Ünay, 1978; *P. canakkalensis* Ünay, 1981; *P. marmarensis* Ünay, 1990; *P. senii* Koliadimou, 1996; *P. incliniformis* (Sarıca and Şen, 2003); *P. sinapensis* (Sarıca and Şen, 2003); *P. berdikensis* (Şen and Sarıca, 2011); *P. complicatus* Şen and Sarıca, 2011.

Distribution Early Miocene–Early Pleistocene of Europe, Turkey and Ukraine.

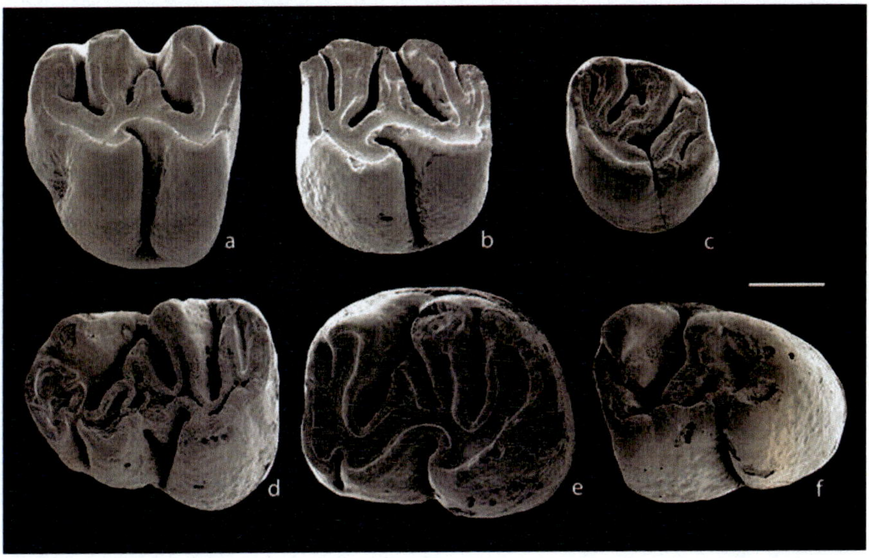

Fig. 14 *Debruijnia kostakii* from Karydia (type locality). **a** M1, **b** M2, **c** M3, **d** m1 (holotype), **e** m2, **f** m3. Images courtesy of H. de Bruijn. All specimens figured as left; c, d, f are reversed. Scale bar equals 1 mm

Remarks Spalacids are a rare element in fossil assemblages. As a result, it is not always easy to identify them to species level. The presence of *Pliospalax* sp. has been reported from Antonios with one fragmented M1 (Vasileiadou and Koufos 2005), in the faunal list for Ano Metochi 3 with no detail on the material (de Bruijn 1989) and from Kessani with one m1 and two m2 (Vasileiadou et al. 2012).

***Pliospalax macoveii* (Simionescu, 1930)**

Nomenclatural and Taxonomical History *Prospalax macoveii* in Simionescu 1930 (new species); *Pliospalax macoveii* in Kormos 1932 (new combination). Şen (1977) considered *Pli. simionescui* a junior synonym of *P. macoveii.*

Type Material Fragmented left mandible with the incisor and m1–2 (holotype), University of Iaşi, Romania (Simionescu 1930: Fig. 33).

Type Locality Malusteni, Romania, Pliocene [MN15].

Distribution Late Miocene–Pliocene of Romania and Ukraine; possibly in Greece and Turkey.

Occurrences in Greece Nea Silata.

Remarks Simionescu did not give a diagnosis or description for his species, but Topachevski (1969:155) gave some diagnostic features: species of medium size, M1–2 without mesoloph, m1 with a free mesolophid only in early stages of wear. Later it fuses with the metaconid, leading to the creation of a longitudinal enamel island, with wear the entoconid fuses with the posterior cingulum and the anterior surface of the upper and lower incisors is smooth. The locality Nea Silata has revealed one m1, one m3, one M1 and two M3 of *Pliospalax* which were described in detail by Vasileiadou (2001), in her unpublished MSc thesis: the M1 lacks a mesoloph and shows a very shallow postero-labial sinus; the labial sinus of the M3 closes in early stages of wear; the mesolophid of the m1 (unworn) is free and long, bending anteriorly. The morphology and the size of the molars are similar to those of *P. macoveii*, but the scantiness of the material allow only for a tentative assignment to *P.* cf. *macoveii* (Vasileiadou et al. 2003).

★*Pliospalax sotirisi* (de Bruijn et al., 1970)

Nomenclatural and Taxonomical History *Spalax sotirisi* in de Bruijn et al. 1970 (new species); *Pliospalax sotirisi* in de Bruijn and van der Meulen 1975 (new combination).

Type Material Partial right mandible with m1–m2, no. 1286 (holotype), Institute of Earth Sciences, Utrecht University, The Netherlands (de Bruijn et al. 1970: Pl. 6.3).

Type Locality Maritsa 1, Rhodes Island, Upper Miocene [Turolian (MN13)].

Distribution Late Miocene of Greece.

Occurrences in Greece Maritsa 1 (Fig. 15a–c), Samos S3 and Mytilinii 1B.

Remarks The diagnosis of de Bruijn et al. (1970) (based on one partial maxilla with Ml–M2, two M2, one M3 and the holotype) mentions that the little-worn m1 and m2 bear two labial and two lingual re-entrant folds, whereas the anterior lingual re-entrant fold in the m2 reaches the anterior border of the tooth. Furthermore, the labial re-entrant folds of the M1 develop into enamel islands at about the same stage of wear. Black et al. (1980) attributed to *P.* cf. *sotirisi* one left m2 and one left m3, possibly from the same individual, from the Upper Miocene locality Samos S3, mentioning that the Samos m2 is similar to that of *P. sotirisi* from the type locality but slightly larger. Additional spalacid material from Samos Island was reported by Vasileiadou and Sylvestrou (2009), from the locality Mytilinii 1B, consisting of a partial skull with the left M1–3. These molars were found to show similar morphology with the molars of *P. sotirisi* from Maritsa 1, in relation to their wear stage; however, the Samos M1 and M2 are narrower than the Rhodes ones (Vasileiadou and Sylvestrou 2009). In any case, all three assemblages (from the type locality and the two from Samos Island) are too small to allow the understanding of the intraspe-

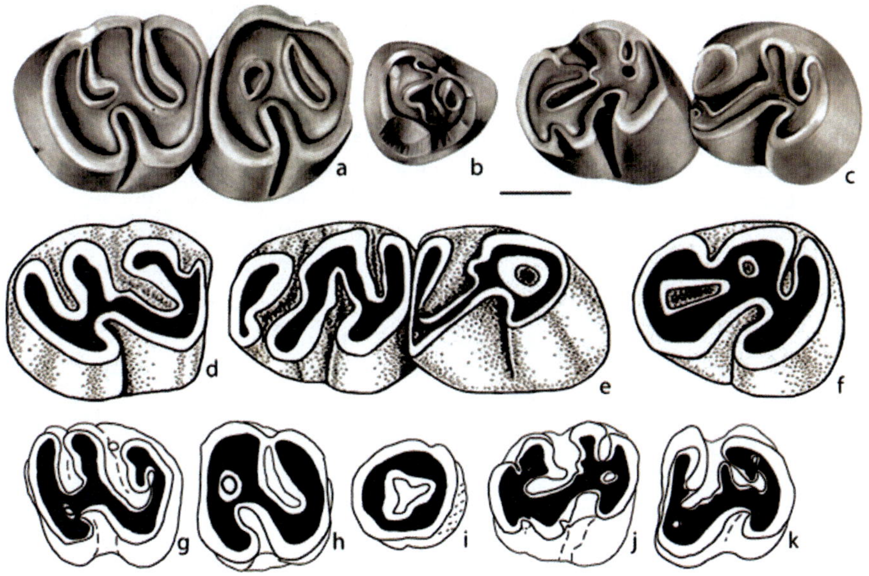

Fig. 15 *Pliospalax sotirisi* from Maritsa 1 (type locality). **a** M1–2, **b** M3, **c** m1–2 (holotype). *Pliospalax tourkobouniensis* from Tourkobounia 1 (type locality). **d** M1, **e** m1–m2 (holotype), **f** m1. Images and drawings courtesy of H. de Bruijn. *Pliospalax senii* from Ravin Voulgarakis (type locality). **g** M1, **h** M2, **i** M3, **j** m1 (holotype), **k** m2. Drawings courtesy of K. Koliadimou. All specimens figured as left; c, g–k are reversed. Scale bar equals 1 mm

cific morphological and metrical variation. Similarities of this species with *P. macoveii* have led de Bruijn and van der Meulen (1975) and Şen and de Bruijn (1977) to tentatively consider the two species synonymous. However, Ünay (1978) noticed differences in the length of the postero-labial and postero-lingual folds and preferred to retain both species; Ünay included *P. sotirisi* as a separate species in her most recent review of the genus (Ünay 1999).

★*Pliospalax tourkobouniensis* de Bruijn and van der Meulen, 1975

Type Material Left mandible with m1–m2, no. 481 (holotype), Institute of Earth Sciences, Utrecht University, The Netherlands (de Bruijn and van der Meulen 1975: Pl. 6.2).

Type Locality Tourkobounia 1, Greece, Pliocene [Villanyian (MN16)].

Distribution Villanyian of Greece and Romania.

Occurrences in Greece Tourkobounia 1 (Fig. 15d–f).

Remarks According to the diagnosis by de Bruijn and van der Meulen (1975), in slightly worn m1 the antero-labial and antero-lingual re-entrant folds are confluent, whereas with moderate wear the former fold disappears and the latter one becomes shallow. Further, the postero-labial re-entrant fold on the m1 does not reach the crown basis. The M1 of the species preserves traces of the posterior cingulum and the mesoloph. *P. tourkobouniensis* differs from *P. macovei* and *P. sotirisi* in having a shorter postero-labial re-entrant fold in the m1 (de Bruijn and van der Meulen 1975). The preservation of the posterior cingulum and the mesoloph in the M1 and the relatively low cheek teeth are archaic characteristics among the spalacines, which oddly enough are present in this Pliocene species (de Bruijn and van der Meulen 1975). Apart from the holotype, the type locality has also revealed two m1, three m2, three m3, three M1 and one M3.

★*Pliospalax senii* Koliadimou, 1996

Type Material Right m1, LGPUT RVL295 (holotype), Museum of Geology-Paleontology-Paleoanthropology, Aristotle University of Thessaloniki (Koliadimou 1996: Fig. 4.16 a).

Type Locality Ravin Voulgarakis, Greece, Pleistocene [Biharian (MNQ20)].

Distribution Early Pleistocene of Greece (only in type locality).

Remarks According to Koliadimou's (1996) diagnosis (translated from Greek), *P. senii* is a medium-sized spalacine; the m1 shows a non-isolated anterior lobe, a postero-labial sinusid reaching the edge of the crown, a very small mesolophid, small postero-lingual and small pre-antero-lingual sinusid; the m2 is characterized by the presence of a large mesolophid; in the M1 there is a small postero-labial sinus, one antero-lingual islet and a cingulum in the form of a cusp in the central-labial sinus; the M3 is reduced in size and without lingual sinus. Apart from the holotype, the material from Ravin Voulgarakis (Fig. 15g–k) includes one m2, one M1, five M2 and two M3. *P. senii* seems to be the youngest *Pliospalax* species.

Spalax Guldenstaedt, 1770

Type Species *Spalax microphthalmus* Guldenstaedt, 1770 (extant).

Other Taxa Included [following Ünay's (1996) concept for the genus, including all younger spalacins usually characterized by two lingual and one labial re-entrant fold on the m1] (extant species found in fossil assemblages) *S. leucodon* Nordmann, 1840; *S. nehringi* (Satunin, 1898); *S. ehrenbergi* Nehring, 1898; (extinct) *S. podolicus* Trouessart, 1897; *S. minor* Topachevski, 1959; *S. odessanus* (Topachevski, 1969); *S. advenus* Kretzoi, 1977; *S. denizliensis* Erten, 2018.

Distribution Pliocene–present of E Europe, W Asia and N Africa.

Remarks *Spalax* contains five extant species of blind, fossorial or subterranean mole rats, living today in E Europe, the Balkans and the E Mediterranean area

(Macdonald 2009). Several Middle–Late Pleistocene *Spalax* assemblages have been reported from Greece (Appendix). Two m1, three m2, one m3, six M1, one M2 and two M3 from the Biharian fauna of Kalymnos-Xi have been attributed to *Spalax* cf. *nehringi* by Kuss and Storch (1978); Koliadimou (1996) suggested the attribution of this material to *Pliospalax*, based on the presence of two labial and two lingual sinusids on the m1 and of an antero-labial islet on the m2. We here retain the initial identification by Kuss and Storch (Appendix). Storch (1975) reported the presence of *Spalax nehringi* in the Middle Pleistocene locality Latomi 1 (three partial skulls, 13 m1, 18 m2, 17 m3, 21 M1, 17 M2, 13 M3 and three incisor fragments). Fossil material of *Spalax leucodon* has been reported by Chatzopoulou (2014) from the Upper Pleistocene localities Loutra Almopias Cave LAC (three m1, 10 m2, two partial m1/2, 17 m3, seven M1, six M2, one partial M1/2, 15 M3) and LAC Ia (three m1, four m2, two m3, one partial mandible with m1–m2, eight M1, eight M2, one M3). Mayhew (1978) reported the presence of *Spalax microphthalmus* in the Upper Pleistocene locality Arnissa. Finally, Sickenberg (1971) attributed one fragmented lower incisor and one fragmented femur from Petralona Cave to ? *Spalax* sp.

***Spalax odessanus* (Topachevski, 1969)**

Nomenclatural and Taxonomical History *Microspalax odessanus* in Topachevski 1969 (new species); *Spalax odessanus* in Savic and Nevo 1990 (new combination).

Type Material Partial skull with dentition, No. 8442 (holotype), Institute of Zoology AN Ukraine (Topachevski 1969: Fig. 58 A).

Type Locality Odessa, Ukraine, Upper Pliocene [MN15].

Distribution Late Pliocene of Ukraine and Greece.

Occurrences in Greece Megalo Emvolon 1.

Remarks This species, according to Topachevski's (1969) diagnosis, is large, with no sharp longitudinal depression in the suture between the nasal bones, a straight line formed by the temporo-nasal and temporo-maxillary sutures, wide palate between the M1s, with a well-developed and wide central ridge, no mesocone on the M1/2, not fully developed mesoconid on the m1 and width of the internal fork covering less than one forth of the total length of the molar's wearing surface, no fusion of the entoconid and metaconid even with advanced wear, leading to the absence of an anterior mark in m1 in all wear stages, the entoconid and posterior collar fused on the m1 in early wear stages, leading to the presence of a posterior mark, and flat and smooth anterior surface of the upper incisor. *S. odessanus* has been reported by de Bruijn (1984) from the Ruscinian fauna of Megalo Emvolon 1 (known also as Karaburun), where it is represented by two partial skulls, one with one mandibular ramus and one with both rami. Both specimens show *S. odessanus* features: large size, very wide incisors, wide rostrum, wide palate, pro-odont upper incisors and robust coronoid process with sharp anterior border (de Bruijn 1984).

Subfamily Cricetodontinae Schaub, 1925

Type Genus *Cricetodon* Lartet, 1851.

***Deperetomys* Mein and Freudenthal, 1971a**

Type Species *Cricetodon sansaniensis hagni* Fahlbusch, 1964.

Other Taxa Included *D. rhodanicus* (Depéret, 1887); *D. intermedius* (de Bruijn et al., 1987); *D. anatolicus* de Bruijn, Fahlbusch, Saraç and Ünay, 1993; *D. magnus* de Bruijn et al., 2013a; *D. calefactus* Marković et al., 2019; *D. saltensis* Marković et al., 2019.

Distribution (early?) Oligocene–middle Miocene of Europe and Turkey.

Remarks The presence of cf. *Deperetomys* sp. in Karydia is reported by Theocharopoulos (2000), but with no details on the material. However, the genus is not included in the new faunal list for the locality (Wessels 2020 personal communication).

***Cricetodon* Lartet, 1851**

Type Species *Cricetodon sansaniensis* Lartet, 1851.

Other Taxa Included *C. causasicus* Argyropulo, 1938; *C. meini* Freudenthal, 1963; *C. lavocati* Freudenthal, 1966; *C. jotae* Mein and Freudenthal, 1971a; *C. albanensis* Mein and Freudenthal, 1971a; *C. aureus* Mein and Freudenthal, 1971b; *C. aguirrei* Sese, 1977; *C. pasalarensis* (Tobien, 1978); *C. candirensis* (Tobien, 1978); *C. cariensis* (Şen and Ünay, 1979); *C. hungaricus* (Kordos, 1986); *C. aliveriensis* Klein Hofmeijer and de Bruijn, 1988; *C. versteegi* de Bruijn et al., 1993; *C. tobieni* de Bruijn et al., 1993; *C. kasapligili* de Bruijn et al., 1993; *C. bolligeri* Rummel, 1995; *C. jumaensis* Rummel, 1995; *C. engesseri* Rummel and Kälin, 2003; *C. orientalis* Bi, 2005; *C. soriae* López Martínez, Cárdaba, Salesa, Hernández Fernández, Cuevas González and Fesharaki, 2006 in Hernández Fernández et al. 2006; *C. klariankae* Hir, 2007; *C. volkeri* Wu, 2009; *C. wanhei* Qiu, 2010; *C. nievei* López-Guerrero et al., 2014; *C. trallesensis* Durgut and Ünay, 2016; *C. fikreti* Durgut and Ünay, 2016; *C. magnesiensis* Durgut and Ünay, 2016; *C. yapintiensis* Durgut and Ünay, 2016; *C. goklerensis* Joniak et al., 2017.

Distribution Miocene of Europe and Asia.

Remarks The presence of *Cricetodon* sp. is reported by de Bonis et al. (1997) for the locality Thymiana A, without details on the material. Skandalos (2017:43), in his master thesis, described one m3 from Karydia as *Cricetodon* sp., as the molar, unlike all other molars from the locality attributed to *C. aliveriensis*, bears an extra ridge in the posterior part that ends freely in the posterosinusid without reaching the lingual border; otherwise, both morphologically and metrically, the molar resembles those of *C. aliveriensis*.

★*Cricetodon aliveriensis* Klein Hofmeijer and de Bruijn, 1988

Type Material Left M1, Nr. 542 (holotype) (Klein Hofmeijer and de Bruijn 1988: Pl. 1.1).

Type Locality Aliveri (South Quarry), Greece, Lower Miocene [MN4].

Distribution Early Miocene of Greece.

Occurrences in Greece Aliveri (Fig. 16) and Karydia.

Remarks According to the diagnosis by Klein Hofmeijer and de Bruijn (1988), this is a small *Cricetodon* species, with an anterocone consisting of two cusps of about equal size, an incomplete anterior protolophule and a complete posterior one on the Ml, either short and low or absent posterior parts of the ectoloph on the Ml–M2, present mesolophid on all the lower molars and a double metalophulid on the ml. Apart from the type locality, from where 87 molars of the species have been described, also Karydia has yielded a large collection of the species including 564 dental elements, first reported by Doukas (2003), and consequently detailed described and studied by Skandalos (2017).

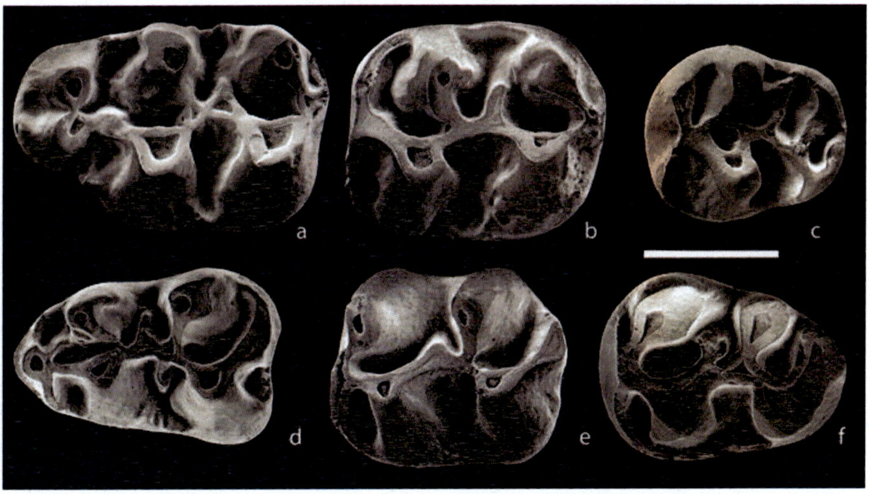

Fig. 16 *Cricetodon aliveriensis* from Aliveri (type locality). **a** M1 (holotype), **b** M2, **c** M3, **d** m1, **e** m2, **f** m3. Images courtesy of H. de Bruijn. All specimens figured as left. Scale bar equals 1 mm

***Cricetodon meini* Freudenthal, 1963**

Type Material Fragment of left maxilla with M1–M3, VxC 309 (holotype) (Mein 1958: Fig. 142; Mein and Freudenthal 1971b: Pl. 1.1).

Type Locality Vieux Collonges, France, middle Miocene [MN5].

Distribution Early–middle Miocene of Europe and Russia.

Occurrences in Greece Komotini, Antonios, Thymiana A and C.

Remarks *C. meini* is smaller than all European *Cricetodon* species except for *C. aliveriensis*. The original diagnosis was given by Freudenthal (1963) [here with adaptations in nomenclature by Mein and Freudenthal (1971b), translated from French]: "Lower molars with an ectomesolophid (may be missing in m2–m3); the hypolophulid of the m1 is transverse; the anterior metalophulid may be missing (i.e., in two of the 10 teeth studied); the posterior metalophulid is always present; upper molars: the posterior ectoloph is absent or very weak; the mesoloph is on average longer than in *C. sansaniensis* Lartet 1851; the M3 is much less reduced than in the latter species." Mein and Freudenthal (1971b) separated the smaller *C. meini* from the larger *C. aureus* in the type locality Vieux Collonges and renewed the diagnosis for the former species. According to this emended diagnosis, the lower molars are two-rooted, most of them bear a posterior or a double connection between the metaconid and the protoconid, next to the posterior connection there is an anterior connection with the anteroconid, the posterior connection is rarely missing and an ectomesolophid is almost always present on the m1 but often missing from the m2 and almost always missing from the m3. On the M1 the anterocone is simple or bifid, on the upper molars the posterior ectoloph is absent or short and the mesoloph is well developed.

This species is present in the faunal list for the localities Thymiana A and C (de Bonis et al. 1997) without further information. *C. meini* from Komotini (Poroton Quarry), represented by 14 molars, has been described by de Bruijn et al. (1993), who mention that the molars are identical to those from the species type locality. The *C. meini* material from Antonios includes 62 molars described by Vasileiadou and Koufos (2005), who mention that the only difference between the Antonios species and that from the type locality is that on the m1 from the type locality the anterior metalophulid is more frequently present than on the m1 from Antonios.

***Byzantinia* de Bruijn, 1976**

Type Species *Byzantinia pikermiensis* de Bruijn, 1976.

Other Taxa Included *B. hellenicus* (Freudenthal, 1970); *B. nikosi* de Bruijn, 1976; *B. sofcaensis* (Tobien, 1978); *B. eskihisarensis* (Tobien, 1978); *B. cariensis* (Şen and Ünay, 1979); *B. bayraktepensis* Ünay, 1980; *B. ozansoyi* Ünay, 1980; *B. darda-*

nellensis Ünay, 1980; *B. orientalis* (Lungu, 1981) [possible synonym of *B. bayraktepensis*, according to Şen 2016]; *B. uenayae* Rummel, 1998.

Distribution Middle–late Miocene of the Balkans and Anatolia.

Remarks De Bruijn (1976) described one unerupted m1 and one M2 from the Turolian fauna of Kalithies (Rhodes Island) as *Byzantinia* sp. A, reporting that this species is different from *B. pikermiensis*, *B. nikosi* and *B. hellenicus* in having higher-crowned cheek teeth. De Bruijn (1989) also reported the presence of *Byzantinia* sp. in the latest Vallesian fauna of Lefkon, without more details and descriptions of the material.

***Byzantinia bayraktepensis* Ünay, 1980**

Type Material Left M3, BA1 no. 22 (holotype) (Ünay 1980: Pl. 1.4).

Type Locality Bayraktepe 1, Turkey, middle Miocene [late Astaracian].

Distribution Middle–late Miocene of Turkey and Greece.

Occurrences in Greece Chryssavgi 1.

Remarks According to Ünay's (1980) original diagnosis: the cheek teeth are relatively large and brachyodont; the upper molars show complete ectolophs; the M1 lacks a lingual spur on the anterolophule; the M3 shows a wider posterior part than in other *Byzantinia* species; the m1 bears either only an anterior metalophulid or a double one. One m1, two m2, one m3, one upper incisor and a partial M1 from the locality Chryssavgi 1 have been attributed to this species by Koliadimou (1996). The molars are brachyodont and of medium size. Comparisons with material from the species type locality confirmed great metrical and morphological similarities (Koliadimou 1996).

★*Byzantinia nikosi* de Bruijn, 1976

Type Material Right M3, no. 170 (holotype) (de Bruijn 1976: Pl. 1.9).

Type Locality Biodrak, Greece, Upper Miocene [late Vallesian (MN10)].

Distribution Middle–late Miocene of Turkey and Greece.

Occurrences in Greece Biodrak (Fig. 17a–f).

Remarks This *Byzantinia* species is of medium size, with brachyodont cheek teeth, complete ectolophs and a metacone that is not completely incorporated in the posteroloph on the upper molars and a not elongated posteriorly M3 with a complete endoloph and a shallow, narrow sinus (de Bruijn 1976). The type locality,

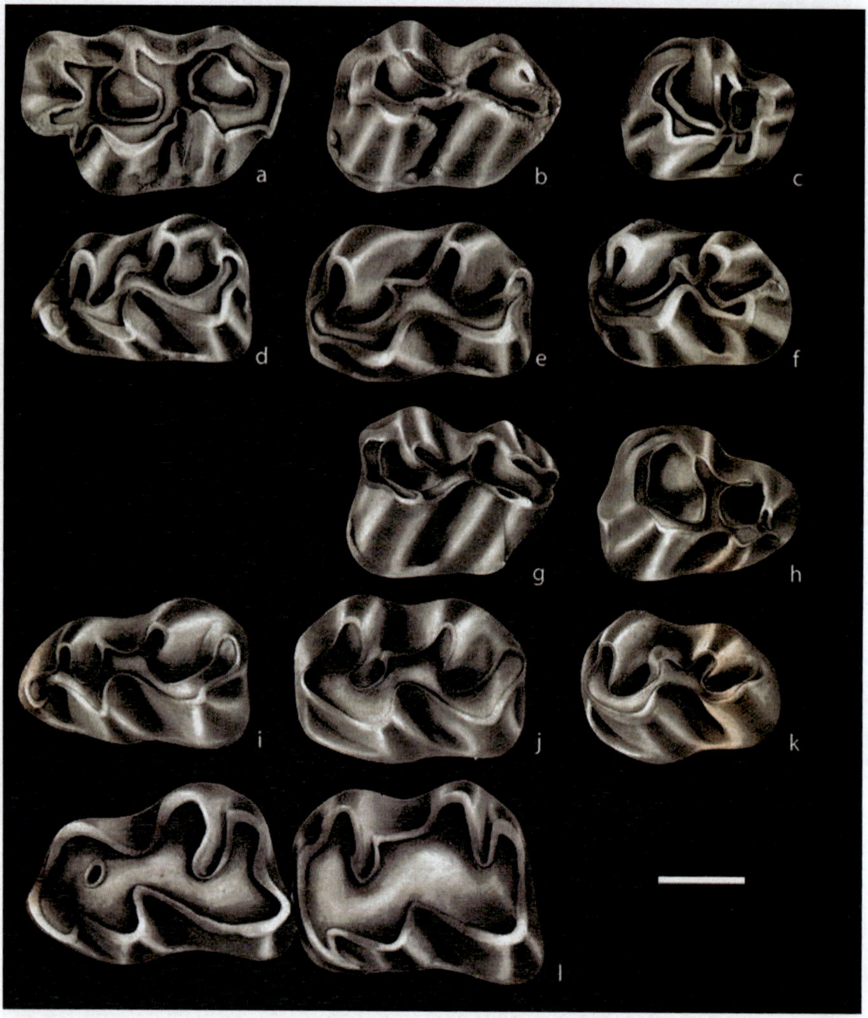

Fig. 17 *Byzantinia nikosi* from Biodrak (type locality). **a** M1 (holotype), **b** M2, **c** M3, **d** m1, **e** m2, **f** m3. *Byzantinia pikermiensis* from Pikermi–Chomateri (type locality). **g** M2, **h** M3 (holotype), **i** m1 **j** m2, **k** m3. *Byzantinia hellenicus* **l** m1–2. Images courtesy of H. de Bruijn. All specimens figured as left; a, c–f, l are reversed. Scale bar equals 1 mm

Biodrak, has yielded 10 molars (two m1, one m2, three m3, one M1, one M2, two M3), including the holotype. *B.* aff. *nikosi* has been listed in the faunal list from the Upper Miocene locality Katheni by de Bruijn and van der Meulen (1979), with no additional information.

★*Byzantinia pikermiensis* de Bruijn, 1976

Type Material Left M3, no. 118 (holotype) (de Bruijn 1976: Pl. 1.6).

Type Locality Pikermi (Chomateri), Greece, Upper Miocene [Turolian (MN12)].

Distribution Late Miocene of Greece and Turkey.

Occurrences in Greece Chomateri (Fig. 17g–k).

Remarks The original diagnosis by de Bruijn (1976) defines this medium-sized brachyodont *Byzantinia* species as having complete ectolophs on the M2–M3, four roots on the M2, a confluent with the posteroloph metaloph that incorporates the metacone on the M2 and a posteriorly elongated M3 with a small hypocone, a complete endoloph and a shallow, wide sinus. The species is represented in the type locality Chomateri by 15 molars (three m1, two m2, two m3, four M2, four M3), which do not include any M1 (de Bruijn 1976). The M1 became known from small collections from the Turkish localities Çorakyerler (Ünay et al. 2006) and Altıntaş (de Bruijn et al. 2013b) and bears a lingual spur of the anterolophule and no funnel.

★***Byzantinia hellenicus* (Freudenthal, 1970)**

Nomenclatural and Taxonomical History *Ruscinomys hellenicus* in Freudenthal 1970 (new species); *Byzantinia hellenicus* in de Bruijn 1976 (new combination).

Type Material Partial right maxilla with M1–M3, AMNH 20766 (holotype), American Museum of Natural History New York (Freudenthal 1970: Fig. 1).

Type Locality Samos (Quarry 4), Greece, Upper Miocene, [middle Turolian (MN12)].

Distribution Late Miocene of Greece and Turkey.

Occurrences in Greece Samos S3 and Samos (Quarry 4; Fig. 17l).

Remarks Freudenthal (1970) gave the diagnosis based on the molars, which are moderately hypsodont, with continuous longitudinal lophs, smaller than those of *R. europaeus* and about similar in size with those of *R. schaubi*. The most distinctive element of the species is the M3, which, in comparison to the M1 and M2, is relatively longer than in any of the (then) known species of *Ruscinomys* or *Cricetodon* s.s. *B. hellenicus* is much bigger than all other *Byzantinia* species. Besides the type material, the species is also represented in Greece by two partial skulls, one left upper jaw with M1–M3, three partial lower jaws with molars, seven m1, seven m2, four m3, six M1, five M2 and one M3 from the locality Samos S3 (Black et al. 1980:371). Based on this material, Black et al. (1980) gave more morphological features: on the M3 "*the metacone is displaced postero–internally so that there is only a faint groove on the posterior face of the tooth between it and the hypocone*"; the m1 and m2 are almost equal in length and the m3 is only slightly shorter; on the m1 there is a well-developed anteroconid connected to the protolophid in its middle, all cusps bend forward and all crests are anteriorly directed, a small short mesolophid closes off the anterior fossettid well bellow the top of the metaconid whereas the posterior fossettid closes near the base of the entoconid, the hypoconid bears a short posterior crest that ends in a distinct hypoconulid and the two buccal and the one lingual valleys are very deep; on the m2 there is a small shelf near the base of the crown on the antero-buccal side, the metalophid projects strongly anteriorly, the mesolophid and posterolophid are short and there is a distinct hypoconulid and the buccal and the lingual valleys are deep as on the m1; the anterior half of the m3 is similar to that of the m2, whereas the posterior one is compressed laterally so that the posterior fossettid forms a narrow but deep notch. One lower jaw of this species from Samos has been erroneously identified as *Cricetodon magnum* by Bakr (1959; Black et al. 1980). Black et al. (1980) noticed that *B. hellenicus* is nearly identical to *B. pikermiensis*, but significantly larger.

Subfamily Cricetinae Fischer von Waldheim, 1817

Type Genus *Cricetus* Leske, 1779.

Remarks A very large number of small- to medium-sized genera have been described to include fossil hamster species from Miocene–Upper Pleistocene Eurasian localities, as well as extant Eurasian species. Very often, the dental morphology of these animals is quite similar, a fact that hampers generic and specific identification and the establishment of phylogenetic relationships. It is very possible that some of the genera will be proven to be synonymous after more careful study of fossil type materials and dental material of recent animals. As de Bruijn et al. (2012) pointed out: "*a revision of the late Neogene to Recent small and medium-sized hamsters based on dental morphology is badly needed*".

Hamster dental elements from the Upper Miocene locality Plakias and the Biharian fauna of Tourkobounia 3 and 4 have been attributed to Cricetinae indet. (de Bruijn et al. 2012 and Reumer and Doukas 1985, respectively). The teeth from Plakias show a unique mixture of primitive and advanced characters that cannot be seen in any cricetine genus known; however de Bruijn et al. (2012) refrain from naming a new taxon due to the scarcity of the material (five molars in total). The material from Tourkobounia 3 and 4 has not been described.

***Democricetodon* Fahlbusch, 1964**

Type Species *Democricetodon crassus* Freudenthal, in Freudenthal and Fahlbusch, 1969.

Other Taxa Included *D. affinis* (Schaub, 1925); *D. brevis* (Schaub, 1925); *D. gaillardi* (Schaub, 1925); *D. vindobonensis* (Schaub and Zapfe, 1953); *D. romieviensis* (Freudenthal, 1963); *D. freisingensis* (Fahlbusch, 1964); *D. gracilis* (Fahlbusch, 1964); *D. mutilus* (Fahlbusch, 1964); *D. franconicus* (Fahlbusch, 1966); *D. hispanicus* Freudenthal, 1967; *D. sulcatus* Freudenthal, 1967; *D. affinis cretensis* de Bruijn and Meulenkamp, 1972; *D. nemoralis* Agusti, 1981; *D. kohatensis* Wessels et al., 1982; *D. hasznosensis* Kordos, 1986; *D. iazygum* Radulescu and Samson, 1988; *D. zarandicus* Radulescu and Samson, 1988; *D. walkeri* Tong and Jaeger, 1993; *D. lindsayi* Qiu, 1996; *D. tongi* Qiu, 1996; *D. kaonou* Mein and Ginsburg, 1997; *D. hanniae* Aguilar et al., 1999; *D. sudrei* Aguilar et al., 1999; *D. anatolicus* Theocharopoulos, 2000; *D. doukasi* Theocharopoulos, 2000; *D. fourensis* Maridet et al., 2000; *D. moralesi* van der Meulen et al., 2003; *D. suensis* Qiu, 2010; *D. haltmari* Joniak et al., 2017.

Distribution Early–late Miocene (MN1–9) of Europe and Asia.

Remarks Most of the *Democricetodon* species show great intraspecific morphological variation, and many have been named based on small collections. These facts have resulted in a great number of species, some of which are considered to be synonymous; for instance, *D. freisingensis* is considered to be a junior synonym of *D. gaillardi* by Theocharopoulos in de Bruijn et al. (2003). The genus is in need of a revision (as in, e.g., Maridet 2003, van der Meulen et al. 2003), but as we do not aim to do that in this work, we here list all species assigned to the genus so far (see Appendix).

One m2 and two m3 of *Democricetodon* have been reported from the Lower Miocene locality Lapsarna by Vasileiadou and Zouros (2012). They show metrical and morphological similarity with *D. gracilis*, *D. franconicus* and *D. doukasi*, but as the assemblage does not include the diagnostic first molars, the authors refrained from giving a specific identification and attributed the material to *Democricetodon* sp. Furthermore, one *Democricetodon* species is reported from the locality Thymiana C (*Democricetodon* n. sp.) and three from Thymiana A (*D. gracilis*, *Democricetodon* n. sp. I and *Democricetodon* n. sp. II) by de Bonis et al. (1997), but there is no information on the material.

***Democricetodon franconicus* (Fahlbusch, 1966)**

Nomenclatural and Taxonomical History *Democricetodon minor franconicus* in Fahlbusch 1966 (new subspecies); *Copemys* (*Democricetodon*) *minor franconicus* in Fahlbusch 1967 (new combination); *Democricetodon franconicus* in Fejfar 1974 (elevation to species level).

Type Material Left m1, Slg. München 1962 XIX 20 (holotype) (Fahlbusch 1966:Taf. 10.21).

Type Locality Erkertshofen 1, Germany, Lower Miocene [MN4].

Distribution Early–middle Miocene of Europe and Turkey.

Occurrences in Greece Aliveri, Karydia and Antonios.

Remarks According to the original diagnosis in German by Fahlbusch (1966) and its translation by van der Meulen et al. (2003), this species shows the following features: mesolophid of the m1–2 and mesoloph of the M1–2 mostly reaching the molar border, semi-circular anterior cingulum and weak anteroconid on the m1, no posterior metalophulid nor mesolophid on the m3; posteriorly directed proto- and metalophule on the M1, double protolophule and single, mostly transverse metalophule in the M2 and rounded M3 with a strongly reduced posterior part.

This species has been described from three Lower Miocene Greek localities: Aliveri, Karydia and Antonios. The large assemblage from Aliveri (255 molars in total) described by Klein Hofmeijer and de Bruijn (1988) comes from the South Quarry. The smaller size of the Aliveri molars and some morphological deviation from the molars in the type locality Erkertshofen 1 (e.g., mesolophs/mesolophids of the Aliveri molars shorter than those of the Erkertshofen 1 ones) have been attributed by Klein Hofmeijer and de Bruijn (1988) to the more "primitive" stage of evolution of the Aliveri species and, consequently, the older age of the Aliveri species compared to that of the Erkertshofen 1 species. Theocharopoulos (2000) tested the homogeneity of the Aliveri material using the "V" coefficient of variation introduced by Freudenthal and Cuenca Bescós (1984) and, due to the results that proved the Aliveri "*D. franconicus*" material to be non-homogenous, allocated most of the molars to *D. gracilis*, leaving only 99 molars to *D. franconicus*. However, he mentioned the great difficulty in allocating each molar to one of the two species due to the similarities in size and morphology and admitted that some of the molars might have been included in the wrong species. Theocharopoulos (2000) also assigned a sample of 72 molars from Karydia to *D. franconicus*, mentioning that the stage of evolution of the Karydia species seems to be similar with that of the Aliveri species, hence more "primitive" and probably older than the type material of *D. franconicus*. The most recent review of the Karydia material led to the verification of the presence of the species (Wessels 2020 personal communication). Finally, Vasileiadou and Koufos (2005) assigned to this species 45 molars from Antonios. The authors mentioned the large size variation within the sample, which is not accompanied, however, by morphological variation.

Democricetodon gracilis **(Fahlbusch, 1964)**

Nomenclatural and Taxonomical History *Democricetodon* (*Democricetodon*) *minor gracilis* in Fahlbusch 1964 (new subspecies); *Democricetodon gracilis* in Fejfar 1974 (elevated to species level).

Type Material Right m1, BSPG 1959 II 247 (holotype), Bavarian State Collection for Paleontology and Geology in Munich (Fahlbusch 1964: Abb. 10a; Taf. 2.9).

Type Locality Sandelzhausen bei Mainburg, Germany, middle Miocene [MN5].

Distribution Early–middle Miocene (MN4–6) of Europe, Russia and Turkey.

Occurrences in Greece Aliveri, Karydia and Thymiana A.

Remarks According to the original diagnosis by Fahlbusch (1964) (in German) and the English translation by Wessels and Reumer (2009), this species is characterized by the smaller size and more elegant appearance than *D. minor*, the somewhat concave crown sides near the labial and lingual sinusids in most lower molars, the very short anteroconid that forms a flat triangle on most m1s, the very short to half long at most mesolophid and the often anteriorly curved metaconid on the lower molars, the mostly undulant labial crown side and the long mesoloph that often reaches the tooth border on the upper molars and the relatively narrow and undivided anterocone on the M1. Wessels and Reumer (2009) studied a very large collection of molars from the type locality (767 molars versus 15 molars that were included in the original description by Fahlbusch 1964) and gave a general description of the species, based on the most common occurrences of some characteristics. According to these authors, the mesoloph is of medium length or long on the M1, long or connected to the border on the M2, the mesolophid is of medium length or short on the m1 and short on the m2; the protolophule is simple on the M1 and double on the M2; the metalophule is posterior on the M1 and anterior or transverse on the M2.

This species has been described from Aliveri and Karydia by Theocharopoulos (2000) and mentioned in the faunal list of Thymiana A by de Bonis et al. (1997) without details on the material. Even though the species had already been reported in Aliveri by Theocharopoulos (2000) (see Remarks of *D. franconicus*), van den Hoek Ostende et al. (2015), in the most recent review of the Aliveri fauna, did not mention it in their faunal list (van den Hoek Ostende et al. 2015:Tab 1). Wessels and Reumer (2009) noticed that only *D. gracilis* from Aliveri has a double protolophule in most M1 (75%), which could be a "primitive" feature and is explained by the fact that the Aliveri *gracilis* is the oldest described until now. The most recent faunal list for Karydia (Wessels 2020 personal communication) does not include the species, but the molars previously assigned to it by Theocharopoulos (2000) are now assigned to *Democricetodon* sp.

***Democricetodon gaillardi* (Schaub, 1925)**

Nomenclatural and Taxonomical History *Cricetodon gaillardi* in Schaub 1925 (new species); *Democricetodon gaillardi* in Fahlbusch 1964 (new combination).

Type Material (Lectotype) fragment of left mandible with m1–m3, SMNS 43551, State Museum of Natural History Stuttgart (Schaub 1925: Pl. 1.9).

Type Locality Steinheim, Germany, middle Miocene [MN7/8].

Distribution Early–middle Miocene of Europe, Russia and Turkey.

Occurrences in Greece Karydia.

Remarks According to Maridet's (2003) summary, the diagnosis by Schaub (1925) includes the following features: the M2 is relatively larger than in *D. affinis*, the upper molars bear a strong mesoloph that ends at the molar border with a small elevation, a cingulum longer than in *D. affinis* that closes the transverse labial valley, posteriorly directed proto- and metaloph but also always preserved anteriorly directed proto- and metaloph ("primitive" lophs); the m1 is somewhat elongated with a very strong anteroconid; the m3 is almost as long as the m2, and the posterior part is reduced in comparison with that in *D. affinis*; the mesolophid is long and ends on the edge of the tooth with a small cuspulid; the labial sinusid is symmetrical on the m1–2 and is directed labially from the anterior arm of the hypoconid. 11 molars from Karydia have been allocated to *D.* cf. *gaillardi* by Theocharopoulos (2000). According to the author, the Karydia molars are smaller than those from the type locality, the anterior protoloph is stronger than the posterior one in all four M2 and the metaloph is more often anteriorly directed (anteriorly directed in three and transverse in one M2) than in the type material. The Karydia material is interpreted by Theocharopoulos (2000) as more "primitive" than that from the type locality. However, these molars are now assigned to *Democricetodon* sp. (Wessels 2020 personal communication).

Democricetodon anatolicus **Theocharopoulos, 2000**

Type Material Left m1, HA 3 252 (holotype) (Theocharopoulos 2000: Pl. 6.2).

Type Locality Harami 3, Turkey, Lower Miocene [MN1].

Distribution Early Miocene of Turkey and Greece.

Occurrences in Greece Karydia.

Remarks Theocharopoulos (2000) gave the species original diagnosis as follows: "*Low–crowned and small* Democricetodon *with blade–like anteroconid that is situated very close to the anterior cusps; a posterior paracone–spur is frequently present in the upper molars; the posterior arm of the protoconid is connected to the metaconid in some m1*". He also mentioned that *D. anatolicus* is smaller than all other *Democricetodon* species. Indeed, also the species described after 2000 are larger than *D. anatolicus*. This species is known only with small assemblages from five Turkish localities (apart from the type locality) and Karydia. The Karydia molars (twelve in total) are somewhat smaller than the ones from the type locality, but morphologically similar to them (Theocharopoulos 2000). Since all known collections are small (the largest being that of Kılçak 3a with 29 molars), the intraspecific metrical and morphological variation is not well-known.

Karydomys **Theocharopoulos, 2000**

Type Species *Karydomys symeonidisi* Theocharopoulos, 2000.

Other Taxa Included *K. zapfei* (Mein and Freudenthal, 1971a); *K. boskosi* Theocharopoulos, 2000; *K. dzerzhinskii* Kordikova and de Bruijn, 2001; *K. wigharti*

Mörs and Kalthoff, 2004; *K. debruijni* Maridet et al., 2011; *K. strati* López-Antoñanzas et al., 2019.

Distribution Early–middle Miocene of Eurasia.

Remarks This genus was created by Theocharopoulos (2000) to include species similar to *Democricetodon* but different in the more inflated cusps, the better developed ectolophs on the upper molars and the very small anteroconid on the m1.

★*Karydomys symeonidisi* Theocharopoulos, 2000

Type Material Right m1, KRD 541 (holotype) (Theocharopoulos 2000: Pl. 11.1).

Type Locality Karydia 1, Greece, Lower Miocene [MN4].

Distribution Early Miocene of Greece (only in Karydia).

Remarks According to the original diagnosis, this species' molars have the following characteristics: they are larger than those of *K. boskosi* and smaller than those of *K. zapfei*; their cusps are conical; the M1 and m1 show "inflated" cusps; the M1 shows a metalophule that ends against the posteroloph; the metalophule may be absent in the M2; the mesoloph on the upper molars and the mesolophid on the lower ones are long and narrow. This species is only known from its type locality, with 27 molars from the type-level Karydia 1 and 32 molars (Fig. 18a–f) from the layer Karydia 2 (Theocharopoulos 2000).

★*Karydomys boskosi* Theocharopoulos, 2000

Type Material Left m1, KRD 2192 (holotype) (Theocharopoulos 2000: Pl. 9.13).

Type Locality Karydia 2, Greece, Lower Miocene [MN4].

Distribution Early Miocene of Greece (only in Karydia).

Remarks This is the smallest *Karydomys* species, with slender M1 and m1, both M1 and M2 bear a metalophule that ends against the posteroloph and the lophs on the upper molars are short and wide (Theocharopoulos 2000). It is known with 19 molars from the type-level Karydia 2 and one molar from the layer Karydia 1 (Fig. 18g–i).

★*Karydomys strati* López-Antoñanzas et al., 2019

Type Material Left M2, UM Thy 0–30 (holotype), University of Montpellier, France (López-Antoñanzas et al. 2019: Fig. 3B, I–L).

Type Locality Thymiana 0, Chios Island, Greece, late Lower Miocene [MN4].

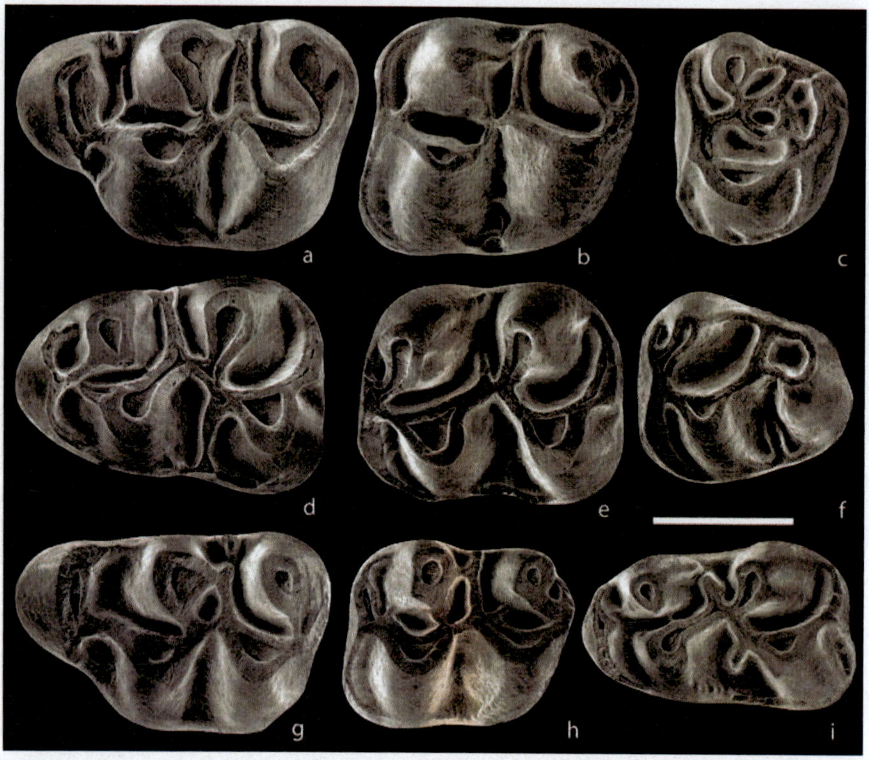

Fig. 18 *Karydomys symeonidisi* from Karydia (type locality). **a** M1, **b** M2, **c** M3, **d** m1 (holotype), **e** m2, **f** m3. *Karydomys boskosi* from Karydia (type locality). **g** M1, **h** M2, **i** m1 (holotype). Images courtesy of C. Doukas. Scale bar equals 1 mm

Distribution Late early Miocene of Greece (only type locality).

Remarks According to the diagnosis, the M1–2 bear a posteriorly directed spur on the paracone and long and narrow mesolophs, and the M1 shows a labial spur on the anterolophule but lacks both anterior protolophule and metalophule, whereas the M2 has double protolophule and metalophule; the lower molars bear an ectomesolophid and long and narrow mesolophids (López-Antoñanzas et al. 2019). The material from the type locality is very limited, consisting only of one M1, three M2, one m1, two m2 and one fragmented m3 (including the holotype). It is thought to be "less evolved" than the other two Greek *Karydomys* species (*K. symeonidisi* and *K. boskosi*) due to the presence of an anterior metalophule on the M2, which is absent in these two species (López-Antoñanzas et al. 2019:43).

Megacricetodon Fahlbusch, 1964

Type Species *Cricetodon gregarious* Schaub, 1925.

Other Taxa Included *M. minor* (Lartet, 1851); *M. bourgeoisi* (Schaub, 1925); *M. ibericus* (Schaub, 1944); *M. lappi* (Mein, 1958); *M. collongensis* (Mein, 1958); *M. primitivus* (Freudenthal, 1963); *M. crusafonti* (Freudenthal, 1963); *M. bavaricus* (Fahlbusch, 1964); *M. schaubi* Fahlbusch, 1964; *M. similis* (Fahlbusch, 1964); *M. minutus* Daxner, 1967; *M. debruijni* Freudenthal, 1968; *M. bezianensis* Bulot, 1980; *M. germanicus* Aguilar, 1980; *M. gersii* Aguilar, 1980; *M. sinensis* Qiu et al., 1981; *M. freudenthali* García Moreno (in Álvarez-Sierra and García Moreno 1986); *M. lopezae* García Moreno (in Álvarez-Sierra and García Moreno 1986); *M. roussillonensis* Aguilar, Calvet and Michaux, 1986; *M. crisiensis* Radulescu and Samson, 1988; *M. aguilari* Lindsay, 1988; *M. sivalensis* Lindsay, 1988; *M. daamsi* Lindsay, 1988; *M. mythikos* Lindsay, 1988; *M. rafaeli* Daams and Freudenthal, 1988; *M. fournasi* Aguilar, 1995; *M. lemartineli* Aguilar, 1995; *M. pusillus* Qiu, 1996; *M. wae* Aguilar, Clauzon and Michaux, 1999; *M. fahlbuschi* Aguilar, Clauzon and Michaux, 1999; *M. lalai* Aguilar, Clauzon and Michaux, 1999; *M. dzhungaricus* (Kordikova and de Bruijn, 2001); *M. robustus* Kälin and Engesser, 2001; *M. andrewsi* Peláez-Campomanes and Daams, 2002; *M. aunayi* Lazzari and Aguilar, 2007; *M. tautavelensis* Lazzari and Aguilar, 2007; *M. yei* Bi et al., 2008; *M. beijiangensis* Maridet et al., 2011; *M. vandermeuleni* Oliver and Peláez-Campomanes, 2013; *M. hellenicus* Oliver and Peláez-Campomanes, 2016.

Distribution Miocene of Europe and Asia.

Remarks The huge number of species attributed to this genus (39) suggests that the genus might be over-split. Usually, the attribution of a *Megacricetodon* assemblage to a species is based on the predominance of certain morphotypes mainly of the two first upper and lower molars. This means that small assemblages cannot be identified with certainty. An assemblage from the Lower Miocene locality Antonios (10 m1, eight m2, three m3, six M1, seven M2, one M3) had initially been identified as *M. primitivus* (Vasileiadou and Koufos 2005), but was later mentioned as *Megacricetodon* sp. by Oliver and Peláez-Campomanes (2016). Additional material from the locality that is available to the authors will soon give more information on the assignment to a species. A single M3 from the locality Rema Patsiaris has been attributed to ?*Megacricetodon* sp. by Sylvestrou (2002).

★*Megacricetodon hellenicus* Oliver and Peláez-Campomanes, 2016

Type Material Right m1, IAUU AL–N–2096 (holotype), Institute of Earth Sciences, Utrecht University, The Netherlands (Oliver and Peláez-Campomanes 2016: Fig. 10).

Type Locality Aliveri (North Quarry), Greece, Lower Miocene [MN4].

Distribution Known only from the type locality.

Remarks According to the diagnosis by Oliver and Peláez-Campomanes (2016), this is a small-sized *Megacricetodon* species with slender crests and mesolophs/-ids generally of medium length. The m1 bears a rounded and simple anteroconid, lower than the four main cusps. The upper molars bear a long and angular entoloph and transverse sinus. The *Megacricetodon* assemblage from Aliveri (eight m1, 11 m2, seven m3, 17 M1, seven M2 and seven M3 in total from South and North Quarry) had been attributed to *M. primitivus* by Klein Hofmeijer and de Bruijn (1988). However, Oliver and Peláez-Campomanes (2016) differentiated the two species through the deeply split anterocone, shorter mesoloph and short entoloph on the M1, a higher percentage of absent ectoloph on the M2, a short entoloph on the M3, the same height of all cusps on the m1, shorter mesolophids on the m1–2 and the shorter lingual anterolophid on the m2 of *M. primitivus* than on the molars of *M. hellenicus*.

Megacricetodon primitivus **(Freudenthal, 1963)**

Nomenclatural and Taxonomical History *Cricetodon minor primitivus* in Freudenthal 1963 (new subspecies); *Democricetodon (Megacricetodon) minor primitivus* in Fahlbusch 1964 (new combination); *Megacricetodon minor primitivus* in Freudenthal 1965 (new combination); *Megacricetodon primitivus* in Bulot 1980 (erected to species).

Type Material Right m1–m3 (holotype), Museum Sabadell 56–142 (Freudenthal 1963:Pl. 1).

Type Locality Valtorres, Spain, Lower Miocene [MN4].

Distribution Early–middle Miocene of W and SE Europe and Turkey.

Occurrences in Greece Thymiana A and C.

Remarks The original diagnosis by Freudenthal (1963), as translated in English and emended by Daams and Freudenthal (1988), mentioned: "*Three out of 36 m1 have an ectomesolophid; in five out of 15 m3 a weak mesolophid or a remnant of the posterior metalophulid is present; in other respects this species agrees with* M. minor". According to the emended diagnosis by Oliver and Peláez-Campomanes (2014), this is a small *Megacricetodon* species showing a rounded and simple anteroconid, a short anterolophulid and usually no ectomesolophid on the m1; the mesoloph is present only rarely on the M3; the M1 always bears a double anterocone, normally with a small platform in front of the splitting groove; a strong lingual mesocingulum connecting the hypocone and the anterocone is present in more than 20% of the M1, whereas the mesoloph is usually short to medium. The presence of *M. primitivus* in the middle Miocene localities Thymiana A and C is mentioned in faunal lists by de Bonis et al. (1997), without additional information. According to Oliver and Peláez-Campomanes (2014), *M. primitivus* is restricted to the Iberian Peninsula and France, and all other assemblages attributed to this species differ in such a degree that they can be considered as different species. These authors have already shown that what was initially identified as *M. primitivus* from Aliveri by Klein Hofmeijer and de Bruijn (1988) belongs to *M. hellenicus*, whereas the material that was described as *M. primitivus* from Antonios by Vasileiadou and Koufos (2005) does not belong to this species. The species might not be present in Greece at all. The newest faunal list for the locality Karydia includes *M.* aff. *primitivus*, but no information on the material is yet available (Wessels 2020 personal communication).

Megacricetodon minor (Lartet, 1851)

Nomenclatural and Taxonomical History *Cricetodon minus* in Lartet 1851 (new species); *Megacricetodon minor* in Freudenthal 1965 (new combination).

Type Material Left m1–m3, Ss 1235 (holotype), Museum Basil, Switzerland.

Type Locality Sansan, France, middle Miocene [MN6].

Distribution Middle–late Miocene (MN5–9) of Europe and Turkey.

Occurrences in Greece Chryssavgi 1.

Remarks The diagnosis by Lartet (1851) only mentions that this species is smaller than a house mouse. This very short diagnosis for the Sansan species has created many controversies about which species Lartet was referring to, as many additional "*Cricetodon*" species started being identified from Sansan. Schaub (1925) gave the first accurate description of the species, according to which, the m1 is long and narrow and bears a simple anteroconid, the m3 bears four cusps, a reduced entoconid and a not reduced posterior part, all lower molars show an anteriorly directed metalophulid and a mesolophid of variable length, the M1 has an asymmetrically split anterocone far from the protocone and posteriorly directed proto- and metalophule, the M2–3 have anteriorly directed proto- and metalophule, the M2 shows traces of a posterior protolophule, the M3 is very short and rounded and the mesoloph of all upper molars is of variable length (from Daams and Freudenthal 1988). For a detailed revision on the nomenclatural and taxonomical history of the species, see Maridet and Şen (2012).

One m1 and one M1 from the middle Miocene locality Chryssavgi 1 have been attributed to this species by Koliadimou (1996). The two molars have been compared with a number of species and were found to be similar in some features and different in some others to *M. minor*, *M. debruijni* and *M. primitivus*. Our opinion is that an assemblage of just two molars is far too small to be attributed to any species and the Chryssavgi 1 taxon should be referred to *Megacricetodon* sp. However, here, we retain the initial identification of Koliadimou.

***Cricetulodon* Hartenberger, 1965**

Type Species *Cricetulodon sabadellensis* Hartenberger, 1965.

Other Taxa Included *C. hartenbergeri* (Freudenthal, 1967); *C. cretensis* (de Bruijn and Meulenkamp, 1972); *C. meini* (Agusti, 1986); *C. lucentensis* (Freudenthal et al., 1991); *C. complicidens* Topachevski and Scorik, 1992; *C. bugesiensis* Freudenthal et al., 1998.

Distribution Late Miocene of Spain and France, possibly Greece, Turkey and Ukraine (Kälin 1999).

Remarks A small cricetid from the late Vallesian fauna of Kastellios 1 was initially attributed to *Cricetulodon* cf. *sabadellensis* by de Bruijn et al. (1971), but after the recovery of a few more molars to *Cricetulodon* sp. by de Bruijn and Zachariasse (1979), who commented that the total material is "*too poor a basis for a reliable assignment*" and that the distinction of *Cricetulodon* from *Kowalskia* (now *Neocricetodon*) is "*hazardous unless adequate material is available*".

★*Cricetulodon cretensis* (de Bruijn and Meulenkamp, 1972)

Nomenclatural and Taxonomical History *Democricetodon affinis cretensis* in de Bruijn and Meulenkamp 1972 (new subspecies); *Cricetulodon cretensis* in de Bruijn et al. 2012 (new combination and referral to the species level).

Type Material Left M1, PL. no. 23 (holotype) (de Bruijn and Meulenkamp 1972: Pl. 1.4).

Type Locality Plakias, Crete, Greece, Upper Miocene [early Vallesian (MN9)].

Distribution Known only from the type locality (Fig. 19).

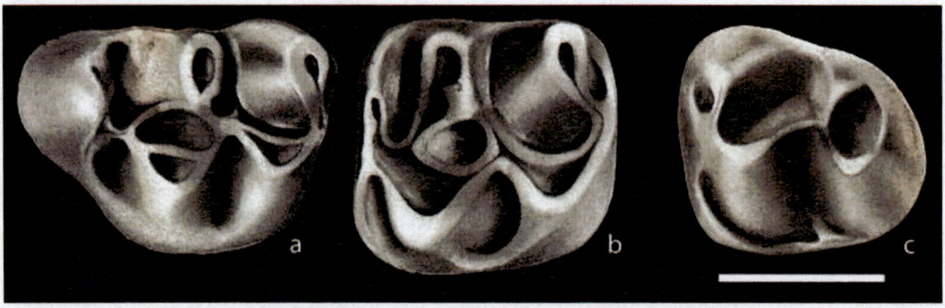

Fig. 19 *Cricetulodon cretensis* from Plakias (type locality). **a** M1 (holotype), **b** M2, **c** m3. Images courtesy of H. de Bruijn. All specimens figured as left; b is reversed. Scale bar equals 1 mm

Remarks The poor material from Plakias was initially described as a *Democricetodon* subspecies, but later tentatively transferred to *Cricetulodon*. De Bruijn et al. (2012) mentioned that "*the material, enigmatic though it may be, is inadequate to properly define an unspecialized hamster species*". According to the same authors, the generic allocation of the small hamster from Plakias is problematic mainly because: "(1) the anterocone of the M1 is narrow, basically undivided and situated labially and (2) the protolophule 2 of the M1–2 is stronger than the protolophule 1 and connects the paracone to the anterior arm of the hypocone instead of to the posterior side of the protocone as in most cricetines". They temporarily allocated the material to *Cricetulodon*, which, occasionally, shows the above characters. Compared with the smallest *Cricetulodon*, *C. hartenbergeri*, the Plakias species has somewhat smaller teeth, a narrower anterocone on the M1 and a stronger paracone–hypocone connection on the M1–2.

***Neocricetodon* Schaub, 1934**
(= *Epicricetodon* Kretzoi, 1951, *Neocricetodon* Kretzoi, 1951, *Kowalskia* Fahlbusch, 1969, *Karstocricetus* Kordos, 1987, according to Freudenthal et al. 1998)

Type Species *Cricetulus grangeri* Young, 1927.

Other Taxa Included (according to Sinitsa and Delinschi, 2016) *N. schaubi* (Kretzoi, 1951); *Neocricetodon ? lavocati* (Hugueney and Mein, 1965); *N. polonicus* (Fahlbusch, 1969); *N. magnus* (Fahlbusch, 1969); *N. intermedius* (Fejfar, 1970); *N. fahlbuschi* (Bachmayer and Wilson, 1970); *N. moldavicus* (Lungu, 1981); *N. occidentalis* (Aguilar, 1982); *N. skofleki* (Kordos, 1987); *N. yinanensis* (Zheng, 1984); *N. nestori* (Engesser, 1989); *N. polgardiensis* (Freudenthal and Kordos, 1989); *N. browni* (Daxner-Höck, 1992); *N. progressus* (Topachevski and Skorik, 1992); *N. similis* (Wu, 1991); *N. hanae* (Qui, 1994); *N. seseae* Aguilar, Calvet and Michaux, 1995; *N. ambarrensis* Freudenthal, Mein and Martín Suárez, 1998.

Distribution Late Miocene–Pliocene (MN10–15) of Europe and Asia.

Remarks De Bruijn (1976) reported the presence of *Neocricetodon* sp. (as *Kowalskia* sp.) in the Turolian fauna of Kalithies. The material includes only one damaged m1 and one complete m3 and, according to the author, may belong to the same species with *Neocricetodon* (*Kowalskia*) specimens from Pikermi (i.e., *N.* aff. *lavocati*). Furthermore, Daams and van de Weerd (1980) identified a poor assemblage from the Lower Pliocene locality Agios Ioannis (Karpathos Island) as *Neocricetodon* sp. (as *Kowalskia* sp.). De Bruijn (1989) mentioned the presence of *Neocricetodon intermedius* (as *Kowalskia intermedia*) in the latest Vallesian faunal list from Lefkon, without details on the material. However, Daxner-Höck (1995) attributed the same material to *Neocricetodon* sp. (as *Kowalskia* sp.), because it is too scanty for a species identification and shows metrical and morphological differences from *N. intermedius*.

***Neocricetodon lavocati* (Hugueney and Mein, 1965)**

Nomenclatural and Taxonomical History *Cricetulus lavocati* in Hugueney and Mein 1965 (new species); *Kowalskia lavocati* in Fahlbusch 1969 (new combination); *Neocricetodon lavocati* in Freudenthal et al. 1998 (new combination).

Type Material Left M3, FSL 65212 (originally identified as an M2) (Hugueney and Mein 1965: Pl. 2.61–65).

Type Locality Lissieu, France, late Upper Miocene.

Occurrences in Greece Chomateri.

Distribution Late Miocene–Pliocene of Europe.

Remarks The species was initially named by Hugueney and Mein (1965) on the basis of four isolated molars from Lissieu. Daxner-Höck et al. (1996), in a revision of the genera *Neocricetodon* and *Kowalskia*, refer to the species as *Kowalskia*? *lavocati*, mentioning that a definite generic attribution was not possible, due to the scarcity of the material from the type locality. Later, Freudenthal et al. (1998) ascribed to this species some more molars from the type locality, but did not describe or illustrate the elements. According to Sinitsa and Delinschi (2016), the generic attribution of the species *lavocati* is still uncertain due to the poor knowledge of the material from the type locality.

De Bruijn (1976) described a small sample (10 molars) from Pikermi (Chomateri) as *Kowalskia* aff. *lavocati*, after comparison with *Cricetulodon* and *Kowalskia* (now *Neocricetodon*) species and recent *Tscherskia triton*. He found the Pikermi material morphologically closer to *K. lavocati* than to any other species; however, he mentions "*the material from Pikermi is, notwithstanding its differences with known species, not described as a new species because the association is considered to be too small to give a proper picture of the variation*" (de Bruijn 1976:371). Ünay et al. (2006) created the species *Allocricetus aylasevimae* from the Upper Miocene locality Çorakyerler (Anatolia, Turkey) and mentioned that the cricetine material from Pikermi (Chomatri) is very similar dimensionally and morphologically to their species and may belong to it. Here, we retain the initial attribution by de Bruijn, since we have not seen the material in order to form our own opinion on its taxonomy.

★*Neocricetodon browni* (Daxner-Höck, 1992)

Nomenclatural and Taxonomical History *Kowalskia browni* in Daxner-Höck 1992 (new species); *Neocricetodon browni* in Freudenthal et al. 1998 (new combination).

Type Material Right Ml, MAA1 1343 (holotype) (Daxner-Höck 1992: Fig. 18.3).

Type Locality Maramena, Greece, Miocene/Pliocene [Turolian/Ruscinian boundary (MN13/14)].

Distribution Late Miocene/early Pliocene of Greece.

Occurrences in Greece Lava 2 and Maramena.

Remarks The material from the type locality was initially mentioned as *Kowalskia fahlbuschi* by de Bruijn (1989), but was later recognized as a new *Kowalskia* species by Daxner-Höck (1992). The diagnosis (translated from German) includes the following features: medium-sized *Kowalskia* species with a slightly elevated tooth crown, due to relatively high and massive cusps that project far above the tooth base and enclose narrow, deep valleys among them. The cusps are symmetrical and not leaning forwards or backwards. The anterocone of the M1 is broad and posteriorly divided into two parts. The M1 usually has 3 roots, rarely 4. The M3 is greatly reduced and varies in size and morphology of the posterior part. The mesoloph of the M1 is almost always long, of the M2 medium to long. A posterior labial sinus is small but mostly present on the M1–2. The ml is long and anteriorly mostly wide, with a 2–3-pointed anteroconid. The mesolophid is always present, on the ml–2 long, on the m3 short to long. An ectomesolophid and various secondary cuspids in the labial and lingual sinusids are possible. A shorter diagnosis was later given by Daxner-Höck (1995), according to which this species shows medium-sized molars with narrow sinuses and sinusids among the high and conical cusps, wide, deeply split posteriorly and rounded and smooth anteriorly anterocone/conid, long or sometimes of medium length (not connected to the metacone) mesoloph/lophid. The species is represented by a large collection from Maramena with (apart from the holotype) 10 fragmented jaws and 155 isolated molars (Daxner-Höck 1992). It is also listed in the faunal list for the late Turolian fauna of Lava 2 by de Bruijn et al. (1999), with no additional information.

Hypsocricetus Daxner-Höck, 1992

Type Species *Hypsocricetus strimonis* Daxner-Höck, 1992.

Distribution Late Turolian–Turolian/Ruscinian boundary of N Greece.

★*Hypsocricetus strimonis* Daxner-Höck, 1992

Type Material Left maxilla with M1–2, MAA1 1812 (holotype) (Daxner-Höck 1992: Fig. 20).

Type Locality Maramena, Greece, Miocene/Plioece [Turolian/Ruscinian boundary (MN13/14)].

Distribution Besides its type locality, this species might be present in Monasteri (Turolian) and Ano Metochi 2–3 (Turolian/Ruscinian boundary).

Remarks Initially mentioned in the faunal lists of the three N Greek localities by de Bruijn (1989) as "*Cricetus*" sp. A, the species was named by Daxner-Höck (1992) based on the rich material from Maramena. Its diagnosis is the same with that of the genus, since the latter is so far monospecific. The molars are medium-sized and slender. The high, pointed cusps project above the base of the sinuses by at least 2/3 of the total crown height and give the tooth a relatively hypsodont appearance. High ridges separate the labial from the lingual sinuses. The ml is narrow and bears a long, divided anteroconid; the gap between the symmetrical anteroconid cuspids usually reaches the anterior margin. Starting from the two anteroconid cuspids, two ridges converge posteriorly on the longitudinal ridge. A mesolophid is absent from the ml, it may occur on the m2, but remains short; on the m3 it is always present and usually of medium length. The M1 is characterized by 3 opposing cusp pairs that, together with their connecting ridges, enclose 3 "rhomboidal funnels" ("Rautentrichter" of Schaub 1930). The Ml has 4 roots. The mesoloph covers half of the sinus's length; it has the function of the anterior metaloph. On the M2–3, the labial anterolophulid dominates; the lingual is usually reduced. The Maramena collection is very large and contains (apart from the holotype) 22 jaw fragments and 538 isolated molars (Daxner-Höck 1995). De Bruijn (1989) placed the medium-sized cricetine from Maramena, Monasteri and Ano Metochi 2,3 in the same species ("*Cricetus*" sp. A), but Daxner-Höck (1995) suggested that the assignment of the Monasteri and the Ano Metochi 2–3 hamster to *H. strimonis* still has to be proven. No information on the material from these two localities has ever been published.

Mesocricetus **Nehring, 1898**

Type Species *Cricetus nigricans* Brandt, 1832 = *M. raddei* (Nehring, 1894).

Other Taxa Included (extant that are also found as fossils in Pleistocene assemblages) *M. auratus* (Waterhouse, 1839); *M. newtoni* (Nehring, 1898); *M. brandti* (Nehring, 1898); *M. aramaeus* Bate, 1943; (extinct) *M. primitivus* de Bruijn et al., 1970.

Distribution Late Miocene (MN13)–present of E Europe and W Asia.

Remarks Most of the extant species of *Mesocricetus* have been identified in Pleistocene Greek localities. Kuss and Storch (1978) described one m1, one m2, two m3 and one M1 of *M.* cf. *aramaeus* from the Biharian fauna of Kalymnos-Xi. However, van der Meulen and van Kolfschoten (1986) reported the presence of *M.* cf. *auratus* at the same locality, without giving details on the material, so it is not certain that the material they studied is that available to Kuss and Storch. *M.* cf. *brandti* has been reported from the Middle Pleistocene locality Latomi 1 (Storch 1975) and *M. newtoni* from the Upper Pleistocene localities Loutra Almopias Cave LAC and LAC Ia (Chatzopoulou 2014) and Arnissa (Mayhew 1978). Furthermore, Theocharopoulos (1991), in his unpublished diploma thesis, reported the presence of one m2 on *Mesocricetus* sp. from the early Ruscinian fauna from Spilia 1.

★*Mesocricetus primitivus* de Bruijn et al., 1970

Type Material Right m2, no. 113 (holotype) (de Bruijn et al. 1970: Pl. 4.10).

Type Locality Maritsa 1, Rhodes Island, Greece, Upper Miocene [Turolian (MN13)].

Distribution Late Miocene (MN13)–Pliocene (MN15) of Greece and Turkey.

Occurrences in Greece Maritsa 1 (Fig. 20b–c, e–f) and Nea Silata (Fig. 20a, d).

Remarks De Bruijn et al. (1970) named their species based on a collection of 14 small *Mesocricetus* molars, with the following diagnostic features: well-developed mesolophids in the m2–3 and relatively strong "*labial branch*" of the anterior cingulum in the M2–3. The authors added that the species differs from all extant *Mesocricetus* species due to its relatively short m2–3, the isolated position of the mesolophid in the m2–3 and the strong lingual cingulum in the M2–3. Van den Hoek Ostende et al. (2015) correctly noted that the "*labial branch of the anterior cingulum*" should be replaced by "*lingual branch of the anterior cingulum*" in the species' diagnosis, a character which is nevertheless mentioned by de Bruijn et al. (1970) in the discussion as an important difference from the extant species.

De Bruijn et al. (1970) did not attribute any M1 or m1 from Maritsa 1 to their new species. However, they described two m1, one M1 and one M3 as "? *Cricetulus* sp.". Based on the similarity of the Maritsa 1 m1 and M1 with molars of *M.* cf.

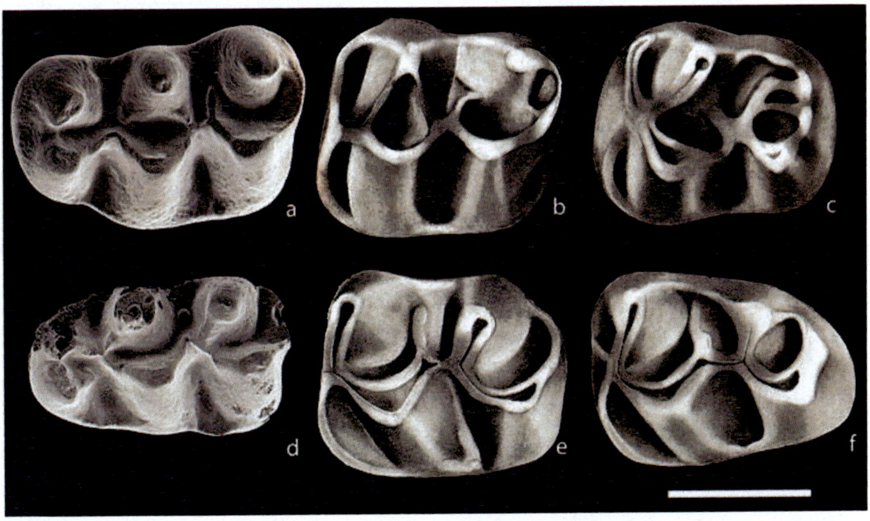

Fig. 20 *Mesocricetus primitivus*. **a** M1, **b** M2, **c** M3, **d** m1, **e** m2 (holotype), **f** m3 [a and d from Nea Silata (KV personal data), b–c and e–f from Maritsa 1 (type locality), courtesy of H. de Bruijn. All specimens figured as left; a, e–f are reversed. Scale bar equals 1 mm

primitivus from Çalta (Turkey), Şen (1977) attributed the Maritsa 1 molars to *M. primitivus*. Later, eight molars from Nea Silata were also attributed to this species, based on comparisons with the type material (Vasileiadou et al. 2003). Finally, one M1, one partial upper molar and three m2 from the locality Kessani were identified as *M.* cf. *primitivus* (Vasileiadou et al. 2012). The Kessani m2 is similar to those from the type locality and from Çalta. However, the Kessani M1 has a stronger anterior protolophule than the Çalta M1 does. Furthermore, the Kessani M1 is longer than the one from Maritsa 1, its anterocone cusps are closer to each other and not as deeply split anteriorly as on the Maritsa 1 one and the connections between the cusps are stronger (Vasileiadou et al. 2012). Due to the small assemblages attributed so far with certainty to this species, the intraspecific morphological and metrical variation is still unknown.

***Cricetulus* Milne-Edwards, 1867**

Type Species *Cricetulus griseus* Milne-Edwards, 1867 (extant).

Other Taxa Included Only extant species belong to the genus: *C. barabensis* (Pallas, 1773); *C. migratorius* (Pallas, 1773); *C. griseus* Milne-Edwards, 1867; *C. longicaudatus* (Milne-Edwards, 1867); *C. kamensis* (Satunin, 1903); *C. lama* Bonhote, 1905; *C. alticola* Thomas, 1917; *C. sokolovi* Orlov and Malygin, 1988 (see following paragraph: *Allocricetus*).

Distribution Pleistocene–present in E Europe and Asia.

Remarks The only species of *Cricetulus* reported from Greek fossil localities is *C. migratorius*. It has been found in the Biharian faunas of Tourkobounia 2 (van der Meulen and Doukas 2001), Volos (van der Meulen and van Kolfschoten 1986), in the Middle Pleistocene Latomi 1 (Storch 1975) and in the Upper Pleistocene Loutra Almopias Cave LAC and LAC Ia (Chatzopoulou 2014), Arnissa (Mayhew 1978) and Kitseli pothole (Tsoukala et al. 2006). *Cricetulus* cf. *migratorius* has been found in the Biharian fauna from Kaiafas (van der Meulen and van Kolfschoten 1986) and Tourkobounia 5 (Reumer and Doukas 1985).

***Allocricetus* Schaub, 1930**

Type Species *Allocricetus bursae* Schaub, 1930.

Other Taxa Included *A. ehiki* Schaub, 1930. We here follow the opinion of Hordijk and de Bruijn (2009: 63) that "*there are a number of other species of small cricetines that possibly belong in* Allocricetus. *However, the group seems to have been over-split on the generic level and needs to be revised. Moreover, a larger part of the fossil material of the small "modern" cricetines available in collections has not been studied. It is therefore impossible at this stage to give a comprehensive list of the species of* Allocricetus."

Distribution Latest Miocene–Late Pleistocene of Europe and Turkey.

Remarks Schaub (1930) established his new genus and separated it from *Cricetulus* on bases of cranial features. He noticed that the dentitions of the two genera are indistinguishable, whereas some differences proposed by Tchernov (1968) are not really functional due to intrageneric variation. Chaline (1972) treats *Allocricetus* as a subgenus of *Cricetulus*. Due to the difficulty in distinguishing *Allocricetus* from *Cricetulus*, Sickenberg (1971) attributed two m1 from the Middle Pleistocene locality Petralona to *Allocricetus* sp./*Cricetulus* sp. Further studies of dentitions of the *Cricetulus* extant species are necessary in order to understand the relationship between the two genera.

Allocricetus ehiki Schaub, 1930

Type Material Schaub (1930) neither designated a holotype for the species, nor did he mention the available material from the type locality. However, he gave measurements of a skull, lower teeth and humeri and figured the reconstruction of a skull with incisor and three molars (Schaub 1930: fig. 18) and one right lower tooth raw with m1–m3 (Schaub 1930: Tab I.10) from the type locality.

Type Locality Kalkberg bei Villány, Hungary, Pliocene [MN16].

Distribution Latest Miocene–Middle Pleistocene of W Europe and Asia.

Occurrences in Greece Komanos 1 low A and B and Komanos 1 high A and B.

Remarks Hordijk and de Bruijn (2009) attributed to this species one m1 and one M1 from Komanos 1 low A and B and one M1 from Komanos 1 high A and B, based on their morphological and metrical similarity with comparative material from the type locality. Four partial mandibles with incomplete dentitions and 42 isolated molars from Maramena have been assigned to *A.* cf. *ehiki* by Daxner-Höck (1992). All this Greek material shows the general dental pattern of a modern cricetine: reduced or absent mesoloph/-id, equally deep sinus/-id with no extra ridge, wide anterocone/-id smooth on the crown anterior side and split in two cusps connected through low ridges anteriorly and posteriorly, usually double connections between the cusp pairs forming funnels between the labial and lingual cusps mainly in the upper molars, weak anterolophulid, anteriorly directed meta- and hypolophid and oblique meta- and hypolophulid on the lower molars.

Cricetinus Zdansky, 1928

Type Species *Cricetinus varians* Zdansky, 1928.

Other Taxa Included *C. europaeus* Kretzoi, 1959; *C. gritzaii* Topachevski and Scorik, 1992; *C. beremendensis* Hir, 1994; *C. janossyi* Hir, 1996; *C. koufosi* Koliadimou, 1996; *C. mesolophidus* Wu and Flynn, 2017.

Distribution Early Pliocene–Middle Pleistocene of Europe and Asia.

Remarks This genus seems to be a rare element in the Plio–Pleistocene Eurasian faunas, known with small collections from a few localities with a great geographical distance among them. Its morphological features present a mixture of characters found also in other genera. More specifically, the m1 shows an undivided or weakly divided anteroconid and a very low single and longitudinal anterolophulid, features also found in *Mesocricetus* (Koliadimou 1996). Furthermore, the upper molars of *Cricetinus* resemble those of *Cricetulus* and *Allocricetus*, with no or weak mesolophs on the M1–2 and rare presence of a posterior metalophule on the M2 (Hir 1997). It is possible, thus, that the records of *Cricetinus* are a few because its elements have been attributed to other genera.

★*Cricetinus koufosi* Koliadimou, 1996

Type Material Left mandible with m1–2, RVL289 (holotype) (Koliadimou 1996:fig. 4.38.a), Museum of Geology-Paleontology-Paleoanthropology, Aristotle University of Thessaloniki.

Type Locality Ravin Voulgarakis, Greece, late Lower Pleistocene [Biharian (MNQ20)].

Distribution Biharian (MNQ19–MNQ20) of Greece.

Occurrences in Greece Marathoussa and Ravin Voulgarakis.

Remarks According to the diagnosis of this middle-sized *Cricetinus* species, given by Koliadimou (1996), it is characterized by the following features (translated from Greek): well-developed pre-anterocone, posterior metalophule and absence of "parametalophule" (an enamel crest between the metalophule and the posteroloph trimming the metacone on the lingual side, according to Hir 1994) on the M1; presence of posterior metalophule and rare parametalophule on the M2; absence of mesoloph on the M1 and presence on the M2–3; low anterolophulid on the m1; absence of mesolophid on the m1–2 and presence of small mesolophid on the m3. The material from the type locality (Fig. 21) includes (apart from the holotype) five m1, four m2, four m3, one partial maxilla with M1–2, 10 M1, eight M2 and four M3. A collection comprising 11 molars from the Lower Pleistocene locality Marathoussa (N Greece) has also been attributed to this species based on comparisons with the type material (Koufos et al. 2001).

***Cricetus* Leske, 1779**

Type Species *Cricetus cricetus* Linnaeus, 1758 (only extant species of genus).

Other Taxa Included *C. major* Woldrich, 1880; *C. runtonensis* Newton, 1909; *C. kormosi* Schaub, 1930; *C. nanus* Schaub, 1930; *C. praeglacialis* Schaub, 1930; *C. lophidens* de Bruijn et al., 1970.

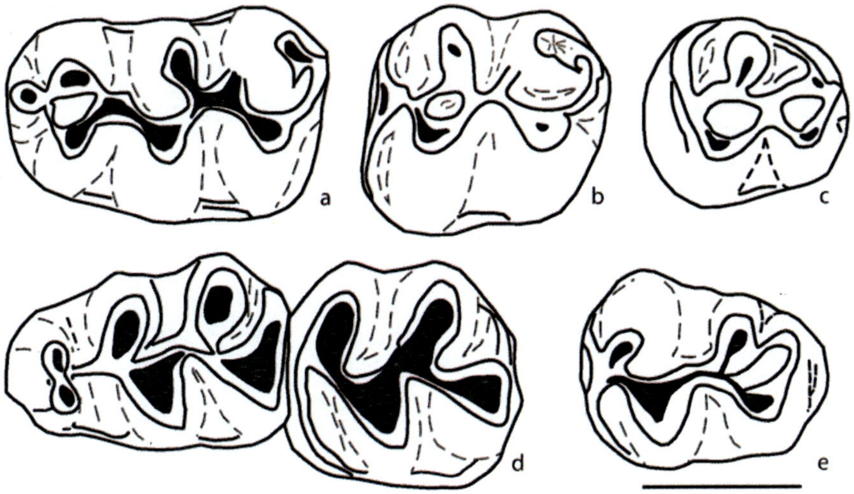

Fig. 21 *Cricetinus koufosi* from Ravin Voulgarakis (type locality). **a** M1, **b** M2, **c** M3, **d** m1–2 (holotype), **e** m3. Drawings courtesy of K. Koliadimou. All specimens figured as left; **a–c** are reversed. Scale bar equals 1 mm

Distribution Late Miocene–present of Europe and Asia.

★***Cricetus lophidens*** **de Bruijn et al., 1970**

Type Material Right M1, no. 16 (holotype) (de Bruijn et al. 1970: Pl. 4.3).

Type Locality Maritsa 1, Rhodes Island, Greece, Upper Miocene [Turolian (MN13)].

Distribution Late Miocene–early Pliocene of Greece and Turkey.

Occurrences in Greece Maritsa 1 (Fig. 22).

Remarks De Bruijn et al. (1970) mentioned the following diagnostic features for this small *Cricetus* species: the molars have the same size with those of *C. nanus* and *C. kormosi* and are high-crowned; well-developed ridges connect the labial cusps, tending to form an ectoloph. A feature separating this species from *C. nanus* and *C. kormosi* is the larger height of the ridges in relation to the cusps on the upper molars, interpreted as a primitive feature (de Bruijn et al. 1970). Another primitive feature, according to the authors, is the rounded anterior margin of the M1. On the M1, the labial cusps are connected through a longitudinal ridge, which is a unique characteristic for *C. lophidens*. Apart from the holotype, the type locality has yielded six m1, 20 m2, five m3, 10 M1, 17 M2 and 10 M3.

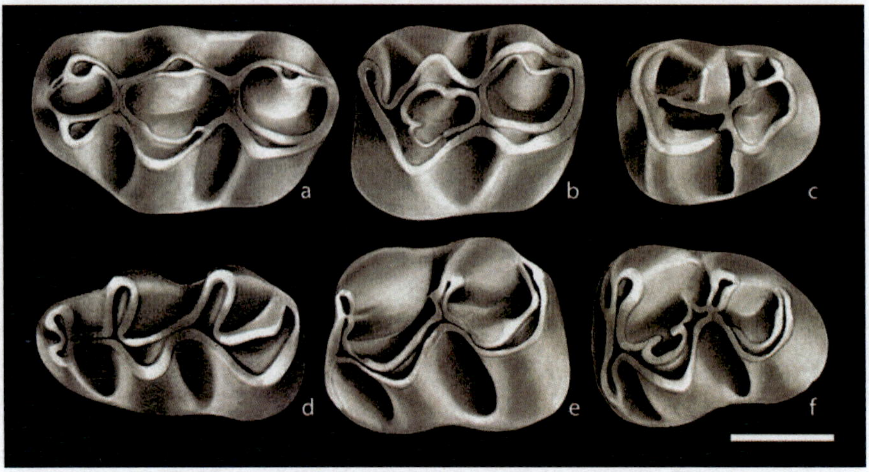

Fig. 22 *Cricetus lophidens* from Maritsa 1 (type locality). **a** M1 (holotype), **b** M2, **c** M3, **d** m1, **e** m2, **f** m3. Images courtesy of H. de Bruijn. All specimens figured as left; **a–c** and **e–f** are reversed. Scale bar equals 1 mm

Subfamily Myocricetodontinae Lavocat, 1961

Type Genus *Myocricetodon* Lavocat, 1952.

Calomyscus Thomas, 1905

Type Species *Calomyscus bailwardi* Thomas, 1905.

Other Taxa Included The genus contains 1–8 extant species that live in Asia; (extinct) *C. minor* de Bruijn et al., 1970; *C. delicatus* Aguilar et al., 1984.

Distribution Late Miocene of Spain, France, Greece and Turkey, present in Asia.

★*Calomyscus minor* de Bruijn et al., 1970

Type Material Partial right maxilla with M1–2, Nr. 786 (holotype) (de Bruijn et al. 1970: Pl. 5.3).

Type Locality Maritsa 1, Rhodes Island, Greece, Upper Miocene [Turolian (MN13)].

Distribution Late Miocene of Greece (and possible presence in Spain).

Occurrences in Greece Maritsa 1.

Remarks This species has similar dentition with the extant *C. bailwardi*, but smaller and with a weaker anterior cingulum on the m1, according to the diagnosis

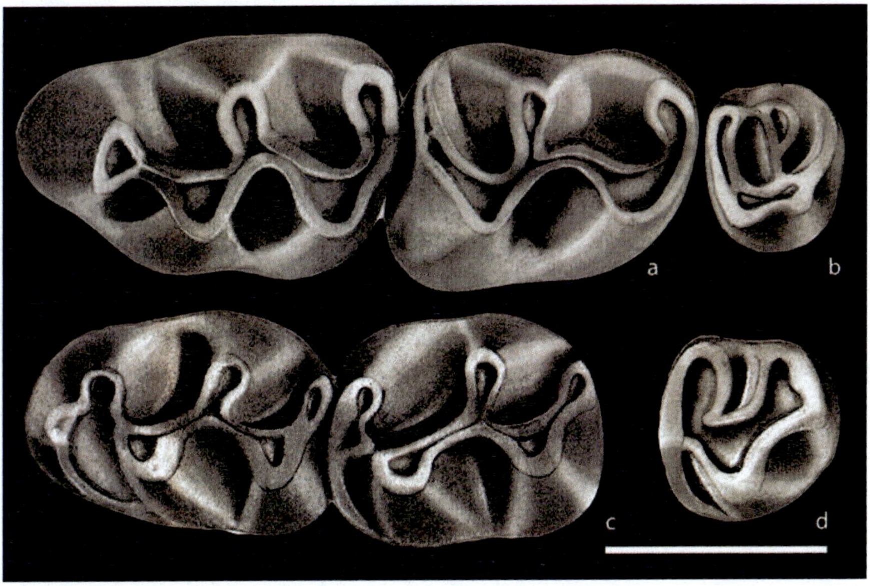

Fig. 23 *Calomyscus minor* from Maritsa 1 (type locality). **a** M1–2 (holotype), **b** M3, **c** m1–2, **d** m3. Images courtesy of H. de Bruijn. All specimens figured as left; **a** is reversed. Scale bar equals 1 mm

by de Bruijn et al. (1970). Its molars are larger than those of *C. delicatus* from the late Miocene of W Europe. The species is well represented in its type locality (Fig. 23), with, apart from the holotype, two maxillary and five mandibular fragments with teeth, 76 m1, 64 m2, 28 m3, 53 M1, 52 M2 and seven M3 (de Bruijn et al. 1970; it is noted by the authors that not all the material collected was taken into account for these counts, as *Calomyscus* elements outnumber all other rodent elements sampled from the locality).

Subfamily Gerbillinae Gray, 1825

Type Genus *Gerbillus* Desmarest, 1804.

Pseudomeriones Schaub, 1934

Type Species *Lophocricetus abbreviatus* Teilhard de Chardin, 1926.

Other Taxa Included *P. tchaltaensis* Şen, 1977; *P. rhodius* Şen, 1977; *P. pythagorasi* Black, Krishtalka and Solounias, 1980; *P. latidens* Şen, 1998; *P. complicidens* Zhang, 1999; *P. megistos* Sylvestrou and Kostopoulos, 2007; *P. hansi* Suata-Alpaslan, 2009.

Distribution Late Miocene–Pliocene of Eurasia.

Remarks Six molars, most of which fragmented, from the Turolian faunas from Tomea Eksi 1 and 2 (Ptolemais basin) have been attributed to *Pseudomeriones* sp. by Hordijk and de Bruijn (2009), who mentioned that the material undoubtedly represents a new species but is too limited to be named with confidence. *Pseudomeriones* sp. is recorded from the Upper Miocene locality Ano Metochi 3 (Wessels 1998, Wessels 1999), previously listed as *P. abbreviatus* by Armour-Brown et al. (1977), de Bruijn and van der Meulen (1979) and Şen (1983) and as *P. rhodius* by de Bruijn (1989). One worn m1 from the Turolian fauna of Kalithies was initially mentioned as *Gerbillus* sp. by de Bruijn (1976), later Black et al. (1980) referred it to *P. pythagorasi* based on its size and, finally, Wessels (1998, 1999) referred it as *Pseudomeriones* sp. – an identification followed herein.

★*Pseudomeriones pythagorasi* Black, Krishtalka and Solounias, 1980

Type Material Right m1, CM 36305 (Black et al. 1980: Fig. 8c).

Type Locality S3 (old collection), Samos Island, Greece, Upper Miocene [middle Turolian (MN12)].

Distribution Late Miocene (Turolian) of Greece and Turkey.

Occurrences in Greece Samos S3 and Mytilinii 1A, B (Fig. 24a–e).

Remarks Black et al. (1980) described this species based on the smaller size of its dental elements in comparison with *P. abbreviatus*, the presence of a stronger posterior arm on the hypoconid of the m1, the more alternate cusps on the M1 and m1 and the wider M1 and m1 than in *P. abbreviatus*. On the m1 of the species, the hypoconid posterior arm is strong and remains so until advanced wear stages (Black et al. 1980). According to Vasileiadou and Sylvestrou (2009), in the *P. pythagorasi* from the new Samos collection (Mytilinii 1A, B), there is a strong labial arm on the m1 that reaches the protoconid base. On the M2 there is a deep and wide, backwards curved lingual sinus, whereas the mesosinus is very shallow and transverse to the longitudinal axis of the teeth. The type locality yielded three m1 and four M1 in total, whereas the new material from Samos consists of one partial mandible with incisor and m1–2, two m1, one m2, two M1 and four M2 (Vasileiadou and Sylvestrou 2009); thus, the present knowledge on the M2 and m2 of this species is based on the new Samos material.

★*Pseudomeriones megistos* Sylvestrou and Kostopoulos, 2007

Type Material Partial right mandible with i, m1–2, PLM–2 (holotype), Museum of Geology-Paleontology-Palaeoanthropology, Aristotle University of Thessaloniki (Sylvestrou and Kostopoulos 2007: Fig. 3).

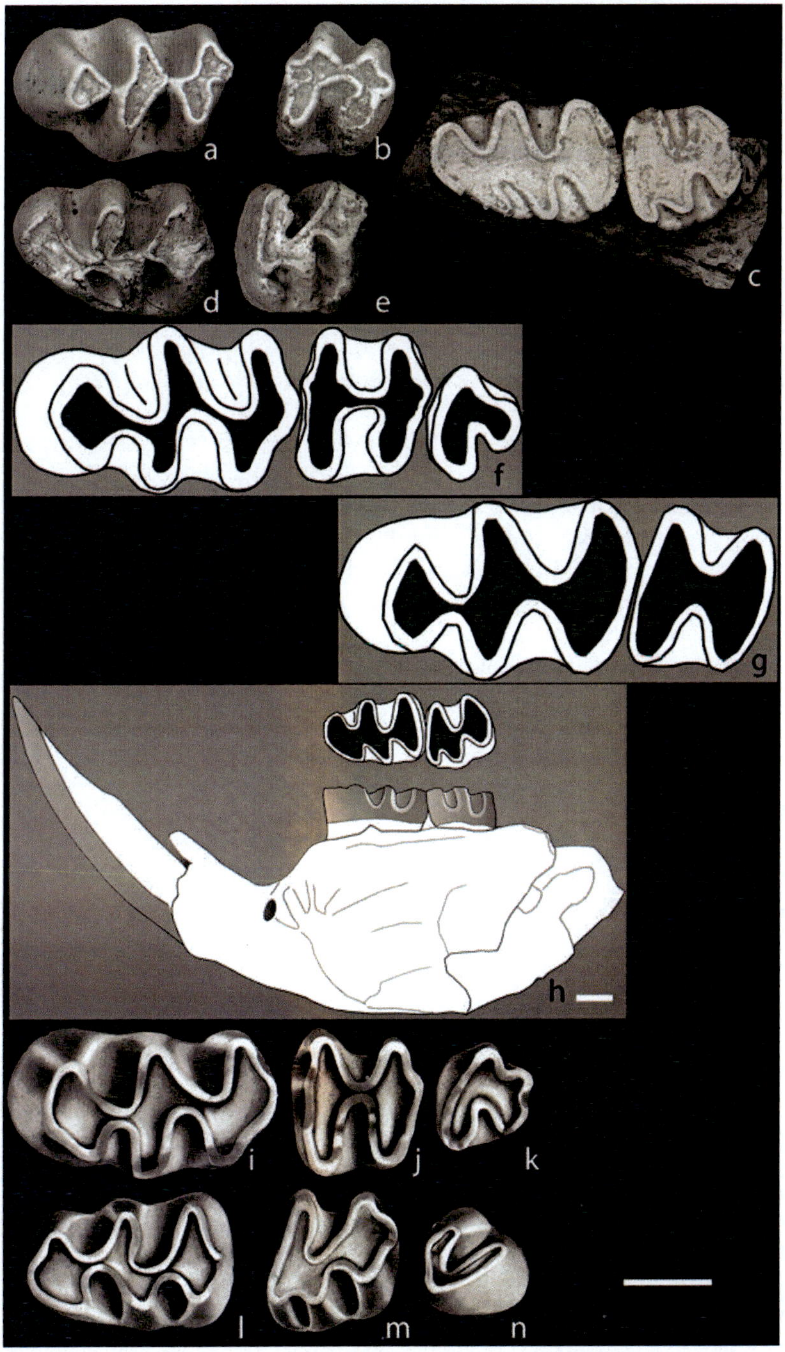

Fig. 24 *Pseudomeriones pythagorasi* from Mytilinii MTLA–B (Samos Island). **a** M1, **b** M2, **c** fragmented mandible with m1–2, **d** m1, **e** m2. Images KV and IS personal data. *Pseudomeriones megistos* from Monasteri. **f** M1–3 and from Paliambela-B (type locality). **g** M1–2, **h** right mandible with i–m2 (holotype). Drawings IS personal data. *Pseudomeriones rhodius* from Maritsa 1 (type locality). **i** M1, **j** M2, **k** M3, **l** m1 holotype, **m** m2, **n** m3. Images courtesy of H. de Bruijn. All specimens figured as left; c–e, g–n are reversed. Scale bar equals 1 mm, common for a–g, i–n

Type Locality Paliambela-B, Katerini Basin, N Greece, Upper Miocene [MN13].

Distribution Late Turolian of Greece (Paliambela-B and Monasteri).

Occurrences in Greece Paliambela-B and Monasteri (Fig. 24f–h).

Remarks This is the largest *Pseudomeriones* species described so far. According to the diagnosis by Sylvestrou and Kostopoulos (2007), the first and second molars of both maxilla and mandible are wide and bear weak to absent posteroloph/id, the anterocone on the M1 is transversally extended, the anteroconid on the m1 is large, simple and crescent-like, the m2 bears well-developed protosinusid and antero-external cuspid and a "U"-shaped and slightly forward directed mesosinusid and the third molars are large and bear no cingula. On the two-rooted M2 the lingual sinus and mesosinus are equally deep and the former is transverse. Apart from the material from the type locality, consisting of the holotype and one partial mandible with m1–m2, one partial maxilla with M1–M2 and one M2, Sylvestrou and Kostopoulos (2007) also attributed to their species a gerbil from Monasteri (Strimon basin, N Greece), previously attributed to *P. rhodius* by de Bruijn (1989) and Wessels (1998). The Monasteri material is richer, consisting of one m1, two m2, four m3, two M1, three M2 and two M3.

★*Pseudomeriones rhodius* Şen, 1977

Nomenclatural and Taxonomical History *P. abbreviatus rhodius* in Şen 1977 (new subspecies); *P. rhodius* in Şen 1983 (raised to species level).

Type Material Right m1, no. 164 (holotype) (de Bruijn et al. 1970: Pl. 5.12).

Type Locality Maritsa 1, Rhodes Island, Greece, Upper Miocene [Turolian (MN13)].

Distribution Late Turolian–earliest Ruscinian (MN13–14) of Greece and Turkey.

Occurrences in Greece Maritsa 1.

Remarks The gerbil material from the type locality Maritsa 1 (Fig. 24i–n), consisting of seven m1, nine m2, five m3, one M1, seven M2 and three M3, was initially attributed to *P. abbreviatus* by de Bruijn et al. (1970). Later, Şen (1977) attributed this material to the new subspecies *P. abbreviatus rhodius*, based on the following differences from *P. abbreviatus* from China: the Maritsa 1 molars are larger and more elongated, the m1 has a triangular anteroconid and its anterior labial and lingual sinusids are equally large, the anterocone of the M1 is wide, the M2 has three roots and the m3 shows a rather complicated morphology (Şen 1983). A poor gerbil material has also been found in the locality Kessani (one m1, two m2, two m3, three M1); it was initially attributed to *P. rhodius* by Syrides et al. (1997) and later to *P.* cf.

rhodius by Vasileiadou et al. (2012), due to differences from the type material. The intraspecific morphological variation within *P. rhodius* is still not adequately known, since the type locality has revealed a small sample (e.g., including only one M1) and even smaller have been reported from other localities.

Meriones **Illiger, 1811**

Type Species *Meriones tamariscinus* Pallas, 1773 (extant).

Other Taxa Included (extant) *M. meridianus* Pallas, 1773; *M. libycus* Lichtenstein, 1823; *M. crassus* Sundevall, 1842; *M. shawii* Duvernoy, 1842; *M. hurrianae* Jerdon, 1867; *M. unguiculatus* Milne-Edwards, 1867; *M. persicus* Blanford, 1875; *M. tristrami* Thomas, 1892; *M. rex* Yerbury and Thomas, 1895; *M. blackleri* Thomas, 1903; *M. grandis* Cabrera, 1907; *M. sacramenti* Thomas, 1922; *M. arimalius* Cheesman and Hinton, 1924; *M. vinogradovi* Heptner, 1931; *M. zarudnyi* Heptner, 1937; *M. dahlia* Shidlovsky, 1962; *M. chengi* Wang, 1964; (extinct) *M. maghrebianus* Tong, 1989; *M. maximus* Tong, 1989.

Distribution The extant *Meriones* is currently distributed from N Africa via the Arabian Peninsula and Anatolia to NW and SW Asia, with 17 species (Macdonald 2009). It is also present in Pleistocene fossil and subfossil occurrences in the same area (e.g., *M. shawii*, *M. tristrami*).

Remarks The genus is presently known in Greece only from Kos Island with the species *M. tristrami*, which is the only European occurrence of the genus (Sozen et al. 2016). The same species has been reported from Pleistocene sediments in a cave of Kalymnos island (Kuss and Storch 1978), with 45 molars.

Subfamily Murinae Illiger, 1811

Type Genus *Mus* Linnaeus, 1758.

Remarks De Bonis and Melentis (1975) described and illustrated a murin m2 from the locality Ravin de Zouaves 5. The tooth is rather large (L= 1.43 mm, W= 1.32 mm), the antero-labial cuspid is medium-sized and connected with the protoconid through wear, there are no accessory cuspids on the labial side and the terminal heel is elongated and very thin; it has the shape of a low ridge. The authors suggest that these characters could place the molar in *Valerymys* Michaux, 1969. However, *Valerymys* was synonymized with *Occitanomys* Michaux, 1969 by Aguilar et al. (1986), who included its type species [*V. ellenbergeri* (Thaler, 1966)] in the latter genus. Later, Mein et al. (1993) transferred some other *Valerymys* species [i.e., *V. vireti* (Schaub, 1938), *V. turoliensis* Michaux, 1969 and *V. oreopitheci* Engesser, 1989] in their new genus *Huerzelerimys*. Since the tooth cannot be located and its morphology is rather general to include it with certainty to a genus, it is here referred to as Murinae indet. Murinae indet. have also been reported from Spilia 3–4 (van der Meulen and van Kolfschoten 1986), Tourkobounia 3, Tourkobounia 4, Tourkobounia 5 (Reumer and Doukas 1985), Choremi 3 (van Vugt et al. 2000) and Dimitra 2 (Syrides and Koliadimou 1994).

***Progonomys* Schaub, 1938**

Type Species *Progonomys cathalai* Schaub, 1938.

Other Taxa Included *P. woelferi* Bachmayer and Wilson, 1970; *P. hispanicus* Michaux, 1971; *P. debruijni* Jacobs, 1978; *P. castilloae* Aguilar and Michaux, 1996; *P. hussaini* Cheema et al., 2000; *P. minus* Şen, 2003; *P. ibrahimi* (Şen, 2003); *P. sinensis* Qiu et al., 2004; *P. shalaensis* Qiu and Li, 2016; *P. morganae* Kimura et al., 2017. Wessels (2009) considered *P. castilloae*, *P. hussaini*, *P. minus*, *P. ibrahimi* and *P. sinensis* to be junior synonyms of *P. cathalai*; Kimura et al. (2017) do not agree with this synonymy and retain them as separate species.

Distribution Late Miocene (MN9–11) of Europe, Asia and N Africa.

Remarks Schaub (1938) did not give a diagnosis for his new genus. Mein et al. (1993), in a revision of the genus, gave the following diagnostic features: slightly larger than the extant *Mus musculus*, long and slender molars, no longitudinal connections between the cusps/-ids; M1 with an almost oval outline, forward-situated tl without t1bis, joint t4–t5 through a high connection, a tendency to fuse with t8 through a low crest that does not form a t7; t6 and t9 usually separated on the upper molars; one single lingual root on the upper molars; m1 with a reduced or absent tma, generally absent anteroconid–metaconid connection with the exception of very worn specimens; two main roots and a small central one on the m1. Based on this diagnosis, they placed only two species in the genus, i.e., *P. cathalai* and *P. woelferi* (and an unnamed species from Sinap Tepe 1).

Wessels (2009) gave a new emended diagnosis for the genus: "*Murinae with long and slender molars, without longitudinal connections between the cusps. Outline of M1 varies between almost elliptical and elliptical with a strong indentation on its antero–lingual part. The t1 and t2 are mostly connected, the t2 is either close to t1 or placed further backwards, a t1bis is absent (>50%) or small. The t4 is united to t5 by a low to medium high connection, the t4 has the tendency to fuse with t8 by a low crest that never forms a t7. Upper molars with t6 and t9 separated in more than 50% of the M1 and M2. m1 has a small or absent tma; the base of the anteroconid and metaconid are close to each other, a short low connection is often present in unworn specimens and almost always present in worn specimens. Upper molars with one single lingual root, m1 with two main roots and a small central one in most specimens*".

Wessels' (2009) study of large *Progonomys* assemblages from Turkey has proven that the *Progonomys* species can be separated mainly based on the size and the general shape of the molars, whereas there is a large intraspecific morphological variation. *P. cathalai* is in general smaller than *P. woelferi*, with a small size overlap. Small *Progonomys* assemblages are, therefore, difficult to identify with certainty to species level. De Bruijn et al. (1971) attributed the small murid samples from the locality Kastellios Hill (Crete, Greece) to *Progonomys*: from layer KA1 to *P.* cf. *woelferi* (16 molars) and from layer KA3 to both *P. cathalai* (13 molars) and *P.* cf. *woelferi* (3 molars). De Bruijn and Zachariasse (1979), after the addition of more material, confirm the presence of *P. woelferi* in the layer KA1 (112 molars in total) and mention the presence of *P. cathalai* also in layer KA2a (3 molars). Unfortunately, the presence of *P.* cf. *woelferi* in layer KA3 is not mentioned at all in this second work. Mein et al. (1993:46) excluded *P. cathalai* from layer KA3 from *Progonomys* (without transferring it to another taxon), because of the absence of a connection between t4 and t8, the absence of a tl2 in the upper molars, the great development of the labial cingula and the absence of a third root in the m1. However, Wöger (2011), in her diploma thesis (Vienna University), after studying *P. woelferi* from its type locality Kohfidisch, gives reasons for not following Mein et al. (1993) in excluding *P. cathalai* from layer KA3 from *Progonomys*. Van Dam (1997) included one of the three molars from layer KA2a in *P. hispanicus* (without mentioning the other 2 specimens); in the same work, he also transferred *P. cathalai* from Biodrak (Greece; de Bruijn 1976) to *P. hispanicus*. Wessels (2009:94) erroneously mentioned in Table 6.1 the presence of *P. cathalai* in layer KA1 (instead of *P. woelferi*) and of *P. woelferi* in Biodrak (instead of *P. cathalai*) and in Table 6.6 (p. 114) of *P.* cf. *cathalai* in Lefkon (instead of *P. cathalai*; de Bruijn 1989). Finally, her Table 6.6 (page 115) shows the presence of *P. woelferi* in levels KA1 and KA3 and of *P. cathalai* in layers KA2a and KA3. Since the low part of the Kastellios Hill Section (including all above-mentioned fossiliferous layers) is correlated with the Chron C4Ar.1r by Zachariasse et al. (2011), it is not wrong to consider that the three fossiliferous layers do not have a big difference in age (9.3–9.1 Ma in Zachariasse et al. 2011) and treat the fauna from the three layers as one local fauna including both *P. woelferi* and *P. cathalai* (de Bruijn 2018 personal communication).

***Progonomys cathalai* Schaub, 1938**

Type Material Right M1, No. 584 (holotype), Natural History Museum, Basel (Schaub 1938: Pl. 1.8).

Type Locality Montredon, niveau Depéret (Hérault), France, Upper Miocene [MN10].

Distribution Late Miocene of Europe, Asia and N Africa.

Occurrences in Greece Kastellios, Ravin de la Pluie, Biodrak and Lefkon.

Remarks Schaub (1938) did not give a diagnosis for the species. Wessels' (2009) study of a very large *P. cathalai* assemblage from the Turkish localities Altıntaş and Kütahya (1247 specimens) contributed greatly to the knowledge of the intraspecific size and morphological variation and the differences of this species with the other

Progonomys species. There is a large variation in the position of the t1, but the cusp is placed backwards less often than in the other species; the t1 and t5 can be connected at their base through wear; the t6–t9 connection is absent in 20% of the M1; the t12 is well developed; a very small tma is present in less than 10% of the m1; the labial cingulum is weak with a medium-sized C1 on the m1 (Wessels 2009:113). The assignment of the small *Progonomys* assemblages from the Greek localities Kastellios (see above Remarks on *Progonomys*), Ravin de la Pluie (one partial right maxilla with M1–3; de Bonis and Melentis 1975) and Biodrak (11 molars; de Bruijn 1976) to this species is mainly based on the small size of the molars. The species is also reported from Lefkon (de Bruijn 1989), without information about the material.

***Progonomys woelferi* Bachmayer and Wilson, 1970**

Type Material Partial maxilla with M1–2, No. 1970/1395 (holotype), Natural History Museum Vienna (Bachmayer and Wilson 1970: Pl. 11.57).

Type Locality Kohfidisch, Austria, Upper Miocene [MN11].

Distribution Late Miocene of Europe (Spain, Austria, Greece).

Occurrences in Greece Kastellios.

Remarks The diagnosis given by Bachmayer and Wilson (1970) mentions that this species has larger molars than those of *Progonomys cathalai* by approximately 15%. The Ml–2 have three roots, but may also show a broad inner root; they are maybe slightly more stephanodont than in *P. cathalai*, as the t6 and t9 may be closer together. The m1 shows no or only a very small anteroconid. *P. woelferi* from Kastellios (de Bruijn et al. 1971; de Bruijn and Zachariasse 1979) shows great morphological similarity with the material from the species' type locality, even though it is on average slightly larger. This size difference, together with the fact that the Kohfidisch upper molars show a better-developed t6–t9 connection than the Kastellios ones, has been attributed to an age difference between the two assemblages: the Kastellios material seems to be more "primitive" than the one from Kohfidisch. This is further confirmed by the ages of the two assemblages, as the Kastellios fauna is dated to 9.3–9.1 Ma (Zachariasse et al. 2011) and the Kohfidisch fauna to 8.6–8.5 Ma (Wöger 2011).

***Parapodemus* Schaub, 1938**

Type Species *Mus gaudryi* Dames, 1883.

Other Taxa Included (following de Bruijn et al. 1999) *P. lugdunensis* Schaub, 1938; *P. barbarae* van de Weerd, 1976; *P. meini* Martín Suárez and Freudenthal, 1993.

Distribution Late Miocene–Pliocene (MN11–15) of Europe and Asia.

Remarks De Bruijn (1976) gave an emended diagnosis for the genus, based on which *Parapodemus* has a dental pattern close to that of *Apodemus*, but the t7 is either absent or weak in the majority of the M1–2. The author clarified that a minimum of 10 M1–2 must be present in an assemblage so as to safely distinguish the two genera. De Bruijn et al. (1999) revised the status of *Parapodemus*, its type species *P. gaudryi* and the holotype of the species, which was considered lost since 1947 but was found in 1999 in the collections of the Geology Department of the University of Athens. They noticed that the grade of evolution for M1–2 is different in large *Parapodemus/Apodemus* assemblages (more M2s than M1s show the "*Apodemus*" morphology with a t7 and less show the "*Parapodemus*" morphology without a t7). They redefined the genus, choosing the use of only the M1s in an assemblage. Thus, the t7 has to be absent in the majority of the M1s in assemblages with at least 10 M1 in order to assign the assemblage to *Parapodemus* (de Bruijn et al. 1999:106).

★*Parapodemus gaudryi* (Dames, 1883)

Nomenclatural and Taxonomical History *Mus* (?*Acomys*) *gaudryi* in Dames 1883 (new species); *Acanthomys gaudryi* in Forsyth Major 1891 (new combination); *Parapodemus gaudryi* in Schaub 1938 (new combination). Considered by de Bruijn (1976) and de Bruijn et al. (1999) as the senior synonym of *Parapodemus schaubi* Papp 1947.

Type Material Mandible with right m2–3 and left m2 (holotype; Dames 1883: Taf. V.2; de Bruijn et al. 1999: Pl. 1.8). The designated neotype of de Bruijn (1976:Pl. 3.8) is set aside upon the re-discovery of the holotype.

Type Locality Pikermi (Megaloremma), Greece, Upper Miocene [Turolian (MN12)].

Distribution Late Miocene, Turolian (MN11–12) of Europe.

Occurrences in Greece Vathylakkos-3, Samos Quarry 4, Chomateri, Megaloremma and Rema Marmara.

Remarks Dames (1883) based his new species on one mandible with complete dentition but damaged in the symphysis area and the posterior part. Consequently, the mandible was further damaged, leading to the current presence of only the right m2–3 and left m2 (for the history of the specimen and its taxonomic situation, see de Bruijn et al. 1999). De Bruijn (1976) attributed a collection from Pikermi (Chomateri) to *P. gaudryi*, based on the presence of a ridge connecting the t4 and t8 instead of a cusp (t7) in most of the M1 (seven out of 12) and in half of the M2 (two out of four). The authors also designated a neotype for the species from Pikermi (Chomateri), since the holotype from Pikermi (Megaloremma) had been lost. De Bruijn et al. (1999) argued that the two localities (Megaloremma and Chomateri), with a distance of about 3.7 km between them, more or less represent the same horizon; thus the designation of a neotype from a locality different than the type locality was justified. Further material from Greece is reported by Schaub (1926), who identified a partial skull with worn dentition from Samos Quarry 4 as *Mus gaudryi*. The locality Rema Marmara in N Greece has also yielded *P. gaudryi* (de Bruijn 1989), but no details on the material has been published. Finally, the presence of *P. schaubi* has been reported from the N Greek locality Vathylakkos-3 without further details on the material (de Bonis and Koufos 1999).

***Parapodemus lugdunensis* Schaub, 1938**

Type Material Right M1, P.L.63 (holotype), Natural History Museum, Basel (Schaub 1938: Fig. 27).

Type Locality Mollon, France, Upper Miocene [Turolian (MN11)].

Distribution Late Miocene, Turolian (MN10–12) of Eurasia.

Occurrences in Greece Lefkon.

Remarks The m1 of the species usually has a more or less symmetrical anteroconid complex, with a well-developed antero-central cuspid almost in half of the population, large c1 and two or three accessory cuspids and no longitudinal spur; the m2 has a distinct antero-labial cuspid and a well-developed c1; the M1 is characterized by high connections among the cusps, the t1 appears compressed on the lingual wall of the t2, the t6 is close to the t9, the t12 is large and there are three roots; on the M2 the t6 is close to and connected to the t9. *Parapodemus* cf. *lugdunensis* is reported from the Upper Miocene (MN10) locality Lefkon by de Bruijn (1989), but without descriptions or details on the assemblage.

***Hansdebruijnia* Storch and Dahlmann, 1995**
(=*Senia* Sarıca-Filoreau, 2002, according to Hordijk and de Bruijn 2009)

Type Species *Occitanomys*? *neutrum* de Bruijn, 1976.

Other Taxa Included *H. pusilla* (Schaub, 1938); *H. perpusilla* Storch and Ni, 2002; *H. amasyaensis* (Sarıca-Filoreau, 2002); *H. erkinsae* Ünay et al., 2006.

Distribution Late Miocene (Turolian, MN11–13) of E Europe and Asia.

Remarks The genus *Hansdebruijnia* was initially named as a subgenus of *Occitanomys* by Storch and Dahlmann (1995) and was later elevated to the generic rank by Storch and Ni (2002).

★*Hansdebruijnia neutrum* (de Bruijn, 1976)

Nomenclatural and Taxonomical History *Occitanomys*? *neutrum* in de Bruijn 1976 (new species); *Hansdebruijnia neutrum* in Storch and Dahlmann 1995 (new combination).

Type Material Right M1, no. 262 (holotype) (de Bruijn 1976: Pl. IV.10).

Type Locality Pikermi (Chomateri), Greece, Upper Miocene [Turolian (MN12)].

Distribution Late Miocene, Turolian (MN11–13) of E Europe and Asia.

Occurrences in Greece Chomateri, Rema Marmara, Kalithies and Maramena.

Remarks De Bruijn (1976) defined this species based on a collection of 15 molars (Fig. 25) from Pikermi (Chomateri) as a "*small* Occitanomys*–like murid*" with similar size to that of *O. sondaari* van de Weerd 1976, without a ridge connecting the t4 and t8 on the M1, without a t1 bis on the M1–2 and with a small antero-central cusp that is lower than the anteroconid cuspids on the m1. The author noticed that this species is different from other *Occitanomys* species in a great degree. He also attributed to *O.? cf. neutrum* four molars from Kalithies, based on the great similarity of the three of them with the respective molars from Chomateri (the M1 from Kalithies

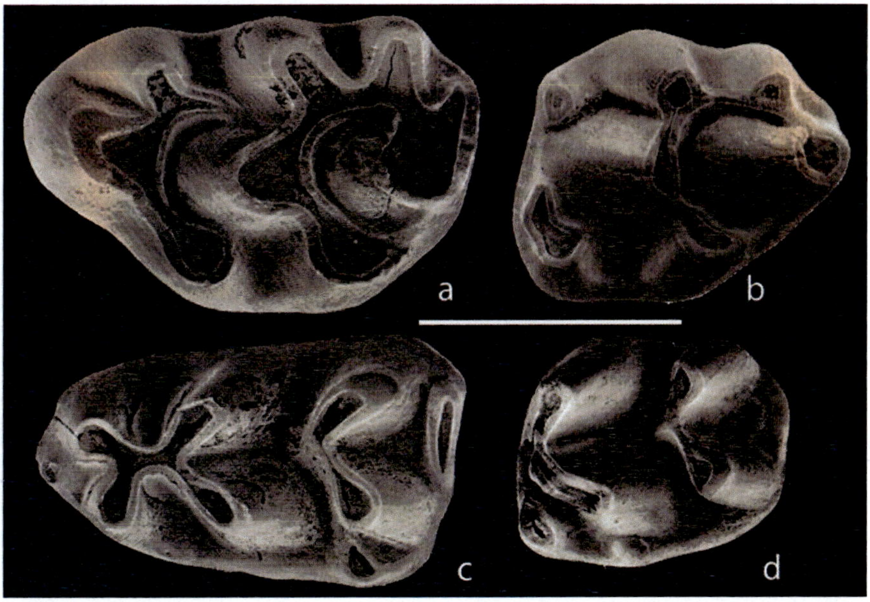

Fig. 25 *Hansdebruijnia neutrum* from Pikermi–Chomateri (type locality). **a** M1 (holotype), **b** M2, **c** m1, **d** m2. Images courtesy of H. de Bruijn. All specimens figured as left; **a** is reversed. Scale bar equals 1 mm

is mentioned as resembling the M1 of *Parapodemus lugdunensis* rather than that of *O.? neutrum*). Storch and Dahlmann (1995) created *Hansdebruijnia* as a subspecies of *Occitanomys* and attributed the rich material from the Turolian/Ruscinian locality Maramena (165 first and second upper and lower molars in total), originally referred to as *O.* cf. *neutrum* by de Bruijn (1989), to *O. (H.) neutrum*, reporting as only difference the presence of a weak to moderate t1bis in about 60% of the Maramena M1 in contrast to the absence of this cuspule from all four M1 from Chomateri. The absence of the t1bis is a diagnostic feature of the species, but it is only based on the four M1 found in Chomateri; the scantiness of the type material cannot verify the intraspecific variability of this feature. The species has also been reported from the locality Rema Marmara by de Bruijn (1989), but there are no details on the assemblage.

"Karnimata" Jacobs, 1978 or Muridae incertae sedis

Type Species *"Karnimata" darwini* Jacobs, 1978.

Taxa Included *"K." hipparionum* (Schlosser, 1924); *"K." provocator* (de Bruijn, 1976); *"K." huxlei* Jacobs, 1978; *"K." minima* Brandy, 1979; *"K." intermedia* Brandy, 1979; *"K." afghanensis* Brandy, 1979; *"K." inflata* Mein, Moissenet and Adrover, 1990.

Distribution Late Miocene (Turolian, MN12–13) of Europe and Asia.

Remarks The type species of *"Karnimata"*, *"K." darwini*, was synonymized with *Progonomys woelferi* Bachmayer and Wilson 1970 by Mein et al. (1993), an opinion also shared by authors such as de Bruijn et al. (1999) and Wessels (2009), but not shared by others such as Storch and Ni (2002) and Kimura et al. (2017). The consequence of the synonymy would be the absence of a proper genus name to include all the species of the genus apart from the type species. Mein et al. (1993) mentioned that the group of species assigned to *"Karnimata"* is not homogenous; thorough study of the available material in order to clarify relationships and generic attribution is still pending.

★*"Karnimata" provocator* (de Bruijn, 1976)

Nomenclatural and Taxonomical History *Occitanomys*? *provocator* in de Bruijn 1976 (new species); *"Karnimata" provocator* in de Bruijn et al. 1999 (new combination).

Type Material Left M1, no. 232 (holotype) (de Bruijn 1976: Pl. IV.1).

Type Locality Pikermi (Chomateri), Greece, Upper Miocene [Turolian (MN12)].

Distribution Late Miocene, Turolian (MN12–13) of Greece and Turkey.

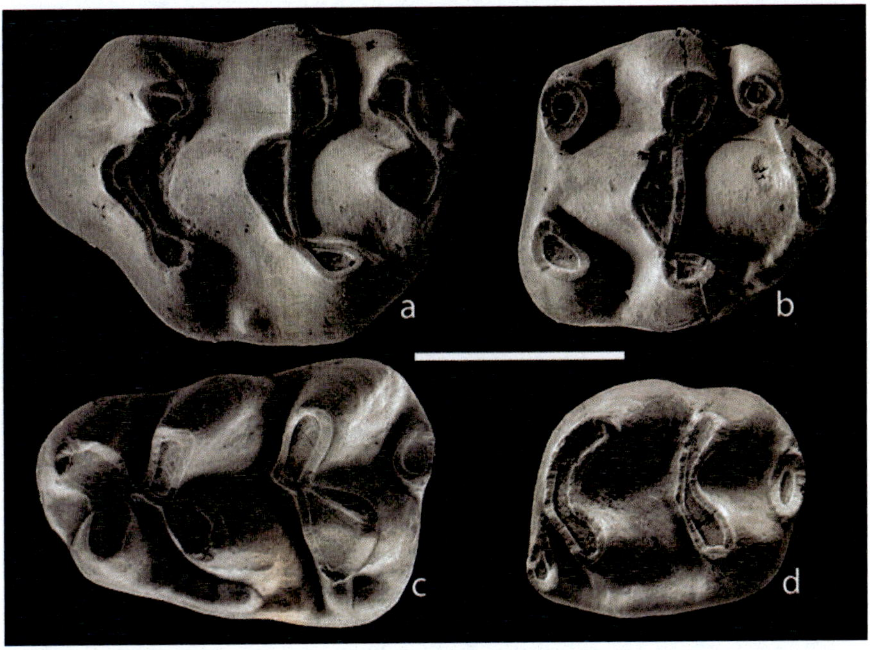

Fig. 26 "*Karnimata*" *provocator* from Pikermi–Chomateri (type locality). **a** M1 (holotype), **b** M2, **c** m1, **d** m2. Images courtesy of H. de Bruijn. All specimens are left. Scale bar equals 1 mm

Occurrences in Greece Samos S3, Mytilinii 1A and Chomateri.

Remarks This species was defined based on 16 first and second upper and lower molars (Fig. 26) from Chomateri, which are similar in size with *Occitanomys adroveri* from Los Mansuetos, with a rounded and isolated t1 and no t1bis on the M1–2, with a t3 without a posterior spur and without a longitudinal ridge on the m1–2 (de Bruijn 1976). One partial skull with the right M1–M2 and the left M1–M3, one isolated M1 and one m1 described by Black et al. (1980), as well as one isolated m1, a partial mandible with the m2–m3, one m2 and two m3 described by Vasileiadou and Sylvestrou (2009), both assemblages from Samos Island (S3 and Mytilinii 1A, respectively), have also been attributed to this species.

Occitanomys Michaux, 1969

Type Species *Occitanomys brailloni* Michaux, 1969.

Other Taxa Included (including the species of *Centralomys* de Giuli, 1989 and *Rhodomys* Martín Suárez and Mein, 1991, according to Hordijk and de Bruijn 2009) *O. adroveri* (Thaler, 1966); *O. sondaari* van de Weerd, 1976; *O. magnus* (Şen, 1977); *O. montheleni* Aguilar et al., 1986; *O. alcalai* Adrover et al., 1988; *O. debruijni* (Şen, Jaeger, Dalfes, Mazin and Bocherens, 1989); *O. benericettii* (de Giuli, 1989); *O. vandami* Suata Alpaslan, 2009.

Distribution Late Miocene (MN10)–middle Pliocene of Europe and Turkey.

Remarks Michaux (1969) established this genus for *brailloni* and *adroveri*, based on their degree of stephanodonty, which is weaker than that of *Stephanomys* and *Castillomys*, but stronger than that of *Apodemus*, and the position of the t1 on the M1, which is posterior. The genus is further characterized by very brachyodont molars; the oldest species show a trace of a longitudinal connection in the lower molars, which becomes stronger in younger species; the m1 lacks a tma, although occasionally a very small cuspule is present; in the upper molars a t12 is poorly developed and in the M1 the t1 is located very far backwards and in younger species it develops a posterior connection with the t5 (Freudenthal and Martín Suárez 1999). It differs from *Progonomys* through the presence of a t6–t9 connection on more than 50% of the M1 (van Dam 1997). Martín Suárez and Mein (1991) created the subgenus *Rhodomys*, including in it the species *debruijni* (type species) and *sondaari*. According to these authors, the subgenus has the following features: the molars are small-sized, the tl on the M1 is positioned very far backwards, often isolated from the t2 and sometimes connected to the t5, on the m1 there is no tma and the longitudinal spur between the entoconid and the protoconid is very weak and the m2 is posteriorly reduced and sometimes bears a c2. Freudenthal and Martín Suárez (1999) included in this subgenus also *O. adroveri* and Suata Alpaslan (2009) her species *O. vandami*. Hordijk and de Bruijn (2009) do not agree with these allocations and consider *Rhodomys* to be a junior synonym of *Occitanomys*, as some diagnostic characters of the subgenus may occur within the range of the intraspecific variation in various *Occitanomys* (Hordijk and de Bruijn 2009:33).

Occitanomys brailloni **Michaux, 1969**

Type Material Left M1, LY8a3 (holotype), Museum of Miquel Crusafont Palaeontological Institute, Sabadell, Spain (Michaux 1969: Pl. 1.4).

Type Locality Layna (Soria), Spain, Lower Pliocene.

Distribution Late Miocene (Turolian, MN12)–Early Pleistocene of Europe and Turkey.

Occurrences in Greece Rema Marmara, Nea Silata, Spilia Kardia, Vorio-1, Ptolemais 1, Vorio 3, Ptolemais 3 and Spilia 3–4.

Remarks This is a rather common species in S Europe including Greece. Michaux (1969) distinguished it from *O. adroveri* through its larger size, the more inflated cusps and the reduced posterior part of the M1–2. Further morphological features of the species are the following: t1 connected with the t5 but not with the t2, well-developed t1bis and a rather high t4–t8 connection on the M1, weak labial cingulum with one or two accessory cuspids on the m1 and very weak to absent labial cingulum with no or one cuspid on the m2. Vasileiadou et al. (2003) assigned to *O. brail-*

loni 167 molars from the locality Nea Silata. A few molars from the localities Kardia (12), Ptolemais 1 (1) and Ptolemais 3 (4) have been attributed to this species by van de Weerd (1979), who mentioned that these molars differ from those from the type locality in the presence of a spur on the t3 of the M1 and of a weak longitudinal ridge on the m1–2. A few molars of *O. brailloni* have been described from Vorio 1 (1) and Vorio 3 (12) by Hordijk and de Bruijn (2009). The species is also present in the faunal lists of Rema Marmara and Spilia 0 and 1 (de Bruijn 1989), as well as of Spilia 3 and 4 and Limni 3 (as cf.) (van der Meulen and van Kolfschoten 1986), with no information on the assemblages.

***Occitanomys adroveri* (Thaler, 1966)**

Nomenclatural and Taxonomical History *Parapodemus adroveri* in Thaler 1966 (new species); *Occitanomys adroveri* in Michaux 1969 (new combination).

Type Material M1–M3, M. 4029 (holotype), Museum of Miquel Crusafont Palaeontological Institute, Sabadell, Spain.

Type Locality Los Mansuetos, Spain, Upper Miocene [Turolian (MN12)].

Distribution Late Miocene (Turolian MN12)–early Pliocene (MN14) of Europe.

Occurrences in Greece Monasteri, Kessani-1, Ano Metochi 3, Spilia 0, Komanos 1 low A and B and Komanos 1 high A and B.

Remarks This species shows a small t1bis, low t4–t5 and t4–t8 connections and a small t12 on the M1, as well as a t1bis confluent with the t1, a small t3 and low t4–t5 and t4–t8 connections on the M2. Hordijk and de Bruijn (2009) noticed minor differences between their samples from Komanos (Ptolemais Basin, N Greece) and the material from the type locality Los Mansuetos, among which is the somewhat stronger t4–t5 connection and the generally weaker t12 in the M1–2, the presence of two fully developed lingual roots on one M2, as well as the slightly better developed longitudinal spur of the last chevron and the better developed terminal heel in the m1–2 from Komanos. The authors commented that "*considering the geographical and stratigraphical distance between these associations this similarity is surprising*" (Hordijk and de Bruijn 2009:33). The Komanos localities have been dated to 5.23 and 5.22 Ma and are lithostratigraphically correlated with the locality Kardia, where *O. brailloni* has been found (Hordijk and de Bruijn 2009). The authors commented that this might mean either that the two species were coeval in the Ptolemais Basin or that the correlation between the localities is wrong. A small collection of seven molars from Kessani (N Greece) has also been assigned to this species by Vasileiadou et al. (2012), whereas the locality Nea Silata (N Greece) has yielded *O. brailloni* (Vasileiadou et al. 2003). Since these two localities are almost coeval, correlated with the time interval 5.4–5.23 Ma (Koufos and Vasileiadou 2015), it appears indeed possible that the two species co-existed in N Greece during that period, but have never been found in the same locality. *O. adroveri* has also been reported from Monasteri, Ano Metochi 3 and Spilia 0 (de Bruijn 1989), with no details on the materials.

★*Occitanomys debruijni* (Şen, Jaeger, Dalfes, Mazin and Bocherens, 1989)

Nomenclatural and Taxonomical History *Castillomys debruijni* in Şen et al. 1989 (new species); *Occitanomys debruijni* in Martín Suárez and Mein 1991 (new combination).

Type Material Right M1, no. 471 (holotytpe) Department of Earth Science Utrecht, The Netherlands (de Bruijn et al. 1970: Pl. 2.3).

Type Locality Maritsa 1, Rhodes Island, Greece, Upper Miocene [Turolian (MN13)].

Distribution Late Miocene (Turolian)–early Pliocene (MN12–14) of Greece and Turkey.

Occurrences in Greece Maritsa 1 and Tomea Eksi 1 and 2.

Remarks This species was named by Şen et al. (1989), based on the rich material from Maritsa 1 (93 molars in total; Fig. 27), which had been identified as *Castillomys*

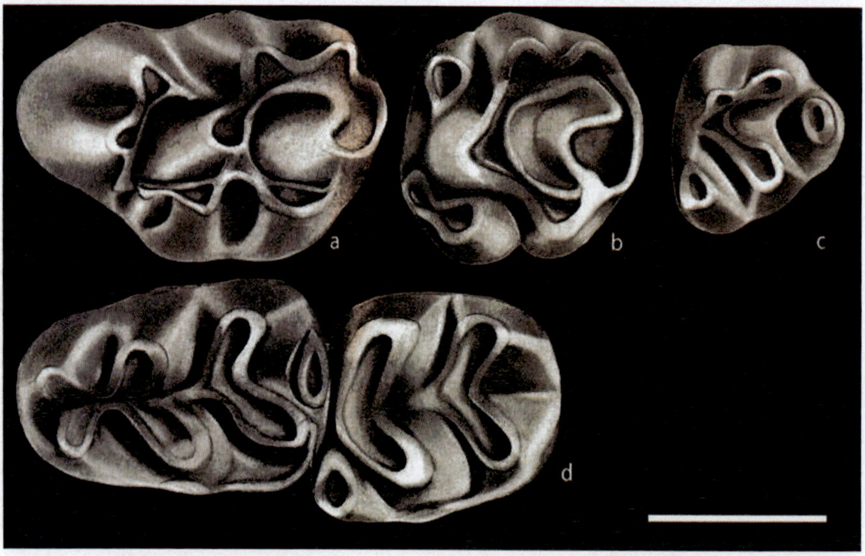

Fig. 27 *Occitanomys debruijni* from Maritsa 1 (type locality). **a** M1 (holotype), **b** M2, **c** M3, **d** m1–2. Images courtesy of H. de Bruijn. All specimens are right but reversed and figured as left. Scale bar equals 1 mm

crusafonti by de Bruijn et al. (1970) and *Castillomys* sp. by van de Weerd (1976), exhibiting the following features: lower molars with weak longitudinal connection, absent or weak connection between the t3 and the t5–t6 on the M1, isolated posterior cusp on the M3 and a size intermediate between those of *C. crusafonti* and *C. magnus*. Martín Suárez and Mein (1991) transferred the species to *Occitanomys* and created the subgenus *Rhodomys* based on it; they gave an emended diagnosis for the species including the following features: large-sized molars, with a large and sometimes isolated t1bis on the M1, a reduced or absent connection between the t3 and t5, an isolated t8 on the M3, a weak longitudinal spur on the lower molars and absence of the c1 from the m2. Hordijk and de Bruijn (2009) assigned to this species two samples from Tomea Eksi 1 (eight molars) and Tomea Eksi 2 (35 molars) (Ptolemais Basin, N Greece). The authors mentioned the striking morphological similarity of their material with the material from the type locality and the presence of only minor differences, e.g., the posterior spur of the t3 is weaker in the M1 and the t1bis is on average less separate from the t1 in the M2 from Tomea Eksi than in the respective molars from Maritsa 1, and the longitudinal spur between the second and third chevron in the m1–2 from Tomea Eksi is weak or absent, while it varies between strong and absent in the molars from Maritsa 1.

***Occitanomys magnus* (Şen, 1977)**

Nomenclatural and Taxonomical History *Castillomys magnus* in Şen 1977 (new species); *Centralomys magnus* in Martín Suárez and Mein 1991 (new combination); *Occitanomys magnus* in Hordijk and de Bruijn 2009 (new combination).

Type Material Right M1, ACA–824 (holotype) (Şen 1977: Pl. 1.1).

Type Locality Çalta, Turkey, Pliocene [Ruscinian (MN15)].

Distribution Late Miocene (middle Turolian)–early Pliocene (late Ruscinian) of Greece and Turkey.

Occurrences in Greece Notio 1.

Remarks The species was created by Şen (1977), based on four m1, five m2, two m3, three M1, three M2 and one M3 from Çalta. The M1 is rather stephanodont, with the t1 and t3 far from the t2 and with posteriorly directed spurs and with a reduced t12; the m1 does not bear an antero-central cuspid and shows strong labial margin and low median ridge between second and third chevron. Hordijk and de Bruijn (2009) assigned one complete and one partial M1 and one complete M2 from Notio 1 to *O. magnus* based on their size and high degree of stephanodonty.

***Micromys* Dehne, 1841**

Type Species *Mus minutus* Pallas, 1771.

Other Taxa Included *M. praeminutus* Kretzoi, 1959; *M. bendai* van de Weerd, 1979; *M. steffensi* van de Weerd, 1979; *M. paricioi* Mein, Moissenet and Adrover, 1983; *M. chalceus* Storch, 1987; *M. cingulatus* Storch and Dahlmann, 1995; *M. caesaris* Minwer-Barakat et al., 2008.

Distribution Late Miocene (Turolian, MN12)–present of Eurasia.

Remarks The molars of the genus are rather brachyodont, with the upper ones showing a large number of roots and M1 showing a wide distance between the t3 and t6 (Freudenthal and Martín Suárez 1999). It is represented by seven extinct species and only one extant, *M. minutus*, with a very wide geographical distribution. Van der Meulen and van Kolfschoten (1986) reported the presence of *M.* cf. *minutus* in the Biharian locality Zeli 2 A and B, with no additional details.

★*Micromys bendai* van de Weerd, 1979

Taxonomic History Considered as the senior synonym of *Micromys kozaniensis* van de Weerd, 1979 and *Micromys telfordi* Wu and Flynn, 1992, according to Hordijk and de Bruijn (2009).

Type Material Left M1, PT1 393 (holotype) (van de Weerd 1979: Pl. 3.4).

Type Locality Ptolemais 1, Greece, Pliocene [early Ruscinian (MN14)].

Distribution Turolian–Ruscinian (MN12–15) of Europe and Asia.

Occurrences in Greece Rema Marmara, Monasteri, Tomea Eksi 1 (Fig. 28a–f), Tomea Eksi 2, Prosilion–Mercurion, Ano Metochi 2, Spilia 0, Vorio 1, Tomea Eksi 3, Ptolemais 1, Vorio 3 and 3A and Komanos 2.

Remarks This medium-sized *Micromys* species was named based on a collection of 56 molars from Ptolemais 1; its upper molars are relatively wide, there is a well-developed anteriorly positioned t7 and a small t12 on the upper molars (van de Weerd 1979). Apart from the type locality, several other localities in the Ptolemais Basin have yielded *M. bendai* (Tomea Eksi 1, Tomea Eksi 2, Prosilion–Mercurion, Vorio 1, Tomea Eksi 3, Vorio 3 and 3A and Komanos 2), described by Hordijk and de Bruijn (2009). Further material has been reported by de Bruijn (1989) from localities in the Strimon Basin (Rema Marmara, Monasteri, Ano Metochi 2–3 and Spilia 0), but without further details, whereas Theocharopoulos (1991), in his unpublished undergraduate thesis, assigned to the species six molars from the locality Spilia 1.

★*Micromys kozaniensis* van de Weerd, 1979
[junior synonym of *Micromys bendai* van de Weerd, 1979, according to Hordijk and de Bruijn 2009]

Type Material Left M1, PT3 306 (holotype) (van de Weerd 1979: PI. 3.8).
Type Locality Ptolemais 3, Greece, Pliocene [late Ruscinian (MN15)].
Remarks Van de Weerd (1979) described the species *M. kozaniensis* based on 47 molars from the locality Ptolemais 3, reporting that this species differs from *M. bendai* through its narrower upper molars with a more posterior position of the t7, a larger t12 and a deep indentation between the t1 and t2 in the contour of the M1. Based on the coexistence of molars similar to both *M. bendai* and *M. kozaniensis* in the Chinese locality Bilike reported by Qiu and Storch (2000), Hordijk and de Bruijn (2009) synonymized the two species.

★*Micromys steffensi* van de Weerd, 1979

Type Material Left M1, KD 244 (holotype) (van de Weerd 1979:Pl. 3.1).

Type Locality Kardia, Greece, Pliocene [early Ruscinian (MN14)].

Distribution Late Miocene–early Pliocene (MN13–14) of Greece and Turkey.

Occurrences in Greece Dytiko-1, Paliambela-A, Kessani-1, Komanos 1 low A and B, Komanos 1 high A and B and Kardia.

Remarks This is the largest *Micromys* species described until now (e.g., see Hordijk and de Bruijn 2009:Fig. 10), based on a rich collection from Kardia (at least 151 molars). The t7 is well developed as a separate cusp and is anteriorly situated on the upper molars, whereas there is a t1bis on all M2 and some M1 (van de Weerd 1979). The author mentioned that the lower molars of the species are similar to those of *Occitanomys*, but differ in the presence of a third small root and the anterocentral cuspid on the m1 and the presence of two labial accessory cuspulids on the m2. Also in the Ptolemais Basin, the species has been known with rich collections (Fig. 28h–l) from Komanos 1 low A and B and Komanos 1 high A and B (Hordijk and de Bruijn 2009). Outside the Ptolemais Basin, the species has been identified, mainly based on the size, with two molars from Kessani (Vasileiadou et al. 2012), with three from Paliambela-A (as *M.* cf. *steffensi* in Sylvestrou 2002), and from Dytiko 1 (Koufos and Vasileiadou 2015 and references therein).

★*Micromys cingulatus* Storch and Dahlmann, 1995

Type Material Left M1, MAA 20/1520 (holotype) (Storch and Dahlmann 1995: Pl. 2.25).

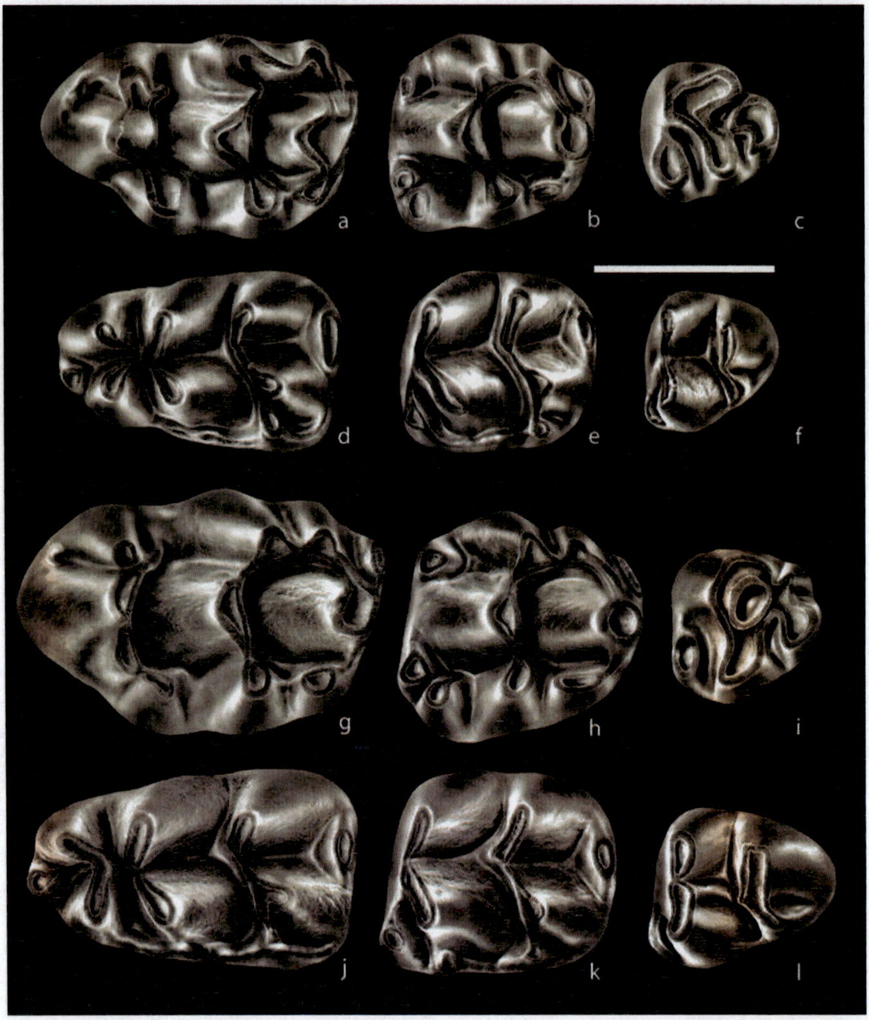

Fig. 28 *Micromys bendai* from Tomeas Eksi 1–3. **a** M1, **b** M2, **c** M3, **d** m1, **e** m2, **f** m3. *Micromys steffensi* from Komanos 1 high B. **g** M1, and from Komanos 1 low A. **h** M2, **i** M3, **j** m1, **k** m2, **l** m3. Images courtesy of K. Hordijk and H. de Bruijn. All specimens figured are reversed. Scale bar equals 1 mm

Type Locality Maramena, Greece, Miocene/Pliocene [Turolian/Ruscinian boundary (MN13/14)].

Distribution Late Miocene–early Pliocene (MN13–14) of Greece and Ukraine.

Occurrences in Greece Maramena.

Remarks This species has been described with rich material from Maramena, based on the following features: molars of medium size; two thirds of the M1–2 lack a t7; the M1–2 bear three roots; on the M2 there is no t1bis and the t6 and t9 are similar in size; the labial cingulum on the m1 is well developed with strong c1 and two to three additional accessory cuspulids in more than 90% of the molars (Storch and Dahlmann 1995). According to Hordijk and de Bruijn (2009), the grouping of this species in *Micromys* is unconvincing, since the presence of a strong labial cingulum on the lower molars is a character shown in *Apodemus*.

***Micromys paricioi* Mein, Moissenet and Adrover, 1983**

Type Material Left M1, CE–4B (holotype), University of Lyon, France (Mein et al. 1983:Pl. 4.1).

Type Locality Celadas 4 b, Spain, Lower Pliocene [Ruscinian (MN14)].

Distribution Late Miocene (late Turolian, MN13)–early Pliocene (early Ruscinian, MN14) of Spain, Greece and Ukraine.

Occurrences in Greece Nea Silata.

Remarks This species was based on a small collection of three molars from the type locality and three additional molars from the coeval locality Peralejos E in the same area (Mein et al. 1983). The molars are very small, of similar size with those of the extant *M. minutus* and the Chinese Turolian species *M. chalceus* (e.g., see Minwer–Barakat et al. 2008:Fig. 3). The upper molars show an elongated ridge between the t4 and t8 instead of a t7, and well-developed t9 and t12, whether the m1 has a large antero-central cuspid and two small isolated cuspulids on the labial cingulum (Mein et al. 1983). Two lower molars from the locality Nea Silata exhibit morphological features and the size of the Spanish species, but due to the presence of a small antero-central cuspid on the m1 and the scarcity of the material, they are identified as *M.* cf. *paricioi* (Vasileiadou et al. 2003).

***Micromys praeminutus* Kretzoi, 1959**

Type Material No holotype has been designated.

Type Locality Csarnóta 2, Hungary, Pliocene [late Ruscinian (MN15)].

Distribution Pliocene of C Europe.

Occurrences in Greece Limni 6.

Remarks This species was insufficiently described by Kretzoi (1959), who did not give a proper diagnosis, descriptions, figures or measurements of the material from the type locality. The only information he gave was that the molars of this species are similar to those of *Apodemus* and differ from them through their smaller size and the absence of accessory cuspules, but the presence of a weak cingulum in their place (presumably on the m1). Later, in a more detailed description of the Csarnóta

2 fauna, he gave figures of M1 and m1 of the species (Kretzoi 1962:Pl.5.5-6). Since then, most Pliocene and Pleistocene fossil *Micromys* is "traditionally" placed in this species, inevitably without comparisons with the material from the type locality. *M. praeminutus*, as currently understood, resembles the extant *M. minutus* in most characters, but differs from it in its significantly larger size (e.g., see van de Weerd 1979). In Greece, the species has been reported only from one Villanyian fauna of Limni 6. De Bruijn and van der Meulen (1979) and van de Weerd (1979) refer to the material as *M. praeminutus*, but van der Meulen and van Kolfschoten (1986) as *M.* cf. *praeminutus* – identification followed herein. The only information published for this material is some measurements of M1 and m1 given by van de Weerd (1979).

***Paraethomys* Petter, 1968**

Type Species *Paraethomys filfilae* Petter, 1968.

Other Taxa Included *P. meini* (Michaux, 1969); *P. jaegeri* Montenat and de Bruijn, 1976; *P. athmeniae* Coiffait and Coiffait, 1981; *P. abaigari* Adrover, Mein and Moissenet, 1988; *P. belmezensis* Castillo Ruiz, 1992.

Distribution Late Miocene–Pleistocene of N Africa, S Europe and W Asia.

Remarks A small sample of two m1, two m2 and three m3 from the locality Chalkoutsi has been provisionally attributed to *Paraethomys*? sp. by de Bruijn (1976), with the comment that the molars resemble both *Paraethomys anomalus* (now *P. meini*) and *Occitanomys provocator*, but the presence of very weak posterior heel on the m1–2, the shape of the m1 and m3 and the location of the wear facets on the m1 remind a *Paraethomys* more than an *Occitanomys* species.

***Paraethomys meini* (Michaux, 1969)**

Nomenclatural and Taxonomical History *Anthracomys meini* in Michaux 1969 (new species); *Paraethomys meini* in Jaeger 1975 (new combination). Considered as the senior synonym of *Occitanomys anomalus* de Bruijn et al. 1970, according to Montenat and de Bruijn (1976) and van de Weerd (1976).

Type Material Right maxilla with M1–3, ST 31 (holotype), University of Montpellier, France (Michaux 1969: Pl. 2.6).

Type Locality Sète, France, Pliocene [late Ruscinian (MN15)].

Distribution Late Miocene–Pliocene of S Europe and N Africa.

Occurrences in Greece Maritsa 1.

Remarks This species shows an isolated t1 on the upper molars and a very reduced labial cingulum on the lower molars (Michaux 1969). It has been reported from Greece only from Maritsa 1, from where the material was initially described as the new species *Occitanomys anomalus* by de Bruijn et al. (1970).

★*Occitanomys anomalus* de Bruijn et al., 1970
[junior synonym of *Paraethomys meini* (Michaux, 1969), according to Montenat and de Bruijn (1976) and van de Weerd (1976)]

Type Material Right M1, no. 339 (holotype) (de Bruijn et al. 1970: Pl. 2.9).
Type Locality Maritsa 1, Rhodes Island, Greece, Upper Miocene [Turolian (MN13)].
Remarks De Bruijn et al. (1970) described the new species *Occitanomys anomalus* from the locality Maritsa 1, based on material consisted of five m1, eight m2, seven m3, eight M1, five M2 and six M3 that showed a well-developed posterior cingulum on the M1–2, a t9 present on the M1 but absent from the M2 and size similar to that of *O. brailloni*. Later, the species was transferred to *Paraethomys* (de Bruijn 1974; Şen 1977). Many authors noticed the morphological and dimensional similarity between *P. anomalus* from Maritsa 1 and *P. meini* from Sète (e.g., Montenat and de Bruijn 1976; van de Weerd 1976) and thus consider the two species synonymous.

***Arvicanthis* Lesson, 1842**

Type Species *Arvicanthis niloticus* (Desmarest, 1822).

Other Taxa Included (only fossil species listed) *A. niloticus irhoudae* Jaeger, 1975; *A. arambourgi* Jaeger, 1975; *A. primaevus* Jaeger, 1976; *A. magnus* Şen et al., 1979; *A. musisii* Mein, 1994; *A. broekhuisi* Hordijk and de Bruijn, 2009.

Distribution Latest Miocene–Late Pleistocene of Africa, Arabian Peninsula, Israel and Greece; extant species are only present in Africa.

Remarks One m2 from the locality Kessani (Thrace, N Greece) has been described by Vasileiadou et al. (2012) as Muridae indet. Some of its characteristics, such as the large angle formed between the labial and lingual cuspids, the lack of connection between the two chevrons and the greatly reduced labial cingulum, are found in some *Arvicanthis* species, such as *A. niloticus* and *A. broekhuisi*, from which it differs in other features. The molar could be described as aff. *Arvicanthis* sp.

★*Arvicanthis broekhuisi* Hordijk and de Bruijn, 2009

Type Material Right M1, TE2–271 (holotype) (Hordijk and de Bruijn 2009: Pl. 2.1).

Type Locality Tomea Eksi 2, Ptolemais Basin, Upper Miocene [Turolian (MN13)].

Distribution Latest Miocene–earliest Pliocene of Greece.

Occurrences in Greece Tomea Eksi 1, Tomea Eksi 2 and Prosilion–Mercurion.

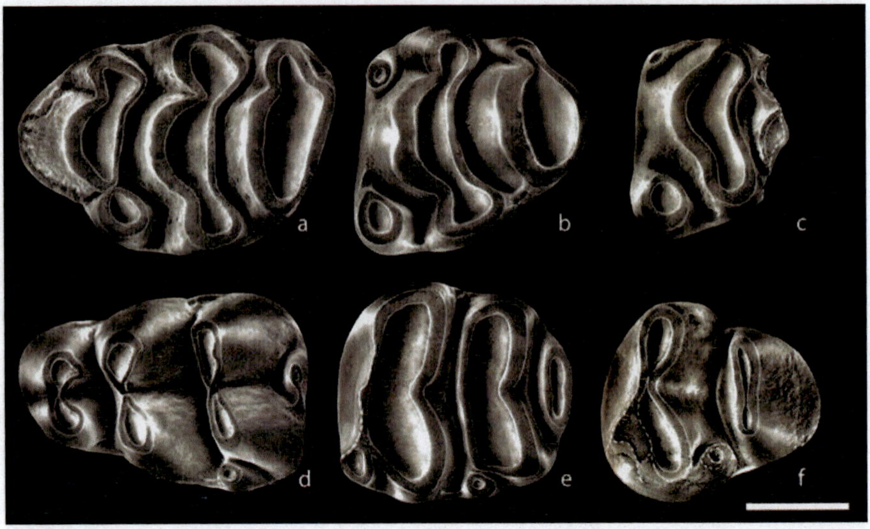

Fig. 29 *Arvicanthis broekhuisi* from Tomea Eksi 2 (type locality). **a** M1 (holotype), **b** M2, **c** M3, **d** m1, **e** m2, **f** m3. Images courtesy of K. Hordijk and H. de Bruijn. All specimens figured as left; c, d, f are reversed. Scale bar equals 1 mm

Remarks The species has molars (Fig. 29) of large size (larger than any other *Arvicanthis* species), five or six roots on the M1 and five on the M2, the t3 is well developed on the M2 and very small on the M3, the antero-labial accessory cusp is absent from the m1, of variable size on the m2 and small on the m3, the c1 is large on the m1 and of variable size on the m2 and the terminal heel is small and oval on the m1 and has the shape of a long transverse ridge on the m2, whereas the m1 is short and wide (Hordijk and de Bruijn 2009). According to the authors of the species, it is the oldest record of *Arvicanthis*, other than *Arvicanthis* sp. from the Lukeino Formation of Kenya, and has the largest and most specialized dentition. These led them to the conclusion that the species is a side-branch of the African *Arvicanthis* and not an ancestor, which reached Greece during the Messinian. The ancestry of the group should be sought for in the late Vallesian or early Turolian faunas of E Africa (Hordijk and de Bruijn 2009). Hordijk and de Bruijn (2009) have also included in this species one M1 and one M2, both slightly damaged, from Tomea Eksi 1, identical to the molars from the type locality. Further, they attributed one M2 and two m2 from Prosilion–Mercurion to *A.* cf. *broekhuisi*, as the m2 is smaller and more square than that from the type locality and the postero-labial accessory cusp is absent from one m2, whereas it is always present in the m2 from Tomea Eksi 2 (Hordijk and de Bruijn 2009: 30).

Pelomys Peters, 1852

Type Species *Mus (Pelomys) fallax* Peters, 1852.

Other Taxa Included (only fossil species listed) *P. europeus* de Bruijn et al., 1970; *P. dietrichi* Jaeger, 1976; *P. orientalis* Şen et al., 1979.

Distribution Latest Miocene of Greece, Pliocene of Algeria and Afghanistan and Pleistocene of Tanzania; present today only in Africa.

★*Pelomys europeus* de Bruijn et al., 1970

Type Material Right M1–2, no 551(holotype) (de Bruijn et al. 1970: Pl. 3.2).

Type Locality Maritsa 1, Rhodes Island, Greece, Upper Miocene [Turolian (MN13)].

Distribution Late Miocene of Greece and Pliocene of Algeria (Amama 3).

Occurrences in Greece Maritsa 1.

Remarks The restricted material from the type locality (Fig. 30) exhibits the following diagnostic characters: teeth with no stephanodonty, first and second lamina of M1 each with three well-isolated cusps that give a very symmetrical pattern, t1 and t4 situated at the same level as t3 and t6, respectively, M1 with median anterocone, t1 larger than t3 on the M2, terminal heel of the m1–2 small and rounded and M3 large (de Bruijn et al. 1970). It is possible that this species does not belong to

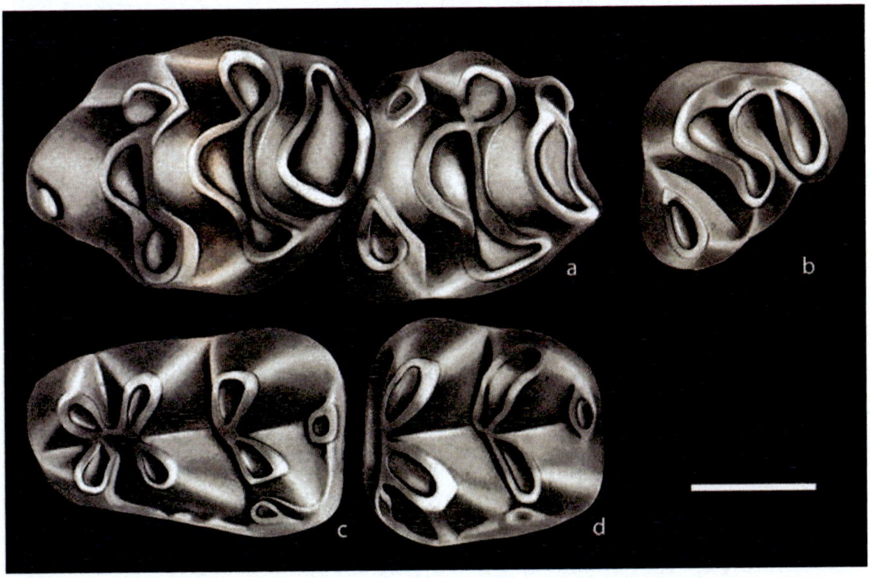

Fig. 30 *Pelomys europeus* from Maritsa 1 (type locality). **a** M1–2 (holotype), **b** M3, **c** m1, **d** m2. Images courtesy of H. de Bruijn. All specimens are right but reversed and figured as left. Scale bar equals 1 mm

Pelomys but to *Rattus*; the absence of upper incisors from Maritsa 1 hampers the certain generic affiliation (de Bruijn et al. 1970:549).

***Thallomys* Thomas, 1920**

Type Species *Mus nigricauda* Thomas, 1882.

Other Taxa Included (only extinct species listed) *T. debruyni* Broom, 1948; *T. quadrilobatus* Jaeger, 1976; *T. laetolilensis* Denys, 1987.

Distribution Pliocene and Pleistocene of Africa, probably Pliocene of Greece; present today only in Africa.

Remarks Van der Meulen and van Kolfschoten (1986) listed *Thallomys*? in the faunal list of the Pliocene locality Damatria (Rhodes Island), with no further details on the material.

***Apodemus* Kaup, 1829**

Type Species *Mus agrarius* Pallas, 1771.

Other Taxa Included (only extinct species listed) *A. atavus* Heller, 1936; *A. orientalis* (Schaub, 1938); *A. levantinus* Bate, 1942; *A. caesareanus* Bate, 1942; *A. dominans* Kretzoi, 1959; *A. jeanteti* Michaux, 1967; *A. mannu* Thaler, 1973; *A. gudrunae* van de Weerd, 1976; *A. gorafensis* Ruiz Bustos et al., 1984; *A. maastrichtiensis* van Kolfschoten, 1985; *A. agustii* Martín Suárez, 1988; *A. etruscus* Engesser, 1989; *A. qiui* Wenyu and Flynn, 1991; *A. zhangwagouensis* Wenyu and Flynn, 1991; *A. lii* Qiu and Storch, 2000; *A. asianicus* Zheng, 2004.

Distribution Late Miocene (late Vallesian)–present in Eurasia.

Remarks We follow Hordijk and de Bruijn (2009) in including in *Apodemus* any homogenous sample of at least 10 M1 the majority of which has a well-formed t7 (exc. *Micromys* and *Rhagapodemus*). There are at least 20 extant *Apodemus* spp., some of which have been reported from Pleistocene Greek fossil localities: *A.* cf. *mystacinus* has been reported from the Biharian fauna of Gerakarou 1 (Koufos et al. 1995), the Lower Pleistocene locality Lagkada (van der Meulen and van Kolfschoten 1986) and the Upper Pleistocene locality Trypiti in Naxos [(originally mentioned by Sondaar (1971) as a large endemic *Apodemus* sp. and later described by van der Geer et al. 2014)]; *A. mystacinus* from the Pleistocene localities Kalymnos-Xi (Kuss and Storch 1978), Ravin Voulgarakis (Koliadimou 1996), Tourkobounia 2 (van der Meulen and Doukas 2001), Kaiafas, Zeli 2, 2A and 2B, Volos (van der Meulen and van Kolfschoten 1986), Varkiza 1 and 2 (van de Weerd 1973), Latomi 1 (Storch 1975), Loutra Almopias Cave LAC and LAC Ia (Chatzopoulou 2014) and Arnissa (Mayhew 1978); *A. sylvaticus/flavicollis* (dental elements of the two species cannot be easily distinguished, and we here chose to group them even if only one of them has been initially identified in one of the following localities) from Marathoussa (N Greece) (Koufos et al. 2001), Ravin Voulgarakis (Koliadimou 1996), Kaiafas, Zeli 2, 2A and 2B, Volos (van der Meulen and van Kolfschoten 1986), Marathousa 1 (Peloponnese) (Doukas et al. 2018), Loutra Almopias Cave LAC and LAC Ia (Chatzopoulou 2014) and Dimitra 2 (Syrides and Koliadimou 1994); *A.* cf. *sylvaticus/flavicollis* from Lagkada (van der Meulen and van Kolfschoten 1986); *A.* cf. *flavicollis* from Tourkobounia 2 (van den Meulen and Doukas 2001); *A.* cf. *sylvaticus* from Choremi 3 (van Vugt et al. 2000); and *A. uralensis* from Loutra Almopias Cave LAC and LAC Ia (Chatzopoulou 2014).

Kuss and Storch (1978) assigned to *A. levantinus* Bate, 1942 three m1, three m2, four m3, two M1, five M2 and two M3 from Kalymnos-Xi, based on similarity in the crown outline, labial cingula of the lower molars and cusp arrangement with the descriptions and illustrations given by Bate. They also attributed three m1, two m2, four m3, two M1, three M2 and one M3 from Kalymnos-Xi to *A. caesareanus* Bate, 1942, based on similarities with the species, such as the bulky and strongly anteriorly pointing anterior root of the M1 and the strongly asymmetrical anteroconid complex that bends labially of the m1. However, later, van der Meulen and van Kolfschoten (1986) assigned all the small-sized molars from the locality to *A. sylvaticus/flavicolis*, followed herein.

The presence of *Apodemus* sp., without details on the material, has been reported from the late Turolian fauna of Lava 2 (de Bruijn et al. 1999), the late Ruscinian fauna of Kastoria 1 (van der Meulen and van Kolfschoten 1986) and the late Biharian fauna from Choremi 4 (van Vugt et al. 2000). Van de Weerd (1973) reported the presence of 22 M1 of *Apodemus* sp. in the Pleistocene locality Varkiza 2; the molars could belong to any of the extant *A. flavicollis*, *A. sylvaticus* and *A. microps*. A large collection consisting of two maxillary and six mandibular fragments, 92 M1, 61 M2, 21 M3, 75 m1, 61 m2 and 36 m3 from Latomi 1 (Chios island) have been assigned to *Apodemus* sp. by Storch (1975); the molars are smaller than those of *A. mystacinus* from the same locality, but the author did not attempt a species determination, as the morphological and metrical features fit those of many extant *Apodemus* species. Finally, five m1, one m2, one m3, six M1 and two M2 from Arnissa have been assigned to *Apodemus* sp. by Mayhew (1978). The author discussed that the molars are smaller than those of *A. mystacinus* and larger than those of *A. microps*, but morphological and metrical variability falls within that of both *A. sylvaticus* and *A. flavicollis*.

***Apodemus gorafensis* Ruiz Bustos et al., 1984**

Type Material Left M1, G–8 (holotype), National Institute of Geology C.S.I.C., Madrid (Ruiz Bustos et al. 1984: Fig. 2.3).

Type Locality Gorafe–A, Granada, Spain, Lower Pliocene [Ruscinian (MN14)].

Distribution Turolian–Ruscinian (MN13–15) of Spain and Greece.

Occurrences in Greece Monasteri, Tomea Eksi 1, Tomea Eksi 2, Kessani-2, Kessani-1, Ano Metochi 2–3, Maramena and Nea Silata.

Remarks This *Apodemus* species is diagnosed through its medium-sized molars, the always present t7 on the M1 and the strong antero-central cuspid on the m1 (Ruiz Bustos et al. 1984). De Bruijn (1989) reported the presence of *A. gudrunae* in the faunal lists for the localities Monasteri and Ano Metochi 2–3, without details on the material, but Storch and Dahlmann (1995) assigned those assemblages to *A. gorafensis*. The latter authors described a large sample of 250 middle-sized first and second upper and lower molars from Maramena and commented that their morphology is very close to that of both *A. gorafensis* and *A. gudrunae* and their size is intermediate between that of the two species; however, the presence of a t7 on all M1–2 assigns the material in the former species (Storch and Dahlmann 1995:122). Vasileiadou (2001) assigned to the species a small sample of five m1, three m2, one m3, two M2 and four M3 from Nea Silata, based on the well-developed and separated antero-central cuspid on the m1 and the strong labial cingulum with one or two large accessory cuspids on the m2 (Vasileiadou et al. 2003). The species has also been found in Kessani (Vasileiadou et al. 2012), represented by six m1, eight m2, one m3, three M1 and four M2. Small samples of *A. gorafensis* have also been found in Tomea Eksi 1 (six molars) and Tomea Eksi 2 (13 molars) (Hordijk and de Bruijn 2009). The authors commented on the overlap in size and the morphological similarity of the material with both *A. gorafensis* and *A. gudrunae*; they assigned it to the former species only based on "the slightly better size match and the greater similarity in the development of the posterior spur of the t3 in the M1 in that species" (Hordijk and de Bruijn 2009:36).

Apodemus dominans **Kretzoi, 1959**

Taxonomical History Senior synonym of *Apodemus occitanus* Pasquier, 1974, according to de Bruijn and van der Meulen (1975).

Type Material Not designated.

Type Locality Csarnóta 2, Hungary, Lower Pliocene [Ruscinian (MN15)].

Distribution Late Miocene (Turolian, MN13)–late Early Pleistocene (Biharian, MNQ18) of Europe and Asia.

Occurrences in Greece Maritsa 1, Ano Metochi 2–3, Maramena, Nea Silata, Spilia 0, 1, Tomea Eksi 3, Limni 3, Vorio 3 and 3A, Kato Afiartis, Apolakkia, Spilia 3–4, Notio 1, Kastoria 1, Tourkobounia 1, Limni 6, Archipolis, Damatria, Rema Aslan, Karnezeika, Kardamena and Kastoria 2.

Remarks Kretzoi (1959) created the species without presenting type material, diagnosis or descriptions of the material from Csarnóta 2. He only mentioned the similarity of the new species with the extant *A. sylvaticus*, from which he separated his species only because it is of older age (Kretzoi 1959:243). Later, in a more detailed description of the Csarnóta 2 fauna, Kretzoi gave figures of the species (Kretzoi 1962:Pl. 4.4–6 and Pl. 5.2, 4), mentioning that it is the most abundant species in the fauna with more than 8000 molars, but again gave no diagnosis or descriptions.

De Bruijn et al. (1970) described a small assemblage from the Turolian fauna of Maritsa 1, consisting of three m1, one m2, two m3, two M1 and two M2, which differ from *A. sylvaticus* in the presence of a strong t12 and three roots on the M1–2. They preferred to assign the material to *A.* cf. *dominans*, admitting that the species is not adequately known (de Bruijn et al. 1970:545). Storch and Dahlmann (1995) described a large assemblage (1221 M1–2 and m1–2) of *A.* cf. *dominans* from the Turolian/Ruscinian fauna of Maramena, which shows similarities with the material from Csarnóta 2, apart from the presence of a fourth root on 15% of the M1 and 26% of the M2 and the presence of a deep vertical groove on the lingual root on the 18% of the M1 and 16% of the M2 (Storch and Dahlmann 1995:123). The Villanyian fauna of Tourkobounia 1 has yielded an assemblage of 172 M1–2 and m1–2 assigned to *A. dominans*, based on the similarity with material from the type locality (de Bruijn and van der Meulen 1975). Both assemblages from Maramena and Tourkobounia 1 show a large morphological and metrical variation, and it cannot be ruled out that they both contain also *A. atavus* dental elements (see following paragraph). The species has also been described from the Turolian/Ruscinian fauna of Nea Silata, where it is present with 16 molars (Vasileiadou 2001, Vasileiadou et al. 2003). Hordijk and de Bruijn (2009) assigned 19 medium-sized molars from Notio 1 to *A. dominans*, based on their similarity with the material from the type locality and their larger size than that of the medium-sized *Apodemus* molars in lower horizons of the Ptolemais Basin (assigned to *A. atavus*; see following paragraph). The authors also mentioned the possibility of the presence of *A. dominans* dental elements in the assemblages from Tomea Eksi 3 and Vorio 3A (Hordijk and de Bruijn 2009:36; see following paragraph). Fifteen molars from Apolakkia-2 have been assigned to *A.* aff. *dominans* because they are larger than the material from the type locality and the separation between the t4 and the t7 is not as deep as in the M1 from the type locality on two out of four M1 (van de Weerd et al. 1982). Identical molars have been found in Karpathos (two m2, two m3, one M1, one M2), which have also been assigned to *A.* aff. *dominans* (Daams and van der Weerd 1980). De Bruijn (1989) listed *A. dominans* in the faunal lists of Ano Metochi 2–3 and Spilia 0 and 1, giving no details on the material. *A. dominans* has been also reported from the localities Spilia 3–4, Kastoria 1, Limni 6 and Kastoria 2; *A.* cf. *dominans* from the localities Limni 3, Rema Aslan, Kardamena, Archipolis and Damatria (van der Meulen and van Kolfschoten 1986) and Karnezeika (Kokotini et al. 2019); and *A.* aff. *dominans* from Agios Ioannis (van der Meulen and van Kolfschoten 1986), with no information on the materials.

***Apodemus atavus* Heller, 1936**

Type Material Partial maxilla with M1–M2 (holotype) (Heller 1936: Tab. X.2).

Type Locality Gundersheim, Germany, Pliocene [Villanyian (MN16)].

Distribution Late Miocene (Turolian, MN13)–late Early Pleistocene (Biharian, MNQ19) of Eurasia.

Occurrences in Greece Kessani-1, 2, Komanos 1 low A and B, Komanos 1 high A and B, Kardia, Tomea Eksi 3, Ptolemais 1, Vorio 3 and 3A, Ptolemais 3 and Notio 1.

Remarks The species shares many morphological characters with *A. dominans*: the presence of a relatively well-developed t12 and three roots on the M1–2 and of a distinct c1 on the m1–2. However, the cheek teeth of *A. atavus* are smaller than those of *A. dominans*, on the m1 the antero-central cuspid is connected with the antero-lingual and antero-labial cuspids in the former species but usually isolated in the latter, there is always a connection between protoconid–metaconid chevron and the anterior complex in the former but not in the latter and on the M1 there is always a strong t7–t4 connection in the former, but the two cusps are usually separated in the latter species. The size difference, in combination with the subtle morphological differences between the two species, have led authors (e.g., Fejfar and Storch 1990:147) to compare this situation with that between the extant species *A. flavicollis* and *A. sylvaticus*: the former species has slightly larger molars, but the morphological differences are not apparent, allowing no certain assignment in one of the two species of Pleistocene middle-sized *Apodemus* assemblages. As Popov (2004:459) mentioned "*As in recent forms, the determination of sympatric fossil species of the genus, based on dental features alone, meets with considerable difficulties*". Indeed, many Turolian–Villanyian middle-sized *Apodemus* assemblages probably contain elements of both *A. dominans* and *A. atavus* (just like *A. flavicollis* and *A. sylvaticus* co-exist in modern ecosystems); it is thus risky to assign, mainly small assemblages, to one species. Therefore, some of the assemblages that have been assigned to *A.* (cf.) *dominans* in the literature might better be assigned to *A. atavus/dominans*. For instance, Vasileiadou et al. (2012) assigned one m1, one m3 and two M1 from Kessani to *A.* cf. *dominans*, admitting that the molars show metrical and morphological similarities also with *A. atavus*; this species is here assigned to *A. atavus/dominans*.

Hordijk and de Bruijn (2009) assigned to *A. atavus* the medium-sized murin dental elements from Komanos 1 (low A and B and high A and B), Tomea Eksi 3, Vorio 3A and Notio 1. The assemblages from Tomea Eksi 3 and Vorio 3A show a rather large size variation, with molars in the size range of *A. dominans*, and the authors commented on the possibility of the co-occurrence of the two species in these assemblages (Hordijk and de Bruijn 2009:36). Van de Weerd (1979) attributed seven molars from Kardia, 26 molars from Ptolemais 1 and nine molars from Ptolemais 3 to *A. dominans*, but Hordijk and de Bruijn (2009) attributed them to *A. atavus* without justification, an identification followed herein.

Apodemus jeanteti Michaux, 1967

Taxonomical History Considered as the senior synonym of *Rhagapodemus primitivus* Aguilar et al., 1995, according to Martín Suárez and Mein (1998).

Type Material Partial right maxilla with M1–2, Sey 12 (holotype), University of Montpellier (Michaux 1967: Fig. 7a).

Type Locality Seynes, France, late Upper Pliocene

Distribution Late Miocene (Turolian, MN13)–Pliocene of S Europe.

Occurrences in Greece Spilia 3–4.

Remarks According to Michaux's (1967) diagnosis, the molars have size similar to that of the extant *A. mystacinus*; the t1, t2 and t3 of the M1 are almost aligned; there is no posterior spur on the t1; the antero-central cuspid of the m1 is either absent or strongly reduced. *A.* cf. *jeanteti* is in the faunal list of the Pliocene locality Spilia 3–4 (van der Meulen and van Kolfschoten 1986), with no details.

Rhagapodemus Kretzoi, 1959

Type Species *Rhagapodemus frequens* Kretzoi, 1959.

Other Taxa Included *R. primaevus* (Hugueney and Mein, 1965); *R. hautimagnensis* Mein and Michaux, 1970; *R. ballesioi* Mein and Michaux, 1970; *R. vandeweerdi* de Bruijn and van der Meulen, 1975; *R. minor* (Brandy, 1978); *R. debruijni* (Kotlia, 1992); *R. azzarolii* Angelone and Kotsakis, 2001.

Distribution Late Miocene–Middle Pleistocene of Europe and Asia.

Remarks *Rhagapodemus* is distinguished from *Apodemus* through its robust semi-hypsodont cheek teeth with "vertical" cusps and the isolated t1 on the M1 (Hordijk and de Bruijn 2006).

Rhagapodemus primaevus (Hugueney and Mein, 1965)

Nomenclatural and Taxonomical History *Apodemus primaevus* in Hugueney and Mein 1965 (new species); *Rhagapodemus primaevus* in Mein 1989 (new combination).

Type Material Left M1, no 65054 (holotype), Département des Sciences de la Terre, Université Claude-Bernard, Lyon (Hugueney and Mein 1965: Pl. 1.18).

Type Locality Lissieu, France, Upper Miocene [MN13].

Distribution Late Miocene–early Pliocene of Europe (Spain, France, Greece).

Occurrences in Greece Kessani, Maramena, Komanos 1 low A and B, Komanos 1 high A and B, Kardia, Vorio 1, Tomea Eksi 3, Ptolemais 1, Vorio 3 and 3A and Ptolemais 3.

Remarks *R. primaevus* is the oldest species of the genus, characterized by relatively small-sized and low-crowned molars, relatively less vertical cusps/-ids, t1 isolated in one third of the upper molars, absence of a posterior spur on the t2 of the M1 and labial cuspids lower than main ones on the lower molars (Martín Suárez and Mein 1998). Hordijk and de Bruijn (2009) assigned to *R. primaevus* the molars from MN14 faunas of the Ptolemais Basin (Komanos 1 low A and B, Komanos 1 high A and B, Vorio 1, Tomea Eksi 3, Vorio 3), as their morphological and metrical characters are within the range for the material from the type locality. *R. hautimagnensis* had been identified in Kardia, Ptolemais 1 and Ptolemais 3 of the Ptolemais Basin, by van de Weerd (1979), but Hordijk and de Bruijn (2009) included the material in *R. primaevus*. A sample of 48 first and second molars from Maramena has been assigned to *R. primaevus* by Storch and Dahlmann (1995). One M1 and one M2 from Kessani have been assigned to this species, based on their size and the anterior position of the t1 (Vasileiadou et al. 2012).

***Rhagapodemus hautimagnensis* Mein and Michaux, 1970**

Type Material Left M1, University Lyon 65350 (holotype) (Mein and Michaux 1970: Pl. 2.1).

Type Locality Hautimagne, France, Lower Pliocene [early Ruscinian (MN14)].

Distribution Late Miocene–Early Pleistocene of Europe.

Occurrences in Greece Monasteri, Ano Metochi 2–3 and Spilia 0 and 1.

Remarks The molars of this species are slightly smaller, and the cusps of the first molar are lower and more oblique than those of *R. frequens* (Mein and Michaux 1970). This species is reported in the faunal lists given for the localities Monasteri and Ano Metochi 2–3 (de Bruijn 1989) and Spilia 0 and 1 (van der Meulen and van Kolfschoten 1986), with no details on the materials. However, the locality Maramena, thought to represent more or less the same horizon with Ano Metochi 2–3 (de Bruijn 1995:88), has yielded *R. primaevus* and not *R. hautimagnensis* (Storch and Dahlmann 1995); it is possible that re-examination of the materials from these localities will prove their assignment also to *R. primaevus*.

★*Rhagapodemus vandeweerdi* de Bruijn and van der Meulen, 1975

Type Material Right M1, no. 231(holotype) (de Bruijn et al. 1970: Pl. 1.3).

Type Locality Maritsa 1, Rhodes Island, Greece, Upper Miocene [Turolian (MN13)].

Distribution Late Miocene of Greece (found only in the type locality).

Remarks The rich material from Maritsa 1 (Fig. 31a–f) was initially attributed to *Apodemus* aff. *jeanteti* (de Bruijn et al. 1970). Later, de Bruijn and van der Meulen (1975) created the species *R. vandeweerdi* diagnosed through its relatively low cheek teeth, the rounded and isolated t1 in early wear stages on the M1, the postero-

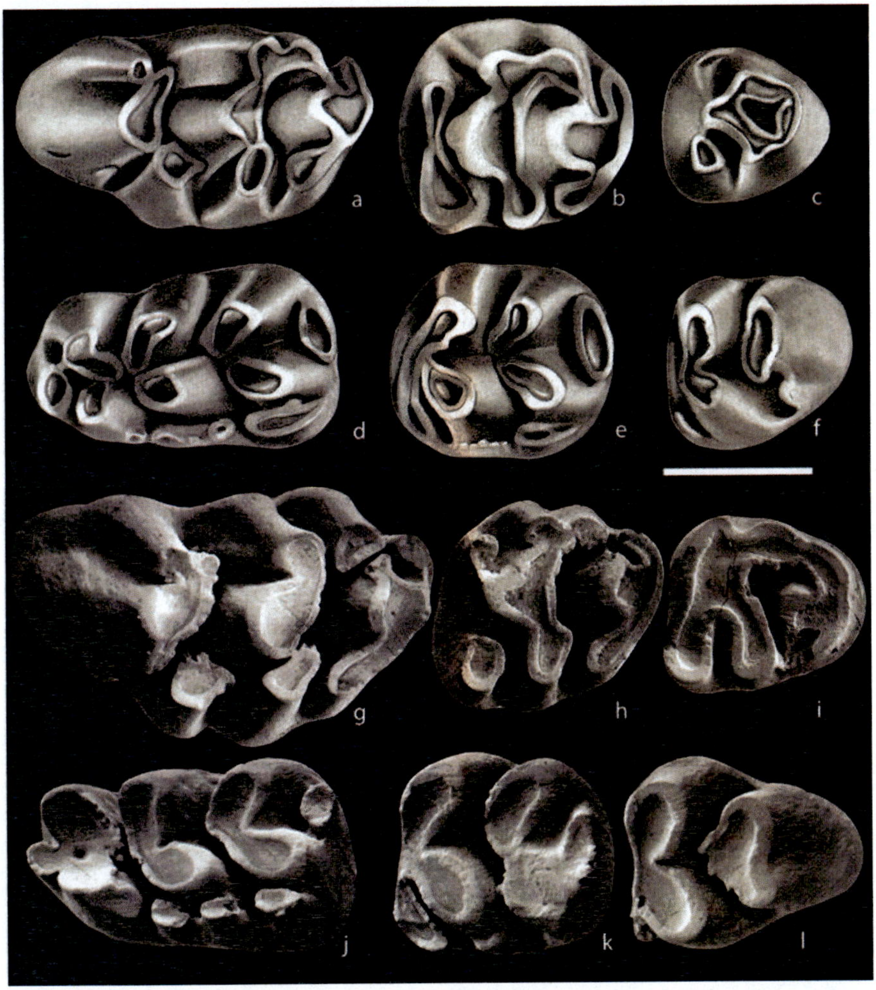

Fig. 31 *Rhagapodemus vandeweerdi* from Maritsa 1 (type locality). **a** M1, **b** M2, **c** M3, **d** m1 (holotype), **e** m2, **f** m3. *Rhagapodemus frequens athensis* from Tourkobounia 1 (type locality). **g** M1 (holotype), **h** M2, **i** M3, **j** m1, **k** m2, **l** m3. Images courtesy of H. de Bruijn. All specimens figured as left; **a**–**g** are reversed. Scale bar equals 1 mm

labially directed spur on the t2 of M1, the presence on all m1 of an antero-central cuspid and the elongated posterior cuspid on the labial cingulum of the m1–2.

***Rhagapodemus frequens* Kretzoi, 1959**

Type Material No holotype has been designated for the species.

Type Locality Csarnóta 2, Hungary, Pliocene [MN15].

Distribution Pliocene–Early Pleistocene of Europe.

Occurrences in Greece Notio 1, Kastoria 1 and Archipolis.

Remarks Kretzoi (1959) created the genus and the species on bases of the similarity with *Apodemus* in the arrangement of the cusps but the higher degree of hypsodonty and an isolated t1 on the M1. Unfortunately, he did not designate a holotype nor did he describe or figure any specimens from the type locality. Later, in a more detailed description of the Csarnóta 2 fauna, Kretzoi gave a figure of one right M1 of the species (Kretzoi 1962:Pl. 4.7). Subsequent authors, based on material from Csarnóta 2 available to them, gave some morphological features of the species: round and isolated t1, fused t2 and t3 on the M1, antero-posteriorly compressed t3 of variable size on the M2, antero-central cusp of variable size, rare connection of the anteroconid complex with the protoconid–metaconid chevron and c1 fused with the hypoconid on the m1 and very weak or absent accessory cusps on the labial cingulum of the m2 (Hordijk and de Bruijn 2009:38). The species is present with 14 molars in the late Ruscinian fauna of Notio 1 (Hordijk and de Bruijn 2009). It has also been reported in Kastoria 1, and *R.* cf. *frequens* has been reported from Archipolis (van der Meulen and van Kolfschoten 1986), without information on the materials.

★*Rhagapodemus frequens athensis* de Bruijn and van der Meulen, 1975

Type Material Right M1, no. 381 (holotype) (de Bruijn and van der Meulen 1975: Pl. 3.7).

Type Locality Tourkobounia 1, Greece, Pliocene [Villanyian (MN16)].

Distribution Pliocene (Villanyian) of Greece.

Occurrences in Greece Tourkobounia 1, Limni 6 and Rema Aslan.

Remarks The molars from Tourkobounia 1 (six m1, eight m2, seven m3, 11 M1, nine M2 and seven M3; Fig. 31g–l) are morphologically very similar to those of *R. frequens frequens*, but they are larger and higher, leading de Bruijn and van der Meulen (1975) to the creation of a new subspecies. The authors reported that the length of the Csarnóta 2 upper molars available to them for comparisons is outside the range of the Tourkobounia ones, whereas the length of the Csarnóta 2 lower molars is close to that of the smallest Tourkobounia specimens. Martín Suárez and Mein (1998), in their revision of various murid genera, elevated *R. frequens athensis* to a species level; however, here we follow the original opinion of de Bruijn and van der Meulen and retain it as a subspecies. *R. frequens athensis* has also been reported from the locality Limni 6, whereas the faunal list for the locality Rema Aslan includes *R.* cf. *frequens athensis* (van der Meulen and van Kolfschoten 1986), both without details on the material.

***Orientalomys* de Bruijn and van der Meulen, 1975**

Type Species *Parapodemus similis* Argyropulo and Pidoplichka, 1939.

Other Taxa Included *O. galaticus* (Şen, 1975); *O. sibiricus* (Erbajeva, 1975); *O. adirganus* Tjutkova, 1993; *O. sinensis* Qiu and Storch, 2000.

Distribution Late Miocene–Early Pleistocene of E Europe and Asia.

Remarks de Bruijn and van der Meulen (1975) included in this genus also the species *Stephanomys pusillus* Schaub, 1938 from Ertemte and *Stephanomys schaubi* Teilhard de Chardin, 1940 from locality 18 near Peking (China); the former one is now referred to *Hansdebruijnia* and the latter one remains in *Stephanomys*. The main characteristics of the genus are the high degree of stephanodonty, the connection of the t1 with the t5 and its separation from the t2 on the M1, the presence of more than three roots on the M1–2 and more than two on the m1–2, the presence of a small and isolated antero-central cuspid on the m1 and the presence of large accessory cusps on the m1–2 (de Bruijn and van der Meulen 1975).

***Orientalomys similis* (Argyropulo and Pidoplichka, 1939)**

Nomenclatural and Taxonomical History *Parapodemus similis* in Argyropulo and Pidoplichka 1939 (new species); *Orientalomys similis* in de Bruijn and van der Meulen 1975 (new combination).

Type Material Left mandible with m1–3, no. 6585 (Argyropulo and Pidoplichka 1939: Figs. b, d, e), lectotype assigned by de Bruijn and van der Meulen (1975).

Type Locality Odessa, Ukraine, Pliocene [late Ruscinian (MN15)].

Distribution Pliocene of Romania, Ukraine, Greece and possibly Turkey.

Occurrences in Greece Tourkobounia 1 and Limni 6.

Remarks The cheek teeth from Tourkobounia 1 (nine m1, four m2, seven m3, 10 M1, six M2 and six M3) are low and show high degree of stephanodonty (de Bruijn and van der Meulen 1975). The species is also reported from the locality Limni 6 (van der Meulen and van Kolfschoten 1986), with no further details.

★*Kritimys* **Kuss and Misonne, 1968**

Type Species *Mus catreus* Bate, 1912.

Other Taxa Included *Kritimys kiridus* (Bate, 1942).

Distribution Endemic genus of the Pleistocene in Crete.

Remarks A detailed report on this endemic genus of Crete is provided by Lyras et al. (volume 2).

Mus **Linnaeus, 1758**

Type Species *Mus musculus* Linnaeus, 1758.

Other Taxa Included The genus includes about 38 extant species (Macdonald 2009), several of which have been found in Pleistocene and Holocene fossil and subfossil assemblages [e.g., *M. musculus* Linnaeus, 1758; *M. minutoides* (A. Smith, 1834); *M. spicilegus* Petényi, 1882; *M. spretus* Lataste, 1883; *M. caroli* Bonhote, 1902; *M. indutus* (Thomas, 1910); *M. shortridgei* (Thomas, 1914); *M. pahari* Thomas, 1916]; (extinct) *M. minotaurus* Bate, 1942; *M. haouzi* (Jaeger, 1975); *M. petteri* Jaeger, 1976; *M. bateae* Mayhew, 1977; *M. aegaeus* Kuss and Storch, 1978; *M. auctor* Jacobs, 1978; *M. elegans* Şen, 1983; *M. jacobsi* Kotlia, 1992; *M. hamidae* (Geraads, 1994); *M. dhailai* Kotlia, 1996; *M. linnaeusi* Patnaik, 1997; *M. ique* Geraads, 1998; *M. flynni* Patnaik, 2001; *M. dulamensis* Kotlia, 2008; *M. narmadaensis* Kotlia, Joshi and Joshi, 2011; *M. denizliensis* Erten, Şen and Sagular, 2015.

Distribution Late Miocene–present, cosmopolitan.

Remarks The late Biharian fauna of Choremi 4 has revealed 39 molars assigned to *M.* cf. *spretus* by van Vugt et al. (2000); the molars are very similar to those of the extant western Mediterranean mouse, but the antero-labial cingulum on the M2 is better developed than on recent material. Van de Weerd (1973) reported the presence of two m1, one m2, two M1 and two M2 of *M. musculus* in the locality Varkiza 2; the material has no differences from recent *M. musculus* from Greece. Two m1 from the Upper Pleistocene locality Loutra Almopias Cave LAC have been attributed to the recent steppe mouse *M. spicilegus* by Chatzopoulou (2014). An assemblage consisting of eight m1, nine m2, three m3, six M1, one M3 and several partial toothless mandibles from the mid-Pleistocene locality Latomi 1 has been assigned to *Mus* sp. by Storch (1975). According to the author, the M1 is relatively large and the M3 relatively small, whereas the lingual root of the M1 is strongly bent forwards. The size and morphology of the molars fit well with those of the recent house mouse, *M. musculus*, but identification to species level of such a small assemblage is not appropriate, according to Storch (1975). *Mus* sp. has been reported from the late Biharian fauna of Choremi 3 (one specimen) (van Vugt et al. 2000). For the endemic Cretan Pleistocene–Holocene species *Mus bateae* Mayhew, 1977 and *Mus minotaurus* Bate, 1942, see Lyras et al. (volume 2).

★*Mus aegaeus* Kuss and Storch, 1978

Type Material Left m2, SMF 76/1331(holotype) (Kuss and Storch 1978: Fig. 42).

Type Locality Kalymnos-Xi, Greece, Lower Pleistocene [Biharian].

Distribution Late Early Pleistocene (Biharian) of Greece (only known from the type locality).

Remarks The species is only based on the holotype, but Kuss and Storch (1978) suggested that its morphological difference from those of other murins of similar age allow the establishment of a new species. This m2 is low-crowned, very large and relatively narrow, of similar size with those of *M. minotaurus*, showing an isolated anterolabial cuspid, a very large protoconid and a small metaconid.

***Rattus* Fischer von Waldheim, 1803**

Type Species *Mus rattus* Linnaeus, 1758.

Other Taxa Included The genus includes around 56 extant species (Macdonald 2009), some of which have also been found in Pleistocene and Holocene fossil and subfossil assemblages [e.g., *R. rattus* (Linnaeus, 1758); *R. norvegicus* (Berkenhout, 1769); *R. pyctoris* (Hodgson, 1845); *R. argentiventer* (Robinson and Kloss, 1916)]; (extinct) *R. haasi* Tchernov, 1968; *R. dobrogicus* Terzea, 1973; *R. casimcensis* Radulescu and Samson, 1973; *R. trinilensis* Musser, 1982.

Distribution Early Pleistocene–present, cosmopolitan.

Remarks Storch (1975) assigned six molars from the mid-Pleistocene locality Latomi 1 to *R. rattus*, based on their morphological and metrical similarity with the modern black rat. He also assigned one M2 to cf. *Rattus* sp.; the molar is larger than those of *R. rattus*, with higher cusps strongly leaning posteriorly and a larger t1, and clearly differs from those of *R. rattus*, *R. turkestanicus*, *R. norvegicus*, *R. haasi* and *Kritimys* (Storch 1975:179–180).

Subfamily Arvicolinae Gray, 1821

Type Genus *Arvicola* Lacépède, 1799.

Remarks Arvicolinae indet. has been reported in the faunal lists for Tourkobounia 3 and Tourkobounia 4 (Reumer and Doukas 1985). Van de Weerd (1973) reported the presence of several unidentifiable fragmentary vole molars in Varkiza 2.

Microtodon Miller, 1927
(= / or including as subgenera _Baranomys_ Kormos, 1933 and _Bjornkurtenia_ Kowalski, 1992, according to de Bruijn 2010)

Type Species *Sigmodon atavus* Schlosser, 1924.

Other Taxa Included *M. loczyi* (Kormos, 1933); *M. longidens* Kowalski, 1960; *M. canterranensis* (Michaux, 1976); *M. komanensis* Hordijk and de Bruijn, 2009.

Distribution Late Turolian–early Pliocene (late Ruscinian) of Eurasia.

Remarks One small M1 from the locality Komanos 1 low A has been attributed to *Microtodon* sp.; it is too small to belong to *M. komanensis* described from the area, whereas its size is similar to that of *M. atavus*, but its prisms are better developed than those of this species; it is within the size range of *M. canterranensis* and *M. longidens* (Hordijk and de Bruijn 2009:52).

★_Microtodon komanensis_ Hordijk and de Bruijn, 2009

Type Material m1, KO1low A–147 (holotype) (Hordijk and de Bruijn 2009: Pl. 14.1).

Type Locality Komanos 1 low A, Greece, Lower Pliocene [early Ruscinian (MN14)].

Distribution Only from the early Pliocene of Greece.

Occurrences in Greece Komanos 1 low A and B, Komanos 1 high A and Vorio 3A.

Remarks The species is based on a small collection (Fig. 32) of six m1, two m2, six m3, two M1, five M2 and six M3 from Komanos 1 low A (Hordijk and de Bruijn 2009). According to the diagnosis by Hordijk and de Bruijn (2009:47), the species has mesodont molars of large size, with clearly alternating re-entrant folds, generally undifferentiated enamel (with rare slight differences in thickness) and almost straight enamel border at the crown base; a small but well-developed anterior cap on the m1, with a small round enamel islet on unworn specimens and an anteriorly curved LRA2 on some m1; confluent dentine fields; primitive m3, with deep and wide LRA2, shallow BRA2, poorly developed or undeveloped LRA1 and posterior lobe; juvenile M3 with usually isolated anterior lobe, moderately worn M2 with connection between anterior lobe and T2 and deeply worn M2 with anterior enamel islet due to the closure of the BRA1; and M3 with a short and wide posterior cap with a deep oval posterior enamel islet and generally closed BRA2 that encloses a small deep round postero-labial enamel islet. The species has also been described from Komanos 1 low B (one M3) and Komanos 1 high A (Fig. 32) (two m1, one m2, three M1, two M2 and two M3). A small assemblage from Vorio 3A (one m1, two m2, two m3, three M1, two M2 and one M3) has been assigned to *M.* aff. *komanensis*; most of the molars show slight differences from the ones from the type locality;

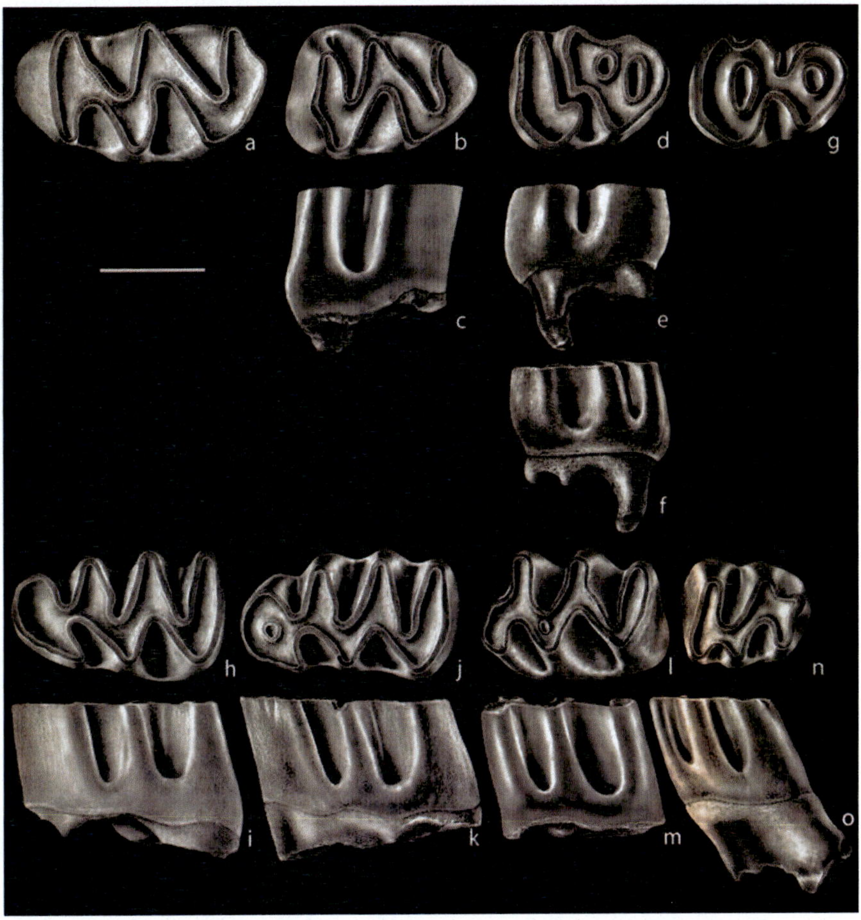

Fig. 32 *Microtodon komanensis* from Komanos 1 low A (type locality). **a** M1, **d–f** M3 (occlusal, lingual and labial view), **g** M3, **h–i** m1 (holotype); occlusal and labial view), **j–k** m1 (occlusal and labial view), **n–o** m3 (occlusal and labial view), and from Komanos 1 high A. **b–c** M2 (occlusal and lingual view), **l–m** m2 (occlusal and labial view). Images courtesy of K. Hordijk and H. de Bruijn. All specimens figured as left; **a**, **b–c**, **h–i** are reversed. Scale bar equals 1 mm

however, the Vorio 3A M3 differs from *M. komanensis* from Komanos: it is slightly higher-crowned, its BRA1 closes earlier with wear and the LRA2 separates the T2 and T3 (Hordijk and de Bruijn 2009:51–52). The authors presume that this material is slightly more evolved compared to *M. komanensis*, but refrain from naming a new species due to the scarcity of the material and its similarity with *M. komanensis*.

***Promimomys* Kretzoi, 1955**
(= *Prosomys* Shotwell, 1956 = *Polonomys* Kretzoi, 1959, according to Hordijk and de Bruijn, 2006)

Type Species *Promimomys cor* Kretzoi, 1955.

Other Taxa Included *P. moldavicus* (Kormos, 1932); *P. mimus* (Shotwell, 1956); *P. insuliferus* Kowalski, 1958; *P. gracilis* Kretzoi, 1959; *P. microdon* Jánossy, 1974; *P. asiaticus* Jin and Zhang, 2005; *P. enginae* Alpaslan, 2015 (see remarks on the status of *Promimomys* in Hordijk and de Bruijn 2009: 52-54).

Distribution Late Miocene (late Turolian)–early Pliocene of Europe and Asia; Late Miocene of N America (with *P. mimus*).

Remarks One heavily damaged M1 from the latest Miocene fauna from Prosilion–Mercurion (Servia Basin) has been provisionally assigned to cf. *Promimomys* sp. by Hordijk and de Bruijn (2009:57), because of its similarity with *P. cor* from Komanos 1.

***Promimomys cor* Kretzoi, 1955**

Type Material Partial left mandible with a heavily worn m1, the anterior root of the m2 and the base of the incisor (holotype) (Kretzoi 1955: Fig. 1).

Type Locality Csarnóta 2, Hungary, Lower Pliocene [early Ruscinian (MN14)].

Distribution Early Pliocene of Europe.

Occurrences in Greece Komanos 1 low B, Komanos 1 high A, Kardia, Vorio 1, and Ptolemais 1.

Remarks According to Kretzoi's (1955:91) original diagnosis (free translation from German), it differs from *P. moldavicus* through the simple and broadly rounded front edge of the anterior cap, more pronounced brachyodonty and larger size. Three Ruscinian faunas from the Ptolemais–Servia Basin have revealed *P. cor*: Komanos 1 low B (three m1, four m2, three m3, three M1, six M2, six M3), Komanos 1 high A (one m1, four m2, four m3, two M1, four M2, two M3) and Vorio 1 (six m1, nine m2, three m3, two M1, six M2, three M3) (Hordijk and de Bruijn 2009). Some of the Vorio 1 M2 show an antero-lingual cingulum at the crown base, whereas all M3 show an LRA3, sometimes connected to the posterior islet, and three thin, well-separated roots. The Komanos molars lack these characters present in the Vorio molars, whereas the anterior roots on some M3 are either fused or reduced to a single root (Hordijk and de Bruijn 2009:56). The rich assemblages from Kardia (nine m1, 19 m2, 20 m3, 15 M1, 13 M2, 24 M3) and Ptolemais 1 (21 m1, 22 m2, 17 m3, 46 M1, 52 M2, 51 M3), previously described as *Promimomys insuliferus* by van der Weerd (1979), have also been attributed to *Promimomys cor* by Hordijk and de Bruijn (2009), after their re-examination. The Ptolemais 1 M2 show a low and small antero-lingual cuspule, and some M3 show an LRA3 and an open posterior enamel islet, features not present in the Kardia molars; the M3 from Kardia have fused anterior roots or a single anterior root, whereas the M3 from Ptolemais 1 more frequently have three separated roots (van der Weerd 1979:153; Hordijk and de Bruijn 2009:56).

***Promimomys insuliferus* Kowalski, 1958**

Type Material Mandible with m1–3, no. MF/22, (holotype) Institute of Systematics and Evolution of Animals of the Polish Academy of Sciences, Krakow (Kowalski 1956: Pl. IV.7 and text–fig. 2a).

Type Locality Podlesice, Poland, Lower Pliocene [early Ruscinian (MN14)].

Distribution Early Pliocene (Ruscinian) of Europe and Turkey.

Occurrences in Greece Spilia 0 and Limni 3.

Remarks According to Kowalski (1958:41), *P. insuliferus* differs from *P. cor.* and *P. moldavicus* through its smaller size and the remarkably long duration of the enamel islet in the anterior cap of the m1. *P. insuliferus* has been included in the faunal list for Spilia 0 (de Bruijn 1989) and *P.* cf. *insuliferus* in the faunal list for Limni 3 (van der Meulen and van Kolfschoten 1986).

***Mimomys* Forsyth Major, 1902**

Type Species *Mimomys pliocaenicus* Forsyth Major, 1902.

Other Taxa Included [including the subgenera *Aratomys* Zazhigin in Gromov and Poljakov, 1977 (=*Kilarcola* Kotlia, 1984), according to Repenning (2003); *Cosomys* Wilson, 1932; *Ogmodontomys* Hibbard, 1941; *Cseria* Kretzoi, 1959; *Ophiomys* Hibbard and Zakrzewski, 1967; *Cromeromys* Zazhigin, 1980; according to Repenning and Grady 1988] *M.* (*Cr.*) *intermedius* (Newton, 1881); *M. savini* Hinton, 1910; *M. reidi* Hinton, 1910; *M. majori* Hinton, 1910; *M. cantianus* Hinton, 1910; *M. pusillus* (Méhely, 1914); *M. franconicus* Heller, 1930; *M.* (*Cs.*) *stehlini* Kormos, 1931; *M. moldavicus* Kormos, 1932; M. (*Oph.*) *parvus* (Wilson, 1933); *M. kormosi* Brunner, 1933; *M. fejervaryi* Kormos, 1934; *M. orientalis* Young, 1935; *M. hassiacus* Heller, 1936; *M.* (*Og.*) *poaphagus* Hibbard, 1941; *M. coelodus* Kretzoi, 1954; *M. kislangensis* Kretzoi, 1954; *M. obtusus* Kretzoi, 1954; *M. arvalinus* Kretzoi, 1954; *M.* (*Og.*) *sawrockensis* Hibbard, 1957; *M. delmi* Brunner, 1958; *M.* (*Oph.*) *taylori* Hibbard, 1959; *M.* (*Cs.*) *gracilis* (Kretzoi, 1959); *M. polonicus* Kowalski, 1960; *M. hajnackensis* Fejfar, 1961; *M. hintoni* Fejfar, 1961; *M. kretzoii* Fejfar, 1961; *M. minor* Fejfar, 1961; *M. proseki* Fejfar, 1961; *M. medasensis* Michaux, 1971; *M. septimanus* Michaux, 1971; *M.* (*Oph.*) *fricki* Hibbard, 1972; *M.* (*Oph.*) *magilli* Hibbard, 1972; *M. monohani* Martin, 1972; *M. blanci* van der Meulen, 1973; *M. antis* Savinov, 1974; *M. silasensis* Jánossy, 1974; *M.* (*Cs.*) *parkeri* Martin, 1975; *M. pitymyoides* Jánossy and van der Meulen, 1975; *M.* (*Cr.*) *tornensis* Jánossy and van der Meulen, 1975; *M. ostramosensis* Jánossy and van der Meulen, 1975; *Mimomys* (*Cr.*) *gansunicus* Zheng, 1976; *M. banchiaonicus* Zheng, 1976; *M. pseudointermedius* Erbajeva, 1976; *M.* (*Ar.*) *multifidus* Zazhigin in Gromov and Poljakov 1977; *M.* (*Oph.*) *mcknighti* (Gustafson, 1978); *M. davakosi* van de Weerd, 1979; *M. antiquus* (Zazhigin, 1980); *Mimomys* (*Cr.*) *irtyshensis* (Zazhigin, 1980); *M. post-*

silasensis Rabeder, 1981; *M. praepliocaenicus* Rabeder, 1981; *M.* (*Cs.*) *carnuntina* Rabeder, 1981; *M.* (*Cs.*) *proopsia* Rabeder, 1981; *M.* (*Cs.*) *opsia* Rabeder, 1981; *M. youhenicus* Xue, 1981; *M. malezi* Rabeder, 1983; *M. clairi* Chaline, 1984; *M.* (*Ar.*) *kashmirensis* (Sahni and Kotlia, 1985); *M. peii* Zheng and Li, 1986; *M. realensis* Esteban and Martinez-Salanova, 1987; *M. hanzhongenicus* Tang and Zong, 1987; *M. haplodentatus* Savinov and Tutkova, 1987; *M.* (*Cr.*) *virginianus* Repenning and Grady, 1988; *M. dakotaensis* Martin, 1989; *M. vandermeuleni* Feifar, Mein and Moissenet, 1990; *M.* (*Oph.*) *panacaensis* Mou, 1997; *M.* (*Cr.*) *hordijki* van Kolfschoten and Tesakov, 1998; *M.* (*Ar.*) *bilikeensis* (Qiu and Storch, 2000); *M.* (*Og.*) *pipecreekensis* Martin et al., 2002.

Distribution Late Miocene (late Turolian)–Middle Pleistocene of Europe, Asia and N America.

Remarks Obviously the genus has been over-split; some of the included species might be proven to be synonymous, e.g., *M. intermedius* = *M. savini* = *M. majori*, according to Agadjanian and Kondrashov (2007). Martin (2003) gave three criteria for including a species in *Mimomys*: highly developed negative enamel differentiation, *Mimomys* schmelzmuster and presence of roots. Based on these, many of the species (see Appendix) could be transported to *Promimomys* (e.g., *M. davakosi*). Since the revision of the genus is beyond the scope of this paper, we list all species described in the literature as *Mimomys* (see Appendix).

Mimomys sp. has been reported from the late Villanyian fauna of Rema Aslan (van der Meulen and van Kolfschoten 1986), the locality Yerakarou (= Gerakarou 1 in Koliadimou 1996) (Zamanis et al. 1980) and the lacustrine deposits of Yialtra (Evia Island) (Katsikatsos et al. 1981), but without details on the assemblages. *Mimomys* sp. is also reported from Alikes, with a complete skull bearing all molars (Athanassiou 1998); the *Mimomys*-type enamel differentiation is recognized in the sample, and the size of the molars indicates a large species (e.g., *M. savini*, *M. pliocaenicus*), but the lack of the diagnostic m1 does not allow further identification. One M2 and one M3 from the Biharian fauna of Marathoussa (N Greece) have been assigned to *Mimomys* sp. (Koufos et al. 2001); they are similar to *M. savini*, but the absence of m1 does not allow specific classification. One M1 from Choremi 2 and one m2 from Choremi 4 have been assigned to *Mimomys* sp.; they are medium-sized and thus too small to belong to *M.* aff. *savini* also present in the localities, and bear roots, cement in the re-entrant angles and *Mimomys*-type enamel differentiation (van Vugt et al. 2000). One M1 and part of a mandible with partially preserved m2 from the Ruscinian fauna of Makrygialos have been identified as ?*Mimomys* sp. (Sylvestrou 2002); the archaic dental features (i.e., three-rooted M1, absence of cement from the M1, relatively low crown height and small size) place the samples in the *Mimomys* primitive forms.

★***Mimomys davakosi*** **van de Weerd, 1979**

Taxonomical History Considered as a senior synonym of *M. vandermeuleni* Fejfar et al. 1990, according to Hordijk and de Bruijn (2009).

Type Material Left m1, PT3 12 (holotype) (van der Weerd 1979: Fig. 6–4)

Type Locality Ptolemais 3, Greece, Lower Pliocene [late Ruscinian (MN15)].

Distribution Early Pliocene (Ruscinian) of Europe, Russia and Turkey.

Occurrences in Greece Tomea Eksi 3, Vorio 3 and 3A, Ptolemais 3, Komanos 2 (Fig. 33).

Remarks Radulescu and Samson (1989) considered *M. davakosi* to be a junior synonym of *M. moldavicus* Kormos 1932; Fejfar et al. (1997:278) commented that *M. moldavicus* is based on "*insufficient material*", but the two species could indeed be synonymous. Hordijk and de Bruijn (2009:58) prefer retaining *M. davakosi* as a separate species pending further study of the type materials. A rich vole tooth collection from the Lower Pliocene locality Ptolemais 3 (20 m1, 28 m2, 27 m3, 25 M1, 27 M2, 30 M3) has been the base for the creation of *M. davakosi* by van der Weerd (1979). The diagnostic dental features are low crown, no cement, low or absent enamel-free areas, closed enamel islet on almost all molars, poorly developed *Mimomys* ridge, prism- and islet-folds and weak fourth lingual re-entrant fold on the m1 (van der Weerd 1979:155). *M. davakosi* is also present in Komanos 2 with one m1, three m2, six m3, two M1 and four M3 and in Vorio 3 and 3a with two m1, one m2, two M1 and two M2 (Hordijk and de Bruijn 2009). Five m1, eight m2, 13 m3, four M1, three M2 and eight M3 from the locality Tomea Eksi 3 have been assigned to *M.* aff. *davakosi*, as "*the material represents an intermediate evolutionary stage between* Promimomys cor *and* Mimomys davakosi" and assignment to either is possible (Hordijk and de Bruijn 2009:61).

Mimomys hajnackensis Fejfar, 1961

Type Material Right m1, No. 65361(holotype) (Fejfar 1961: Abb. 2a, c).

Type Locality Hajnáčka, Slovakia, Upper Pliocene [early Villanyian (MN16)].

Distribution Late Pliocene (early Villanyian) of Europe.

Occurrences in Greece Tourkobounia 1, Limni 6 and Damatria.

Remarks The species has been considered a junior synonym of *M. hassiacus* Heller 1936 by several authors (e.g., Mörs et al. 1998:147, Sabol et al. 2006:267), but others retain it as a separate species (e.g., Mayhew et al. 2008) (see Mörs et al. 1998:149 for synopsis on the synonymy between the species and Mayhew et al. 2008:182 for discussion on taxonomy of large *Mimomys* species). Here we decided to maintain it as a separate species, as we have not studied material from the type localities of the two species. *M.* cf. *hajnackensis* has been reported from the early Villanyian faunas of Damatria (Rhodes Island) and Limni 6 (Evia Island) (van der

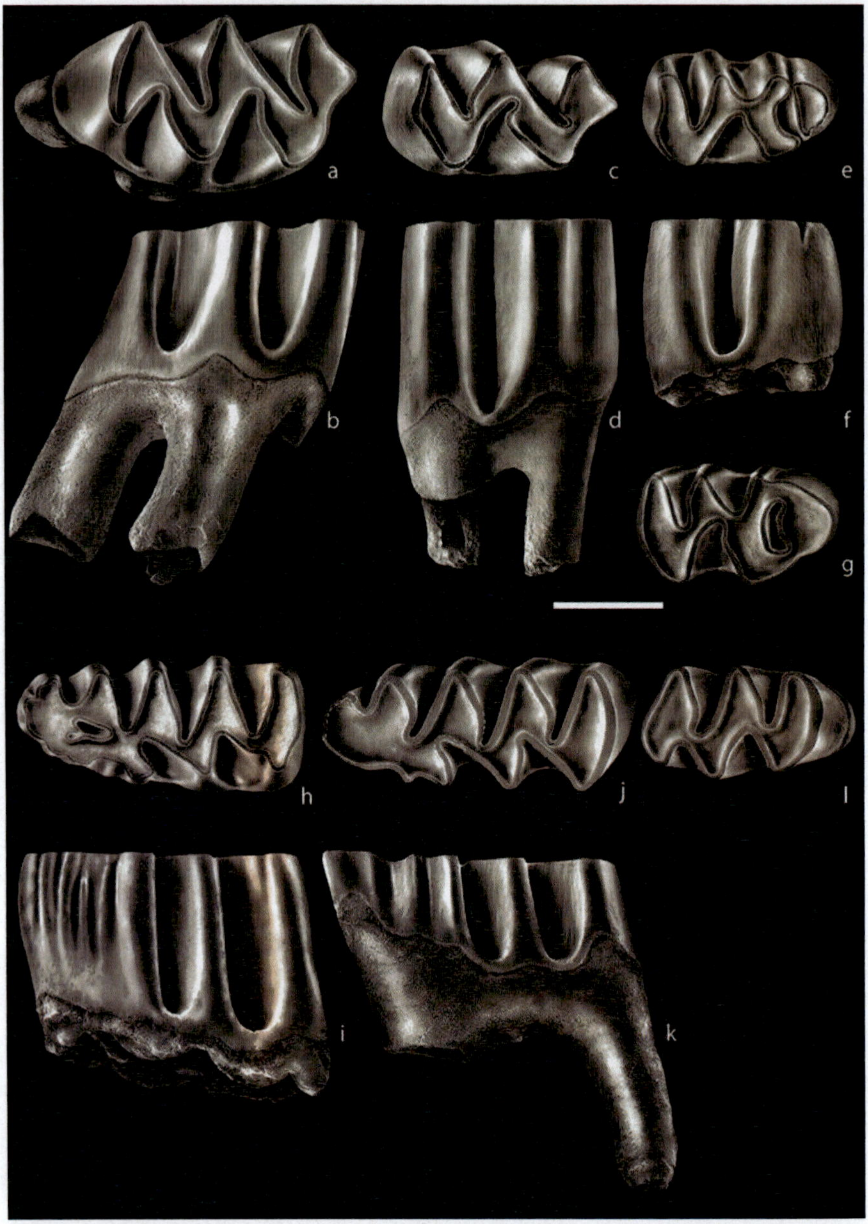

Fig. 33 *Mimomys davakosi* from Vorio 3 and 3A. **a–b** M1 (occlusal and lingual view), **c–d** M2 (occlusal and lingual view), **h–i** m1 (occlusal and labial view), **j–k** m1 (occlusal and labial view), and from Komanos 2. **e–f** M3 (occlusal and lingual view), **g** M3, **l** m3. Images courtesy of K. Hordijk and H. de Bruijn. All specimens figured as left; c–d, e–f, g, j–k, l are reversed. Scale bar equals 1 mm

Meulen and van Kolfschoten 1986). The same material from Limni 6 had been earlier reported as *M.* cf. *polonicus* by de Bruijn and van der Meulen (1979), but no information on the assemblage is available in either publication. Benda et al. (1977) mentioned that one M1 with features similar to those of both *M. stehlini* and *M. polonicus* has been found in "*an exposure of the Damatria formation… in the airport exposure*", but it is not known whether this is the same locality and the same material reported by van der Meulen and van Kolfschoten (1986). Two M1 and two M3 from the Villanyian fauna of Tourkobounia 1 were described by de Bruijn and van der Meulen (1975) and were assigned to *Mimomys* sp., due to the absence of the diagnostic m3; later, the same material was referred to *M.* cf. *septimanus* by de Bruijn and van der Meulen (1979) and to *M. hajnackensis* by van der Meulen and van Kolfschoten (1986), without justification.

Mimomys gracilis (Kretzoi, 1959)

Nomenclatural and Taxonomical History *Cseria gracilis* in Kretzoi 1959 (new species); *Mimomys gracilis* in Kowalski 1960 (new combination).

Type Material Left m1(holotype) (Kretzoi 1962: text–fig. 3).

Type Locality Csarnóta 2, Hungary, Pliocene [late Ruscinian (MN15)].

Distribution Pliocene of Europe and Turkey.

Occurrences in Greece Limni 6 and Archipolis.

Remarks The species was initially described by Kretzoi (1959) as the type species of *Cseria*. He did not assign a holotype, nor did he give a diagnosis, only a short comparison with *Mimomys*: the length of the m1 is 2.4–2.5 mm; the molars are shorter than those of *Mimomys* and without cement; the root is long and forms at an early stage; the enamel is thick and not differentiated. *M. gracilis* had been included in the faunal list for the locality Kastoria 1 by de Bruijn and van der Meulen (1979), but, later, van der Meulen and van Kolfschoten (1986) included two other arvicoline species in the Kastoria 1 faunal list and not *M. gracilis*. *M.* cf. *gracilis* has been included in the faunal list for Limni 6 (van der Meulen and van Kolfschoten 1986), whereas in the faunal list given by de Bruijn and van der Meulen (1979) the same material had been referred to *M.* cf. *septimanus*. *M.* cf. *gracilis* has also been reported from the locality Archipolis (Rhodes Island) by Doukas and Papayianni (2016), whereas the only arvicolid in the faunal list for this locality previously reported by van der Meulen and van Kolfschoten (1986) was *M.* cf. *stehlini*.

Mimomys reidi Hinton, 1910

Type Material Left m1, GMS 48637 (Mayhew and Stuart 1986:Fig. 13.6 a, b; according to Mayhew and Stuart 1986:442 and 458). The holotype had been previously figured as *Arvicola intermedius* by Newton (1882: Pl. XIII.8).

Type Locality "Weybourne Crag", Trimingham, Great Britain, Upper Pliocene.

Distribution Late Pliocene–Middle Pleistocene of Europe.

Occurrences in Greece Kardamena and Kastoria 2.

Remarks The species has been re-diagnosed by Mayhew and Stuart (1986:458), based not only on the type material from the type locality but also on additional material attributed to this species from eastern England. According to their revised diagnosis, it is a medium-sized *Mimomys*, with enamel thicker on the convex sides of the triangles and small to moderate amounts of cement, with moderately high-crowned molars; the enamel islet in the m1 disappears at an early wear stage, the triangles are confluent and the *Mimomys* ridge is persisting for at least half the height of the crown. Further, on the m2, the T1 is separated from the posterior lobe, the T1 and T2 are partly confluent, the T2 is separated from the anterior region and there are three dentine fields. The m3 also has three dentine fields and the T1 and T2 are broadly confluent. The M1 has two or three roots, the dentine fields are not completely separated and the LSA2 is prominent and rounded, with low enamel-free area. The M2 is two-rooted and the T2 and T3 are partly confluent. The M3 is also two-rooted, the enamel islet is present till moderate wear stage, whereas the anterior lobe and the T2 are weakly separated from the posterior lobe. *M. reidi* is in the faunal list for the locality Kastoria 2 and *M.* cf. *reidi* in the faunal list for the locality Kardamena (Kos Island) (van der Meulen and van Kolfschoten 1986), but no additional information is available.

Mimomys pliocaenicus **Forsyth Major, 1902**

Type Material Left mandible with i, m1, m2 (holotype) (Forsyth Major 1902: Fig. 13.8–9).

Type Locality Upper Val d'Arno, Italy, Upper Pliocene.

Distribution Late Pliocene–Early Pleistocene of Europe.

Occurrences in Greece Kastoria 2.

Remarks Mayhew and Stuart (1986:444) gave the diagnosis of the species: large *Mimomys*, abundant crown cement, enamel thicker on the convex sides of the triangles; m1 and M3 with isolated enamel islet before root formation and present during a long ontogenetic period; M1–M3 two- or three-rooted. *M. pliocaenicus* has been reported from the locality Kastoria 2 by van der Meulen and van Kolfschoten (1986), without further information.

Mimomys ostramosensis Jánossy and van der Meulen, 1975

Type Material A viscerocranium with complete right dentition and the left M1–2, together with the left mandible with m1–3 of the same individual, no. V.73.2 (holotype), Hungarian Natural History Museum, Budapest (Jánossy and van der Meulen 1975:Pl. I.1a–b).

Type Locality Osztramos 3, Hungary, Lower Pleistocene [late Villanyian (MN17)].

Distribution Late Pliocene–Middle Pleistocene of Europe and Turkey.

Occurrences in Greece Lagkada (Kos Island).

Remarks The diagnosis states that the species is very large and phylogenetically advanced, with enamel islets on the m1 and M3 and *Mimomys* ridge on the m1 present only in unworn and slightly worn specimens (Jánossy and van der Meulen 1975). *M. ostramosensis* is in the faunal list of the locality Lagkada (Kos Island) (van der Meulen and van Kolfschoten 1986), without further information.

Mimomys savini Hinton, 1910

Type Material Right m1, B.M. No. M. 6986b (holotype) (Savin Collection) (figured by Forsyth Major 1902: text–fig.15, figs. 22, 22a, 22b as *Mimomys intermedius*)

Type Locality West Runton Upper Freshwater Bed, England, Lower Pleistocene [Biharian].

Distribution Late Early Pleistocene (Biharian) of Europe.

Occurrences in Greece Kaifas, Zeli 2, Zeli 2A and B, Ravin Voulgarakis, Kalamoto-1 and Choremi 1, 2, 3 and 4.

Remarks The specific features given by Hinton (1926:365) are (among others) medium size, well-differentiated enamel and re-entrant folds partly filled with cement, and on the m1 the enamel islet is present in early wear stages (before the formation of roots). One partial left mandible with m1–2 and the incisor from the locality Kalamoto has been assigned to *M. savini* by Tsoukala and Chatzopoulou (2005); the molars are large, with cement in the synclines, no enamel islet and restricted communication between the T4 and T5 on the m1. *M. savini* has also been described from Ravin Voulgarakis, where a rich assemblage of 15 m1, 16 m2, 10 m3, eight M1, 15 M2, 15 M3 and a mandible with m1–m2 has been found; the material is characterized by the presence of two roots on all molars, *Mimomys*-type enamel-thickness differentiation and presence of an enamel islet in the first wear stages of the m1 (Koliadimou 1996). *M. savini* has been also reported from the localities Kaiafas, Zeli 2 and Zeli 2A and B (van der Meulen and van Kolfschoten 1986), with no information on the available material. The late Biharian faunas of

Choremi 1–4 (Megalopolis Basin) have yielded several molars (30 in Choremi 1, 30 in Choremi 2, 43 in Choremi 3, 36 in Choremi 4) assigned to *M.* aff. *savini*: the molars are high-crowned and larger than the ones from the type locality (hence the open identification), bear roots, crown cement, negative enamel differentiation, do not show a *Mimomys* islet on the M3 and m1, and one molar bears a *Mimomys* ridge that disappears close to the crown base (van Vugt et al. 2000:74–75).

Dolomys Nehring, 1898

Type Species *Dolomys milleri* Nehring, 1898.

Other Taxa Included *D. occitanus* (Thaler, 1955); *D. nehringi* Kretzoi, 1959; *D. odessanus* Nesin, 1987; *D. adroveri* Fejfar, Mein and Moissenet, 1990.

Distribution Pliocene–Early Pleistocene of Europe.

Remarks One worn and damaged m1 found in the locality Apolakkia–2 (Rhodes Island) has been assigned to *Dolomys* sp.; it shows a shallow but well-developed fourth labial fold in the anteroconid, and there is no enamel islet or a trace of a *Mimomys* ridge (van de Weerd et al. 1982).

Dolomys occitanus (Thaler, 1955)

Nomenclatural and Taxonomical History *Mimomys occitanus* in Thaler 1955 (new species); *Dolomys occitanus* in Maul 1996 (new combination).

Type Material No holotype was designated by Thaler, but Chaline (1974:343) designated a neotype: left m1, SE–91, University of Montpellier [Chaline 1974: fig. 8 (9); fig. 9 (5)].

Type Locality Grotte 1, Sète, France, Pliocene [MN15].

Distribution Pliocene (late Ruscinian–Villanyian) of Europe and Turkey.

Occurrences in Greece Apolakkia–2, Spilia 3–4 and Kastoria 1.

Remarks According to Thaler's (1955:434) diagnosis, early root development, small degree of hypsodonty and three roots on the M3 characterize this species. Chaline (1974:343) also added that there is no cement in the synclines. The presence of *D. occitanus* is mentioned from the late Ruscinian fauna of Apolakkia–2 with "*some rodent remains belonging to* Mimomys occitanus" by Benda et al. (1977:121). Van de Weerd et al. (1982) described mostly fragmentary material from the same site (possibly the same material found by Benda et al. 1977): five anterior parts of m1, one M1, one M2, one complete and three partial M3. The length of the anteroconid, the height of the enamel-free areas and the open labial fold in front of the *Mimomys* ridge of the Apolakkia–2 m1 show similarities with the Sète m1; however, the M3 from Apolakkia–2 is different from that from Sète in having a short anterior labial fold without enamel islet (van de Weerd et al. 1982:100). *D. occitanus* has also been reported in the faunal list from Spilia 3–4 (de Bruijn and van der Meulen 1979) and in the faunal list from Kastoria 1 (van der Meulen and van Kolfschoten 1986), with no further information.

Dolomys nehringi Kretzoi, 1959

Type Material No holotype or neotype has ever been designated; Kretzoi figured one right m1 and one left M3 from the type locality (Kretzoi 1962:text–fig. 4.a–b).

Type Locality Csarnóta 2, Hungary, Pliocene [late Ruscinian].

Distribution Pliocene of Europe.

Occurrences in Greece Notio 1.

Remarks This vole from Csarnóta–2, much smaller than *D. milleri*, was assigned to a new species without a lot of information and detailed descriptions by Kretzoi (1959:242–243). A small assemblage from Notio 1 (two partial m1, two complete and two partial m3, one partial M1 and one partial M3) has been assigned to *D.* cf. *nehringi*, as the size and morphology of the molars appear to be close to those of the material from Csarnóta 2 (Hordijk and de Bruijn 2009:61–62).

Propliomys Kretzoi, 1959

Type Species *Dolomys hungaricus* Kormos, 1934.

Distribution Pliocene of Europe and Turkey.

Propliomys hungaricus (Kormos, 1934)

Nomenclatural and Taxonomical History *Dolomys hungaricus* in Kormos 1934 (new species); *Propliomys hungaricus* in Kretzoi 1959 (new combination).

Type Material Mandible with m1–m3, no. 3799 (holotype) (Kormos 1934: fig. 46)

Type Locality Csarnóta 2, Hungary, Pliocene [late Ruscinian (MN15)].

Distribution Pliocene (late early Ruscinian–early Villanyian) of Europe and Turkey.

Occurrences in Greece Notio 1 and Kastoria 1.

Remarks The length of the m1 is between 2.8 and 3.2 mm; the enamel thickness is even; there is variation in the shape of the anterior cap of the m1 (Kretzoi 1959).

Some more features characterizing the species are as follows: the anteroconid is elongated; the alternating synclines are asymmetrical; the BRA3 is deep; the lingual triangles are larger than the buccal ones (Fejfar et al. 2011:26). A small assemblage from Notio 1, consisting of one complete and five partial m1, five m2, six m3, three M1 (one damaged), six complete (three damaged) and two fragmentary M2 and 12 complete (four damaged) and one fragmentary M3, has been referred to *P. hungaricus*; the dental features of the samples correspond very well to the material from Csarnóta 2 (Hordijk and de Bruijn 2009:62–63). The species is also in the faunal list of the locality Kastoria 1 (as *Pliomys hungaricus*; van der Meulen and van Kolfschoten 1986), with no information of the assemblage.

***Pliomys* Méhely, 1914**

Type Species *Pliomys episcopalis* Méhely, 1914.

Other Taxa Included *P. coronensis* (Méhely, 1914); *P. lenki* (Heller, 1930); *P. dalmaticus* Kormos, 1931; *P. progressus* Kretzoi, 1938; *P. simplicior* Kretzoi, 1956; *P. proavius* Heller, 1958/59; *P. zimmermanni* (Brunner, 1958); *P. soergeli* (Brunner, 1958); *P. kowalskii* Schevtschenko, 1965; *P. ucrainicus* (Topacevski, and Scorik, 1967); *P. posterior* Jánossy, 1969; *P. chalinei* Jeannet, 1974; *P. graecus* de Bruijn and van der Meulen, 1975; *P. hollitzeri* Rabeder, 1981; *P. jalpugensis* Nesin, 1983; *P. schernfeldensis* Carls and Rabeder, 1988; *P. destinatus* Tesakov, 2005.

Distribution Pliocene–Late Pleistocene of Europe, Pliocene of Ukraine, Kazakhstan, Early Pleistocene of N America.

★*Pliomys graecus* de Bruijn and van der Meulen, 1975

Type Material Partial right mandible with incisor and m1–m2, no. 695 (holotype) (de Bruijn and van der Meulen 1975: Pl. 2.1)

Type Locality Tourkobounia 1, Greece, Upper Pliocene [Villanyian (MN16)].

Distribution Late Pliocene–Middle Pleistocene of Greece.

Occurrences in Greece Tourkobounia 1 and Limni 6.

Remarks De Bruijn and van der Meulen (1975) created the species based on a rich material (Fig. 34) from Tourkobounia 1, consisting of 13 m1, 18 m2, 19 m3, 16 M1, 24 M2 and 13 M3. According to the diagnosis, the m1 has a short, wide AC and very low enamel-free areas, the M2 usually has an incipient LRA1 and the posterior part of the M3 is strongly reduced (de Bruijn and van der Meulen 1975:323). *P. graecus* has also been found in Limni 6 (van der Meulen and van Kolfschoten 1986), with no additional information. The Limni 6 material had been previously listed as *Pliomys hungaricus* by de Bruijn and van der Meulen (1979).

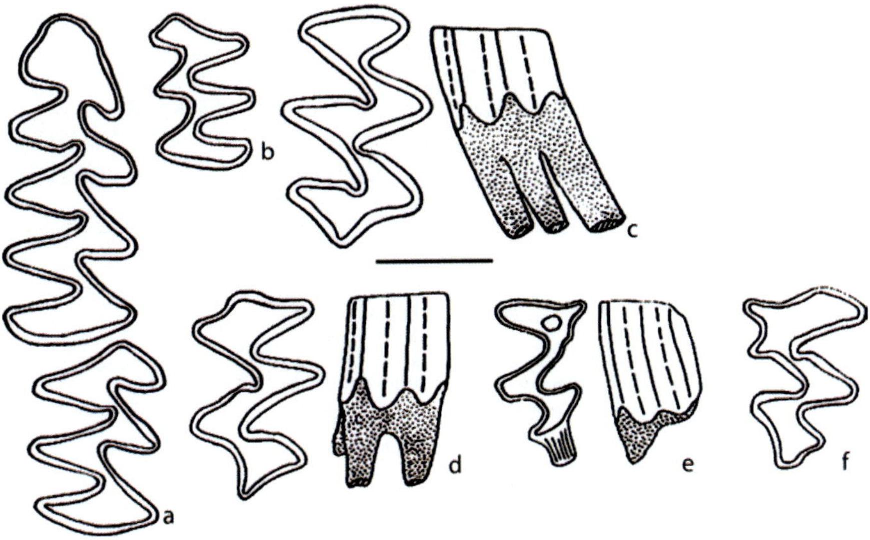

Fig. 34 *Pliomys graecus* from Tourkobounia 1 (type locality). **a** right m1–m2 (holotype), **b** left m3, **c** right M1 (occlusal and lingual view), **d** left M2 (occlusal and lingual view), **e** left M3 (juvenile; occlusal and lingual view), **f** right M3 (occlusal and lingual view). Drawings courtesy of H. de Bruijn. Scale bar equals 1 mm for occlusal views and 2 mm for lingual views

Pliomys episcopalis Méhely, 1914

Taxonomical History Considered as a senior synonym of *Dolomys episcopalis bolkayi* Kormos, 1931, according to Bona et al. (2008).

Type Material Méhely did not designate a holotype; Hinton (1926:342) designated a right mandibular ramus (Méhely 1914: taf. V.10) as the lectotype.

Type Locality Somlyóberg, Püspökfürdő, Hungary, Lower Pleistocene.

Distribution Early–Middle Pleistocene of Europe and Turkey.

Occurrences in Greece Megalopolis (Thoknia), Kastoria 2, Kaiafas, Zeli 2A and B and Choremi 3.

Remarks There is no cement in the synclines, and the thick enamel shows positive differentiation; two roots on the m1; the molar consists of the PL, five well-developed triangles and a small T6; there is a very shallow BRA4 and the AC is rounded anteriorly (Popov 2017). Van der Meulen and van Kolfschoten (1986) mentioned the presence of *P. episcopalis* in the Biharian faunas of Kastoria 2, Kaiafas and Zeli 2A and B, with no information on the assemblages. Benda et al. (1987) referred to *P.* cf. *bolkayi* (here reported as *P.* cf. *episcopalis*) a small collection of three m1, one m2, three M1, four M2 and five M3 from the locality Megalopolis

(Thoknia). The molars have two roots, positive enamel-thickness differentiation, there is no crown cement, the enamel-free areas are high and their size is larger than that of *P. episcopalis* (Benda et al. 1987:133). Four large, high-crowned and rooted molars with positive enamel differentiation and with no cement in the re-entrant angles from Choremi 3 have been assigned to *P.* aff. *episcopalis* by van Vugt et al. (2000); they are larger than the molars from the type locality, which might be an endemic character, according to the authors.

***Borsodia* Jánossy and van der Meulen, 1975**

Type Species *Mimomys newtoni* Forsyth Major, 1902.

Other Taxa Included *B. fejervaryi* (Kormos, 1934); *B. chinensis* (Kormos, 1934); *B. praehungarica* (Schevtschenko, 1965); *B. arankoides* (Alexandrova, 1976); *B. angensis* (Adamenko, 1977); *B. lenensis* (Adamenko, 1977); *B. prolaguroides* (Zazhigin, 1980); *B. parvisinuosa* Rabeder, 1981; *B. aequisinuosa* Rabeder, 1981; *B. altisinuosa* Rabeder, 1981; *B. klochnevi* Erbajeva, 1998. [Some of the described species might be synonymous, e.g., Tesakov (1993) considered the species *B. parvisinuosa*, *B. aequisinuosa* and *B. altisinuosa* to be junior synonyms of *B. praehungarica*.]

Distribution Late Pliocene (early Villanyian)–Early Pleistocene (early Biharian) of Europe, N China, Mongolia and Russia.

Remarks The poor material from the locality Gerakarou 1, consisting of one m2, one m3 and one M1, has been tentatively assigned to this genus (cf. *Borsodia* sp.) by Koliadimou (1996), due to the absence of the diagnostic teeth (m1, M3). The features of the molars are absence of cement, strong hypsodonty, tendency of root formation, no enamel-thickness differentiation (Koliadimou 1996).

***Borsodia newtoni* (Forsyth Major, 1902)**

Nomenclatural and Taxonomical History *Mimomys newtoni* in Forsyth Major 1902 (new species); *Mimomys* (*Borsodia*) *newtoni* in Jánossy and van der Meulen 1975 (new subgenus); *Borsodia newtoni* in Carls and Rabeder 1988 (elevated to genus). Considered as a senior synonym of *Mimomys newtoni hungaricus* Kormos, 1938 according to Mayhew and Stuart (1986), and *Mimomys simplex* Kretzoi, 1954 according to Mayhew (2012).

Type Material Left m1, BMNH no. M 6967a, (holotype) Savin Collection no. 430 (Forsyth Major 1902: text–fig. 13.7; Mayhew and Stuart 1986: Fig. 12.8a & b).

Type Locality Forest Bed or Weybourne Crag of East Runton, Norfolk Coast, England, Middle Pleistocene.

Distribution Early Pliocene–Middle Pleistocene of Europe, Russia, Ukraine and Kazakhstan.

Occurrences in Greece Kastoria 2.

Remarks Mayhew and Stuart (1986:453) gave the diagnosis of the species: small size, absence of crown cement, closed dentine fields, positive enamel differentiation (*Microtus*-type), two-rooted molars, juvenile m1 with a *Mimomys* ridge, m1 without enamel islet and M3 with enamel islet. The species is reported in the faunal list for Kastoria 2 (as *Mimomys newtoni*) by van der Meulen and van Kolfschoten (1986), without any details on the material.

***Lagurodon* Kretzoi, 1956**

Type Species *Lagurus arankae* Kretzoi, 1954.

Distribution Early Pleistocene (Biharian) of Europe, Ukraine, Russia and Kazakhstan.

Remarks *Lagurodon* sp. has been reported in the faunal lists of the localities Kaiafas and Zeli 2A and B by van der Meulen and van Kolfschoten (1986), without any detail on the material.

***Lagurodon arankae* (Kretzoi, 1954)**

Nomenclatural and Taxonomical History *Lagurus arankae* in Kretzoi 1954 (new species); *Lagurodon arankae* in Kretzoi 1956 (new combination).

Type Material Left m1(holotype) (Kretzoi 1954: Fig. 2c).

Type Locality Kisláng, Hungary, Upper Pliocene–Lower Pleistocene (3.5–1.7 Ma).

Distribution Late Pliocene–Early Pleistocene (Biharian) of Eurasia.

Occurrences in Greece Tourkobounia 2, Marathoussa, Alikes, Zeli 2, Ravin Voulgarakis and Apollonia-1.

Remarks Van der Meulen and Doukas (2001:265) assigned an assemblage of one partial mandible with m1–m2, 16 m1, 10 m2, seven M1, two M2 and four M3 from Tourkobounia 2 to *Lagurodon arankae* (as *Lagurus arankae*, following Rabeder 1981:84): the molars are characterized by the absence of crown cement and roots, presence of wide re-entrant angles and narrow alternating triangles and *Microtus*-type enamel-thickness differentiation. Ravin Voulgarakis has yielded the richest *L. arankae* material in Greece, consisting of 160 m1, 85 partial m1, 206 m2, 144 m3, two partial mandibles with m1–m2, 140 M1, 151 M2, 174 M3 and 20 partial M3 (Koliadimou 1996). The main features that classified the material to this species are the absence of roots and crown cement, the *Microtus*-type enamel-thickness differentiation, the presence on the m1 of four buccal and four lingual anticlines and of four enamel-free areas, the presence of strongly developed T6 and BRA4 on the m1, the presence of a small anticline on the LRA2 and the small oval PC on the M3 (Koliadimou 1996). A poor sample of one m1, one m2 and one m3 from Apollonia-1 has also been assigned to *L. arankae* (Koliadimou 1996). A quite rich material of 17 m1, nine m2, eight m3, seven M1, seven M2 and nine M3 from Marathoussa (N Greece) has been indentified as *L. arankae* (Koufos et al. 2001). The species also appears in the faunal list for Zeli 2, (van der Meulen and van Kolfschoten 1986), with no additional information on the material. Finally, a fragmentary left mandible with m1–3 from the locality Alikes has been tentatively assigned to the species (*L.* cf. *arankae*) by Athanassiou (1998), due to poor preservation of the anteroconid on the m1.

***Prolagurus* Kormos, 1938**

Type Species *Lagurus pannonicus* Kormos, 1930.

Other Taxa Included *P. posterius* Zazhigin, 1969; *P. ternopolitanus* (Topachevski, 1973).

Distribution Early–early Middle Pleistocene of Eurasia.

Prolagurus pannonicus **(Kormos, 1930)**

Nomenclatural and Taxonomical History *Lagurus pannonicus* in Kormos 1930 (new species); *Prolagurus pannonicus* in Kormos 1938 (new combination). Also includes the subspecies *P. pannonicus transylvanicus* Terzea, 1989, according to Tesakov et al. (2019).

Type Material Skull fragment with M1–M2, GIH Ob 4617 Vt50 (holotype), Hungarian Geological Institute (Kormos 1930: text–figs. 1, 3–8).

Type Locality Somlyóberg, Püspökfürdő, Romania, terminal Lower Pleistocene.

Distribution Early–early Middle Pleistocene of Europe.

Occurrences in Greece Volos.

Remarks The molars of the species lack roots and crown cement, whereas the enamel-thickness differentiation is of *Microtus*-type; the anteroconid complex is separated from the rest of the m1, and the rounded anterior cap is separated from the anteroconid triangles. *P. pannonicus* is in the faunal list of the locality Volos (van der Meulen and van Kolfschoten 1986), with no details on the material.

Lagurus Gloger, 1842

Type Species *Mus lagurus* Pallas, 1773 (extant).

Other Taxa Included Extinct: *L. transiens* Jánossy, 1962.

Distribution Pleistocene–present of Eurasia.

Remarks Pallas (1773:704) described *Lagurus lagurus*, the only extant species of the genus. Fossil material of the extant species has been found in Latomi 1 (Chios Island), with 26 m1 and 26 M3 and 14 more or less complete mandibles (Storch 1975:180). A poor assemblage of *L. lagurus*, consisting of two m2, one m3, two M2 and one M3, has also been described from the Upper Pleistocene locality Arnissa by Mayhew (1978:312).

Myodes Pallas, 1811 (= *Clethrionomys* Tilesius, 1850, sensu Carleton et al. 2014)

Type Species *Mus rutilus* Pallas, 1779 (extant).

Other Taxa Included Extant: *M. glareolus* (Schreber, 1780); *M. gapperi* (Vigors, 1830); *M. rufocanus* (Sundevall, 1846); *M. californicus* (Merriam, 1890); *M. andersoni* (Thomas, 1905); *M. smithii* (Thomas, 1905); *M. centralis* (Miller, 1906); *M. regulus* (Thomas, 1907); *M. shanseius* (Thomas, 1908); *M. imaizumii* (Jameson, 1961); *M. rex* (Imaizumi, 1971). Extinct: *M. kretzoii* (Kowalski, 1958); *M. intrans* (Kretzoi, 1958); *M. hintonianus* (Kretzoi, 1958); *M. sokolovi* (Topachevski, 1965).

Distribution Late Pliocene–present of Europe, Asia and N America.

Remarks One m1, three m3, three M1, one partial M1, one M2, two M3 and one partial M3 from Loutra Almopias Cave LAC and five m1, one partial m1, two m2, one m3, one partial mandible with m1–2, three M1, two M2 and one partial M3 from Loutra Almopias Cave LAC Ia have been assigned to the extant *M. glareolus* (Chatzopoulou 2014; as *Clethrionomys glareolus*); the molars are rooted, with crown cement, even enamel thickness around the anti- and synclines, rounded edges of the triangles and confluent dentine fields. One m1 and one M3 of *M. glareolus* have also been found in the Upper Pleistocene locality Arnissa (Mayhew 1978:317, as *Clethrionomys glareolus*). *Myodes* sp. has been reported from the locality Kastoria 2 (van der Meulen and van Kolfschoten 1986; as *Clethrionomys* sp.), without details on the material.

Kalymnomys von Koenigswald, Fejfar and Tchernov, 1992

Type Species *Jordanomys major* Kuss and Storch, 1978.

Other Taxa Included *K. haasi* (Tchernov, 1968).

Distribution Late Pliocene–Early Pleistocene of Bulgaria, Greece, Turkey and Israel.

Remarks The presence of *Kalymnomys* sp. in the Villafranchian fauna of Karnezeika has been reported by Kokotini et al. (2019).

★*Kalymnomys major* (Kuss and Storch, 1978)

Nomenclatural and Taxonomical History *Jordanomys major* in Kuss and Storch 1978 (new species); *Kalymnomys major* in von Koenigswald et al. 1992 (new combination).

Type Material Partial left mandible with m1–m3, SMF 76/1491 (holotype) (Kuss and Storch 1978: Fig. 44).

Type Locality Kalymnos-Xi, Greece, Lower Pleistocene [Biharian].

Distribution Early Pleistocene of Bulgaria, Greece and Turkey.

Occurrences in Greece Lagkada, Purgos, Kaiafas, Tourkobounia 2 and Kalymnos-Xi.

Remarks The material from the type locality Kalymnos-Xi consists of 47 m1, 39 m2, 26 m3, 22 M1, 32 M2 and 24 M3 (apart from the holotype) (Kuss and Storch 1978). According to the diagnosis, a large percentage of the m1 bears a *Mimomys* ridge (60% in the type locality). Further, the molars have neither crown cement nor roots; the enamel is either evenly thick along the triangles or shows a slight *Microtus*-type differentiation; the slender and pointed triangles alternate, but the dentine fields remain confluent; on the m1 the anterior cap is elongated or rounded and the lingual re-entrant valleys are symmetrical and rounded (Kuss and Storch 1978:222). One m1 and one M3 of *Kalymnomys* have also been described from the locality Tourkobounia 2 (Fig. 35); even though the length of the m1 falls within the ranges of both *K. major* and *K. haasi*, the material has been assigned to the former species due to the presence of the *Mimomys* ridge (van der Meulen and Doukas 2001:265). *K. major* has also been reported from the localities Pyrgos and Kaiafas and *K.* cf. *major* from Lagkada by van der Meulen and van Kolfschoten (1986), with no details on the material.

***Kislangia* Kretzoi, 1954**

Type Species *Mimomys rex* Kormos, 1934.

Other Taxa Included *K. capettai* (Michaux, 1971); *K. regulus* (Rabeder, 1981); *K. praerex (*Rabeder, 1981); *K. clairi* (Chaline, 1984); *K. ischus* (Esteban and Martinez Salonova, 1987); *K. gusii* Agustí, Galobart and Martín Suárez, 1993.

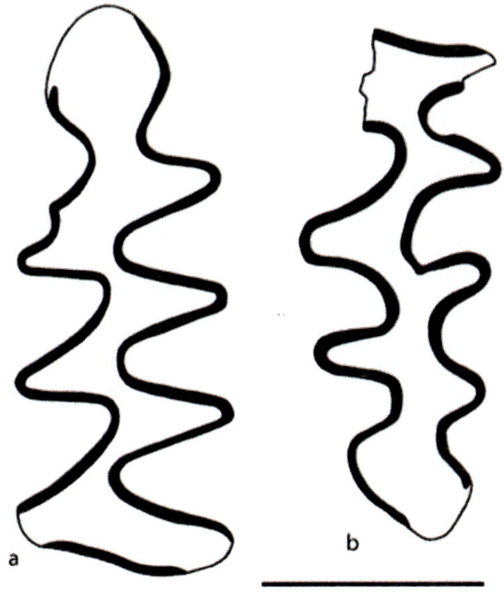

Fig. 35 *Kalymnomys major* from Tourkobounia 2. **a** m1, **b** M3. Drawings courtesy of C. Doukas. All specimens are left. Scale bar equals 1 mm

Distribution Late Pliocene–Early Pleistocene of Europe.

Remarks Three m1 of *Kislangia* sp. have been described from the Biharian fauna of Ravin Voulgarakis; they are rather robust, bearing two roots and cement, showing a *Mimomys*-type enamel-thickness differentiation, consisting of the anterior cap, five triangles and the posterior lobe, whereas an incipient T7 has started forming, bearing no enamel islet and a very small *Mimomys* ridge, with closed dentine fields on the posterior lobe and the T1–T3 and a very wide dentine field in the anteroconid complex (Koliadimou 1996). *Kislangia* sp. has also been reported from the Villafranchian fauna of Karnezeika (Kokotini et al. 2019).

Kislangia rex (Kormos, 1934)

Taxonomical History Considered a senior synonym of *Kislangia ondatrina* Kretzoi, 1954 and *Kislangia kadici* Kretzoi, 1954, according to Mayhew (2012).

Type Material Mandible with m1–m3, No. 3802/4 (holotype), Hungarian Natural History Museum (Kormos 1934: Fig. 49–upper).

Type Locality Kalkberg, Hungary, Lower Pleistocene [Villanyian (MN17)].

Distribution Late Pliocene–Early Pleistocene of Europe.

Occurrences in Greece Thoknia (Megalopolis).

Remarks *K. rex* has extremely wide molars with abundant crown cement and relatively thin enamel with a *Mimomys*-type differentiation. Van der Meulen and van Kolfschoten (1986) reported the presence of *K.* cf. *rex* in the fauna from Megalopolis (Thoknia). Benda et al. (1987) described (probably) the same material, consisting of four m1, nine m2, one m3, five M1, two M2 and eight M3 as *Mimomys* (*Kislangia*) *rex*. The dental features are abundant crown cement in the synclines, the dentine fields of the triangles and loops usually separated, high enamel-free areas, deep reentrant angles, *Mimomys* ridge on two but barely present on the third m1 and absence of the *Mimomys* islet (Benda et al. 1987:132).

***Pitymimomys* Tesakov, 1998**

Type Species *Mimomys pitymyoides* Jánossy and van der Meulen, 1975.

Other Taxa Included *P. baschkiricus* (Suchov, 1970); *P. altenburgensis* (Rabeder, 1981); *P. stranzendorfensis* (Rabeder, 1981); *P. stenokorys* (Rabeder, 1981); *P. jota* (Rabeder, 1981); P. *inceptor* Tesakov, 2003; *P. koenigswaldi* Erbajeva et al., 2006.

Distribution Late Pliocene–Early Pleistocene of Europe, Russia and Turkey.

***Pitymimomys pitymyoides* (Jánossy and van der Meulen, 1975)**

Nomenclatural and Taxonomical History *Mimomys pitymyoides* in Jánossy and van der Meulen 1975 (new species); *Pitymimomys pitymyoides* in Tesakov 1998 (new combination).

Type Material A nearly complete skull with the occipital region somewhat damaged, with the left mandible of the same individual, both with complete dentition, inv. no. V 73.4 (holotype), Natural History Museum of Budapest (Jánossy and van der Meulen 1975: Pl. II.8–9).

Type Locality Osztramos 3, Hungary, Lower Pleistocene [late Villanyian (MN17)].

Distribution Early Pleistocene (late Villanyian) of Europe, Russia and Turkey.

Occurrences in Greece Kastoria 2.

Remarks According to the diagnosis by Jánossy and van der Meulen (1975), this is a small, advanced *Mimomys* species, with confluent T2 and T3 on the m1 and the M2. Additional features were given by Mayhew and Stuart (1986), based on material from eastern England: uniform enamel thickness, sparse or absent crown cement, long and narrow anterior cap on the m1, with a *Mimomys* ridge and narrow pointed salient angles. The species has been reported from the Lower Pleistocene locality Kastoria 2 by van der Meulen and van Kolfschoten (1986; as *Mimomys pitymyoides*), without details on the available material.

***Microtus* Schrank, 1798**

Type Species *Mus arvalis* Pallas, 1778 (= *Microtus terrestris* Schrank, 1798).

Other Taxa Included The genus comprises today 61 species (Macdonald 2009) and eight subgenera, many of which have been reported from fossil and subfossil localities. Extinct species of Eurasia (in the subgenera *Allophaiomys* Kormos, 1933, *Tibericola* von Koenigswald, Fejfar and Tchernov, 1992, *Terricola* Fatio, 1867, *Tyrrhenicola* Forsyth Major, 1905, *Stenocranius* Katschenko, 1901, *Pallasiinus* Kretzoi, 1964, *Microtus*): *M.* (*M.*) *nivaloides* Forsyth Major, 1902; *M.* (*Tyr.*) *henseli* (Major, 1905); *M.* (*S.*) *gregaloides* (Hinton, 1923); *M.* (*M.*) *arvalinus* Hinton, 1923; *M.* (*P.*) *ratticepoides* (Hinton, 1923); *M.* (*A.*) *pliocaenicus* (Kormos, 1933); *M.* (*Ter.*) *arvalidens* (Kretzoi, 1958); *M.* (*A.*) *deucalion* (Kretzoi, 1969); *M.* (*S.*) *hintoni* (Kretzoi, 1941); *M.* (*A.*) *ruffoi* (Pasa, 1947); *M.* (*Tib.*) *jordanica* (Haas, 1966); *M.* (*A.*) *nutiensis* (Chaline, 1972); *M.* (*A.*) *pitymyoides* (Chaline, 1972); *M.* (*A.*) *bourgondiae* (Chaline, 1972); *M.* (*A.*) *praehintoni* (Rabeder, 1981); *M.* (*A.*) *superpliocaenicus* Rabeder, 1981; *M.* (*Ter.*) *tarentinus* Brunet-Lecomte, 1988; *M.* (*Tib.*) *vandermeuleni* (Agustí, 1991); *M.* (*Ter.*) *grafi* Brunet-Lecomte, Nadachowski and Chaline, 1992; *M.* (*P.*) *protoeconomus* Rekovets, 1994; *M.* (*Ter.*) *atapuerquensis* Gil, 1996; *M.* (*A.*) *eoratticeps* Paunovic and Rabeder, 1996; *M.* (*A.*) *lavocati* Laplana and Cuenca-Bescós, 2000; *M.* (*A.*) *valerotensis* Laplana et al., 2000; *M.* (*Tib.*) *sakaryaensis* (Ünay et al., 2001); *M.* (*Tib.*) *eleniae* van der Meulen and Doukas, 2001; *M.* (*Tyr.*) *sondaari* Marcolini et al., 2006.

Distribution Late Pliocene–present in Eurasia and N America.

Remarks Many extant species have been found in fossil and subfossil localities with Middle–Late Pleistocene ages. *Microtus* (*Microtus*) *guentheri* (Levant or Günther's vole) has been found in the Middle Pleistocene locality Latomi 1, represented by 226 m1, 288 M3 and 39 more or less complete lower jaws with teeth (Storch 1975), and in the Upper Pleistocene locality Arnissa, represented by nine m1 (Mayhew 1978). The European common vole *Microtus* (*Microtus*) *arvalis* has been described from Latomi 1 (115 m1, 66 M3, five more or less complete toothed low jaws; Storch 1975), from Arnissa (102 m1; Mayhew 1978) and from the Upper Pleistocene Kitseli pothole (one low jaw with m1–m3, three M3; Tsoukala et al. 2006), whereas 10 m1 and six M3 from the Middle Pleistocene locality Marathousa 1 (Peloponnese) (Doukas et al. 2018) and six fragmentary teeth from the Upper Pleistocene locality Dimitra 2 (Syrides and Koliadimou 1994) have been attributed to *Microtus* (*Microtus*) cf. *arvalis*. *Microtus* (*Microtus*) *arvalis*/*agrestis* has been described from the Upper Pleistocene localities Loutra Almopias Cave LAC (one partial mandible with m1–m2, one partial mandible with m1, 81 m1, one M2, one partial maxilla with M3 and 72 M3) and LAC Ia (11 partial mandibles with m1–m2, nine partial mandibles with m1 and 141 m1), as the molars of the two extant species cannot be easily separated based on morphology and size (Chatzopoulou 2014). 31 m1 and 49 M3 from Loutra Almopias Cave LAC and 28 m1 and a partial lower jaw with m1 from LAC Ia have been attributed to *Microtus* (*Pitymys*) cf. *subterraneus* (alpine pine vole) by Chatzopoulou (2014). One complete and one fragmentary M3 from Kitseli pothole have been assigned to cf. *Microtus* (*Pitymys*) *subterraneus*/*multiplex* (Tsoukala et al. 2006).

Remains of *Microtus* sp. have been found in the MNQ20 faunas of Zeli 2 and 2A and B (van der Meulen and van Kolfschoten 1986), but no details on the material are known. Sickenberg (1971) assigned 20 molars from the Petralona Cave to *Microtus* sp.; they show dimensions and morphology close to those of the extant *M. guentheri* and *M. hartingi* and also of the extinct *M. arvalinus*, whereas a few small specimens in the collection could even belong to *Pitymys* (Sickenberg 1971:237). Kuss and Storch (1978) reported the presence of one complete m1, one fragmentary m1 and one complete M1 in Kalymnos-Xi and attributed the first to *Allophaiomys* sp. 1 and the other two to *Allophaiomys* sp. 2 (see below paragraph on *Microtus* (*Tibericola*) *jordanica* for allocation of the latter species), whereas van der Meulen and van Kolfschoten (1986) reported the presence of *Microtus* sp. in Kalymnos, without giving details on the material (probably they refer to the material from Kalymnos-Xi already published by Kuss and Storch). A small number of rootless molars with abundant cement (one partial m1, one m2, two m3, one M1, two M2, one M3) from Varkiza 1 has been reported as *Microtus* sp.; the m1 could belong to a species of *Pitymys*, whereas the M3 is too large to be a *Microtus*, but its morphology does not remind an *Arvicola* species (van de Weerd 1973:164). Mayhew (1978) described two partial mandibles, 14 m1 and seven M3 of *Microtus* (*Pitymys*) sp. from the Upper Pleistocene locality Arnissa: the T4 and T5 of the m1 and triangles 1 and 2 of the M3 are broadly confluent. The great morphological variability of the shape of the anteroconid complex of the m1 is, according to Mayhew, an indication that a second *Pitymys* species might be present. Finally, Mayhew (1978) assigned a large M3 from Arnissa to *Microtus* sp.; the molar has the size of *Arvicola* but has broadly confluent first and second triangles, which is a characteristic of *Pitymys*. Mayhew suggested that this molar belongs to the same species with the large M3 described from Varkiza 1 by van de Weerd (Mayhew 1978:315).

Microtus (*Allophaiomys*) *pliocaenicus* (Kormos, 1933)

Nomenclatural and Taxonomical History *Allophaiomys pliocaenicus* in Kormos 1933 (new species); *Microtus* (*Allophaiomys*) *pliocaenicus* in Chaline 1966 (new combination). Considered a senior synonym of *A. laguroides* Kormos, 1933, according to Ch. IV.4.6 (van der Meulen 1973:97).

Type Material Left m1, V. 61. 1491 (holotype), Hungarian Natural History Museum, Budapest (van der Meulen 1973: Pl. VIII.19).

Type Locality Betfia 2, Romania, Lower Pleistocene.

Distribution Early Pleistocene of Eurasia and N America.

Occurrences in Greece Marathoussa.

Remarks *A. pliocaenicus* is characterized by a simple and short anteroconid complex in the ml, showing no tendency to subdivide; the three parameters defined by van der Meulen (1973:56–57) have the following values: M–A/L<44.5 (length of AC/maximum length of occlusal surface), M–B/W≥23.0 (shortest distance between BRA3 and LRA4/width of posterior part of AC) and M–C/W≥20.0 (shortest distance between BRA3 and LRA3/width of posterior part of AC) (van der Meulen 1973:97). A collection of 37 m1, 19 m2, 23 m3, 24 M1, 28 M2 and 29 M3 from the Biharian fauna of Marathoussa (N Greece) has been assigned to *M. (A.) pliocaenicus* (Koufos et al. 2001:71–78): 90% of the m1 AC2 belong to the "arvicolid–hintonid" type as defined by van der Meulen (1973:54), the mean A/L ratio is 43.6 and 91% of the M3 belong to the simplex–Group as defined by Rabeder (1981:201–204).

Microtus (Allophaiomys) pitymyoides (Chaline, 1972)

Type Material Right m1, B. 2545 (holotype) (Chaline 1972: Fig. 30.11).

Type Locality Bourgade (Hérault), France, Lower Pleistocene [Biharian].

Distribution Early Pleistocene of France, Slovakia and Greece.

Occurrences in Greece Ravin Voulgarakis.

Remarks According to the diagnosis given by Chaline (1972:101), the m1 presents the morphotype "pliocaenicus", associated with "pre–hintoni", "hintoni" (isolated AC, well-developed T4, incipient T7), "pitymyo–gregaloid" (isolated AC, developed T7), "microto–gregaloid", "arvaloid" and "pitymyo–arvalidens" (as figured by Chaline 1972:fig. 30); the length of the m1 ranges between 2.27 and 2.98, its mean value being 2.629; the height/length ratio ranges between 0.92 and 1.4; and its mean value is 1.07. Koliadimou (1996) identified a large collection of 43 m1, one partial low jaw with m1, 39 m2, 12 m3, 38 M1, 31 M2 and 34 M3 from Ravin Voulgarakis as *M. (Allophaiomys) pitymyoides*, based on the similarities with the material from the type locality.

Microtus (Allophaiomys) ruffoi (Pasa, 1947)

Type Material One adult palate with M2-M3 is figured in Pasa 1947 (Fig. 10.1) as well as several isolated teeth that represent the type material (Pasa 1947: Fig. 10.2-8) but without designation of a holotype.

Type Locality Soave Cava Sud, Italy, Lower Pleistocene.

Distribution Early Pleistocene of Spain, Italy and Greece.

Occurrences in Greece Kaiafas.

Remarks This species has a large AC2 and deep LRA4 and BRA3 (van der Meulen and Zagwijn 1974). Van der Meulen and van Kolfschoten (1986) have reported the presence of *M.* cf. *ruffoi* in the faunal list of the Lower Pleistocene locality Kaiafas, but details on the material are unknown.

***Microtus* (*Tibericola*) *jordanica* (Haas, 1966)**

Nomenclatural and Taxonomical History *Arvicola* (?) *jordanica* in Haas 1966 (new species); *Tyrrhenicola jordanica* in Kretzoi 1969 (new combination); ? *Chionomys jordanica* in Nadachowski 1991 (new combination); *Microtus* (*Tibericola*) *jordanica* in von Koenigswald et al. 1992 (new combination).

Type Material Partial right mandible with m1 and m2 (holotype) (Haas 1966: Pl. XIV.1).

Type Locality Ubeidiya, Israel, Lower Pleistocene.

Distribution Early Pleistocene of Israel, Turkey and Greece.

Occurrences in Greece Kalymnos-Xi.

Remarks *M.* (*Tib.*) *jordanica* has large and rootless molars with abundant cement in the synclines; the length of the m1 ranges between 2.88 and 4.12 mm., with a mean value of 3.51 mm.; the m1 consists of the posterior lobe, five triangles and the anterior cap; the T1–T4 are isolated whereas the T5 is confluent with the AC; the posterior lobe is tilted towards the labial side of the molar (von Koenigswald et al. 1992:3, 5). The large *Microtus* (*Allophaiomys*) sp. 2 reported from Kalymnos-Xi, with one fragmented m1 and one complete M1 (Kuss and Storch 1978:224), was compared with *Microtus* (*Tibericola*) *jordanica* from Ubeidiya (Israel) by von Koenigswald et al. (1992:13) and was found to belong to this species.

★*Microtus* (*Tibericola*) *eleniae* van der Meulen and Doukas, 2001

Type Material Left m1, TB2 709 (holotype) (van der Meulen and Doukas 2001: Fig. 1, 2).

Type Locality Tourkobounia 2, Greece, Lower Pleistocene [Biharian].

Distribution Early Pleistocene of Greece (only in type locality).

Remarks According to the species diagnosis, the molars (Fig. 36) have small size (mean length of m1= 2.9 mm.), the AC is of "hintonid" morphotype, the T4 and T5 are confluent on the m1 and the T2 and T3 are usually confluent on the M3 (van der Meulen and Doukas 2001:266). The enamel thickness is slightly differentiated; the enamel is reduced in the inner tips of the re-entrant angles; the crown cement in the re-entrant angles is abundant. 25 of 40 m1 show "hintonid" (broadly confluent T4–

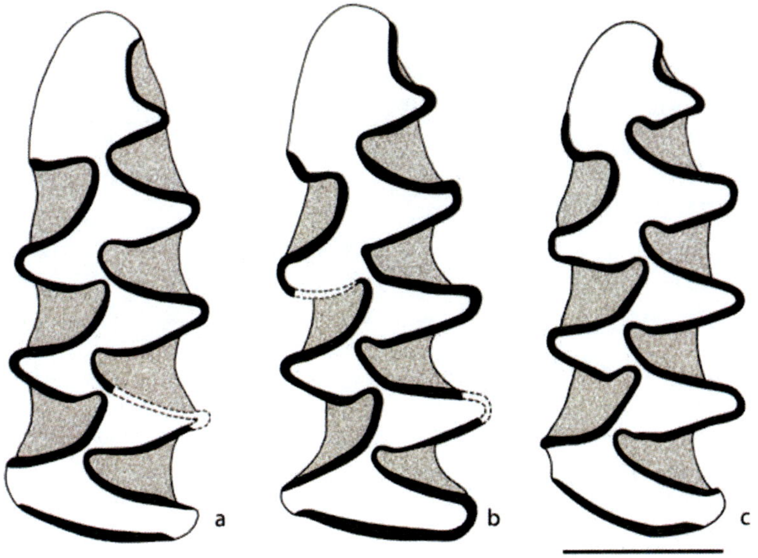

Fig. 36 *Microtus* (*Tibericola*) *eleniae* from Tourkobounia 2 (type locality). **a–c** m1, **b** (holotype). Drawings courtesy of C. Doukas. All specimens are left. Scale bar equals 1 mm

T5 and a more or less separated AC), five of 40 m1 show the "nivaloid", seven "hintonid–nivaloid", two "arvicolid–hintonid" and one "ratticepoid" type; the PC of the M3 is elongated and shows no buccal salient angle, the T2 and T3 are broadly confluent in five, slightly confluent in four and separated in three of 12 M3 (van der Meulen and Doukas 2001:266–267).

Microtus (*Terricola*) *arvalidens* (Kretzoi, 1958)

Nomenclatural and Taxonomical History *Microtus* (*Pitymys*) *arvalidens* in Kretzoi 1958 (new species); *Microtus* (*Terricola*) *arvalidens* in Rekovets 1994 (new combination). Considered a senior synonym of *Pitymys arvaloides* Hinton, 1923, according to Kretzoi (1958).

Type Material A right m1, M.6987d (holotype) Natural History Museum of London (Forsyth Major 1902:106, Fig. 28; Hinton 1923)

Type Locality West Runton Norfolk, Middle Pleistocene.

Distribution Early–Middle Pleistocene of Europe and Turkey.

Occurrences in Greece Volos.

Remarks The anterior cap of the m1 is mostly rounded; the T1–T3 are separated, the T4–T5 form one broad dentine field separated anteriorly and posteriorly, the T6

and T7 are almost opposite each other and are broadly confluent with the AC, less confluent (usual case), or almost separated from the AC; the M3 shows a well-developed T5 and an incipient LRA5 (Rekovets and Nadachowski 1995). *M.* (*Ter.*) *arvalidens* is reported in the faunal list of the Biharian fauna of Volos [as *Microtus* (*Pitymys*) *arvalidens*; van der Meulen and van Kolfschoten 1986], without details.

Chionomys **Miller, 1908**

Type Species *Arvicola nivalis* Martins, 1842 (extant).

Other Taxa Included Extant: *C. gud* (Satunin, 1909); *C. roberti* (Thomas, 1906).

Distribution Middle Pleistocene–present of S and E Europe and SW Asia.

Remarks The oldest fossils that can be with certainty assigned to the European snow vole *C. nivalis* have been found in the Middle Pleistocene locality Latomi 1 (Chios Island) and described by Storch (1975). The material includes one partial skull with teeth, 17 partial low jaws with teeth, 49 isolated m1 and 30 isolated M3; it shows the typical features of the extant species, but the molars are larger than those of recent molars from west Anatolia (Storch 1975:182). Mayhew (1978) attributed 169 m1 from the Upper Pleistocene locality Arnissa to *C. nivalis*. The majority of the molars (80%) consist of the PL, five triangles and a triangular AC closing at the posterior part due to the two opposing synclines LRA4 and BRA 3; the M3 attributed to the species (unknown number) have a simple morphology, with two lingual and two buccal re-entrant angles (Mayhew 1978:315–316).

C. nivalis has also been described from the Upper Pleistocene localities Loutra Almopias Cave LAC (182 m1, one partial mandible with m1 and 142 M3) and LAC Ia (two partial mandibles with m1–3, 12 partial mandibles with m1–2, nine partial mandibles with m1 and 176 m1): the robust molars have thick enamel and no roots; the majority of the m1 (82% from LAC and 90% from LAC Ia) show equally developed T6 and T7, broadly confluent with the AC, thus giving a triangular-pointed shape to the ACC; the M3 show three lingual and three buccal re-entrant angles with cement (Chatzopoulou 2014). Tsoukala et al. (2006:279) assigned one M3 from Kitseli Pothole to *C.* cf. *nivalis*: there are no roots, the enamel is thick, the two lingual and two buccal synclines are filled with cement and the PC is confluent with the T4 and T5 as the BRA3 and LRA4 are very shallow, whereas the T2–T3–T4 are less confluent among them.

Dinaromys **Kretzoi, 1955**

Type Species *Microtus* (*Chionomys*) *bogdanovi* Martino, 1921 (extant).

Other Taxa Included (extinct) *D. dalmatinus* (Kormos, 1931); *D. topachevskii* Nesin and Skorik, 1989; *D. allegranzii* Sala, 1996.

Distribution Pleistocene–present in Carpathians, Balkans, Ukraine and Uzbekistan.

Remarks One complete and one partial m1, as well as one M3 from Varkiza 1, have been assigned to *D.* cf. *bogdanovi* by van de Weerd (1973). The molars are two-rooted, with *Mimomys*-type enamel-thickness differentiation and some cement. The uncertain determination is due to the absence on the M3 of a small buccal fold, present on the M3 of the extant Balkan snow vole *D. bogdanovi* (van de Weerd 1973:163).

Arvicola Lacépède, 1799

Type Species *Mus terrestris* Linnaeus, 1758 (extant) (including *Mus amphibius* Linnaeus, 1758, according to Macdonald 2009).

Other Taxa Included Extant: *A. sapidus* Miller, 1908. Extinct: *A. mosbachensis* (Schmidtgen, 1911); *A kalmankensis* Zazhigin, 1980; *A. jacobaeus* Cuenca-Bescós et al., 2010.

Distribution Middle Pleistocene–present in Eurasia.

Remarks Several middle–early Late Pleistocene *Arvicola* species have been occasionally described, mainly from C and W Europe. All these species were later included in *A. cantianus* (Hinton, 1910) (Mayhew 1978; von Koenigswald and van Kolfschoten 1996; Santel and von Koenigswald 1998; Maul et al. 2000). It is common, thus, that any middle–early Late Pleistocene *Arvicola* with *Mimomys*-type enamel-thickness differentiation and no roots is assigned to *A. cantianus* (e.g., Escudé et al. 2008). Maul et al. (2000:136) suggested that the name *A. cantianus* should be used only for the type material, as it consists only of two partial m1, one M1 and one M2, and all other *Arvicola* with negative enamel-thickness differentiation should be referred to *A. mosbachensis*. More recently, Marcolini et al. (2011:481) considered *A. cantianus* a nomen dubium, suggesting that this name should not be used in taxonomy.

Mayhew (1978:324) described five m1, one M1 and four M3 from the Upper Pleistocene locality Arnissa as *A. cantiana/terrestris*, following the suggestion of von Koenigswald (1973:164, Fig. 2) for the *Arvicola* assemblages that show neither *Mimomys* (like in *A. cantianus*) nor *Microtus* (like in *A. terrestris*) type of enamel-thickness differentiation, but undifferentiated enamel along the prisms. Many authors have followed this suggestion and have assigned middle–early Late Pleistocene *Arvicola* to *A. cantiana/terrestris* or *A. cantianus/terrestris* (e.g., van Kolfschoten 1985; von Koenigswald and van Kolfschoten 1996; Desclaux et al. 2000; Marquet and Séronie-Vivien 2016), mainly in W Europe, whereas in E Europe other *Arvicola* species have been proposed to represent this transitional form: *A. chosaricus* Alexandrova 1976 and *A. hunasensis* Carls 1986 (Carls et al. 1988; Kalthoff et al. 2007). Here we retain the initial assignment of the Arnissa material to *A. cantiana/terrestris*, as the taxonomical problem with these species has not been solved yet.

The extant European water vole (or northern water vole) *A. terrestris* has been described from the Upper Pleistocene localities Loutra Almopias Cave LAC and LAC Ia, represented by one partial mandible with m1, 10 m1, four m2, four m3, eight M1, three M2 and seven M3 from the former and two partial mandibles with m1, 48 m1, six m2, 13 m3, 26 M1, 14 M2 and 22 M3 from the latter locality (Chatzopoulou 2014). One fragmentary m1, one M1 and one M3 from the locality Dimitra 2 have been attributed to *A.* cf. *terrestris* (Syrides and Koliadimou 1994).

***Arvicola mosbachensis* (Schmidtgen, 1911)**

Nomenclatural and Taxonomical History *Microtus mosbachensis* in Schmidtgen 1911 (new species); *Arvicola mosbachensis* in Heller 1933 (new combination).

Type Material Left mandible with m1–m2, SMF 74/4835 (holotype), Senckenberg Research Institute and Natural History Museum Frankfurt am Main (Maul et al. 2000: Pl. 1.1).

Type Locality Mosbach 2, Germany, Middle Pleistocene (Cromer lntergl. III or IV).

Distribution Middle–early Late Pleistocene of Eurasia.

Occurrences in Greece Marathousa 1 (Peloponnese) and Latomi 1

Remarks Maul et al. (2000) studied a large number of molars from the type locality (102 m1 and 23 M3): molars are smaller than those of the extant *A. terrestris*, high-crowned and bear cement, 2% of the m1 and 10% of the M3 show incipient roots, 27% of the m1 bear a *Mimomys* fold and the mean SDQ is 133.34 (range 117.6–159.27); the M3 bears three anticlines and two synclines on each side, and usually the T3 and T4 are more confluent than in *A. terrestris*. The Middle Pleistocene locality Marathousa 1 (Peloponnese) has yielded 47 m1, 35 M3, one mandible with m1–m3, nine toothless mandible fragments and a small number of other molars with no diagnostic features attributed to *A. mosbachensis* by Doukas et al. (2018); the m1 show a negative enamel-thickness differentiation (mean SDQ=122, range 102–149), and the anterior cap of the m1 consists of the opposite confluent T4 and T5 and a mostly rounded and less often angular AC2, whereas the AL1 of the M3 is elongated and shows concave sides with thin enamel, the T2 is closed, the T3 and T4 are mostly confluent and the PC is relatively rounded. Storch (1975) identified nine m1 and 11 M3 from the locality Latomi 1 in Chios Island as *Arvicola praeceptor*; the teeth show negative enamel-thickness differentiation, typical for early *Arvicola* species, and are larger than those of *A. mosbachensis* (Storch 1975:180, 182). We here include the Chios material to *A. mosbachensis*, following the above-mentioned literature, without, however, having seen the material.

Family Eomyidae Depéret and Douxami, 1902

Type Genus *Eomys* Schlosser, 1884.

Remarks The Turolian fauna of Kalithies (Rhodes Island) has yielded one M2 with a very simple morphology, bearing the two main lophs and a short mesoloph, which has been assigned to Eomyidae indet. (de Bruijn 1976).

Subfamily Eomyinae Winge, 1887

Type Genus *Eomys* Schlosser, 1884.

Pseudotheridomys Schlosser, 1925–1926

Type Species *Theridomys? parvulus* Schlosser, 1884.

Other Taxa Included Eurasian species: *P. schaubi* Lavocat, 1951; *P. pusillus* Fahlbusch, 1969; *P. lacombai* Álvarez-Sierra, 1987; *P. fejfari* Álvarez-Sierra and Daams, 1987; *P. bernensis* Engesser, 1990; *P. rolfoi* Engesser, 1990; *P. asiaticus* Wang and Emry, 1991; *P. yanshini* Lopatin, 2000. North American species: *P. hesperus* Wilson, 1960; *P. pagei* Shotwell, 1967; *P. cuyamensis* Lindsay, 1974.

Distribution Late Oligocene–middle Miocene of Eurasia and N America.

Pseudotheridomys parvulus (Schlosser, 1884)

Nomenclatural and Taxonomical History *Theridomys? parvulus* in Schlosser 1884 (new species); *Pseudotheridomys parvulus* in Schlosser 1925–1926 (new combination). Senior synonym of *P. bouziguensis* Escarguel and Aguilar, 1997, according to Marković et al. (2016).

Type Material Left mandibular fragment embedded in the original sediment, with its lingual side exposed, very severely fractured, with strongly worn p4–m1 (m2–3 are no longer present), Slg. München 1881 IX 79 (Schlosser 1925–1926: Pl. 11.7 and Fahlbusch 1968:Pl. 14.1). Schlosser did not assign a holotype for his species; Stehlin and Schaub (1951:357) referred to the mandible figured by Schlosser (1884) in Plate 3, figures 26 and 37 as a holotype; Fahlbusch (1968) clarified that the specimen can only be a lectotype.

Type Locality Haslach, Germany, Lower Miocene [MN2].

Distribution Early Miocene of Europe and Turkey.

Occurrences in Greece Aliveri and Karydia.

Remarks The molars of *P. parvulus* are characterized by the presence of five, more or less parallel and equal in length, smooth and continuous ridges; this feature characterizes the morphotype A suggested by Fahlbusch (1983:215). His morphotype B is characterized by the absence of a mesoloph/-id (character of the genus *Ligerimys*). As discussed by Marković et al. (2016), in large assemblages of *P. parvulus*, there are random specimens that show four ridges and can be considered to be within the morphological variation of *P. parvulus*. Rarely, a short additional ridge is present between the meta- and mesolophid on the lower molars (Engesser 1999). The Lower Miocene locality Aliveri has yielded a very rich collection of 681 *P. parvulus* isolated cheek teeth (Álvarez–Sierra et al. 1987). The authors compared the Aliveri material with *P. parvulus* from Bouzigues (France), Alcocer 3B (Spain) and Jungingen (Germany) and found it to be of the same size, whereas they noticed some minor differences: the Aliveri assemblage has slightly less reduced anterolophid on the p4 and ml–2 than the Bouzigues assemblage, slightly less reduced mesoloph and anterolophid than the Alcocer 3B assemblage and a slightly wider P4 than the Jungingen assemblage. All Aliveri molars show five ridges on their occlusal surface (morphotype A). Theocharopoulos (2000) and later Doukas (2003) reported the presence of *Ligerimys* sp. in the Lower Miocene locality Karydia. However, Duncan (2012), in the frames of her Master thesis, studied the Karydia material thoroughly and assigned it to *P.* aff. *parvulus*. She gave more morphological characteristics for the two morphotypes of Fahlbusch and mentioned that 93% of the Karydia specimens belong to morphotype A and 7% to morphotype B (Duncan 2012:30). The presence of this small percentage, as well as the larger size of the cheek teeth from Karydia in comparison with those from Aliveri, led Duncan to her uncertain assignment to the species. These features, according to Fahlbusch (1983), can be observed in assemblages at the early stage of the *Pseudotheridomys* to *Ligerimys* transition. Based only on this, Karydia can be considered younger that Aliveri (Duncan 2012:44).

***Eomyops* Engesser, 1979**

Type Species *Eomys catalaunicus* Hartenberger, 1966.

Other Taxa Included *E. bodvanus* (Jánossy, 1972); *E. oppligeri* Engesser, 1990; *E. hebeiseni* Kälin, 1997; *E. noeliae* Ruiz-Sanchez et al., 2009.

Distribution Middle Miocene (MN5)–Early Pleistocene (MNQ 17) of Europe.

Remarks De Bruijn (1976) assigned one unerupted p4 from Biodrak to *Leptodontomys* sp., noticing the "*unusual lophodont dental pattern for a* Leptodontomys" and mentioning the resemblance of this specimen with specimens of *L.* cf. *catalaunicus* from Anwil (de Bruijn 1976:380). After his publication, Engesser (1979) created the genus *Eomyops* to include all the European species of *Leptodontomys*, which he considered a N American genus. Since then, there are both views in the available literature (see the historical review of Kälin 1997:629

and the remarks in de Bruijn et al. 2012:69–70 for details), with authors assigning European species to *Leptodontomys* (e.g., de Bruijn 1995, de Bruijn et al. 1999) and others who assign them to *Eomyops* (e.g., Engesser 1999, Daxner-Höck 2005). Here we follow the view of Engesser (1999) and assign them to *Eomyops*. The premolar from Biodrak is here therefore assigned to *Eomyops* sp.

***Eomyops catalaunicus* (Hartenberger, 1966)**

Nomenclatural and Taxonomical History *Eomys catalaunicus* in Hartenberger 1966 (new species); *Leptodontomys catalaunicus* in Hugueney and Mein 1968 (new combination); *Eomyops catalaunicus* in Engesser 1979 (new combination).

Type Material Left M1/2, CL 301 (holotype), Museum of Sabadell (Hartenberger 1966: Fig. 4a).

Type Locality Can Llobateres, Spain, Upper Miocene [Vallesian (MN9)].

Distribution Middle Miocene–early Pliocene of Europe and Turkey.

Occurrences in Greece Plakias, Lava 2 and Maramena.

Remarks The cheek teeth of *E. catalaunicus* do not differ much in morphology from the other *Eomyops* species, but they are somewhat larger than those of *E. oppligeri* and smaller than those of *E. hebeiseni* and *E. bodvanus* (Daxner-Höck 2005). Hartenberger (1966) described them as small and bunodont, with a lingual anteroloph on the M1–2. Ruiz-Sánchez et al. (2009) reported that their new species (*E. noeliae*) shows two parallel ridges on the enamel along the lower incisors, "*like the type species of the genus,* E. catalaunicus", and that this character should be used to distinguish between the European *Eomyops* and the N American *Leptodontomys* (which had also been the view of Engesser 1979). De Bruijn et al. (2012) do not agree with this view, as most assemblages include no incisor or include incisors that cannot be clearly associated with the eomyid. The locality Plakias has revealed eight p4, 18 m1/2, four m3, seven P4, 12 M1/2 and one M3 assigned to *E.* aff. *catalaunicus* by de Bruijn et al. (2012). Even though the size and morphology of the teeth are very similar to those of the species, the authors assign the material only with reservation to *E. catalaunicus* because there are no incisors with ridges associated with the eomyid material. De Bruijn et al. (1999) have listed *Eomyops* cf. *catalaunicus* (as *Leptodontomys* cf. *catalaunicus*) for the locality Lava 2, without details of the material. The only material from Greece assigned to *E. catalaunicus* with certainty comes from the locality Maramena, where one M1 and one M2 have been found, which show a "striking similarity" with the molars from the type locality of the species (de Bruijn 1995: 98).

***Keramidomys* Hartenberger, 1966**

Type Species *Keramidomys pertesunatoi* Hartenberger, 1966.

Other Taxa Included *K. carpathicus* (Schaub and Zapfe, 1953); *K. thaleri* Hugueney and Mein, 1968; *K. mohleri* Engesser, 1972; *K. anwilensis* Engesser, 1992; *K. reductus* Bolliger, 1992; *K. fahlbuschi* Qiu, 1996; *K. ermannorum* Daxner-Höck and Höck, 2009. [*K. thaleri* is considered by several authors as a junior synonym of *K. carpathicus* (e.g., Fahlbusch 1975, Bolliger 1992, de Bruijn 1995), as it is difficult to assign small assemblages in one of the two species. Other authors retain *K. thaleri* as a distinct species (e.g., Engesser 1999, Prieto 2010).]

Distribution Middle Miocene–early Pliocene (MN5–15) of Eurasia.

Remarks Van der Meulen and van Kolfschoten (1986) reported the presence of *Keramidomys* sp. in Kastoria 1, without giving information on the material.

***Keramidomys carpathicus* (Schaub and Zapfe, 1953)**

Nomenclatural and Taxonomical History *Pseudotheridomys carpathicus* in Schaub and Zapfe 1953 (new species); *Keramidomys carpathicus* in Hugueney and Mein 1968 (new combination). Considered as the senior synonym of *K. octaviae* Baudelot 1972 by de Bruijn (1995).

Type Material Partial left maxilla with P4–M1(holotype) (Schaub and Zapfe 1953: Fig. 1).

Type Locality Neudorf, Slovakia, middle Miocene [MN6].

Distribution Middle Miocene (MN5)–early Pliocene (MN14) of Europe and Turkey.

Occurrences in Greece Maritsa 1, Monasteri and Maramena.

Remarks Since the genus seems to have undergone little morphological changes during its long existence, the differences among the *Keramidomys* species are rather subtle, with all of them showing a lophodont tooth pattern with the cusps entirely incorporated in the ridges (Engesser 1999). *K. carpathicus* shows a more distinct reduction of the third molars, in comparison with other species. According to de Bruijn (1995), there is a large variation in the cheek tooth morphology within large assemblages, and thus small assemblages of *Keramidomys* cannot be clearly assigned to a species. A small collection of cheek teeth (one d4, two p4, two m3, one D4, three P4, six M1/2 and one M3) from the locality Maritsa 1 has been assigned to *K. carpathicus* by de Bruijn et al. (1970). The authors found no morphological difference between the Maritsa 1 material and that from the type locality; however the elements differ in size, but the authors mention that this could be due to differences in the method of measuring the teeth (de Bruijn et al. 1970:569). *K.* cf. *carpathicus* and *K.* aff. *carpathicus* had been listed for the localities Monasteri and Maramena, respectively, by de Bruijn (1989). The material from Maramena was later thoroughly studied by de Bruijn (1995) and assigned to *K.* cf. *carpathicus*. It consists of three d4, six p4, 11 m1/2, three m3, two D4, four P4, 11 M1/2 and two M3. There is a large metrical and morphological variation within the sample. Overall, the Maramena molars are close in size to those from the species type locality, but the premolars are larger (de Bruijn 1995). Some characters of the Maramena cheek teeth seem to be "more primitive" than those of the Neudorf ones (e.g., anteroloph and protoloph of P4 not completely fused, longitudinal ridge of M1–2 not interrupted in about half of the specimens), and others seem to be "more derived" (e.g., reduced mesoloph/-id and reduced dental pattern on the third molars). The author concluded that there is no clear trend within the genus (de Bruijn 1995:100). Later, Daxner-Höck and Höck (2009) established a new *Keramidomys* species, *K. ermannorum*, based on material from the early Vallesian fauna of Richardhof–Golfplatz in Austria and included in it material from other Austrian Vallesian–Turolian faunas, as well as that from Maramena. Their species shows a combination of the morphological characteristics of *K. thaleri*, *K. carpathicus* and *K. mohleri*, but differs from them in its trend to develop lophodonty, its deeper sinus/-id, its plane occlusal surface, the presence of labial and lingual connections of the lophs/-ids, respectively and its wider lower molars, whereas its cheek teeth are in size intermediate between the larger ones of *K. mohleri* and the smaller ones of *K. thaleri* and *K. carpathicus* (Daxner-Höck and Höck 2009:386). The authors noticed the wide variability of dental characters within the assemblages, but also morphological and metrical trends, such as an increase in size and degree of lophodonty, enlargement and deepening of the sinus/-id and flattening of the occlusal surface (Daxner-Höck and Höck 2009:388). Since the Maramena material has not been studied for the purposes of the present report, we here retain the initial assignment by de Bruijn (1995).

Family Ctenodactylidae Zittel, 1893

Subfamily Ctenodactylinae Alston, 1876

Type Genus *Ctenodactylus* Gray, 1830.

Sayimys Wood, 1937

Type Species *Sayimys perplexus* Wood, 1937.

Other Taxa Included *S. sivalensis* (Hinton, 1933); *S. obliquidens* Bohlin, 1946; *S. intermedius* (Şen and Thomas, 1979); *S. badauni* Vasishat, 1985; *S. chinjiensis* Baskin, 1996; *S. baskini* López-Antoñanzas and Şen, 2003; *S. assarrarensis* López-Antoñanzas and Şen, 2004; *S. giganteus* López-Antoñanzas, Şen and Saraç, 2004; *Sayimys* n.sp. was described from Libya by Wessels et al. (2003), but the species has not been officially named yet.

Distribution Early Miocene–Pliocene of N Africa, Asia and Greece.

Remarks The Lower Miocene locality Antonios has yielded one d4 and one m1/2, both fragmented, assigned to *Sayimys* sp.: the two teeth show morphological and metrical similarities with both *S. intermedius* and *S. sivalensis*, including the absence of a metalophulid II from the d4; however, the absence of the anterior part of both teeth and the failure to study the taxonomically important anteroconid area do not allow a specific attribution (Vasileiadou and Koufos 2005).

***Sayimys intermedius* (Şen and Thomas, 1979)**

Nomenclatural and Taxonomical History *Metasayimys intermedius* in Şen and Thomas 1979 (new species); *Sayimys intermedius* in de Bruijn et al. 1989 (new combination). Considered a senior synonym of *S. minor* de Bruijn et al., 1981, according to López–Antoñanzas and Şen (2004).

Type Material Left mandible with d4–m2, AJ 545 (holotype) (Şen and Thomas 1979: fig. 1A–B).

Type Locality Al Jadidah, Saudi Arabia, middle Miocene.

Distribution Early–middle Miocene of Saudi Arabia, Pakistan, Greece and possibly Turkey [*S.* cf. *intermedius* in Paşalar (Peláez–Campomanes and Daams 2002)].

Occurrences in Greece Thymiana A and C (Chios Island).

Remarks Şen and Thomas (1979) gave the following diagnostic features for their species: the masseteric crest of the mandible is strong; a groove runs all along the anterior enamel of the incisor; the dp4 bears a small anterior cuspid attached to the protoconid–metaconid loph; on the lower molars the talonid is as wide as the trigonid; the synclines and cusps are arranged transversally and not obliquely as in *S. obliquidens*; there is no cement and no pseudomesolophid on the m2. Some additional features of the species were reported by de Bruijn et al. (1989): absence of a metalophulid II from the d4; d4 anteroconid more often isolated; the mean length of the m3 is smaller than that of the m1/2. A rather rich collection of ctenodactylid isolated dental elements (seven d4, two p4, 12 m1–3, three D3, four D4, eight M1–2 and one M3) with similar features has been recovered from two horizons (A and C) of the middle Miocene locality Thymiana (Chios Island).

The material has been assigned to *S. intermedius* by López–Antoñanzas et al. (2005), but more recently Hartman et al. (2019) referred it to as *Sayimys* aff. *giganteus* due mainly to its larger d4 and long m3.

Family Hystricidae Burnett, 1830

Type Genus *Hystrix* Linnaeus, 1758.

Remarks The presence of Hystricidae indet. in the Pliocene locality Tourkobounia 1 has been reported by Reumer and Doukas (1985) without details on the material.

Hystrix Linnaeus, 1758

Type Species *Hystrix cristata* Linnaeus, 1758 (extant).

Other Taxa Included (extant) *H. brachyura* Linnaeus, 1758; *H. indica* Kerr, 1792; *H. javanica* (F. Cuvier, 1823); *H. africaeaustralis* Peters, 1852; *H. crassispinis* (Günther, 1877); *H. pumila* (Günther, 1879); *H sumatrae* (Lyon, 1907); (extinct) *H. primigenia* (Wagner, 1848); *H. refossa* Gervais, 1852; *H. sivalensis* Lydekker, 1878; *H. lagrelii* Lönnberg, 1924; *H. kiangsenensis* Wang, 1931; *H. parvae* (Kretzoi, 1951); *H. makapanensis* Greenwood, 1958; *H. trofimovi* Shevireva, 1986; *H. leakeyi* Denys, 1987; *H. aryanensis* Şen, 2001; *H. depereti* Şen, 2001; *H. gansuensis* Wang and Qiu, 2002; *H. lufengensis* Wang and Qiu, 2005; *H. paukensis* Nishioka et al., 2011. According to van Weers (1994), *H. hirsutirostris* Brandt, 1835 = *H. indica*; according to van Weers (2005), *H. vinogradovi* Argyropulo, 1941 = *H. brachyura*, *H. vanbreei* van Weers, 1994 = *H. lagrelli.*

Distribution Late Miocene–present in Eurasia and Africa. Most of the extant species are also found in Pliocene and Pleistocene fossil localities.

Remarks van der Meulen and van Kolfschoten (1986) listed *Hystrix* sp. in the Pliocene locality Damatria. Furthermore, one m1 of *Hystrix* sp. has been found in the Middle Pleistocene layers of the Petralona cave (Sickenberg 1971). Athanassiou (2018) has included cf. *Hystrix* in the faunal list of the Lower Pleistocene locality Sesklon; this report is based on the presence of a fragmentary lower jaw (Athanassiou, personal communication). Symeonidis (1992) had already reported the presence of a large rodent lower incisor in Sesklon, which he attributed to *Castor* sp. (see above paragraph). Since the incisors of *Castor* and *Hystrix* differ neither in size nor in morphology, and the overall fauna from Sesklon does not support the presence of a local water body, it is thought that Symeonidis' identification was erroneous and the incisor also belongs to *Hystrix* (Athanassiou, personal communication).

★*Hystrix primigenia* (Wagner, 1848)

Nomenclatural and Taxonomical History *Lamprodon primigenius* in Wagner 1848 (new species); *Hystrix primigenia* in Gaudry and Lartet 1856 (new combination). Considered the senior synonym of *Castor atticus* Roth and Wagner, 1854, according to Gaudry and Lartet (1856); of *H. bessarabica* Rjabinin, 1918, according to Sulimski (1960); and of *H. trofimovi* Shevyreva, 1986 according to van Weers (1994).

Type Material A lower left incisor (holotype) (Wagner 1848: Tab. 4.7–8).

Type Locality Pikermi, Greece, Upper Miocene [Turolian (MN12)].

Distribution Late Miocene (early–middle Turolian)–Pliocene (late Ruscinian) of Eurasia.

Occurrences in Greece Samos S3, Chomateri, Pikermi, Halmyropotamos, Alifakas and Dytiko 3.

Remarks The species *Lamprodon primigenius* (Wagner 1848) was based on one lower left incisor from Pikermi. The locality later yielded two isolated molars for which the species *Castor atticus* was created by Roth and Wagner (1854). The name *Hystrix primigenia* was established by Gaudry and Lartet (1856) to include the material previously reported. Wagner (1857), recognizing the validity of *H. primigenia*, added to the already known material one skull from Pikermi. Gaudry (1862–1867) described two additional cranial remains with dentition from the type locality. More recently, one mandible and one isolated tooth of the species from the locality Pikermi (Chomateri) have been figured, but without details on the material (Bachmayer et al. 1982).

There is no original diagnosis of the species. Some characters of the species were later given by several authors. The cheek teeth are very similar to those of other species, such as *H. parvae* and *H. depereti*; the features that can better distinguish this species are the size and the degree of hypsodonty of the cheek teeth, the depth of the hypoflexus/id on the molars and the development of the third molars (Şen 2001). *H. primigenia* is on average larger than *H. parvae* and smaller than *H. depereti*, the teeth are rather low-crowned, the hypoflexus/id are as deep as the half of the crown height and the posterior part of the third molar is well developed, giving a square shape to the tooth (Şen 2001; Şen and Purabrishemi 2010). Furthermore, the length of the M1 is smaller than that of the M2 (Şen 2001).

Apart from the material from Pikermi, a few other specimens of this species have been unearthed in Greece. Melentis and Schneider (1966) described one partial left maxilla with the M1–2 of *H. primigenia* from the Upper Miocene locality Alifakas (Thessaly). The Turolian fauna of Halmyropotamos has yielded both semi-mandibles of one *H. primigenia* individual, the right one with the m2–3 and the alveoli of the p4 and the m1 and the left one with the m1–2 and the alveoli of the p4 (Melentis 1967). Furthermore, the species is represented by one almost complete skull and one left mandible with roots of the p4–m2 and part of the m3 in the late Turolian fauna from Samos–S3 (Bakr 1959; Black et al. 1980). Finally, one skull with complete dentition of *H. primigenia* has been described from the late Turolian fauna from Dytiko-3 (Fig. 37) by de Bonis et al. (1992).

★*Castor atticus* **Roth and Wagner, 1854**
[junior synonym of *Hystrix primigenia* (Wagner, 1848), according to Gaudry and Lartet (1856)]

Type Material No holotype has been designated; the only available material is two lower right molars (Roth and Wagner 1854: Pl. 4.3).
Type Locality Pikermi, Greece, Upper Miocene [Turolian (MN12)].
Remarks Roth and Wagner (1854) erected the species *Castor atticus* based on two isolated lower right molars from Pikermi. One of the two is worn down and does not show any clear features on the occlusal surface. They have three separate roots. The authors compared them with molars from the then known species of *Castor*, *Trogontherium*, *Chalicomys*, *Palaeomys* and *Stenofiber* and found differences in the size and shape, in "the very oblique direction of the enamel–duplicature penetrating from the outside of the tooth and the more complex nature of the enamel figures on the occlusal surface". They created a new species for this material, commenting, however, on the unfortunate situation of justifying its existence only through the presence of two teeth. Consequently, Gaudry and Lartet (1856) established the name *Hystrix primigenia* to include the incisor described by Wagner (1848) and the molars of Roth and Wagner (1854). Wagner (1857:129–130) described a skull from Pikermi and agreed with Gaudry and Lartet that the incisor previously described as *Lamprodon primigenius* and the two molars previously described as *Castor atticus* actually belong to the same species, *H. primigenia*.

Hystrix refossa **Gervais, 1852**

Taxonomical History Considered by van Weers (2005) as the senior synonym of *H. major* Gervais, 1859; *H. crassidens* Lydekker, 1886; *H. etrusca* Bosco, 1899; *H. makapensis* Greenwood, 1958; *H. angressi* Frenkel, 1970; *H. gigantea* van Weers, 1985; and *H. magna* Pei, 1987.

Type Material One partial mandible with the p4 and the four alveoli of the m1 (holotype) Muséum National, Laboratoire de Paléontologie, Paris (van Weers 1994: Pl. 1.a–b).

Type Locality Les Etouaires, Perrier, Puy de Dôme, France, Lower Pleistocene.

Distribution Early late Pliocene–Late Pleistocene of Eurasia.

Occurrences in Greece Milia-5, Gerakarou 1 and Alikes.

Remarks The cheek teeth of the species are slender, with rounded occlusal surface and complicated synclinid I (Koliadimou and Koufos 1991). The emended diagnosis by van Weers (1994) mentions that the species has very large-sized and strongly

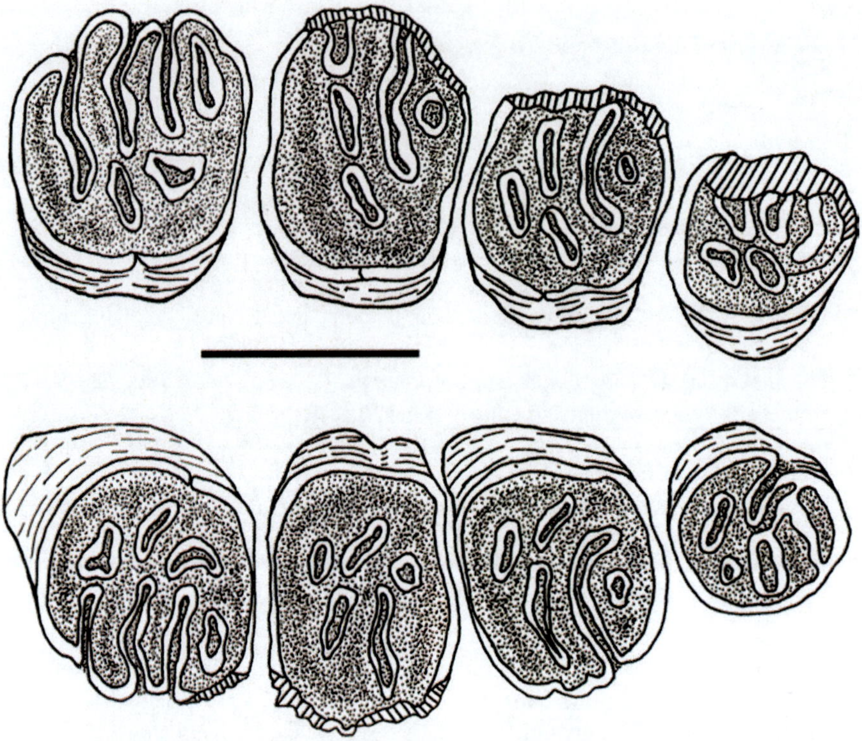

Fig. 37 *Hystrix primigenia* from Dytiko-3. Upper dentition of the skull DKO–41. Drawings courtesy of GD. Koufos. Scale bar equals 1 cm

hypsodont teeth. The same author also distinguishes *H. refossa* from *H. primigenia* mainly through the degree of hypsodonty: the ratio tooth height/occlusal surface length is nearly always higher than 1.5 and maximum 2.6 for the former and rarely higher than 1.5 and maximum 1.6 for the latter species. One partial left mandible with i–m2, one upper incisor, two P4 and two M3 from the Biharian fauna from Gerakarou 1 have been assigned to *H. major*, with the note that they are also very similar to *H. refossa*, but more material of the latter species is needed for its better knowledge (Koliadimou and Koufos 1991). The species is also described from the Biharian fauna of Alikes, with one partial right mandible with d4–m1 (Athanassiou 1998). Lazaridis et al. (2019) descripted from Milia-5 a right mandible fragment with part of the first lower incisor and the alveolus of the fourth premolar as *H. refossa*. The Milia record appears to be the oldest known of the species so far, dated at MN16a (early Villafrancian, early late Pliocene).

6 Concluding Remarks

The rodent record of Greece is rather rich during the Neogene and Pleistocene, with fossils revealed in 68 and 39 localities, respectively. Unfortunately it is impossible to establish any knowledge for the pre-Neogene rodent faunas, since there is only one Late Oligocene fauna from Kyprinos, a fact that has to do with the absence of appropriate fossiliferous sediments from the area. Of the eight families represented, the castorids, dipodids, eomyids, ctenodactylids and hystricids are very rare, with the eomyids disappearing after the Pliocene, following the pattern of the European family evolution. The glirids and sciurids show greater diversity during the Miocene, with eight and six genera, respectively, in the Early–Middle Miocene and seven genera each in the Late Miocene, and their genus numbers start declining after the Late Miocene, apparently due to the change of the environment to less humid and more open. The murids are by far the best represented rodents in the Greek fossil record, with 4 genera in the Early Miocene, 12 in the Early–Middle Miocene, 24 in the Late Miocene (including the Turolian/Ruscinian boundary), 18 in the Pliocene, 19 in the Early Pleistocene and 14 in the middle–Late Pleistocene. Four new glirid species, one new sciurid genus (*Aliveria*), seven new sciurid species, six new murid genera (*Heramys*, *Byzantinia*, *Karydomys*, *Hypsocricetus*, *Hansdebruijnia*, *Kalymnomys*) and 42 new murid species and one subspecies have been described from Greek fossil localities, proving the great importance of the Greek rodent fossil record in the knowledge of the Eurasian rodent evolution.

Acknowledgments This chapter is dedicated to the late Kostakis Theocharopoulos, an esteemed friend and colleague. We are deeply obliged to Dr. Wilma Wessels for access to important literature and the original images and drawings, as well as for having offered the Karydia faunal list. We are also grateful to Dr. Hans de Bruijn, Emeritus Prof. George Koufos, Dr. Constantine Doukas, Dr. Kalliopi Koliadimou, Dr. Katerina Chatzopoulou, Dr. Athanassios Athanassiou and Dr. Kees Hordijk for giving permission to use original images and drawings. All the hard and tedious work of scanning original plates by Frans Verdaasdonk and Dr. Wilma Wessels is very much appreciated. Without all the intense and lifelong work on fossil rodents, in general, and on the Greek fossil rodents, in particular, of Hans de Bruijn, most of the knowledge presented here would not have been established. For this and also for all the help he offered and knowledge he shared with both authors during the past 24 years, we both want to express our gratitude and dedicate this chapter to him. We greatly appreciated fruitful comments and suggestions by Dr. Wilma Wessels and Dr. Sevket Şen.

Editorial Note The works in this book (lagomorphs, rodents, insectivores) are the last of Katerina, who left us recently after a long battle with cancer. Katerina deeply believed in this collective project and worked with passion and dedication for its implementation, giving even her very last strength. Unfortunately, she could not deal with the constructive comments and suggestions of the reviewers, a task undertaken by the co-authors of the articles and her friends in the same scientific field.

Appendix

List of Greek localities with known occurrences of fossil rodents arranged in approximate geochronological order. Locality numbers refer to the collection numbers of the PaleoBiology Database (PBDB). Type localities are marked with boldface

Localities[PBDB No]	Age (MNQ)	Taxon
Pili[204953]	Sub-recent	*Dryomys nitedula*[1] *Myomimus roachi*[1]
Loutra Almopias Cave Ia[204954]	Late Pleistocene, Younger Dryas (13,000–14,500 BP)	*Dryomys nitedula*[2] *Glis glis*[2] *Muscardinus avellanarius*[2] *Spermophilus citellus*[2] *Spalax leucodon*[2] *Mesocricetus newtoni*[2] *Cricetulus migratorius*[2] *Apodemus mystacinus*[2] *Apodemus sylvaticus/flavicollis*[2] *Apodemus uralensis*[2] *Myodes glareolus*[2] *Microtus* (*Microtus*) *arvalis/agrestis*[2] *Microtus (Pitymys)* cf. *subterraneus*[2] *Chionomys nivalis*[2] *Arvicola terrestris*[2]
Kitseli Pothole[204955]	Late Pleistocene	*Cricetulus migratorius*[3] *Microtus* (*Microtus*) *arvalis*[3] cf. *Microtus (Pitymys) subterraneus/multiplex*[3] *Chionomys* cf. *nivalis*[3]
Trypiti[203848]	Late Pleistocene	*Apodemus* cf. *mystacinus*[4,5]
Dimitra 2[204956]	Late Pleistocene	*Spermophilus citelloides*[6] *Apodemus sylvaticus/flavicollis*[6] Murinae indet.[6] *Microtus* (*Microtus*) cf. *arvalis*[6] *Arvicola* cf. *terrestris*[6]

Localities[PBDB No]	Age (MNQ)	Taxon
Arnissa[204957]	Late Pleistocene [penultimate interglacial (Eemian) age]	*Dryomys nitedula*[7] *Spermophilus citellus*[7] *Spalax microphthalmus*[7] *Mesocricetus newtoni*[7] *Cricetulus migratorius*[7] *Apodemus mystacinus*[7] *Apodemus* sp.[7] *Lagurus lagurus*[7] *Myodes glareolus*[7] *Microtus* (*Microtus*) *guentheri*[7] *Microtus* (*Microtus*) *arvalis*[7] *Microtus (Pitymys)* sp.[7] *Microtus* sp.[7] *Chionomys nivalis*[7] *Arvicola cantiana/terrestris*[7] *Sicista subtilis*[7]
Loutra Almopias Cave LAC[203847]	Late Pleistocene, middle Würm (42,500 BP)	*Dryomys nitedula*[2] *Glis glis*[2] *Muscardinus avellanarius*[2] *Spermophilus citellus*[2] *Spalax leucodon*[2] *Mesocricetus newtoni*[2] *Cricetulus migratorius*[2] *Apodemus mystacinus*[2] *Apodemus sylvaticus/ flavicollis*[2] *Apodemus uralensis*[2] *Mus spicilegus*[2] *Myodes glareolus*[2] *Microtus* (*Microtus*) *arvalis/ agrestis*[2] *Microtus (Pitymys)* cf. *subterraneus*[2] *Chionomys nivalis*[2] *Arvicola terrestris*[2] *Sicista subtilis*[2]
Petralona Cave[183123]	Middle Pleistocene	*Myomimus personatus*[8] *Spermophilus* sp.[8] ? *Spalax* sp.[8] *Allocricetus* sp./*Cricetulus* sp.[8] *Microtus* sp.[8] *Hystrix* sp.[8]

Localities[PBDB No]	Age (MNQ)	Taxon
Latomi 1[204077]	Middle Pleistocene	*Dryomys nitedula*[9] *Myomimus roachi*[9] *Spermophilus* cf. *xanthoprymnus*[9] *Spalax nehringi*[9] *Mesocricetus* cf. *brandti*[9] *Cricetulus migratorius*[9] *Apodemus mystacinus*[9] *Apodemus* sp.[9] *Mus* sp.[9] *Rattus rattus*[9] cf. *Rattus* sp.[9] *Lagurus lagurus*[9] *Microtus* (*Microtus*) *guentheri*[9] *Microtus* (*Microtus*) *arvalis*[9] *Chionomys nivalis*[9] *Arvicola mosbachensis*[9] *Sicista subtilis*[9] *Allactaga* cf. *euphratica*[9]
Varkiza 2[204167]	Middle or Late Pleistocene	*Myomimus personatus*[10] *Apodemus mystacinus*[10] *Apodemus* sp.[10] *Mus musculus*[10] Arvicolinae indet.[10]
Varkiza 1[204166]	Middle or Late Pleistocene	*Apodemus mystacinus*[10] Arvicolinae indet.[10] *Microtus* (*Pitymys*) sp.[10] *Dinaromys* cf. *bogdanovi*[10]
Marathousa 1[34765, 187637, 199937-40]	Middle Pleistocene, Galerian (~500–400 ky)	*Apodemus sylvaticus/flavicollis*[11,12] *Microtus* sp. (cf. *M. arvalis*)[11,12] *Arvicola mosbachensis*[11,12] *Castor fiber*[11,12] ?*Allactaga* indet.[11,12]
Choremi 4[200082]	Middle Pleistocene, late Biharian (~580–620 ky)	*Apodemus* sp.[13] *Mus* cf. *spretus*[13] *Mimomys* sp.[13] *Mimomys* aff. *savini*[13] *Castor fiber*[13]
Choremi 3[200081]	Middle Pleistocene, late Biharian (~710–730 ky)	*Sciurus* cf. *vulgaris*[13] *Apodemus* cf. *sylvaticus*[13] Murinae indet.[13] *Mus* sp.[13] *Mimomys* aff. *savini*[13] *Pliomys* aff. *episcopalis*[13] *Castor fiber*[13]

Localities[PBDB No]	Age (MNQ)	Taxon
Choremi 2[200080]	Middle Pleistocene, late Biharian (~840–880 ky)	*Mimomys* sp.[13] *Mimomys* aff. *savini*[13] *Castor fiber*[13]
Choremi 1[200079]	Middle Pleistocene, late Biharian (~840–880 ky)	*Mimomys* aff. *savini*[13]
Volos[34808]	Pleistocene (MNQ21)	*Cricetulus migratorius*[14] *Apodemus mystacinus*[14] *Apodemus sylvaticus/ flavicollis*[14] *Prolagurus pannonicus*[14] *Microtus* (*Terricola*) *arvalidens*[14]
Kalamoto-1[200083]	Pleistocene (MNQ20)	*Mimomys savini*[15]
Apollonia-1[34784]	Pleistocene (MNQ20)	*Lagurodon arankae*[16,17]
Ravin Voulgarakis[34783]	**Pleistocene (MNQ20)**	*Glis* aff. *glis*[17] *Myomimus* sp.[17] *Spermophilus nogaici*[17] ***Pliospalax senii*[17]** ***Cricetinus koufosi*[17]** *Apodemus mystacinus*[17] *Apodemus sylvaticus/ flavicollis*[17] *Mimomys savini*[17] *Lagurodon arankae*[17] *Kislangia* sp.[17] *Microtus (Allophaiomys) pitymyoides*[17] *Sicista subtilis*[17]
Zeli 2A+B[34786]	Pleistocene (MNQ20)	*Micromys* cf. *minutus*[14] *Apodemus mystacinus*[14] *Apodemus sylvaticus/ flavicollis*[14] *Mimomys savini*[14] *Pliomys episcopalis*[14] *Lagurodon* sp.[14] *Microtus* sp.[14]
Zeli 2[34785]	Pleistocene (MNQ20)	*Apodemus mystacinus*[14] *Apodemus sylvaticus/ flavicollis*[14] *Mimomys savini*[14] *Lagurodon arankae*[14] *Microtus* sp.[14]
Alikes[34782]	Early Pleistocene, Biharian (?MNQ19)	*Mimomys* sp.[18] *Lagurodon* cf. *arankae*[18] *Hystrix refossa*[18]

Localities[PBDB No]	Age (MNQ)	Taxon
Kalymnos-Xi[34770]	**Early Pleistocene, Biharian (MNQ19)**	*Myomimus roachi*[14,19,20] *Sciurus* sp.[14,19,20] *Spalax* cf. *nehringi*[14,19,20] *Meriones tristrami*[14,19,20] *Mesocricetus* cf. *auratus*[14,19,20] *Apodemus mystacinus*[14,19,20] *Apodemus sylvaticus/flavicollis*[14,19,20] ***Mus aegaeus***[14,19,20] ***Kalymnomys major***[20] *Microtus* (*Allophaiomys*) sp.[14,19,20] *Microtus* (*Tibericola*) *jordanica*[14,19,20]
Marathoussa[34765]	**Early Pleistocene, Biharian (MNQ19)**	*Spermophilus* sp.[21] ***Cricetinus koufosi***[17,21] *Apodemus sylvaticus/flavicolis*[21] *Mimomys* sp.[21] *Lagurodon arankae*[21] *Microtus* (*Allophaiomys*) *pliocaenicus*[21] *Sicista* cf. *subtilis*[21]
Tourkobounia 5[34592]	Early Pleistocene, Biharian	Gliridae indet.[22] Sciuridae indet.[22] *Cricetulus* cf. *migratorius*[22] Murinae indet.[22]
Tourkobounia 4[34592]	Early Pleistocene, Biharian	Gliridae indet.[22] Cricetinae indet.[22] Murinae indet.[22] Arvicolinae indet.[22]
Tourkobounia 3[34592]	Early Pleistocene, Biharian	Sciuridae indet.[22] Cricetinae indet.[22] Murinae indet.[22] Arvicolinae indet.[22]
Tourkobounia 2[34767]	**Early Pleistocene, Biharian (MNQ19)**	*Eliomys quercinus*[23] *Glis* sp.[23] *Myomimus roachi*[23] *Sciurus* cf. *anomalus*[23] *Cricetulus migratorius*[23] *Apodemus mystacinus*[23] *Apodemus* cf. *flavicollis*[23] *Lagurodon arankae*[23] *Kalymnomys major*[23] ***Microtus* (*Tibericola*) *eleniae***[23]

Localities[PBDB No]	Age (MNQ)	Taxon
Kaiafas[34766]	Early Pleistocene, Biharian (MNQ19)	*Glis* n. sp.[14] *Glis sackdillingensis*[14] *Sciurus* sp.[14] *Cricetulus* cf. *migratorius*[14] *Apodemus mystacinus*[14] *Apodemus sylvaticus/ flavicollis*[14] *Mimomys savini*[14] *Pliomys episcopalis*[14] *Lagurodon* sp.[14] *Kalymnomys major*[14] *Microtus* (*Allophaiomys*) cf. *ruffoi*[14]
Pyrgos[34655]	Early Pleistocene, Biharian (?MNQ18)	*Kalymnomys major*[14]
Lagkada[205041]	Early Pleistocene, Biharian (MNQ18)	*Myomimus roachi*[14] *Apodemus* cf. *mystacinus*[14] *Apodemus* cf. *sylvaticus/ flavicollis*[14] *Mimomys ostramosensis*[14] *Kalymnomys* cf. *major*[14]
Kastoria 2[34653]	Early Pleistocene, Biharian (MNQ18)	*Myomimus roachi*[14] *Apodemus dominans*[14] *Mimomys reidi*[14] *Mimomys pliocaenicus*[14] *Pliomys episcopalis*[14] *Borsodia newtoni*[14] *Myodes* sp.[14] *Pitymimomys pitymyoides*[14]
Gerakarou 1[34617]	Early Pleistocene, Biharian (MNQ18)	*Apodemus* cf. *mystacinus*[16] *Mimomys* sp.[16,24,25] cf. *Borsodia* sp.[16] *Hystrix refossa*[16,25]
Kardamena[204662]	Early Pleistocene, Biharian	*Myomimus* sp.[14] *Apodemus* cf. *dominans*[14] *Mimomys* cf. *reidi*[14]
Thoknia[205042]	Early Pleistocene, (lower part of MNQ18 in Benda et al. 1987)	*Mimomys (Kislangia) rex*[14,26] *Pliomys* cf. *episcopalis*[14,26]
Sesklo (upper level)[34614]	Early Pleistocene, late Villanyian (MNQ17)	cf. *Hystrix*[27]
Yialtra[204804]	Early Pleistocene	*Mimomys* sp.[28]
Karnezeika[202122]	Early Pleistocene, middle-late Villafranchian (2.1–1.9 Ma, Iliopoulos personal communication August 2019)	Gliridae indet.[29] *Apodemus* cf. *dominans*[29] *Kalymnomys* sp.[29] *Kislangia* sp.[29]
Rema Aslan[205075]	Early Pleistocene, late Villanyian	*Myomimus roachi*[14] *Apodemus* cf. *dominans*[14] *Rhagapodemus* cf. *frequens athensis*[14] *Mimomys* sp.[14]

Localities[PBDB No]	Age (MNQ)	Taxon
Damatria[34591]	Pliocene, Villanyian (MN16)	*Myomimus* sp.[14] *Thallomys*? sp.[14] *Apodemus* cf. *dominans*[14] *Mimomys* cf. *hajnackensis*[14] *Hystrix* sp.[14]
Archipolis[205077]	Pliocene, Villanyian (MN 16a in Doukas and Papayianni 2016)	*Apodemus* cf. *dominans*[14,30] *Rhagapodemus* cf. *frequens*[14,30] *Mimomys* cf. *gracilis*[14,30]
Milia-5[185859]	Pliocene, Villanyian (MN16a)	*Hystrix refossa*[31]
Limni 6[34590]	Pliocene, Villanyian (MN16)	*Dryomys* sp.[14,32] *Eliomys* sp.[14,32] *Myomimus* cf. *roachi*[14,32] *Micromys* cf. *praeminutus*[14,28,32] *Apodemus dominans*[14,28,32] *Rhagapodemus frequens athensis*[14,28,32] *Orientalomys similis*[14,28,32] *Mimomys* cf. *hajnackensis*[14,32] *Mimomys* cf. *gracilis*[14,32] *Pliomys graecus*[14,28,32]
Tourkobounia 1[34589]	**Pliocene, Villanyian (MN16)**	*Dryomys* sp.[1,14,33] *Glis minor*[1,14,33] *Dryomimus* aff. *eliomyoides*[1,14,33] *Sciurus* cf. *anomalus*[1,14,33] ***Pliospalax tourkobouniensis***[1,14,33] *Apodemus dominans*[1,14,33] ***Rhagapodemus frequens athensis***[1,14,33] ***Orientalomys** similis*[1,14,33] *Mimomys* cf. *hajnackensis*[1,14,33] ***Pliomys graecus***[1,14,33] Hystricidae indet.[14]
Kastoria 1[34588]	Pliocene, late Ruscinian (MN15 in Koufos 2013)	*Dryomimus* cf. *eliomyoides*[14] *Keramidomys* sp.[14] *Apodemus* sp.[14] *Apodemus dominans*[14] *Rhagapodemus frequens*[14] *Dolomys occitanus*[14] *Propliomys hungaricus*[14]

Localities[PBDB No]	Age (MNQ)	Taxon
Notio 1[205081]	Pliocene, late Ruscinian (MN15, ~3.97–3.92 Ma)	*Glis minor*[34] *Myomimus* aff. *maritsensis*[34] *Sciurus anomalus*[34] *Prospalax priscus*[34] *Occitanomys magnus*[34] *Apodemus dominans*[34] *Apodemus atavus*[34] *Rhagapodemus frequens*[34] *Dolomys* cf. *nehringi*[34] *Propliomys hungaricus*[34] Castoridae indet.[34]
Spilia 3–4[31900, 31907]	Pliocene, late Ruscinian (MN15)	*Myomimus maritsensis*[14] Murinae indet.[14] *Occitanomys brailloni*[14] *Apodemus dominans*[14] *Apodemus* cf. *jeanteti*[14] *Dolomys occitanus*[14]
Komanos 2[205081]	Pliocene, late Ruscinian (MN15, 4.81–4.77 Ma)	*Micromys bendai*[34] *Mimomys davakosi*[34]
Ptolemais 3[35088]	**Pliocene, late Ruscinian (MN15, ~4.95–4.77 Ma)**	*Occitanomys brailloni*[34,35] *Micromys bendai* (= ***Micromys kozaniensis***) [34,35] *Apodemus atavus*[34,35] *Rhagapodemus primaevus*[34,35] ***Mimomys davakosi***[34,35] *Castor fiber*[34,35]
Megalo Emvolon 1[36579]	Pliocene, late Ruscinian, MN 15	*Spalax odessanus*[36]
Apolakkia[184242]	Pliocene, late Ruscinian (MN15)	*Apodemus* aff. *dominans*[14,37] *Dolomys* sp.[14,37] *Dolomys occitanus*[14,37] *Castor fiber*[14,37]
Kato Afiartis[205291]	Pliocene, early Ruscinian	*Muscardinus* sp.[38] Pteromyini indet.[38] *Neocricetodon* sp.[38] *Apodemus* aff. *dominans*[38]
Makrygialos[205292]	Pliocene, Ruscinian (older than MN15–16)	cf. *Mimomys* sp.[39]
Vorio 3 and 3A[205086]	Pliocene, early Ruscinian (MN14, ~4.95–4.85 Ma)	*Pliopetaurista dehneli*[34] *Occitanomys brailloni*[34] *Micromys bendai*[34] *Apodemus dominans*[34] *Apodemus atavus*[34] *Rhagapodemus primaevus*[34] *Microtodon* aff. *komanensis*[34] *Mimomys davakosi*[34] Castoridae indet.[34]

Localities[PBDB No]	Age (MNQ)	Taxon
Ptolemais 1[35086]	**Pliocene, early Ruscinian (MN14, ~5.04–4.95 Ma)**	*Hylopetes* sp.[34,35] *Occitanomys brailloni*[34,35] ***Micromys bendai*[34,35]** *Apodemus atavus*[34,35] *Rhagapodemus primaevus*[34,35] *Promimomys cor*[34,35] *Castor fiber*[34,35]
Limni 3[205045]	Pliocene, early Ruscinian (MN14)	*Occitanomys* cf. *brailloni*[14] *Apodemus* cf. *dominans*[14] *Promimomys* cf. *insuliferus*[14]
Tomea Eksi 3[205088]	Pliocene, early Ruscinian (MN14, ~4.97–4.93 Ma)	*Micromys bendai*[34] *Apodemus dominans*[34] *Apodemus atavus*[34] *Rhagapodemus primaevus*[34] *Mimomys* aff. *davakosi*[34]
Vorio 1[205087]	Pliocene, early Ruscinian (MN14, ~5.04 Ma)	*Occitanomys brailloni*[34] *Micromys bendai*[34] *Rhagapodemus primaevus*[34] *Promimomys cor*[34]
Kardia[35087]	**Pliocene, early Ruscinian (MN14, ~5.23–5.22 Ma)**	*Occitanomys brailloni*[34,35] ***Micromys steffensi*[34,35]** *Apodemus atavus*[34,35] *Rhagapodemus primaevus*[34,35] *Promimomys cor*[34,35] *Castor fiber*[34,35]
Komanos 1 high A and B[205083]	Pliocene, early Ruscinian (MN14, ~5.22 Ma)	*Pliopetaurista dehneli*[34] *Allocricetus ehiki*[34] *Occitanomys adroveri*[34] *Micromys steffensi*[34] *Apodemus atavus*[34] *Rhagapodemus primaevus*[34] *Promimomys cor*[34] *Microtodon komanensis*[34] Castoridae indet.[34]
Komanos 1 low A and B[205084]	**Pliocene, early Ruscinian (MN14, ~5.23 Ma)**	*Glis minor*[34] *Tamias* sp.[34] *Allocricetus ehiki*[34] *Occitanomys adroveri*[34] *Micromys steffensi*[34] *Apodemus atavus*[34] *Rhagapodemus primaevus*[34] *Promimomys cor*[34] ***Microtodon komanensis*[34]** *Microtodon* sp.[34]

Localities[PBDB No]	Age (MNQ)	Taxon
Spilia 0[31931], 1[31229]	Pliocene, early Ruscinian (MN14, ~5.23–4.95 Ma)	*Mesocricetus* sp.[14,40,41] *Occitanomys brailloni*[14,40,41] *Occitanomys adroveri*[14,40,41] *Micromys bendai*[14,40,41] *Apodemus dominans*[14,40,41] *Rhagapodemus hautimagnensis*[14,40,41] *Promimomys insuliferus*[14,40,41]
Nea Silata[191612]	Miocene/Pliocene, Turolian/ Ruscinian, (MN13/14, ~5.4–5.23 Ma)	*Myomimus maritsensis*[42,43] *Pliopetaurista dehneli*[42,43] *Spermophilinus* cf. *turolensis*[42,43] *Pliospalax* cf. *macoveii*[42,43] *Mesocricetus primitivus*[42,43] *Occitanomys brailloni*[42,43] *Micromys* cf. *paricioi*[42,43] *Apodemus gorafensis*[42,43] *Apodemus dominans*[42,43]
Maramena[32189]	**Miocene/Pliocene, Turolian/ Ruscinian, (MN13/14, ~5.4–5.23 Ma)**	*Glis* cf. *minor*[44,47] *Muscardinus pliocaenicus*[44,47] *Myomimus maritsensis*[44,47] ***Hylopetes macedoniensis***[44] *Pliopetaurista dehneli*[46,47] *Miopetaurista thaleri*[46,47] ***Tamias atsali***[46,47] *Spermophilinus turolensis* *Eomyops catalaunicus*[46,47] *Keramidomys* cf. *carpathicus*[46,47] ***Neocricetodon browni***[45,47] ***Hypsocricetus strimonis***[45,47] *Allocricetus* cf. *ehiki*[44,47] *Hansdebruijnia neutrum* ***Micromys cingulatus***[47,48] *Apodemus gorafensis*[47,48] *Apodemus* cf. *dominans*[47,48] *Rhagapodemus primaevus*[47,48] Castoridae indet.[47]

Localities[PBDB No]	Age (MNQ)	Taxon
Ano Metochi 2–3[31924, 31928]	Miocene/Pliocene, Turolian/ Ruscinian, (MN13/14, ~5.4–5.23 Ma)	*Myomimus* cf. *maritsensis*[40,45,48,49] *Pliopetaurista dehneli*[40,45,48,49] *Tamias atsali*[40,45,48,49] *Pliospalax* sp.[40,45,48,49] *Pseudomeriones* sp.[40,45,48,49] *Hypsocricetus strimonis*[45,48,49] *Occitanomys adroveri*[40,45,48,49] *Micromys bendai*[40,45,48,49] *Apodemus gorafensis*[40,45,48,49] *Apodemus dominans*[40,45,48,49] *Rhagapodemus hautimagnensis*[40,45,48,49]
Kessani-1[205296]	Miocene/Pliocene, Turolian/ Ruscinian, (MN13/14, ~5.4–5.23 Ma in Koufos and Vasileiadou 2015)	*Myomimus maritsensis*[50,51] *Pliopetaurista dehneli*[50,51] *Pliospalax* sp.[50,51] *Pseudomeriones* cf. *rhodius*[50,51] *Mesocricetus* cf. *primitivus*[51] *Occitanomys adroveri*[51] *Micromys steffensi*[51] aff. *Arvicanthis* sp.[51] *Apodemus gorafensis*[51] *Apodemus atavus/ dominans*[50] *Rhagapodemus primaevus*[51]
Kessani-2[205297]	Miocene/Pliocene, Turolian/ Ruscinian, (MN13/14, ~5.4–5.23 Ma in Koufos and Vasileiadou 2015)	*Pliopetaurista dehneli*[50,51] *Pseudomeriones* cf. *rhodius*[51] *Mesocricetus* cf. *primitivus*[51] aff. *Arvicanthis* sp.[51] *Apodemus gorafensis*[51] *Apodemus atavus/ dominans*[50,51] *Rhagapodemus primaevus*[50,51]
Prosilion–Mercurion[205093]	Late Miocene, late Turolian (MN13)	*Micromys bendai*[34] *Arvicanthis* cf. *broekhuisi*[34] cf. *Promimomys* sp.[34]
Tomea Eksi 2[205089]	**Late Miocene, late Turolian (MN13, 5.40–5.30 Ma)**	*Pseudomeriones* sp.[34] *Occitanomys debruijni*[34] *Micromys bendai*[34] ***Arvicanthis broekhuisi*[34]** *Apodemus gorafensis*[34]
Tomea Eksi 1[205090]	Late Miocene, late Turolian, MN 13 (5.43–5.32 Ma)	*Pseudomeriones* sp.[34] *Occitanomys debruijni*[34] *Micromys bendai*[34] *Arvicanthis broekhuisi*[34] *Apodemus gorafensis*[34]

Localities[PBDB No]	Age (MNQ)	Taxon
Monasteri[31940]	Late Miocene, late Turolian (uppermost MN13, ~5.96–5.33 Ma)	*Myomimus maritsensis*[40,45,48,52] *Hylopetes* sp.[40,45,48,52] *Keramidomys* cf. *carpathicus*[40,45,48,52] *Pseudomeriones megistos*[40,45,48,52] *Hypsocricetus strimonis*[45,48,52] *Occitanomys adroveri*[40,45,48,52] *Micromys bendai*[40,45,48,52] *Apodemus gorafensis*[45,48,52] *Rhagapodemus hautimagnensis*[45,48,52]
Paliambela-B[205333]	**Late Miocene, late Turolian (uppermost MN13, ~5.96–5.33 Ma)**	***Pseudomeriones megistos*[52]**
Paliambela-A[205335]	Late Miocene, late Turolian (uppermost MN13, ~5.96–5.33 Ma)	*Micromys* cf. *steffensi*[39]
Lava 2[205341]	Late Miocene, late Turolian (MN13, ~6.8 Ma)	*Glis* cf. *minor*[53] *Muscardinus* cf. *vireti*[53] *Pliopetaurista* sp. (larger than *dehneli*)[53] *Hylopetes* cf. *hungaricus*[53] *Blackia miocaenica*[53] *Spermophilinus* sp. (larger than *turolensis*)[53] *Eomyops* cf. *catalaunicus*[53] *Prospalax* aff. *petteri*[53] *Neocricetodon browni*[53] *Apodemus* sp.[53]
Dytiko-3[32376]	Late Miocene, late Turolian (MN13, 7.0–6.0 Ma)	*Hystrix primigenia*[54]
Dytiko-1[32374]	Late Miocene, late Turolian (MN13, 7.0–6.0 Ma)	*Micromys steffensi*[55]
Maritsa 1[68046]	**Late Miocene, late Turolian (MN13)**	*Eliomys* aff. *intermedius*[56] ***Myomimus maritsensis*[56]** ***Atlantoxerus rhodius*[56]** ***Spermophilinus giganteus*[56]** *Keramidomys carpathicus*[56] ***Pliospalax sotirisi*[33,56]** ***Calomyscus minor*[56]** *Pseudomeriones rhodius*[57,58]* ***Mesocricetus primitivus*[56]** ***Cricetus lophidens*[56]** ***Occitanomys debruijni*[58]** *Paraethomys meini*[56] (= ***Occitanomys anomalus*[56]**) ***Pelomys europeus*[56]** *Apodemus* cf. *dominans*[56] ***Rhagapodemus vandeweerdi*[33]**

Localities[PBDB No]	Age (MNQ)	Taxon
Chalkoutsi[205344]	Late Miocene, Turolian (MN11–13)	? *Paraethomys* sp.[59]
Kalithies[205343]	Late Miocene, Turolian (MN11–13)	Eomyidae indet.[59–61] *Byzantinia* sp.[59–61] *Pseudomeriones* sp.[59–61] *Neocricetodon* sp.[59–61] *Hansdebruijnia* cf. *neutrum*[59–61]
Alifakas[207134]	Late Miocene, Turolian (MN11–13)	*Hystrix primigenia*[62]
Rema Marmara[32190]	Late Miocene, middle/late Turolian (MN 12; tentatively)	*Parapodemus gaudryi*[40] *Hansdebruijnia neutrum*[40] *Occitanomys brailloni*[40] *Micromys bendai*[40]
Halmyropotamos[202213]	Late Miocene, Turolian (MN12)	*Hystrix primigenia*[63]
Pikermi[182754]	**Late Miocene, end of middle Turolian (MN12, ~7.1 Ma)**	***Hystrix primigenia*[64,65] (= *Castor atticus*)**
Pikermi–Megaloremma	**Late Miocene, Turolian (MN12)**	***Parapodemus gaudryi*[66]**
Chomateri[195562]	**Late Miocene, Turolian (MN12)**	*Muscardinus* sp.[59] *Myomimus* cf. *dehmi*[59] ***Byzantinia pikermiensis*[59]** *Neocricetodon* aff. *lavocati*[59] *Parapodemus gaudryi*[59] ***Hansdebruijnia neutrum*[59]** **"*Karnimata*" *provocator*[59]** *Hystrix primigenia*[67]
Mytilinii 1B[202216]	Late Miocene, Turolian (MN12, 7.17–7.13 Ma)	*Spermophilinus* cf. *bredai*[68] *Pliospalax* cf. *sotirisi*[68] *Pseudomeriones pythagorasi*[68]
Mytilinii 1A[202215]	Late Miocene, Turolian (MN12, 7.17–7.13 Ma)	*Spermophilinus* cf. *bredai*[68] *Pseudomeriones pythagorasi*[68] "*Karnimata*" *provocator*[68]
Samos Quarry 4[95689]	**Late Miocene, middle Turolian (MN12, slightly younger than 7.3 Ma)**	***Byzantinia hellenicus*[60,70]** *Parapodemus gaudryi*[60,69]
Samos S3[205351]	**Late Miocene, middle Turolian (MN12, near the fossil level MYT, ~7.3 Ma)**	*Spermophilinus* cf. *bredai*[60] *Pliospalax* cf. *sotirisi*[60] *Byzantinia hellenicus*[60] ***Pseudomeriones pythagorasi*[60]** "*Karnimata*" *provocator*[60] *Hystrix primigenia*[60]
Vathylakkos-3[182750]	Late Miocene, middle Turolian (MN12, ~7.3 Ma)	*Parapodemus gaudryi*[71]
Ravin des Zouaves 5[195489]	Late Miocene, latest Vallesian (MN11, 8.23–8.07 Ma)	Murinae indet.[72]

Localities[PBDB No]	Age (MNQ)	Taxon
Lefkon[31223]	Late Miocene, latest Vallesian (MN10)	*Myomimus dehmi*[40,45] *Pliopetaurista bressana*[40,45] *Byzantinia* sp.[40,45] *Neocricetodon* sp.[40,45] *Progonomys cathalai*[40,45] *Parapodemus* cf. *lugdunensis*[40,45] Castoridae indet.[40,45]
Biodrak[205342]	**Late Miocene, latest Vallesian (MN10)**	*Miodyromys multicrestatus*[59] *Eomyops* sp.[59] ***Byzantinia nikosi*[59]** *Progonomys cathalai*[59]
Ravin de la Pluie[191070]	Late Miocene, late Vallesian (MN10, 9.31–9.23 Ma)	*Spermophilinus* sp.[72,73] *Progonomys cathalai*[72,73]
Kastellios (KA1, 2a, 3) 205360–205362, 205365	Late Miocene, late Vallesian (MN10, 9.3–9.1 Ma)	*Muscardinus* cf. *hispanicus*[74,75] *Spermophilinus bredai*[74,75] *Cricetulodon* sp.[74,75] *Progonomys cathalai*[74,75] *Progonomys woelferi*[74,75]
Plakias[183151]	**Late Miocene, early Vallesian (lower part of MN 9, ~9.9 Ma or 9.311–9.6 Ma)**	*Glirulus (Paraglirulus) werenfelsi*[76,77] Gliridae indet.[76,77] cf. *Blackia* sp.[76,77] cf. *Hylopetes* sp.[76,77] *Eomyops* aff. *catalaunicus*[76,77] *Eumyarion leemanni*[76,77] Cricetinae indet.[76,77] ***Cricetulodon cretensis*[76,77]**
Serrae	Late Miocene, Vallesian–Turolian (MN9–13)	*Chalicomys jaegeri*[78]
Katheni[207010]	Late Miocene, Vallesian–Turolian (MN9–13)	*Myomimus dehmi*[32] *Byzantinia* aff. *nikosi*[32]
Chryssavgi 1[205366]	Middle Miocene, late Astaracian (MN7+8)	*Myomimus* sp.[17] *Megacricetodon minor*[17] *Byzantinia bayraktepensis*[17]
Rema Patsiaris[205388]	Middle Miocene	cf. *Megacricetodon* sp.[39]
Thymiana C[205390]	Middle Miocene, Orleanian (MN5, ~15.5 Ma, interval of Chrons C5Bn.1n - C5Cr in Koufos 2006)	Sciuridae indet.[79] *Megacricetodon primitivus*[79] *Cricetodon meini*[79] *Democricetodon* sp. 1[79] *Sayimys intermedius*[79,80]

Localities[PBDB No]	Age (MNQ)	Taxon
Thymiana A[205389]	Middle Miocene, Orleanian (MN5, ~15.5 Ma, interval of Chrons C5Bn.1n - C5Cr in Koufos 2006)	*Microdyromys* sp.[79] *Myomimus* sp.[79] *Heramys* sp.[79] *Megacricetodon primitivus*[79] *Cricetodon meini*[79] *Democricetodon* n. sp. I[79] *Democricetodon* n. sp. II[79] *Democricetodon gracilis*[79] *Cricetodon* sp.[79] *Sayimys intermedius*[79,80]
Antonios[73861]	Early Miocene, Orleanian (MN4/5)	cf. *Microdyromys* sp.[81] *Palaeosciurus sutteri*[81] *Spermophilinus besana*[81] *Pliospalax* sp.[81] *Megacricetodon* sp.[81,82] *Cricetodon meini*[81] *Democricetodon franconicus*[81] *Sayimys* sp.[81]
Komotini[68085]	Early Miocene, Orleanian (?MN4)	*Tamias* sp.[32,83] Sciuridae indet.[32,83] *Cricetodon meini*[32,83]
Thymiana 0	Late Early Miocene (MN4)	*Karydomys strati*[84]
Moschopotamos[208601]	Early Miocene, Orleanian (MN4)	*Aliveria* cf. *brinkerinki*[85]
Karydia[69435]	**Early Miocene, Orleanian (MN4)**	*Glirulus (Glirulus) diremptus*[86–89] *Glirulus (Paraglirulus) agelakisi*[86–89] *Glirudinus gracilis*[86–89] *Glis galitopouli*[86–89] *Seorsumuscardinus alpinus*[86–89] *Miodyromys* cf. *praecox*[86–89] *Aliveria luteijni*[86–89] *Blackia miocaenica*[86–89] *Palaeosciurus* aff. *fissurae*[86–89] *Pseudotheridomys* aff. *parvulus*[86–89] *Mirrabella tuberosa*[87] *Eumyarion* aff. *latior*[87] *Anomalomys minor*[86–89] ***Debruijnia kostakii***[89] *Cricetodon aliveriensis*[87] *Democricetodon franconicus*[86–89] *Democricetodon gracilis*[86–89] *Democricetodon* cf. *gaillardi*[86–89] *Democricetodon anatolicus*[86–89] ***Karydomys symeonidisi***[86–89] ***Karydomys boskosi***[86–89] cf. *Deperetomys* sp.[86–89] *Miopetaurista* sp.[90]

Localities[PBDB No]	Age (MNQ)	Taxon
Aliveri[68032]	**Early Miocene, Orleanian (MN4, 18–17.5 Ma)**	*Bransatoglis complicatus*[92,98] *Glirulus (Glirulus) diremptus*[92] ***Glirulus (Paraglirulus) agelakisi*[92]** *Microdyromys* sp.[92] ***Glirudinus euryodon*[92]** ***Glis galitopouli*[92]** ***Aliveria brinkerinki*[91]** ***Aliveria luteyni*[91]** *Miopetaurista dehmi*[91] *Blackia miocaenica*[91] *Palaeosciurus* aff. *fissurae*[91,99] ***Tamias eviensis*[91]** *Pseudotheridomys parvulus*[96] ***Mirrabella tuberosa*[95]** *Eumyarion latior/ weinfurteri*[97] ***Anomalomys aliveriensis*[97]** ***Heramys eviensis*[93]** ***Megacricetodon hellenicus*[82,97]** ***Cricetodon aliveriensis*[93]** *Democricetodon franconicus*[86,97] *Democricetodon gracilis*[86,99]
Lapsarna[186558]	Early Miocene, Orleanian (MN4, older than 18.4±0.5 Ma)	*Glirulus (Glirulus) diremptus*[100,101] *Eumyarion* aff. *latior/ weinfurteri*[100,101] *Democricetodon* sp.[100,101]
Kyprinos[196839]	**Late Oligocene**	*Bransatoglis sjeni*[102] *Paracricetodon* aff. *kavakderensis*[102] *Pseudocricetodon moguntiacus* cf. *orientalis*[102] ***Mirrabella hansoulii*[102]** *Kerosinia* aff. *variabilis*[102]

MN Mammal Neogene Zone, *MNQ* Mammal Quaternary Zone

[a]The assignment of the Monasteri and the Ano Metochi 2,3 hamster to *H. strimonis* still has to be proven; see text.

[1] Daams 1981, [2] Chatzopoulou 2014, [3] Tsoukala et al. 2006, [4] Sondaar 1971, [5] van der Geer et al. 2014, [6] Syrides and Koliadimou 1994, [7] Mayhew 1978, [8] Sickenberg 1971, [9] Storch 1975, [10] van de Weerd 1973, [11] Konidaris et al. 2018, [12] Doukas et al. 2018, [13] van Vugt et al. 2000, [14] van der Meulen and van Kolfschoten 1986, [15] Tsoukala and Chatzopoulou 2005, [16] Koufos et al. 1995, [17] Koliadimou 1996, [18] Athanassiou 1998, [19] Kuss and Storch 1978, [20] von Koenigswald et al. 1992, [21] Koufos et al. 2001, [22] Reumer and Doukas 1985, [23] van der Meulen and Doukas 2001, [24] Zamanis et al. 1980, [25] Koliadimou and Koufos 1991, [26] Benda et al. 1987, [27] Athanassiou 2018, [28] Katsikatsos et al. 1981, [29] Kokotini et al. 2019, [30] Doukas and Papayianni 2016, [31] Lazaridis et al. 2019, [32] de Bruijn and van der Meulen 1979, [33] de Bruijn and van der Meulen 1975, [34] Hordijk and de Bruijn 2009, [35] van de Weerd 1979, [36] de Bruijn 1984, [37] van de Weerd et al. 1982, [38] Daams and van de Weerd 1980, [39] Sylvestrou 2002, [40] de Bruijn 1989, [41] Theocharopoulos 1991, [42] Vasileiadou 2001, [43] Vasileiadou et al. 2003, [44] Bouwens and de Bruijn 1986, [45] Daxner-Höck 1995, [46] de Bruijn 1995,

[47] Schmidt-Kittler et al. 1995, [48] Storch and Dahlmann 1995, [49] Wessels 1998, [50] Syrides et al. 1997, [51] Vasileiadou et al. 2012, [52] Sylvestrou and Kostopoulos 2007, [53] de Bruijn et al. 1999, [54] de Bonis et al. 1992, [55] Koufos and Vasileiadou 2015, [56] de Bruijn et al. 1970, [57] Şen 1977, [58] Şen et al. 1989, [59] de Bruijn 1976, [60] Black et al. 1980, [61] Wessels 1999, [62] Melentis and Schneider 1966, [63] Melentis 1967, [64] Wagner 1848, [65] Gaudry and Lartet 1856, [66] Dames 1883, [67] Bachmayer et al. 1982, [68] Vasileiadou and Sylvestrou 2009, [69] Schaub 1926, [70] Freudenthal 1970, [71] de Bonis and Koufos 1999, [72] de Bonis and Melentis 1975, [73] Aguilar et al. 1995, [74] de Bruijn et al. 1971, [75] de Bruijn and Zachariasse 1979, [76] de Bruijn and Meulenkamp 1972, [77] de Bruijn et al. 2012, [78] Melentis 1966, [79] de Bonis et al. 1997, [80] López-Antoñanzas et al. 2005, [81] Vasileiadou and Koufos 2005, [82] Oliver and Peláez-Campomanes 2016, [83] de Bruijn et al. 1993, [84] López-Antoñanzas et al. 2019, [85] Benda and Steffens 1981, [86] Theocharopoulos 2000, [87] Doukas 2003, [88] Duncan 2012, [89] de Bruijn 2016, [90] Wessels 2020 personal communication, [91] de Bruijn et al. 1980, [92] van der Meulen and de Bruijn 1982, [93] Klein Hofmeijer and de Bruijn 1985, [94] Ziegler and Fahlbusch 1986, [95] de Bruijn et al. 1987, [96] Alvarez-Sierra et al. 1987, [97] Klein Hofmeijer and de Bruijn 1988, [98] Ünay 1994, [99] van den Hoek Ostende et al. 2015, [100] Vasileiadou and Zouros 2012, [101] Vasileiadou et al. 2017, [102] Doukas and Theocharopoulos 1999

References

Adkins RM, Gelke EL, Rowe D, Honeycutt RL (2001) Molecular phylogeny and divergence time estimates for major rodent groups: evidence from multiple genes. Mol Biol Evol 18(5):777–791

Adrover R, Mein P, Moissenet E (1976) Mise en évidence du Pliocène moyen continental dans le nord du fossé de Teruel (Espagne): le gisement de Villalba–Alta. C R d'activités annuelles. Association régionale pour le développement des recherches de paléontologie et de préhistoire et des Amis du Muséum 14:11–14

Agadjanian AK, Kondrashov PE (2007) Mollusks and small mammals from the Kuznetsovka locality, Pleistocene of the Oka–Don Plain. Paleontol J 41(4):395–406

Aguilar JP (1974) Les rongeurs du Miocène inférieur en Bas-Languedoc et les corrélations entre échelles stratigraphiques marine et continentale. Geobios 7(4):345–398

Aguilar J, Calvet M, Michaux J (1986) Description des rongeurs Pliocènes de la faune du Mont–Hélène (Pyrénées–Orientales), nouveau jalon entre les faunes de Perpignan (Serrat–d'en–Vacquer) et de Sète. Palaeovertebrata 16(3):127–144

Aguilar JP, de Bonis L, Denys C, Koufos GD (1995) Découverte d'un *Spermophilinus* (Rodentia, Mammalia) dans le gisement Vallésien du Ravin de la Pluie (Macédoine, Gréce). Ann Géol Pays Helléniques 36:645–648

Agustí J (1981) Roedores Miomorfos del Neógeno de Cataluña. Thesis, Univ de Barcelona

Álvarez-Sierra MÁ, Daams R, van der Meulen AJ (1987) The mammals from the lower Miocene of Aliveri (Island of Evia, Greece). Part 7. The Eomyidae. Proc Kon Neder Akad Weten Ser B 90(1):47–56

Argyropulo AI, Pidoplichka IG (1939) Recovery of a representative of Murinae (Glires, Mammalia) in Tertiary deposits of the USSR. Doklady Akad Nauk SSSR 23:209–212

Armour-Brown A, de Bruijn H, Maniati C, Siatos G, Niesen P (1977) The Geology of the Neogene sediments north of Serrai and the use of rodent faunas for biostratigraphic control. Proc 6th Colloquium Geology Aegean Region 6(2):615–622

Athanassiou A (1998) Contribution to the study of the fossil mammals of Thessaly. National and Kapodistrian University of Athens, Gaia 5, p 354

Athanassiou A (2018) A Villafranchian Hipparion–Bearing mammal fauna from Sésklo (E. Thessaly, Greece): implications for the question of *Hipparion–Equus* sympatry in Europe. Quaternary 1(2):12. https://doi.org/10.3390/quat1020012

Bachmayer F, Wilson RW (1970) Die Fauna der altpliozänen Höhlen–und Spaltenfüllungen bei Kohfidisch, Burgenland (Österreich) Small Mammals (Insectivora, Chiroptera, Lagomorpha, Rodentia) from the Kohfidisch Fissures of Burgenland, Austria. Ann Naturhist Mus Wien:533–587

Bachmayer F, Symeonidis N, Zapfe H (1982) Die Ausgrabungen in Pikermi–Chomateri bei Athen. Eine Dokumentation. Ann Naturhist Mus Wien 84(A):7–12

Bakr A (1959) Rodentia and Carnivora from the Pontian of Samos: In The American Museum of Natural History. Doctoral dissertation, Harvard University

Benda L, de Bruijn H (1982) Biostratigraphic correlations in the Eastern Mediterranean Neogene. Part 7. Calibration of Sporomorph– and rodent–associations in the Aliveri–Kymi basin (Island of Euboea, Greece). Newsl Stratigr 11:128–135

Benda L, Steffens P (1981) Aufbau und Alter des Neogens von Katerini (Griechenland). Geol Jb B 42:93–103

Benda L, Meulenkamp JE, van de Weerd A (1977) Biostratigraphic correlations in the Eastern Mediterranean Neogene. 3. Correlation between mammal, sporomorph and marine microfossil assemblages from the Upper Cenozoic of Rhodes, Greece. Newsl Stratigr 6(2):117–130

Benda L, van der Meulen AJ, Meyer K-J, van der Weerd A (1987) Biostratigraphic correlations in the Eastern Mediterranean Neogene. 8. Calibration of Sporomorph– and Rodent–Associations form the Megalopolis Basin (Peloponnesus, Greece). Newsl Stratigr 17(3):129–141

Black CC (1966) Tertiary Sciuridae (Mammalia: Rodentia) from Bavaria. Mitt Bayerischen Staatss Paläontol Hist Geol 6:51–63

Black CC, Krishtalka L, Solounias N (1980) Mammalian fossils of Samos and Pikermi, Part 1. The Turolian rodents and insectivores of Samos. Ann Carnegie Mus 49:359–378

Blanga-Kanfi S, Miranda H, Penn O, Pupko T, RW DB, Huchon D (2009) Rodent phylogeny revised: analysis of six nuclear genes from all major rodent clades. BMC Evolut Biol 9(1):71

Bolliger T (1992) Kleinsäugerstratigraphie in der miozänen Hörnlischüttung (Ostschweiz). In Gregor H-J, Unger HJ (eds) Kleinsäuger aus der Miozänmolasse der Ostschweiz. Documenta naturae, München 75, pp 1–296

Bona F, Sala B, Tintori A (2008) Early Toringian small mammals fauna from Fontana Marella cave (Varese, Lombardy, North Italy). Riv It Paleontol Stratigrafia 114:133–144

Bosma AA, de Bruijn H (1982) Eocene and Oligocene Gliridae (Rodentia, Mammalia) from the Island of Wight, England. Part 2. *Gliravus minor* n. sp., *Gliravus daamsi* n. sp. and *Bransatoglis bahloi* n. sp. Proc Kon Ned Akad Wet B 85:365–380

Bosma AA, de Bruijn H, Wessels W (2013) Late Miocene Sciuridae (Mammalia, Rodentia) from Anatolia, Turkey. J Vertebr Paleontol 33(4):924–942

Bosma AA, de Bruijn H, Wessels W (2019) Early and middle Miocene Sciuridae (Mammalia, Rodentia) from Anatolia, Turkey. J Vertebr Paleontol. https://doi.org/10.1080/02724634.2018.1537281

Bouwens P, de Bruijn H (1986) The flying squirrels *Hylopetes* and *Petinomys* and their fossil record. Proc Kon Neder Akad Weten Ser B 89:113–123

Bulot C (1980) Nouvelle description de deux espèces du genre *Megacricetodon* (Cricetidae, Rodentia) du Miocène de Bézian (Zone de La Romieu). Bull Mus Nat Hist Natur Paris Sci Terre 2:3–16

Carleton MD, Gardner AL, Pavlinov IY, Musser GG (2014) The valid generic name for red-backed voles (Muroidea: Cricetidae: Arvicolinae): restatement of the case for *Myodes* Pallas, 1811. J Mammal 95(5):943–959

Carls N, Rabeder G (1988) Arvicolids (Rodentia, Mammalia) from the earlier Pleistocene of Schernfeld (Bavaria). Beit Palaeontol 14:79–139

Carls N, Groiss JT, Rabeder G (1988) Die mittelpleistozäne Höhlenfüllung von Hunas, Fränkische Alb. Chronologische und paläoklimatologische Ergebnisse. Beit Palaontol 14:239–249

Casanovas-Vilar I, Alba DM (2011) The never–ending problem of Miocene beaver taxonomy. Acta Palaeontol Polonica 56(1):217–221

Casanovas-Vilar I, Alba DM, Almécija S, Robles JM, Galindo J, Moyà-Solà S (2008) Taxonomy and paleobiology of the genus *Chalicomys* Kaup, 1832 (Rodentia, Castoridae), with the description of a new species from Abocador de Can Mata (Vallès–Penedès Basin, Catalonia, Spain). J Vertebr Paleontol 28:851–862

Chaline J (1966) Les Lagomorphes et les Rongeurs. Glires Linné 1758. In Faunes et flores préhistoriques de l'Europe occidentale. Boubée et cie, Atlas Préhist 3, Paris, pp 397–440

Chaline J (1972) Les rongeurs du Pléistocène moyen et supérieur de France: (systématique, biostratigraphie, paléoclimatologie) Cahiers de Paleontol. CNRS, Paris

Chaline J (1974) Un nouveau critère d'étude des *Mimomys*, et les rapports de *Mimomys occitanus–Mimomys stehlini* et de *Mimomys polonicus* (Arvicolidae, Rodentia). Acta Zool Cracoviensia 19:337–355

Chatzopoulou K (2014) The micromammals of the Quaternary deposits from Cave A, Loutra Almopias (Pella, Northern Greece): Stratigraphy–Taphonomy–Paleoenvironment. PhD thesis, Aristotle University of Thessaloniki, Greece

Cicha I, Fahlbusch V, Fejfar O (1972) Die biostratigraphische Korrelation einiger jungtertiärer Wirbeltierfaunen Mitteleuropas. N Jahr Geol Paläontol Abh 140:129–145

Coiffait B, Coiffait PE, Jaeger J-J (1985) Découverte en Afrique du Nord des genres *Stephanomys* et *Castillomys* (Muridae) dans un nouveau gisement de microvertébres néogènes d'Algérie orientale: Argoub Kemellal. Proc Kon Neder Akad Weten Ser B 88:167–183

Cuenca Bescós G (1988) Revisión de los Sciuridae del Aragoniense y del Rambliense en la fosa de Calatayud–Montalban. Scripta Geol 87:1–116

Daams R (1974) *Pseudodryomys* (Gliridae, Rodentia, Mammalia) from Early and Middle Miocene Deposits in Spain. Kon Ned Akad Weten Proc Ser B 77(3):201–240

Daams R (1976) Miocene rodents (Mammalia) from Cetina de Aragon (Prov. Zaragoza) and Buñol (Prov. Valencia), Spain. Proc Kon Ned Akad Wet B 79(3):152–182

Daams R (1981) The dental pattern of the Dormice *Dryomys*, *Myomimus*, *Microdyromys* and *Peridyromys*. Utrecht Micropaleontol Bull Sp Pub Utrecht 3:1–115

Daams R (1985) Glirinae (Gliridae, Rodentia) from the type area of the Aragonian and adjacent areas (Prov. of Teruel and Zaragoza, Spain). Scripta Geol 77:1–20

Daams R (1999) Family Gliridae. In: Rössner GE, Heissig K (eds) The Miocene land mammals of Europe. Verlag Dr F. Pfeil, München, pp 301–318

Daams R, de Bruijn H (1995) A classification of the Gliridae (Rodentia) on the basis of dental morphology. Hystrix 6(1–2):3–50

Daams R, Freudenthal M (1988) Cricetidae (Rodentia) from the type–Aragonian; the genus *Megacricetodon*. In Freudenthal M (ed) Biostratigraphy and paleoecology of the Neogene micromammalian faunas from the Calatayud–Teruel Basin (Spain). Scripta Geol 1:39–132

Daams R, van der Weerd AV (1980) Early Pliocene small mammals from the Aegean island of Karpathos (Greece) and their paleogeographic significance. Geologie en Mijnbouw 59(4):327–331

Dahlmann T (2001) Die Kleinsäuger der unter–pliozänen Fundstelle Wölfersheim in der Wetterau (Mammalia: Lipotyphla, Chiroptera, Rodentia). Courier Forschungsinstitut Senckenberg 227:1–129

Dangremond A (2012) Age or paleoenvironment? What is the cause of the differences between two Early Miocene small–mammal assemblages from Kaplangı (Turkey)? Master thesis, Utrecht University

Daxner-Höck G (1992) Die Cricetinae aus dem Obermiozän von Maramena (Mazedonien, Nordgriechenland). Paläontol Zeit 66(3–4):331–367

Daxner-Höck G (1995) The vertebrate locality Maramena (Macedonia, Greece) at the Turolian–Ruscinian boundary (Neogene). 9. Some glirids and cricetids from Maramena and other late Miocene localities in Northern Greece. Mün Geowissen Abh 28:103–120

Daxner-Höck G (2005) Eomyidae and Gliridae from Rudabánya. Palaeontographia Italica 90:149–161

Daxner-Höck G, Höck E (2009) New data on Eomyidae and Gliridae (Rodentia, Mammalia) from the late Miocene of Austria. Ann Naturhist Mus Wien, p:375–444

Daxner-Höck G, Mein P (1975) Taxonomische Probleme um das Genus *Miopetaurista* Kretzoi, 1962 (Fam. Sciuridae). Paläontol Zeit 49(1):75–77

Daxner-Höck G, Fahlbusch V, Kordos L, Wu W (1996) The late Neogene cricetid genera *Neocricetodon* and *Kowalskia*. In: Bernor RL, Fahlbusch V, Mittmann H-W (eds) The evolution of western Eurasian Neogene mammal faunas. Columbia University Press, New York, pp 220–226

de Bonis L, Koufos GD (1999) The Miocene large mammal succession in Greece. In: Agusti J, Rook L, Andrews P (eds) Hominoid evolution and climatic change in Europe, The evolution of Neogene terrestrial ecosystems in Europe, vol 1. Cambridge University Press, London, pp 205–223

de Bonis L, Melentis J (1975) Première découverte de Muridés (Mammalia, Rodentia) dans le Miocène de la région de Thessalonique. Précision sur l'âge géologique des Dryopithécinés de Macédonien. C R Acad Sci Paris 280:1233–1236

de Bonis L, Bouvrain G, Geraads D, Koufos G (1992) A skull of *Hystrix primigenia* from the late Miocene of Macedonia (Greece). N Jb Geol Palaontol Mh H 2:75–87

de Bonis L, Koufos GD, Sen S (1997) The Sanitheres (Mammalia, Suoidea) from the Middle Miocene of Chios Island, Aegean Sea, Greece. Rev Paléobiol Genève 16(1):259–270

de Bruijn H (1966) On the mammalian fauna of the *Hipparion*–beds in the Calatayud–Teruel basin (Prov. Zaragoza, Spain). Part II. The Gliridae (Rodentia). Proc Kon Neder Akad Weten Ser B 69(3):367–387

de Bruijn H (1974) The Ruscinian rodent succession in Southern Spain and its implications for the biostratigraphic correlation of Europe and North Africa. Senkenbergiana Lethaea 55:435–443

de Bruijn H (1976) Vallesian and Turolian rodents from Biotia, Attica and Rhodes (Greece). Proc Kon Neder Akad Weten Ser B 79(5):361–389

de Bruijn H (1984) Remains of the mole–rat *Microspalax odessanus* Topachevski from Karaburun (Greece, Macedonia) and the family Spalacidae. Proc Kon Neder Akad Weten Ser B 87:417–425

de Bruijn H (1989) Smaller mammals from the Upper Miocene and Lower Pliocene of the Strimon basin, Greece; I Rodentia and Lagomorpha. Boll Soc Paleontol Italiana 28(2/3):189–195

de Bruijn H (1995) The vertebrate locality Maramena (Macedonia, Greece) at the Turolian–Ruscinian boundary (Neogene). 8. Sciuridae, Petauristidae and Eomyidae (Rodentia, Mammalia). Münchner Geowissen Abh A 28:87–102

de Bruijn H (1998) Vertebrates from the Early Miocene lignite deposits of the opencast mine Oberdorf (Western Styrian Basin, Austria): 6. Rodentia I (Mammalia). Ann Naturhist Mus Wien 99A:99–137

de Bruijn H (2009) The *Eumyarion* (Mammalia, Rodentia, Muridae) assemblage from Sandelzhausen (Miocene, Southern Germany): a test on homogeneity. Palaeontol Zeit 83(1):77–83

de Bruijn H (2010) A tentative classification of the Eurasian Paleogene and Neogene Muridae above the species level. In Fortelius M (coord) Neogene of the Old World Database of Fossil Mammals (NOW), University of Helsinki. http://www.helsinki.fi/science/now/pdf/Muridae_tentative.pdf

de Bruijn H (2016) A new stage in the evolution of the mole rats (Rodentia, Spalacinae) from the Early Miocene of northern Greece. Hist Biol 29(5):571–575

de Bruijn H, Mein P (1968) On the mammalian fauna of the *Hipparion*–beds in the Calatayud–Teruel Basin (Prov. Zaragoza, Spain). Part V. The Sciurinae. Proc Kon Neder Akad Weten Ser B 71:73–90

de Bruijn H, Meulenkamp JE (1972) Late Miocene rodents from the Pandanassa formation (prov. Rethymnon), Crete. Greece. Proc Kon Neder Akad Weten Ser B 75(1):54–60

de Bruijn H, Saraç G (1991) Early Miocene rodent faunas from the eastern Mediterranean area. 1. The genus *Eumyarion*. Proc Kon Neder Akad Weten Ser B 94(1):1–36 de Bruijn H, van der Meulen AJ (1975) The early Pleistocene rodents from Tourkobounia–1 (Athens, Greece). Proc Kon Neder Akad Weten Ser B 78(4):314–338

de Bruijn H, van der Meulen AJ (1979) A review of the Neogene rodent succession in Greece. Ann Géol Pays Helléniques 1:207–217

de Bruijn H, Zachariasse WJ (1979) The correlation of marine and continental biozones of Kastellios hill reconsidered. Ann Géol Pays Helléniques 1:219–226

de Bruijn H, Dawson MR, Mein P (1970) Upper Pliocene Rodentia, Lagomorpha and Insectivora (Mammalia) from the Isle of Rhodes (Greece). Proc Kon Neder Akad Weten Ser B 73(5):535–584

de Bruijn H, Sondaar P, Zachariasse WJ (1971) Mammalia and Foraminifera from the Neogene of Kastellios hill (Crete): correlations of continental and marine biozones. Proc Kon Neder Akad Weten Ser B 74(5):1–22

de Bruijn H, van der Meulen AJ, Katsikatsos G (1980) The mammals from the lower Miocene of Aliveri (Island of Evia, Greece) 1. The Sciuridae. Proc Kon Neder Akad Weten Ser B 83(3):241–261

de Bruijn H, Ünay E, Saraç G, Hofmeijer GK (1987) An unusual new eucricetodontine from the Lower Miocene of the Eastern Mediterranean. Proc Kon Neder Akad Weten Ser B 90(2):119–132

de Bruijn H, Boon E, Hussain ST, Drooger CW (1989) Evolutionary trends in *Sayimys* (Ctenodactylidae, Rodentia) from the Lower Manchar Formation (Sind, Pakistan). Proc Kon Neder Akad Weten Ser B 92(3):191–214

de Bruijn H, Fahlbusch V, Saraç G, Ünay E (1993) Early Miocene rodent faunas from the eastern Mediterranean area. Part III. The genera *Deperetomys* and *Cricetodon* with a discussion of the evolutionary history of the Cricetodontini. Proc Kon Ned Akad Weten 96(2):151–216

de Bruijn H, Saraç G, van den Hoek Ostende LW, Roussiakis S (1999) The status of the genus name *Parapodemus* SCHAUB, 1938; new data bearing on an old controversy. In Reumer JWF, de Vos J (eds) Elephants have a Snorkel! Papers in honour of Paul Y. Sondaar. Deinsea 7:95–112

de Bruijn H, van den Hoek OL, Kristkoiz-Boon E, Rummel M, Theocharopoulos C, Ünay E (2003a) Rodents, lagomorphs and insectivores, from the middle Miocene hominoid locality of Çandir (Turkey). Courier–Forschungsinst Senckenberg 240:51–87

de Bruijn H, Ünay E, Yïlmaz A, Saraç G (2003b) A rodent assemblage from the Eo/Oligocene boundary interval near Süngülü, Lesser Caucasus, Turkey. Coloquios Paleontol 1:47–76

de Bruijn H, van den Hoek Ostende LW, Donovan SK (2007) *Mirrabella*, a new name for the genus *Mirabella* de Bruijn et al., 1987 (Mammalia), preoccupied by *Mirabella* Emeljanov, 1982 (Insecta). Contrib Zool 76(4):279–280

de Bruijn H, Doukas CS, van den Hoek Ostende LW, Zachariasse WJ (2012) New finds of rodents and insectivores from the Upper Miocene at Plakias (Crete, Greece). Swiss J Palaeontol 131(1):61–75

de Bruijn H, Marković Z, Wessels W (2013a) Late Oligocene rodents from Banovići (Bosnia and Herzegovina). Palaeodiversity 6:63–105

de Bruijn H, Ünay E, Hordijk K (2013b) A review of the Neogene succession of the Muridae and Dipodidae from Anatolia with special reference to taxa known from Asia and/or Europe. In: Fossil Mammals of Asia: Neogene Biostratigraphy and Chronology. Columbia University Press, New York, pp 564–580

de Bruijn H, Bosma AA, Wessels W (2015) Are the Rhizomyinae and the Spalacinae closely related? Contradistinctive conclusions between genetics and palaeontology. Palaeobiodiv Palaeoenviron 95(3):257–269

de Giuli C (1989) The rodents of the Brisighella latest Miocene. Boll Soc Paleontol Italiana 28(2–3):197–212

Dehm R (1950) Die Nagetiere aus dem Mittel–Miocän (Burdigalium) von Wintershof–West bei Eichstätt in Bayern. N Jahr Mineral Geol Paläontol 91:312–428

Depéret CJJ (1892) La faune de mammifères miocènes de la Grive–Saint–Alban (Isère) et de quelques autres localités du bassin du Rhône: documents nouveaux et révision générale. Arch Mus d'Hist Natur Lyon 5:1–15

Desclaux E, Abbassi M, Marquet J-C, Chaline J, van Kolfschoten T (2000) Distribution and evolution of *Arvicola* Lacépède, 1799 (Mammalia, Rodentia) in France and Liguria (Italy) during the Middle and the Upper Pleistocene. Acta Zool Cracoviensa 43:107–125

Doukas CS (2003) The MN4 faunas of Aliveri and Karydia (Greece). Coloquios Paleontol 1:127–132

Doukas CS, Papayianni K (2016) Small mammals in the Plio/Pleistocene sediments of Greece. In: Paleoanthropology of the Balkans and Anatolia. Springer, Dordrecht, pp 291–302

Doukas CS, Theocharopoulos CD (1999) Smaller mammals from the Oligocene of Kyprinos (Thrace, N. Greece). In Reumer JWF, de Vos J (eds) Elephants have a Snorkel! Papers in honour of Paul Y. Sondaar. Deinsea 7:133–146

Doukas C, van Kolfschoten T, Papayianni K, Panagopoulou E, Harvati K (2018) The small mammal fauna from the Palaeolithic site Marathousa 1 (Greece). Quat Int 497:95–107

Duncan NE (2012) The early Miocene small mammal fauna from Karydiá, Greece: biostratigraphic and palaeogeographic implications. MSc thesis, Department of Earth Science, Faculty of Geosciences, Utrecht University

Engesser B (1972) Die obermiozäne Säugetierefauna von Anwil (Baselland). Tätigkeitsberichte Naturforschenden Gesells Baselland 28:37–363

Engesser B (1979) Relationships of some insectivores and rodents from the Miocene of North America and Europe. Bull Carnegie Mus Natur Hist 14:1–68

Engesser B (1999) Family Eomyidae. In: Rössner GE, Heissig K (eds) The Miocene land mammals of Europe. Verlag Dr F. Pfeil, München, pp 319–335

Escudé E, Montuire S, Desclaux E, Quéré JP, Renvoisé E, Jeannet M (2008) Reappraisal of 'chronospecies' and the use of *Arvicola* (Rodentia, Mammalia) for biochronology. J Archaeol Sci 35(7):1867–1879

Fahlbusch V (1964) Die Cricetiden (Mamm.) der Oberen Süßwasser–Molasse Bayerns. Abh Bayerischen Akad Wissen Math–Naturwissen Kl 118:1–136

Fahlbusch V (1966) Cricetidae (Rodentia, Mamm.) aus der mittelmiozänen Spaltenfüllung Erkertshofen bei Eichstätt. Mitt Bayerischen Staats Paläontol Hist Geol 6:109–131

Fahlbusch V (1967) Die Beziehungen zwischen einigen Cricetiden (Mamm., Rodentia) des nordamerikanischen und europäischen Jungtertiärs. Paläontol Zeit 41(3):154–164

Fahlbusch V (1968) Neue Eomyidae (Rodentia, Mamm.) aus einer aquitanen Spaltenfüllung von Weißenburg in Bayern. Mitt Bayer Staatss Paläontol Hist Geol 8:219–245

Fahlbusch V (1969) Pliozäne und Pleistozäne Cricetinae (Rodentia, Mammalia) aus Polen. Acta Zool Cracov 14(5):99–137

Fahlbusch V (1975) Die Eomyiden (Rodentia, Mammalia) der Oberen Süsswasser-Molasse Bayerns. Mitt Bayerischen Staatss Paläontol Hist Geol 15:63–90

Fahlbusch V (1983) Mikroevolution – Makroevolution – Punktualismus. Ein Diskussionsbeitrag am Beispiel miozäner Eomyiden (Mammalia, Rodentia). Paläontol Zeit 57(3/4):213–230

Fejfar O (1972) Ein neuer Vertreterder Gattung *Anomalomys* Gaillard, 1900 (Rodentia, Mammalia) aus dem europäischen Miozän (Karpat). N Jb Geol Paläontol Abh 141(2):168–193

Fejfar O (1974) Die Eomyiden und Cricetiden (Rodentia, Mammalia) des Miozäns der Tschechoslowakei. Palaeontographica 146:100–180

Fejfar O, Kaiser TM (2005) Insect bone-modification and paleoecology of Oligocene mammal-bearing sites in the Doupov Mountains, north Western Bohemia. Palaeontol Electronica 8(1, 8A):1–11

Fejfar O, Storch G (1990) Eine pliozäne (ober–ruscinische) Kleinsäugerfauna aus Gundersheim, Rheinhessen. 1. Nagetiere: Mammalia, Rodentia. Senckenbergiana lethaea 71(1/2):139–184

Fejfar O, Heinrich W-D, Pevzner MA, Vangengeim EA (1997) Late Cenozoic sequence of mammalian sites in Eurasia: an updated correlation. Palaeogeogr Palaeoclimatol Palaeoecol 133:259–288

Fejfar O, Heinrich W-D, Kordos L, Maul LC (2011) Microtoid cricetids and the early history of arvicolids (Mammalia, Rodentia). Palaeontol Electronica 14(3, 27A):1–38
Forsyth Major CJ (1891) Considérations nouvelles sur la faune des Vertébrés du Miocène supérieur dans l'île de Samos. C R Hebd l'Acad Sci Paris 113:608–610
Forsyth Major CJ (1899) Some rodents from the Middle Miocene lacustrine deposits of Oeningen, southern Germany. Geol Mag 4:363–373
Forsyth Major CJ (1902) Exhibit of some jaws and teeth of Pliocene voles (*Mimomys* gen.nov.). Proc Zool Soc London 1902:102–107
Freudenthal M (1963) Entwicklungsstufen der Miozänen Cricetodontinae (Mammalia, Rodentia) Mittelspaniens und ihre stratigrafische Bedeutung. Beaufortia 119(10):51–157
Freudenthal M (1965) Betrachtungen über die Gattung *Cricetodon*. Proc Kon Neder Akad Weten Ser B 68(5):293–305
Freudenthal M (1970) A new *Ruscinomys* (Mammalia, Rodentia) from the late Tertiary (Pikermian) of Samos, Greece. Ame Mus Novitates 2402:1–10
Freudenthal M, Cuenca Bescos G (1984) Size variation of fossil rodent populations. Scripta Geol 76:1–28
Freudenthal M, Martín Suárez E (1999) Family Muridae. In: Rössner GE, Heissig K (eds) The Miocene land mammals of Europe, Verlag Dr F. München, Pfeil, pp 401–409
Freudenthal M, Hugueney M, Moissenet E (1994) The genus *Pseudocricetodon* (Cricetidae, Mammalia) in the Upper Oligocene of the province of Teruel (Spain). Scripta Geol 104:57–114
Freudenthal M, Mein P, Suárez EM (1998) Revision of Late Miocene and Pliocene Cricetinae (Rodentia, Mammalia) from Spain and France. Treballs del Museu de Geologia de Barcelona 7:11–93
Freudenthal M, Martín Suárez E, Carballo NH, Rodriguez-Fernandez LR, González FM (2010) Rodents from the Lower Oligocene of the Bierzo Basin (Leon, Spain). N Jahr Geol Paläontol Abh 257(3):317–340
Friant M (1953) Une faune du Quaternaire ancien en France Méditerranéenne (Sète, Herault). Ann Soc Géol Nord 73:161–170
García-Moreno E, López Martínez N (1986) *Ramys*, a new genus of Gliridae (Rodentia) from the lower Vallesian of Spain. Proc Kon Neder Akad Weten B 89(4):337–355
Gaudry A (1862–1867) Animaux Fossiles et Géologie de l'Attique. Savy, Paris
Gaudry A, Lartet E (1856) Résultats des recherches paléontologiques entreprises dans l'Attique sous les auspices de l'Académie. C r hebd séances Acad Sci 43:271–274
Haas G (1966) On the vertebrate fauna of the Lower Pleistocene site Ubeidiya. Israel Acad Sci Human:1–68
Hartenberger JL (1965) Les Cricetidae (Rodentia) de Can Llobateres (Néogène d' Espagne). Bull Soc Géol France 7:487–498
Hartenberger J-L (1966) Les rongeurs du Vallésien (Miocène supérieur) de Can Llobateres (Sabadell, Espagne): Gliridae et Eomyidae. Bull Soc Géol France 7(8):596–604
Heller F (1930) Eine Forest–Bed–Fauna aus der Sackdillinger Höhle (Oberpfalz). N Jahr Mineral Geol Paläontol Abh–Abt B 63:247–298
Heller F (1933) Ein Nachtrag zur Forest–Bed–Fauna aus der Sackdillinger Höhle (Oberpfalz). Centralblatt Mineral Geol Paläontol Abt B 1:60–68
Heller F (1936) Eine oberpliocäne Wirbeltierfauna aus Rheinhessen. N Jahr Mineral Geol Paläontol Beil Bd B 76:99–160
Hinton AC (1923) Diagnoses of species of *Pitymys* and *Microtus* occurring in the Upper Freshwater bed of West Runton, Norfolk. Ann Mag Nat Hist 12(70):541–542
Hinton AC (1926) Monograph of the Voles & Lemmings (Microtinae) living and extinct. British Museum, Natural History
Hir J (1994) *Cricetinus beremendensis* sp. n. (Rodentia, Mammalia) from the Pliocene fauna of Beremend 15.(S Hungary). Fragmenta Mineral Palaeontol 17:71–89
Hir J (1997) A short scetch of the evolution and stratigraphy of the Plio–Pleistocene cricetids (Rodentia, Mammalia) in Hungary. Folia Hist Natur Mus Matraensis 22:43–49

Hordijk K, de Bruijn H (2009) The succession of rodent faunas from the Mio/Pliocene lacustrine deposits of the Florina–Ptolemais–Servia Basin (Greece). Hellenic J Geosci 44:21–103

Hugueney M (1999) Family Castoridae. In: Rössner GE, Heissig K (eds) The Miocene Land Mammals of Europe. Verlag Dr. Friedrich Pfeil, München, pp 281–300

Hugueney M, Mein P (1965) Lagomorphes et rongeurs du Néogène de Lissieu (Rhône). Travaux Lab Géol Faculté Sci Lyon 12:109–123

Hugueney M, Mein P (1968) Les Eomyides (Mammalia, Rodentia) néogènes de la Region lyonnaise. Geobios 1:187–204

Jaeger JJ (1975) Les rongeurs du Miocene Moyen et superieur du Maghreb. Academie de Montpellier, These

Jánossy D, van der Meulen AJ (1975) On *Mimomys* (Rodentia) from Osztramos–3, North Hungary. Proc Kon Neder Akad Weten Ser B 78(5):381–391

Kälin D (1997) *Eomyops hebeiseni* n. sp., a new large Eomyidae (Rodentia, Mammalia) of the Upper Freshwater Molasse of Switzerland. Eclogae Geol Helvetiae 90(3):629–637

Kälin D (1999) Tribe Cricetini. In: Rössner GE, Heissig K (eds) Land Mammals of Europe. Verlag Dr. Friedrich Pfeil, München, pp 373–385

Kalthoff DC, Mörs T, Tesakov A (2007) Late Pleistocene small mammals from the Wannenköpfe volcanoes (Neuwied Basin, western Germany) with remarks on the stratigraphic range of *Arvicola terrestris*. Geobios 40(5):609–623

Katsikatsos G, de Bruijn H, van der Meulen AJ (1981) The Neogene of the Island of Euboea (Evia), a review. Geol en Mijnbouw 60:509–516

Kaup J (1832) Beschreibung dreyer Gattungen urweltlicher Nager des Zoologischen Museums zu Darmstadt welche von den jetzt lebenden Genera verschieden sind. Leipzig, Isis von Oken:p992–p996

Kimura Y, Flynn LJ, Jacobs LL (2017) Early Late Miocene murine rodents from the upper part of the Nagri formation, Siwalik Group, Pakistan, with a new fossil calibration point for the tribe Apodemurini (*Apodemus/Tokudaia*). Fossil Imprint 73(1–2):197–212

Klein Hofmeijer G, de Bruijn H (1985) The mammals from the Lower Miocene of Aliveri (Island of Evia, Greece). IV: The Spalacidae and Anomalomyidae. Proc Kon Neder Akad Weten Ser B 88(2):185–198

Klein Hofmeijer G, de Bruijn H (1988) The mammals from the Lower Miocene of Aliveri (Island of Evia, Greece). Part 8: the Cricetidae. Proc Kon Neder Akad Weten Ser B 91(2):185–204

Kokotini M, Kargopoulos N, Iliopoulos G, Roussiakis S, Skandalos P, Michailidis D, Svorligkou G, Kampouridis P, Theodorou G (2019) Karnezeika (Argolis, Peloponnese): preliminary data concerning a new Villafranchian locality of Southern Greece. Bull Geol Soc Greece Sp Pub 7:104

Koliadimou K (1996) Palaeontological and biostratigraphical study of the Neogene/Quaternary micromammals of Mygdonia basin (Macedonia, Greece). PhD thesis, Aristotle Univiversity of Thessaloniki Greece

Koliadimou K, Koufos GD (1991) The Hystricidae from the Pleistocene of Macedonia (Greece) and a review of the European representatives of the family. Proc. 5th Congr. Geol. Soc. Greece. Bull Geol Soc Greece 25(2):453–471

Konidaris GE, Athanassiou A, Tourloukis V, Thompson N, Giusti D, Panagopoulou E, Harvati K (2018) The skeleton of a straight–tusked elephant (*Palaeoloxodon antiquus*) and other large mammals from the Middle Pleistocene butchering locality Marathousa 1 (Megalopolis Basin, Greece): preliminary results. Quat Int 497:65–84

Kormos T (1916) Die Säugetiere der Felsnische Pilisszántó in systematischer zoogeographischer und phylogenetischer Hinsicht. Mitteilungen aus dem Jahrbuch der Konigl. ungar. Geologischen Reichsanstalt, Budapest 23:365–458

Kormos T (1930) Diagnosen neuer Saugetiere aus der ober–Pliozänen fauna des Somlyóberges bei Püspökfürdő. Ann Mus Nat Hungarici 27:237–246

Kormos T (1932) Neue pliozäne Nagetiere aus der Moldau. Paläontol Zeit 14(3):193–199

Kormos T (1934) Neue Insektenfresser, Fledermaiuse und Nager aus dem Oberpliozan der Villanyer Gegend. Földtani Közlöny 64:296–328

Kormos T (1938) *Mimomys newtoni* F. Major und *Lagurus pannonicus* Kormos, zwei gleichzeitige verwandte Wuhlmause von verschiedener phylogenetischer Entwicklung. Anz Ungarns Akad Wissen math–naturw Kl 57:353–379

Koufos GD, Vasileiadou K (2015) Miocene/Pliocene mammal faunas of southern Balkans: implications for biostratigraphy and palaeoecology. Palaeobiodiv Palaeoenviron 95(3):285–303

Koufos GD, Syrides GE, Kostopoulos DS, Koliadimou KK (1995) Preliminary results about the stratigraphy and the palaeoenvironment of Mygdonia Basin, Macedonia, Greece. Geobios 28:243–249

Koufos GD, Vassiliadou A, Koliadimou K, Syrides G (2001) Early Pleistocene small mammals from Marathousa, a new locality in the Mygdonia basin, Macedonia, Greece. Deinsea 8:49–102

Kowalski K (1956) Insectivores, bats and rodents from the early Pleistocene bone breccias of Podlesice near Kroczyce (Poland). Acta Palaeontol Polonica 1(4):331–394

Kowalski K (1958) An early Pleistocene fauna of small mammals from the Kadzielnia hill in Kielce (Poland). Acta Palaeontol Polonica 3(1):1–47

Kowalski K (1960) Cricetidae and Microtidae (Rodentia) from the Pliocene of Węże (Poland). Acta Zool Cracoviensia 5(11):447–506

Kowalski K (1963) The Pliocene and Pleistocene Gliridae (Mammalia, Rodentia) from Poland. Acta Zool Cracoviensia 14:533–567

Kretzoi M (1954) Bericht über die calabrische (Villafranchische) Fauna von Kisláng, Kom. Fejér. Állami Földtani Intézet Évijjelentése az 1953(1):239–265

Kretzoi M (1955) *Promimomys cor* n.g.n.sp, ein alterttimlicher Arvicolide aus dem ungarischen Unterpliozan. Acta Geol Acad Sci Hungaricae 3:89–94

Kretzoi M (1958) New names for arvicolid homonyms. Ann Hist–Natur Mus Nat Hungarici 50:55–58

Kretzoi M (1959) Insectivoren, Nagetiere und Lagomorphen der Jüngstpliozänen Fauna von Csarnóta im Villányer Gebirge (Südungarn). Vertebr Hungarica Mus Hist–Natur Hungarici 1:237–246

Kretzoi M (1962) Fauna und Faunenhorizont von Csarnóta. Ann Rep Hungarian Geol Inst 1959:297–395

Kretzoi M (1969) Skizze einer Arvicoliden Phylogenie. Vertebr Hungarica Mus Hist–Natur Hungarici 11:155–193

Kuss SE, Storch G (1978) Eine Säugetierfauna (Mammalia: Artiodactyla, Rodentia) des älteren Pleistozäns von der Insel Kalymnos (Dodekanés, Griechenland). N Jahr Geol Paläontol Monats:206–227

Lartet E (1851) Notice sur la colline de Sansan, suivi d'une récapitulation des diverses espèces d'animaux vertébrés fossiles, trouvés soit à Sansan, soit dans d'autres gisements du terrain Miocène dans le bassin sous–Pyrénéen, 45. Portes, Auch

Lazaridis G, Tsoukala E, Maul LC (2019) The earliest Hystrix refossa: a new early Villafranchian record from Milia (Grevena, Macedonia, Greece). Hystrix 30(1):12–18

López-Antoñanzas R, Şen S (2004) Ctenodactylids from the lower and middle Miocene of Saudi Arabia. Palaeontol 47(6):1477–1494

López-Antoñanzas R, Şen S, Koufos GD (2005) A ctenodactylid rodent (Mammalia: Rodentia) from the Middle Miocene of Chios Island (Greece). Geobios 38(1):113–126

López-Antoñanzas R, Peláez-Campomanes P, Prieto J, Knoll F (2019) New species of *Karydomys* (Rodentia) from the Miocene of Chios Island (Greece) and phylogenetic relationships of this rare democricetodontine genus. Papers Palaeontol 5(1):33–45

Lyras GA, Athanassiou A, van der Geer AAE (volume 2) The fossil record of insular endemic mammals from Greece. In: Vlachos E (ed) The Fossil Vertebrates of Greece, vol 2 – Laurasiatherians, artiodactyles, perissodactyles, carnivorans, and island endemics. Springer, Cham

Macdonald DW (ed) (2009) The encyclopedia of mammals. Oxford University Press, The Brown Reference Group Ltd, Oxford

Mahboubi S (2014) Les rongeurs du Miocène supérieur et terminal d'Afrique nord–occidentale:biochronologie, magnétostratigraphie, biogéographie et paléoenvironnements. Thèse, Université de Poitiers

Marcolini F, Piras P, Kotsakis T, Claude J, Michaux J, Ventura J, Cubo J (2011) Phylogenetic signal and functional significance of incisor enamel microstructure in *Arvicola* (Arvicolinae, Rodentia, Mammalia). C R Palevol 10:479–487

Maridet O (2003) Révision du genre *Democricetodon* (Mammalia, Rodentia, Cricetinae) et dynamique des faunes de rongeurs du Néogène d'Europe occidentale: évolution, paléobiodiversité et paléobiogéographie. Doctoral dissertation, Université Claude Bernard–Lyon I

Maridet O, Şen S (2012) Les Cricetidae (Rodentia) de Sansan. In Mammifères de Sansan. Mémo Mus Paris 203:29–65

Marković Z, de Bruijn H, Wessels W (2016) A revision of the new rodent collections from the Early Miocene of Sibnica, Serbia. Life on the shore–geological and paleontological research in the Neogene of Sibnica and vicinity (Levač basin, Central Serbia), pp 63–117

Marquet JC, Séronie-Vivien MR (2016) Les rongeurs du sondage Séronie–Vivien de la grotte d'âge Pléistocène moyen de Pradayrol à Caniac–du–Causse (Lot). Quat 27:111–131

Martin RA (2003) The status of *Mimomys* in North America revisited. Deinsea 10:399–406

Martín Suárez E, Mein P (1991) Revision of the genus *Castillomys* (Muridae, Rodentia). Scripta Geol 96:47–81

Martín Suárez E, Mein P (1998) Revision of the genera *Parapodemus*, *Apodemus*, *Rhagamys* and *Rhagapodemus* (Rodentia, Mammalia). Geobios 31(1):87–97

Maul L (1996) A discussion of the referral of *Mimomys occitanus* THALER, 1955 (Rodentia: Arvicolidae) to the genus *Mimomys*. Acta Zool Cracoviensia 39(1):343–348

Maul LC, Rekovets L, Heinrich WD, Keller T, Storch G (2000) *Arvicola mosbachensis* (Schmidtgen 1911) of Mosbach 2: a basic sample for the early evolution of the genus and a reference for further biostratigraphical studies. Senckenbergiana lethaea 80(1):129–147

Mayhew DF (1978) Late Pleistocene small mammals from Arnissa (Macedonia, Greece). Proc Kon Neder Akad Weten Ser B 81(3):302–321

Mayhew DF (2012) Revision of the fossil vole assemblage (Mammalia, Rodentia, Arvicolidae) from Pleistocene deposits at Kisláng, Hungary. Palaeontol 55(1):11–29

Mayhew DF, Stuart AJ (1986) Stratigraphic and taxonomic revision of the fossil vole remains (Rodentia, Microtinae) from the Lower Pleistocene deposits of Eastern England. Phil Trans R Soc London Ser B Biol Sci 312(1157):431–485

Mayhew DF, Dieleman FE, Boele J, Verhaard L, van den Hoek Ostende LW (2008) *Mimomys hajnackensis* from the Pliocene of the Netherlands. Netherlands J Geosci 87(2):181–188

Mayr H (1979) Gebissmorphologische Untersuchungen an miozänen Gliriden (Mammalia, Rodentia) Süddeutschlands. Thesis University Munich

McKenna CM, Bell KS (eds) (1997) Classification of Mammals. Above the species level. Columbia University Press, New York

Méhely L (1908) *Prospalax priscus* (Nhrg), die pliocäne Stammform der heutigen *Spalax*–Arten. Ann Hist Natur Mus Nat Hungarici 6:305–316

Méhely LV (1914) Fibrinae Hungariae. Die tertären und quartären wurzelzänigen Wühlmäuse Ungarns. Ann Nat Mus Hungary 12:155–243

Mein P (1958) Les mammifères de la faune sidérolithique du Vieux–Collonges. Nouv Arch Mus Hist Nat Lyon 5:1–122

Mein P (1970) Les Sciuroptères (Mammalia, Rodentia) néogènes d'Europe occidentale. Géobios 3(3):7–77

Mein P (1983) Composition quantitative des faunes de mammifères du Miocène moyen et supérieur de la régión Lyonnaise. R.C.M.N.S. Interim–Coll. Mediterranean Neogene Continental Paleoenvironments and Paleoclimatic evolution, p 75–79

Mein P (1989) Continental faunas at the Miocene-Pliocene boundary. La gisement karstique turolien de Lissieu (Rhône). International Workshop "Continental faunas at the Miocene/Pliocene boundary", Proceedings, Faenza 25

Mein P, Freudenthal M (1971a) Une nouvelle classification des Cricetidae (Mammalia, Rodentia) du Tertiaire de l'Europe. Scripta Geol 2:1–37

Mein P, Freudenthal M (1971b) Les Cricetidae (Mammalia, Rodentia) du Néogène Moyen de Vieux–Collonges (Rhône, France). Partie 1: Le genre *Cricetodon* Lartet, 1851. Scripta Geol 5:1–51

Mein P, Michaux J (1970) Un nouveau stade dans l'évolution des rongeurs pliocènes de l'Europe sud–occidentale. C R Séances l'Acad Sci Paris 270:2780–2783

Mein P, Moissenet E, Adrover R (1983) L'extension et l'âge des formations continentales pliocènes du fossé de Teruel (Espagne). C R l'Acad Sci Paris 296:1603–1610

Mein P, Martín Suárez E, Agustí J (1993) *Progonomys* Schaub, 1938 and *Huerzelerimys* gen. nov. (Rodentia); their evolution in Western Europe. Scripta Geol 103:41–64

Melentis JK (1966) *Steneofiber jaegeri* aus Ligniten von Serrae und die Datierung der Fundschichten. Ann Géol Pays Helléniques 17:289–297

Melentis JK (1967) Studien über fossile Vertebraten Griechenlands. 19. Die Pikermifauna von Halmyropotamos (Euboa, Griechenland). 1. Teil: Odontologie und Kraniologie. Ann Géol Pays Helléniques 19:283–411

Melentis JK, Schneider H (1966) Studien über fossile Vertebraten Griechenlands. 17. Eine neue Pikermifauna in der Nähe der Ortschaft Alifaka in Thessalien (Griechenland). Ann Géol Pays Helléniques 17:267–288

Meng J, Wyss A (2005) Glires. In: Rose KD, Archibald JD (eds) The rise of placental mammals: origins and relationships of the major extant clades. Johns Hopkins University Press, Baltimore, pp 145–158

Michaux J (1967) Origine du dessin dentaire *Apodemus* (Rodentia, Mammalia). C R Acad Sci Paris sér D 264:711–714

Michaux J (1969) Muridae (Rodentia) du Pliocène supérieur d'Espagne et du Midi de la France. Palaeovertebrata 3:1–25

Minwer-Barakat R, García-Alix A, Martín Suárez E, Freudenthal M (2008) *Micromys caesaris*, a new murid (Rodentia, Mammalia) from the late Pliocene of the Guadix Basin, Southeastern Spain. J Paleontol 82(2):436–441

Montenat C, de Bruijn H (1976) Ruscinian rodent faunule from La Juliana (Murcia)–its implication for correlation of continental and marine biozones. Proc Kon Neder Akad Weten Ser B 79(4):245–255

Mörs T, von Koenigswald W, von der Hocht F (1998) Rodents (Mammalia) from the late Pliocene Reuver Clay of Hambach (Lower Rhine Embayment, Germany). In van Kolfschoten T, Gibbard P (eds), The Dawn of the Quaternary. Mededelingen Nederlands Instituut voor Toegepaste Geowetens 60:135–160

Nadachowski A (1991) Systematics, geographic variation, and evolution of snow voles (*Chionomys*) based on dental characters. Acta Theriologica 36(1–2):1–45

Nehring A (1897) Mehrere neue *Spalax*–Arten. Sitzungsberichte der Gesellschaft Naturforschender Freunde zu Berlin:163–183

Newton ET (1882) The Vertebrata of the Forest Bed Series of Norfolk and Suffolk. Mem Geol Surv UK, London

Nicoara I (2011) Upper Turolian Sciuroidea (Rodentia, Mammalia) from the Republic of Moldova. Acta Palaeontol Romaniae 7:257–265

Oliver A, Peláez-Campomanes P (2014) Evolutionary patterns of early and middle Aragonian (Miocene) of *Megacricetodon* (Rodentia, Mammalia) from Spain. Palaeontographica Abt A:85–135

Oliver A, Peláez-Campomanes P (2016) Early Miocene evolution of the rodent *Megacricetodon* in Europe and its palaeobiogeographical implications. Acta Palaeontol Polonica 61(1):211–219

Pasa A (1947) I mammiferi di alcune antiche brecce veronesi. Mem Mus Civico Storia Nat Verona 1:1–111

Peláez-Campomanes P, Daams R (2002) Middle Miocene rodents from Paşalar, Anatolia, Turkey. Acta Palaeontol Polonica 47(1):125–132

Popov VV (2004) Pliocene small mammals (Mammalia, Lipotyphla, Chiroptera, Lagomorpha, Rodentia) from Muselievo (North Bulgaria). Geodiversitas 26:403–491
Popov VV (2017) Early Pleistocene small mammals (Eulipotyphla, Chiroptera, Lagomorpha and Rodentia) from Futjova Cave, North Bulgaria. Acta Zool Bulgarica 69(2):263–282
Prieto J (2009) Comparison of the dormice (Gliridae, Mammalia) *Seorsumuscardinus alpinus* de Bruijn, 1998 and *Heissigia bolligeri* PRIETO & BÖHME, 2007. N Jb Geol Paläontol Abh 252:377–379
Prieto J (2010) Note on the morphological variability of *Keramidomys thaleri* (Eomyidae, Mammalia) from Puttenhausen (North Alpine Foreland Basin, Germany). Zitteliana:103–109
Qiu Z, Storch G (2000) The early Pliocene micromammalian fauna of Bilike, Inner Mongolia, China (Mammalia: Lipotyphla, Chiroptera, Rodentia, Lagomorpha). Senckenbergiana lethaea 80(1):173–229
Rabeder G (1981) Die arvicoliden (Rodentia, Mammalia) aus dem älteren Pleistozän von Niederösterreich. Beitr Paläont 8:1–373
Radulescu C, Samson P (1989) Contributions to the knowledge of the mammalian faunas from Malusteni and Beresti (Romania). Order Rodentia, Family Arvicolidae. Travaux de l'Institut de Spéologie « Emile Racovitza ». Bucarest 28:43–56
Ramos-Lara N, Koprowski JL, Kryštufek B, Hoffmann IE (2014) *Spermophilus citellus* (Rodentia: Sciuridae). Mammalian Species 46(913):71–87
Rekovets L (1994) Small Mammals of the Athropogene of the Southern Part of East Europe. Naukova Dumka, Kiev
Rekovets L, Nadachowski A (1995) Pleistocene voles (Arvicolidae) of the Ukraine. Paleont Evol 28–29:145–245
Repenning CA (2003) Chapter 17: *Mimomys* in North America. Bull Am Mus Natur Hist:469–512
Repenning CA, Grady F (1988) The microtine rodents of the Cheetah Room fauna, Hamilton Cave, West Virginia, and the spontaneous origin of *Synaptomys*. US Geol Surv Bull 1853:1–32
Reumer JWF, Doukas CS (1985) Early Pleistocene Insectivora (Mammalia) from Tourkobounia (Athens, Greece). Proc Kon Neder Akad Weten Ser B 88:111–121
Roth JR, Wagner JA (1854) Die fossilen knochenueberreste von Pikermi in Griechenland. Kön Bayerische Akad Wiss Abth, München
Ruiz Bustos A, Sesé C, Dabrio CJ, Peña Ruano JA, Padial JM (1984) Geología y fauna de micromamíferos del nuevo yacimiento del Plioceno inferior de Gorafe–A (Depresión de Guadix–Baza, Granada). Estudios Geol 40:231–241
Ruiz-Sánchez FJ, Calatayud BL, Freudenthal M (2009) *Eomyops noeliae* sp. nov., a new Eomyidae (Mammalia, Rodentia) from the Aragonian of Spain. C R Palevol 8:375–384
Ruiz-Sánchez FJ, Murelaga X, Freudenthal M, Larrasoana JC, Furió M, Garcés M, González-Pardos M, Suarez-Hernando O (2013) Micromammalian faunas from the middle Miocene (middle Aragonian) of the Tudela formation (Ebro Basin, Spain). Bull Geosci 88:131–152
Ryu SH, Kwak MJ, Hwang UW (2013) Complete mitochondrial genome of the Eurasian flying squirrel *Pteromys volans* (Sciuromorpha, Sciuridae) and revision of rodent phylogeny. Molecular biology reports 40(2):1917–1926
Sabol M, Konečný V, Vass D, Kováčová M, Ďurišová A, Túnyi I (2006) Early Late Pliocene site of Hajnáčka I (Southern Slovakia) – geology, palaeovolcanic evolution, fossil assemblages and palaeoenvironment. Courier Forschungsinst Senckenberg 256:261–274
Santel W, von Koenigswald W (1998) Preliminary report on the Middle Pleistocene small mammal fauna from Yarimburgaz Cave in Turkish Thrace. Eiszeitalter Ggw 48:162–169
Savic IR, Nevo E (1990) The Spalacidae: evolutionary history, speciation and population biology. Progress Clinical Biol Res 335:129–153
Schaub S (1925) Die hamsterartigen Nagetiere des Tertiärs und ihre lebenden Verwandten. Abh Schweiz Paläontol Gesell 45:1–114
Schaub S (1926) Über *Mus gaudryi* Dames aus dem Pontien von Samos. Eclog Geol Helvetiae 19:768–776
Schaub S (1930) Quartäre und jungtertiäre Hamster. Abh Schweiz paläontol Ges 49:1–49

Schaub S (1938) Tertiare und Quartare Murinae. Abh Schweiz Pal Gesellsch 61:1–39
Schaub S, Zapfe H (1953) Die Fauna der miozänen Spaltenfüllung von Neudorf an der March (ČSR): Simplicidentata. Sitz österr Akad Wiss math–naturwiss Kl Abt 1(162):81–215
Schlosser M (1884) Die Nager des europäischen Tertiärs nebst Betrachtungen über die Organisation und die geschichtliche Entwicklung der Nager überhaupt. Palaeontographica 31:19–162
Schlosser M (1890) Die Affen, Lemuren, Chiropteren, Insectivoren, Marsupialier, Creodonten und Carnivoren des europäischen Tertiärs. Beitr Pal Geol Österr 8:387–492
Schlosser M (1925–1926) Die Säugetierfauna von Peublanc (Dép. Allier). Soc Sei Natur Croatica 38–39:372–394
Schmidt-Kittler N, de Bruijn H, Doukas C (1995) The Vertebrate locality Maramena (Macedonia, Greece) at the Turolian–Ruscinian boundary (Neogene). 1. General Introduction. Münchner Geowiss Abh A 28:9–18
Schötz M (1993) Zwei Hamsterfaunen (Rodentia, Mammalia) aus der niederbayerischen Molasse. Mitt Bayerischen Staats Paläontol Hist Geol 33:155–193
Şen S (1977) La faune de Rongeurs pliocènes de Çalta (Ankara, Turquie). Bull Mus Nat d'hist natur Sci Terre 61:89–171
Şen S (1983) Rongeurs et lagomorphes du gisement pliocène de Pul–e Charkhi, bassin de Kabul, Afghanistan. Bull Mus Nat d'Hist Natur Paris C, p:33–74
Şen S (2001) Early Pliocene porcupine (Mammalia, Rodentia) from Perpignan, France: A new systematic study. Geodiversitas 23:303–312
Şen S (2003) Muridae and Gerbillidae (Rodentia). In: Fortelius M, Kappelman J, Sen S, Bernor R (eds) Geology and Paleontology of the Miocene Sinap Formation. Columbia University Press, Turkey, pp 125–140
Şen S (2016) Rodents, lagomorphs and insectivores. In Şen S (ed) Late Miocene mammal locality of Küçükçekmece. European Turkey. Geodiversitas 38(2):299–314
Şen S, de Bruijn H (1977) Comparaison des rongeurs du Néogène terminal de Maritsa, Çalta et Tourkobounia–1, Méditerranée orientale. Proc Kon Ned Akad Weten Ser B 80(3):221–227
Şen S, Purabrishemi Z (2010) First porcupine fossils (Mammalia, Rodentia) from the late Miocene of NW Iran, with notes on late Miocene–Pliocene dispersal of porcupines. Paläontol Zeit 84(2):239–248
Şen S, Thomas H (1979) Découverte de Rongeurs dans le Miocène moyen de la Formation Hofuf (Province du Hasa, Arabie Saoudite). C R somm Soc Géol France 21:34–37
Şen S, Jaeger J-J, Dalfes N, Mazin J-M, Bocherens H (1989) Découverte d'une faune de petits mammifères pliocènes en Anatolie occidentale. C R Acad Sci Paris 309:1729–1734
Şen S, Seyitoğlu G, Karadenizli L, Kazancı N, Varol B, Araz H (1998) Mammalian biochronology of Neogene deposits and its correlation with the lithostratigraphy in the Çankiri–Çorum Basin, central Anatolia, Turkey. Eclogae Geol Helvetiae 91:307–320
Sickenberg O (1971) Revision der Wirbeltier fauna der Höhle Petralona (Griech. Mazedonien). Ann Géol Pays Helléniques 23:230–264
Simionescu I (1930) Vertebratele Pliocene de la Măluşteni (Covurlui). Academia Română, Publicaţiile Fondului "Vasile Adamachi" 9(49):1–69
Sinitsa MV, Delinschi A (2016) The earliest member of *Neocricetodon* (Rodentia: Cricetidae): a re–description of *N. moldavicus* from Eastern Europe, and its bearing on the evolution of the genus. J Paleontol 90(4):771–784
Sinitsa MV, Pogodina NV (2019) The evolution of early *Spermophilus* in eastern Europe and the antiquity of the Old World ground squirrels. Acta Palaeontol Polonica 64:643–667
Skandalos P (2017) Study of *Cricetodon* (Rodentia, Mammalia) from the Early Miocene locality of Karydia (Rhodope, North Greece). Master thesis, National Kapodistrian Univ Athens
Sondaar PY (1971) Palaeozoogeography of the Pleistocene mammals from the Aegean. In Strid A (ed) Evolution in the Aegean. Opera Botanica 30:65–70
Sozen M, Bukhnikashvili A, Shenbrot G, Scott D, Amori G, Kryštufek B, Yigit N, Mitsain G (2016) *Meriones tristrami* (errata version published in 2017). The IUCN Red List of

Threatened Species 2016: e.T13170A115110701. https://doi.org/10.2305/IUCN.UK.2016-3.RLTS.T13170A22432669.en. Downloaded on 07 May 2020
Stehlin HG, Schaub S (1951) Die Trigonodontie der simplicidentaten Nager. Schweiz Paläontol Abh 67:1–385
Storch G (1975) Eine mittelpleistozäne Nager–Fauna von der Insel Chios, Ägäis (Mammalia: Rodentia). Senckenbergiana Biol 56(4/6):165–189
Storch G, Dahlmann T (1995) The vertebrate locality Maramena (Macedonia, Greece) at the Turolian–Ruscinian boundary (Neogene). 10. Murinae (Rodentia, Mammalia). Münchner Geowiss Abh 28:121–132
Storch G, Ni X (2002) New Late Miocene murids from China (Mammalia, Rodentia). Geobios 35(4):515–521
Suata-Alpaslan F (2009) *Occitanomys* (*Rhodomys*) *vandami* nov. sp. (Muridae, Mammalia) from the early Pliocene/Early Ruscinian fauna of Iğdeli (Turkey). Open Geol J 3:64–69
Sulimski A (1960) *Hystrix primigenia* (Wagner) in the Pliocene fauna from Weze. Acta Palaeontol Polonica 5(3):319–335
Sulimski A (1964) Pliocene Lagomorpha and Rodentia from Weze 1 (Poland): study on the Tertiary bone breccia fauna from Weze near Dzialoszyn in Poland. Acta Palaeontol Polonica 9:149–261
Sylvestrou I (2002) Stratigraphic, palaeontological and palaeogeographical study of the Neogene–Quaternary deposits of Katerini basin, Northern Greece. PhD thesis, Department of Geology, Aristotle University of Thessaloniki, Greece
Sylvestrou IA, Kostopoulos DS (2007) *Pseudomeriones megistos* nov. sp. (Gerbillinae, Mammalia) from the latest Miocene of Northern Greece and its phylogenetic relationships. Geobios 40(6):833–848
Symeonidis NK (1992) Fossil mammals of Lower Pleistocene (Villafranchian) age from the Sesklo Basin (Volos). Ann Géol Pays Helléniques 35:1–21
Syrides G, Koliadimou K (1994) A new Pleistocene locality from the Strymon basin (Macedonia, Greece) with continental mollusks and micromammals. Bull Geol Soc Greece 30:331–339
Syrides GE, Koliadimou KK, Koufos GD (1997) New Neogene molluscan and mammalian sites from Thrace, Greece. C R Acad Sci 324(5):427–433
Tchernov E (1968) A Pleistocene faunule from a karst fissure filling near Jerusalem, Israel. Verhandl Naturf Ges Basel 79(2):161–185
Tesakov AS (1993) Evolution of *Borsodia* (Arvicolidae, Mammalia) in the Villanyian and in the early Biharian. Quat Int 19:41–45
Tesakov AS (1998) Voles of the Tegelen fauna. Mededelingen Nederlands Instituut voor Toegepaste Geowetens TNO 60:71–134
Tesakov AS, Simakova AN, Frolov PD, Sytchevskaya EK, Syromyatnikova EV, Foronova IV, Shalaeva EA, Trifonov VG (2019) Early–Middle Pleistocene environmental and biotic transition in NW Armenia, southern Caucasus. Palaeontol Electronica 22(2):1–39
Thaler L (1955) Sur l' âge pliocène de la faune des grottes du Lazaret (Sète, Hérault). C R hebd séances l'Acad sci 240(1):433–435
Thaler L (1966) Les rongeurs fossiles du Bas–Languedoc dans leurs rapports avec l'histoire des faunes et la stratigraphie du Tertiaire d'Europe. Mém Mus Nat d'Hist Natur nouv sér C 42:3–284
Theocharopoulos KD (1991) Contribution in the study of micromammals of the Lower Pleistocene from the Strimonas basin (Serres, N. Greece). Diploma–thesis, University of Athens, School of Geology
Theocharopoulos CD (2000) Late Oligocene–middle Miocene *Democricetodon*, *Spanocricetodon* and *Karydomys* n. gen. from the eastern Mediterranean area. Gaia 8:1–92
Topachevski VA (1957) New species of souslik from the Upper Pleistocene deposits of Azov region, Ukraine. Proc Acad Sci USSR 2:204–208
Topachevski VA (1969) Fauna of the USSR, mammals, mole rats, Spalacidae. Academy of Sciences of the USSR, Fauna of the USSR, Mammals, New Delhi, Amerind Publishing Co Pvt. Ltd Vol 99:1–150

Tsoukala E, Chatzopoulou K (2005) A new early Pleistocene (latest Villafranchian) site with mammals in Kalamotó (Mygdonia Basin, Macedonia, Greece). Preliminary Report. Mitt Komm Quartärforsch Österr Akad Wiss 14:213–233

Tsoukala E, Bartsiokas A, Chatzopoulou K, Lazaridis G (2006) Quaternary mammalian remains from the Kitseli Pothole (Alea, Nemea, Peloponnese). Sci Ann School Geol Aristotle Univ Thessaloniki 98:273–284

Ünay E (1978) *Pliospalax primitivus* n. sp. and *Anomalomys gaudryi* Gaillard from the *Anchitherium* fauna of Sariçay (Turkey). Bull Geol Soc Turkey 21:119–132

Ünay E (1980) The Cricetodontini (Rodentia) from the Bayraktepe section (Çanakkale, Turkey). Proc Kon Neder Akad Weten Ser B 83:399–418

Ünay E (1994) Early Miocene rodent faunas from the eastern Mediterranean area: Part IV. The Gliridae. Proc Kon Neder Akad Weten 97(4):445–490

Ünay E (1996) On fossil Spalacidae (Rodentia) In Bernor RL, Fahlbusch V, Mittmann HW (eds) The Evolution of Western Eurasian Neogene Mammal Faunas. Columbia University Press, New York, pp 246–252

Ünay E (1999) Family Spalacidae. In Rössner GE, Heissig K (eds) The Miocene land mammals of Europe, Verlag Dr F. Pfeil München, p 421–426

Ünay E, de Bruijn H, Suata-Alpaslan F (2006) Rodents from the Upper Miocene hominoid locality Çorakyerler (Anatolia). Beitr Paläont 30:453–467

Ünay-Bayraktar E (1989) Rodents from the middle Oligocene of Turkish Thrace. Utrecht Micropaleontol Bull Sp Pub 5:1–119

van Dam JA (1997) The small mammals from the upper Miocene of the Teruel–Alfambra region (Spain): paleobiology and paleoclimatic reconstructions. Geol Ultraiectina 156:1–204

van de Weerd A (1973) Rodentia from two Pleistocene fissure fillings near Athens. Proc Kon Neder Akad Weten Ser B 76:148–166

van de Weerd A (1976) Rodent faunas of the Mio–Pliocene continental sediments of the Teruel–Alfambra region, Spain. Utrecht Micropaleontol Bull Sp Pub 2:218

van de Weerd A (1979) Early Ruscinian rodents and lagomorphs (Mammalia) from the lignites near Ptolemais (Macedonia, Greece). Proc Kon Neder Akad Weten 82:127–170

van de Weerd A, Reumer JW, de Vos J (1982) Pliocene mammals from the Apolakkia Formation (Rhodes, Greece). Proc Kon Neder Akad Weten Ser B 85:89–112

van de Weerd A, de Brujin H, Marković Z, Wessels W (2018) Paracricetodontinae (Mammalia, Rodentia) from the Late Eocene and Early Oligocene of south-east Serbia. Palaeobio Palaeoenv 98:489–508

van den Hoek Ostende LW, Mayda S, Oliver A, Madern A, Hernández-Ballarín V, Peláez-Campomanes P (2015) Aliveri revisited, a biogeographical appraisal of the early Miocene mammals from the eastern Mediterranean. Palaeobiodiv Palaeoenviron 95(3):271–284

van der Geer AA, Lyras GA, van den Hoek Ostende LW, de Vos J, Drinia H (2014) A dwarf elephant and a rock mouse on Naxos (Cyclades, Greece) with a revision of the palaeozoogeography of the Cycladic Islands (Greece) during the Pleistocene. Palaeogeogr, Palaeoclimatol, Palaeoecol 404:133–144

van der Meulen AJ (1973) Middle Pleistocene smaller mammals from the Monte Peglia (Orvieto, Italy) with special reference to the phylogeny of *Microtus* (Arvicolidae, Rodentia). Quaternaria 17:1–144

van der Meulen AJ, de Bruijn H (1982) The mammals from the Lower Miocene of Aliveri (Island of Evia, Greece). Part 2. The Gliridae. Proc Kon Neder Akad Weten Ser B 85:485–524

van der Meulen AJ, Doukas CS (2001) The Early Biharian rodent fauna from Tourkobounia 2. Lynx ns 32:255–277

van der Meulen A, van Kolfschoten T (1986) Review of the late Turolian to early Biharian mammal faunas from Greece and Turkey. Mem Soc Geol Italiana 31:201–211

van der Meulen AJ, Zagwijn WH (1974) *Microtus* (*Allophaiomys*) *pliocaenicus* from the lower Pleistocene near Brielle, the Netherlands. Rijksmus Geol Min:21–30 van der Meulen AJ, Peláez-Campomanes P, Daams R (2003) Revision of medium-sized Cricetidae from the Miocene of the Daroca–Villafeliche area in the Calatayud–Teruel basin (Zaragoza, Spain). Coloquios Paleontol 1:385–441

van Kolfschoten T (1985) The Middle Pleistocene (Saalian) and Late Pleistocene (Weichselian) mammal faunas from Maastricht–Belvedere, Southern Limburg, The Netherlands. Meded Rijks Geol Dienst 39:45–74

van Vugt N, de Bruijn H, van Kolfschoten T, Langereis CG, Okuda M (2000) Magneto– and cyclostratigraphy and mammal–fauna's of the Pleistocene lacustrine Megalopolis Basin, Peloponnesos, Greece. Geol Ultrajectina 189:69–92

van Weers DJ (1994) The porcupine *Hystrix refossa* Gervais, 1852 from the Plio–Pleistocene of Europe, with notes on other fossil and extant species of the genus *Hystrix*. Scripta Geol 106:35–52

van Weers DJ (2005) A taxonomic revision of the Pleistocene *Hystrix* (Hystricidae, Rodentia) from Eurasia with notes on the evolution of the family. Contrib Zool 74:301–312

Vasileiadou K (2001) Study of the Late Miocene/Early Pliocene micromammalian fossil assemblage from Nea Silata, Chalkidiki, North Greece. Master thesis, Aristotle University of Thessaloniki.

Vasileiadou K, Koufos GD (2005) The micromammals from the early/middle Miocene locality of Antonios, Chalkidiki, Greece. Ann Paléontol 91:197–225

Vasileiadou K, Sylvestrou IA (2009) The Late Miocene mammal faunas of the Mytilinii Basin, Samos Island, Greece: new collection. 4. Micromammals. Beit Paläontol Wien 31:37–55

Vasileiadou K, Zouros N (2012) Early Miocene micromammals from the Lesvos Petrified Forest (Greece): preliminary results. Palaeobiodiv Palaeoenviron 92(2):249–264

Vasileiadou KV, Koufos GD, Syrides GE (2003) Silata, a new locality with micromammals from the Miocene/Pliocene boundary of the Chalkidiki peninsula, Macedonia, Greece. Deinsea 10(1):549–562

Vasileiadou K, Konidaris G, Koufos GD (2012) New data on the micromammalian locality of Kessani (Thrace, Greece) at the Mio–Pliocene boundary. Palaeobiodiv Palaeoenviron 92(2):211–237

Vasileiadou K, Böhme M, Neubauer TA, Georgalis GL, Syrides GE, Papadopoulou L, Zouros N (2017) Early Miocene gastropod and ectothermic vertebrate remains from the Lesvos Petrified Forest (Greece). PalZ 91(4):541–564

von Koenigswald W (1973) Veränderungen in der Kleinsäugerfauna von Mitteleuropa zwischen Cromer und Eem (Pleistozän). E&G–Quat Sci J 23(1):159–167

von Koenigswald W, van Kolfschoten T (1996) The *Mimomys–Arvicola* boundary and the enamel thickness quotient (SDQ) of *Arvicola* as stratigraphic markers in the Middle Pleistocene. The Early Middle Pleistocene in Europe, p:211–226

von Koenigswald W, Fejfar O, Tchernov E (1992) Revision einiger alt– und mittelpleistozaner Arvicoliden (Rodentia, Mammalia) aus dem ostlichen Mittelmeergebiet (Ubeidiya, Jerusalem und Kalymnos–Xi). N Jahr Geol Paläontol Abh 184(1):1–23

von Meyer H (1848) Mitteilungen an Professor Bronn gerichtet. N Jahr Mineral Geol Paläontol, p:465–473

Wagner A (1848) Urweltliche SäugthierUeberreste aus Griechenland. Abh mathphys Cl Kön Bayer Akad Wiss 5:333–510

Wagner A (1857) Neue Beitrage zur Kenntnis der fossilen SäugthierUeberreste von Pikermi. Abh mathphys Cl Kön Bayer Akad Wiss 8:109–158

Wessels W (1998) Gerbillinae from the Miocene and Pliocene of Europe. Mitt Bayer Staatss Paläontol hist Geol 38:187–207

Wessels W (1999) Family Gerbillidae. In: Rössner G, Heissig K (eds) The Miocene Land Mammals of Europe. Pfeil Verlag, München, pp 395–400

Wessels W (2009) Miocene rodent evolution and migration. Muroidea from Pakistan, Turkey and North Africa. Geologica Ultraiectina Univ Utrecht 307:1–289

Wessels W, Reumer BM (2009) *Democricetodon* and *Megacricetodon* (Mammalia, Cricetidae) from the Miocene of Sandelzhausen, Southern Germany. Paläontol Zeit 83(1):187–205
Wessels W, Fejfar O, Peláez-Campomanes P, van der Meulen A, de Bruijn H (2003) Miocene small mammals from Jebel Zelten, Libya. Coloquios Paleontol 1:699–715
Wilson DE, Reeder DM (2005) Mammal species of the world: a taxonomic and geographic reference 3rd edition. MD, Johns Hopkins University Press, Baltimore
Wöger JA (2011) A Revision of the Murinae (Rodentia, Mammalia) of Kohfidisch (Upper–Miocene, Burgenland, Austria). Doctoral dissertation, University of Vienna
Wu W (1993) Neue Gliridae (Rodentia, Mammalia) aus untermiozänen (orleanischen) Spaltenfüllungen Süddeutschlands. Institut für Paläontologie und historische Geologie
Zachariasse WJ, van Hinsbergen DJ, Fortuin AR (2011) Formation and fragmentation of a late Miocene supradetachment basin in central Crete: implications for exhumation mechanisms of high–pressure rocks in the Aegean forearc. Basin Res 23(6):678–701
Zamanis A, Faugères L, de Bonis L, Fountoulis D, Simeakis C, Panayotis H, Dimitrakopoulos R, Mercier J-L (1980) Découverte d'une faune de Mammifères du Quaternaire ancient dans les formations du lac Langhada (Macédoine centrale, Grèce). Implications néotectoniques. C R Acad Sci Paris 291:813–816
Ziegler R, Fahlbusch V (1986) Kleinsäuger–Faunen aus der basalen Oberen Süßwasser–Molasse Niederbayerns. Zitteliana A 14:3–58

The Fossil Record of Hares, Rabbits, and Pikas (Mammalia: Lagomorpha) in Greece

Katerina Vasileiadou and Ioanna Sylvestrou

1 Introduction

The order Lagomorpha Brandt, 1855, comprises two extant families: Ochotonidae Thomas, 1897 (pikas), and Leporidae Fischer de Waldheim, 1817 (rabbits and hares). Lagomorphs for long are classified as a suborder of Rodentia; it was in 1912 that they were separated to form their own order, mainly due to the presence of a second pair of nonfunctional upper incisors, absent in rodents (Gidley 1912; see also Montuire 2001 for details). The first upper pair of incisors is ever-growing (not the second upper ones) and there is a diastema between the two pairs of incisors and the premolars. The dental formula is 2,0,3,2/1,0,2,3 for extant ochotonids and 2,0,3,3/1,0,2,3 for extant leporids, but it may slightly vary in fossil taxa.

Ochotonids are represented today by only one genus, *Ochotona* Link, 1795, with more than 30 species, living mainly in rocky, alpine meadow, steppe, or semi-desert environments across the western part of North America and in Asia north of the Himalayas, from the Middle East and the Urals to the Bering Sea (MacDonald 2009). Leporids are represented today by 11 genera with more than 60 species, living worldwide (native in Eurasia, Africa, and North America and introduced to South America and Oceania) in a great variety of environments, from desert to tropical rain forest and arctic tundra (MacDonald 2009).

The extant Greek fauna includes only two species of Leporidae: *Lepus europaeus* (European or brown hare) and *Oryctolagus cuniculus* (European rabbit). Nevertheless the fossil record indicates the presence in the area of numerous

K. Vasileiadou (Deceased)
Natural History Museum of the Lesvos Petrified Forest, Mytilene, Lesvos, Greece

I. Sylvestrou (✉)
School of Geology, Aristotle University of Thessaloniki, Thessaloniki, Greece
e-mail: sylvest@geo.auth.gr

E. Vlachos (ed.), *Fossil Vertebrates of Greece Vol. 1*,
https://doi.org/10.1007/978-3-030-68398-6_16

representatives of both lagomorph families since the early Miocene. Although diagnostic dental elements (p3 and P2) are few in several cases and definite specific determinations maybe difficult, the Greek lagomorph fossil record is fairly complete. The aim of this work is to present the diversity of the Greek lagomorphs through time, providing taxonomical and chronological information. Fossil Greek occurrences of taxa surviving today are not treated separately and take part of the genus remarks. Taxa originally described from Greece are marked with a star.

2 Historical Overview

No mention to fossil lagomorphs in Greece can be found in literature before the beginning of the 1970s. Lagomorph fossil remains were reported for the first time by de Bruijn et al. (1970) from the Upper Miocene locality Maritsa 1 (Rhodes Island), in which one ochotonid and one leporid were present. Later in the same decade, two more species were described from the Turolian faunal complex of Pikermi-Chomateri by López Martínez (1976), two species from the Upper Pleistocene locality Arnissa by Mayhew (1978), and one from the Pliocene locality Ptolemais 1 by van de Weerd (1979). During the following two decades, and as the attempts to find micromammalian fossil localities in Greece were intensified,[1] a few lagomorph remains were usually found alongside large assemblages of rodents. These investigations led to the production of several publications describing Neogene lagomorph remains (e.g., Katsikatsos et al. 1981; van de Weerd et al. 1982; Reumer and Doukas 1985; López Martínez 1986; de Bruijn 1989, 1995; de Bruijn et al. 1999).

In the years that followed, doctoral, postgraduate, and undergraduate studies carried out in the Geology Department of the Aristotle University of Thessaloniki led to the discovery of additional lagomorph remains and the production of a series of publications describing them (e.g., Koliadimou 1996; Vasileiadou 2001; Sylvestrou 2002; Konidaris 2005; Chatzopoulou 2014).

The study of the Greek fossil lagomorphs, even though hampered by the absence of large assemblages, has led to the creation of one new ochotonid genus, *Albertona* López Martínez, 1986, and one new ochotonid and one new leporid species: *Albertona balkanica* López Martínez, 1986, from Aliveri and *Trischizolagus maritsae* de Bruijn et al., 1970, from Maritsa 1.

[1] Mainly by a Dutch-Greek collaboration between the Faculty of Earth Sciences of the University of Utrecht, the Geology Department of the University of Athens, and the Greek Institute of Geology and Mineral Exploration

3 Evolution and Phylogenetic Relationships

The order Lagomorpha originated in Asia likely around the Paleocene–Eocene boundary (Ge et al. 2013). Eocene stem lagomorphs, representing several distinct lineages with a mosaic of ochotonid and leporid features, show that the two living families split most probably during the early Eocene from different roots and at different times (Averianov and Lopatin 2005; Rose et al. 2008).

According to Ge et al. (2013), apart from the extant *Ochotona*, the family Ochotonidae includes 31 extinct genera with around 150 species, but opinions vary among scholars. *Sinolagomus* from the Oligocene of China and Mongolia appears as the earliest member of the family (Erbajeva et al. 2011; Erbajeva 2016). The attribution of the even earlier *Desmatolagus* to the family is strongly debated (e.g., Fostowicz-Frelik and Meng 2013; Ge et al. 2013). From mid-Oligocene onward, the ochotonid diversity increased, whereas the family showed great diversity and geographical dispersal during the Miocene (Angelone 2008). However, in the Miocene–Pliocene transition, a significant diversity reduction and geographical restriction that continued in the Pleistocene took place, with only 3–4 genera surviving, leading to the current monogeneric status of the family (Ge et al. 2013). It seems that their peak took place in a period during which the temperatures were high and the distribution of the C3 plants (forests) increased, whereas when the climatic conditions deteriorated near the end of the Miocene and the C4 grasses started expanding, the ochotonid diversity dropped dramatically (Ge et al. 2013). During their evolution, they showed specific evolutionary trends in their dental morphology: a shift toward hypsodonty and disappearance of roots, increase in the complexity of the P3/p3 enamel structure, increase in the depth of the P4–M2 hypostria and widening of the p4–m2 talonids (Erbajeva 1994).

As for ochotonids the number of extinct genera of Leporidae is disputable (about 34 genera and 130 species in Ge et al. 2013). The earliest members of the family (i.e., Archaeolaginae) are known from the early Oligocene, but leporid diversity increased from the middle Miocene onward and accelerated at the end of Miocene to Pliocene (White 1991; Dawson 2008; Lopez-Martinez 2008). During the Pliocene/Pleistocene did they thrive and expand geographically, to decrease again in genera number during the Holocene (possibly due to extreme climate conditions and human impact). It seems that their diversity followed a pattern opposite of that of the ochotonids, since the genera numbers increased and spread geographically mainly in the period of cool and dry climatic conditions and predominance of C4 plants in open grasslands, near the Miocene/Pliocene transition (Ge et al. 2013).

The monophyly of Lagomorpha is undoubted and is also supported by molecular data (Graur et al. 1996). Based on morphological characters (including the presence of ever-growing incisors with enamel only on the anterior side), lagomorphs and rodents have been traditionally grouped together in the clade Glires as sister groups. Although this relationship was repeatedly questioned (e.g., Graur et al. 1996; Li et al. 1987; Meng and Wyss 2005), the monophyly of Glires is today generally accepted (Fostowicz-Frelik 2017 and references therein).

In the fossil record, lagomorphs are mainly represented by dental elements. Consequently, the dental characters in both extinct and extant genera have been mainly used for the creation of cladograms depicting the relationships within the lagomorphs (e.g., McKenna 1982). Based on these cladograms, Leporidae and Ochotonidae are monophyletic clades (McKenna 1982).

4 Distribution

Lagomorph fossil remains have been reported from 36 localities in Greece, dated from the early Miocene to the Late Pleistocene (Fig. 1, Appendix): 15 are correlated with the Miocene, 6 with the Miocene/Pliocene boundary, 6 with the Pliocene, and 9 with the Pleistocene. The majority (23) is located in Northern Greece (Macedonia and Thrace), four in Attica, one in Central Greece, and eight in insular Greece (Evia, Chios, Rhodes, Karpathos, and Crete). The most diverse lagomorph assemblages are those of the Pliocene locality Megalo Emvolon 1 and of the latest Pleistocene locality Loutra Almopias Cave LAC Ia, with three lagomorph species in each (Appendix).

5 Systematic Paleontology

Lagomorpha Brandt, 1855

Family Ochotonidae Thomas, 1897

Type Genus *Ochotona* Link, 1795.

***Prolagus* Pomel, 1853**

Type Species *Lagomys sansaniensis* Lartet, 1851.

Other Taxa Included At least 26 extinct species of *Prolagus* are considered as valid today: *P. oeningensis* (König, 1825); *P. sardus* (Wagner, 1829); *P. calpensis* Major, 1905; *P. vasconiensis* Viret, 1930; *P. bilobus* Heller, 1936; *P. savagei* Berzi, 1967; *P. crusafonti* López Martínez in López Martínez and Thaler, 1975; *P. michauxi* López Martínez in López Martínez and Thaler, 1975; *P. ibericus* López Martínez in López Martínez and Thaler, 1975; *P. figaro* López Martínez in López Martínez and Thaler, 1975; *P. depereti* (López Martínez in López Martínez and Thaler, 1975); *P. schnaitheimensis* Tobien, 1975; *P. tobieni* López Martínez, 1977; *P. major* López Martínez, 1977; *P. praevasconiensis* Ringeade, 1978; *P. apriceni-*

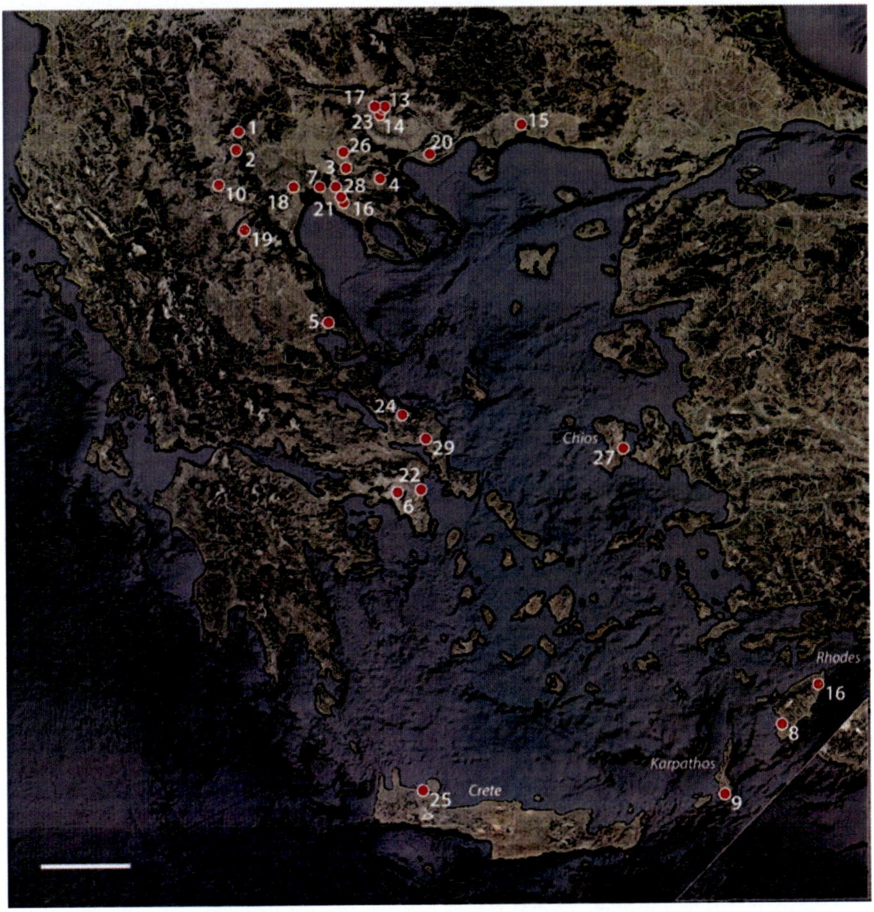

Fig. 1 Map of Greece showing the geographic distribution of the most important localities with lagomorph fossils. **1**, Loutra Almopias Cave (localities LAC and LAC Ia); **2**, Arnissa; **3**, Apollonia 1; **4**, Ravin Voulgarakis; **5**, Alikes; **6**, Tourkobounia localities 2, 3, 5; **7**, Megalo Emvolon 1; **8**, Apolakkia-2; **9**, Agios Ioannis; **10**, Ptolemais 1; **11**, Spilia localities 0, 1; **12**, Nea Silata; **13**, Maramena; **14**, Ano Metochi localities 2, 3; **15**, Kessani localities 1, 2; **16**, Maritsa 1; **17**, Monasteri; **18**, Paliambela-A; **19**, Lava 2; **20**, Rema Marmara; **21**, Elaiochoria; **22**, Chomateri; **23**, Lefkon; **24**, Katheni; **25**, Vrysses; **26**, Chryssavgi 1; **27**, Thymiana localities A, C; **28**, Antonios; **29**, Aliveri. See Appendix for more information. Image exported from Google Earth Pro © 2019, map data from US Dept. of State Geographer, SIO, NOAA, US Navy, NGA, GEBCO, image from Landsat/ Copernicus. Scale bar equals 80 km, North faces upward

cus Mazza, 1987; *P. imperialis* Mazza, 1987; *P. sorbinii* Masini, 1989; *P. aguilari* López Martínez, 1997; *P. caucasicus* Averianov and Tesakov, 1998; *P. italicus* Angelone, 2008; *P. lopezmartinezae* Hordijk and van der Meulen, 2010; *P. osmolskae* Fostowicz-Frelik, 2010; *P. vargasensis* Hordijk and van der Meulen, 2010; *P. pannonicus* Angelone and Čermák, 2014; *P. latiuncinatus* Angelone and Čermák, 2014.

Distribution Early Miocene to Holocene in Europe, Turkey, and NW Africa.

Remarks One P2 of *Prolagus* sp. from Antonios has been described by Vasileiadou and Koufos (2005): slightly deeper paraflexus than mesoflexus, both filled with cement; large mesial hypercone; small postcone; anterior side covered by cement. It shows similarities with *P. oeningensis* from Paşalar described by de Bruijn et al. (2003), but the material is too scanty to allow specific identification. The presence of *Prolagus* sp. is also mentioned from several Upper Miocene to Pliocene Greek localities: Katheni (Evia Island; Katsikatsos et al. 1981), Lefkon and Monasteri in the Strimon Basin (de Bruijn 1989), and Paliambela-A (with one broken p3, two broken P2, one P4, and one broken P3; Sylvestrou 2002; Sylvestrou and Kostopoulos 2007). Van der Made (1996) mentioned the presence of one left P3 of *Prolagus* sp. at Vrysses (Crete) (collected by de Bruijn).

The scanty material (one d3, two p3, two I1, two P2, one P3, and one P4) from the middle Miocene locality Chryssavgi 1 does not allow specific determination. However, these dental elements are distinguished from all known species of *Prolagus*, and, for that reason, Koliadimou (1996) proposed that they belong to a new species of *Prolagus*, but she refrained from naming it due to the limited material. The distinctive features she gave for this species mentioned: small to medium size; the width of the p3 is small in relation to its length, the anteroconid is large and rhomboid, with one main buccal fold and numerous small ones; the precone of the P2 is elongated and covers the paraflexus; slightly folded lagiloph on the P3.

Angelone (2007:417–418) mentioned that some of the late Miocene to Pliocene *Prolagus* assemblages from Greece (i.e., from the localities Kessani, Karpathos, Pikermi, and Maramena; see following paragraphs for details) seem to exhibit unusual features compared to the Western European *Prolagus* species and their revision might show a late Miocene to Pliocene speciation event in the area of the Aegean. Until such a revision takes place, we here refer each of these assemblages to the species originally attributed to.

***Prolagus crusafonti* López Martínez in López Martínez and Thaler, 1975**

Type Material IPS.L-1 (holotype), left mandible with p3–m2, Miquel Crusafont Catalan Institute of Paleontology (López Martínez and Thaler 1975:Pl. 1.5a–d)

Type Locality Can Ponsic, Spain, Upper Miocene [Vallesian (MN9)].

Distribution Late Miocene of SW Europe, Greece, and SW Russia.

Remarks According to López Martínez and Thaler (1975:854), *P. crusafonti* is smaller than *P. oeningensis*; the P2 is small with three cusps and variation on the hypercone; the upper molars show enamel fossettes; the lower teeth are wider than in *P. oeningensis*; the crochet is absent or very small on half of the population; the anteroconid is equal in surface to the metaconid and much more developed than in *P. oeningensis*. López Martínez (1976) described three fragmented mandibles, one d4, 23 p3, six P2, one P3, one P4, and several upper and lower cheek teeth from Pikermi as *P.* cf. *crusafonti*. The morphology of the p3 shows a wide variation: there are specimens with tiny crochet ("advanced" feature) and others with a large one ("primitive" feature), specimens with small and rounded anteroconid ("primitive" feature) and others with a large one ("advanced" feature), specimens with a short and straight protoconulid ("primitive" feature) and others with a curved one ("advanced" feature) (López Martínez 1976:237). This combination of advanced and primitive features is—according to the author—characteristic for the late Miocene *P. crusafonti*, but the P2 shows a weakly developed mesial hyperloph, less however, than in *P. michauxi*, and the enamel of the precone is always present (López Martínez 1976:237, 239). Tesakov and Averianov (2002:82) preferred assigning the material from Pikermi to *P. sorbinii*, due to the presence of large enamel fossettes and a short hypostria on the P4.

***Prolagus michauxi* López Martínez in López Martínez and Thaler, 1975**

Type Material CB.UMSL-1001 (holotype), skull, University of Montpellier (López Martínez and Thaler 1975:Pl. 2.1–2).

Type Locality Sète, France, Lower Pliocene [late Ruscinian (MN15)].

Distribution Late Miocene to Pliocene (MN12–MN16) of S Europe, W Turkey, Algeria, and Morocco.

Remarks A species larger than *P. oeningensis* but smaller than *P. sardus*. The diagnosis includes the following dental features (Fig. 2): P2 evolved like in *P. sardus*, but the maxilla bears only one alveolus, the upper molars have primitive characters and often bear enamel islets like in *P. oeningensis*; short muzzle, very large premolar foramen, wide and prominent zygomatic arches; large p3, common presence of a crochet and a robust protoconulid (López Martínez and Thaler 1975). The material from the Turolian/Ruscinian fauna of Kessani (Thrace), consisting of two d3, seven p3, one p4/m1, two P2, 11 P3, six P4, and five M1/2, has been described as *P. michauxi* by Konidaris (2005) and Vasileiadou et al. (2012); all p3 have a crochet, whereas on the P2 there is a long mesial hyperloph that covers the paraflexus, centrocone, and mesoflexus. Syrides et al. (1997) had previously referred the material to *P.* aff. *michauxi* in the faunal list given for Kessani. A very rich material of *P. michauxi* (eight d3, three d4, 56 p3, 24 p4/m1, nine m2, one partial mandible with p4 and m2, 44 P2, 56 P3, 57 P4, 51 M1/2) has been found in Nea Silata (Vasileiadou 2001); the anteroconid of the p3 is large and elongated, oblique to the longitudinal axis of the premolar, the protoconulid is narrower and longer than the protoconid, the metaconid is rounded and large, the endoconid and hypoconid are elongated and large, the crochet is absent (32.6%), small (55.8%), or large (11.6%) (Vasileiadou et al. 2003). The species is also present in Ptolemais 1 with two p3 (one damaged), one P2, one P3, and two P4 (van de Weerd 1979). One P3 and one P2 from the Lower Pliocene locality Agios Ioannis (Karpathos Island) have been attributed to *P.* cf. *michauxi* (Daams and van der Weerd 1980).

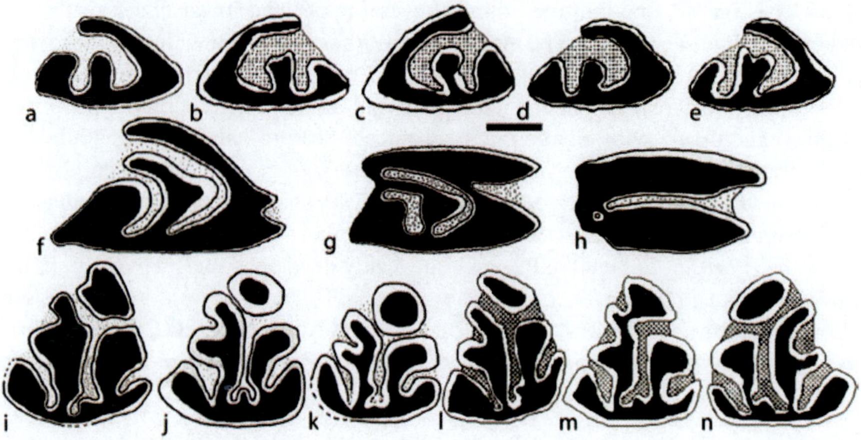

Fig. 2 *Prolagus michauxi* from Kessani (a, f, g, h, i, j, k; Vasileiadou et al. 2012) and Nea Silata (b, c, d, e, l, m, n; Vasileiadou 2001). **a** right P2, **b** left P2, **c** left P2, **d** right P2, **e** right P2, **f** right P3, **g** right P4, **h** right M1/2, **i** left p3, **j** left p3, **k** left p3, **l** left p3, **m** left p3, **n** right p3. Scale bar equals 1 mm

Prolagus sorbinii Masini, 1989

Type Material V.7026 (holotype), articulated skeleton with skull and mandible with complete dentition of one individual, Museum of Natural History of Verona (Masini 1989:Pl. 1–2, Pl. 3, Figs. 1, 2, and 4).

Type Locality Monte Castellaro, Italy, Upper Miocene [Messinian].

Distribution Late Miocene to Pliocene (MN12–MN15) of South Europe.

Remarks According to the diagnosis given by Masini (1989), the species is slightly larger than *P. michauxi*, the muzzle is elongated, the incisive foramen tends to split in anterior and posterior parts, the premolar foramen is large, the P2 is evolved with a well-developed anteroflexus, the p3 has either no or very small crochet and no protoisthmus, and the upper molars have no or reduced enamel islet and a not much extended labially hypoflexus. De Bruijn (1995) assigned the ochotonid material from Maramena (six d3, three d4, 17 p3, two D2, two D3/4, 11 P2, 14 P3, nine P4) to *P.* cf. *sorbinii*, stating, however, that the length of the snout that is the main characteristic that distinguishes *P. sorbinii* from *P. michauxi* cannot be observed in the available material. He based his identification on the average greater similarity of the P2–4 from Maramena with the premolars of *P. sorbinii* than with those of *P. michauxi*. *P.* cf. *sorbinii* is also present in three other localities in the Strimon Basin: Rema Marmara, Ano Metochi 3, and Spilia 0 (de Bruijn 1995:135; initially reported as *P. michauxi* in de Bruijn 1989). Masini (1989:305) stated that "the major distinctive features of *P. sorbinii* lie in its skull morphology," but later on Angelone (2007) provided several distinctive dental features differentiating the three taxa,

P. crusafonti, *P michauxi*, and *P. sorbinii*. Hence, several records of Greek *Prolagus* may need to be revised.

Albertona López Martínez, 1986

Type Species *Albertona balkanica* López Martínez, 1986.

Other Taxa Included *A. aegeensis* Ünay and Göktaş 1999.

Distribution Early Miocene (Orleanian) of Turkey, Greece, and Serbia.

★*Albertona balkanica* López Martínez, 1986

Type Material Left lower jaw with p3–4 and m2, no. 1 (holotype), Department of Earth Sciences of the University of Utrecht, the Netherlands (López Martínez 1986:Pl. 1.1a–b).

Type Locality Aliveri, Greece, Lower Miocene [MN4; van den Hoek Ostende et al. 2015].

Distribution The species is also known from the early Miocene of Serbia.

Remarks López Martínez (1986) created the genus and the species based on a very rich material from Aliveri consisting of five lower incisors, 10 d3, seven d4, 45 p3, and four upper incisors, 24 P2, 60 P3, 24 P4, 63 M1, and 33 M2. The dental diagnostic features of both the genus and the species are as follows: dental formula is 2,0,3,2/1,0,2,3; the teeth are rootless, and all premolars show a very thick cement layer; the P4 is molariform; the P2 shows two flexa, and its posterior and lingual sides form a right angle; the P3 does not show a mesoflexus but shows an anterolingual constriction in the protoloph; the d3 has a small anteroconid; the p3 bears neither an anteroconid nor a protoflexid, it bears a centro- and a mesoflexid and sometimes a small anterostylid fused to the protoconid; the lower molariform teeth show a protoflexid (López Martínez 1986). Based on a cladogram of the Old World Oligocene to Miocene ochotonids, an *Albertona-Alloptox* sister group is proposed by López Martínez (1986:191) that may imply an ancestor-descendant relationship, which, however, is challenged by later studies (e.g., Marković 2010; Angelone and Hír 2012).

Alloptox Dawson, 1961

Type Species *Ochotona gobiensis* Young, 1932.

Other Taxa Included There are at least five valid species of *Alloptox: A. anatoliensis* Ünay and Şen, 1976; *A. minor* Li, 1978; A. *guangheensis* Cao et al., 1990; *A. xichuanensis* Liu and Zheng, 1997; *A. katinkae* Angelone and Hír, 2012. According to Liu and Zheng (1997) *A. chinghaiensis* Qiu et al., 1981, is synonym of *A. anatoliensis* and *A. sihongensis* Wu, 1995, of *A. minor*.

Distribution Early to middle Miocene of Eurasia and North Africa.

Remarks *Alloptox* sp. has been reported by de Bonis et al. (1997) from the middle Miocene localities Thymiana A and C (Chios Island).

***Alloptox anatoliensis* Ünay and Şen, 1976**

Type Material AKÇ-1 (holotype), partial left mandible with p3–m2 (Ünay and Şen 1976:Fig. 1a).

Type Locality Çandir, Turkey, middle Miocene [MN5].

Distribution Early to middle Miocene of Greece and Turkey.

Remarks According to the diagnosis, the p3 is triangular, with massive trigonid, rounded lingual surface and angular but slightly concave buccal one, with deep and narrow anterobuccal groove, with only one and deep lingual groove (paraflexid in Şen 1990) that reaches the middle of the talonid posterolingually, with a clearly concave lingual surface of the talonid (entoconid in Şen 1990), with a well-developed anteriorly and posteriorly lingual part of the talonid (entoconid in Şen 1990) and a protruding buccal part of the talonid (hypoconid in Şen 1990) (Ünay and Şen 1976). In the available material from Çandir, the metaflexid on the p3 is a very shallow fold without cement (Ünay and Şen 1976:Figs. 1a and 2a), whereas the much larger *A. anatoliensis* sample from Çandir 2 includes some p3 with a well-developed metaflexid (de Bruijn et al. 2003). *A. anatoliensis* from Paşalar shows a large variety: ten p3 exhibit a well-developed metaflexid with cement, ten p3 a shallow fold without cement, and 34 p3 show no trace of a metaflexid (Şen 1990). As Vasileiadou and Koufos (2005) pointed out, the presence of only one lingual groove (paraflexid), thus the absence of a metaflexid, cannot be diagnostic for *A. anatoliensis*. De Bruijn et al. (2003) mentioned that the size of both *A. minor* and *A. chinghaiensis* is within the size variation of *A. anatoliensis*; thus the validity of the two species is doubtful (Liu and Zheng (1997) had already established the synonymy of *A. chinghaiensis* with *A. anatoliensis*). Vasileiadou and Koufos (2005) assigned two p3, one P2, three P3, and 12 partial molariform teeth from Antonios to *A. anatoliensis*; there is a shallow metaflexid on one of the two p3. Angelone and Hír (2012:3) referred to the Antonios species as "*A. anatoliensis*" due to the peculiar dental characters. The middle Miocene locality Chryssavgi 1 has yielded one P3 and three molars, assigned to *Alloptox* sp. by Koliadimou (1996); as the author mentioned, the material cannot be attributed to a species due to the absence of diagnostic dental elements (p3). However, Koufos (2006) added *A.* aff. *anatoliensis* in the faunal list for Chryssavgi 1; apparently, he was led to this assignment due to the similarity of the specimens with specimens of *A. anatoliensis* from Paşalar (as reported by Koliadimou 1996).

***Ochotona* Link, 1795**

Type Species *Ochotona minor* Link, 1795 (extant) (= *Lepus dauuricus* Pallas, 1776).

Other Taxa Included The genus includes 30 extant species living in Asia and North America (MacDonald 2006), some of which can be found in late Pleistocene-Holocene fossil and sub-fossil assemblages (e.g., *O. pusilla* Pallas, 1768). It also includes at least 41 extinct taxa: *O. eximia* Khomenko, 1914; *O. lagreli* Schlosser, 1924; *O. ursui* Simionescu, 1930; *O. antiqua* Argyropulo and Pidoplichko, 1939; *O. minor* Bohlin, 1942; *O. spanglei* Shotwell, 1956; *O. tologoica* Habaeva, 1958; *O. pseudopusilla* Gureev and Schevtschenko, 1964; *O. gureevi* (Erbajeva, 1966); *O. dodogolica* Erbajeva, 1966; *O. transcaucasica* (Vekua, 1967); *O. azerica* Gadzhiev and Aliev, 1969; *O. whartoni* Guthrie and Matthews, 1971; *O. gromovi* Erbajeva, 1976; *O. intermedia* Erbajeva, 1976; *O. guizhongensis* Ji et al., 1980; *O. polonica* Sych, 1980; *O. kalfense* (Lungu, 1981); *O. sibirica* (Erbajeva, 1983); *O. zazhigini* Erbajeva, 1983; *O. bazarovi* Erbajeva, 1983; *O. zasuchini* Erbajeva, 1983; *O. nihewanica* Qiu, 1985; *O. filippovi* Erbajeva, 1999; *O. valerotae* Erbajeva et al., 2001; *O. agadjianiani* Melik-Adamian, 2003; *O. ozansoyi* Şen, 2003; *O. dehmi* Erbajeva, 2005; *O. horaceki* Čermák, 2005; *O. plicodenta* Erbajeva and Zheng, 2005; *O. magna* Erbajeva and Zheng, 2005; *O. youngi* Erbajeva and Zheng, 2005; *O. lingtaica* Erbajeva and Zheng, 2005; *O. gracilis* Erbajeva and Zheng, 2005; *O. zhangi* Erbajeva and Zheng, 2005; *O. tedfordi* Erbajeva et al., 2006; *O. chowmincheni* Erbajeva et al., 2006; *O. gudrunae* Erbajeva et al., 2006; *O. zabiensis* Fostovicz-Frelik, 2008; *O. kormosi* Fostovicz-Frelik et al., 2011; *O. mediterranensis* Suata-Alpaslan, 2012.

Distribution The extant species live in Eastern Europe, Asia, and North America. The genus occurs from the Miocene and in Central and Western Europe is recorded till the Late Pleistocene.

Remarks A few isolated dental elements (one p3, one M2, two upper molariform teeth, and four upper incisors) from Maritsa 1 (Rhodes Island) of a rather small species, with one anterolingual, one anterobuccal, and one posterobuccal fold on the p3, a well-developed process protruding posteriorly from the talon on the M2, a long hypostria on the upper molariform teeth, and a deep groove medially of the center of the anterior upper incisor, have been assigned to *Ochotona* sp. (de Bruijn et al. 1970). The remains from the Upper Pleistocene locality Arnissa (a partial left mandible with the m3, a partial right maxilla with the alveoli of the P4–M2, one P3, one M1, and one M2) are similar in size and morphology to the extant steppe pika *O. pusilla* currently restricted to the steppes of S Russia and N Kazakhstan but common in Europe during the cool phases of the Late Pleistocene (Mayhew 1978, Erbajeva 1988). The same species has been described from Loutra Almopias Cave LAC I and LAC Ia (one lower incisor, four d3, five d4, two p3, four p4/m1/2, 12 upper incisors, one D2, one D3, three D4, three P3, three P4/M1, five M2; Fig. 3); the material shows morphological features of the steppe pika: p3 with small anteroconid, wide posteroconid, shallow para- and anteroflexid (Chatzopoulou 2014).

Ochotonoma Şen, 1998

Type Species *Ochotonoma anatolica* Şen, 1998.

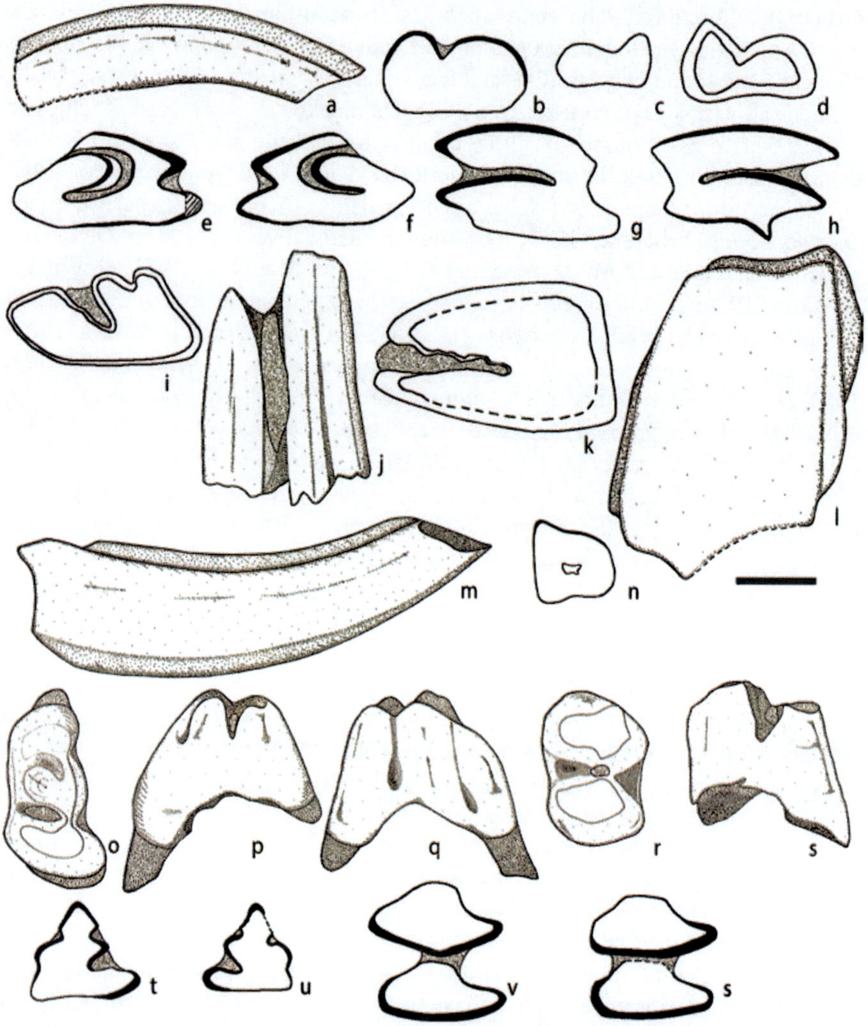

Fig. 3 *Ochotona pusilla* from Loutra Almopias Cave LAC. **a** left I1 in proximal view, **b** same in occlusal view, **c–d** right I1 in **c** occlusal and **d** root view, **e** right P3, **f** left P3, **g** left P4/M1, **h** right M2, **i–j** right D2 in **i** root and **j** anterior view, **k–l** right D4 in **k** root and **l** anterior view, **m–n** left i1 in **m** buccal and **n** occlusal view, **o–q** left d3 in **o** occlusal, **p** buccal, and **q** lingual view, **r–s** left d4 in **r** occlusal and **s** buccal view, **t** right p3, **u** left p3, **v** right p4/m1/2, **w** right p4/m1/2. Drawings courtesy of K. Chatzopoulou. Scale bar equals 1 mm

Other Taxa Included There are at least three valid species of *Ochotonoma* and some invalid ones: *O. csarnotana* (Kretzoi, 1959), *O. primitiva* (Zheng and Li, 1982), and *O. ortalicensis* (Ünay and de Bruijn, 1998). *O. anatolica* and *O. ortalicensis* are regarded as junior synonyms of *O. csarnotana* by Čermák (2007).

Distribution Late Turolian to early Villányian of Eastern Europe and Asia.

Remarks A few dental elements (one complete and one partial p3, one P2, two P3, one P4, one M1) of a small pika from Apolakkia-2 (Rhodes Island) were initially assigned to *Ochotonoides* sp., due to the presence of three labial and one lingual folds on the p3 (van de Weerd et al. 1982). Later, Şen (1998) noticed that the Apolakkia-2 species is markedly smaller than any of the *Ochotonoides* species, its p3 has a less folded anteroconid and smooth enamel along the proto- and paraflexid, and its P2 is shorter than those of other *Ochotonoides* species. Comparisons with the *Ochotonoma* material from Çalta revealed greater similarities with this genus (Şen 1998:375). Both Şen (1998) and later Čermák (2007, 2010) suggested that the material from Rhodes could represent a new species of *Ochotonoma*, due to its unique paraflexid morphology and the orientation of the strongly developed lingual part of the anteroconid; however, no new species can be named due to the limited material available.

Family Leporidae Fischer de Waldheim, 1817

Type Genus *Lepus* Linnaeus, 1758.

Remarks Unidentifiable leporid dental elements have been found in several Greek sites: three molariform teeth in Nea Silata (Vasileiadou et al. 2003), one d4 and one upper molariform tooth in Kessani (Vasileiadou et al. 2012), one lower and two upper molariform teeth and one P2 in Ravin Voulgarakis, and one partial left mandible with p4–m3 in Apollonia 1 (Koliadimou 1996), whereas Leporidae indet. is in the faunal list for Tourkobounia 3 and 4 without further details on the available material (Reumer and Doukas 1985).

Alilepus Dice, 1931

Type Species *Lepus annectens* Schlosser, 1924.

Nomenclatural and Taxonomical history Dice (1929) established *Allolagus* designating *Lepus annectens* Schlosser, 1924, as the type species. However, the same generic name had already been used by Ognev (1929) with *Lepus mandshuricus* Radde, 1861, as type species. For that reason, the new replacement name *Alilepus* was erected by Dice (1931:159) (see Čermák et al. 2015).

Other Taxa Included Eurasian *Alilepus* includes at least ten extinct valid species: *A. laskarewi* (Khomenko, 1914); *A. hungaricus* Kormos, 1934; *A. ucrainicus* Gureev, 1964; *A. turolensis* López Martínez, 1977; *A. longisinuosus* Qiu and Han, 1986; *A. zhoukoudianensis* Cheng et al., 1995; *A. lii* Jin, 2004; *A. elongatus* Winkler et al., 2011; *A. meini* Angelone and Rook, 2011; *A. parvus* Wu and Flynn, 2017.

Distribution Late Miocene to late Pliocene of North America; late Miocene to Early Pleistocene of Eurasia; late Miocene to Pliocene of Africa.

Remarks One fragmentary p3 and one d3 of *Alilepus* sp. are known from Pikermi (López Martínez 1976): the isthmus between trigonid and talonid on the p3 is narrow, typical for *Alilepus*; the Pikermi p3 is similar in size and morphology with the one of *A. lascarewi* from Taraclia (Romania) and similar morphologically with that of *A. annectens* from Ertemte (Inner Mongolia) but smaller.

***Alilepus turolensis* López Martínez, 1977**

Type Material ARQ–64 (holotype), left p3, University of Montpellier (López Martínez 1977:fig. 4.3).

Type Locality El Arquillo (Rambla de Valdecebro 2), Spain, Upper Miocene [late Turolian (MN13)].

Distribution Late Miocene of S Europe (Spain, Italy, Greece) and Turkey.

Remarks The diagnostic features are as follows: large size, P2 similar to that of *A. annectens* but with a weak metaflexus, hypoflexus is deeper than that of *A. annectens*, in advanced stages of wear the posterior edge loses its crenulation; the p3 is larger than that of *A. annectens* and *A. laskarewi*, with proto- and hypoflexid equally short, but much deeper mesoflexid, so that the isthmus connecting the trigonid and the talonid is much narrower and displaced buccally than in *A. annectens*; unlike in *Trischizolagus*, there are no lingual or anterior synclines (López Martínez 1977:20–21). *A. turolensis* is present with a quite rich material (six d3, five d4, one p3, one p4/m2, two m3, one D2, three D3, one D4, three P2, six P3–M2; Fig. 4) at Maramena (Strimon Basin). The material is of interest because it is the first record of *A. turolensis* in the eastern Mediterranean area, whereas it is much richer than the type material from El Arquillo (de Bruijn 1995). The species is also reported in the faunal list of Spilia 0 and 1 (de Bruijn 1989) and likely Elaiochoria 2 (Hulva et al. 2007) but without further details.

***Pliopentalagus* Gureev and Konkova in Gureev, 1964**

Type Species *Pliopentalagus moldaviensis* Gureev and Konkova in Gureev, 1964.

Other Taxa Included *P. dietrichi* (Fejfar, 1961); *P. progressivus* Liu and Zheng, 1997; *P. huainanensis* Jin, 2004; *P. dajushanensis* Tomida and Jin, 2009; *P. anhuiensis* Tomida and Jin, 2009. The North American species *P.? agilis* (Russel and Harris, 1986) was originally attributed to *Aztlanolagus* Russel and Harris, 1986, but Tomida and Jin (2009) transferred it to *Pliopentalagus*. Its reference to the latter genus is treated with caution by some authors (e.g., Čermák and Wagner 2013).

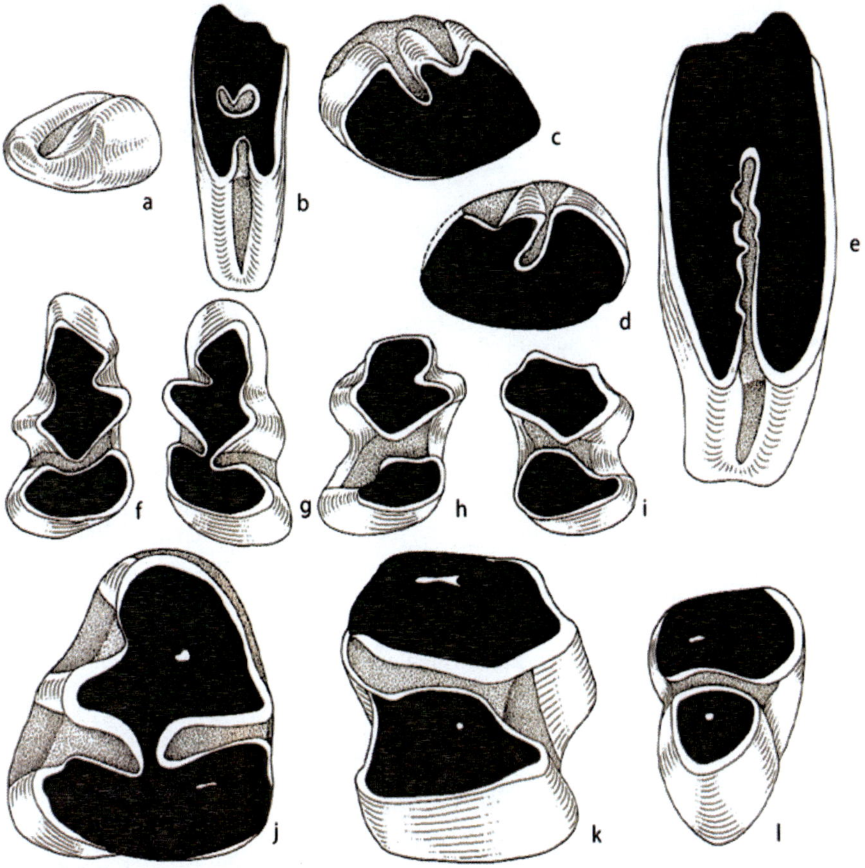

Fig. 4 *Alilepus turolensis* from Maramena. **a** right D2, **b** left D3, **c** left P2, **d** right P2, **e** P4/M2, **f** left d3, **g** right d3, **h** left d4, **i** right d4, **j** left p3, **k** right p4/m2, **l** left m3. Drawings courtesy of H. de Bruijn

Distribution Latest Miocene to late Pliocene/Early Pleistocene of China; early Pliocene of C and SE Europe. Late Pliocene to Late Pleistocene of North America.

Remarks *Pliopentalagus* sp. is included in the fauna list for the locality Lava 2 without any additional information on the material (de Bruijn et al. 1999).

***Trischizolagus* Radulesco and Samson, 1967**

Type Species *Trischizolagus dumitrescuae* Radulesco and Samson, 1967.

Other Taxa Included *T. crusafonti* (Janvier and Montenat, 1971); *T. maritsae* de Brujin, 1970; *T. gambariani* (Melik-Adamyan, 1986); *T. nihewanensis* (Cai, 1989); *T.? raynali* (Geraads, 1994); *T. mirificus* Qiu and Storch, 2000.

Distribution Late Miocene (late Turolian MN13) to Early Pleistocene of Eurasia and possibly in Morocco.

***Trischizolagus dumitrescuae* Radulesco and Samson, 1967**

Type Material IS 6001 (holotype), partial right mandible with p3–m1.

Type Locality Mălusteni, Iaşy Region, Romania, Lower Pliocene.

Distribution Late Miocene to early Pliocene of Eurasia.

Remarks *T. dumitrescuae* is large, with a mesoflexid that transforms into an enamel islet with wear, well-developed hypoflexid and a tetralobed and elongated trigonid on the p3, and one- to threefold on the anterior wall of the P2 (Koufos and Koliadimou 1993; Averianov 1995). The material from the Pliocene locality Megalo Emvolon 1 (= Karaburun), consisting of two partial maxillae with all cheek teeth, one partial left mandible with p4–m2, one partial left mandible with p3–m3, and one left lower incisor (Fig. 5), has clear morphological and metrical similarities with the type material from Mălusteni and is thus attributed to *T. dumitrescuae* (Koufos and Koliadimou 1993).

★***Trischizolagus maritsae* De Bruijn, Dawson and Mein, 1970**

Type Material No. 1301, right jaw with p3–m3 (holotype), Department of Earth Sciences of the University of Utrecht, the Netherlands (de Bruijn et al. 1970:Pl. 10.5).

Type Locality Maritsa 1, Rhodes Island, Greece, Upper Miocene [Turolian (MN13); Hordijk and de Bruijn 2009].

Distribution Latest Miocene (late Turolian) to early Pliocene of S Europe and Afghanistan.

Remarks De Bruijn et al. (1970) created this species with the following diagnostic dental features: P2 with one persistent anterior fold, upper molariform teeth with only slightly folded walls of the hypostria, p3 with antero-external anterior and antero-internal folds on the trigonid and a postero-external fold between the trigonid and the talonid, presence of a postero-internal fold (mesoflexid) on unworn p3 and absence or presence of a transversely short postero-internal fold on worn p3 (Fig. 6). The type locality has yielded, apart from the holotype, six p3, two lower molariform teeth, one upper incisor, two P2, 11 upper molariform teeth, and 13 deciduous teeth. *T. maritsae* suggested as smaller than *T. dumitrescuae* and has a

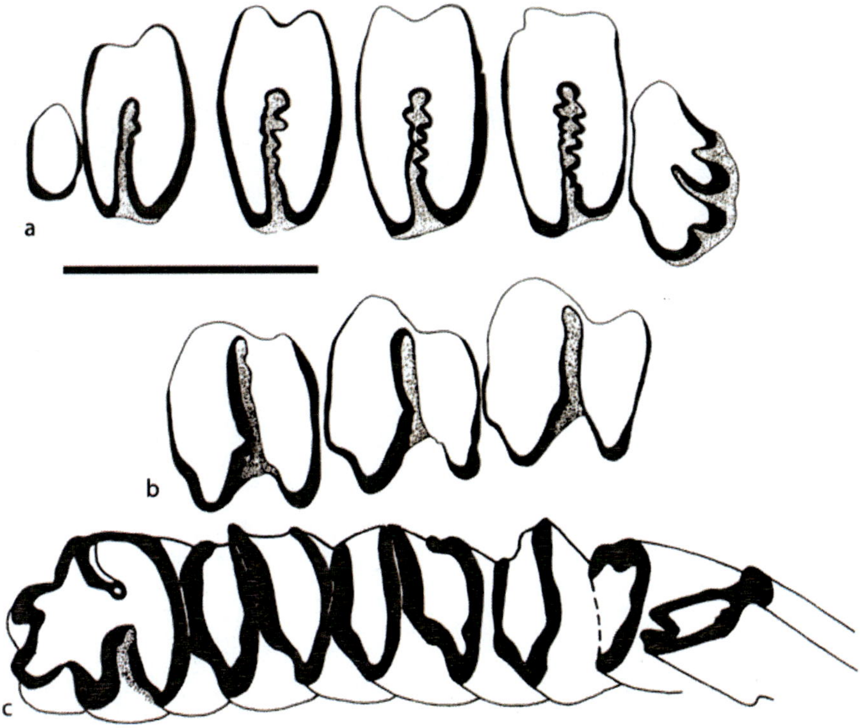

Fig. 5 *Trischizolagus dumitrescuae* from Megalo Emvolon 1. **a** right maxillary fragment with P2–M3, **b** left mandibular fragment with p4–m2, **c** left mandibular fragment with p3–m3. Drawings courtesy of K. Koliadimou. Scale bar equals 5 mm

weaker postero-internal fold on the p3, a feature interpreted as a primitive character for *Trischizolagus* (de Bruijn et al. 1970:574–575). One leporid partial left maxilla with P4–M2 from Megalo Emvolon 1, referred by Koufos and Koliadimou (1993) to as *T.* cf. *maritsae*, along with previous reports of the taxon from the same site (Steffens et al. 1979) needs further investigation.

***Oryctolagus* Lilljeborg, 1873**

Type Species *Lepus cuniculus* Linnaeus, 1758.

Other Taxa Included (extinct) *O. lacosti* (Pomel, 1853); *O. laynensis* López Martínez, 1977; *O. burgi* Nocchi and Sala, 1998; *O. giberti* de Marfà, 2008.

Distribution Pliocene to Present of Europe.

***Oryctolagus laynensis* López Martínez, 1977**

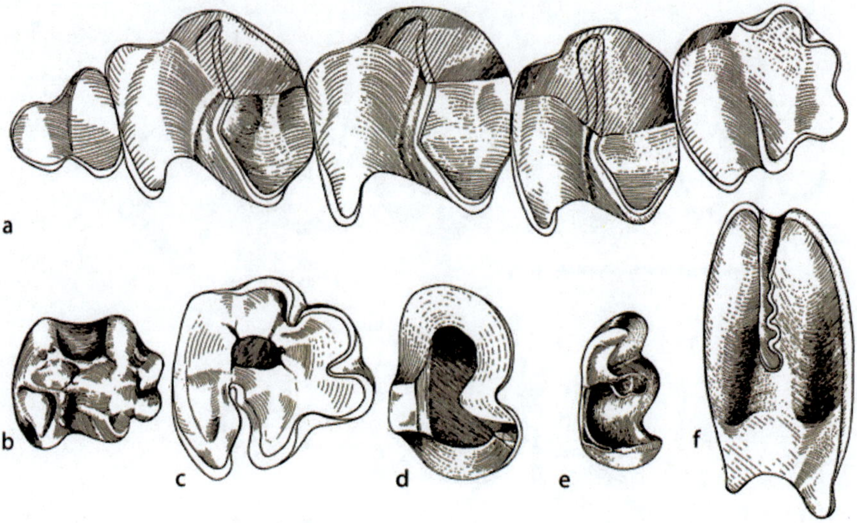

Fig. 6 *Trischizolagus maritsae* from Maritsa 1 (type locality). **a** right mandibular fragment with unworn p3–m3 (holotype), **b** right p3 (unworn), **c** right p3, **d** left I1, **e** right P2 (unworn), **f** left upper molariform tooth. Drawings courtesy of H. de Bruijn

Type Material LAY 3071 (holotype), partial skull with teeth, Lucas Mallada Institute (López Martínez 1977:Pl. 1.1).

Type Locality Layna, Soria, Spain, Lower Pliocene [Ruscinian (MN15)].

Distribution Pliocene to Early Pleistocene of Europe (Spain, France, Greece).

Remarks The dental diagnostic features include the following: size similar to that of the extant subspecies *O. c. cuniculus* and larger than *O. c. alqirus*; the lower dental elements similar to those of *O. cuniculus*; the p3 has a large labial anteroconid, similar in size to the protoconid; the m2 and M2 have size similar to that of the m1 and M1, as in *Lepus*; the m3 and M3 are better developed than in *Oryctolagus* and *Lepus*; the upper teeth have an hypoflexus that extents in less than half the crown width, either smooth or crenulated only in the anterior part; the P2 has a very weak hypoflexus, a lagicone equal to or smaller than the mesial hypercone (López Martínez 1977). *Oryctolagus* cf. *laynensis* is reported from Megalo Emvolon 1 (Steffens et al. 1979), but no details are known.

Oryctolagus lacosti (Pomel, 1853)

Nomenclatural History Pomel did not give description, holotype or figures for *O. lacosti*; de Marfa and Mein (2007) designated a lectotype and gave a diagnosis.

Type Material FSL-211646 (lectotype), palate with both upper dental rows (P2-M2), Univ. Claude-Bernard, Lyon, France (de Marfa and Mein 2007:Fig. 2a).

Type Locality Perrier-Étouaires, Auvergne France, Upper Pliocene [early Villafranchian (MN16)].

Distribution Late Pliocene to Middle Pleistocene of Europe.

Remarks Following the lectotype designation, *O. lacosti* is diagnosed as a large rabbit (size similar to that of modern *Lepus*), the ratio of the palatal bridge length to the choanas width is above 0.75, the mental foramen of the mandible is large and very close to the p2, the p3 shows a deep anteroflexid that penetrates vertically the tooth anterior border (de Marfa and Mein 2007). A rich material of *O. lacosti* cranial and postcranial elements has been found in Alikes (Athanassiou 1998; Fig. 7). The dimensions of the Alikes elements are similar to those of *O. lacosti* from various W European localities, and the morphology of the p3 (the hypoflexid runs almost across the entire tooth, whereas the protoflexid is well developed and the posterior wall of the trigonid is crenulated) is similar with that of the p3 from Saint-Vallier attributed to *O. lacosti* by Viret (1954) (Athanassiou 1998).

Lepus Linnaeus, 1758

Type Species *Lepus timidus* Linnaeus, 1758 (extant mountain hare).

Other Taxa Included The genus *Lepus* includes 32 extant species (MacDonald 2009), some of which have been found in Pleistocene-Holocene fossil and subfossil assemblages (e.g., *L. timidus*, *L. europaeus* Pallas, 1778). Extinct species: *L. valdarnensis* Weithofer, 1889; *L. wongi* Young, 1927; *L. terraerubrae* Kretzoi, 1956; *L. tanaiticus* Gureev, 1964; *L. qinhuangdaoensis* Wang et al., 2010; *L. teilhardi* Zhang, 2010; *L. ziboensis* Zhang, 2010.

Distribution Widespread today in Africa, Eurasia, North America, and the Japanese archipelago. Early *Lepus* has been found in the Early Pleistocene of Eurasia.

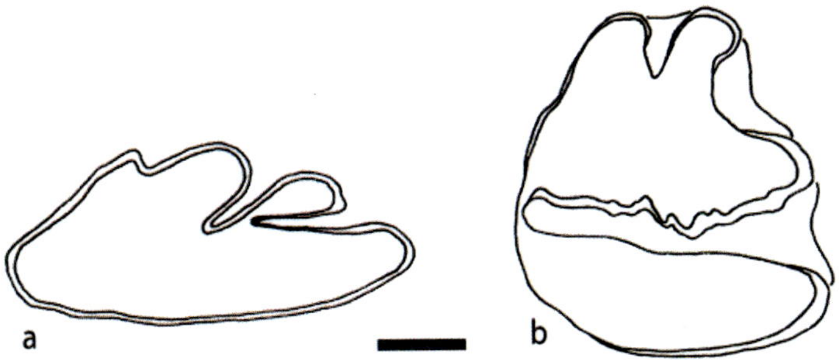

Fig. 7 *Oryctolagus lacosti* from Alikes. **a** right P2, **b** right p3, Drawings courtesy of A. Athanassiou. Scale bar equals 1 mm

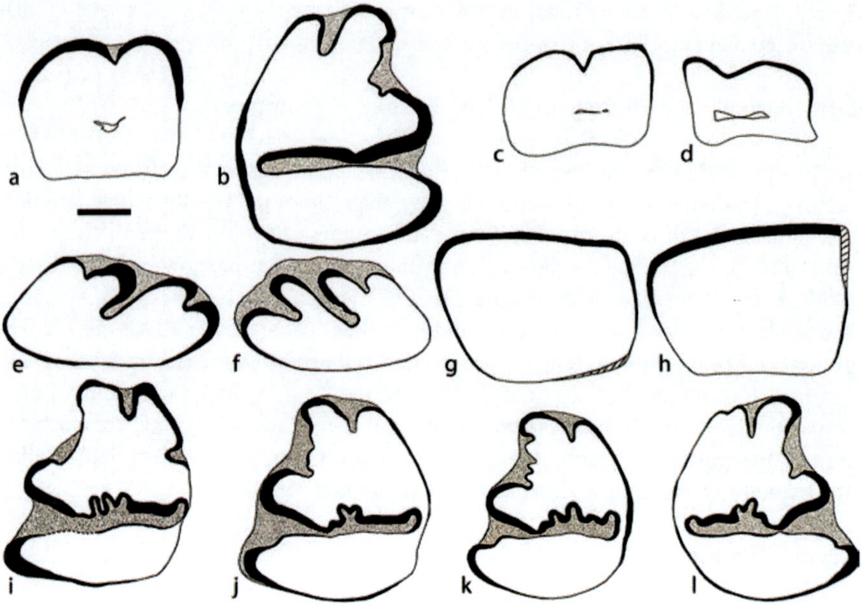

Fig. 8 *Lepus timidus* from Loutra Almopias Cave LAC. **a** right I1, **b** right p3. *Lepus europaeus* from Loutra Almopias Cave LAC. **c** right I1, **d** left I1, **e** right P2, **f** left P2, **g** right i1, **h** left i1, **i** left p3, **j** left p3, **k** left p3, **l** right p3. Drawings courtesy of K. Chatzopoulou. Scale bar equals 1 mm

Remarks Three partial upper cheek teeth from Arnissa have been attributed to *Lepus* sp. by Mayhew (1978). *Lepus* sp. is also present in the faunal list of Tourkobounia 5 (Reumer and Doukas 1985). Chatzopoulou (2014) assigned six lower incisors, six p3, two upper incisors, and two P2 from Loutra Almopias Cave LAC and LAC Ia (Fig. 8) to *Lepus europaeus* (including *L.* cf. *europaeus*). The locality Loutra Almopias Cave LAC Ia also yielded two p3, one upper incisor, and one P2 of *L. timidus* (including *L.* cf. *timidus*) (Chatzopoulou 2014). The rest of the leporid material from Loutra Almopias Cave LAC and LAC Ia (three P2, two partial maxillae, one humerus, and one partial tibia), which shows no diagnostic features, has been assigned to *Lepus* sp. (Chatzopoulou 2014).

Lepus terraerubrae **Kretzoi, 1956**

Type Material See Kretzoi 1956:167.

Type Locality Beremend 9, Hungary, early Middle Pleistocene.

Distribution Middle Pleistocene of Europe.

Remarks Several leporid dental elements (two semi-adult i1, seven p3, three semi-adult I1, 13 P2, and several molariform cheek teeth) from Tourkobounia 2 have been referred to as *L. terraerubrae* (van der Meulen and Doukas 2001), but as the taxonomic status of the species is questionable (Čermák, pers. com), further analysis is needed.

6 Concluding Remarks

The lagomorph record of Greece is far from complete, as remains of the order have been found only in a few micromammalian localities and, most times, the material does not include diagnostic dental elements. However, one new genus, *Albertona*, and two new species, *Albertona balkanica* and *Trischizolagus maritsae*, have been described from Aliveri and Maritsa 1, respectively. Furthermore, the *Ochotona* record in Maritsa 1 (Rhodes Island) represents the most southern European occurrence of the genus during the Pliocene. The distribution of fossil lagomorphs in Greece vaguely follows the general pattern in the evolution of the order: during the early to middle Miocene, it is represented only by ochotonids (*Albertona*, *Alloptox*, *Prolagus*), usually with one or, rarely, two species per locality. Leporids appear in the Greek fossil record during the middle to late Turolian, with *Alilepus*, *Trischizolagus*, and possibly *Pliopentalagus*. However, and in contrast with the European fossil record, Upper Miocene to Pliocene Greek localities reveal either only ochotonids or one ochotonid and one leporid. This situation clearly changes in the Pleistocene: the faunas include only leporids, and only in the Late Pleistocene the ochotonids make one small return with the extant steppe pika *O. pusilla*.

Acknowledgments We are indebted to Dr. Wilma Wessels for access to important literature and to the original plates. We also want to express our gratitude to Dr. Hans de Bruijn, to Emeritus Prof. George Koufos and Dr. Kalliopi Koliadimou, to Dr. Katerina Chatzopoulou, and to Dr. Athanassios Athanassiou for permission to use original figures from Maritsa 1 and Maramena, Megalo Emvolon, Loutra Almopias Cave, and Alikes, respectively. All the hard and tedious work of scanning original plates by Frans Verdaasdonk and Dr. Wilma Wessels is very much appreciated. Thanks are also due to Dr. Chiara Angelone and Dr. Stanislav Čermák for their fruitful comments and suggestions.

Editorial Note The works in this book (lagomorphs, rodents, insectivores) are the last of Katerina, who left us recently after a long battle with cancer. Katerina deeply believed in this collective project and worked with passion and dedication for its implementation, giving even her very last strength. Unfortunately, she could not deal with the constructive comments and suggestions of the reviewers, a task undertaken by the co-authors of the articles and her friends in the same scientific field.

Appendix

Occurrences of hares, rabbits, and pikas in Greece. Locality numbers refer to the collection numbers of the PaleoBiology Database (PBDB). Type localities are marked with boldface

Locality[PBDB No]	Age (MNQ/MN)	Identification
Loutra Almopias Cave LAC Ia[204954]	Younger Dryas (13,000–14,500 BP)	*Ochotona pusilla*[1], *Lepus europaeus*[1], *Lepus timidus*[1]
Arnissa[204957]	Late Pleistocene [penultimate interglacial (Eemian) age]	*Ochotona pusilla*[2], *Lepus* sp.[2]
Loutra Almopias Cave LAC[203847]	Middle Würm (42,500 BP)	*Ochotona pusilla*[1], *Lepus europaeus*[1]
Apollonia 1[34784]	Pleistocene (MNQ20)	Leporidae indet.[3]
Ravin Voulgarakis[34783]	Pleistocene (MNQ20)	Leporidae indet.[3]
Alikes[34782]	Early Pleistocene, Biharian (?MNQ19)	*Oryctolagus lacosti*[4]
Tourkobounia-5[34592]	Early Pleistocene, Biharian	*Lepus* sp.[5]
Tourkobounia-3[34592]	Early Pleistocene, Biharian	Leporidae indet.[5]
Tourkobounia-2[34767]	Early Pleistocene, Biharian (MNQ19)	*Lepus terraerubrae*[6]
Megalo Emvolon 1[36579]	Pliocene, late Ruscinian (MN15)	*Trischizolagus* cf. *maritsae*[7], *Trischizolagus dumitrescuae*[7], *Oryctolagus* cf. *laynensis*[8]
Apolakkia-2[34607]	Pliocene, late Ruscinian (MN15)	*Ochotonoma* sp.[9–12]
Agios Ioannis (Karpathos Island)[205291]	Pliocene, early Ruscinian	*Prolagus* cf. *michauxi*[13]
Ptolemais 1[35086]	Pliocene, uppermost part of the early Ruscinian (MN14, ~5.04–4.95 Ma in Hordijk and de Bruijn 2009)	*Prolagus michauxi*[14]
Spilia 0[31931]	Pliocene, early Ruscinian (MN14, ~5.23–4.95 Ma in Koufos and Vasileiadou 2015)	*Prolagus michauxi*[15], *Alilepus turolensis*[15]
Spilia 1[31229]	Pliocene, early Ruscinian (MN14, ~5.23–4.95 Ma in Koufos and Vasileiadou 2015)	*Prolagus michauxi*[15], *Alilepus turolensis*[15]
Nea Silata[191612]	Miocene/Pliocene, Turolian/Ruscinian, (MN13/14, ~5.4–5.23 Ma in Koufos and Vasileiadou 2015)	*Prolagus michauxi*[16–17], Leporidae indet.[16–17]
Maramena[32189]	Miocene/Pliocene, Turolian/Ruscinian, (MN13/14, ~5.4–5.23 Ma in Koufos and Vasileiadou 2015)	*Prolagus* cf. *sorbini*[18], *Alilepus turolensis*[18]
Ano Metochi 2[31924]	Miocene/Pliocene, Turolian/Ruscinian, (MN13/14, ~5.4–5.23 Ma in Koufos and Vasileiadou 2015)	*Prolagus michauxi*[15]
Ano Metochi 3[31928]	Miocene/Pliocene, Turolian/Ruscinian, (MN13/14, ~5.4–5.23 Ma in Koufos and Vasileiadou 2015)	*Prolagus michauxi*[15]

Locality[PBDB No]	Age (MNQ/MN)	Identification
Kessani-1[205296]	Miocene/Pliocene, Turolian/Ruscinian, (MN13/14, ~5.4–5.23 Ma in Koufos and Vasileiadou 2015)	*Prolagus michauxi*[19–21], Leporidae indet.1[19–21]
Kessani-2[205297]	Miocene/Pliocene, Turolian/Ruscinian, (MN13/14, ~5.4–5.23 Ma in Koufos and Vasileiadou 2015)	*Prolagus michauxi*[19–21], Leporidae indet.[19–21]
Maritsa 1[68046]	**Late Miocene, Turolian (MN13 in Hordijk and de Bruijn 2009)**	*Ochotona* sp.[22], ***Trischizolagus maritsae***[22]
Monasteri[31940]	Late Miocene, Turolian, uppermost (MN13, ~5.96–5.33 Ma in Koufos and Vasileiadou 2015)	*Prolagus* sp.[15]
Paliambela-A[205335]	Late Miocene, Turolian, uppermost (MN13, ~5.96–5.33 Ma in Koufos and Vasileiadou 2015)	*Prolagus* sp.[23–24]
Lava 2[205341]	Late Miocene, late Turolian (MN13, ~6.8 Ma)	*Pliopentalagus* sp.[25]
Rema Marmara[32190]	Late Miocene, middle/late Turolian, (MN12; tentatively)	*Prolagus michauxi*[15]
Elaiochoria	Late Miocene, Turolian	cf. *Alilepus turolensis*[27]
Pikermi - Chomateri[68214]	Late Miocene, Turolian (MN12)	*Prolagus* cf. *crusafonti*[26], *Alilepus* sp.[26]
Lefkon[31223]	Late Miocene, latest Vallesian (MN10)	*Prolagus* sp.[15]
Katheni[207011]	Late Miocene, Vallesian	*Prolagus* sp.[28]
Vryses[207012]	Late Miocene, Vallesian or Turolian	*Prolagus* sp.[29]
Chryssavgi 1[205366]	Middle Miocene, late Astaracian (MN7&8 in Koufos 2006)	*Alloptox* aff. *anatoliensis*[3,8], *Prolagus* n. sp.[3]
Thymiana C[205390]	Middle Miocene, Orleanian (MN5, ~15.5 Ma, interval of Chrons C5Bn.1n– C5Cr in Koufos 2006)	*Alloptox* sp.[30]
Thymiana A[205389]	Middle Miocene, Orleanian (MN5, ~15.5 Ma, interval of Chrons C5Bn.1n– C5Cr in Koufos 2006)	*Alloptox* sp.[30]
Antonios[73861]	Early Miocene, Orleanian (MN4/5)	*Alloptox anatoliensis*[31], *Prolagus* sp.[31]
Aliveri[68032]	**Early Miocene, Orleanian (MN4, 18–17.5 Ma in van den Hoek Ostende et al. 2015)**	***Albertona balkanica***[32]

MN Mammal Neogene Zone, *MNQ* Mammal Quaternary Zone

[1]Chatzopoulou (2014), [2]Mayhew (1978), [3]Koliadimou (1996), [4]Athanassiou (1998), [5]Reumer and Doukas (1985), [6]van der Meulen and Doukas (2001), [7]Koufos and Koliadimou (1993), [8]Koufos (2006), [9]van de Weerd et al. (1982), [10]Şen (1998), [11]Čermák (2007), [12]Čermák (2010), [13]Daams and van de Weerd (1980), [14]van de Weerd (1979), [15]de Bruijn (1989), [16]Vasileiadou (2001), [17]Vasileiadou et al. (2003), [18]de Bruijn (1995), [19]Syrides et al. (1997), [20]Konidaris (2005), [21]Vasileiadou et al. (2012), [22]de Bruijn et al. (1970), [23]Sylvestrou (2002), [24]Sylvestrou and Kostopoulos (2007), [25]de Bruijn et al. (1999), [26]López Martínez (1976), [27]Hulva et al. (2007), [28]Katsikatsos et al. (1981), [29]van der Made (1996), [30]de Bonis et al. (1997), [31]Vasileiadou and Koufos (2005), [32]López Martínez (1986)

References

Angelone C (2007) Messinian *Prolagus* (Ochotonidae, Lagomorpha) of Italy. Geobios 40:407–421

Angelone C (2008) Family Ochotonidae (Lagomorpha) and its application in biochronology: some case studies from the Plio-quaternary of Eurasia. Quat Int 179:5–8

Angelone C, Hír J (2012) *Alloptox katinkae* sp. nov. (Lagomorpha: Ochotonidae), westernmost Eurasian record of the genus from the early Middle Miocene vertebrate fauna of Litke 2 (N Hungary). Neues Jahrb Geol Palaontol Abh 264:1–10

Athanassiou A (1998) Contribution to the study of the fossil mammals of Thessaly, vol 5. National and Kapodistrian University of Athens, Editions of the Department of Geology/Gaia, Athens, p 354

Averianov A (1995) Osteology and adaptations of the Early Pliocene rabbit *Trischizolagus dumitrescuae* (Lagomorpha: Leporidae). J Vertebr Paleontol 15:375–386

Averianov A, Lopatin A (2005) Eocene Lagomorphs (Mammalia) of Asia: 1. Aktashmys (Strenulagidae fam. nov.). Paleontol J 39:308–317

Čermák S (2007) New finds of *Ochotonoma csarnotana* (Lagomorpha, Ochotonidae) from the Pliocene of Hungary: a new look on the species. N Jahr Geol Paläontol Abh 246:247–256

Čermák S (2010) The late Miocene and Pliocene Ochotoninae (Lagomorpha, Mammalia) of Europe – the present state of knowledge. In: Nowakowskli D (ed) Morphology and systematics of fossil vertebrates. DN Publisher, Wroclaw, pp 9–28

Čermák S, Wagner J (2013) The Pliocene record of *Trischizolagus* and *Pliopentalagus* (Leporidae, Lagomorpha, Mammalia) in Central Europe, with comments on taxonomy and evolutionary history of Leporinae. N Jahr Geol Paläontol Abh 268:97–111

Čermák S, Angelone C, Sinitsa MV (2015) New Late Miocene *Alilepus* (Lagomorpha, Mammalia) from Eastern Europe – a new light on the evolution of the earliest old world Leporidae. Bull Geosci 90:431–451

Chatzopoulou A (2014) The micromammals of the quaternary deposits from cave A, Loutra Almopias (Pella, Northern Greece). Stratigraphy-Taphonomy-Paleoenvironment. PhD thesis, Aristotle University of Thessaloniki, School of Geology

Daams R, van der Weerd AV (1980) Early Pliocene small mammals from the Aegean island of Karpathos (Greece) and their paleogeographic significance. Geol en Mijnbouw 59:327–331

Dawson MR (2008) 17 Lagomorpha. In: Janis CM, Gunnell GF, Uhen MD (eds) Evolution of tertiary mammals of North America volume 2: small mammals, Xenarthrans, and Marine Mammals. Cambridge University Press, Cambridge, pp 293–310

de Bonis L, Koufos GD, Şen S (1997) The Sanithera (Mammalia, Suoidea) from the Middle Miocene of Chios Island, Aegean Sea, Greece. Rev Paléobiol Genève 16:259–270

de Bruijn H (1989) Smaller mammals from the Upper Miocene and Lower Pliocene of the Strimon basin, Greece. Part 1. Rodentia and lagomorpha. Boll Soc Paleontol It 28:189–195

de Bruijn H (1995) The vertebrate locality Maramena (Macedonia, Greece) at the Turolian -Ruscinian boundary (Neogene). 11. Lagomorpha (Mammalia). Münchner Geowiss Abh 28:133–136

de Bruijn H, Dawson MR, Mein P (1970) Upper Pliocene Rodentia, Lagomorpha and Insectivora (Mammalia) from the isle of Rhodes (Greece). Proc K Ned Akad Wet B 73:535–584

de Bruijn H, Saraç G, van der Hoek Ostende LW Roussiakis S (1999) The status of the genus name *Parapodemus* SCHAUB, 1938; new data bearing on an old controversy. In: Reumer JWF, de Vos J (eds) Elephants have a Snorkel! Papers in Honour of Paul Y. Sondaar, Deinsea 7:95–112

de Bruijn H, van der Hoek Ostende L, Kristkoiz-Boon E, Rummel M, Theocharopoulos C, Ünay E (2003) Rodents, lagomorphs and insectivores, from the Middle Miocene hominoid locality of Candir (Turkey). CFS Cour Forschinst Senckenb 240:51–87

de Marfa R, Mein P (2007) Révision d' *Oryctolagus lacosti* (Lagomorpha, Mammalia) du Pliocène supérieur de Perrier (Auvergne, France). C R Palevol 6:327–334

Dice LR (1929) The phylogeny of the Leporidae, with description of a new genus. J Mammal 10:340–344

Dice LR (1931) *Alilepus*, a new name to replace *Allolagus* Dice, preoccupied, and notes on several species of fossil hares. J Mammal 12:159–160

Erbajeva MA (1988) Cenozoic pikas (taxonomy, systematics, phylogeny). Nauka Press, Moscow. (In Russian)

Erbajeva MA (1994) Phylogeny and evolution of Ochotonidae with emphasis on Asian ochotonids. In: Tomida Y, Li CK, Setoguchi K (eds) Rodent and lagomorph families of asian origin and diversification. Nat Sci Mus Monogr Tokio 8:1–13

Erbajeva MA (2016) The Ochotonids of Eurasia: biochronology and taxonomic diversity. Biol Bull 43:729–735

Erbajeva MA, Mead JI, Alexeeva NV, Angelone C, Swift SL (2011) Taxonomic diversity of Late Cenozoic Asian and North American ochotonids (an overview). Palaeontol Elect 14:9

Fostowicz-Frelik L (2017) Convergent and parallel evolution in early Glires (Mammalia). In: Pontarotti P (ed) Evolutionary biology: self/nonself evolution, species and complex traits evolution, methods and concepts. Springer, Cham. https://doi.org/10.1007/978-3-319-61569-1_11

Fostowicz-Frelik L, Meng J (2013) Comparative morphology of premolar foramen in lagomorphs (Mammalia: Glires) and its functional and phylogenetic implications. PLoS One 8(11):e79794. https://doi.org/10.1371/journal.pone.0079794

Ge D, Wen Z, Xia L, Zhang Z, Erbajeva M, Huang C, Yang Q (2013) Evolutionary history of lagomorphs in response to global environmental change. PLoS One 8:e59668

Gidley JW (1912) The lagomorphs an independent order. Science 36(922):258–256

Graur D, Duret L, Gouy M (1996) Phylogenetic position of the order Lagomorpha (rabbits, hares and allies). Nature 379:333–335

Hordijk K, de Bruijn H (2009) The succession of rodent faunas from the Mio/Pliocene lacustrine deposits of the Florina-Ptolemais-Servia Basin (Greece). Hellenic. J Geosci 44:21–103

Hordijk K, van der Meulen A (2010) Systematics of immigrant species of *Prolagus* (Ochotonidae, Lagomorpha, Mammalia) from the late Early and Middle Miocene of north–eastern Central Spain, Chapter 3. Geol Ultraiectina 333:87–116

Hulva P, Horáček I, Benda P (2007) Molecules, morphometrics and new fossils provide an integrated view of the evolutionary history of Rhinopomatidae (Mammalia: Chiroptera). BMC Evol Biol 7:165

Katsikatsos G, de Bruijn H, van der Meulen AJ (1981) The Neogene of the island of Euboea (Evia), a review. Geol Mijnb 60:509–516

Koliadimou KK (1996) Palaeontological and Biostratigraphical study of the Neogene/Quaternary micromammals from Mygdonia basin. PhD thesis, Aristotle University of Thessaloniki, School of Geology

Konidaris G (2005) Contribution to the study of the early Pliocene lagomorphs in the Xanthi Basin. Diploma Thesis, Aristotle University of Thessaloniki

Koufos GD (2006) The Neogene mammal localities of Greece: faunas, chronology and biostratigraphy. Hellenic J Geosci 41:183–214

Koufos GD, Koliadimou K (1993) Two lagomorphs from the Pliocene of Macedonia (Greece). Bull Geol Soc Greece 28:117–129

Li C-K, Wilson RW, Dawson MR, Krishtalka L (1987) The origin of rodents and lagomorphs. In: Mammalogy C (ed) Genoways HH. Springer, Boston, pp 97–108

Liu L, Zheng S (1997) Note on the Late Cenozoic lagomorphs of Danjiang reservoir area in Hubei and Henan. Vert PalAsiat 35:130–144. (in Chinese)

López Martínez N (1976) Lagomorpha from the Turolian of Pikermi (Greece). Proc K Ned Akad Wet B 79:235–244

López Martínez N (1977) Nuevos lagomorfos (Mammalia) del Neogeno y Cuaternario Español. Trabajos sobre Neógeno-Cuaternario 8:7–45

López Martínez N (1986) The mammals from the lower Miocene of Aliveri (Island of Evia, Greece). 6. The ochotonid lagomorphs *Albertona balkanica* n. gen. n. sp. and its relationships. Proc K Ned Akad Wet B 89:177–194

López Martínez N (2008) The lagomorph fossil record and the origin of the European rabbit. In: Alves PC, Ferrand N, Hackländer K (eds) Lagomorph biology evolution, ecology, and conservation. Springer, Berlin/Heidelberg, pp 27–46

López Martínez N, Thaler L (1975) Biogéographie, évolution et compléments à la systématique du groupe d' Ochotonides *Piezodus-Prolagus* (Mammalia, Lagomorpha). Bull Soc Géol Fr 5:850–866

MacDonald DW (2009) The encyclopedia of mammals. Oxford University Press, The Brown Reference Group Ltd, Oxford

Marković Z (2010) *Albertona balkanica* from Early Miocene of Snegotin (Serbia) – implications for the distribution and evolution of Miocene Ochotonidae. Ann Paléontol 96:25–32

Masini F (1989) *Prolagus sorbinii* n. sp., a new ochotonid (Mammalia, Lagomorpha) from the Messinian of Italy. Boll Soc Paleontol It 28:295–306

Mayhew DF (1978) Late Pleistocene small mammals from Arnissa (Macedonia, Greece). Proc K Ned Akad Wet B 81:302–321

McKenna MC (1982) Lagomorph interrelationships. Geobios 15:213–223

Meng J, Wyss AR (2005) Glires (Lagomorpha, Rodentia). In: Rose KD, Archibald JD (eds) The rise of placental mammals. Johns Hopkins University Press, Baltimore, pp 145–158

Montuire S (2001) Lagomorpha (Rabbits, Hares and Pikas). eLS. https://doi.org/10.1038/npg.els.0001567

Ognev SI (1929) Zur Systematic der russischen Hasen. Zool Anz 84:68–83

Reumer JWF, Doukas CS (1985) Early Pleistocene Insectivora (Mammalia) from Tourkobounia (Athens, Greece). Proc K Ned Akad Wet B 88:111–121

Rose KD, DeLeon VB, Missiaen P, Rana RS, Sahni A, Singh L, Smith T (2008) Early Eocene lagomorph (Mammalia) from Western India and the early diversification of Lagomorpha. Proc Royal Soc B 275:1203–1208

Şen S (1990) Middle Miocene lagomorphs from Paşalar, Turkey. J Hum Evol 19:455–461

Şen S (1998) Pliocene vertebrate locality of Çalta, Ankara, Turkey. 4. Rodentia and Lagomorpha. Geodiversitas 20:359–378

Steffens P, de Brujin H, Meulenkamp JE, Benda I (1979) Field guide to the Neogene of Northern Greece (Thessaloniki area and Strimon basin). Publ Dept Geol Pal Univ Athens A 35:1–14

Sylvestrou I, (2002) Stratigraphic, palaeontological and palaeogeographical study of the Neogene-Quaternary deposits of Katerini basin, Northern Greece. PhD thesis, Department of Geology, Aristotle University of Thessaloniki

Sylvestrou IA, Kostopoulos DS (2007) *Pseudomeriones megistos* nov. sp. (Gerbillinae, Mammalia) from the latest Miocene of Northern Greece and its phylogenetic relationships. Geobios 40:833–848

Syrides GE, Koliadimou K, Koufos GD (1997) New Neogene molluscan and mammalian sites from Thrace, Greece. CR Acad Sci Ser IIa Sci Terre Planets 324:427–433

Tesakov A, Averianov A (2002) *Prolagus* (Lagomorpha, Prolagidae) from the Pliocene of Moldova and Ukraine. Paleontol J 36:80–86

Tomida Y, Jin C-Z (2009) Two new species of *Pliopentalagus* (Leporidae, Lagomorpha) from the Pliocene of Anhui Province, China, with a revision of *Pl. huainanensis*. Vertebr PalAsiatica 47:53–71

Ünay E, Şen S (1976) Une nouvelle espèce d' *Alloptox* (Lagomorpha, Mammalia) dans Ie Tortonien d' Anatolie. Bull Miner Res Explor Inst Turkey 85:145–119

van de Weerd A (1979) Early Ruscinian rodents and lagomorphs (Mammalia) from the lignites near Ptolemais (Macedonia, Greece). Proc K Ned Akad Wet B 82:127–170

van de Weerd A, Reumer JWF, de Vos J (1982) Pliocene mammals from the Apolakkia formation (Rhodes, Greece). Proc K Ned Akad Wet B 85:89–112

van den Hoek Ostende LW, Mayda S, Oliver A, Madern A, Hernández-Ballarín V, Peláez-Campomanes P (2015) Aliveri revisited, a biogeographical appraisal of the early Miocene mammals from the eastern Mediterranean. Palaeobiodiv Palaeoenviron 95:271–284 van der Made J (1996) Chapter 6. Pre-pleistocene land mammals from Crete. In: Reese DS (ed) Pleistocene and Holocene fauna of Crete and its first settlers. Monogr World Archaeol 28:69–79

van der Meulen AJ, Doukas CS (2001) The Early Biharian rodent fauna from Tourkoubounia 2. Lynx ns 32:255–277

Vasileiadou K (2001) Study of the Late Miocene/Early Pliocene micromammalian fossil assemblage from Nea Silata, Chalkidiki, North Greece. Master thesis, Department of Geology, Aristotle University of Thessaloniki (in Greek)

Vasileiadou K, Koufos GD (2005) The micromammals from the Early/Middle Miocene locality of Antonios, Chalkidiki, Greece. Ann Paléontol 91:197–225

Vasileiadou K, Koufos GD, Syrides G (2003) Silata, a new locality with micromammals from the Miocene/Pliocene boundary of the Chalkidiki Peninsula, Macedonia, Greece. In: Reumer JWF, Wessels W (eds) Distribution and migration of tertiary mammals in Eurasia. A volume in honour of Hans de Bruijn. Deinsea 10:549–562

Vasileiadou K, Konidaris GE, Koufos GD (2012) New data on the micromammalian locality of Kessani (Thrace, Greece) at the Mio-Pliocene boundary. Palaeobiodivers Palaeoenviron 92:211–237

Viret J (1954) Le lœss à bancs durcis de Saint-Vallier (Drôme) et sa faune de mammifères villafranchiens. Nouv Arch Mus Hist Nat Lyon 4:1–200

White JA (1991) North American Leporinae (Mammalia: Lagomorpha) from Late Miocene (Clarendonian) to latest Pliocene (Blancan). J Vertebr Paleontol 11:67–89

The Fossil Record of the Old World Monkeys (Mammalia: Primates: Cercopithecidae) in Greece

George D. Koufos

1 Introduction

The family Cercopithecidae or "Old World monkeys" are known from Oligocene up to now with representatives distributed in the Old World, from the Iberian Peninsula (Gibraltar) to Eastern Asia (Japan) and Africa. In Europe, the family is known mainly from the fossil record, although an extant group of macaques still survives in Gibraltar. The family includes 22 genera and 133 living species including baboons, macaques, mandrills, guenons, colobus, etc. Jablonski (2002) refers to 19 extant and 25 extinct genera; only twelve of the living genera are present in the fossil record. They are living in a variety of environments from rain forests to savanna type ones and high mountainous terrains; one macaque species is adapted to the cold and snowy regions of northern Japan. All species have tail and their size varies from medium to large; their length (body and head) varies from 34 to 37 cm up to 70 cm and their body weight from ~1.3 to 50 kg.

The cercopithecids have a robust skull with strong sagittal crest; in some species (e.g., baboons) the muzzle is elongated. The elongated palate reaches beyond the M3. The tympanic bullae are small with eustachian tube. The dentition consists of 32 teeth (dental formula 2, 1, 2, 3/2, 1, 2, 3); the first incisors are wide and shovel like; the upper canines are relatively large; the upper premolars are D-shaped with straight buccal face and bear a high paracone (buccal cusp) and a variable in size protocone (lingual cusp); the molars bear four cusps(−ids) on their occlusal surface linked by transverse loph(−ids) and forming three foveae; a fifth cuspid (hypoconulid) is often present in the m3, situated on the distal shelf; this bunolophodont feature of the molars represents an adaptation to their diet (leaves and fruits). The Old World monkeys are separated into two subfamilies, Colobinae and Cercopithecinae.

G. D. Koufos (✉)
Department of Geology, Laboratory of Geology and Paleontology, Aristotle University of Thessaloniki, Thessaloniki, Greece
e-mail: koufos@geo.auth.gr

E. Vlachos (ed.), *Fossil Vertebrates of Greece Vol. 1*,
https://doi.org/10.1007/978-3-030-68398-6_17

The subfamily Victoriapithecinae, which was previously included in Cercopithecidae and contains only fossil forms, is now upgraded to the separate family Victoriapithecidae. All information of the introduction were taken from Szalay and Delson (1979) and Myers (2000). More data for the fossil Old World monkeys are given in Delson (1973), Szalay and Delson (1979), Benefit and McCrossin (2002), and Jablonski (2002).

2 Historical Overview

The fossil record of the cercopithecids is quite rich and widespread in the late Miocene to Early Pleistocene deposits of Greece. The most common taxon is the colobine monkey *Mesopithecus* discovered in several Greek fossiliferous sites. It was originally recognized from Pikermi, about 20 km from Athens. Two Bavarian soldiers, who were in Greece, following the first king Otto, were hunting in the area of Pikermi in 1838, and they found a block with some fossil bones. Some calcite crystals, deposited inside the bone, aroused their interest by thinking that they are diamonds. The block was carried in Munich and given to Prof. Wagner, who recognized the fossils. Among them there was the maxilla of a monkey, which then was described by him (Wagner 1839). Earlier in 1835, Finlay and Lindermayer found some fossils in Pikermi and gave them to the Physiographic Society of Greece, but nobody took care for them.

The discovery of Pikermi brought many palaeontologists in Athens who worked there collecting fossils for their museums and institutes (Wagner, Gaudry, Woodward, Abel, etc.). The Pikermi collection of *Mesopithecus* is the richest including numerous remains corresponding to ~100 individuals; they are housed in various European museums and institutes. Quite later, the French geologist Arambourg discovered a number of localities in the lower Axios Valley, near Thessaloniki and collected numerous fossils, including *Mesopithecus*; all the material was described later by Arambourg and Piveteau (1929). During the new excavations in Axios valley (1973–present), new remains of *Mesopithecus* have been found in the various fossiliferous horizons of the area (Bonis et al. 1990). *Mesopithecus* is also recognized in several localities of northern and central Greece. The Greek material of *Mesopithecus* has been studied by various authors since the middle of the nineteenth century. Besides the initial publication of Wagner (1839), *Mesopithecus* material has been studied by several other palaeontologists, including Gaudry (1862–67); Delson (1973), Bonis et al. (1990), Zapfe (1991), Koufos (2009a), Tsoukala and Bartziokas (2008), and Lazaridis et al. (2018). The cercopithecids *Dolichopithecus*, *Paradolichopithecus*, and *Macaca* have been also discovered in Greece, enriching the cercopithecid fossil record. *Dolichopithecus* was discovered in Greece at the beginning of the 1990s (Koufos et al. 1991) and later was recognized in Western Macedonia (Doukas and de Bruijn 2002). The first occurrence of *Paradolichopithecus* in Greece was traced in Lesvos Island (de Vos et al. 2002) and recently it has been recognized in Western Macedonia (Kostopoulos et al. 2018) as well. The genus *Macaca* is only known from a fissure filling near Athens by a few dental remains (Symeonidis and Zapfe 1976).

3 Phylogenetic Relationships

The extensive study of the living Old World monkeys gave important information for their evolution and ecology. On the other hand, the rich African fossil record provided a great amount of morphological and genetic information. All these data provided a good knowledge for their origin and evolutionary history. The ancestral group of the Old World monkeys is the family Victoriapithecidae (this family was considered for a long time as a subfamily of the Cercopithecidae). Victoriapithecids are known from the middle Miocene of Africa, more exactly between 19.0 and 12.5 Ma, although there is evidence for their presence in the uppermost Oligocene at ~25.0 Ma as well (Frost 2017). The earliest known evidence of Cercopithecidae is traced from Tugen Hills, Kenya, dated at 12.5 Ma (Rossi et al. 2013). The first known evidence for the presence of the cercopithecids in Europe is traced in the upper part of late Miocene. Most authors refer a single tooth of *Mesopithecus* from the Vallesian (MN9) locality of Wissberg, Germany, as the oldest known evidence of the cercopithecids in Europe. However, the stratigraphic position of this specimen is doubtful and probably is the result of a mixing of the material (Andrews et al. 1996). In Asia, the known evidence suggests that the cercopithecids appeared at the middle Turolian with *Mesopithecus*. This genus is present in the old collection of Mecquenem (1924), housed in the Museum national d'Histoire naturelle of Paris, dated mainly at the middle Turolian (MN12) without excluding the possibility to be extended in the upper part of early Turolian (Kostopoulos and Bernor 2011); however the *Mesopithecus* mandible from Maragheh has some similarities to *M. delsoni* from Axios Valley and it is quite possible to originate from older levels (Bonis et al. 1990). The known data suggest that the first presence of cercopithecids in Eurasia is certainly documented from the early Turolian of Greece; *Mesopithecus* is present in the locality Nikiti 2, dated at the beginning of Turolian (MN11) between 8.7 and 8.2 Ma (Koufos 2016). *Mesopithecus* survived in Europe until the early Pliocene. At the same time, various forms related to *Dolichopithecus* appeared in Asia and Europe. These forms are related to the modern *Rhinopithecus* or to some African forms.

Molecular studies suggest that the divergence of the recent Cercopithecinae from Colobinae is estimated between 17.6 Ma and 14.5 Ma and the origin of crown Cercopithecinae is around the beginning of late Miocene or 11.5 Ma (Ting 2008; Perelman et al. 2011). However, the oldest known cercopithecines appeared at ~7.4 Ma in East Africa (McDougall and Feibel 2003; Leakey et al. 2003). The cercopithecines are quite rare in the fossil record and the first evidence of their presence out of Africa is a single tooth of a guenon (Cercopithecini sp. indet.) from Abu Dhabi, Arabia, dated from 8.5 to 6.0 Ma (Gilbert et al. 2014). After a long gap in the fossil record, the first cercopithecines migrated to Eurasia with the genus *Macaca*

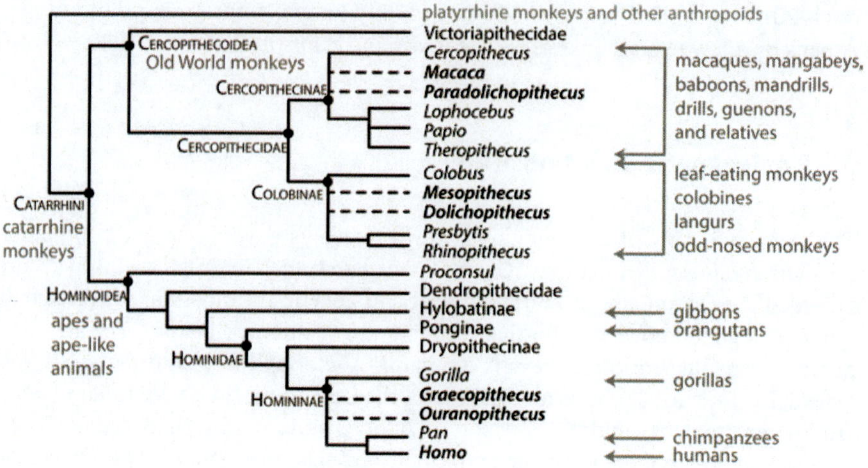

Fig. 1 Cladogram of extant and fossil Old World monkeys. (Modified from Fleagle 2013)

that occurred in Eurasia at ~6.0–5.0 Ma and is recognized in Spain and Italy (Delson et al. 2000; Alba et al. 2014).

As mentioned above, the Old World monkeys are abundant at present, the number of which exceeds remarkably that of the fossil ones. The fossil record of the Old World monkeys is rich in the early–middle Miocene Victoriapithecidae, whereas there is a rich amount of fossils from late Miocene to Pleistocene. Besides their relative abundance and the extensive studies on the modern monkeys, the phylogenetic relationships between the extant and extinct taxa are poorly known. Since the 1970s, several studies and attempts have been proposed to determine the relationships between them, but the results are doubtful. A recent cladogram for the relationships of the Old World monkeys (Fig. 1) is proposed by Fleagle (2013), which explains their relationships based on the available data, at that time.

4 Distribution

The fossil cercopithecids were widely distributed in the Old World with several fossils from various regions. In Greece, they have also been common, especially in the late Miocene deposits (Fig. 2). The colobine *Mesopithecus* is the first cercopithecid, recognized in Greece with *M. pentelicus* in Pikermi and then it was discovered in several fossiliferous sites. The oldest occurrence of the genus is traced at the beginning of the Turolian from Nikiti 2, dated between 8.7 and 8.2 Ma (Koufos 2016). The dimensions of the available metapodials, from Nikiti 2, are closer to those of *M. delsoni*, but the poor material cannot allow a definite attribution. The species *M. delsoni* is only recognized from Ravin des Zouaves 5 in Axios Valley (GPTS age: ~8.2 Ma; Sen et al. 2000). This species is larger than the typical *M. pentelicus*

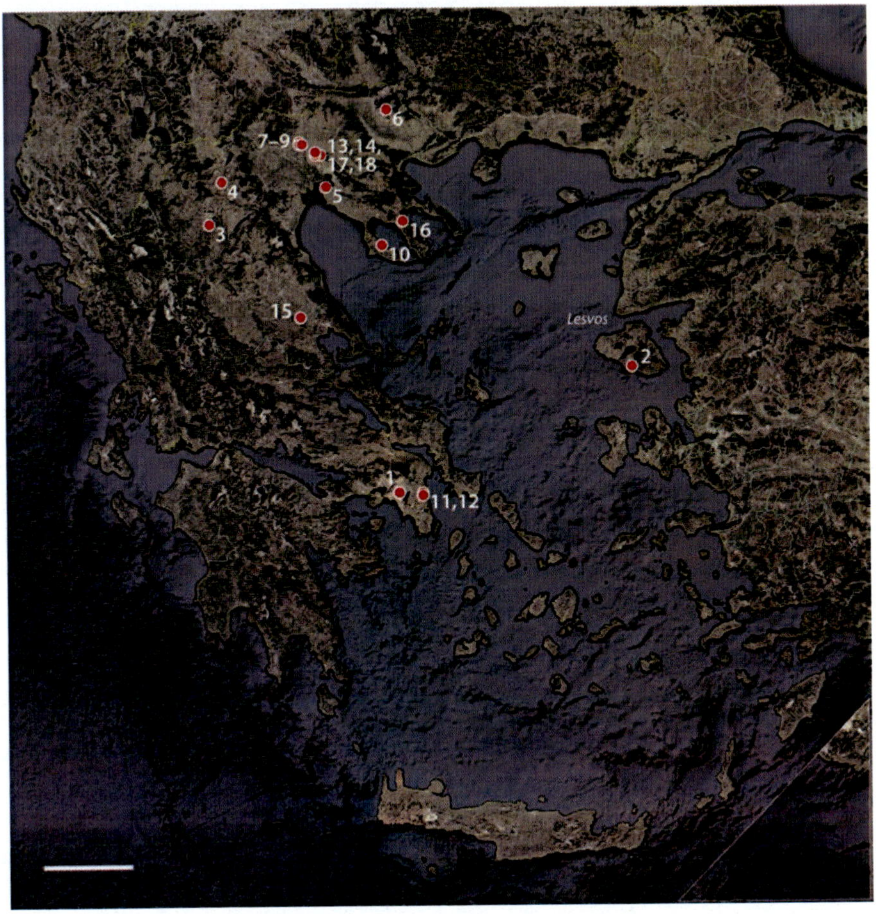

Fig. 2 Map of Greece showing the geographic distribution of the most important localities with Old World monkey fossils. **1**, Tourkovounia 2; **2**, Vatera-F; **3**, Dafnero 3; **4**, Ptolemais basin; **5**, Megalo Emvolon 1; **6**, Maramena; **7**, Dytiko 3; **8**, Dytiko 2; **9**, Dytiko 1; **10**, Kryopigi; **11**, Chomateres; **12**, Pikermi; **13**, Vathylakkos 3; **14**, Vathylakkos 2; **15**, Perivolaki; **16**, Nikiti 2; **17**, Ravin des Zouaves 5; **18**, Ravin X. See Appendix for more information. Image exported from Google Earth Pro © 2019, map data from US Dept. of State Geographer, SIO, NOAA, U.S. Navy, NGA, GEBCO, image from Landsat/Copernicus. Scale bar equals 80 km, north faces upward

from Pikermi (age of the Pikermi fossiliferous beds 7.25–7.10 Ma; Böhme et al. 2017). In the meantime, *Mesopithecus* is represented by an intermediate morphotype between *M. delsoni* and *M. pentelicus*; this morphotype is known from Vathylakkos 1 and 3 in Axios valley (GPTS age: ~7.3 Ma; Sen et al. 2000 and Koufos 2006) and from Perivolaki in Thessaly (GPTS age: 7.3–7.1 Ma; Koufos et al. 2006). During late Turolian, MN13, *M. pentelicus*, and *M. monspessulanus* were recognized by some remains from the three Dytiko localities in Axios Valley (late Turolian, MN13, 7.0–6.0 Ma; Koufos and Vasileiadou 2015; Koufos 2019a, b).

Besides these occurrences, *M. pentelicus* is recognized from the localities Kryopigi, dated at 7.3–6.4 Ma (Lazaridis 2015), and Maramena, dated at the Miocene/Pliocene boundary (Kullmer and Doukas 1995). In conclusion, *Mesopithecus* presents a continuous presence in the Turolian deposits of Greece from 8.7 Ma to 5.3 Ma (see below). During this period, *Mesopithecus* size decreases gradually and reaches its smaller size in the Pliocene with *M. monspessulanus*. The latter taxon is mainly known from Western Europe, but it is traced in the Pliocene of neighboring Bulgaria together with *Dolichopithecus ruscinensis* (Delson et al. 2005). *Dolichopithecus ruscinensis* has been recognized from two Greek Pliocene localities of Greece. *Paradolichopithecus* has been originally described from Lesvos Island and was later discovered in Western Macedonia, while *Macaca* is only found near Athens. The geographic and temporal distribution of the Greek cercopithecids is given in the Appendix and Fig. 2.

5 Systematic Paleontology

Cercopithecidae Gray, 1821
Colobinae Jerdon, 1867
***Mesopithecus* Wagner, 1839**

Type Species *Mesopithecus pentelicus* Wagner, 1839.

Remarks Two metapodials from Nikiti 2 in Chalkidiki have been described as *Mesopithecus* sp.; their dimensions are similar to the large-sized *M. delsoni* (Koufos 2016). Among the Arambourg collection of *Mesopithecus* from Axios Valley, housed in the Museum national d'Histoire naturelle of Paris, there is a mandibular fragment with symphysis from Ravin-X, which has similarities with *M. delsoni*, but the poor material cannot allow a definite determination and thus it is referred to as *M.* cf. *delsoni* (Koufos 2009a). The Greek *Mesopithecus* sample includes some maxillary, mandibular, and postcranial remains from Vathylakkos 2, 3 in Axios valley and Perivolaki, in Thessaly, housed in the Museum national d'Histoire naturelle of Paris (old Arambourg collection) and in the Laboratory of Geology and Paleontology, Aristotle University of Thessaloniki (new collection of Axios and the Perivolaki collection) (Figs. 3, 4, 5 and 6). This material represents a morphotype intermediate between *M. delsoni* and *M. pentelicus*, referred to as *M. delsoni/M. pentelicus*; a similar morphotype has also been recognized in Bulgaria (Koufos et al. 2003; Koufos 2009a, b). This morphotype is characterized by intermediate size between *M. delsoni* and *M. pentelicus*, relatively shallow mandibular corpus similar to *M. pentelicus*, high internal cusp (protocone) in the P3 and P4, flattened anterior symphysis with symphyseal constriction, slightly inclined backward alveolar plane, large fossa genioglossa, strong honing facet in the p3, large hypoconulid with distal groove in the m3, similar dental dimensions with *M. pentelicus*, and longer bones than *M. pentelicus* (Koufos 2009a). Some material from the Miocene/Pliocene

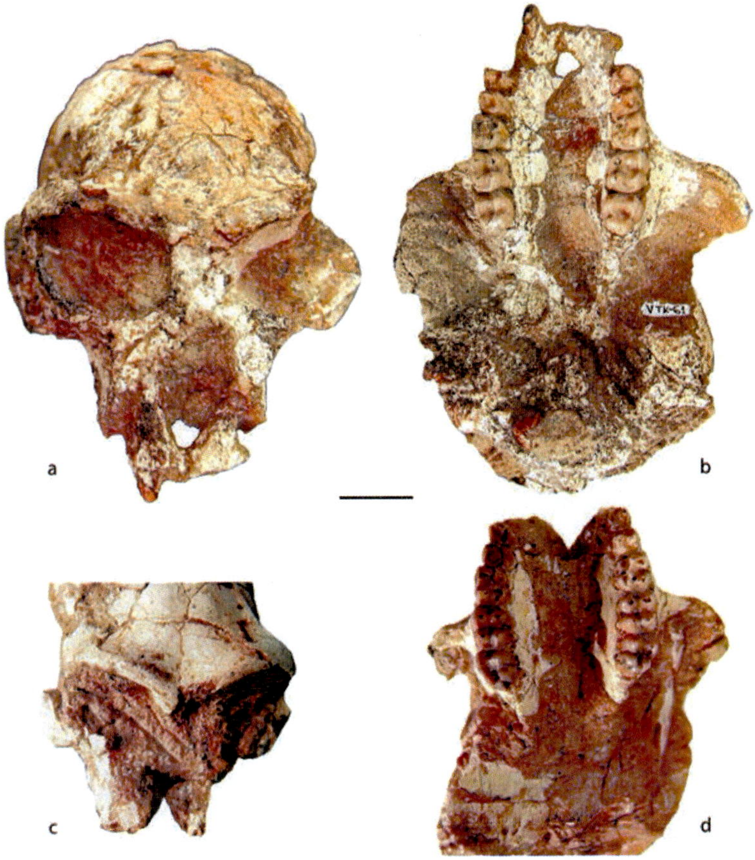

Fig. 3 Selected fossils of the intermediate cercopithecid morphotype between *Mesopithecus delsoni* and *Mesopithecus pentelicus* from Vathylakkos 2, Axios valley, late Miocene (MN12). Cranium of a male individual in (**a**) frontal and (**b**) ventral views. Cranium of a female individual in (**c**) frontal and (**d**) ventral views. Scale bar equals 1 cm

locality of Maramena in Serres Basin is also reported as *M. pentelicus* by Kullmer and Doukas (1995), but needs revision.

★*Mesopithecus delsoni* Bonis et al., 1990

Type Specimen LGPUT-RZO-159 (holotype), mandible of male adult individual with both tooth rows, housed in the Laboratory of Geology and Paleontology, Aristotle University of Thessaloniki (Fig. 4).

Type Locality Ravin des Zouaves 5, Axios Valley, Macedonia, late Miocene, MN11.

Distribution It is certainly known only from the type locality, but it is probably present in Ravin-X of Axios Valley and Nikiti 2 in Chalkidiki (Koufos 2009a, 2016).

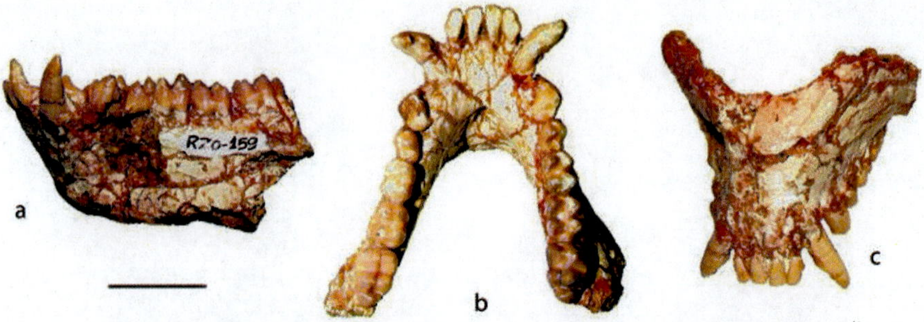

Fig. 4 The holotype of *Mesopithecus delsoni* from the late Miocene (MN11) of Ravin des Zouaves 5, Axios Valley. Mandible of a male individual in (**a**) buccal, (**b**) occlusal, and (**c**) external symphysial views. Scale bar equals 1 cm

The single known mandible from Maragheh, Iran (Mecquenem 1924) is probably similar to this species (Bonis et al. 1990).

Remarks The taxon is known from two male mandibles (Fig. 4) and a mandibular fragment of a female individual, characterized by large size (larger than *M. pentelicus*), deep mandibular corpus, flattened anterior symphysis, strong symphysial constriction, slightly inclined alveolar plane, large fossa genioglossa, thick inferior transverse torus, large honing facet on the p3, and well-developed and bicuspid talonid on the m3 (Koufos 2009a, b). These characters separate *M. delsoni* from the typical *M. pentelicus* from Pikermi. Some authors consider *M. delsoni* as a synonym of *M. pentelicus* or they refer this as a subspecies of the later taxon. However, its morphology and size are quite different from *M. pentelicus* and allow their separation (Bonis et al. 1990; Koufos 2009a, b).

★*Mesopithecus pentelicus* Wagner, 1839

Nomenclatural and Taxonomical History *Mesopithecus pentelicus* in Wagner (1839) (initial identification, new genus and species); the species is referred under various generic and specific names, which are all synonyms of the original name.

Type Specimen BSPM-PIK-AS II.11 (holotype) maxillary fragment with M1–M3 (Wagner 1839), housed in the Bayerische Staatssammlung für Paläontologie und Geologie, München, Germany.

Type Locality Pikermi, Attica, Greece, late Miocene, MN12.

Distribution In Greece, *M. pentelicus* is recognized from Chomateres in Attica, Dytiko 1, 3 in Axios Valley and Kryopigi in Chalkidiki. In Eurasia, the species is known from Spain, Italy, Germany, Hungary, FYROM, Bulgaria, Ukraine, Iran, Afghanistan, Pakistan, and China.

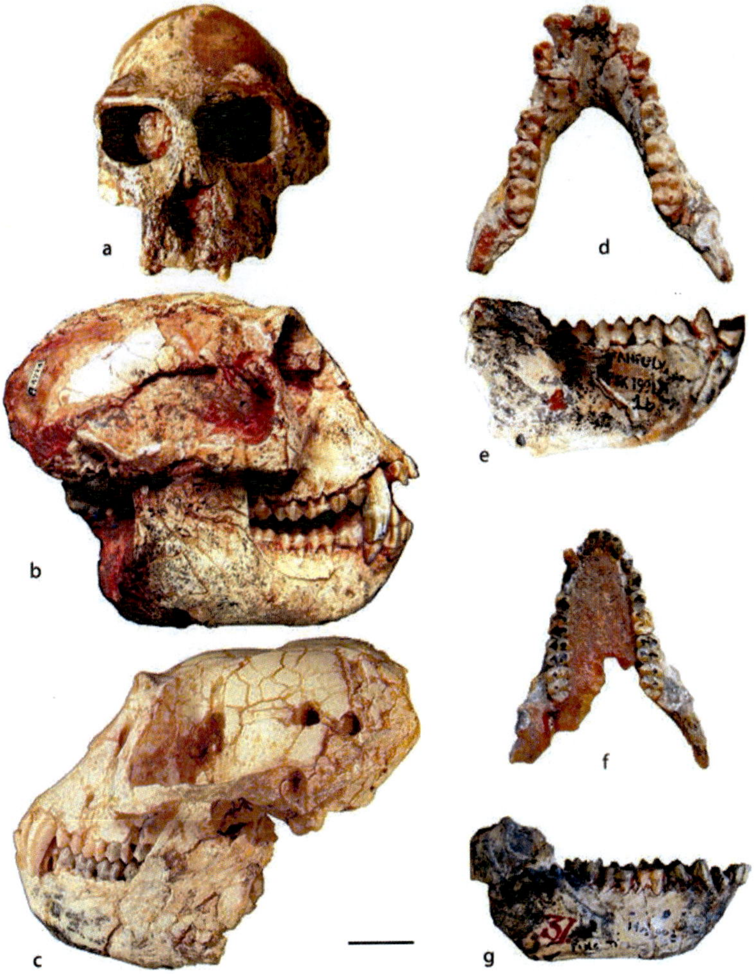

Fig. 5 Selected specimens of *Mesopithecus pentelicus* from the late Miocene of Greece. (**a**) Cranium of a male individual in frontal view, (**b**) cranium of a male individual with associated mandible in right lateral view, (**c**) cranium and associated mandible in left lateral view (photo by G. Lazaridis). Mandible of a male individual in (**d**) occlusal and (**e**) buccal views. Mandible of a female individual in (**f**) occlusal and (**g**) buccal views. All specimens are from the late Miocene of Pikermi (MN12), except **c**, which is from the late Miocene (MN12–13) of Kryopigi (Chalkidiki). Scale bar equals 1 cm

Remarks Based on the Pikermi collection, the species is medium sized for a colobine monkey, has a short face, sexual dimorphism expressed in the cranium, canines and postcranial bones, absent or very small sagittal crest in the females, enlarged mandibular angle, shallow mandibular corpus with straight inferior margin between the p4 and m3, anteriorly convex symphysis without symphysial constriction, small and deeply inclined planum alveolare, absent or weak fossa genioglossa, small

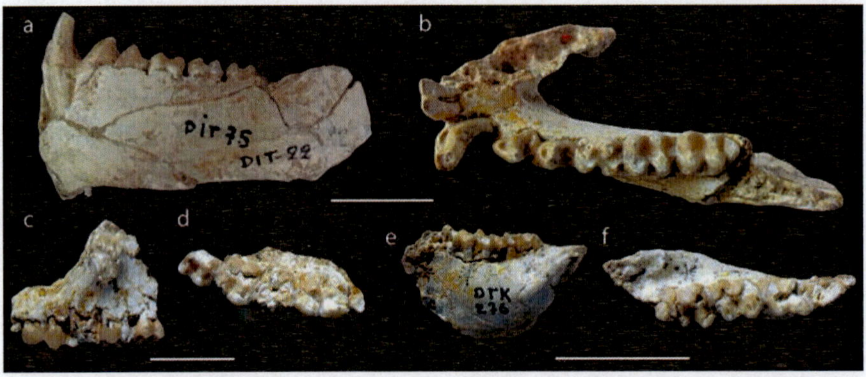

Fig. 6 Selected specimens of *Mesopithecus monspessulanus* from the late Miocene of Greece. Mandible of male individual in (**a**) buccal and (**b**) occlusal views, associated maxillary fragment in (**c**) lateral and (**d**) occlusal views, and associated mandibular fragment in (**e**) buccal and (**f**) occlusal views. Scale bars equal 2 cm for **a**, **b**, **c**, **e** and 1 cm for **d**, **f**

lingual cusp in the P3 and P4, small honing facet on the p3, and small hypoconulid on the m3 (Koufos 2009a). The Dytiko *M. pentelicus* has, at average, somewhat smaller dimensions, and narrower lower molars, which are more derived characters appeared in *M. monspessulanus* (Bonis et al. 1990; Koufos 2019a, b).

Mesopithecus monspessulanus Gervais, 1849

Nomenclatural and Taxonomical History *Mesopithecus monspessulanus* in Gervais (1849) (initial identification); *Mesopithecus monspessulanus* in Trouessart (1879) (new combination). The species have been described under various generic and specific names till now, more details in Szalay and Delson (1979).

Type Specimen The type material described by Gervais (1849) includes three molars, two male lower canines, and the proximal fragment of ulna, but he did not designate a type specimen (see details in Delson 1973 and Szalay and Delson 1979).

Type Locality Montpellier (freshwater marls of the foundation of Palais de Justice), France, early Pliocene, MN15 (Delson 1973).

Distribution The species is known from France, Germany, Hungary, Romania, Bulgaria, and Italy. In Greece, it is known from Dytiko 1 and Dytiko 2 in Axios valley.

Remarks The presence of a small-sized *Mesopithecus* in Dytiko material (Fig. 6) was early recognized by a mandibular fragment from Dytiko 2, which has smaller size than the typical *M. pentelicus* from Pikermi, being closer to *M. monspessulanus* and referred to as *M.* cf. *monspessulanus* (Bonis et al. 1990). A reconsideration of the Dytiko material indicated the presence of this small-sized form in Dytiko 1 as well. The revision of the material indicated the presence of both species *M. pentelicus* and *M. monspessulanus* in the Dytiko sample. The Dytiko *M. monspessulanus* represents the more primitive form of the species (it is slightly larger than the Pliocene typical form) and the earliest appearance of the species in Europe (Koufos 2019a, b).

***Dolichopithecus* Depéret, 1889**

Type Species *Dolichopithecus ruscinensis* Depéret, 1889.

***Dolichopithecus ruscinensis* Depéret, 1889**

Nomenclatural and Taxonomical History *Semnopithecus monspessulanus* in Depéret (1886) (initial identification); *Dolichopithecus ruscinensis* in (Depéret 1889) (new species).

Type Specimen MNHN-PER-001 (lectotype), female partial cranium figured by Depéret (1889: pl. 1, Figs. 2, 2a) and designated by Delson (1973).

Type Locality Perpignan fossil localities, especially Serrat d'en Vacquer, France, early Pliocene (Delson 1973).

Distribution Besides Greece, the species is recognized in Spain, France, Germany, Hungary, Bulgaria, and Romania. In Greece, it is known from Megalon Emvolon, near Thessaloniki and from Ptolemais Basin in Western Macedonia.

Remarks The species is known from the two mandibular fragments of the same individual from Megalon Emvolon (Fig. 7) and two isolated teeth found in Ptolemais lignitic deposits. The Megalon Emvolon specimen displays moderate size for a colobine monkey, constant mandibular depth, strong sexual dimorphism in the canines, strong metaconid and protoconid on the relatively narrow p4, larger mesial breadth of the m3 than the distal one of the m2, m3's hypoconulid between the midline and the line joining the buccal cuspids of the tooth, no clear fifth cuspid on the m3's talonid but there is a cingular elevation indicating its development (Koufos et al. 1991).

Cercopithecinae Gray, 1821
***Paradolichopithecus* Necrasov et al., 1961**

Type Species *Dolichopithecus arvernensis* Depéret, 1929.

Remarks The genus is known from the Early Pleistocene Greek locality Dafnero 3 by an almost complete cranium, housed in the Laboratory of Geology and Paleontology, Aristotle University of Thessaloniki (Fig. 8a). The cranium belongs to a female individual and it is characterized by moderately long face, elongated and narrow nasal bones, heart-shaped and longer than wide nasal cavity, rounded-trapezoidal orbits, small interorbital distance, shallow and straight zygomatic arches, small lacrymal fossa completely engulfed by the lacrymal bone, moderately

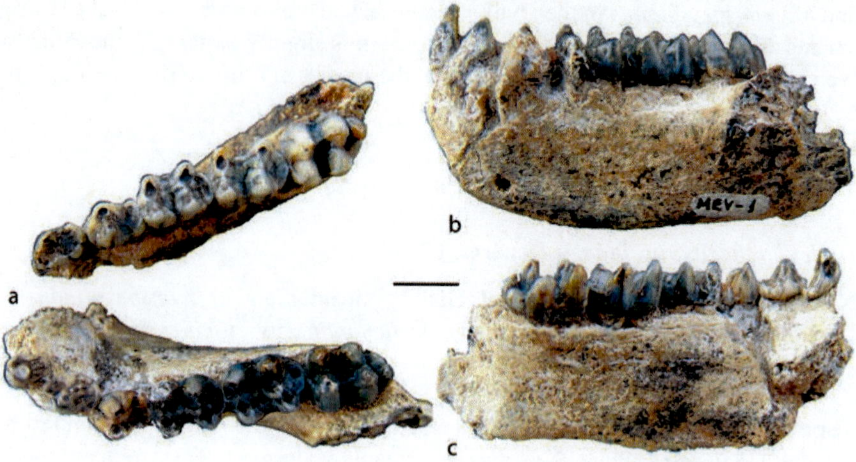

Fig. 7 Selected specimens of *Dolichopithecus ruscinensis* from the Pliocene (MN15) of Megalon Emvolon, near Thessaloniki. Mandible of a female individual in (**a**) occlusal, (**b**) buccal, and (**c**) lingual views. Scale bar equals 1 cm

thick supraorbital torus, small and rounded auditory bullae, small occipital condyles, oval foramen magnum, presence of maxillary sinus, small premolar row relative to the molar one, large protocone on the P3, and strongly flare molars (Kostopoulos et al. 2018). The Dafnero skull with a mosaic of outer and inner cranial features resembles both the Chinese *Procynocephalus* and the Eurasian *Paradolichopithecus*, supporting the idea of their synonymy, but as the known material is scarce, it is reported as *P.* aff. *arvernensis*, at the moment.

Paradolichopithecus arvernensis (Depéret, 1929)

Nomenclatural and Taxonomical History *Dolichopithecus arvernensis* in Depéret (1929) (initial identification, new species); *Paradolichopithecus geticus* in Necrasov et al. (1961) (new genus); *Paradolichopithecus arvernensis* in Delson (1973) (new combination).

Type Species Holotype, cranium and associated mandible, housed in the collections of the University of Lyon.

Type Locality Sénèze, France, Early Pleistocene, MN17.

Distribution In Greece, it is known from the Early Pleistocene locality Vatera, Lesvos Island. Besides Greece, the species is also known from Spain, France, and Romania.

Remarks The known Greek sample of the taxon includes two mandibles and some postcranial remains (Fig. 8b), housed in the museum of the village Vrissa in Lesvos

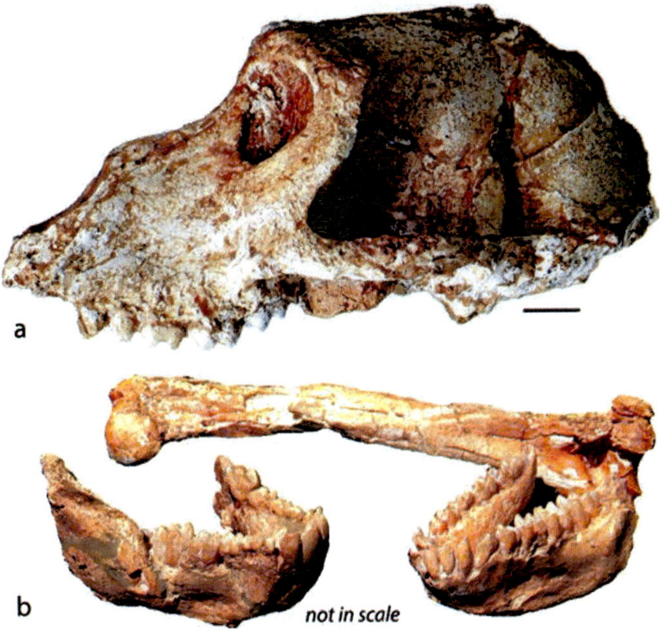

Fig. 8 Selected specimens of *Paradolichopithecus* from Greece. (**a**) *Paradolichopithecus* aff. *arvernensis* from the Early Pleistocene (MN17) of Dafnero 3 (Grevena basin), cranium of female individual in left lateral view. Scale bar equals 1 cm. (**b**) *Paradolichopithecus arvernensis* from the Early Pleistocene (MN17) of Vatera (Lesvos Island), two mandibles and a humerus (not in scale). Photo provided by G. Lyras

Island (van der Geer and Sondaar 2002; van der Geer and Dermitzakis 2008). The main characters of the Greek *Paradolichopithecus* are the large size, clear sexual dimorphism in the canines, large anterior lower teeth, larger labial than lingual cuspid on the p4, elongated trigonid of the m3, low relief and presence of a large sixth cuspid in the m3 (van der Geer and Dermitzakis 2008).

Macaca Lacépède, 1799

Type Species *Simia inuus* Linnaeus, 1766.

Remarks In Greece, the genus is known from the Early Pleistocene locality Tourkobounia 2 (fissure filling), near Athens. The known material includes four isolated teeth (two M1, one M2, and one m1), found during washing process for micromammals. According to the original describers, their morphological features differ slightly from those of *M. florentina* and attributed them to this species with caution (Symeonidis and Zapfe 1976). However, as the available material is scarce it is better to refer them as *Macaca* sp.

Acknowledgments I would like to thank D. Youlatos for comments that improved this manuscript.

Appendix

List of Greek localities containing old world monkey fossils. Type localities are marked with bold. Locality numbers refer to the collection numbers of the PaleoBiology Database (PBDB)

Localities[PBDB No]	Age (MN; GPTS in Ma)	Taxon
Tourkovounia 2[34767]	Latest Villafranchian (?MNQ19)	*Macaca* sp.[1]
Vatera F [183341]	Middle Villafranchian (MN17)	*Paradolichopithecus arvernensis*[2]
Dafnero 3[193513]	Middle Villafranchian (MN17)	*Paradolichopithecus* aff. *arvernensis*[3]
Ptolemais Basin	Ruscinian (MN14–15)	*Dolichopithecus ruscinensis*[4]
Megalo Emvolon 1[214080]	Ruscinian (MN15)	*Dolichopithecus ruscinensis*[5]
Maramena[32189]	Miocene/Pliocene (MN13/14)	*Mesopithecus* sp.[6]
Dytiko 3[32376]	Late Turolian (MN13)	*Mesopithecus pentelicus*[7]
Dytiko 2[32375]	Late Turolian (MN13)	*Mesopithecus monspessulanus*[7] *Mesopithecus pentelicus*[7]
Dytiko 1[32374]	Late Turolian (MN13)	*Mesopithecus monspessulanus*[7] *Mesopithecus pentelicus*[7]
Kryopigi[157582]	Middle-late Turolian (MN13; 7.3–6.4)	*Mesopithecus pentelicus*[8]
Chomateres[195562]	Middle Turolian (MN12)	*Mesopithecus pentelicus*[9]
Pikermi[182754]	**Middle Turolian (MN12; 7.3–7.1)**	***Mesopithecus pentelicus***[10]
Vathylakkos 3[182750]	Middle Turolian (MN12; ~7.3)	*Mesopithecus* sp.[11,*]
Vathylakkos 2[202703]	Middle Turolian (MN12; ~7.3)	*Mesopithecus* sp.[12,13,*]
Perivolaki[194879]	Middle Turolian (MN12; 7.3–7.1)	*Mesopithecus* sp.[14,*]
Nikiti 2[73869]	Early Turolian (MN11; 8.7–8.2)	*Mesopithecus* sp. [15,**]
Ravin des Zouaves 5[195489]	**Early Turolian (MN11; ~8.2)**	***Mesopithecus delsoni***[16]
Ravin X[182745]	?Early Turolian	*Mesopithecus* cf. *delsoni*[17]

GPTS Geomagnetic Polarity Time Scale
[*]Intemediate morphotype between *M. delsoni and M. pentelicus*
[**]Size closer to *M. delsoni*
[1]Symeonidis and Zapfe (1976), [2]de Vos et al. (2002), [3]Kostopoulos et al. (2018), [4]Doukas and de Bruijn (2002), [5]Koufos et al. (1991), [6]Kullmer and Doukas (1995), [7]Koufos (2019a, b), [8]Tsoukala and Bartziokas (2008), [9]Zapfe (1991), [10]Wagner (1839), [11]Arambourg and Piveteau (1929), [12]Koufos et al. (1991), [13]Koufos et al. (2004), [14]Koufos (2006), [15]Koufos (2016), [16]Bonis et al. (1990); [17]Koufos (2009a, b)

References

Alba DM, Delson E, Carnevale G, Colombero S, Delfino M, Giuntelli P, Pavia M, Pavia G (2014) First joint record of *Mesopithecus* and cf. *Macaca* in the Miocene of Europe. J Hum Evol 67:1–18

Andrews P, Harrison E, Delson E, Bernor RL, Martin L (1996) Distribution and biochronology of European and Southwest Asian Miocene Catarrhines. In: Bernor RL, Fahlbusch V, Mittmann

H-W (eds) The evolution of Western Eurasian Neogene mammal faunas. Columbia University Press, pp 168–207

Arambourg CJ, Piveteau J (1929) Les Vertébrés du Pontien de Salonique. Ann Paleontol 18:59–138

Benefit BR, McCrossin ML (2002) The Victoriapithecidae, Cercopithecoidea. In: Hartwig W (ed) The primate fossil record. Cambridge University Press, Cambridge, pp 254–299

Böhme M, Spassov N, Ebner M, Geraads D, Hristova L, Kirscher U, Kötter S, Linnemann U, Prieto J, Roussiakis S, Theodorou G, Uhlig G, Winklhofer M (2017) Messinian age and savannah environment of the possible hominin *Graecopithecus* from Europe. PLoS One 12:e0177347

de Bonis L, Bouvrain G, Geraads D, Koufos GD (1990) New remains of *Mesopithecus* (Primates, Cercopithecidae) from the late Miocene of Macedonia with the description of a new species. J Vert Paleontol 10:473–483

de Mecquenem R (1924) Contribution a l'étude des fossils de Maragha (suite). Ann Paleontol 14:135–160

De Vos J, van der Made J, Athanassiou A, Lyras G, Sondaar PY, Dermitzakis MD (2002) Preliminary note on the late Pliocene fauna from Vatera (Lesvos, Greece). Ann Geol Pays Helleniques 39:37–70

Delson E (1973) Fossil colobine monkeys of the circum Mediterranean region and the evolutionary history of the Cercopithecidae (Primates, Mammalia). PhD thesis, Columbia University, New York

Delson E, Terranova CJ, Jungers WL, Sargis EJ, Jablonski NG (2000) Body mass in Cercopithecidae (Primates, Mammalia): estimation and scaling in extinct and extant taxa. Anthropol Pap Am Mus Nat Hist 83:1–159

Delson E, Thomas H, Spassov N (2005) Fossil Old World monkeys (Primates: Cercopithecidae) from the Pliocene of Dorkovo, Bulgaria. Geodiversitas 27:159–166

Depéret C (1889) Sur le *Dolichopithecus ruscinensis*, nouveau singe fossile du Pliocene du Roussillon. C R Acad Sci Paris 109:982–983

Doukas CS, De Bruijn H (2002) A new occurrence of *Dolichopithecus* (Mammalia, Primates) in N. Greece. Ann Geol Pays Helleniques 39:295–297

Fleagle J (2013) Primate adaptation and evolution, 3rd edn. Academic, San Diego

Frost SR (2017) Evolution of the Cercopithecidae. In: Fuentes A (ed) The international encyclopedia of primatology. Wiley, Chichester, pp 1–3

Gaudry A (1862–67) Animaux fossils et géologie de l'Attique. Savy, Paris

Gervais P (1849) Zoologie et Paleontologie francaises, ed. I, Bertrand, Paris

Gilbert CC, Bibi F, Hill A, Beech MG (2014) Early guenon from the late Miocene Baynunah Formation, Abu Dhabi, with implications for cercopithecoid biogeography and evolution. Proc Natl Acad Sci U S A 111:10119–10124

Jablonski N (2002) Fossil Old World monkeys: the late Neogene radiation. In: Hartwig WC (ed) The primate fossil record. Cambridge University Press, Cambridge, pp 254–299

Kostopoulos DS, Bernor RL (2011) The Maragheh bovids (Mammalia, Artiodactyla): systematic revision and biostratigraphic-zoogeographic interpretation. Geodiversitas 33:649–708

Kostopoulos DS, Guy F, Kynigopoulou Z, Koufos GD, Valentin X, Merceron G (2018) A 2Ma old baboon-like monkey from Northern Greece and new evidence to support the *Paradolichopithecus – Procynocephalus* synonymy (Primates: Cercopithecidae). J Hum Evol 121:178–192

Koufos GD (2006) Primates. In: Koufos GD (ed.), The Late Miocene mammal locality of Perivolaki, Thessaly, Greece. Palaeontographica Abt A 276:23–37

Koufos GD (2009a) The Neogene cercopithecids (Mammalia, Primates) of Greece. Geodiversitas 31:817–850

Koufos GD (2009b) The genus *Mesopithecus* (Primates, Cercopithecidae) in the late Miocene of Greece. Boll Soc Paleontol Ital 48:157–166

Koufos GD, Syrides GE, Koliadimou KK (1991) A Pliocene primate from Macedonia (Greece). J Hum Evol 21:283–294

Koufos GD, Spassov N, Kovatchev D (2003) Study of *Mesopithecus* from the late Miocene of Bulgaria. Palaeontographica Abt A 269:39–91

Koufos GD, Bonis Lde, Kostopoulos DS, Viriot L, Vlachou T (2004) *Mesopithecus* (Primates, Cercopithecoidea) from the Turolian locality of Vathylakkos 2 (Macedonia, Greece). Palaeont Z 78:213–228

Koufos GD, Sen S, Kostopoulos DS, Sylvestrou IA, Vlachou TD (2006) Chronology. In: Koufos GD (ed) The late Miocene vertebrate locality of Perivolaki, Thessaly, Greece. Palaeontographica, Abt. A 276:185–200

Koufos GD, Vasileiadou K (2015) Miocene/Pliocene mammal faunas of southern Balkans: implications for biostratigraphy and palaeoecology. Palaeobiodivers Palaeoenviron 95:285–303

Koufos GD (2016) Primates. In: Koufos GD, Kostopoulos DS (eds), Palaeontology of the Upper Miocene vertebrate localities of Nikiti (Chalkidiki Peninsula, Macedonia, Greece). Geobios 49:45–51

Koufos GD (2019a) First evidence of *Mesopithecus monspessulanus* (Mammalia: Cercopithecidae) in the late Miocene of Macedonia, Greece. Bull Geol Soc Greece. Sp Pub 7 Ext Abs GSG2019-068

Koufos GD (2019b) Late Turolian *Mesopithecus* (Mammalia: Cercopithecidae) from Axios valley (Macedonia, Greece): earliest presence of *M. monspessulanus* in Europe. C R Palevol 18:1057–1072

Kullmer O, Doukas C (1995) The deciduous dentition of *Mesopithecus pentelicus* Wagner (Primates, Mammalia). In: Schmidt-Kittler N (ed.), The vertebrate locality of Maramena (Macedonia, Greece) at the Turolian-Ruscinian boundary (Neogene). Münch Geowissen Abh 28:65–74

Lazaridis G (2015) Study of the Late Miocene Vertebrate Locality of Kryopigi and other Localities of Kassandra Peninsula, Chalkidiki (Greece). Systematics, Taphonomy, Paleoecology, Biochronology. Ph.D. Dissertation, Aristotle University of Thessaloniki

Lazaridis G, Tsoukala E, Rae TC, Gomez-Olivencia A, Nagel D, Bartsiokas A. (2018) *Mesopithecus pentelicus* from the Turolian locality of Kryopigi (Kassandra, Chalkidiki, Greece). J Hum Evo 121:128–146

Leakey MG, Teaford MF, Ward CV (2003) Cercopithecidae from Lothagam. In: Leakey MG, Harris JM (eds) Lothagam: the dawn of humanity in eastern Africa. Columbia University Press, New York, pp 201–248

McDougall I, Feibel CS (2003) Numerical age control for the Miocene-Pliocene succession at Lothagam, a hominoid-bearing sequence in the northern Kenya Rift. In: Leakey MG, Harris JM (eds) Lothagam: the dawn of humanity in eastern Africa. Columbia University Press, New York, pp 43–64

Myers P (2000) Cercopithecidae. Animal Diversity Web. http://animaldiversity.org/accounts/Cercopithecidae/. Accessed 13 Feb 2018

Perelman P, Johnson WE, Roos C, Seuánez HN, Horvath JE, Moreira MAM, Kessing B, Pontius J, Roelke M, Rumpler Y, Schneider MPC, Silva A, O'Brien SJ, Pecon-Slattery J (2011) A molecular phylogeny of living primates. PLoS Genet 7(3):e1001342

Rossi JB, Gilbert CC, Hill A (2013) Early Cercopithecid Monkeys from the Tugen Hills, Kenya. Proc Natl Acad Sci U S A 110:5818–5822

Sen S, Koufos GD, Kondopoulou D, de Bonis L (2000) Magnetostratigraphy of the late Miocene continental deposits of the lower Axios valley, Macedonia, Greece. In: Koufos GD, Ioakim C (eds) Mediterranean Neogene cyclostratigraphy in marine-continental deposits. Bull Geol Soc Greece 9:197–206

Symeonidis N, Zapfe H (1976) Primatenzahne (Cercopithecidae) aus einer Pleistozänen Spaltenfullung im Steibruch Tourkobounia, Athens. Ann Geol Pays Helleniques 28:207–214

Szalay F, Delson E (1979) Evolutionary history of the primates. Academic, New York

Ting N (2008) Mitochondrial relationships and divergence dates of the African colobines: evidence of Miocene origins for the living colobus monkeys. J Hum Evol 55:312–325

Trouessart EL (1879) Catalogue des mammifered vivants et fossiles. Rev Mag Zool 7:223–230

Tsoukala E, Bartziokas A (2008) New *Mesopithecus pentelicus* specimens from Kryopigi, Macedonia, Greece. J Hum Evol 54:448–451

Van der Geer AAE, Dermitzakis MD (2008) Dental eruption sequence in the Pliocene papionin *Paradolichopithecus arvernensis* (Mammalia, Primates) from Greece. J Vertebr Palaeontol 28:1238–1244

Van der Geer AAE, Sondaar PY (2002) The postcranial elements of *Paradolichopithecus arvernensis* (Primates, Cercopithecidae, Papionini) from Lesvos, Greece. Ann Geol Pays Helleniques 39A:71–86

Wagner A (1839) Fossile Überreste von einem Affenschädel und andern Saugethierreste aus Griechenland. Gelehrte Anzeigen Bayer Akad Wiss 38:301–312

Zapfe H (1991) *Mesopithecus pentelicus* Wagner aus dem Turolium von Pikermi bei Athens, Odontologie und Osteologie. Neue Denkschriften Naturhist Mus Wien 5:1–203

The Fossil Record of the Miocene Hominoids (Mammalia: Primates: Hominoidea) in Greece

George D. Koufos

1 Introduction

The hominoids represent a group of catarrhine primates, including humans, their fossil ancestors, and the anthropoid apes. The fossil hominoids are known from Africa, Asia, and Europe and constitute a group that is assumed to include the common ancestor of humans and extant great apes. Therefore, each fossil remain of this group is of special importance for human evolution. The hominoids were widespread in the Old World expanding from Spain to China/Thailand and from Germany to Namibia. Temporarily, extended from the Oligocene of Africa (*Aegyptopithecus, Kamoyapithecus*) to the Pleistocene of China (*Gigantopithecus*), the bulk of their fossil remains come from the Miocene, and thus this group is often referred informally as "Miocene hominoids". At present, hominoids are absent in Europe, but they are present in Africa and Asia; gorillas and chimpanzees are living in Africa and orangutans in Asia.

The hominoids are relatively rare (~13 genera) in the late Oligocene and early middle Miocene (from 25.0 to ~15.0 Ma) deposits, restricted in Africa and Arabia with the genera *Proconsul* and *Ekembo*, being the most common and better known. These primitive hominoids were quadrupedal, arboreal, fruit, and/or leaf-eaters and characterized by a quite variable body mass, from ~3 to ~80 kg (Pilbeam 2002). During the late middle Miocene, the hard-object eater, *Kenyapithecus* from Kenya seems to have migrated in Eurasia. The late Miocene hominoids of Africa are poorly represented in relation to those of Eurasia. However, some important hominoids like *Sahelanthropus* and *Orrorin* have been discovered from the Late Miocene deposits of Tchad (7.0–6.0 Ma) and Kenya (~6.0 Ma), respectively (Senut et al. 2001, Brunet et al. 2002). In the middle Miocene, the story of hominoid evolution moved to

G. D. Koufos (✉)
Department of Geology, Laboratory of Geology and Paleontology, Aristotle University of Thessaloniki, Thessaloniki, Greece
e-mail: koufos@geo.auth.gr

E. Vlachos (ed.), *Fossil Vertebrates of Greece Vol. 1*,
https://doi.org/10.1007/978-3-030-68398-6_18

Eurasia. The first known African immigrants to Eurasia were *Griphopithecus* and *Kenyapithecus* found in the middle Miocene (Andrews et al. 1996, Kelley et al. 2008). The hominoids are strongly diversified in Europe with the occurrence of several taxa (>10 genera), which survived until the latest Miocene. Among them, *Pierolapithecus* and *Hispanopithecus* from Spain, *Dryopithecus* from France and Germany, *Oreopithecus* from Italy, *Rudapithecus* from Hungary, and *Ouranopithecus* from Greece are important and directly relevant for understanding the evolution of the human clade. The main goal of this chapter is to summarize the fossil record of the Miocene hominoids from Greece, whereas the Greek hominins are presented in another chapter (Harvati this volume).

2 Historical Overview

In Greece the first hominoid fossils, though discovered in 1944, remained unknown since the beginning of the 1970s. They were found in the fossiliferous site, named Pyrgos Vassilissis near Athens. The collected material includes some mammal remains among which there is an almost toothless mandible (see below). The fossils were discovered, during the constructions in the garden by the German army in the World War II and were carried to Germany and given to the University of Erlangen. Prof. Freyberg studied them and attributed the primate mandible to *Mesopithecus pentelicus* (Freyberg 1951). Quite later, von Koenigswald (1972) described the mandible, as belonging to a new hominoid, named *Graecopithecus freybergi*. Recently, the Pyrgos Vassilissis fauna was revised, and there is an updated faunal list and age, recognizing *Graecopithecus* as a separate taxon (Böhme et al. 2017, Fuss et al. 2017). Since the 1970s, several remains of *Ouranopithecus* have been found in Axios Valley and Chalkidiki Peninsula, Macedonia, Greece (Bonis et al. 1974, 1990; Koufos 1993, 1995 and references therein).

3 Phylogenetic Relationships

The origin of the hominoids should be found among the Oligocene primates, but there are different opinions about the possible candidate. The Oligocene *Aegyptopithecus* (35.0–32.0 Ma) from Fayum in Egypt, though it shares some derived features with *Proconsul* (Simons 1965), there are not clear evidence linking the two taxa, and there is a long age gap (~10.0 Ma) between them. The Oligocene *Kamoyapithecus* from Kenya (~25.0 Ma) with primitive morphology and fragmentary material fails to fall among the Hominoidea in a quantitative cladistic analysis (Begun 2014). Stevens et al. (2013) proposed *Rukwapithecus* from Tanzania (25.2 Ma) as the oldest ape and classified it among the nyanzapithecines. Begun (2014) did not agree with this hypothesis because he considered a possible synonymy with *Kamoyapithecus* and that it is premature to consider nyanzapithecines and

Rukwapithecus as more closely related to the middle Miocene and later hominoids than *Proconsul*. The early Miocene African primate taxa (e.g., *Morotopithecus, Limnopithecus, Kalepithecus*) could be related with hominoids, but they cannot provide certain evidences because of their limited known material and its preservation.

The African genus *Proconsul* (early Miocene, 21.0–19.0 Ma; Begun (2014), although the existing controversies, preserves a set of characters linking it with the Miocene hominoids and with the extant great apes and humans. Besides *Proconsul*, other proconsulids, like *Afropithecus, Morotopithecus*, etc., share some derived features with the later and extant hominoids, but most of them are poorly represented in the fossil record. However, the ancestor of the early/middle Miocene hominoids should be sought among this group as *Proconsul*-like taxa seem to share various derived characters with the later hominoids together with some primitive ones. The African hominoids migrated to Eurasia at about 16.0 Ma; during that time, the physical connection of Africa and Eurasia was completed, while three main migration waves have been recognized at that time, including the arrival of the hominoids in Eurasia (Rögl 1999). In Europe the oldest hominoids are known from Slovakia, Austria, and Germany, and in Asia from Turkey with the genus *Griphopithecus*, dated in the middle Miocene (Andrews et al. 1996, Casanovas-Vilar et al. 2011). An evidence for the linking of African and Eurasian hominoids is the recognition of the African genus *Kenyapithecus* in the middle Miocene Turkish locality Paşalar (Kelley et al. 2008). The hominoids after their arrival in Eurasia dispersed, diversified, and survived until the end of the Miocene, represented by several taxa in the fossil record. The Asian Miocene hominoids (e.g., *Ankarapithecus, Sivapithecus, Lufengpithecus, Khoratpithecus*) are considered as ancestral forms of the modern orangutans. In Europe, the Miocene hominoids are strongly diversified and represented by several taxa (e.g., *Pierolapithecus, Hispanopithecus, Dryopithecus, Rudapithecus, Ouranopithecus, Oreopithecus*), the relationships of which are discussed. *Ouranopithecus* shares several derived characters with the australopiths and human clade and proposed as a potential common ancestor of this clade and gorillas. Among the latest Miocene hominoids, the African genera *Sahelanthropus* and *Orrorin* are interesting. The femur of *Orrorin* shares some derived features with australopiths and *Homo*, and it is considered as closer to the last taxon (Pickford et al. 2002). On the other hand, the cranium of *Sahelanthropus* preserves some derived features linking it with the hominid clade being closer to the common ancestor of chimpanzee and *Homo* (Brunet et al. 2002).

Several cladograms have been proposed for the evolutionary relationships among humans and extinct and extant hominoids, which changed rapidly as new material discovered especially during the last decades. A simplified version of a recent cladogram proposed by McNulty (2010) is given in Fig. 1, highlighting the position of the Greek taxa. *Ouranopithecus* seems to be the best candidate, among the ancestral hominines, for the African great apes and human clade. Some authors (Benefit and McCrossin 1995; Andrews 1992; Dean and Delson 1992; Andrews et al. 1996; McNulty 2005; Clarke 2012) follow this opinion. On the other hand, Begun (1994, 2002, 2009), although he gives a slight precedence to *Dryopithecus*,

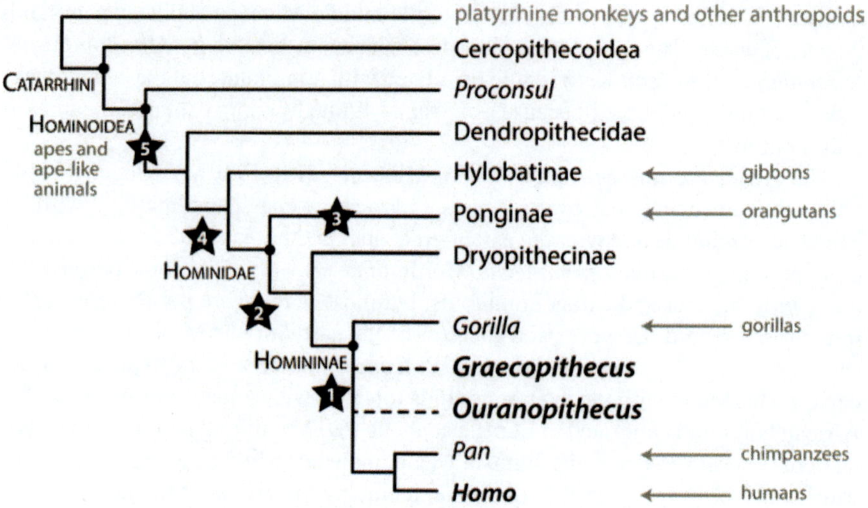

Fig. 1 Simplified cladogram showing the evolutionary relationships of the Miocene hominoids compared to the modern ones and *Homo*, highlighting the positions of the taxa from Greece (modified from McNulty 2010). Stars indicate the major splitting events between ape lineages. **1**: ancestral African apes, including *Ouranopithecus*, *Graecopithecus*, *Nakalipithecus*, and *Chororapithecus*; **2**, European apes, including *Dryopithecus*, *Oreopithecus*, *Pierolapithecus*, and *Anoiapithecus*; **3**, Asian apes, including *Sivapithecus*, *Gigantopithecus*, *Ankarapithecus*, *Lufengpithecus*, *Khoratpithecus*, and *Yuanmoupithecus*; **4**, first modern apes, including *Griphopithecus*, *Kenyapithecus*, *Nacholapithecus*, *Otavipithecus*, and *Equatorius*; **5**, stem hominoids, including *Kamoyapithecus*, *Dendropithecus*, *Micropithecus*, *Simiolus*, *Proconsul*, *Heliopithecus*, *Afropithecus*, *Morotopithecus*, *Nyanzapithecus*, *Mabokopithecus*, *Limnopithecus*, *Turkanapithecus*, *Rangwapithecus*, *Kalepithecus*, and ?*Samburupithecus*

mentions that *O. macedoniensis* has strong similarities with the dryopithecines, proposing them as a sister group of the hominins and African apes. In addition, he considers that the absence of postcranial is a disadvantage for *Ouranopithecus* as its locomotion is unknown. However, the two available phalanges of *Ouranopithecus* indicate that it is closer to terrestrial forms (Bonis and Koufos 2014). On the contrary, Dean and Delson (1992) consider *O. macedoniensis*, in facial morphology, closer to *Gorilla* than to the dryopithecines. The Greek hominoid and the Asian *Lufengpithecus* were synonymized with *Ankarapithecus*, and together with several Eurasian, dryopithecines can be considered as ancestral forms of the pongines as they share affinities with *Pongo* (Köhler et al. 2001). The recent 3D geometric morphometric study of *O. macedoniensis* cranium from Xirochori suggests that it is clearly separated from *Pongo*, being closer to *Gorilla* (Ioannidou et al. 2019a, b).

However, the relatively poor fossil record of the Miocene hominoids and the unequal geographic and chronological distribution of their finds as well as the absence of a rich fossil record for gorillas and chimpanzees are disadvantages to build reasonable hypotheses for the ancestor of African great apes and human clade.

4 Distribution

Both genera *Ouranopithecus* and *Graecopithecus* are certainly known from Greece, the first from three localities in Northern Greece and the other from one locality in Athens (Fig. 2 and Appendix). *Ouranopithecus* is also reported outside Greece by the species *O. turkae* that has been described from the early Turolian Turkish locality Çorakyerler (Guleç et al. 2007). The taxonomic value of this species debated, considering that the Çorakyerler hominoid belongs to a different genus (Begun 2009). The second Greek hominoid *Graecopithecus* is probably known from Bulgaria by an isolated P4 from the locality Azmaka (~7.2 Ma) referred to as cf. *Graecopithecus* sp. (Fuss et al. 2017).

5 Systematic Paleontology

Hominidae Gray, 1825
***Ouranopithecus* Bonis and Melentis, 1977**

Type Species *Dryopithecus macedoniensis* Bonis et al., 1974.

★Ouranopithecus macedoniensis (Bonis and Melentis, 1977)

Nomenclatural and Taxonomical History *Dryopithecus macedoniensis* in Bonis et al., 1974 (initial identification, new species); *Ouranopithecus macedoniensis* in Bonis and Melentis, 1977 (new genus).

Type Species LGPUT-RPl-54 (holotype), mandible of a female sub-adult individual with both tooth rows (Fig. 2d), described and figured by Bonis et al. (1974: p.1393; Fig. 1), and housed in the Laboratory of Geology and Paleontology, University of Thessaloniki.

Type Locality Ravin de la Pluie, Axios Valley, Greece, late Miocene, MN10.

Distribution Besides the type locality, *Ouranopithecus* is known from the Greek late Miocene localities Xirochori 1 of Axios Valley and Nikiti 1 in Chalkidiki (Figs. 3, 4 and 5). A hominoid, under the name *Ouranopithecus turkae*, is referred from Turkey (Guleç et al. 2007), but its attribution is debated (Begun 2009).

Remarks During the beginning of the 1970s, a new series of excavations started in the late Miocene fossiliferous sites of Axios Valley (Macedonia, Greece), which were known since the beginning of the twentieth century, when C. Arambourg collected a quite rich mammal fauna from several localities of the area (Arambourg and Piveteau 1929). However, the presence of the fossils in the area is earlier referred by Bourcart (1919). During the new field campaign in Axios Valley, the prospecting of

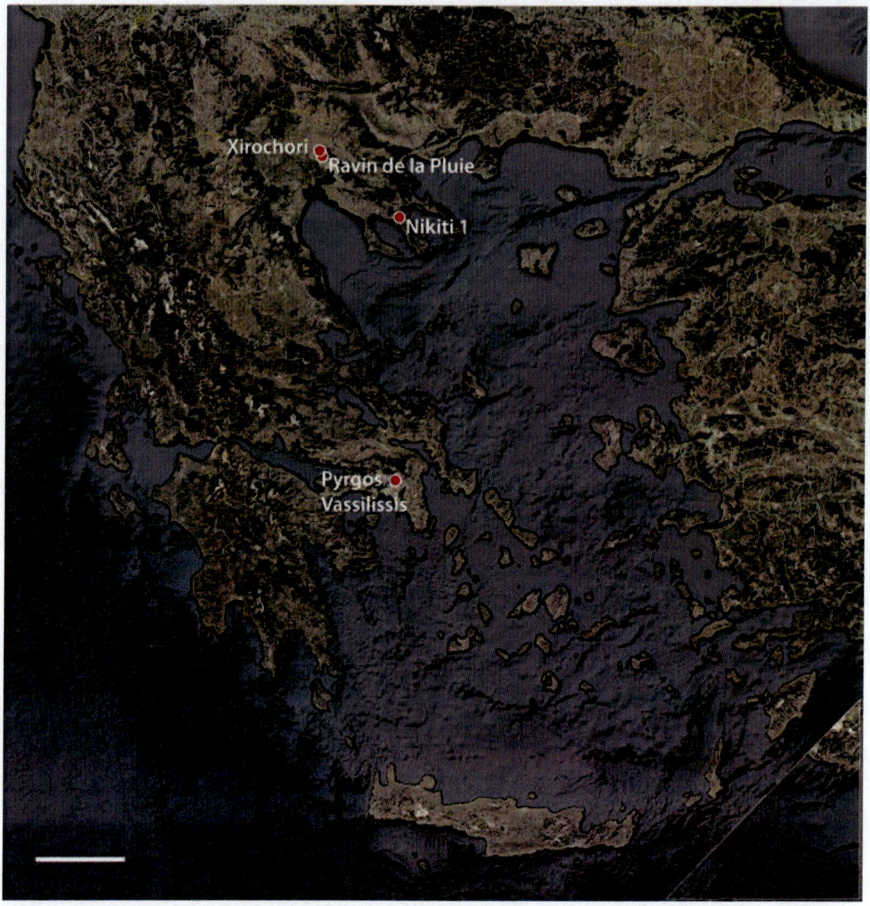

Fig. 2 Map of Greece showing the geographic distribution of the most important localities with Miocene hominoid fossils. See Appendix for more information. Image exported from Google Earth Pro © 2019, map data from US Dept. of State Geographer, SIO, NOAA, U.S. Navy, NGA, GEBCO, image from Landsat/Copernicus. Scale bar equals 80 km, North faces upward

the area led to the discovery of a new locality, named Ravin de la Pluie, which provided a rich mammal fauna including the hominoid *O. macedoniensis* (Bonis et al. 1974). The site was discovered after a heavy rain, and therefore, the locality was named Ravin de la Pluie (= ravine of rain). Among the first collected fossils from the new locality was a primate mandible of a "Miocene hominoid". The founders considered it as a gift which came with the rain from the sky (=*ουρανός* in Greek), and thus the name *Ouranopithecus* (= ape from the sky) was given to the new genus.

In 1989 the author, prospecting the area of Axios Valley, discovered a new fossiliferous site, named Xirochori 1, and among the first collected material was a partial cranium of *O.macedoniensis* (Bonis et al. 1990; Fig. 4). One year later, some mammal remains, including the hominoid *O.macedoniensis*, have been found near the village of Nikiti, Chalkidiki Peninsula (Koufos et al. 1991).

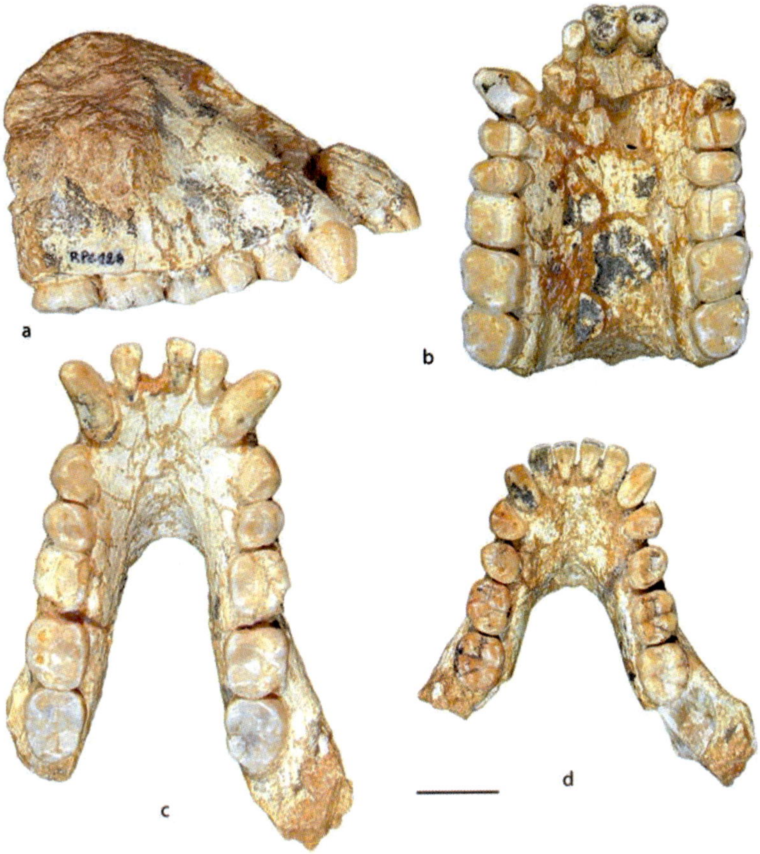

Fig. 3 Selected specimens of *Ouranopithecus macedoniensis* from the late Miocene (MN10) of its type locality, Ravin de la Pluie (Axios Valley). (**a–b**) Maxilla of a male individual in (**a**) lateral, and (**b**) occlusal views, (**c**) mandible of a male individual in occlusal view, (**d**) mandible of a female individual (holotype) in occlusal view. Scale bar equals 2 cm

The taxonomic status of *Ouranopithecus* was discussed for a long time as it was synonymized with *Graecopithecus* (Andrews et al. 1996 and references therein). However, the comparison of the single known badly preserved mandible of *Graecopithecus* with the *Ouranopithecus* material suggests that the last taxon is clearly different and older than *Graecopithecus* (Koufos and Bonis 2005 and references therein). Recently, Fuss et al. (2017) proposed some characters distinguishing the two taxa, while magnetochronology suggests an estimated age of 7.175 Ma for *Graecopithecus*, quite younger than that of *Ouranopithecus* dated from 9.6–8.7 Ma. The majority of *Ouranopithecus* fossils originate from Ravin de la Pluie, corresponding to more than 25 juvenile, adults, and old male and female individuals.

O. macedoniensis is characterized by large body size (in the range of variation of gorillas), strong sexual dimorphism expressed in the overall size of the dentition and especially in the size and morphology of the canines, short and vertical premaxilla,

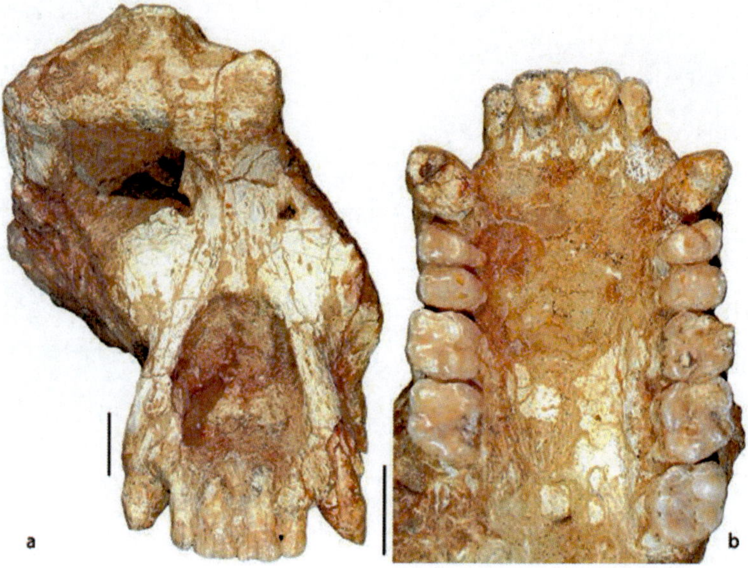

Fig. 4 The partial cranium of a male individual of *Ouranopithecus macedoniensis* from the late Miocene (MN10) of Xirochori 1 (Axios Valley) in (**a**) frontal, and (**b**) occlusal views. Scale bars equal 2 cm

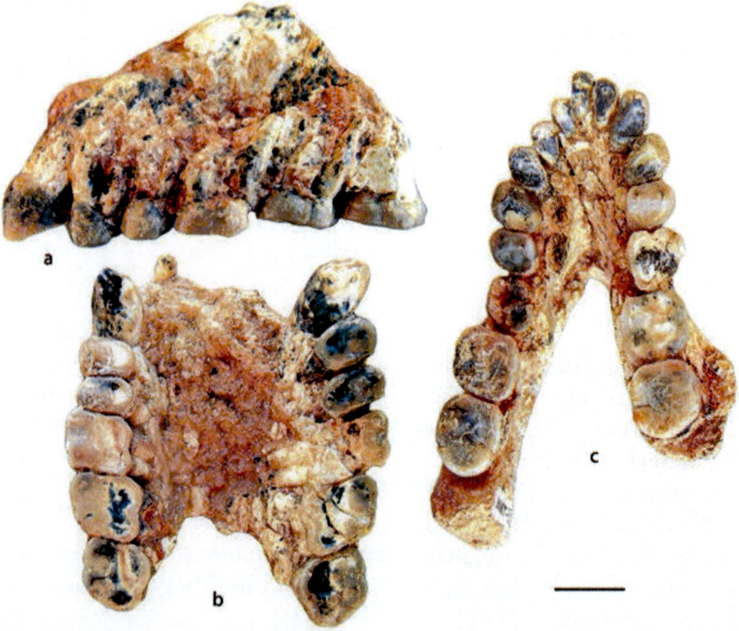

Fig. 5 Selected specimens of *Ouranopithecus macedoniensis* from the late Miocene (MN10) of Nikiti 1 (Chalkidiki Peninsula). (**a–b**) maxilla of a female individual in (**a**) lateral, and (**b**) occlusal views, (**c**) mandible of a female individual in occlusal view. Scale bar equals 2 cm

moderate supraorbital torus with a glabellar depression, large interorbital distance, relatively short nasals, small and low quadrangular orbits, subvertical nasal aperture margins, primitive nasoalveolar area with a large fossa incisiva, strong and low-rooted zygomatic arches, narrow and convex mandibular condyle, low and thick horizontal ramus, long planum alveolare, well-developed fossa genioglossa, strong superior and inferior torus, shovel-like and strongly heteromorphic upper incisors, relatively reduced canines, relatively symmetric upper premolars, low cusps in the molars, more symmetric p3 without honing facet, thick enamel, and absent or very weak cingulum. The study of the associated rich mammal fauna from Ravin de la Pluie and Nikiti 1 and its comparison with the modern ones from known environments, and the dental microwear of the hominoid and bovids suggest that *Ouranopithecus* was living in an open, savannah-like environment with low vegetation (bushes, shrubs, small trees) and thick herbaceous layer (Bonis et al. 1992, 1999, Koufos 2006, Merceron et al. 2005, 2007, 2016). It was a hard-object feeder, feeding on roots, tubers, nuts, and graminoids (Merceron et al. 2005), without excluding the possibility of feeding on soft-objects during the rain period, when it was possible to find fresh fruits and shoots.

***Graecopithecus* von Koenigswald, 1972**

Type Species *Graecopithecus freybergi* von Koenigswald, 1972.

★*Graecopithecus freybergi* von Koenigswald, 1972

Nomenclatural and Taxonomical History *Mesopithecus pentelicus* in Freyberg 1951 (initial identification); *Graecopithecus freybergi* in von Koenigswald, 1972 (new genus and species).

Type Specimen RIM 438/387 (holotype), fragmentary mandible preserving the roots of the right P4 and M1 and the worn M2 (Fig. 6), housed in the collection of B. von Freyberg, Erlangen (von Koenigswald 1972).

Type Locality Pyrgos Vassilissis, Attica, late Miocene, MN12.

Distribution It is only known from its type locality and possibly from Bulgaria; a single P4 from the locality Azmaka was described as cf. *Graecopithecus* sp. (Fuss et al. 2017), representing a possible presence of the genus outside Greece.

Remarks The Pyrgos Vassilissis mandible has been originally attributed to *Mesopithecus pentelicus*, and quite later, it was described as a new genus and species of hominoids. For a long time, there was a debate concerning its synonymy with the other Greek hominoid *O.macedoniensis*. Some authors synonymized them, and others considered that they are different proposing to keep the name *G. freybergi* only for the Pyrgos mandible until finding more evidences for its taxonomic value (Koufos and Bonis 2005 and ref. therein; see also above).

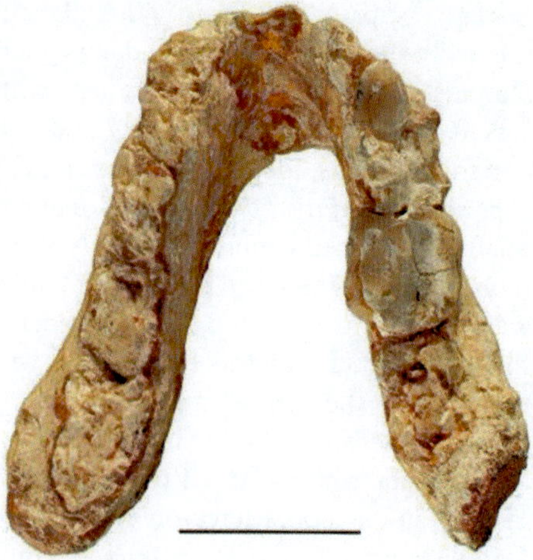

Fig. 6 The only known specimen of *Graecopithecus freybergi* from the late Miocene (MN12) of Pyrgos Vassilissis (Attica), a mandible in occlusal view, photo by J. Fuss. Scale bar equals 2 cm

Acknowledgments Many thanks to J. Fuss for providing the photo of *Graecopithecus freybergi*. I would also like to thank D. Youlatos for comments that improved this manuscript.

Appendix

List of Greek fossil localities containing Miocene hominoid fossils. Type localities are marked with bold. Locality numbers refer to the collection numbers of the PaleoBiology Database (PBDB)

Localities[PBDB No]	Age (MN; GPTS in Ma)	Taxon
Pyrgos Vassilissis[195555]	**Middle Turolian (MN12; ~7.175)**	***Graecopithecus freybergi*[1]**
Nikiti 1[202729]	Terminal Vallesian (MN10; 9.3–8.7)	*Ouranopithecus macedoniensis*[2]
Ravin de la Pluie[191070]	**Late Vallesian (MN10; ~9.3)**	***Ouranopithecus macedoniensis*[3]**
Xirochori 1[195490]	Late Vallesian (MN10; ~9.6)	*Ouranopithecus macedoniensis*[4]

GPTS Geomagnetic Polarity Time Scale

[1]von Koenigswald (1972), [2]Koufos et al. (1991), [3]Bonis et al. (1974), [4]Bonis et al. (1990)

References

Andrews P (1992) Evolution and environment in the hominoidea. Nature 360:641–646

Andrews P, Harrison T, Delson E, Bernor RL, Martin L (1996) Distribution and biochronology of European and Southwest Asian Miocene catarrhines. In: Bernor RL, Fahlbusch V, Mittmann H-W (eds) The evolution of western Eurasian Neogene mammal faunas. Columbia University Press, New York, pp 168–207

Arambourg CJ, Piveteau J (1929) Les Vertébrés du Pontien de Salonique. Ann Paleontol 18:59–138

Begun DR (1994) Relations among the great apes and humans: new interpretations based on the fossil great ape Dryopithecus. Am J Phys Anthr 37(S19):11–63

Begun DR (2002) European hominoids. In: Hartwig W (ed) The primate fossil record. Cambridge University Press, Cambridge, pp 339–368

Begun DR (2009) Dryopithecins, Darwin, de Bonis, and the European origin of the African apes and human clade. Geodiversitas 31:789–816

Begun DR (2014) Fossil record of Miocene hominoids. In: Henke W, Tattershall I (eds) Handbook of palaeoenthropology. Springer, Berlin/Heidelberg/New York, pp 1–66

Benefit BR, McCrossin ML (1995) Miocene hominoids and hominid origins. Ann Rev Anthropol 24:237–256

Böhme M, Spassov N, Ebner M, Geraads D, Hristova L, Kirscher U, Kötter S, Linnemann U, Prieto J, Roussiakis S, Theodorou G, Uhlig G, Winklhofer M (2017) Messinian age and savannah environment of the possible hominin *Graecopithecus* from Europe. PLoS ONE 12:e0177347

Bonis L, Koufos GD (2014) First discovery of postcranial bones of *Ouranopithecus macedoniensis* (Primates, Hominoidea) from the late Miocene of Macedonia (Greece). J Hum Evol 74:21–36

Bonis L, Bouvrain G, Geraads D, Melentis JK (1974) Première découverte d'un primate hominoïde dans le Miocène supérieur de Macédoine (Grèce). C R Acad Sci Paris 278:3063–3066

Bonis L, Bouvrain G, Geraads D, Koufos GD (1990) New hominid skull material from the late Miocene of Macedonia in Northern Greece. Nature 345:712–714

Bonis L, Bouvrain G, Geraads D, Koufos GD (1992) Diversity and palaeoecology of Greek late Miocene mammalian faunas. Palaeogeogr Palaeoclimatol Palaeoecol 19:99–121

Bonis L, Bouvrain G, Koufos GD (1999) Palaeoenvironments of the hominoid primate *Ouranopithecus* in the late Miocene deposits of Macedonia, Greece. In: Agusti J, Rook L, Andrews P (eds) Hominoid Evolution and climatic change in Europe. Cambridge University Press, London, pp 205–237

Bourcart J (1919) Note préliminaire sur les terrains sédimentaires de la région de Salonique. C R Som Soc Geol Fr 8:77–79

Brunet M, Guy F, Pilbeam D, Mackaye HT, Likius A, Ahounta D, Beauvilain A, Blondel C, Bocherens H, Boisserie JR, De Bonis L (2002) A new hominid from the Upper Miocene of Chad, Central Africa. Nature 418:145

Casanovas-Vilar I, Alba DM, Garcés M, Robles JM, Moyà-Solà S (2011) Updated chronology for the Miocene hominoid radiation in Western Eurasia. Proc Nat Acad Sci 108:5554–5559

Clarke R (2012) The history of research in human evolution in Africa and what lessons have been learned. In: Sanz N (ed) Human origin sites and the World Heritage Convention in Africa. World Heritage Papers 33:43–68

Dean D, Delson E (1992) Second gorilla or third chimp? Nature 359:676–677

Freyberg B von (1951) Die Pikermi-fauna von tour la Reine (Attica). Ann Geol Pays Helleniques 3:7–10

Fuss J, Spassov N, Begun DR, Böhme M (2017) Potential hominin affinities of *Graecopithecus* from the Late Miocene of Europe. PLoS ONE 12(5):e0177127

Güleç ES, Sevim A, Pehlevan C, Kaya F (2007) A new great ape from the late Miocene of Turkey. Anthropol Sci 115(2):153–158

Harvati K (this volume) The hominin fossil from Greece. In: Vlachos E (ed) The Fossil Vertebrates of Greece Vol. 1 – Basal vertebrates, amphibians, reptiles, afrotherians, glires, and primates. Springer, Cham

Ioannidou M, Koufos GD, Bonis Lde, Harvati K (2019a) 3D geometric morphometrics analysis of mandibular fragments of *Ouranopithecus macedoniensis* (late Miocene, Central Macedonia, Greece). Am J Phys Anthrop https://doi.org/101/1002/ajpa.23900

Ioannidou M, Koufos, GD, Bonis L de, Harvati K (2019b) A new three-dimensional geometric morphometrics analysis of the *Ouranopithecus macedoniensis* cranium (Late Miocene, Central Macedonia, Greece). Am J Phys Anthropol 170:295–307

Kelley J, Andrews P, Alpagut B (2008) A new hominoid species from the Middle Miocene site of Paşalar, Turkey. J Hum Evol 54:455–479

Köhler M, Moya-Sola S, Alba, DM (2001) Eurasian hominoid evolution in the light of recent *Dryopithecus* findings. In: Bonis Lde, Koufos GD, Andrews, P. (eds) Hominoid evolution and climatic change in Europe: Volume 2: Phylogeny of the neogene hominoid primates of Eurasia, Cambridge University Press, Cambridge, pp. 192-212

Koufos GD (1993) Mandible of *Ouranopithecus macedoniensis* (Hominidae, Primates) from a new late Miocene locality of Macedonia (Greece). Am J Phys Anthropol 91(2):225–234

Koufos GD (1995) The first female maxilla of the hominoid *Ouranopithecus macedoniensis* from the late Miocene of Macedonia, Greece. J Hum Evol 29(4):385–399

Koufos GD (2006) Palaeoecology and chronology of the Vallesian (late Miocene) in the Eastern Mediterranean region. Palaeogeogr Palaeoclimatol Palaeoecol 234:127–145

Koufos GD, Bonis L (2005) The late Miocene hominoids *Ouranopithecus* and *Graecopithecus*. Implications about their relationships and taxonomy. Ann Paleontol 9:227–240

Koufos GD, Syrides GE, Koliadimou KK, Kostopoulos DS (1991) Un nouveau gisement de Vertébrés avec hominoide dans le Miocene supérieur de Macédoine (Gréce). C R Acad Sci Paris 313:691–696

McNulty KP (2005) A geometric morphometric assessment of the hominoid supraorbital region: affinities of the Eurasian Miocene hominoids *Dryopithecus*, *Graecopithecus*, and *Sivapithecus*. In: Slice D (ed) Modern morphometrics in physical anthropology. Kluwer, New York, pp 349–373

McNulty KP (2010) Apes and Tricksters: The evolution and diversification of humans' closest relatives. Evol Educ Outreach 3:322–332

Merceron G, Blondel C, de Bonis L, Koufos GD, Viriot L (2005) A new method of dental microwear analysis: application to extant Primates and *Ouranopithecus macedoniensis* (Late Miocene of Greece). Palaios 20:551–561

Merceron G, Blondel C, Viriot L, Koufos GD, Bonis L de (2007) Dental microwear analysis of bovids from the Vallesian (late Miocene) of Axios valley in Greece: reconstruction of the habitat of *Ouranopithecus macedoniensis* (Primates, Hominoidea). Geodiversitas 29:421–433

Merceron G, Novello A, Scott RS (2016) Paleoenvironments inferred from phytoliths and dental microwear texture analyses of meso-herbivores. In: Koufos GD, Kostopoulos DS (eds) Palaeontology of the upper Miocene vertebrate localities of Nikiti (Chalkidiki Peninsula, Macedonia, Greece). Geobios 49:135–146

Pickford M, Senut B, Gommery D, Treil J (2002) Bipedalism in *Orrorin tugenensis* revealed by its femora. C R Palevol 1:191–203

Pilbeam DR (2002) Perspectives on the Miocene Hominoidea. In: Hartwig WC (ed) The primate fossil record. Cambridge University Press, Cambridge, pp 303–310

Rögl F (1999) Circum-Mediterranean Miocene paleogeography. In: Rössner G, Heissig K (eds) The land mammals of Europe. Verlag Dr F. Pfeil, München, pp 39–48

Senut B, Pickford M, Gommery D, Mein P, Chcboi K, Coppens Y (2001) First hominid from the Miocene (Lukcino Formation, Kenya). C R Acad Sci Paris 332:137–144

Simons EL (1965) New fossil apes from Egypt and the initial differentiation of Hominoidea. Nature 205:135–139

Stevens NJ, Seiffert ER, O'connor PM, Roberts EM, Schmitz MD, Krause C, Gorscak E, Ngasala S, Hieronymus TL, Temu J (2013) Palaeontological evidence for an Oligocene divergence between Old World monkeys and apes. Nature 497:611

Von Koenigswald GHR (1972) Ein Unterkiefer eines fossilen Hominoiden aus dem Unterpliozan Griechenlands. Proc K Ned Akad Wet 75:385–394

The Hominin Fossil Record from Greece

Katerina Harvati

1 Introduction

The study of the human fossil record is by nature an interdisciplinary endeavor. Human paleontology differs from other paleontology subfields not only due to the rarity of human fossils but because it is tightly intertwined with the study of Paleolithic archaeology and early prehistory, as the human fossil record cannot be properly understood without consideration of the behavioral record testified by material cultural remains produced by early humans, most commonly represented by lithic artifacts. Although material cultural remains do not appear until several million years after the earliest purported human ancestor, *Sahelanthropus tchadensis* dating to 7–6 Ma (Brunet et al. 2005)—with the Lomekwian lithic industry possibly representing the earliest such evidence at ca. 3.3 Ma (Harmand et al. 2015)—they take on increasing importance in later periods and become an inextricable part of the human fossil record, especially from the Middle Pleistocene onward. Modern paleoanthropology is not only tightly intertwined with Paleolithic archaeology but relies on an interdisciplinary approach involving a number of disciplines, including genetics, paleontology, zooarchaeology, geoarchaeology, taphonomy, geoarchaeology, and dating methods, as well as primatology and cultural anthropology.

Paleoanthropology as a discipline is not well established in Greece, and, until recently, there had been relatively little systematic research in human evolution and in the Paleolithic era. Previous work has mainly focused on single important specimens, most notably the Petralona cranium (e.g., Kurtén and Poulianos 1977; Poulianos 1981, 1982). As a result, the known record suffers from important

K. Harvati (✉)
Paleoanthropology, Senckenberg Centre for Human Evolution and Palaeoenvironments, Institute for Archaeological Sciences, Department of Geosciences, Eberhard Karls Universität Tübingen, Tübingen, Germany
e-mail: katerina.harvati@ifu.uni-tuebingen.de

E. Vlachos (ed.), *Fossil Vertebrates of Greece Vol. 1*,
https://doi.org/10.1007/978-3-030-68398-6_19

shortcomings, the most serious of which is the lack of context for many of the most important specimens, as few have been recovered from controlled excavations. Despite this, the known human fossil record from Greece comprises a number of fossil human individuals representing different taxa. Recently, new research has focused both on systematic interdisciplinary fieldwork—aiming to discover new Paleolithic/paleoanthropological sites preserving evidence of early human presence and activity in Greece (e.g., Panagopoulou et al. 2015; Galanidou et al. 2016; Harvati et al. 2018; Thompson et al. 2018; Tourloukis and Harvati 2018)—and on the state-of-the-art reconstruction and analysis of the existing human fossil record, aiming to clarify their taxonomy, chronology, and phylogenetic relationships (e.g., Harvati 2009; Harvati et al. 2011, 2013, 2019; Röding et al. in press). This work has uncovered the earliest currently known evidence of human occupation of the region at the site Marathousa 1, Megalopolis basin, dated by multiple dating approaches to ca. 400–500 ka (Blackwell et al. 2018; Jacobs et al. 2018; Tourloukis et al. 2018), as well as the earliest known possible evidence of early *Homo sapiens* in Eurasia at ca. 210 ka at Apidima, Mani (Harvati et al. 2019). These results highlight the great potential of Greece and SE Europe to shed light on human evolution, dispersals, and adaptations.

2 Phylogenetic Relationships

The human fossil record of Greece spans part of the Middle and the Late Pleistocene and, similarly to the rest of Europe, evidences the species *Homo heidelbergensis*, *Homo neanderthalensis*, and *Homo sapiens*. *H. sapiens* and *H. neanderthalensis* are morphologically distinct taxa (e.g., Harvati et al. 2004) differentiated by a suite of respectively derived features, as well as, possibly, behavioral characteristics (see, e.g., Harvati 2015). They are considered sister species, who are known to have hybridized repeatedly when their ranges overlapped, as occurred during the last out of Africa expansion of modern humans ca. 70–50 ka (e.g., Green et al. 2010; Fu et al. 2015; Villanea and Schraiber 2019). Their relationship to *H. heidelbergensis* is not completely resolved, partly due to the difficulty in the definition of the latter taxon. The holotype of this species is the Mauer specimen, a mandible without associated cranial remains, making the definition of its hypodigm problematic. Cranial remains from Arago, France, were assigned to *H. heidelbergensis* on the basis of similarities between the Arago and Mauer mandibles, and further Middle Pleistocene fossil crania from Europe, Africa, and possibly Asia have been attributed to this taxon, originally on the basis of similarities to the Arago material (e.g., Harvati 2007 and references therein). No agreement exists over the integrity of this sample—in the past sometimes loosely referred to as the grade "archaic *Homo sapiens*" (Stringer et al. 1979), a term that is no longer in use. The question of the hypodigm and geographic/temporal range of *Homo heidelbergensis* is debated, as is its phylogenetic position (see, e.g., Hublin 2009; Stringer 2012; Roksandic et al. 2018). Generally the European members of this group (*H. heidelbergensis* sensu stricto) are considered ancestral to Neanderthals, possibly a chronospecies of the Neanderthal lineage. Similarly, the morphologically similar African specimens are sometimes assigned to *H. rhodesiensis* and considered ancestral to *H. sapiens*. However, some authors consider *H. heidelbergensis* (sensu lato) as a wide-ranging species occurring in both Africa and Europe and the last common ancestor of both *H. neanderthalensis* and *H. sapiens* (e.g., Rightmire 2008).

The position of the little known species *H. antecessor* from Spain, dated to ca. 800 ka, is crucial in this discussion. Unlike *H. heidelbergensis*, *H. antecessor* shows some modern human-like facial features and has also been proposed as the last common ancestor of *H. sapiens* and *H. neanderthalensis* (e.g., Bermúdez de Castro et al. 1997). Under this scheme, the *H. heidelbergensis* morphology might already be derived in the direction of *H. neanderthalensis* and therefore not ancestral to *H. sapiens*. The late chronology proposed for specimens such as Petralona (Grün 1996) and Broken Hill (Grün et al. 2020), the type specimen of *H. rhodesiensis*, further complicates this picture. The date of ca. 300 ka for both Broken Hill, which shows no *H. sapiens*-derived features (Grün et al. 2020), and the early *H. sapiens* specimens from Jebel Irhoud (Hublin et al. 2017) suggests either a late survival of *heidelbergensis*-like populations or an exclusion of this specimen and other similar samples from *H. sapiens* ancestry. Similarly, the proposed late date of Petralona at ca. 250–300 ka (Grün 1996), although highly uncertain, suggests a possible late survival of *H. heidelbergensis* in Greece.

The origin of *H. neanderthalensis* is relatively well understood, with European Middle Pleistocene specimens generally considered ancestral to this taxon. The latter exhibit "incipient" (e.g., Dean et al. 1998) or in some cases well-developed (Martinón-Torres et al. 2012) Neanderthal-derived features, as well as Neanderthal-like nuclear DNA (Meyer et al. 2016). The origin of *H. sapiens*, on the other hand, as well as the timing of the split between the two lineages, is less clear. An emerging consensus sees the origin of modern humans to have occurred earlier than previously thought, dating back to 300 ka based on both the fossil record (Hublin et al. 2017) and genetic data (Schlebusch et al. 2017; Lipson et al. 2020). The origin of *H. sapiens* is also now considered to have involved multiple, geographically and morphologically diverse, Middle Pleistocene African populations in an "African Multiregionalism" scenario (e.g., Scerri et al. 2018). It is currently estimated that the *sapiens* and Neanderthal lineages split between 550 and 765 ka on the basis of paleogenetic evidence (Meyer et al. 2016).

3 Distribution

With the exception of the Petralona Cave (N. Greece), all other occurrences of hominins from Greece come from sites in the Peloponnese, S. Greece (Fig. 1 and Table 1). Also, with the exception of the dubious location of the isolated find from Megalopolis (see below), all other occurrences come from caves.

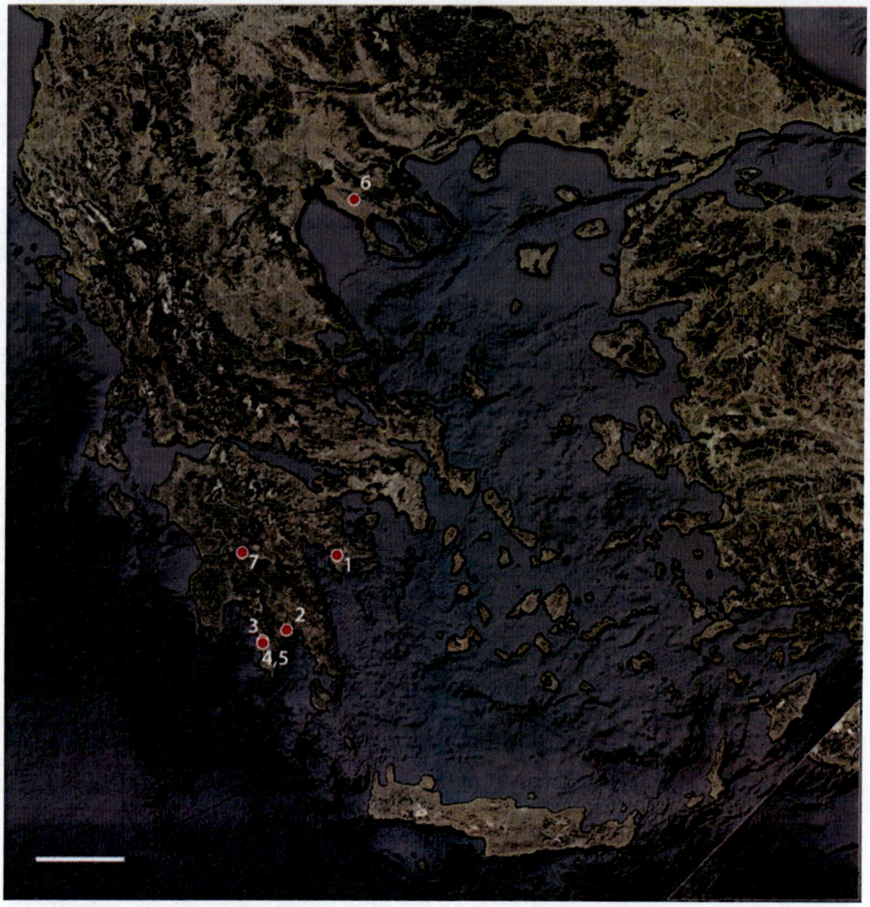

Fig. 1 Map of the occurrences of fossil hominins from Greece. See Appendix for further information. **1**, Franchthi Cave; **2**, Lakonis I; **3**, Kalamakia Cave; **4**, Apidima Cave Γ; **5**, Apidima Cave A; **6**, Petralona Cave; **7**, Megalopolis lignite. See Table 1 for more information. Image exported from Google Earth Pro © 2019, map data from US Dept. of State Geographer, SIO, NOAA, US Navy, NGA, GEBCO, image from Landsat/Copernicus. Scale bar equals 80 km, North faces upward

4 Systematic Paleontology

Hominidae Gray, 1825
Homininae Gray, 1825
***Homo* Linnaeus,** 1758

Type Species *Homo sapiens* Linnaeus, 1758.

Remarks An isolated human upper M3 of indeterminate, possible Middle Pleistocene age was collected together with a large fossil mammal assemblage

Table 1 List of Greek localities containing hominin fossils. Locality numbers refer to the collection numbers of the PaleoBiology Database (PBDB)

Localities[PBDB No]	Age	Taxon	References
Franchthi Cave[182709]	Latest Pleistocene to Early Holocene	*Homo sapiens*	1
Lakonis I Cave[207778]	Late Pleistocene (ca. 44–38 ka BP)	*Homo neanderthalensis*	1–5
Kalamakia Cave[184245]	Late Pleistocene (ca. 100–39 ka BP)	*Homo neanderthalensis*	5, 6
Apidima Cave Γ[32082]	? Upper Paleolithic	*Homo sapiens*	1, 7–9
Apidima Cave A[32059]	Middle Pleistocene*	*Homo neanderthalensis*, early *Homo sapiens*	1, 10–14
Petralona Cave[183123]	Middle Pleistocene	*Homo heidelbergensis*	1, 15, 16
Megalopolis paleolake basin	? early Middle Pleistocene	*Homo* sp.	5, 17

*The absolute direct dates of the hominin remains are ca. 170 ka BP for *H. neanderthalensis* and ca. 210 ka BP for early *H. sapiens*. Although the two specimens were found in close proximity, they were originally deposited in different parts of the cave and eroded into their final position in a fissure close to the cave ceiling at a later date. The surrounding matrix dates to ca. 150 ka and is younger than both specimens, as would be expected (see Harvati et al. 2019)

[1]Harvati et al. (2009), [2]Harvati et al. (2003), [3]Panagopoulou et al. (2002–2004), [4]Elefanti et al. (2009), [5]Harvati (2016), [6]Harvati et al. (2013), [7]Pitsios (1985), [8]Pitsios (1995), [9]Darlas (1995), [10]Harvati and Delson (1999), [11]Pitsios (2002), [12]Harvati et al. (2011), [13]Harvati et al. (2019), [14]Bräuer et al. (2020), [15]Stringer (1974), [16]Stringer et al. (1979), [17]Sickenberg (1975)

during survey (1962–1965) of the Megalopolis basin by the Geological Society of Hannover (Sickenberg 1975). The fossil assemblage was assigned to the early Middle Pleistocene by Sickenberg (1975). However neither the exact location of the finds nor its actual chronology can be ascertained, and concerns have been expressed that it might represent an intrusive *H. sapiens* (Sickenberg 1975).

The classification of the Megalopolis UM3 is hindered by the high variability of this element, which complicates the assessment of its affinities, as well as by the strong abrasion of its crown surface. The latter appears to be at least in part the result of acid etching by Xirotiris et al. (1979), which has left the specimen damaged and with little remaining informative morphology on its crown surface. A preliminary comparative analysis of crown dimensions showed that the Megalopolis specimen is relatively small, overlapping with values reported in the literature for recent *H. sapiens*, but also with the lower end of the range of variation of *H. neanderthalensis*, *H. heidelbergensis*, as well as *H. erectus* (Harvati 2016). Surprisingly, its crown shape index, calculated from its buccolingual dimensions, was more similar to African early *Homo* than to later *Homo* species, as represented in that small dataset (Harvati 2016). More recently, Röding et al. (in press) assessed the Megalopolis crown shape using outline semilandmarks analysis. Results tentatively suggest affinities with the Neanderthal lineage, although additional analysis, perhaps of internal tissue proportions, would be needed to confirm this finding.

Interestingly, recent field research in the Megalopolis basin, undertaken under the auspices of the Ephoreia of Paleoanthropology and Speleology and directed by Dr. Panagopoulou in collaboration with the author and her team, have confirmed the presence of humans in the Middle Pleistocene Megalopolis basin on the basis of archaeological evidence. This consists of lithic and bone artifacts associated with faunal remains exhibiting cutmarks and other human modifications at the sites Marathousa 1, dated to ca. 400–500 ka, and Choremi 3, dated to ca. 260 ka (Harvati et al. 2018 and articles therein; Panagopoulou et al. 2018; Thompson et al. 2018). Marathousa 1, a Lower Paleolithic elephant butchering site, is currently the oldest known chronometrically dated archaeological site in Greece (Panagopoulou et al. 2018; Harvati et al. 2018 and references therein). It is therefore conceivable that the Megalopolis M3 represents these populations. Ongoing research continues to explore exposed early Middle Pleistocene sediments in this region, aiming to push back even further in time the earliest known appearance of *Homo* in Greece.

Homo heidelbergensis **Schoetensack, 1908**

Nomenclatural and Taxonomic History The nomen *H. heidelbergensis* fell into disuse after the revision of the human evolutionary tree according to the principles of the New Synthesis (e.g., Mayr 1963). Specimens from the Middle and early Late Pleistocene, including Neanderthals, were at the time considered as representing either *H. erectus* or early forms of *H. sapiens* and often collectively referred to as "archaic *H. sapiens*," a grade (Stringer et al. 1979). This term is no longer in use. Current classification schemes reflect the common recognition of Neanderthals and modern humans as distinct lineages, with *H. heidelbergensis* possibly representing their last common ancestor (e.g., Rightmire 2008; Stringer 2012—but see "Phylogenetic Relationships").

Type Specimen M.1, a mandible with dentition, now stored in the collections of the Institute of Geology and Palaeontology, University of Heidelberg, Germany. This is the so-called Mauer mandible, found in 1907 and later designated as the type specimen of *H. heidelbergensis* by Schoetensack (1908).

Type Locality Grafenrain mine, Mauer, near Heidelberg, Germany; early Middle Pleistocene, ca. 500 ka (e.g., Harvati 2007 and references therein; Stringer 2012).

Distribution The geographic distribution of this species is disputed. Outside its type locality, several sites in W. Eurasia (e.g., Arago, Petralona, Steinheim, Sima de los Huesos, Boxgrove) and Africa (Broken Hill, Bodo, Elandsfontein, Ndutu) are proposed, with additional possible occurrences (see Stringer 2016 and references therein). Nevertheless, the question of whether this taxon comprises only European specimens (*H. heidelbergensis* sensu stricto) or can be expanded to include African and possibly Asian material (*H. heidelbergensis* sensu lato) is debated (see "Phylogenetic Relationships"). The European *H. heidelbergensis* is commonly considered as part of the *H. neanderthalensis* lineage, although a primitive status is also proposed.

Remarks The only broadly accepted occurrence of *Homo heidelbergensis* in Greece comes from Petralona Cave, Northern Greece. Discovered in 1960 (Kokkoros and Kanellis 1960; Poulianos 1981, 1982), the Petralona cranium (Fig. 2) is in excellent state of preservation and perhaps one of the most complete cranial specimens in the fossil human record of Europe. It is housed in the collections of the Museum of Geology, Palaeontology, and Palaeoanthropology of the School of Geology, Aristotle University of Thessaloniki, Greece.

The Petralona cranium is thought to be of Middle Pleistocene age with a proposed date of ca. 250–300 ka, although there is a high degree of uncertainty about its chronological placement (see Harvati et al. 2009; Grün 1996). Along with other early European *H. heidelbergensis*, it has been described as showing incipient Neanderthal facial characteristics (e.g., Dean et al. 1998). However, it also shows strong overall similarities with African penecontemporary specimens, particularly with the Broken Hill (Kabwe) cranium (e.g., Stringer 1974; Stringer et al. 1979; Harvati 2009; but see Freidline et al. 2012).

Given their Middle Pleistocene chronology and the Lower Paleolithic character of their lithic assemblages, the sites recently identified in the Megalopolis basin (Marathousa 1, Choremi 3) possibly also represent the remnants of activity of *H. heidelbergensis* populations in the Peloponnese. Although no clear affinities with this taxon could be observed for the Megalopolis molar (above), its tentative assignment to the Neanderthal lineage (Röding et al. in press) and possible Middle Pleistocene chronology also make it a potential candidate for this taxon.

Homo neanderthalensis **King, 1864**

Nomenclatural and Taxonomic History After discovery of the type specimen from the Neander Valley in 1856, the antiquity and status of these remains as human fossils were fiercely debated, due to their vast philosophical implications for human

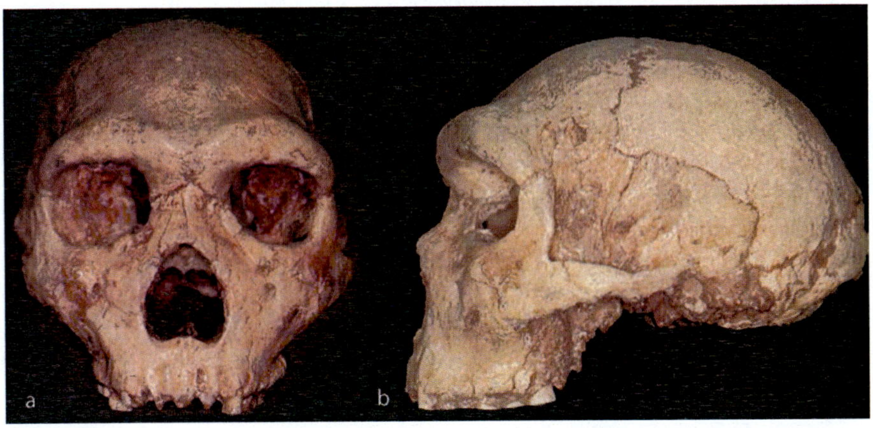

Fig. 2 The Petralona cranium, in (**a**) frontal and (**b**) lateral views. Photographs copyright and courtesy of E. Delson

origins in light of the nascent development of evolutionary theory. Pathological conditions were proposed as alternative reasons for their peculiar cranial and postcranial anatomy. This debate was only settled with the eventual discovery of additional skeletons of similar morphology associated with lithic artifacts and extinct faunal assemblages from other sites in Europe. *Homo neanderthalensis* was named by King (1864) nearly a decade after discovery (see Harvati 2015 and references therein). The validity of this species has been questioned, with the dominant view on Neanderthals alternating between two extremes throughout the twentieth century: on the one hand, *H. neanderthalensis* has been considered a distinct species with no ancestral relationship to *H. sapiens*; on the other hand, it has been seen as a subspecies or even population of *H. sapiens* and possibly ancestral to modern Europeans (see Harvati 2015 and references therein). The validity of *H. neanderthalensis* as a distinct lineage is currently generally recognized on the basis of their anatomical, genetic, and possibly behavioral differences (see Harvati 2015 and references therein). However, paleogenetic evidence has indicated repeated, if limited, interbreeding with *H. sapiens*, leading to a small percentage of Neanderthal ancestry for all non-African modern humans (e.g., Green et al. 2010; Sankararaman et al. 2014). In light of extensive interbreeding among closely related primate and other mammalian species and genera (e.g., Zinner et al. 2011; Ackermann et al. 2019), *H. neanderthalensis* can be interpreted as the sister species of *H. sapiens*, with the two lineages not having achieved complete reproductive isolation at the time of the major dispersal of modern humans out of Africa 70–50 ka when their ranges overlapped.

Type Specimen Skull cap and a partial skeleton, stored in the collections of LVR-Landesmuseum (Bonn, Germany; King (1864:Pl. I)).

Type Locality Cave or fissure intersecting the southern rocky side of the deep narrow Neanderthal valley, river Düssel, near Hochdal, Germany (King 1864); Late Pleistocene, ca. 40 ka (Schmitz et al. 2002).

Distribution Members of *H. neanderthalensis* are known from across Western Eurasia, including Europe, the Near East (e.g., Israel, Syria, Iraq), reaching as far East as Uzbekistan, and the Altai (Siberia; see Harvati 2015 and references therein). In Greece, fossil remains of *H. neanderthalensis* are known from the Late Pleistocene of Lakonis and Kalamakia and the late Middle Pleistocene of Apidima, all in the Peloponnese (see below). However, Middle Paleolithic archaeological assemblages, generally associated with Neanderthals in Europe, are the most common Paleolithic finds in Greece and occur in most of mainland Greece, as well as on several Ionian and Aegean islands, having also been reported from Crete (Tourloukis and Harvati 2018). These remains suggest that much of present-day Greece was inhabited by Neanderthal populations in the Late Pleistocene.

Remarks *Homo neanderthalensis* is represented in Greece by fossil assemblages from three sites, all of them located in the Mani Peninsula, Southern Peloponnese.

Two of these assemblages derive from excavated contexts and date broadly to the Late Pleistocene (at Lakonis and Kalamakia), while one was extracted from breccia deposits and has been dated to the late Middle Pleistocene (at Apidima).

Lakonis Cave Specimen LKH1 (a left lower third molar) derives from the upper layers of the Lakonis cave, situated on the eastern coast of the Mani Peninsula, near Gytheion. The site was excavated from 1997 to 2011 by the Ephoreia of Paleoanthropology and Speleology (Greek Ministry of Culture). It preserves a rich Middle Paleolithic lithic assemblage through most of the stratigraphic sequence; however, LKH1—the only human fossil recovered at Lakonis—was found in the top-most layer, which yielded an Initial Upper Paleolithic stone tool assemblage and dates to ca. 48–42 ka cal (AMS ^{14}C on charcoal; Harvati et al. 2003, 2009; Panagopoulou et al. 2002–2004; Elefanti et al. 2009; Harvati 2016). LKH1 shows features strongly aligning it with *H. neanderthalensis*, including a large anterior fovea, complex root morphology, a relatively enlarged pulp cavity, and a midtrigonid crest. The latter trait is expressed at very high frequencies in Neanderthals but almost never on modern human lower third molars (Harvati et al. 2003). Also aligning it with *H. neanderthalensis* are its relatively high enamel secretion rates and relatively thin enamel (Smith et al. 2009).

Kalamakia Cave Kalamakia cave is located on the western coast of the Mani Peninsula. Thirteen fragments of human remains were excavated from several layers of Unit IV, and one (KAL3) from the uppermost layer of Unit III, during work by the Ephoreia of Paleoanthropology and Speleology (Greek Ministry of Culture) and the Muséum national d'Histoire naturelle (Paris) between 1993 and 2006. The fossil fragments represent at least eight individuals, two of them juveniles, and date roughly to between 100,000 (U/Th radiometric dating of a marine shell at the Institut de Paléontologie Humaine in Paris, IPH Kal 9304: 109.000 + 14.000/−13.000) and > 39,000 radiocarbon years (^{14}C AMS dating on charcoal at Gif-sur-Yvette in France, GifA 94592) (Harvati et al. 2013; Harvati 2016). The site also yielded Mousterian lithics with Levallois elements (Harvati et al. 2013).

The fossil human assemblage comprises ten isolated teeth KAL2 (left upper third premolar), KAL3 (left upper third? molar), KAL5 (upper fourth premolar), KAL6 (left lower fourth premolar), KAL8 (right upper second? molar), KAL9 (right lower fourth premolar), KAL10 (right upper lateral incisor), KAL11 (left upper central incisor), and KAL12 (left upper deciduous lateral incisor), KAL13 (left? upper deciduous central incisor); one cranial fragment KAL1 (occipital? fragment); and three postcranial elements KAL4 (subadult lumbar vertebra, L3 or L4), KAL7 (fibular shaft fragment), and KAL14 (left navicular). Although not all are taxonomically informative, several elements deriving from throughout the stratigraphic sequence show diagnostic Neanderthal anatomy. Among the dental remains, the two lower premolars and the two upper incisors display combinations of crown features that are observed at very high frequencies among Neanderthals but not in modern humans (Harvati et al. 2013). The two upper incisors (KAL10 and KAL11) show shoveling, lingual tubercles and labial convexity, a combination considered as derived for the *H. neanderthalensis* lineage (e.g., Bailey 2007; Martinón-Torres et al. 2012). The two lower premolars (KAL6 and KAL9) exhibit multiple lingual cusps, a transverse crest and an asymmetric crown, also a combination of traits considered unique for *H. neanderthalensis* (e.g., Bailey 2007; Martinón-Torres et al. 2012). The crown dimensions of the upper third premolar (KAL2) and the upper third molar (KAL3) place them closer to Neanderthals than to early modern humans (Harvati et al. 2013). Finally, the dimensions of the navicular KAL14 also align it with *H. neanderthalensis*, rather than with *H. sapiens* (Harvati et al. 2013). None of elements in the Kalamakia human skeletal assemblage show modern human-derived features.

Apidima Cave A Apidima is a complex of five caves on the western coast of the Mani Peninsula. It was investigated between 1978 and 1985 by the Museum of Anthropology, National and Kapodistrian University of Athens Medical School (Pitsios 1995; Harvati and Delson 1999; Harvati et al. 2009, 2019). Cave A has produced two human fossil crania, found encased in a block of matrix attached to the cave walls and close to the ceiling (Pitsios 1985, 1995, 1999; Harvati and Delson 1999; Harvati et al. 2009, 2011).

Of these two specimens, Apidima 2 (also referred to as LAO 1/S2; Fig. 3) is assigned to *Homo neanderthalensis* (Harvati et al. 2009, 2011, 2019). It preserves most of the cranium, with the exception of the occipital bone, and exhibits a suite of Neanderthal-derived features in both its facial and neurocranial morphology: a continuous, doubly arched, thick and rounded supraorbital torus with no break between the glabellar, orbital, and lateral regions; an anterior position of the nasal root; an inflated infraorbital region; a bi-level morphology of the inferior nasal margin; a rounded "*en bombe*" cranial profile in posterior view; as well as some ancestral features, such as a low and elongated neurocranium (Harvati and Delson 1999;

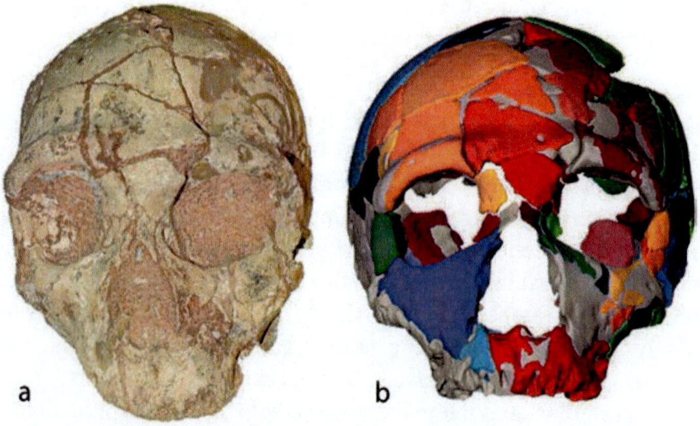

Fig. 3 The Apidima 2 cranium, in (**a**) frontal and (**b**) virtual reconstruction of the frontal view. (Copyright K. Harvati)

Pitsios 2002; Harvati et al. 2009, 2019; Bräuer et al. 2020). Furthermore, it metrically aligns with Neanderthals in both linear (Harvati et al. 2011; Bräuer et al. 2020) and 3D measurements of its facial and neurocranial shape (Harvati et al. 2019).

Apidima 2 was dated directly with the U-series method by Bartsiokas et al. (2017) and later again by Harvati et al. (2019), yielding an age of ca. 170 ka (172 ± 11; Harvati et al. 2019). The same method was also applied to the breccia matrix encasing the specimens, yielding an age of ca. 150 ka, which represents the age of the sediment solidification (Harvati et al. 2019). It is important to note that, although excavation of Cave A purportedly yielded Middle Paleolithic lithic remains (see Harvati and Delson 1999) and a mixed fauna (Tsoukala 1999), these materials derive from the cave floor sediments of Cave A and are not associated with, but rather significantly postdate, the breccia sediments from the ceiling of Cave A and their contents (see Harvati et al. 2009, 2011, 2019).

Pleistocene *Homo sapiens* Linnaeus, 1758

Remarks Early modern human remains, as well as sites dating to the Upper Paleolithic period, are very rare in Greece. Currently, the most important site yielding such fossils is Apidima, on the western coast of the Mani Peninsula. Additional, fragmentary Late Pleistocene human remains have been recovered from Franchthi Cave, also in the Peloponnese. *H. sapiens* is considered to have evolved in Africa as early as ca. 300 ka or earlier on the basis of both fossil and genetic data (Hublin et al. 2017; Schlebusch et al. 2017; Lipson et al. 2020), from multiple morphologically and geographically diverse populations (e.g., Scerri et al. 2018). This taxon dispersed into Eurasia and the rest of the world between 70 and 50 ka, in large part replacing *H. neanderthalensis* as well as other Late Pleistocene taxa in Eurasia by 40–35 ka. Limited interbreeding among the lineages is thought to have occurred at this time, leading to small contributions of genetic ancestry from archaic humans (e.g., on average 2–4% for Neanderthals) to all modern non-Africans. Early members of *H. sapiens* are also known to have dispersed out of Africa in at least one earlier migration, with sites in the Levant dating as early as ca. 185 ka (Mislya Cave, Israel; Hershkovitz et al. 2018). The recent evidence from Apidima Cave, Greece, suggests that this early dispersal event occurred earlier and dispersed further than previously suspected, reaching South East Europe (Harvati et al. 2019), consistent with genetic evidence for interbreeding between Neanderthals and modern humans before 200 ka (Posth et al. 2017; Chen et al. 2020).

Apidima Cave A As mentioned above, Apidima Cave A has yielded two human fossil crania, found encased in a block of matrix attached to the cave walls and close to the ceiling during investigations by the Museum of Anthropology, Medical School, National and Kapodistrian University of Athens (Pitsios 1985, 1995, 1999; Harvati and Delson 1999; Harvati et al. 2009, 2011). Of the two fossil crania, Apidima 1 (also referred to as LAO 1/S1; Fig. 4) was identified as an early *Homo sapiens* by Harvati and colleagues (Harvati et al. 2019), on the basis of its modern human-derived morphology in combination with primitive traits and the absence of Neanderthal-derived features.

Apidima 1 preserves most of the posterior part of the cranium and part of the cranial base. It is undistorted, allowing virtual reconstruction by mirror-imaging. Although found in close proximity to Apidima 2, a Neanderthal (see above), Apidima 1 shows a typical modern human-like rounded posterior cranium, a *H. sapiens*-derived feature. Furthermore, it shows no Neanderthal characteristics. It lacks a Neanderthal-like rounded "*en bombe*" profile in posterior view, a Neanderthal-like occipital convexity and lambdoid flattening (occipital "bun"), or Neanderthal-like occipital torus and associated suprainiac fossa. A small, faint depression above inion does not conform to the Neanderthal suprainiac fossa condition, being much smaller and without clear borders. Its internal configuration, as evaluated on the basis of a CT scan, differs from that described for Neanderthals in that the diploic layer does not appear affected (Harvati et al. 2020). The Apidima 1 depression is therefore similar to those described for African Middle-Late Pleistocene specimens (Bosman et al. 2020). Apidima 1 also exhibits primitive morphology: the parietal walls are nearly parallel and the occipital plane low, both plesiomorphic conditions. However, unlike many Middle Pleistocene specimens, Apidima 1 does not show a steep angulation of the occipital bone or an occipital torus. Principal components analyses of the overall shape of its neurocranium and midsagittal profile (Fig. 5) placed Apidima 1 with modern humans, to the exclusion of Neanderthals or Middle Pleistocene Europeans/Africans. Its overall shape, as measured by Procrustes distance, was closest to modern human specimens. Apidima 1 was classified as *H. sapiens* with posterior probability >0.9 in all analyses performed by Harvati et al. (2019).

Apidima 1 was dated directly by Harvati et al. (2019) using laser ablation U-series, yielding a date of ca. 210 ka (211 ± 16), indicating that the population that this individual represents pre-dated the Neanderthals known from the same region (Apidima 2, Kalamakia, and Lakonis individuals). These authors therefore interpreted it as the earliest known representative of an early dispersal of early *Homo sapiens* out of Africa (Harvati et al. 2019; Delson 2019), also documented by fossil

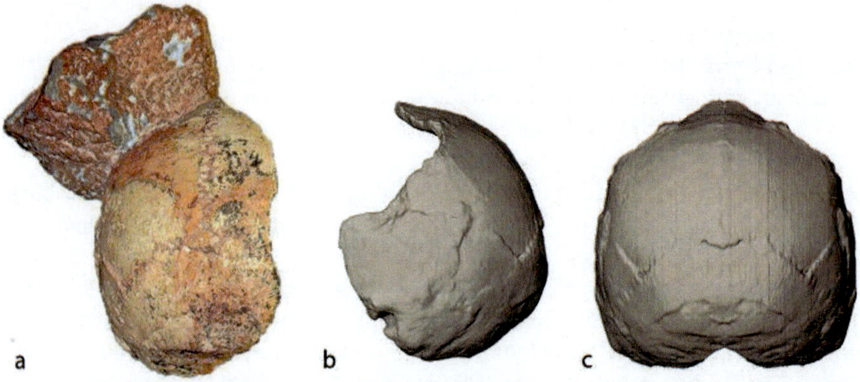

Fig. 4 The Apidima 1 cranium, in (**a**) posterior, (**b**) virtual reconstruction of the lateral, and (**c**) virtual reconstruction of the posterior view. (Copyright K. Harvati)

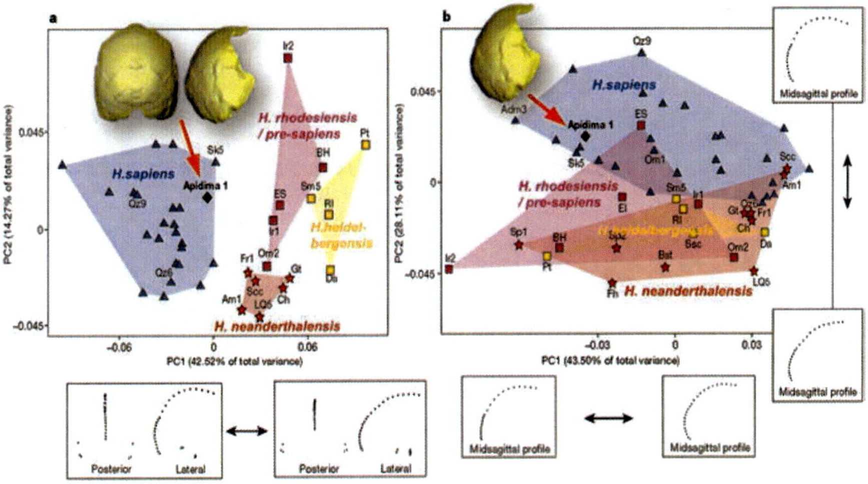

Fig. 5 Principal component analysis of the shape of the Apidima 1 neurocranium (**left**) and midsagittal profile (**right**), based on 3-D landmarks and semilandmarks, with a comparative sample of Pleistocene *H. sapiens* (blue triangles), *H. neanderthalensis* (red stars), *H. heidelbergensis* sensu stricto (or pre-Neanderthals, yellow squares), and African Middle Pleistocene specimens (*H. rhodesiensis* or pre-*sapiens*; magenta squares). Petralona: Pt. Apidima 1 (black diamond) is projected into the PCA, which was calculated on the comparative sample. Both analyses classified Apidima 1 as *H. sapiens* (neurocranium: posterior probability 100%; midsagittal profile posterior probability 93%). (Adapted from Harvati et al. (2019))

human evidence from Israel (Misliya, Skhul, and Qafzeh sites; Hershkovitz et al. 2018; Stringer and Galway-Witham 2018), and the first such specimen found in Europe (see Delson 2019).

Some authors have disagreed with this interpretation. De Lumley et al. (2020) consider this specimen to represent an early Neanderthal, despite the lack of Neanderthal-derived features. These authors present only a few and undiagnostic linear measurements for this specimen, with no statistical analysis to support their attribution. Furthermore, while they attempted to assess the internal configuration of the Apidima 1 suprainiac depression (see Fig. 6), they misidentified its position, likely in part due to the poor resolution of the medical CT scan used (de Lumley et al. 2020). Rosas and Bastir (2020) also did not assess the morphology (either external or internal) of the Apidima 1 suprainiac depression, although they used this feature as the only basis on which to support their conclusion that this specimen represents an "atypical" Neanderthal. Unlike the comparative shape analysis conducted by Harvati et al. (2019), Rosas and Bastir's (2020) analysis was limited to the occipital bone and did not consider the entire available morphology of the specimen, nor did it include a representative comparative sample of early *H. sapiens*. Even so, they did not find any morphological similarities with Neanderthals for Apidima 1, instead placing it close to *H. erectus* and Pleistocene *H. sapiens* in their

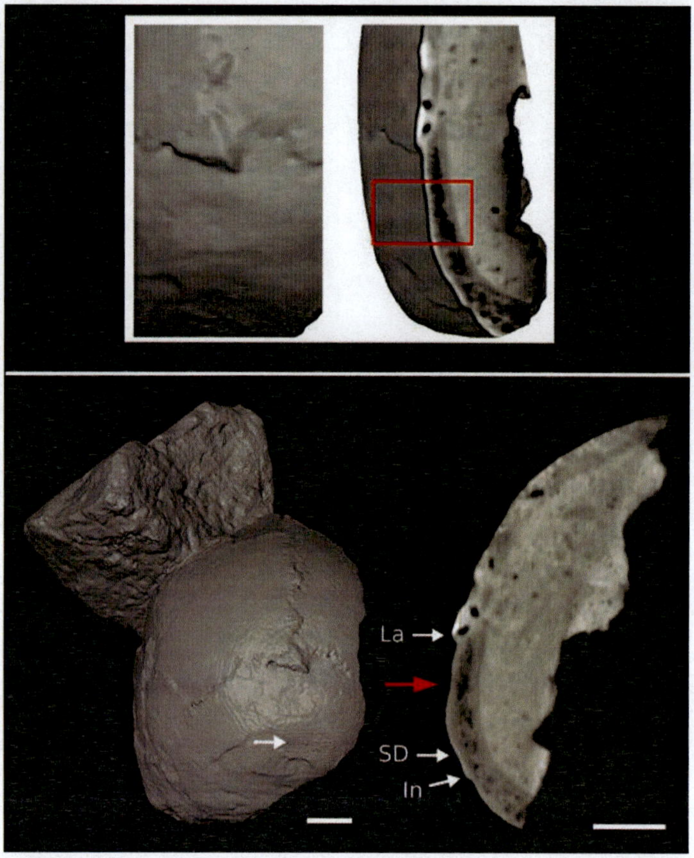

Fig. 6 Apidima 1 occipital region. Top: CT scan images of Apidima 1 illustrating what is presented as a Neanderthal-like suprainiac fossa by De Lumley et al. (2020), enclosed in the red box (Image adapted from De Lumley et al. (2020)). Bottom: posterior view of the CT scan used in Harvati et al. (2019) and sagittal section illustrating the internal morphology of the suprainiac depression. La, lambda; In, inion; SD, suprainiac depression. Red arrow points to the area misidentified as the suprainiac depression by De Lumley et al. (2020). Note the nonstandard, oblique angle of the CT section presented by De Lumley et al. (2020; top), which does not give a clear image of the internal configuration of the bone. Our CT section (bottom) illustrates that there is no thinning of the diploic layer at the supriniac depression of Apidima 1, contra De Lumley et al. (2020), and unlike the condition described for Neanderthals (Balzeau and Rougier 2013)

PCA plot (Rosas and Bastir 2020). It is worth noting that, to the author's knowledge, none of these researchers have examined the Apidima fossils, relying instead on low-resolution medical CT scans obtained in the early 2000s. Further research is currently ongoing to help shed additional light on the phylogenetic affinities and archaeological context of this important specimen.

Apidima Cave Γ Of the five caves constituting the Apidima Cave complex in western Mani, Cave Γ has yielded modern human skeletal remains presumed to date to the Upper Paleolithic. Cave Γ is situated >13 m above Cave A and stratigraphically above the breccia cone encasing the Apidima 1 and 2 fossil human crania (see Harvati et al. 2019). An incomplete skeleton (LAO 1/S3) from Cave Γ has been proposed to represent an early Upper Paleolithic burial (Pitsios 1985, 1995) and was reportedly found associated with lithics described as possibly Aurignacian (Darlas 1995). However, neither the Cave Γ skeleton nor the artifacts or faunal remains recovered there have been described in detail or dated. Similarly, isolated teeth reported from this cave may represent additional individuals, possibly not associated with the archaeological material or the purported burial (Pitsios 1985, 1995). The chronology and cultural affiliation of these remains are therefore as yet uncertain (Harvati et al. 2009; Harvati 2016). Recent direct dating of human skeletal remains from Cave Γ using the U-series method found a terminal Pleistocene minimum age for these individuals, thus confirming their Pleistocene age (Harvati et al. in press). However, further research is needed to establish an upper age limit and to investigate their possible association with the described Upper Paleolithic assemblage.

Franchthi Cave Fragmentary human remains dating to the terminal Pleistocene have been recovered during excavation at Franchthi Cave, southern Argolid. These include a mandibular fragment, three isolated teeth, and two postcranial fragments, all of them representatives of *H. sapiens* (Harvati et al. 2009). These remains are associated with Epigravettian cultural remains.

5 Summary

Although paleoanthropology in Greece is at its infancy, the currently known Greek human fossil record is surprisingly rich for the state of research. As expected for a region acting as both refugium and dispersal corridor throughout the Pleistocene, it is also quite diverse. Major shortcomings are the lack of controlled excavations and therefore lack of context and chronological control for most of the human fossils known, although recent research is helping to remedy this situation through renewed attempts at direct dating (e.g., Bartsiokas et al. 2017; Harvati et al. 2019) and detailed analyses of their affinities (e.g., Harvati 2009, 2016; Harvati et al. 2011, 2013, 2019; Röding et al. in press). This work has already led to breakthroughs in our understanding of human evolution in Greece, as with the identification of an early modern human expansion reaching as far as Europe, something that was not suspected before (Harvati et al. 2019; Delson 2019). Further efforts in this direction can be expected to further clarify the record. At the same time, renewed efforts at systematic, targeted identification and investigation of paleoanthropological sites in the field should be pursued, both to add to this important record and to further clarify the context of human evolution in Greece.

Acknowledgments This research has been funded by the European Research Council Grants ERC StG PaGE (283503) and ERC CoG CROSSROADS (724703), awarded to K. Harvati. I am grateful to Dr. Abel Bosman for providing the images for Fig. 6 (bottom panel) and to Dr. Eric Delson for the photographs of the Petralona cranium shown in Fig. 2. Finally, I thank Eric Delson and an anonymous reviewer for their helpful comments and suggestions.

References

Ackermann RR, Arnold ML, Baiz MD, Cahill JA, Cortés-Ortiz L, Evans BJ, Grant BR, Grant PR, Hallgrimsson B, Humphreys RA, Jolly CJ, Malukiewicz J, Percival CJ, Ritzman TB, Roos C, Roseman CC, Schroeder L, Smith FH, Warren KA, Wayne RK, Zimmer D (2019) Hybridization in human evolution: insights from other organisms. Evol Anthropol 28:189–209

Bailey S (2007) The evolution of non-metric dental variation in Europe. Mitteilungen der Gesells Urgeschichte 15:9–30

Balzeau A, Rougier H (2013) New information on the modifications of the Neandertal suprainiac fossa during growth and development and on its etiology. Am J Phys Anthropol 151:38–48

Bartsiokas A, Arsuaga JL, Aubert M, Grün R (2017) U-series dating and classification of the Apidima 2 hominin from Mani Peninsula, Southern Greece. J Hum Evol 109:22–29

Bermúdez de Castro JM, Arsuaga JL, Carbonell E, Rosas A et al (1997) A hominid from the lower Pleistocene of Atapuerca, Spain: possible ancestor to Neanderthals and modern humans. Science 276:1392–1395

Blackwell BAB, Sakhrani N, Singh I, Gopalkrishna KK, Tourloukis V, Panagopoulou E, Karkanas P, Blickstein JIB, Skinner AR, Florentin JA, Harvati K (2018) ESR dating ungulate teeth and molluscs from the Paleolithic site Marathousa 1, Megalopolis Basin, Greece. Quaternary 2:22

Bosman A, Reyes-Centeno H, Harvati K (2020) A virtual assessment of the suprainiac depressions on the Eyasi I (Tanzania) and Aduma ADU-VP-1/3 (Ethiopia) Pleistocene hominin crania. J Hum Evol 145:102815

Bräuer G, Pitsios T, Säring D, von Harling M, Jessen F, Kroll A, Groden C (2020) Virtual reconstruction and comparative analyses of the middle Pleistocene Apidima 2 cranium (Greece). Anat Rec 303:374–392

Brunet M, Guy F, Pilbeam D, Lieberman DE, Likius A, Mackaye HT, Ponce de León MS, Zollikofer CPE, Vignaud P (2005) New material of the earliest hominid from the upper Miocene of Chad. Nature 434:752–755

Chen L, Wolf AB, Fu W, Li L, Akey JM (2020) Identifying and interpreting apparent Neanderthal ancestry in African individuals. Cell 180:677–687

Darlas A (1995) [Τα λίθινα εργαλεία του σκελετού ΛΑΟ 1/Σ 3 (Απήδημα – Μάνη)]. Acta Anthropol 1:59–62

De Lumley MA, Guipert G, de Lumley H, Protopapa N, Pitsios T (2020) Apidima 1 and Apidima 2: two neandertal skulls in the Peloponnese, Greece. L'Anthropol 144:1–41

Dean D, Hublin J-J, Holloway R, Ziegler R (1998) On the phylogenetic position of the pre-Neanderthal specimen from Reilingen, Germany. J Hum Evol 34:485–508

Delson E (2019) An early modern human outside Africa. Nature 571:487–488

Elefanti P, Panagopoulou E, Karkanas P (2009) The transition from the Middle to the Upper Palaeolithic in the southern Balkans: the evidence from the Lakonis I cave, Greece. Eurasian Prehist 5:85–96

Freidline SE, Gunz P, Harvati K, Hublin J-J (2012) Middle Pleistocene human facial morphology in an evolutionary and developmental context. J Hum Evol 63:723–740

Fu Q, Hajdinjak M, Moldovan OT, Constantin S, Mallick S, Skoglund P, Patterson N, Rohland N, Lazaridis I, Nickel B, Viola B (2015) An early modern human from Romania with a recent Neanderthal ancestor. Nature 524:216–219

Galanidou N, Athanassas C, Cole J, Iliopoulos G, Katerinopoulos A, Magganas A, McNabb J (2016) The Acheulian site at Rodafnidia, Lisvori, on Lesbos, Greece: 2010–2012. In: Harvati K, Roksandic M (eds) Paleoanthropology of the Balkans and Anatolia: human evolution and its context. Springer, Dordrecht, pp 119–138

Gray J E (1825) An outline of an attempt at the disposition of Mammalia into Tribes and Families, with a list of genera apparently appertaining to each Tribe, Annals of Philosophy, new series 10, 337–344

Green RE, Krause J, Briggs AW, Maricic T, Stenzel U, Kircher M, Patterson N, Li H, Zhai W, Fritz MHY, Hansen NF (2010) A draft sequence of the neandertal genome. Science 328:710–722

Grün R (1996) A re-analysis of electron spin resonance dating results associated with the Petralona hominid. J Hum Evol 30:227–241

Grün R, Pike A, McDermott F, Eggins S, Mortimer G, Aubert M, Kinsley L, Joannes-Boyau R, Rumsey M, Denys C, Brink J, Clark T, Stringer C (2020) Dating the skull from Broken Hill, Zambia, and its position in human evolution. Nature 580:372–375

Harmand S, Lewis JE, Feibel CS, Lepre CJ, Prat S, Lenoble A, Böes X, Quinn RL, Brenet M, Arroyo A, Taylor N, Clément S, Daver G, Brugal JP, Leakey L, Mortlock RA, Wright JD, Lokorodi S, Kirwa C, Kent DV, Roche H (2015) 3.3-million-year-old stone tools from Lomekwi, West Turkana, Kenya. Nature 521:310–315

Harvati K (2007) 100 years of *H. heidelbergensis* – life and times of a controversial taxon. Mitt Gesells Urgeschichte 16:85–94

Harvati K (2009) Petralona: link between Africa and Europe? In: Schepartz L, Bourbou C, Fox S (eds) New directions in the skeletal biology of Greece, Athens, Wiener laboratory series. American School of Classical Studies at Athens, Princeton, pp 31–49

Harvati K (2015) Neanderthals and their contemporaries. In: Henke W, Tattersall I (eds) Handbook of paleoanthropology, 2nd edn. Springer, Berlin, Heidelberg, pp 2243–2279

Harvati K (2016) Paleoanthropology in Greece: recent findings and interpretations. In: Harvati K, Roksandic M (eds) Paleoanthropology of the Balkans and Anatolia: human evolution and its context, Vertebrate paleobiology and paleoanthropology series. Springer, Dordrecht, pp 3–14

Harvati K, Delson E (1999) Conference report: paleoanthropology of the Mani Peninsula (Greece). J Hum Evol 36:343–348

Harvati K, Panagopoulou E, Karkanas P (2003) First Neanderthal remains from Greece: the evidence from Lakonis. J Hum Evol 45:465–473

Harvati K, Frost SR, McNulty KP (2004) Neanderthal taxonomy reconsidered: implications of 3D primate models of intra- and inter-specific differences. Proc Natl Acad Sci U S A 101:1147–1152

Harvati K, Panagopoulou E, Runnels C (2009) The paleoanthropology of Greece. Evol Anthropol 18:131–143

Harvati K, Stringer C, Karkanas P (2011) Multivariate analysis and classification of the Apidima 2 cranium from Mani, southern Greece. J Hum Evol 60:246–250

Harvati K, Darlas A, Bailey SE, Rein TR, El Zaatari S, Fiorenza L, Kullmer O, Psathi E (2013) New Neanderthal remains from Mani Peninsula, Southern Greece: the Kalamakia Middle Paleolithic cave site. J Hum Evol 64:486–499

Harvati K, Konidaris G, Tourloukis V (eds) (2018) Special issue human evolution at the gates of Europe. Quat Int Sp Issue Vol 497 Part A, pp 1–240

Harvati K, Röding C, Bosman A, Karakostis FA, Grün R, Stringer C, Karkanas P, Thompson NC, Koutoulidis V, Moulopoulos LA, Gorgoulis VG, Kouloukoussa M (2019) Apidima Cave fossils provide earliest evidence of *Homo sapiens* in Eurasia. Nature 571:500–504

Harvati K, Röding C, Bosman A, Karakostis FA, Stringer C, Gorgoulis VG, Kouloukoussa M (2020) Apidima 1 and the early dispersal of *Homo sapiens* out of Africa Abstract presented at the European Society for the Study of Human Evolution 2020 meeting

Harvati K, Grün R, Duval M, Zhao J, Karakostis A, Tourloukis V, Gorgoulis V, Kouloukoussa M (In press) Direct U-series dating of the Apidima C human remains. In: Reyes-Centeno H, Harvati K (eds) Ancient connections in Eurasia. Kerns Verlag, Tübingen

Hershkovitz I, Weber G W, Quam R, Duval M, Rainer Grün R, Kinsley L, Ayalon A, Bar-Matthews M, Valladas H, Mercier N, Arsuaga J L, Martinón-Torres M, Bermúdez de Castro J M, Fornai C, Martín-Francés L, Sarig R, May H, Krenn V A, Slon V, Rodríguez L, García R, Lorenzo C, Carretero J M, Frumkin A, Shahack-Gross R, Bar-Yosef Mayer D E, Cui Y, Wu X, Peled N, Groman-Yaroslavski I, Weissbrod L, Yeshurun R, Tsatskin A, Zaidner Y, Weinstein-Evron M (2018) The earliest modern humans outside Africa. Science 359:456–459

Hublin J-J (2009) The origin of Neandertals. Proc Natl Acad Sci U S A 106:16022–16027

Hublin J-J, Ben-Ncer A, Bailey S, Freideline S, Neubauer S, Skinner MM, Bergmann I, Le Cabec A, Benazzi S, Harvati K, Gunz P (2017) New fossils from Jebel Irhoud (Morocco) and the pan-African origin of *Homo sapiens*. Nature 546:289–292

Jacobs Y, Li B, Karkanas P, Tourloukis V, Thompson NC, Panagopoulou E, Harvati K (2018) Optical dating of K-feldspar grains from middle Pleistocene lacustrine sediment at Marathousa 1 (Greece). Quat Int 497:170–177

King W (1864) The reputed fossil man of the Neanderthal. Q J Sci 1:88–97

Kokkoros P, Kanellis A (1960) Découverte d'un crâne d'homme paléolithique dans la péninsule Chalchidique. L'Anthropol 64:438–446

Kurtén B, Poulianos AN (1977) New stratigraphic and faunal material from Petralona cave, with special reference to the Carnivora. Anthropos (Athens) 4:47–130

Lipson M, Ribot I, Mallick S, Rohland N, Olalde I, Adamski N, Broomandkhoshbacht N, Lawson AM, López S, Oppenheimer J, Stewardson K, Asombang RN, Bocherens H, Bradman N, Culleton BJ, Cornelissen E, Crevecoeur I, de Maret P, Fomine FLM, Lavachery P, Mindzie CM, Orban R, Sawchuk E, Semal P, Thomas MG, Van Neer W, Veeramah KR, Kennett DJ, Patterson N, Hellenthal G, Lalueza-Fox C, MacEachern S, Prendergast ME, Reich D (2020) Ancient west African foragers in the context of African population history. Nature 577:665–670

Linnaeus C (1758) Systema Naturae per Regna Tria Naturae, Secundum Classes, Ordines, Genera, Species, cum Characteribus, Differentiis, Synonymis, Locis. Editio Decima 1:1–824

Martinón-Torres M, Bermúdez de Castro JM, Gómez-Robles A, Prado-Símon L, Arsuaga JL (2012) Morphological description and comparison of the dental remains from Atapuerca-Sima de los Huesos site (Spain). J Hum Evol 62:7–58

Mayr E (1963) The taxonomic evaluation of fossil hominids. In Classification and Human Evolution. Edited by S. L. Washburn. Chicago: Aldine Press, pp. 332–346

Meyer M, Arsuaga JL, Filippo C, Nagel S, Aximu-Petri A, Nickel B, Martínez I, Gracia A, Bermúdez de Castro JM, Carbonell E, Viola B, Kelso J, Prüfer K, Pääbo S (2016) Nuclear DNA sequences from the middle Pleistocene Sima de los Huesos hominins. Nature 531:504–507

Panagopoulou E, Karkanas P, Tsartsidou G, Kotjabopoulou E, Harvati K, Ntinou M (2002–2004) Late Pleistocene archaeological and fossil human evidence from Laconia Cave, southern Greece. J Field Archaeol 29:323–349

Panagopoulou E, Tourloukis V, Thompson N, Athanassiou A, Tsartsidou G, Konidaris GE, Giusti D, Karkanas P, Harvati K (2015) Marathousa 1: a new Middle Pleistocene archaeological site from Greece. Antiquity 343: Project Gallery

Panagopoulou E, Tourloukis V, Thompson N, Konidaris G, Athanassiou A, Giusti D, Tsartsidou G, Karkanas P, Harvati K (2018) The Lower Palaeolithic site of Marathousa 1, Megalopolis, Greece: overview of the evidence. Quat Int 497:33–46

Pitsios TK (1985) Παλαιοανθρωπολογικές έρευνες στη θέση «Απήδημα» της Μέσα Μάνης. Αρχαιολογία 15:26–33

Pitsios TK (1995) Paleoanthropological research at the cave site of Apidima, Laconia, Greece. Acta Anthropol 1:1–180

Pitsios TK (1999) Paleoanthropological research at the cave site of Apidima and the surrounding region (South Peloponnese, Greece). Anthropol Anz 57:1–11

Pitsios TK (2002) The fossil hominid findings from the cave site of Apidima, South Peloponnese Greece. Coll Anthropol 26(Suppl):158

Posth C, Wißing C, Kitagawa K, Pagani L, van Holstein L, Racimo F, Wehrberger K, Conard NJ, Kind CJ, Bocherens H, Krause J (2017) Deeply divergent archaic mitochondrial genome provides lower time boundary for African gene flow into Neanderthals. Nat Commun 8:16046
Poulianos AN (1981) Pre-sapiens man in Greece. Curr Anthropol 22:187–288
Poulianos AN (1982) [Το Σπήλαιο του Αρχανθρώπου των Πετραλώνων: Η ανάσα της σπηλιάς. Βιβλιοθήκη Ανθρωπολογικής Εταιρίας Ελλάδος, Αριθμός 3. Αθήνα / Πετράλωνα. Γιάννης Ρέκος και Σια: Θεσσαλονίκη]
Rightmire GP (2008) Homo in the middle Pleistocene: hypodigms, variation, and species recognition. Evol Anthropol 17:8–21
Röding C, Zastrow J, Scherf H, Doukas C, Harvati K (in press) Crown outline analysis of the hominin upper third molar from the megalopolis basin, Peloponnese, Greece. In: Reyes-Centeno H, Harvati K (eds) Ancient connections in Eurasia. Kerns Verlag, Tübingen
Roksandic M, Radović P, Lindal J (2018) Revising the hypodigm of *Homo heidelbergensis*: a view from the eastern Mediterranean. Quat Int 466:66–81
Rosas A, Bastir M (2020) An assessment of the late middle Pleistocene occipital from Apidima 1 skull (Greece). L'Anthropol 144:1–10
Sankararaman S, Mallick S, Dannemann M, Prüfer K, Kelso J, Pääbo S, Patterson N, Reich D (2014) The genomic landscape of Neanderthal ancestry in present-day humans. Nature 507:354–357
Scerri EML, Thomas MG, Manica A, Gunz P, Stock JT, Stringer C, Grove M, Groucutt HS, Timmermann A, Rightmire GP, d'Errico F, Tryon CA, Drake NA, Brooks AS, Dennell RW, Durbin R, Henn BM, Lee-Thorp J, Demenocal P, Petraglia MD, Thompson JC, Scally A, Chikhi L (2018) Did our species evolve in subdivided populations across Africa, and why does it matter? Trends Ecol Evol 33:582–594
Schlebusch CM, Malmström H, Günther T, Sjödin P, Coutinho A, Edlund H, Munters AR, Vicente M, Steyn M, Soodyall H, Lombard M, Jakobsson M (2017) Southern African ancient genomes estimate modern human divergence to 350,000 to 260,000 years ago. Science 358:652–655
Schmitz RW, Serre D, BonanI G, Feine S, Hillgruber F, Krainitzki H, Pääbo S, Smith FH (2002) The Neandertal type site revisited: interdisciplinary investigations of skeletal remains from the Neander Valley, Germany. Proc Natl Acad Sci U S A 99:13342–13347
Schoetensack O (1908) Der Unterkiefer des *Homo heidelbergensis* aus den Sanden von Mauer bei Heidelberg. W. Engelmann, Leipzig, pp 1–67
Sickenberg O (1975) Eine Säugertierfauna des tieferen Bihariums aus den Becken von Megalopolis (Peloponnes, Griechenland). Ann Géol Pays Héllén 27:26–71
Smith TM, Harvati K, Olejniczak AJ, Reid DJ, Hublin JJ, Panagopoulou E (2009) Dental development and enamel thickness in the Lakonis Neanderthal molar. Am J Phys Anthropol 138:112–118
Stringer CB (1974) A multivariate study of the Petralona skull. J Hum Evol 3:397–404
Stringer CB (2012) The status of *Homo heidelbergensis* (Schoetensack 1908). Evol Anthropol 21:101–107
Stringer C (2016) The origin and evolution of *Homo sapiens*. Philos Trans R Soc B 371:20150237
Stringer C, Galway-Witham J (2018) When did modern humans leave Africa?. Science 359 (6374):389–390
Stringer CB, Howell FC, Melentis J (1979) The significance of the fossil hominid skull from Petralona, Greece. J Archaeol Sci 6:235–253
Thompson N, Tourloukis V, Panagopoulou E, Harvati K (2018) In search of Pleistocene remains at the gates of Europe: directed surface survey of the megalopolis basin. Quat Int 497:22–32
Tourloukis V, Harvati K (2018) The Palaeolithic record of Greece: a synthesis of the evidence and a research agenda for the future. Quat Int 466:48–65
Tourloukis V, Muttoni G, Karkanas P, Monesi E, Scardia G, Panagopoulou E, Harvati K (2018) Magnetostratigraphic and chronostratigraphic constraints on the Marathousa 1 Lower Palaeolithic site and the middle Pleistocene deposits of the Megalopolis basin, Greece. Quat Int 497:154–169

Tsoukala E (1999) Quaternary large mammals from the Apidima caves (Lakonia, S Peloponnese, Greece). Beitr Paläontol 24:207–229

Villanea FA, Schraiber JG (2019) Multiple episodes of interbreeding between Neanderthal and modern humans. Nat Ecol Evol 3:39–44

Xirotiris N, Henke W, Symeonidis N (1979) Der M^3 von Megalopolis – ein Betrag zu seiner morphologischen Kennzeichnung. Z Morphol Anthropol 70:117–122

Zinner D, Arnold ML, Roos C (2011) The strange blood: natural hybridization in primates. Evol Anthropol 20:96–103

Index

E. Vlachos (ed.), *Fossil Vertebrates of Greece Vol. 1*,
https://doi.org/10.1007/978-3-030-68398-6

D

H

M

N

U

V